MATERIALS SCIENCE AND TECHNOLOGIES

FIBER-REINFORCED COMPOSITES

MATERIALS SCIENCE AND TECHNOLOGIES

Additional books in this series can be found on Nova's website
under the Series tab.

Additional E-books in this series can be found on Nova's website
under the E-book tab.

MATERIALS SCIENCE AND TECHNOLOGIES

FIBER-REINFORCED COMPOSITES

QINGZHENG CHENG
EDITOR

Nova Science Publishers, Inc.
New York

Copyright © 2012 by Nova Science Publishers, Inc.

All rights reserved. No part of this book may be reproduced, stored in a retrieval system or transmitted in any form or by any means: electronic, electrostatic, magnetic, tape, mechanical photocopying, recording or otherwise without the written permission of the Publisher.

For permission to use material from this book please contact us:
Telephone 631-231-7269; Fax 631-231-8175
Web Site: http://www.novapublishers.com

NOTICE TO THE READER

The Publisher has taken reasonable care in the preparation of this book, but makes no expressed or implied warranty of any kind and assumes no responsibility for any errors or omissions. No liability is assumed for incidental or consequential damages in connection with or arising out of information contained in this book. The Publisher shall not be liable for any special, consequential, or exemplary damages resulting, in whole or in part, from the readers' use of, or reliance upon, this material. Any parts of this book based on government reports are so indicated and copyright is claimed for those parts to the extent applicable to compilations of such works.

Independent verification should be sought for any data, advice or recommendations contained in this book. In addition, no responsibility is assumed by the publisher for any injury and/or damage to persons or property arising from any methods, products, instructions, ideas or otherwise contained in this publication.

This publication is designed to provide accurate and authoritative information with regard to the subject matter covered herein. It is sold with the clear understanding that the Publisher is not engaged in rendering legal or any other professional services. If legal or any other expert assistance is required, the services of a competent person should be sought. FROM A DECLARATION OF PARTICIPANTS JOINTLY ADOPTED BY A COMMITTEE OF THE AMERICAN BAR ASSOCIATION AND A COMMITTEE OF PUBLISHERS.

Additional color graphics may be available in the e-book version of this book.

Library of Congress Cataloging-in-Publication Data

Fiber-reinforced composites / editor, Qingzheng (George) Cheng.
   p. cm.
 Includes index.
 ISBN 978-1-61470-303-7 (hardcover)
 1. Fibrous composites. I. Cheng, Qingzheng.
 TA418.9.C6F47 2011
 620.1'18--dc23

                    2011022198

Published by Nova Science Publishers, Inc. † New York

CONTENTS

Preface		vii
Chapter 1	Fiber-Reinforced Composites *Quazi T. H. Shubhra, A. K. M. M. Alam, M. D. H. Beg and M. Saha*	1
Chapter 2	Natural Fiber-Reinforced Composites *Di Bella Guido, Fiore Vincenzo and Valenza Antonino*	57
Chapter 3	Polyvinyl Alcohol Nanocomposites Reinforced with Cellulosic Nanofibers Isolated from Juvenile Poplar *Jingxin Wang, Qingzheng Cheng, Adebola Adebayo and* *Stephen Difazio*	91
Chapter 4	Physical, Flammability and Mechanical Properties of Polymer Eco- Nanocomposites *A. Alhuthali, H. Alamri and I. M. Low*	105
Chapter 5	Static and Dynamic Behaviours of Composite Laminate *R. Zitoune, A. Mir, B. Hachemane, F. Boyer, C. Bouvet,* *F. Collombet and B. Bezzazi*	125
Chapter 6	Performance Properties and Micro-Structural Quantification of Cellulose Fibre-Polypropylene Composites *Marianne Lenes, Gary Chinga-Carrasco and Øyvind Gregersen*	143
Chapter 7	The In-Plane Shear Behaviour of a Carbon Fabric Reinforced Thermoplastic under Quasi-Static and Fatigue Loading Conditions *Ives De Baere, Wim Van Paepegem and Joris Degrieck*	159
Chapter 8	On Toughening of Epoxies and Their Carbon Fibre-Reinforced Composites *Jin Zhang, Tong Lin and Xungai Wang*	217
Chapter 9	Plasma Surface Modification of Glassy Carbon Plates, Carbon Fibres and UHMWPE Fibres for Adhesion Improvement	251

	Yukihiro Kusano, Henrik Mortensen, Steluta Teodoru, Joanna M. Drews, Tom L. Andersen, Bent F. Sørensen, Noemi Rozlosnik, Stergios Goutianos, Peter Kingshott, Susanta Mitra, Frank Leipold, Kristoffer Almdal, Poul K. Michelsen and Henrik Bindslev	
Chapter 10	Adhesion Improvement of Glass Fibre-Reinforced Polyester Composite by Atmospheric Pressure Plasma Treatment *Yukihiro Kusano, Tom L. Andersen, Bent F. Sørensen, Noemi Rozlosnik, Henrik Mortensen, Steluta Teodoru, Peter Kingshott, Frank Leipold, Shailendra Vikram Singh, Jørgen Bilde-Sørensen, Povl Brøndsted, Poul K. Michelsen, Henrik Bindslev, Alexander Bardenshtein and Niels Krebs*	297
Chapter 11	Ultra-High Temperature Ceramics Reinforced with SiC Short Fiber or Whisker *Laura Silvestroni and Diletta Sciti*	319
Chapter 12	Carpet Waste Fibers Applications in Civil Engineering *Reza Jamshidi Chenari and Hamed Shahbazi Fashtali*	369
Chapter 13	Machining of Fibre-Reinforced Composites *Luís Miguel Durão*	387
Chapter 14	ANN Applications in Machining of Fiber-Reinforced Composites *Devi K. Kalla, Jamal Sheikh-Ahmad and Janet Twomey*	439
Chapter 15	Initial versus Final Fracture of Fiber-Reinforced Composites, Analyzed via Acoustic Emission *Nadia Ereifej and Mohammad Al-Rababah*	467
Chapter 16	Analysis of Dynamic Loading of Long Fiber-Reinforced Composites *Carlos Santiuste, Xavier Soldani, Jorge López-Puente and Henar Miguélez*	497
Chapter 17	A Micromechanical Model and Reinforcing Distribution Optimisation in Fibre-Reinforced Materials *Roberto Brighenti*	527
Chapter 18	Application of FRP Composites for Rehabilitation in Civil Engineering: Experiments and Advanced Compuational Methods *B. R. Kim, S. K. Ha, S. Na and H. K. Lee*	571
Chapter 19	Flexible Fiber Composites for Space Structures *Leri Datashvili and Horst Baier*	605
Index		641

PREFACE

Fiber-Reinforced Composite (FRC) is a system material consisting of fibers as reinforcements in discontinuous or dispersed phase and matrix as the continuous phase with distinct interphase or interface regions between them. Other additives (modifiers, fillers) may be added to reduce cost, improve workability, and/or impart desired properties. The fibers carry the loads along their longitudinal directions, while the matrix transfers stresses between the reinforcing fibers and protect the fibers from mechanical and/or environmental damages. Compared to the neat matrix material, FRC has highly favorable mechanical properties, and their strength-to-weight ratios are superior. Other advantages of FRCs may include non-corrosiveness, good bonding properties, ease of repair, and the potential for laboratory fabrication. FRCs have been used in many areas, such as space structures, sports equipment, automobiles, and construction materials. In addition, FRCs have potential for use in dentistry and are expected to gain increasing applications in the future. This book includes both review and research papers in different FRC areas from contributors around the world. In Chapter 1, the contributors reviewed different fiber and matrix types, as well as FRC processes and applications. Other Chapters include the traditional knowledge and the new development of different studies of natural fiber including cellulosic nanofibers (Chapter 2 to 6), carbon fiber (Chapter 7 to 10), and other fibers (Chapter 11 to 12) reinforced composites, FRC machining (Chapter 13 to 14), FRC testing methods (Chapter 15), FRC modeling (Chapter 16 to 17), and FRC applications (Chapter 18 to 19).

Chapter 1 – A typical composite material is a system of materials consisting of two or more materials on a macroscopic scale. It is generally composed of reinforcement (fibers, particles, fillers, etc.) embedded in a matrix (polymers, metals, ceramics, etc.). The matrix holds the reinforcement to form the desired shape while the reinforced material improves the overall properties of the matrix. Composites are favored because they are made of distinct elements with the target to produce materials which have better properties. Fiber-reinforced composites (FRCs) are composed of fibers and a matrix with distinct interfaces (boundaries) between them. Fibers can be incorporated into a matrix either in continuous lengths or in discontinuous (short) lengths. In FRCs, both fibers and matrix retain their chemical and physical identities. But interestingly they produce a combination of properties that cannot be achieved with either of the constituents acting alone. In FRCs, fibers are the reinforcement and the main source of strength (principal load-carrying members) while the matrix keeps them in the desired location and orientation and protect them from environmental damages. Glass, carbon and aramid fibers are widely used synthetic fibers for the preparation of FRCs although nylon, orlon, etc are used to some extent. On the other hand, both plant-based (jute, cotton, etc) and animal based natural fibers (silk, wool, etc) are widely used with various polymer matrices. But both these fibers have advantages and drawbacks. Synthetic fibers-reinforced composites normally show better properties than that made by natural fibers, but they are not environment friendly. Sometimes, fillers or modifiers are added during the preparation of FRCs to facilitate manufacturing process, to reduce cost, improve workability and to impart desired properties. FRCs find widespread applications in aerospace, spacecraft, outdoor deck floorings, railings, window and door frames, fencing, landscaping timbers, benches, indoor furniture, sports equipments such as body parts of race cars and some automobiles, bicycle frames, baseball bats, exercise equipment, etc. Different types of matrix materials can be used for FRCs like thermoplastic and thermosetting resin matrix (epoxy, polyester, polypropylene, etc), metal matrix (aluminum, copper, lead, etc) and also ceramics matrix. In this chapter, metal matrix composites (MMCs), ceramic matrix composites (CMCs), carbon matrix composites and polymer matrix composites (PMCs) are discussed with their properties, fabrication and applications. Especial emphasis was given on mechanical properties, thermal properties and surface morphology of fiber reinforced polymer matrix composites.

Chapter 2 – Over the last decade, composites of polymers reinforced with natural fibers have received increasing attention, both from the academic world and from several industries. There is a wide range of natural fibers which can be applied as reinforcements or fillers thanks to their properties and availability; i.e. flax, hemp, jute, kenaf and sisal. Natural fibers are mainly attractive for the following reasons: specific properties, price, health advantages and recyclability. Particularly, industry is getting more and more interested in environment-friendly products and therefore the research on natural fiber based on composite materials is gaining importance. Some of the benefits linked to the usage of such natural composites are their low density and good specific properties. Furthermore, they are renewable and have a CO_2-neutral life cycle, in contrast with their synthetic opponents (i.e. glass and carbon). Despite such good premises, these fibers have also some negative characteristics: they are highly hydrophilic and their properties may vary in time. The natural fibers have a complex structure of elementary fibers, consisting of cellulose, hemicellulose, pectin, lignin and others and thus they should not be considered as monofilament fibers. Mechanical, physical and even chemical properties of these fibers are strongly harvest-dependent, influenced by climate, location, soil characteristics, and weather circumstances. They are affected also by fiber processing (i.e. retting, scutching, bleaching, spinning) and by their incorporation into composites: handling, impregnation and consolidation may introduce supplementary changes. Obviously, such variability complicates the prediction and the evaluation of the composite properties. Another known problem in natural fiber-reinforced composites is the poor interface quality between the fibers and the polymer matrix; chemical pre-treatments are often applied in order to enhance the adhesion between these components. In the last few years several new components based on natural fiber composites have been mainly developed by industries in the automotive field; i.e. in 2003 around 43,000 tonnes of natural fiber have been used by the European automotive industry as composite reinforcement. This interest is due to the excellent tensile mechanical properties of fiber obtained from plants such as flax, hemp and the stinging nettle. Other main applications of the natural fibers are in the following fields: nautical and biomedical ones. The aim of this work is to present a review on the main natural fibers investigated by researchers and to report the results of some original studies performed by the authors.

Chapter 3 – In this Chapter, cellulosic nanofibers (CNFs) were isolated from juvenile poplar (P. trichocarpa) (PT) and juvenile hybrid poplar (HP) using acid hydrolysis. A film casting method was used to fabricate biodegradable nanocomposites from CNFs and a polyvinyl alcohol (PVA) solution. CNFs were characterized using atomic force microscopy and dynamic light scattering. Most nanofibers were bundles of single nanocrystals and the average diameter of the nanofibers was about 21 nm. The mechanical properties and thermal degradation behavior of the nanocomposites were analyzed by tensile test and thermogravimetric analysis, respectively. The elastic modulus of PVA was significantly improved by the CNFs, while the strength of the composites was comparable with that of pure PVA. No significant differences of tensile mechanical properties of PVA nanocomposites reinforced with CNFs isolated from PT and HP were detected. The thermal degradation behavior of PVA was improved by CNFs.

Chapter 4 – When compared to their synthetic counterparts, natural fibres represent an environmentally friendly alternative by virtue of several attractive attributes that include lower density, lower cost, non-toxicity, ease of processing, renewability and recyclability. In addition, the use of natural fibres in polymer matrix composites has the potential to produce materials with higher specific strength and specific modulus due to their low density. In this chapter, epoxy and vinyl-ester resin matrix composites reinforced with recycled cellulose fibre (RCF) have been fabricated and characterised. Nano-composites from both polymers reinforced with different concentration (1%, 5% and 10%) of nano-clay platelets (30B) and halloysite nano-tubes (HNTs) have been synthesized. The mechanical properties, such as flexural strength, impact toughness, fracture toughness and the effect of water absorption on impact toughness were evaluated. Porosity and flammability properties have been investigated. Results indicated that flexural strength decreased for the majority of samples due to the poor dispersion of nano-fillers and the formation of micro-voids within the samples. In contrast, impact toughness and fracture toughness were improved for all reinforced samples. Water absorption resulted in enhanced impact toughness as a result of local plasticization at the crack-tip. Addition of nanoclay increased the porosities but improved the flammability of all nanocomposites. The effects of RCF and nanoclay additions on the physical and mechanical properties have been discussed in terms of the observed microstructures.

Chapter 5 – The use of composites materials made of natural fibres in buildings and automobiles increases more and more. However, the mechanical properties and the mechanical behaviours of these materials depend of the type of natural fibre used as well as of the type of manufacturing process used for obtaining the final part. In this chapter, the authors are interested in the characterization of the mechanical and thermo mechanical behaviour of composite laminate made of jute/epoxy. For the mechanical characterization, standard static tests (tensile, shear and bending) are carried out for the identification of the mechanical proprieties. Thanks to the digital image correlation technique, the authors observe that the mechanical proprieties have important variabilities during the static tests. The impact tests show that these variabilities have few impact on the first force responsible of the first damage during the impact tests. Nevertheless, the maximum forces obtained during impact tests are affected by the variabilities of this material.

Chapter 6 – This study focuses on the effects of cellulose content and cellulose morphology on the structural and mechanical properties of PP/cellulose fibre composites. A detailed structural characterisation was performed, including SEM-BEI analysis of fibre spatial distribution and X-μCT analysis of fibre orientation. The results indicated that the fibres were well-dispersed and oriented mostly along the main direction of the PP/cellulose test specimen. Based on light microscopy it was found that the cellulose fibres acted as nucleators in the PP matrix. By use of DSC it was shown that addition of cellulose fibres increased both the crystallinity and Tc. The increased crystallinity is mainly a result of the transcrystallinity formation. The tensile properties of the composites were affected by the fibre dimensions and the fibre concentration. The long fibres yielded the highest increase of Young's modulus. Elongation at break was mostly affected by the fibre concentration. The results thus indicate that there is a major potential for improving the mechanical properties of fibre-reinforced composites, provided an adequate concentration of fibres is applied.

Chapter 7 – In this chapter, the nonlinear shear stress-strain relationship of a carbon fabric-reinforced polyphenylene sulphide is investigated, both under quasi-static and fatigue loading conditions. In order to determine the in-plane shear properties of a fibre reinforced composite, there are various possibilities. One of them is the standard three-rail shear test, as described in "ASTM D 4255/D 4255M The standard test method for in-plane shear properties of polymer matrix composite materials by the rail shear method". This set-up, however, requires drilling holes through the specimen. In this study, a new design based on friction and geometrical gripping, without the need of drilling holes through the composite specimen is presented. Next, this setup is used and compared to $[(+45°,-45°)]_{4s}$ tensile tests to study the shear behaviour of the material under consideration.

For the fatigue behaviour, fatigue tests with $R = 0$ were conducted, to assess the behaviour of the newly designed grips under fatigue loading conditions on one hand and to investigate the in-plane shear behaviour on the other hand. With respect to the grips, excellent results were obtained; the specimen fails under shear loading conditions in the loaded area. With respect to the material, an increase in permanent deformation and a decrease in shear stiffness until a certain point in time is visible, after which a drastic increase in deformation and temperature, higher than the softening temperature of the matrix occurred. Also, the maximum value of the shear stress for fatigue with $R = 0$ has a large influence on the fatigue lifetime. These findings were also assessed with the $[(+45°,-45°)]_{4s}$ fatigue tensile tests, with very good correspondence.

For the quasi-static behaviour, a previously derived material model with four material constants is considered. Using quasi-static and hysteresis tests, both with the rail shear setup and with $[(+45°,-45°)]_{4s}$ tensile tests, the necessary data for determining the material constants are obtained. Then, the material constants are optimised by comparing finite element simulations with the data derived from the experiments. The conducted experiments are simulated and the results are compared with the experiments, with excellent correspondence.

Chapter 8 – Epoxy resins were introduced commercially in the late 1940s and have been used widely in industrial applications, including fibre reinforced polymer matrix composites, adhesives and coatings. While epoxies have numerous advantages, they suffer from brittleness as a result of their high crosslink polymer structure. The present work is a brief review of the toughening methods used previously to improve the crack growth resistance of epoxies and their fibre reinforced composites. Toughening agents discussed include rubbers, inorganic fillers, thermoplastics and hyperbranched polymers. The chapter provides the state-of-the-art knowledge in this field as well as indicates the future directions towards improving the fracture toughness of epoxies.

Chapter 9 – Carbon fibres and ultra-high-molecular-weight polyethylene (UHMWPE) fibres are extensively used for improving mechanical properties of fibre-reinforced polymer (FRP) composites. Strong adhesion between the fibre surfaces and the polymer matrix is one of the key issues for improving the longitudinal tensile strength of FRP composites. However, adhesion to these fibres is often difficult or weak because of their low surface energy and poor reactivity. Therefore, adhesion improvement of the fibre surfaces is of considerable industrial importance, which is enhanced by adding chemical functional groups and micro- or nano-scale roughness at the surfaces. Plasma treatment is attractive for this application because of its environmental compatibility and high treatment efficiencies without affecting the textural characteristics of the bulk material. Such a plasma is generally obtained at low pressures, but can also be generated at atmospheric pressure.

In the present work, low and atmospheric pressure plasmas are used to modify the surfaces of glassy carbon plates, carbon fibres and UHMWPE fibres for adhesion improvement. Here, a glassy carbon plate, consisting of amorphous-like sp2 hybridized carbon material, can be an ideal model specimen for fundamental studies of adhesive properties of carbon fibres due to the structural similarity and easier handling than carbon fibres. The plasma treatment typically improved wettability of the surface, increased the oxygen containing polar functional groups at the surface, and enhanced the surface roughness. The results of the surface characterization were compared with adhesion properties with epoxy resin.

Chapter 10 – Glass fibre reinforced polyester (GFRP) composites exhibit high strength-weight ratio and corrosion resistance, and are therefore used for a variety of applications in civil engineering, aerospace and automobile industry as well as medical and dental applications. Adhesives are often required for joining these materials to components fabricated from similar GFRPs or other types of materials. GFRPs usually have smooth surfaces composed mainly of the polymer matrix materials with low surface energies. In addition, they often have a wide range of contaminants on the surfaces, which can form a weak boundary layer in a bond. Therefore, an adhesive joint usually requires careful surface preparation. Among several surface modification techniques for adhesion improvement, atmospheric pressure plasma treatment is promising, since it shows high treatment efficiency, and is applicable to a variety of materials without affecting the bulk properties and easily up-scalable with low equipment cost. Its example includes dielectric barrier discharge (DBD) generated between electrodes covered with dielectrics, and gliding arc generated between divergent electrodes and extended by a high speed gas flow. In addition it is reported that atmospheric pressure plasma treatment can be highly enhanced by high-power ultrasonic irradiation onto the surface during the treatment.

In the present work, GFRP plates are treated with a DBD in air, helium and gliding arc in air for adhesion improvement. The effect of ultrasonic irradiation on surface modification is investigated during helium DBD treatment. The treated surfaces are characterized and the adhesive properties with vinylester resin are tested. It is found that the treatment increased the density of oxygen-containing polar functional groups and polar components of surface energy. The ultrasonic irradiation during the treatment enhanced the treatment efficiency.

Chapter 11 – The IV and V group transition metals— borides, carbides and nitrides— are widely known as ultra-high temperature ceramics (UHTCs), thanks to their high melting point above 2500°C. These ceramics possess outstanding physical and engineeristic properties, such as high hardness and strength, low electrical resistivity and good chemical inertness.

The materials more deeply investigated are the ZrB2-based ones, in view of the lower density, compared to Hf and Ta composites, and the better oxidation resistance of the borides compared to the carbides and nitrides.

So far, the research activity has mainly focused on the achievement of high strength UHTCs usually based on ZrB2 and HfB2 composites and SiC particles in amounts of 20-30vol%. These ceramics can in fact reach strength up to 1 GPa and even higher, through careful tailoring of processing parameters. However, these strong ceramics still have very low fracture toughness, with values ranging from 2.5 to 4 MPam1/2 for ZrB2-based materials. The demand to design and realize reinforced UHTC materials comes from the need to overcome this brittleness, which makes the scale-up of big components a technological issue. The basic concept is the improvement of the fracture toughness through the introduction of elongated reinforcing phases, like SiC fibers (macro-reinforcement), or whiskers (micro-reinforcement) into UHTC matrices.

A number of critical issues arise when elongated secondary phases are added to the matrix: it is difficult to obtain a homogeneous dispersion, fracture of the reinforcing phase occurs during milling and it is problematic to achieve dense ceramics without deteriorate the reinforcing phase.

For example, whiskers tend to degenerate into particles if the sintering temperature is too high and then lose their reinforcing action. For carbon or silicon carbide fibers there is a strong interface reaction with the ZrB2 matrix, which should be instead avoided to promote significant pull out.

This work presents the latest results on ZrB2-based materials reinforced with SiC fibers or whiskers. Several sintering additives or sintering techniques are experienced in order to preserve the integrity of the reinforcing element, thus to exploit at best their toughening action. The microstructure of the dense materials is studied by SEM and TEM techniques to investigate the effect of the sintering additive and the interface between matrix and reinforcing phase.

The mechanical properties are compared to those of the reference material to assess the real variation obtained by the introduction of these reinforcing agents. Following the inspection of the crack/microstructure interaction, the experimental fracture toughness is quantitatively compared to theoretical models. The addition of such reinforcements can improve the fracture toughness, from 3.7 to 5.5-6.3 MPam1/2 upon a proper choice of sintering additive and right amount of reinforcing phase. The main mechanisms invoked for explaining such improvements are crack deflection and crack bridging. The values of high temperature strength and fracture toughness up to 1500 °C are also presented and discussed. Mention is also made to the effect of the introduction of SiC short fibers to other UHTCs, namely to HfB2 and TaC.

Chapter 12 – Soil reinforced with fiber acts as a composite material and fiber inclusion has a significant effect on soil permeability. With regard to the higher void ratio of carpet fibers, it is expected at the first glance that an increase in the void ratio of the reinforced soil would have resulted in an increase in permeability of the mixture. However in contrary, the present chapter indicates that the more the carpet fiber content, the lower the permeability of sand-fiber composite.

This chapter investigates the effect of fiber inclusion on the permeability and hydraulic properties of coarse grained soils. For this purpose a series of constant head permeability tests have been carried out in order to show the effects.

A new system of phase relationships was introduced to calculate the degree of compaction in sand portion of the composite. It was also tried to maintain a constant relative density so as to isolate the effect of fiber inclusion on the model. It was found that carpet fiber inclusion in particular media will render a reduction of hydraulic conductivity of the reinforced soil and this was shown to be due to the clogging of the porous spaces inherent within the carpet fibers with sand fine grains.

Chapter 13 – Although composites components are produced to near-net shape, finishing operations like drilling, in order to fulfil requirements related with tolerances or assembly needs, are usually required. It is known that machining operations in composites can be carried out with conventional tools and machinery with proper adaptations. However, due to their inhomogeneity, this operation can lead to different damages, like delamination, intralaminar cracks, fibre pull out or thermal damage. The most frequent and noticeable evidence of these damages is the existence of a border around the machined hole in consequence of the drilling process. From these damages, delamination is considered the most serious as it can affect the mechanical properties of the composites parts, hence, lower reliability.

Production of higher quality holes, with dimensional precision, acceptable surface quality and damage minimization, is a challenge to everyone related with composites industry. However, the knowledge on this subject is yet limited. So, it means that a better understanding of the damage mechanisms is needed until composites machining reach the same level of confidence that can be found about metal machining. Another difficulty in getting that level of knowledge may be related with the composites nature itself, as their inhomogeneity results in properties dependence on fibre and matrix materials, fibre volume fraction, ply lay-up sequence, orientation and other factors. Hence, conclusions that are valid for a certain type of fibre reinforced composite with some fibre orientation may not be valid if changed from unidirectional to cross-ply or other, just to quote an example.

The purpose of this chapter is to be organised as follows. The first part is dedicated to a review of the principles of conventional composites machining, including orthogonal cut and fibre orientation influence. It will include a state-of-the-art of recent advances on dedicated tools and equipment. Materials for tools will also be evaluated. In the end of this part, a brief review of non-conventional machining will be presented. The second part will be dedicated to machining related damage, focusing on delamination. The concept of critical thrust force for delamination onset and existent analytical models for its determination will be presented. As delamination involves fracture related mechanisms, fracture modes will be reviewed and relevance of interlaminar fracture toughness discussed. The third part will deal with damage evaluation criteria. Different numerical ratios are normally referred to when damage extension is to be compared. Those criteria are based on the availability of non-destructive inspection methods, from tool maker's microscope to computerized tomography. These and other techniques will be presented. Results of damage extension will be related with main drilling parameters - cutting speed or feed rate. Finally, a brief reference to the mechanical tests normally used for damage assessment, like bearing test, and their correlation with damage extension will be the object of the fourth part.

Chapter 14 – Composite materials are inhomogeneous in nature and the study of their behavior during secondary manufacturing is rather difficult due to this complex nature. During machining of fiber reinforced composites, the basic mechanisms of chip formation and the resulting effects on machinability are greatly influenced by the material composition and architecture. The chip formation mode, type of chip produced, cutting force and surface quality are influenced by fiber type, fiber orientation, and cutting tool geometry. The advancement of the cutting tool against alternating hard fibers and soft matrix phases generates cyclic cutting forces and nonuniform surface characteristics. In addition, brittle failure of the fibers and the fiber-binder interfaces induce surface and subsurface cracks and delamination. Thus, the experimental study of machining of composites often involves the analysis and modeling of complex phenomena relying on noisy data that is highly nonlinear and lacking normality. Such data may not be easily analyzed by rigid mathematical models.

Artificial Neural Network (ANN) methods have emerged as a powerful and flexible approximating and predictive tool that is capable of self-learning and adapting to complex data sets. The advantages of ANN over mathematical methods are that it requires no explicit mathematical solutions or restrictive assumptions, and it is faster than other algorithms because of the use of parallel computing. ANN has been successfully applied in the analysis and modeling of machining fiber reinforced composites. This includes the prediction of tool wear, surface roughness, cutting forces and delamination. ANN has also been utilized in devising process control techniques for preventing and minimizing delamination. Special types of ANN structures and training methods were developed specifically to address the type of data obtained from composites machining. Significant improvements in the model predictive capabilities were achieved when ANN methods are compared to multiple regression methods. In this chapter the authors review recent advancements in neural network techniques and their application in machining FRP composites. Future trends of this modeling technology are also discussed.

Chapter 15 – The aim of this chapter was to compare fracture resistance of composite resin-based materials using three-point bending in conjunction with acoustic emission (AE) analysis. Samples were prepared of dimensions of (25 x 2 x 2 mm), according to ISO 4049 standard. Non-reinforced composite samples (NRC) were prepared using Synergy composite (Coltene, Altstätten-Switzerland), while in unidirectional (UFRC) and multidirectional fiber-reinforced groups (MFRC), samples were prepared by incorporating unidirectional EverStick and multidirectional fibers (Stick Tech Ltd Oy, Turku, Finland), respectively. Eight samples were prepared per group, placed at a universal testing machine and loaded at a crosshead speed of 0.2 mm/min. An AE system (Physical Acoustics Corporation, New Jersey, USA) was used to detect audible signals produced while the specimens were loaded. Flexural strength (σ) was calculated at initial and final fracture. One-Way ANOVA was used to detect differences in fracture strength and AE amplitudes between the groups. Paired t-tests were used to detect differences between the flexural strength at initial and final failure of each group. Correlation and regression analyses were used to detect the correlations between initial and final failure strength in each group.

Two major sets of acoustic signals were detected demarcating initial and final failure. Initial fracture strengths and AE amplitudes were significantly lower than those at final fracture in all groups ($p<0.05$). Initial fracture strength of UFRC (171.4 MPa) was significantly higher than that of MFRC (125.1 MPa) and NRC (88.3 MPa). Final fracture strength of UFRC was also significantly higher (197.7 MPa) than the rest of the groups; MFRC (150.5 MPa) and NRC (108.8 MPa). Initial and final fracture strengths were significantly correlated ($r= 0.983$). It was concluded that fibre-reinforcement with unidirectional or multidirectional glass fibres improves the fracture resistance of composite resin materials and monitoring acoustic signals released during fracture can reveal significant information regarding the fracture process.

Chapter 16 – Structural applications of LFRP composites could involve dynamic loading during the service life of the component; for instance, the impact of foreign objects such as birds against components of aircrafts during service. On the other hand, LFRP composite components are usually made to the final size of the desired product however some machining operations are needed, in general, to achieve work-piece requirements. Cutting operations of composite can be considered a dynamic problem and usually involve damage in the component.

Experimental work is needed to understand phenomena involved during dynamic loading states and it is also required to validate models. Numerical modeling based on Finite Element Analysis (FEA) is a powerful tool used to simulate dynamic loading of LFRP composites, allowing the analysis of different parameters influent on the process. The use of numerical models validated with experimental results is of crucial importance for designing, giving an alternative to costly and complex experiments. This paper is focused on the analysis of dynamic loading of LFRP composites including both impact and machining studies. Previous work of the authors is reviewed and summarized in the paper together with other contributions obtained from the scientific literature.

Chapter 17 – The use of composite materials has known a crescent interest in the last decades in many application fields due to their desirable mechanical characteristics such as high specific strength and stiffness, high fracture and fatigue resistance, high wear resistance, high damping durability performance, low thermal coefficient, and so on. For the above mentioned reasons composite materials replace or strategically compliment other traditional structural materials.

The extensive use of advanced materials such as composite materials, requires to describe, with an appropriate accuracy, their overall mechanical behaviour to correctly assess the safety level of structural components in the design process. In the present chapter the problem of the description of the macroscopic mechanical characteristics of such a class of materials is considered. The behaviour of fibre-reinforced composites is examined through the formulation of a micromechanical-based model. The macro constitutive equations for such a class of materials, composed by a matrix phase, for which an elastic-plastic behaviour is eventually allowed, and a fibre-reinforcing phase, is obtained through a micro mechanical model which takes into account the possibility of an imperfect bond between the matrix and the fibres. A two-parameters mechanical model, obtained from energetic considerations and by considering the evolution of the shear stress distribution along a single fibre during the loading process, is formulated to determine the entity of the debonding and its mechanical influence, from a macroscopic point of view, on the composite material. The detailed aspects of the mechanical model are presented and discussed and its implementation in a 2D FE code is finally illustrated.

In the second part of the chapter the problem of the optimal content distribution of fibres in a fibre-reinforced composites, in order to maximise or minimise a given objective function and by assuming some suitable constraints, is investigated by using a biological-based procedure known as Genetic Algorithm (GA). The evolution process simulated by the developed algorithm, can be performed in order to get the maximisation or minimisation of some mechanical desired performance of the structure (stiffness, compliance, peak stress, etc) by keeping constant the total fibres content (optimal constrained problem). The proposed mechanical model with the optimisation algorithm is finally used in some numerical simulations in order to assess its reliability in material design composition with respect to some expected optimal performance, and quantitative comparisons – in term of the improvements with respect to classical homogeneously distributed fibres situations – is finally illustrated and quantified.

Chapter 18 – FRP (Fiber–Reinforced Polymer) composite is an advanced composite created from fibers with high stiffness and strength placed in a resin matrix. Up to date, FRP has revolutionized the civil engineering society as it rapidly gained recognition as a superior construction material. In the field of civil engineering, majority of FRP application is confined to strengthening and repairing of structures (Zhao and Zhang, 2007). In addition, FRP composite materials have been used to construct transportation infrastructures in numerous countries (USA, Germany, England, etc.). Technology advancement of FRP composite in civil engineering is growing as birth to newer materials and knowledge improve its uses with the increasing number of FRP composites replacing conventional construction materials such as steel and concrete in civil engineering (Karbhari and Zhao, 2000; Van Den Einde et al., 2003). In this chapter, current technologies for FRP composites in civil engineering will be reviewed, such as new materials, manufacturing methods and advanced FRP technologies that exist today.

Chapter 19 – Flexible fiber composites are those in which fibers reinforce (highly) flexible matrix materials like urethane, silicone or similar. In this chapter basics of flexible fiber composites are given focusing on carbon fiber-reinforced silicone (CFRS) composites. Specific methods are addressed for manufacturing, testing, and analysis of unidirectional CFRS laminate as well as CFRS of both biaxially and triaxially woven fabric (TWF). A method of homogenization of material properties of TWF flexible composites is given in detail. For this a derivation procedure is developed and discussed for a 9 x 9 ABD stiffness matrix of an equivalent laminate, which is based on micromechanical finite element analysis of a repeated unit cell of the composite. Using an effective beam finite element model, 3D deformability of the cell is fully accounted for. The resulting properties include tensile and flexural as well as transverse shear stiffnesses, which are particularly needed for characterizing the flexible fiber composites. It is concluded that TWF composites behave as transversely (quasi) isotropic materials according to the calculated homogenized properties. After addressing possible space applications like large deployable reflectors and large membrane structures, a focus is given on shape variable space structures using the flexible fiber composites as reflecting morphing skins. The prospective of the use of CFRS in shape morphing structures opens new challenging application direction for this material.

In: Fiber-Reinforced Composites
Editor: Qingzheng Cheng

ISBN: 978-1-61470-303-7
© 2012 Nova Science Publishers, Inc.

Chapter 1

FIBER-REINFORCED COMPOSITES

Quazi T. H. Shubhra[1,2], A. K. M. M. Alam[2,3], M. D. H. Beg[3] and M. Saha[1]*

[1]Department of Applied Chemistry and Chemical Engineering,
University of Dhaka, Dhaka 1000, Bangladesh
[2]Nuclear and Radiation Chemistry Division,
Institute of Nuclear Science and Technology,
Bangladesh Atomic Energy Commission, Dhaka 1207, Bangladesh
[3]Faculty of Chemical and Natural Resources Engineering,
Universiti Malaysia Pahang, 26300 Gambang, Kuantan, Pahang, Malaysia

ABSTRACT

A typical composite material is a system of materials consisting of two or more materials on a macroscopic scale. It is generally composed of reinforcement (fibers, particles, fillers, etc) embedded in a matrix (polymers, metals, ceramics, etc). The matrix holds the reinforcement to form the desired shape while the reinforced material improves the overall properties of the matrix. Composites are favored because they are made of distinct elements with the target to produce materials which have better properties. Fiber-reinforced composites (FRCs) are composed of fibers and a matrix with distinct interfaces (boundaries) between them. Fibers can be incorporated into a matrix either in continuous lengths or in discontinuous (short) lengths. In FRCs, both fibers and matrix retain their chemical and physical identities. But interestingly they produce a combination of properties that cannot be achieved with either of the constituents acting alone. In FRCs, fibers are the reinforcement and the main source of strength (principal load-carrying members) while the matrix keeps them in the desired location and orientation and protect them from environmental damages. Glass, carbon and aramid fibers are widely used synthetic fibers for the preparation of FRCs although nylon, orlon, etc are used to some extent. On the other hand, both plant-based (jute, cotton, etc) and animal based natural fibers (silk, wool, etc) are widely used with various polymer matrices. But both these fibers have advantages and drawbacks. Synthetic fibers reinforced composites normally show better properties than that made by natural fibers, but they are not

---

[*] Email: dr.manoranjansaha@gmail.com (D. Saha); shubhro.du@gmail.com (Quazi T. H. Shubhra)

environment friendly. Sometimes, fillers or modifiers are added during the preparation of FRCs to facilitate manufacturing process, to reduce cost, improve workability and to impart desired properties. FRCs find widespread application in aerospace, spacecraft, outdoor deck floorings, railings, window and door frames, fencing, landscaping timbers, benches, indoor furniture, sports equipments such as body parts of race cars and some automobiles, bicycle frames, baseball bats, exercise equipment, etc. Different types of matrix materials can be used for FRCs like thermoplastic and thermosetting resin matrix (epoxy, polyester, polypropylene, etc), metal matrix (aluminum, copper, lead, etc) and also ceramics matrix. In this chapter, metal matrix composites (MMCs), ceramic matrix composites (CMCs), carbon matrix composites and polymer matrix composites (PMCs) are discussed with their properties, fabrication and applications. Especial emphasis was given on mechanical properties, thermal properties and surface morphology of fiber reinforced polymer matrix composites.

1. INTRODUCTION

A typical composite material is a system of materials consisting of two or more materials (mixed and bonded) on a macroscopic scale with two or more distinct phases having recognizable interfaces between them [1]. Proper combinations of materials into composite materials give rise to properties which transcend those of the constituents, as a result of the principle of combined action. Composites are produced to optimize various material properties like mechanical (mainly strength), chemical and/or physical properties. Since the early 1960s, there has been an increasing demand for very strong and stiffer, yet lighter, materials in aeronautic, energy, civil engineering and in various constructional applications. Unfortunately, no monolithic engineering material available is able to satisfy them. This need and demand certainly led to the concept of combining different materials in an integral composite structure.

Composites were first used near the 1500s B.C. when the early Egyptians and Mesopotamian settlers used a mixture of mud and straw to create strong and durable buildings. Straw continued to provide reinforcement to ancient composite products including pottery and boats. The Mongols invented the first composite bow in 1200 AD. A combination of wood, bone, and "animal glue" were used in bows and were pressed and wrapped with birch bark. These bows were not only powerful, but also accurate. Composite Mongolian bows provided Genghis Khan with military dominance. Because of the composite technology, this weapon was the most powerful weapon on earth until the invention of gunpowder.

The modern era of composites began after the development of plastics by scientists. Until then, the only source of glues and binders were natural resins derived from plants and animals. In the early 1900s, various plastics like polyester, polystyrene, vinyl and phenolic were developed and became superior to natural resins. But plastics alone could not provide sufficient strength for structural applications. Reinforcement was needed to improve the strength, rigidity and other properties. In 1935, Owens Corning introduced the first glass fiber, fiberglass. Fiberglass was made almost by accident in 1930, when an engineer became intrigued by a fiber that was formed during the process of applying lettering to a glass milk bottle. Fiberglass, when combined with plastic polymers results incredibly strong and lightweight structure. This is the beginning of the fiber-reinforced polymers (FRPs) industries, as we know it today.

Many of the greatest advancements in composite materials were incubated by war. Like the development of the composite bow by Mongols, World War II brought the FRP industry from the laboratory into actual production. For lightweight applications in military aircraft, scientists started using composites. It was observed that fiberglass composites were transparent to radio frequencies and the material was soon preferred for using in sheltering electronic radar equipment (Radomes). Later, boats were made from composites and the first commercial boat hull was introduced in 1946. By 1948 several thousand commercial boats had been made using fiberglass reinforced polyester composites. At that time Brandt Goldsworthy who is often referred to as the "grandfather of composites," developed new manufacturing processes for composites. He was the first person who used fiberglass in a surfboard which revolutionized the sport. Goldsworthy also invented a manufacturing process known as pultrusion and products being manufactured using this process include pipes, ladder rails, tool handles, arrow shafts, armor, train floors, medical devices, etc. In the 1970s the composite industries began to mature due to the development of better plastic resins and improved reinforcing fibers. DuPont developed an aramid fiber known as Kevlar and due to its high tenacity, this fiber has become the standard in armor. Another important fiber named "Carbon fiber" was also developed around that time. This fiber has some special advantages and replacing other fibers for composite manufacture as the new material of choice. The composites industry is now growing with its focus around renewable energy and being used extensively in wind turbines. Wind turbine blades are now made of composites.

In future, development of nanocomposites will bring additional improvement in the properties of composites offering a wider range of applications. Composites are on the path towards being more environmental friendly. Composites will continue to make the world stronger, lighter and more durable.

Composites usually consist of a reinforcing material embedded in various matrices (binder). The effective method to increase the strength and to improve the overall properties of composites is to incorporate dispersed phases into the matrix which can be an either polymer or engineering materials such as ceramics or metals. Hence, metal matrix composites , ceramic matrix composites and polymer matrix composites are obtained. Besides, hybrid composites, metal/ceramic/polymer composites and carbon matrix composites can also be obtained. MMC and CMC composites are developed to withstand high temperature applications. MMCs are also used in heat dissipation/electronic transmission applications due to the conductive nature of metals (electrically and thermally).

In fiber-reinforced composites, fibers embedded in or bonded to a matrix with distinct interfaces (boundaries) between them. The matrix holds the fibers to form the desired shape while the reinforced fibers improve the overall mechanical properties of the matrix. In such composites, fibers are the main source of strength and principal load-carrying members while the matrix acts as a load transfer medium between them and keeps them in the desired location and orientation and protect them from environmental damages. The most common form in which fiber-reinforced composites are used in structural applications is called a laminate. Laminates are made by stacking number of thin layers of fibers and matrix and consolidating them into the desired thickness. Fiber orientation in each layer as well as the stacking sequence of various layers in a composite laminate can be controlled easily which helps to generate a wide range of physical and mechanical properties for the composite laminate.

Natural fibers such as flax, hemp, silk, jute, sisal, kenaf, cotton, etc are being used to reinforce matrices mainly thermoplastics and thermosets by many researchers. The principal fibers in commercial use are various types of glass, carbon or aramid although other fibers, such as boron, silicon carbide and aluminum oxide are used in limited quantities. All these fibers can be incorporated into a matrix either in continuous lengths or in discontinuous (short) lengths. Both these fibers have some advantages and disadvantages. Natural fibers have low abrasion resistance, low density, high toughness, acceptable specific strength properties, good thermal properties, enhanced energy recovery, biodegradability, etc. [2]. Natural fibers produce composites offering advantages like environmental friendliness, renewability of the fibers, good sound abatement capability and improved fuel efficiency resulted from the reduced weight of the components [3]. On the other hand, synthetic fibers reinforced composites attracted much attention due to their high strength, better durability and moisture resistance properties [4].

Today, fiber reinforced composites are in used in a variety of structures, ranging from spacecraft and aircraft to buildings and bridges. FRCs find applications in construction industries, decking, window and door frames [5], sports equipments such as bicycle frames, baseball bats, exercise equipment, etc. They are also suited for many automotive applications [6]. Boron and all carbon/graphite composites are used in military aircraft due to their better performance; very high modulus carbon/graphite composites are used in spacecraft with high positional accuracy requirements for optical sensors; fiberglass composites are used in tanks and piping due to corrosion resistance. This wide use of composites has been facilitated by the introduction of new materials, improvements in manufacturing processes and developments of new analytical and testing methods.

2. REINFORCING AGENT

Reinforcing agents are special classes of fillers that have better 'reinforcement' properties than the matrix and after incorporation into the matrix, influence the matrix properties. Usually composites consist of a reinforcing material embedded in a matrix. In the broadest sense, composites are the result of a combination of substantial volume fractions of high strength, high stiffness reinforcing components with a lower modulus matrix. Generally, the reinforcing component in a composite structure can be divided into two classes:

a) Discontinuous: It can be in the form of dispersions/particles, flakes, whiskers and discontinuous short fibers with different aspect ratios.
b) Continuous: It can be long fibers and sheets.

In practice, the most commonly employed reinforcing component is used in particulate or in fibrous form. On the basis of these two types of reinforcing agents, sometimes composites are divided into two groups: particulate reinforced composites and fiber reinforced composites. In addition to these two types of composites, existence of another group of composite system should be noted which is called "laminar composites (or simply laminates)". In laminar composites, the reinforcing agents are in the form of sheets bonded together and are often impregnated with more than one continuous phase in the system [7].

Particulate (or particle) reinforcing agents have aspect ratios around '1' and possess similar length and breadths. They cover particulates of both regular shapes (such as spheres) and irregular shapes that may have extensive convolution and porosity. Particulate reinforcing agents are the most common and the cheapest reinforcing agent. They are divided into two groups: organic (e.g., powdered cellulose, powdered rubber, a wide variety of starches and particulate carbon [8]) and inorganic (e.g., Silica and some simple metal oxides like alumina, glass etc.).

Fibers are stronger and stiffer materials than any other form of reinforcements which is the main reason for the overwhelming attraction for fibrous reinforcements. Fibers are the important class of reinforcements, as they satisfy the desired conditions with strengthening the matrix constituent influencing and enhancing their properties as desired. Generally fiber means a continuous single material whose length is at least 200 times of its width or diameter and filaments are endless or continuous fibers. Fiber reinforcing agents can be either natural or synthetic. Fibers can be used in as woven, non-woven fabrics, yarn or rovings.

2.1. Fibers

Fiber is a class of material that is continuous filaments or is in discrete elongated pieces, similar to lengths of thread. They can be spun into filaments, rope or string. Fibers are of two types: natural and synthetic. Figure 1 shows general classification of fibers.

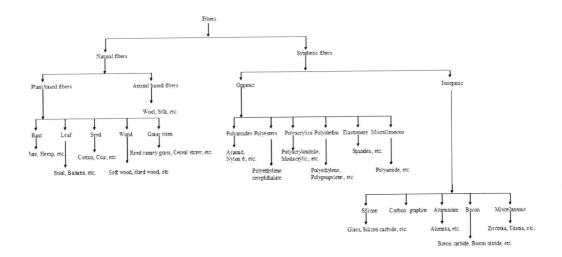

Figure 1. General classification of fibers.

All of the different fibers used in composites have different properties and so affect the properties of the composite in different ways. Table 1 shows properties of different fibers [9]. The mechanical properties of most reinforcing fibers are considerably higher than those of un-reinforced resin systems.

Table 1. Properties of different fibers [9]

Parameter	Unit	Flax	Jute	Hemp	Cotton	Kenaf	Ramie	Sisal	E-Glass
Diameter	m	11-33	200	15–50	-	200	40–80	50–200	5–25
Fineness	dtex	1-7	2–3	2–6	1-4	5–6	5–13	-	-
Length	mm	10-40	1–5	15–28	10-60	2–6	60–260	1–5	–
Maximum stress	cN/tex	30–62	30–34	35–70	25-50	34.5–50	40–70	30-45	-
Tensile strength	MPa	345–1035	393–773	690–1000	330-585	930	400–1050	511–635	1800
Young's modulus	GPa	27.6–45.0	26.5	50.0	4.5-12.6	53.0	61.5	9.4–15.8	69.0–73.0
Density	gcm^{-3}	1.43–1.52	1.44–1.50	1.47–1.50	1.5-1.54	1.5	1.5–1.6	1.16–1.5	2.5
Maximum strain	%	2.7–3.2	1.5–1.8	1.0–1.6	7-8	1.6	3.6–3.8	2.0–2.5	2.5–3.0
Specific tensile strength	km	73.8	52.5	69.3	39.2	63.2	71.4	43.2	73.4
Specific stiffness	km	3.21	1.80	3.47	0.85	3.60	4.18	1.07	2.98

The mechanical properties of the fiber/resin composite are therefore dominated by the contribution of the fiber to the composite. The four main factors that govern the fiber's contribution are:

a) The basic mechanical properties of the fiber itself.
b) The surface interaction of fiber and resin (the 'interface').
c) The amount of fiber in the composite ('fiber volume fraction').
d) The orientation of the fibers in the composite.

2.1.1. Natural Fibers

Natural fiber-reinforced polymers have created interest in the recent years in the automotive industry. Besides, other applications of natural fiber composites include door inner panel, seat back, roof inner panel, and so on. Natural fibers are being used in automotive industry due to the following reasons:

a) Natural fibers are comparatively of very low cost than glass and carbon fibers.
b) They are biodegradable and environment friendly. Unlike glass and carbon fibers, the energy consumption to produce them is very small.
c) The density of natural fibers is in the range of 1.25-1.5 g/cm^3 whereas that of E-glass fibers and carbon fibers is 2.54 g/cm^3 and 1.8–2.1 g/cm^3 respectively.
d) Natural fiber composites provide higher acoustic damping than glass or carbon fiber composites that makes natural fibers more suitable for noise attenuation which is an increasingly important requirement in automotive applications.

e) The modulus–weight ratio of some natural fibers is greater than that of E-glass fibers indicating that they can be very competitive with E-glass fibers in stiffness-critical designs.

However, there are several limitations of natural fibers such as-

a) The strength of natural fibers is relatively low.
b) They absorb moisture.
c) They start to degrade at a comparatively lower temperature than synthetic fibers like glass or carbon fibers.

The two main sources of natural fibers are plants and animals. The main component of animal based fibers is various proteins [10]: examples include mohair, wool, silk, alpaca, angora, etc. The components of plant fibers are cellulose microfibrils dispersed in an amorphous matrix of lignin and hemicellulose [11]: examples include cotton, jute, flax, ramie, sisal, hemp, etc.

2.1.1.1. Plant Based Natural Fibers

Plant fibers are generally comprised mainly of cellulose: examples include cotton, jute, flax, ramie, sisal and hemp. These fibers are abundantly available throughout the world. Cellulose fibers, depending on the part of the plant from which they are taken, can be classified as [12]:

Grasses and reeds: These types of fibers are collected from the stems of monocotyledonous plants such as bamboo and sugar cane.

Leaf fibers: Leaf fibers are fibers that run lengthwise through the leaves of most monocotyledonous plants such as abaca, sisal, henequen, esparto, etc. These fibers are referred to as 'hard fibers'.

Bast fibers: Bast fibers (bundles) are those fibers that come from the inner bark (phloem or bast) of the stems of dicotyledonous plants. Common examples are hemp, jute, flax, kenaf, etc.

Seed and fruit hairs: These are fibers that come from seed-hairs and flosses. Common examples of these fibers are cotton and coconut.

Wood fibers: Wood fibers are those fibers that come from the xylem of angiosperm (hardwood) and gymnosperm (softwood) trees. Examples include yellow poplar, pine, maple, spruce, etc. Wood fiber is the most abundantly used cellulosic fiber due to their extensive use in pulp and paper industries.

2.1.1.1.1. Jute

Jute is the common name given to the fibers extracted from the stems of plants belonging to the genus *Corchorus* and family *Tiliaceae*. Jute fibers can be extracted from the bast of stems of plants. Favorable conditions for cultivation of jute are found in the deltas of the great rivers of the tropics and subtropics where alluvial soils and irrigation are combined with long day lengths to provide an opportunity for considerable vegetative growth before flowering. Optimum growing temperature of jute is 18 to 33°C with a minimum annual moisture requirement of 250 mm in a soil pH between 6.6 and 7.0. Jute is not uniform in the chemical composition as found by the multi-cellular structure. Jute has three principle constituents:

cellulose, hemicellulose and lignin. The approximate composition of raw jute materials is cellulose: 62-65%, hemicelluloses: 15-16% and lignin: 14-20%. The cellulose, hemicelluloses and lignin all exist in the form of long chain molecules. In addition to these major components, it contains minor constituents such as fats and waxes, inorganic (mineral) matters, nitrogenous matter, vitamins, pectinious materials and traces of pigments like– carotene and xanthophylls. The main structural element of jute fiber is cellulose which is held entirely within the cell units where the lignin is and hemicellulose are distributed throughout the entire body of the fiber serving as a cementing material. Cellulose is the key component of the jute fiber. Basically it is the major component of the vegetable kingdom. Cellulose content varies from plant to plant. The term cellulose applies only to the pure plant cell materials consisting of high polymer of the hydro glucose unit as monomer. The degree of polymerization usually varies from several hundred to several thousands. Cellulose is a $1{\rightarrow}4$-linked linear polymer of β-D-glucopyranose as shown in Figure 2.

These monomers are linked together by elimination of one molecule of water between the hydroxyl groups attached to carbon number 1 of one glucose molecule and the number 4 carbon atom of another glucose molecule. Since the condensation reaction to make the cellulose polymer eliminates a molecule of water per glucose, the monomers are often referred to as anhydroglucose units or as glucose residues. Jute fibers have some disadvantages such as high moisture sorption, poor dimensional stability, intrinsic polarity, low thermal resistance, anisotropic fiber resistance, variability, etc.

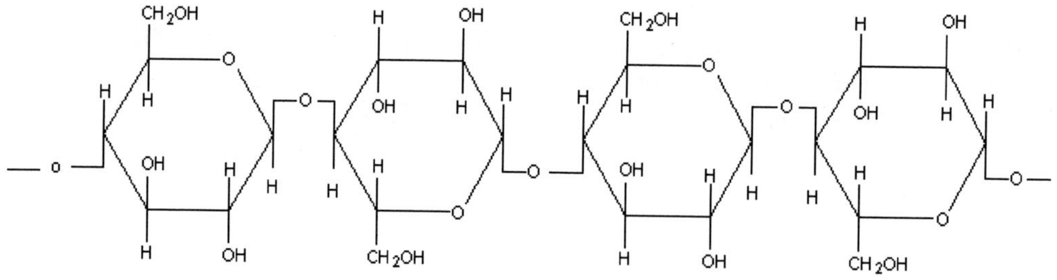

Figure 2. Segment of cellulose chain.

2.1.1.1.2. Cotton

Cotton fibers are seed hairs from plants of the order *Malvales*, family *Malvaceae*, tribe *Gossypieae* and genus *Gossypium*. Each cotton fiber is a single, elongated, complete cell developing in the surface layer of cells of the cottonseed. The mature cotton fiber is actually a dead, hollow, dried cell wall. In the dried out fiber, the tubular structure is collapsed and twisted to give cotton fiber convolutions which differentiate cotton fibers from all other forms of seed hairs and are partially responsible for some of the unique characteristics of this fiber. Raw cotton fiber, after ginning and mechanical cleaning, is approximately 95% cellulose. The structure of cotton cellulose is nothing but a linear polymer of β-D-glucopyranose. Noncellulosic constituents of cotton fibers are proteins, amino acids, other nitrogen-containing compounds, wax, pectic substances, organic acids, sugars, inorganic salts and a very small amount of pigments. These noncellulosic constituents are located principally in the cuticle, in the primary cell wall and in the lumen. Cotton is the only significant textile fiber whose strength increases with humidity while most others are weakened by increased moisture. For most cotton, elongation at break is in the range 6–9%, initial Young's moduli for cotton ranges from about 80 g/den for Sea Island cottons down to 40 g/den for Asian cottons. Tenacity of the Upland cottons ranges from approximately 20 to 30 gf/tex. Cotton being a strong fiber is used in composite fabrication especially with polymeric matrices.

2.1.1.1.3. Flax

Flax is among the oldest crop plants cultivated for the purpose of oil and fiber. It belongs to the genus *Linum* and family *Linaceae*. It is an annual herbaceous plant with shallow root system. Flax has been cultivated for nearly 10,000 years and its fibers, i.e., linen, are reported to be the oldest textile known, dating back to early Mesopotamian times. The flax plant consists of the root, stalk and branches carrying the seed capsules. Flax is dicotyledonous. Only the central portion of the plant (up to 75% of the plant height) can be used to produce fibers. Flax cell wall generally consists of ~70−75% cellulose, 15% hemicelluloses, 10−15% pectic material, about 2% waxes and 2% lignin. Color of the flax fibers varies from light blond to gray. Flax fibers are very strong and flexible. It is a particularly inextensible fiber. Within the small degree of extension, flax is an elastic fiber. Flax fibers lose strength gradually on exposure to sunlight. Flax has good resistance to alkaline solutions and can withstand dilute weak acids. Hot dilute acids or cold concentrated acids attack it. Due to many labor-intensive production steps, flax fibers are quite expensive. Flax is used as reinforcement for high value-added composite products especially for interior automotive components.

2.1.1.1.4. Hemp

Hemp (*Cannabis sativa*) is another famous bast fiber. Hemp plants have impressive growth rates so that they quickly cover the ground and therefore, suppress weeds and some soil-borne pathogens and need almost no or minimal herbicides, pesticides, fungicides and fertilizers. Hemp can be whitish to yellow in color. It has good tensile strength and is water-resistant. Hemp fibers are coarser as compared to flax. The fibers have an excellent moisture resistance. These fibers have high tenacity but low elongation at break. Hemp fibers are used in a wide range of products including specialty paper, textiles, construction materials, plastics and composites, etc.

2.1.1.1.5. Kenaf

Kenaf is the name given to a fiber similar to jute and obtained from the stems of plants belonging to the genus *Hibiscus*, family *Malvaceae*, especially the species *H. cannabinus* L. Kenaf is a pale-colored fiber and this fiber contains less noncellulosic materials than jute. Kenaf fibers are coarse, brittle and they are difficult to process. They have a breaking strength similar to that of low-grade jute and are weakened only slightly when wet. Now, various new applications for kenaf products are emerging, including those for paper products, building materials, absorbents, etc. Kenaf fibers are used as a reinforcing fiber in thermoplastic composites because of its superior toughness and high aspect ratio in comparison with other fibers.

2.1.1.2. Animal Based Natural Fibers

Animal based natural fibers can also be used as alternatives for producing biodegradable and environment friendly composite materials for bio-engineering and bio-medical applications. Animal based fibers are mainly composed of various proteins, like wool, spider and silkworm silk.

2.1.1.2.1. Silk

Silk is a solidified viscous fluid excreted from a special gland by a number of insects and spiders. It is a protein fiber. It is composed of amino acids. Silk protein is a kind of protein like collagen, elastin, keratin, fibroin, sporgin, etc. Insects mainly belong to two families, viz., Saturnidae and Bombycidae which spins silk fibers. There are two main types of silkworms: 'mulberry silks' (*Bombyx mori*) also called 'cultivated silk' and 'wild silk' of which Tussah silk is the most important representative. Silk proteins are natural polymers and are biodegradable with reactive functional groups that open up the possibility to be cross-linked with many other polymers. The mechanical properties of silk fibers consist of a combination of high strength, extensibility and compressibility [10]. The mature silkworm builds its cocoon by extruding a viscous fluid from two large glands in the body of the silkworm. This solution is extruded through two ducts in the head of the silkworm into a common spinneret. The viscous part (fibroin) is covered by another secretion (sericin) which flows from two other symmetrically placed glands. These two components are cemented together by emerging into the air, coagulating and producing a firm continuous filament. As a consequence of this spinning process, the fiber has two main parts called sericin and fibroin. Sericin, called silk gum is a minor component of the fiber and it also has some impurities such as waxes, fats and pigments. Sericin is a yellow, brittle and inelastic substance. It acts as an adhesive for the twin fibroin filaments and conceals the unique luster of fibroin. Sericin dissolves in a hot soap solution. The greatest sericin content is present in the outer layer of a cocoon whereas the least sericin proportion is present in the innermost layer of a cocoon. Fibroin is the principal water insoluble protein. Fibroin has a highly oriented and crystalline structure. Microfibrils are packed together to form the fibril bundle and several fibril bundles produce a single strand. Fibroin has high proportions of alanine, glycine and serine. A small amount of cysteine residue gives a very small amount of sulfur in the fiber. Fibroin contains only a very small amount of amino acids which have acid side chains. The amount of basic side groups in fibroin is relatively low. Because of this, the amount of acids and alkali that can be absorbed by silk are relatively lower than those absorbed by wool.

2.1.1.2.2. Wool

Wool is a natural protein fiber that grows from the follicles of the sheep's skin. Like human hair, wool is also composed of keratin-type protein. Chemically these proteins contain five elements: carbon, hydrogen, oxygen, nitrogen and sulfur. These five elements are combined into 19 amino acids linked together in ladder-like polypeptide chains. Wool fibers consist of two types of cells: cuticle (the outer layer composed of scales) and cortex (the main part of the fiber, composes about 90% of the fiber mass) cells. In coarse fibers, a central medulla also exists. Each cuticle cell consists of three layers with different cystine and isodipeptide contents: epicuticle, exocuticle and endocuticle. On the other hand, the cortex consists of spindle shaped cells. The medulla in coarser wool fibers consists of hollow cells with a skeleton of amorphous proteins and fine filaments. Wool fiber is a hygroscopic fiber which means that it takes up moisture readily. Wool can absorb moisture almost one-third of its own weight. Wool has moderate abrasion resistance, good resilience, medium density, lower electrical conductivity and poor dimensional stability.

2.1.2. Synthetic Fibers

Synthetic fibers are usually more uniform in size, more economical to use and behave in a more predictable manner. For engineering applications the most commonly employed significant fibers are synthetic fibers which includes glass fibers, metallic fibers, organically derived synthetic fibers, etc. Most strong and stiff fibers (e.g., ceramic, fibers of glass, graphite-carbon, boron carbide and silicone carbide) are usually difficult to use as structural materials in bulk. However, embedding such materials in a ductile matrix (such as a polymer or metal) enables them to behave as a stronger, stiffer and tougher material.

2.1.2.1. Carbon Fibers

Carbon fibers are generally manufactured from two types of precursors (starting materials) which are: textile precursors and pitch precursors. The most common textile precursor is polyacrylonitrile (PAN). The molecular structure of PAN contains highly polar CN groups that are randomly arranged on either side of the chain. Pitch is a by-product of petroleum refining or coal coking and comparatively cheaper than PAN. The carbon atoms in pitch are arranged in low molecular weight aromatic ring patterns. Structurally, carbon fibers contain a blend of amorphous carbon and graphitic carbon. Carbon fibers are typically carbonized between 1,200–1,400°C and contain 92%–95% carbon. After carbonization, tensile strength values of 3,000 MPa and moduli of 250 GPa or even higher are usually obtained. Tensile moduli can even be improved up to 350 GPa, at the expense of some drop in strength by the post treatment. Their high tensile modulus results due to graphitic form in which carbon atoms are arranged in a crystallographic structure of parallel planes. In each plane, carbon atoms are arranged at the corners of interconnecting regular hexagons. Although the bond between the planes is much weaker (van der Waals-type), strong covalent bonds exist between the carbon atoms in each plane. This results in highly anisotropic physical and mechanical properties for the carbon fiber. There are different types of carbon fibers depending on the origin of precursor, such as (a) PAN-based, (b) isotropic pitch based, (c) anisotropic pitch-based, (d) rayon-based, or (e) gas phase grown. Depending on mechanical properties, carbon fibers are classified into four categories: (a) PAN-based high modulus (HM)-low strain to failure- type (b) PAN-based high tensile (HT)-high strain to failure-type and (c) PAN-based intermediate modulus (IM) type and (d) mesophase (pitch)-based. Among the advantages of carbon fibers, their exceptionally high tensile strength-weight ratios as well as tensile modulus-weight ratios are of special interest. Carbon fibers also possess very low coefficient of linear thermal expansion, high fatigue strengths and high thermal conductivity. Main disadvantages of carbon fibers are their low strain-to-failure, low impact resistance and high electrical conductivity that may cause "shorting" in unprotected electrical machinery. Their high cost has so far excluded this fiber from widespread commercial applications. They are used mostly in the aerospace industry, where weight saving is considered more critical than cost.

2.1.2.2. Glass Fibers

Glass fibers are the most common of all synthetic fibers for fiber-reinforced composites. Glass fibers are normally divided into three classes: E-glass, S-glass and C-glass. The E-glass is designated for electrical use, the S-glass for high strength and C-glass is for high corrosion resistance. Of the three fibers, the E-glass is the most common reinforcement material used in civil structures. It is produced from lime-alumina-borosilicate which can be easily obtained from abundance of raw materials like sand. Main chemical constituents of E-glass are SiO_2, Al_2O_3, CaO, MgO, B_2O_3 and Na_2O, etc. The fibers are drawn into very fine filaments with varying diameters. Glass fiber's strength and modulus can degrade with increasing temperature. Although the glass material creeps under a sustained load, it can be designed to perform satisfactorily. The fiber itself is regarded as an isotropic material and has a lower thermal expansion coefficient than that of steel. The average tensile strength of freshly drawn glass fibers may exceed 3.45 GPa. The principal advantages of glass fibers are low cost, high strength, high chemical resistance and excellent insulating properties. The disadvantages are relatively low tensile modulus and high density (among the commercial fibers), sensitivity to abrasion during handling which frequently decreases its tensile strength, relatively low fatigue resistance, and high hardness (which causes excessive wear on molding dies and cutting tools).

2.1.2.3. Aramid Fibers

These are synthetic organic fibers consisting of highly crystalline aromatic polyamides. The aramid fibers have excellent fatigue and creep resistance. Although there are several commercial grades of aramid fibers available, the most common one used in structural applications is Kevlar 49. When a solution of Kevlar is extruded into fiber form, a structure with an exceptional degree of alignment of straight polymer chains parallel to fiber axis develops. Weak hydrogen bonds between hydrogen and oxygen atoms in adjacent molecules hold them together in the transverse direction. The resulting filament is highly anisotropic with much better physical and mechanical properties including higher strength and modulus in the longitudinal direction than in the radial direction. Aramid fibers are used as reinforcement in many marine and aerospace applications where high tensile strength, lightweight and resistance to impact damage are important. Aramid fibers have a negative coefficient of thermal expansion in the longitudinal direction which is used in designing low thermal expansion composite panels. The major disadvantages of aramid fiber-reinforced composites are their low compressive strengths and difficulty in cutting or machining.

2.1.2.4. Ceramic Fibers

Ceramic fibers are polycrystalline refractory materials. These fibers are composed of various metal oxides, metal nitrides, metal carbides and their mixtures. Ceramic fibers are available in different fiber lengths and as fabrics. Continuous alumina ceramic fibers offer good compressive strength rather than tensile strength. It's important property is it's high melting point which is about 2,000°C and the composite can be successfully used at temperature up to about 1,000°C. Magnesium and aluminum matrices frequently use alumina fiber as they do not damage the fiber even in the liquid state. Alumina silicate ceramic fibers with significant quantities of boria as continuous filament yarns are produced in braided and woven forms. These fibers are the most advanced high performance fibers available commercially and are known as Sumitomo and Nextel ceramic fibers. A considerable amount of zirconium oxide exists in Nextel ceramic fiber and is called zirconiasilica ceramic fiber. Other ceramic fibers include silicone nitride, silicone carbidethoria, aluminium nitride, boron nitride, potassium titanate, high silica, quartz and a recently introduced polymeric material, polycarbosilane. Silicone carbide fibers are low cost fibers and are used mostly for ceramic matrix composites and metal matrix composites.

2.1.2.5. Metallic Fibers

Metallic fibers can be used in the form of whiskers, metal wool and filament length fibers. The term whisker applies to single crystals with fibrous or fibrillar characteristics having higher tensile strength and moduli as compared with polycrystalline continuous or discontinuous counterparts. As reinforcement, metal fibers have many advantages. They are easily produced using several fabrication processes and are more ductile, apart from being not too sensitive to surface damage. They also possess high strengths and temperature resistance. However, their weight and the tendency to react each other through alloying mechanisms are major disadvantages. Better flexural properties are observed in some metal fibers reinforced plastic composites that also offer improved strength and weight, than glass fibers. Metals in wire or filament form exhibit considerable high strength. Metal wires, of the continuous version, also reinforce plastics like polyethylene and epoxy. Such combinations ensure high strength, lightweight and good fatigue resistance. Besides, continuous metal fibers are easily handled, unlike glass fibers. The main disadvantage of wires is their high densities. Tungsten wire is used in some nickel and cobalt-based super alloys, while steel wire is commonly used for concrete and tyres.

2.1.2.6. Boron Fibers

Boron fiber is a synthetic fiber with extremely high tensile modulus which is in the range of 379–414 GPa. Boron fibers are manufactured by chemical vapor deposition (CVD) of boron onto a heated substrate (either a tungsten wire or a carbon monofilament). By the reaction of boron chloride with hydrogen, vapor of boron is produced:

$$2BCl_3 + 3H_2 = 2B + 6HCl$$

The most common substrate used in the production of boron fibers is tungsten wire. It is continuously pulled through a reaction chamber in which boron is deposited on its surface at 1,100°C–1,300°C. During boron deposition, the tungsten substrate is converted into tungsten boride by diffusion and reaction of boron with tungsten. The major disadvantage of boron fiber is its high cost which is even higher than that of many forms of carbon fibers. For this reason, its use is at present restricted to a few aerospace applications. Boron fiber is primarily used in space and aerospace applications.

3. MATRIX

As stated earlier, the matrix holds the reinforcement in a composite. Matrix supports the fibers and keeps them together in the composite material. If load is applied to composite, matrix transfers that to the fibers, thereby keeping the fibers in their position and chosen orientation.

The matrix also gives the composite environmental resistance. Some important functions of the matrix in FRCs are: (1) to hold the fibers in place in the structure, (2) to protect the fibers from environment, (3) to help to distribute or transfer loads, (4) to enhance transverse properties of a laminate, (5) to improve impact and fracture resistance, (6) to avoid propagation of crack growth through the fibers by providing alternate failure path along the interface between the fibers and the matrix, (7) to carry interlaminar shear, etc.

Matrix must have some desired properties which are important for a composite structure. Some of these properties are:

a) Matrix must wet the fibers with bonding.
b) Minimization of moisture absorption.
c) Easily processable into the desired composite shape.
d) Must flow to penetrate the fiber bundles completely and eliminate voids.
e) Must have reasonable strength, modulus and elongation (elongation should be greater than fiber).
f) Have low shrinkage and coefficient of thermal expansion.
g) Must be elastic to transfer load to fibers.
h) Have dimensional stability to maintain its shape.

There are many matrix choices available and different types have different impact on the processing techniques, physical and mechanical properties and environmental resistance of the finished composites.

In selecting matrix material, some factors may be taken into consideration like: (1) the matrix must be easy to use in the selected fabrication process, (2) the resultant composite should be cost effective, (3) the matrix is be able to withstand service conditions, viz., temperature, humidity, exposure to ultra-violet environment, exposure to chemical atmosphere, abrasion by dust particles, etc.

Matrix can be classified in different ways. Generally it can be divided into three classes: (1) ceramic, (2) metal and (3) organic.

4. DIFFERENT TYPES OF FRCS

Fiber-reinforced composite materials can be classified based on matrix and fibers. Based on fiber source, FRCs may be natural fiber-reinforced and synthetic fiber reinforced. Based on fiber length, they can be continuous fiber-reinforced and discontinuous fiber-reinforced. But FRCs are generally classified based on matrix component. Thus according to the types of matrices stated earlier, composites are of three types: (1) Ceramic Matrix Composites (CMCs), (2) metal Matrix Composites (MMCs) and (3) Organic Matrix Composites (OMCs). Organic matrix is subdivided into two classes, namely polymer matrix and carbon matrix. All these types of composites are discussed in the next section.

4.1. Metal Matrix Composites (MMCs)

A metal matrix composite is a material that consists of a metallic base with a reinforcing constituent which is usually non-metallic and is commonly a ceramic. Metal matrix composites are a new range of advanced materials used in applications where conventional materials and alloys are not suitable for use. Metal matrix has the advantage over polymeric matrix in applications requiring a long-term resistance to severe environments, for example high temperature [13]. The properties of MMCs depend on the nature of the two components. For example, the reinforcement of light metals opens up the possibility of application of these materials in areas where weight reduction is given the first priority. The precondition here is the improvement of the component properties. Metals have some excellent characteristics for which they are chosen as the matrix material in MMC structures. The advantages of MMCs are:

a) Fire resistance.
b) No outgassing.
c) No moisture absorption.
d) Higher temperature capability.
e) Better radiation resistance.
f) Higher transverse stiffness and strength.
g) Higher electrical and thermal conductivities.
h) They can be fabricated with conventional metal working equipment.

However, MMCs have the following disadvantages:

a) MMCs usually have high fabrication costs.
b) Most metals are heavy. So, MMCs comparatively possess heavy weight.

Metal matrix composites can be classified in various ways. One classification is based on the type of reinforcement. According to reinforcement, MMCs are of two different types:

a) Continuous reinforcement composites in which reinforcements are continuous fibers or filaments (continuous fibers in a composite are usually called filaments).
b) Discontinuous reinforcement composites in which reinforcements are short fibers, whiskers or particles.

The two most commonly used metal matrices are based on aluminum and titanium. These two metals have comparatively low densities and are available in a variety of alloy forms. Magnesium is lighter but it is less suitable for many applications due to its great affinity towards oxygen which promotes atmospheric corrosion. Although Beryllium is the lightest of all structural metals and has a tensile modulus higher than that of steel, it suffers from extreme brittleness which is the reason for its exclusion as a potential matrix material. Aluminum and its alloys have been used widely as matrix material in metal matrix composites. Aluminum alloys, such as 201, 6061, and 1100, have been used for their higher tensile strength–weight ratios. When carbon fiber is used with aluminum alloys, carbon reacts with aluminum to form aluminum carbide (Al_4C_3) at typical fabrication temperatures of

500°C or higher which severely degrades the mechanical properties of the composite. A more common reinforcement for aluminum alloys is SiC.

Graphite/aluminum and graphite/magnesium systems were developed more recently which show significant potential in structural applications. The low coefficient of thermal expansion of these two material systems makes them particularly attractive in space structures. Other MMC systems of interest include alumina/magnesium, beryllium/titanium, boron/stainless steel/aluminum, silicon carbide/aluminum, boron/titanium/aluminum, borsic/aluminum, borsic/titanium, copper/graphite, graphite/lead, graphite/aluminum, tungsten/nickel, etc.

4.1.1. Manufacturing Process

Metal matrix composite materials can be produced by many different techniques.

The focus of the selection of suitable process depends on type, quantity and distribution of the reinforcement components (particles and fibers), the matrix alloy and the application. By altering the manufacturing method, the processing and the finishing, as well as by the form of the reinforcement components it is possible to obtain different characteristic profiles, although the same composition and amounts of the components are involved. MMCs are produced by means of processes other than conventional metal alloying. MMC production processes are described below.

4.1.1.1. Continuously Reinforced MMCs

The most common method of manufacturing continuous fiber-reinforced MMCs is vacuum hot pressing (VHP). The starting material (precursor) can be made in different ways. Sheets are made by cutting preformed tapes and placed into a vacuum bag and pressed by placing between the plates of a hydraulic press followed by cold press. Consolidation of layers in VHP takes place through diffusion bonding, plastic deformation and creep of the metal matrix [14] and affected by the basic process parameters: temperature, pressure and holding time.

4.1.1.2. Discontinuously Reinforced MMCs

4.1.1.2.1. Powder Metallurgy

Powder metallurgy [15] involves blending of atomized metal powders with whiskers or particulates. The blended mixture is cold compacted in a graphite die, outgassed and vacuum hot-pressed to form a cylindrical billet. Most billets are extruded to rods or rectangular bars using suitable dies. By hot extrusion of these rods and bars, various structural shapes can be fabricated and by hot rolling on conventional rolling mills, sheets and plates can be produced.

4.1.1.2.2. Casting or Liquid Metal Infiltration

The conventional casting methods can be adopted for producing MMCs which will reduce production costs. In this process, the liquid metal is poured over a fiber preform and forced to infiltrate the preform either by gravity or by the application of moderate suitable pressure. But for liquid metal infiltration, flowing liquid must have a low viscosity and the liquid metal must wet the fiber surface. Liquid metal after infiltration is allowed to cool slowly in the mold.

4.1.1.2.3. Compocasting

Compocasting is a modified form of rheocasting to produce near net-shape MMC parts [16]. The process of casting liquid metal slurry of fine spheroidal solids floating in the liquid is called rheocasting. If particulates, whiskers or short fibers are mixed with the slurry before casting, the process is called compocasting. This process allows uniform distribution of reinforcement in the matrix. In this process, slurry viscosity affects the bonding. Good bonding results when the slurry viscosity is reduced which can be obtained by increasing the shear rate as well as increasing the slurry temperature.

4.1.1.2.4. Squeeze Casting

Squeeze casting involves solidification of liquid metal under pressure [17]. At first preheated fiber perform is placed in an open die cavity which is mounted in the bottom platen of a hydraulic press. Then, measured quantity of liquid metal is poured over the preform, the die cavity is closed and pressure is applied to force the liquid metal into the preform. In this process, high pressure is maintained throughout solidification. After solidification, the pressure is released. It is a net-shape metal casting process and differs from the more familiar process of pressure die-casting. Squeeze-cast metals generally have higher tensile strengths as well as greater strains-to-failure than the gravity-cast or die-cast metals. Variables that may influence the quality of squeeze-cast MMCs are: infiltration speed, the temperatures of the molten metal and the fiber preform, and the final squeeze pressure.

4.1.2. Use of MMCs

MMC systems are being used for superconducting electrical applications. These include hydrofoils, missiles, Navy weapon systems, assault bridges and battle tanks, and spacecraft components such as deployable antennas and booms.

4.2. Ceramic Matrix Composites (CMCs)

Ceramics are solid materials which generally exhibit very strong ionic bonding and in few cases covalent bonding. Ceramics are known for their high temperature stability, high modulus, high hardness, high thermal shock resistance, high corrosion resistance and low density. The main characteristics of ceramics for which they are chosen as the matrix material are:

a) They have low densities.
b) Elastic modulus values of ceramics are very high.
c) They can withstand very high temperature (>2000°C); hence they can be used in advanced heat engines.

On the other hand, the major disadvantages of ceramic matrix materials are their brittleness and low resistance to crack propagation which makes them easily susceptible to flaws. That's why, existence of even minor surface flaws, scratches or internal defects (pores, microcracks) can result in disaster. The primary reason for reinforcing a ceramic matrix is to increase its fracture toughness. Moreover they usually lack uniformity in properties. They also have low thermal and mechanical shock resistances, as well as low tensile strengths. CMCs are non-brittle refractory materials designed for applications in severe environments (often combining high temperatures, high stress levels and corrosive atmospheres). Naturally, ceramic matrices are preferred for high temperature applications.

The selection of matrix materials for ceramic composites is strongly influenced by thermal stability and processing considerations. The properties of matrix materials commonly used in ceramic composites are shown in Table 2 [18].

Table 2. The properties of different ceramic matrix materials [18]

Materials	Young's modulus (GPa)	Poisson's ratio	Modulus of rupture (MPa)	Fracture toughness (MPa m$^{1/2}$)	Density (gcm^{-3})	Thermal expansion (10^{-6}/°C)	Melting point (°C)
TiO$_2$	283	0.28	83	2.53	4.25	9.36	1849
LAS	117	0.24	138	2.42	2.61	5.76	-
Al$_2$O$_3$	345	0.26	483	3.52	3.97	8.64	2050
ZrO$_2$ PS	207	0.23	648	8.46	5.75	7.92	2760
ZrO$_2$ FS	207	0.23	248	2.75	5.56	13.5	-
Pyrex	48	0.2	55	0.08	2.23	3.24	1252
Mullite	145	0.25	186	2.20	3.30	5.76	1850
SiO$_2$	76	0.16	-	0.77	2.20	0.54	1610
SiC Sn	331	0.19	386	4.94	3.21	4.32	1980
SiC HP	414	0.19	462	4.94	3.21	4.32	1980
B$_4$C	290	-	310	-	2.41	3.06	2350
BeO	359	0.24	234	-	3.00	5.76	2350
WC	669	0.20	-	-	15.80	4.50	2870
TiB$_2$	552	0.2	896	6.92	4.62	8.10	2900
TiC	427	0.19	248	-	4.92	8.46	3140
TaC	283	0.24	200	-	14.50	6.66	3880
Si$_3$N$_4$ SN	310	0.24	496	5.60	3.18	3.06	1870
Si$_3$N$_4$ RB	165	0.24	303	3.41	-	-	-
Si$_3$N$_4$ HP	310	0.24	827	5.60	3.19	3.06	1870

Ceramic matrix materials can be classified in different ways. Structural ceramics used as matrix materials can be categorized into two main groups:

a) Oxides: Common oxide ceramics are Alumina (Al$_2$O$_3$) and mullite (Al$_2$O$_3$–SiO$_2$).
b) Nonoxides: Common nonoxide ceramics are silicon carbide (SiC), silicon nitride (Si$_3$N$_4$), boron carbide (B$_4$C) and aluminum nitride (AlN).

According to another classification, ceramic matrix materials can be categorized in the following four main groups:

a) Glass ceramics (e.g., lithium aluminosilicate),
b) Oxides (e.g., alumina and mullite),
c) Nitrides (e.g., silicone nitride), and
d) Carbides (e.g., silicone carbide).

Silicon carbide matrix can withstand high temperature and found wider applications, particularly where high modulus is desired. Silicone nitride matrices are specifically used for the production of ceramic matrix composite systems where tough, strong, oxidation resistant and very high temperature or high heat flux resistant materials are needed, e.g., for advanced heat engines. For these kinds of applications, high temperature resistant fibers are being employed (30% aligned SiC, i.e., 'SiCS6' fibers) in ceramic matrices. AlN is of interest because of its high thermal conductivity and Si_3N_4 is considered for applications requiring high strength.

The reinforcements used in ceramic matrix composites are those fibers that can withstand relatively high temperatures required for the production of ceramics, without significant damage. Other desired requirements are long-term high-temperature stability, creep resistance and oxidation stability. The importance of each of these demands depends on the type of application. Organic, polymeric fiber materials cannot be used in CMCs because of their degradation at temperatures below 500°C. Conventional glass fibers having melting or softening points below 700°C cannot be used for this purpose. Possible candidates for the reinforcement of ceramic materials are polycrystalline or amorphous inorganic fibers or carbon fibers. Commonly used fibers for CMCs are SiC, Si_3N_4, AlN and other ceramic fibers. Of these, SiC has been the most commonly used reinforcement because of its thermal stability and compatibility with a broad range of both oxide and nonoxide ceramic matrices. Carbon fibers can also be used under certain conditions in CMCs. Although carbon fibers degrade in an oxidizing atmosphere above 450°C, they are stable under non-oxidizing conditions up to temperatures of 2,800°C.

The forms in which the reinforcement is used in ceramic matrix composites include whiskers (with length to diameter ratio as high as 500), particulates and both monofilament and multifilament continuous fibers.

4.2.1. Manufacturing Process

Ceramic matrix composites are manufactured by the following two different processes:

4.2.1.1. Powder Consolidation Process

Powder consolidation process is a two-step process. In first step a "green" compact is made and in the second step the compact is hot pressed into the final shape. Temperature of hot pressing ranges from 1,200°C to 1,600°C which consolidate the compact into a dense material with as little porosity as possible [19].

Green compact can be made in various ways for example by infiltrating a fiber preform with either matrix slurry or a colloidal sol prepared from alkoxides and other organic precursors. Consolidation of green compact is possible either by vacuum hot pressing (VHP) in a press or by hot isotactic pressing (HIP) inside an autoclave which will transform the green compact into a dense composite. According to Rice and Lewis, the strength and toughness of hot-pressed ceramic matrix composites depend strongly on the processing conditions which include time, temperature and pressure. To reduce porosity in the composite, the hot-pressing temperature must be higher than that of hot pressing of the matrix alone. Hot pressing temperature being high may cause fiber degradation and adverse fiber–matrix reactions that can be reduced with a fiber coating as well as by using an inert atmosphere (such as argon) during hot pressing.

HIP process is advantageous over VHP process and has the following advantages:

1. The processing temperature is usually lower in HIP due to application of pressure in all directions instead of uniaxially as in VHP.
2. HIP is capable of producing more intricate net-shape structures with higher density and greater uniformity whereas VHP can produce limited simple shapes, such as rods, plates or blocks.

4.2.1.1.2. Chemical Processes

Of the chemical processes, commonly used processes are chemical vapor infiltration (CVI) and polymer pyrolysis [19-20]. Polymer pyrolysis is a two-step process. The first step involves making a polymer-impregnated preform via standard polymer impregnation techniques. In the second step, pyrolysis of the polymer at high temperatures yields the ceramic matrix. But this process has some disadvantages and the principal disadvantage is that it results a highly porous matrix which is due to shrinkage cracks originating from the polymerization process during pyrolysis. One method of reducing the polymer shrinkage is to mix it with fine ceramic fillers. In CVI, the infiltration of the reinforcing phase is carried out by a gas that decomposes to form the solid matrix phase. The temperature range for CVI is 1,000°C–1,200°C. For example, SiC matrix can be deposited on SiC fibers by passing methyltrichlorosilane over the SiC fiber preform in the presence of hydrogen at 1,200°C. The overall chemistry of this process follows the simple scheme shown below:

$$CH_3SiCl_3 + H_2 \rightarrow SiC + 3\ HCl + H_2$$

CVI process has some advantages such as-

a)   This process can be used to get a variety of shapes.
b)   The matrix distribution in the composite is uniform.
c)   In this process, fibers undergo less mechanical damage.

But the control of porosity can be a problem of this process.

4.2.3. Use of CMCs

Light weight, high performance brake systems based on carbon/carbon (C/C) materials have some disadvantages such as low friction coefficient at temperatures below 450°C and high wear of brake pads and discs at elevated operating temperatures. This has been improved by introducing silicon carbide (SiC) into the carbon fiber reinforced carbon base material. The process yields the much improved SiC Matrix in the disc structure. Friction properties of the resulting ceramic matrix composite   outperform the carbon-carbon composite material in many ways. A growing range of brake disc, rotor and clutch products for cars, trucks and motorcycles are commercially available. Vehicle components are based on a Silicon Carbide-Carbon based tough fiber-reinforced ceramic materials especially developed for friction applications, such as: brake discs, pads, clutch facing/plates, etc. In friction applications, components are preferred to have advantages like: (1) high and stable friction coefficient, (2) low wear and consistent friction properties, (3) excellent smooth performance in wet conditions and (4) light weight with long life. CMCs have these advantages. In mechanical and industrial applications, CMCs are preferred due to excellent heat/oxidation resistance, good thermal conductivity, good fire resistance and shock resistance, stability of mechanical properties at room and high temperature, excellent corrosion and wear resistance in harsh environments, etc.

4.3. Carbon Matrix Composites

Carbon matrix composites are composed of a carbon matrix into which reinforcing carbon fibers are embedded. They are commonly referred to as carbon–carbon (C–C) composites. Fibers in the C–C composites can be either continuous or discontinuous. Continuous fibers are selected for structural applications. Some mechanical and thermal properties of continuous carbon fiber-reinforced C–C composites containing carbon fibers having strengths up to 2.5 GPa and elastic modulus in the range of 350–450 Gpa are shown in Table 3 [19].

Carbon fiber reinforced carbon is a high strength composite material and it is also resistant to high temperature in a non-oxidizing atmosphere. Such composites are thermally stable up to 3,000°C in a nonoxidative environment. But they oxidize and degrade in presence of oxygen, even at 400°C–500°C. That's why they need to be protected by a surface coating or must be chemically modified to prevent oxidation. Such a material was first used under extreme thermal and mechanical loads in space technology.

Table 3. Properties of Carbon–Carbon Composites [19]

Material	Fiber volume fraction (V_f, %)	Tensile Strength at 23°C (MPa)	Shear Strength at 23°C (MPa)	Compressive Strength at 23°C (MPa)	Thermal Conductivity[a] (W/m°C)
C–C with 1D Unidirectional continuous	65	650–1000 (x) 2 (z)	7-14 (xy)	620 (x)	125 (x) 10 (z)
C–C with 2D Fabric	31 (x) 30 (y)	300–350 (x) 2.8–5 (z)	7–14 (xy)	150 (x)	95(x) 4 (z)
C–C with 3D Woven orthogonal fibers	13 (x) 13 (y) 21 (z)	170 (x) 300 (z)	21–27 (xy)	140 (z)	57(x) 80 (z)

[a] At 800°C.

4.3.1. Manufacturing Process

The starting material for C-C composites is a carbon fiber preform which may contain unidirectional fibers, bi- or multidirectional fabrics, or a three dimensional structure of carbon fibers. C–C composites are produced by two basic fabrication methods: (1) liquid infiltration and (2) chemical vapor deposition (CVD).

In the liquid infiltration process, the preform is infiltrated with a liquid [19]. The used liquid on heating, carbonizes to yield at least 50wt% carbon. Two types of liquids are used in the liquid infiltration method: (1) pitch and (2) a thermoset resin. Pitch is made from coaltar or petroleum. Thermoset epoxy, polyester, etc resins are available in market.

In the CVD process, a hydrocarbon gas is infiltrated into the preform under the application of high temperature and pressure [19]. The chemical breakdown of the hydrocarbon gas produces the carbon matrix. Hydrocarbon gas used is methane, propane, benzene, etc. Thermal decomposition of gas results formation of a layer of pyrolitic carbon which deposits on the hot preform surface.

Advantages of C-C composites: C-C composites have the following advantages:

a) Low specific weight.
b) Resistance to thermal shock.
c) Resistance to high temperatures.
d) High heat absorption capacity.
e) Exceptional frictional properties at high energy levels.
f) High resistance to damage.
g) Chemical inertness.

Disadvantages: The main disadvantage of carbon-carbon composite is the lack of resistance to oxidation at temperatures in excess of 500°C if C-C composites are not coated or modified chemically. The economic problem is its long manufacturing time and high production cost. Although tensile and compressive properties of C–C composites are generally very high, the shear strength and modulus are very low. This is one of the limitations of C–C composites. Another point to be noted is that the failure strain of the matrix in C–C composites is lower than that of the fibers.

4.3.2. Use of C-C Composites

The main application of C-C composite is as braking material. It is used as aircraft brakes, brake system for high-speed trains and racing cars. Its application as braking material is due to high-energy absorption capacity and low specific weight. Some other examples of its use include heavy duty clutches, tools for high temperature production of alloys like titanium, etc. Light weight, high performance brake systems based on carbon-carbon (C-C) materials for brake discs and pads have been known in the aviation industry and car or motorcycle racing field for quite some time. It is also known that this system has some disadvantages such as low friction coefficient at temperatures below 450°C and high wear of brake pads and discs at elevated operating temperature which can be improved by introducing silicon carbide (SiC) into the carbon fiber reinforced carbon base material.

4.4. Polymer Matrix Composites (PMCs)

Polymer matrix composites are very popular due to their low cost, simple fabrication methods, lightweight and desirable mechanical properties. But the main drawbacks of PMCs are their low thermal resistance and high coefficient of thermal expansion. PMCs are the most developed composite materials group and they have found widespread applications. Polymers are long chain molecules containing one or more repeating units of atoms joined together by covalent bonds. Strong fibers reinforced PMCs are characterized by some properties such as high strength, high stiffness, high fracture toughness, good corrosion resistance, abrasion resistance, etc. In fiber reinforced PMCs, fibers may be arranged in different forms such as unidirectional, rovings, woven, short, etc. Properties of fiber-reinforced PMCs depend on: (1) properties of the fibers, (2) orientation of the fibers, (3) concentration of fibers and (4) properties of the polymer matrix.

Two different types of polymers are used as matrix materials for composite fabrication: (1) thermoset polymer and (2) thermoplastic polymer.

4.4.1. Thermoset Polymer

Thermoset polymers are generally stronger than thermoplastic materials. Molecules of thermoset polymers are chemically joined together by cross-links, forming a rigid, three-dimensional network structure. Cross-links are formed during the polymerization reaction which is also called the curing reaction. After cross-linking, melting of thermoset polymers is not possible by the application of heat. There are many advantages of thermosets. Some of them are:

a) Production is simpler, well-established processing history. Resin molders can be used allowing the matrix and the reinforcing agent to be cured under mild processing conditions.
b) Wetting and adhesion to reinforcements are very good.
c) Thermosets have a highly cross-linked chain structure which does not allow the chains to slide and rotate easily. This chain structure gives thermosets high strength and good stiffness properties.
d) They have better high temperature (H/T) properties.
e) Better general creep behavior, the links between the chains restricting the relative displacements of the macromolecules.
f) Overall, thermosets have better economics than thermoplastic polymers.

But thermosets also have disadvantages. Some disadvantages of the thermosets are as follow:

a) The chemical reaction of cross-linking takes a considerable time which lengthens the production cycles.
b) Material cost is high.
c) They are brittle and possibility of cracks is high.
d) Repairing of thermosets is very difficult.
e) They have poor recycling capabilities. The wastes are not reusable as virgin matter because of the irreversibility of the hardening reaction.

The most common thermoset polymer matrix materials are polyesters (unsaturated), epoxies, bismaleimids (BMI), vinyl ester, polyimide, polyurethane (PUR), silicone, etc. Brief description on two different types of thermoset (polyester and epoxy) matrices is given below:

4.4.1.1. Polyester (Unsaturated)

Polyester resins are unsaturated resins and they are produced by the reaction of dibasic organic acids and polyhydric alcohols. Typical polyhydric alcohols used are glycols like ethylene glycol and typical acids used are phthalic acid, maleic acid, etc. The chemical reactions forming unsaturated polyester and cross-linked polyester resin are shown in Figure 3a and 3b respectively.

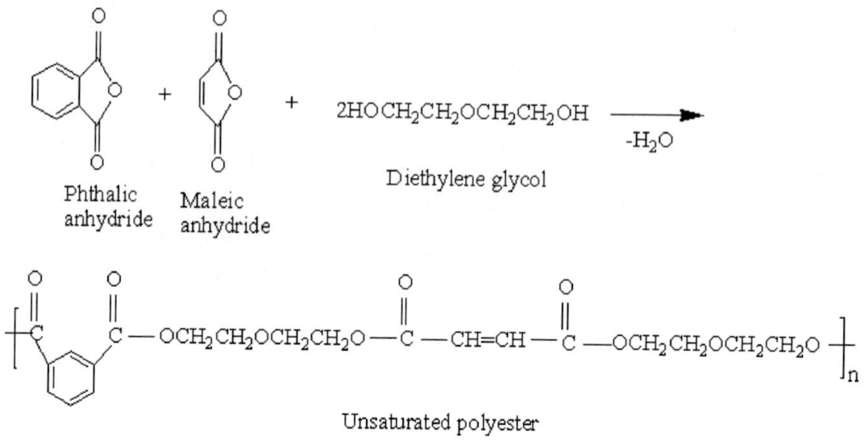

(a)

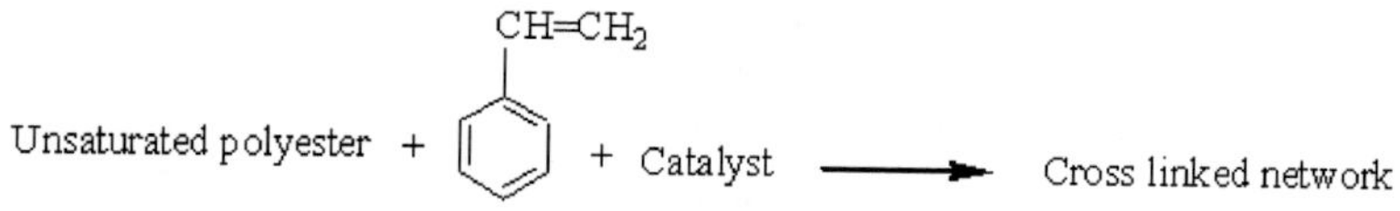

(b)

Figure 3. The chemical reactions (a) forming unsaturated polyester and (b) cross-linked polyester resin.

The polyester is formed by the condensation reaction (3a) and the ester thus formed contains an alcohol group and an acid group which can then react further, forming a polymer. To make the polymer stronger and more resistant to heat, it is then 'cured' by reacting with a monomer such as styrene (3b) in the presence of a peroxide catalyst which results cross-linking in the polyester chains to form a harder, more durable, thermosetting polymer.

The curing reaction for polyester resins is initiated by adding small quantities of a catalyst. In order to be used in the fabrication of composites, the liquid resin must be mixed with a catalyst such as organic peroxide or an aliphatic azo compound to initiate curing reaction. Catalyst concentrations generally range from 1 to 2 percent by original weight of resin. Within certain limits, the higher the catalyst concentration, the faster the cross-linking reaction. With the application of heat (in the temperature range of 107°C–163°C), the decomposition of catalyst results free radicals which react (mostly) with the styrene molecules and break their C=C bonds. At the unsaturation points of polyester molecules, styrene radicals join and form cross-links between them. The resulting material is a solid polyester resin.

Polyesters are extensively used with glass fibers. They are inexpensive, are somewhat resistant to environmental exposure and are lightweight. They are the most widely employed class of thermosetting resins used for automotive. Their poor impact and hot/wet mechanical properties, limited shelf life and high curing shrinkages avoid their high performance applications.

4.4.1.2. Epoxy

Low molecular weight liquid organic resins containing a number of epoxide groups are the starting materials for epoxy matrix; for example diglycidyl ether of bisphenol A (DGEBA) is a common starting material. Molecules that contain the epoxide group (Figure 4a) as part of their structure are called epoxides. Structure of DGEBA is shown in Figure 4b.

To improve the impact strength of the cured epoxy matrix, diluents are often mixed with the starting liquid. Polymerization (curing) reaction transforms the liquid resin to solid state. To initiate the polymerization, small amounts of a reactive curing agent such as diethylene triamine (DETA) is added to the liquid mix just before incorporating the fibers. As the reaction continues, three-dimensional network structure is formed by cross linkages.

Due to inherently polar nature, they show excellent adhesion to a wide variety of fibers for which epoxy matrix is favored for fiber reinforced composites. They show good strength, dimensional stability, excellent mechanical properties and a better moisture resistance. Their maximum application temperature is slightly higher (175°C). A large number of different types and different formulations are available for epoxies. Lots of the high performances PMCs have epoxies as matrices. But they are more expensive than polyesters and have lower shrinkage on curing.

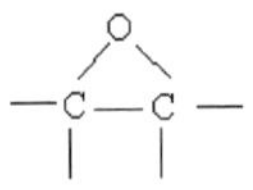

(a)

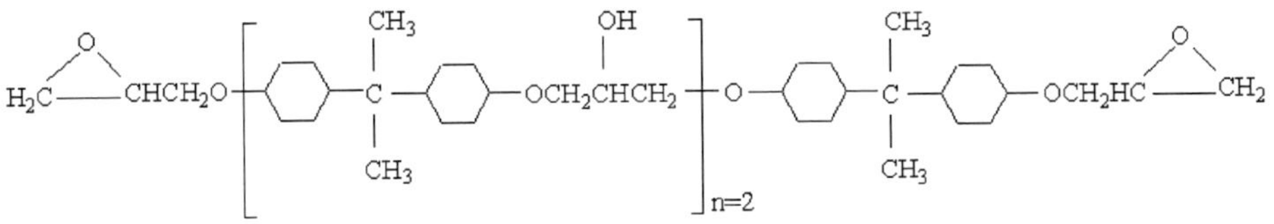

(b)

Figure 4. Structure of (a) a generic epoxide (b) a molecule of diglycidyl ether of bisphenol A (DGEBA) epoxy resin.

4.4.2. Thermoplastic Polymer

In a thermoplastic polymer, chains are associated through weak secondary bonds or intermolecular forces, such as van der Waals bonds and hydrogen bonds. When heat is applied to solid thermoplastic polymers, these secondary bonds can be temporarily broken and the relative movement of molecules to each other or flow to a new configuration is possible if pressure is applied on them. Upon cooling, these molecules can be frozen in their new configuration with restoring of the secondary bonds resulting in a new solid shape. Thus, a thermoplastic polymer becomes soft on heating and can be reshaped (or post formed) as many times as desired. Thermoplastics have higher coefficients of thermal expansion (CTE) compared to thermosets. There are many advantages of thermoplastics. Some of them are:

a) They have low production cycle time. The processing cycles are very short because of the absence of the chemical reaction (e.g. cross-linking).
b) Cost of processing is lower than that of thermosets.
c) They have high toughness and impact resistance.
d) They are flexible and exhibit good mechanical properties.
e) Design flexibility and ease of molding complex parts. Conventional plastics processing techniques, such as injection molding, extrusion and blow molding, etc are used for processing of thermoplastics. So, manufacturing of PMCs is very easy for thermoplastic polymer matrices.
f) They have high reparability.
g) Recycling of composites is possible with thermoplastics.
h) They have good chemical resistance.

But thermoplastics also have disadvantages. Some disadvantages of the thermoplastics are as follow:

a) Glass transition temperature (Tg) of thermoplastic polymers is not too high and they soften upon heating at the characteristic glass transition temperature. So, the application temperatures are limited for PMCs.
b) The fiber orientation in the composites is random and the property modification is not as high as is observed in the thermoset composites.
c) With the increase in temperature, modulus retention decreases. This is due to the absence of chemical links between the macromolecules.
d) For the same reason, the creep and relaxation behaviors are not as good as that in thermosets.
e) It requires high processing temperatures and pressures.

The most commonly used thermoplastic matrix materials are vinylic polymers (polyvinyl chloride (PVC)), polyolefinics (polyethylene, polypropylene), polyacetals, polyamides (PA), polyphenylenes (polyphenylene sulphide (PPS)), polyetheretherketone (PEEK), polysulphone, etc.

4.4.2.1. Polylactic Acid (PLA)

PLA is a rigid biodegradable thermoplastic polymer which is extensively used as a matrix material alone or sometimes with plasticizers like polyethylene glycol (PEG). PLAs can be semicrystalline or totally amorphous, depending on the stereopurity of the polymer backbone. PLA is a unique polymer that in many ways behaves like PET and polyolefin like polypropylene (PP). It is a polymer with the broadest range of applications because of its ability to be stress crystallized, thermally crystallized, impact modified, filled, copolymerized, and processed in most polymer processing equipment. It can be formed into transparent fibers, films or injection molded into blow-moldable preforms for bottles, like PET.

Most of the available biodegradable polymers are expensive including PLA and if used alone will result expensive final product. To reduce cost and improve some properties, use of plasticizer would be a wiser choice. Plasticizers are widely used in plastic industry to reduce brittleness, increases toughness, impart flexibility and tear resistance and impact resistance of the polymer. Plasticizers are also capable of reducing glass transition temperature of glassy polymers. Plasticizers used for PLA are poly (1,3-butylene glycol adipate), poly (1,3-butanediol), triacetin, acetyl trin-butyl citrate, glycerol, citrate ester, PEG, etc.

PLAs belong to the family of aliphatic polyester and they are commonly made from α–hydroxy acids. PLA can be produced from lactic acid (2-hydroxy propionic acid) through fermentation of renewable resources such as cornstarch and sugarcane. Lactic acid is a hydroxy acid with an asymmetric carbon atom and exists in two optically active configurations; L (−) lactic acid and D (+) lactic acid (Figure 5). Lactic acid is a three-carbon chiral acid and is most commonly found in the L (−) form. Two different routes of polymerization can be followed to produce PLA from lactic acid monomers which are shown in Figure 6 [22].

Figure 5. Different isomeric forms of lactic acid: (a) L (−) lactic acid and (b) D (+) lactic acid.

PLA has been extensively used for ages in biomedical fields for applications such as sutures, drug delivery and orthopedic implants. Mechanical properties of PLA are similar to those of commodity plastics such as PP and PE. It also exhibits inherent brittle behavior similar to polystyrene (PS). For amorphous PLAs, the upper temperature range for most commercial applications are determined by glass transition temperature (Tg) whereas for semicrystalline PLAs, both the Tg (~58°C) and melting point (Tm, 130°–230°C, depending on structure) are important for determining the use temperatures across various applications. PLA shows high elastic modulus and tensile strength in the range of 3.2 to 3.7 GPa and 55 to 70 MPa, respectively. But their disadvantages are low impact toughness and elongation at break. PLA is biodegradable. Degradation of PLA occurs by two step degradation processes which are: (a) degradation by hydrolysis followed by (b) bacterial attack on the fragmented residue to biomass, carbon monoxides, carbon dioxide and water.

4.4.2.2. Polypropylene (PP)

Polypropylene (PP) is a thermoplastic polymer and can be made by polymerizing propylene molecules. It is derived from three major sources. Globally, most propylene monomer comes from the steam cracking process using naphtha, a valuable fraction of crude oil. Usually, naphtha cracker's target product is ethylene monomer.

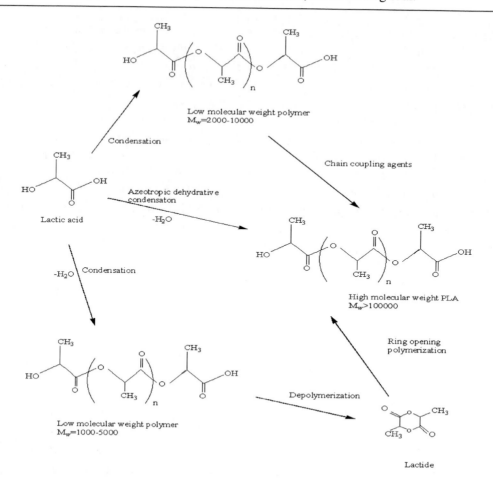

Figure 6. Routes of polymerization to produce PLA from lactic acid monomers.

Propylene is a byproduct of the cracking process produced at various ratios depending on the crude oil feedstock. Many cracking processes have a propylene plant intimately connected to effectively collect the propylene that comes from naphtha cracking. The second largest production of propylene comes from the gasoline refining process. Finally, and most recently, a new process by which propane is dehydrogenated to propylene monomer has been used to produce propylene.

The melting of polypropylene occurs as a range. Most commercial PP has an intermediate level of crystallinity, between 40-60%. Perfectly isotactic PP has a melting point of 171°C (340°F).

Depending on atactic material and crystallinity, commercial isotactic PP has a melting point that ranges from 160 to 166°C (320 to 331°F) whereas, syndiotactic PP with a crystallinity of 30% has a melting point of 130°C (266°F). Shubhra et. al., used commercial grade PP for composite fabrication and he found that tensile strength, tensile modulus, bending strength, bending modulus and impact strength of PP was 22 MPa, 545 MPa, 27 MPa, 2050 MPa and 4.2 kJ/m^2, respectively [10]. Some mechanical and thermal properties of commercial PP is shown in Table 4 [21].

Fiber-Reinforced Composites

Table 4. Some mechanical and thermal properties of commercial PP [21]

Property	Test method	Homopolymer		
Melt flow index (MFI)	(a)	3.0	0.7	0.2
Tensile strength	(b)	500 (lb in^{-2})	4400 (lb in^{-2})	4200 (lb in^{-2})
		34 (MN/m^2)	30 (MN/m^2)	29 (MN/m^2)
Elongation at break (%)	(b)	350	115	175
Flexural modulus	-	190000 (lb in^{-2})	170000 (lb in^{-2})	160000 (lb in^{-2})
		1310 (MN/m^2)	1170 (MN/m^2)	1100 (MN/m^2)
Brittleness temperature (°C)	I.CI./ASTM D746	+15	0	0
Vicat softening oint (°C)	BS 2782	145-150	148	148
Rockwell hardness	-	95	90	90
Impact strength (ft lb)	-	10	25	34

a) Standard polyethylene grader: load 2.16 kg at 230°C.
b) Straining rate 18 in/min.
c) Falling weight test on 14 in diameter moulded bowls at 20°C.

4.4.3. Manufacturing Process

PMCs are manufactured by various processes using open or closed molds. Some processes are discontinuous and some are continuous. Most used processes are hand lay-up, spray lay-up, filament winding, vacuum bag molding, autoclave molding, injection molding, compression molding, pultrusion, etc.

4.4.3.1. Hand Lay-Up Process

The hand lay-up (also called wet lay-up) is one of the oldest and most commonly used methods for manufacturing composite parts such as boat hulls and building panels, etc [23]. Different fibers can be arranged in different directions to enhance the overall performance of the laminated composite material. At first, the mold is cleaned and a mold releasing agent is applied on it. Often a hard wax or a film of poly vinyl alcohol deposited from solution is used as mold releasing agent. A gel coat of resin containing pigment (if required) and curing additives is brushed evenly over the cleaned mold surface. This will form a pure resin outer surface to the molding. After the gel coat has become stiff, successive alternate layers of fiber reinforcement (in the form of woven, knitted, stitched or bonded fabrics) and resin are applied. Resins are impregnated into fibers by hands. The fiber layers are fully wetted and impregnated with the resin by rollers, or brushes used, with an increasing use of nip-roller type impregnators for forcing resin into the fabrics by means of rotating rollers and a bath of resin. If required, a final resin-only sealing layer can be applied in this process. After the laminate becomes fully hardened, it is stripped from the mold and trimmed to size, usually with a power saw. Some of the advantages and disadvantages of hand lay-up process are as follows:

Advantages

a) Low tooling cost.
b) Flexibility of design.
c) Sandwich constructions are possible.
d) Semi-skilled workers are needed.
e) Large and complex items can be produced.
f) Design changes are easily effected.

Disadvantages

a) The waste factor can be high in this process.
b) This is a low volume process.
c) Only one molded surface is obtained, the other being rough.
d) Quality of product is related to the skill of the operator.
e) Resins need to be low in viscosity to be workable by hand.
f) Longer cure times required.

4.4.3.2. Spray Lay-Up Process

Processing steps of this process are almost similar to those in hand lay-up. In this process, the mold is cleaned and a mold releasing agent is applied on the mold. The release agent is then applied to the mold and a layer of gelcoat is applied. The gelcoat is left for sufficient times, until it hardens. After the gelcoat hardens, a spray gun is used to deposit the fiber resin mixture onto the surface of the mold. The spray gun chops the incoming fibers to a predetermined length and impels it through the resin/catalyst mixture. Resin/catalyst mixing can take place inside the gun or after releasing from the gun. Airless spray guns are widely used because they provide more controlled spray patterns and reduced emission of volatiles. Hydraulic pressure is used in airless spray system for dispensing the resin through special nozzles that break up the resin stream into small droplets which then become saturated with the reinforcements. Air-atomized spray gun system dispenses the resin using pressurized air. After spraying the materials on the mold, brushes or rollers are used to remove entrapped air ensuring good fiber wetting. Fabric layers or continuous strand mats may be added into the laminate, depending on performance requirements. Curing of resin depends on the resin formulation and is done at room temperature. After the laminate becomes fully hardened, it is removed from the mold. Using this process, lightly loaded structural panels, e.g. caravan bodies, bathtubs, shower trays, truck fairings, etc are produced.

Advantages

a) Low tooling cost, it is a very economical process for making small to large parts.
b) It is suitable for small to medium volume parts.

Disadvantages

a) Only short fibers are incorporated which severely limits the mechanical properties of the product.
b) Produced composites tend to be very resin-rich and therefore excessively heavy.

4.4.3.3. Filament Winding

Filament winding is a mechanically automated process in which a band of continuous reinforcements (fibers) or mono-filaments is wrapped around a rotating mandrel and cured to produce closed-form hollow parts [24]. The winding operation is achieved by the use of specially designed machines. This process is well suited for parts which are curved in shape. The filaments are wrapped around the mandrel as adjacent bands of repeating patterns. These cover the mandrel surface to produce one complete layer. The winding is continued with successive layers. In wet winding, the fibers are passed through a resin bath before the winding on the mandrel. The filament winding process is named dry if it utilizes prepreg material. In post-impregnation process, the reinforcements are wound on the mandrel without any resin application. After completion of the winding, the filament wound structure is cured at room temperature or in an oven. The mandrel determines accurate internal geometry for the component. It is generally the only major tool and removed after the curing. For winding low cost routine parts, low cost mandrel materials such as cardboard or wood can be used whereas for critical parts requiring close tolerances, expensive mandrels designed for long term use may be required. For high temperature cure, graphite mandrels with low thermal expansion may be advantageous. So, the selection of mandrel is important for this process.

Advantages

a) Inexpensive process.
b) Excellent mechanical properties of products due to the use of continuous fibers.
c) Continuous fibers are used in the loading direction: directional strength can be easily satisfied with the change in winding angle and winding pattern.
d) Filament winding has a high repeatability.
e) High fiber volume percentages are used which will result high strength.
f) Huge parts can be manufactured.

Disadvantages

a) Reverse curvature parts cannot be produced.
b) Complicated shapes require critical mandrel designs and extra cost.
c) Poor surface quality is obtained than that obtained by autoclave production, and generally surface machining is required.
d) Mandrels are expensive and indispensable.

4.4.3.4. Vacuum Bag Molding

This is basically an extension of the wet lay-up process. In this process, pressure is applied to the laminate once laid-up in order to improve its consolidation. This process uses a flexible film or rubber bag to cover the part lay-up. The bag allows the evacuation of air from the part to apply atmospheric pressure. The air under the bag is extracted by a vacuum pump and thus up to one atmosphere of pressure can be applied to the laminate which facilitates consolidation process. Vacuum helps to eliminate entrapped air. The bag used in this method has two fold objectives: (1) it provides a means for removing volatile products during cure and (2) it provides a means for the application of one atmosphere of pressure. Two types of vacuum bags are commonly used: expandable bag or reusable bag. This process is used to laminate curved and irregular shaped workpieces. Boats, racecar components, etc are produced by this process. Typically, polyurethane or vinyl materials are used to make the bag.

Advantages

a) The vacuum bag reduces the amount of volatiles emitted during cure.
b) Better fiber wet-out due to pressure and resin flow throughout structural fibers.
c) Lower void contents.

Disadvantages

a) Skilled operators required.
b) The extra process adds cost both in labor and in disposable bagging materials.

4.4.3.5. Autoclave Molding

In this process a two-sided mold is used. Lower side is a rigid mold and upper side is a flexible membrane. In most cases, reinforcements are pre-impregnated with the resin in the form of prepreg. Sometimes a resin film is placed upon the lower mold and dry reinforcement is placed above that mold. The upper mold is installed and vacuum is applied to the mold cavity. The assembly is placed into an autoclave. An autoclave system is a pressure vessel. The commonly used autoclave is capable of applying high pressure (15 MPa) and temperature (700°C). The autoclaves are generally programmable and temperature/pressure history can be automated. Autoclave method is the most common method used in the aerospace industry [25].

4.4.3.6. Injection Molding

Injection molding refers to a process that generally involves forcing or injecting a fluid plastic material into a closed mold of desired shape. The molding compound is fed into injection chamber through the feed hopper. In the injection chamber, the molding compound is heated wherein it changes into liquid form. It is forced into the injection mold by the plunger. This method is normally used for high-volume and low-cost component manufacturing.

Both thermoplastic and thermoset are subjected to injection molding [26]. A thermoplastic material is first melted and then forced through an orifice into the mold which is kept relatively cool. The material solidifies in the mold and can be removed from the mold after solidification. But in thermoset injection molding, high temperature is required for solidification. Therefore, a reaction material is forced into a generally warm mold in which the material further polymerizes into a solid part. This method is suitable for high-volume and low-cost component manufacturing. But the method is limited to materials with short fibers.

4.4.3.7. Compression Molding

Compression molding is one of the oldest manufacturing techniques that use large presses to compress the prepreg material between two matched steel dies [27]. The recent development of high strength, fast cure, sheet molding compounds and advancement in press technology has made this process very popular for mass production of composite parts. In comparison with the injection molding process, better physical and mechanical properties can be obtained in compression molding. This process utilizes large tonnage presses wherein the curing occurs between two matched steel dies under pressure and high temperature. Compression molding basically involves the pressing of a deformable materials charged between the two heated mold and its transformation into a solid product under the effect of the elevated mold temperature. After placing the laminate to be cured, the cavity is closed. The molds are heated to a high temperature which causes the charge viscosity to be reduced. With increasing mold pressure, the charge flows towards the cavity extremities, forcing air out of the cavity. High pressure helps to eliminate the problem of development of voids. The primary advantage of the compression molding is its ability of producing large number of parts with little dimensional variations. Various shapes, sizes and complexity can be achieved by compression molding. This process has high tooling cost and not practical for low volume production. It is very important to control the cure time otherwise cracking, blistering or warping may occur.

4.4.3.8. Pultrusion

Pultrusion is an automated process that integrates the reinforcement impregnation and the composite consolidation in the same process and used to create shapes by pulling rovings through a shaped and heated die [28]. The process consists of several successive units: the creel, the resin bath, the forming die, the heated curing die, the pullers and the cut-off saw. The process is based on the transformation of the liquid resin in the solid final composite while it passes through the pultrusion die. The creel is the beginning of this process and is the material storage system from which the fibers and mat or fabric are drawn in the required sequence to match the design requirements of the structural shape. A resin impregnation bath is used to facilitate the impregnation of the resin into the fibers. The use of pre-impregnated fibers eliminates the resin bath. Usually, different heating zones are present along the die and these are dependent on several factors, such as the type of resin, the pulling speed and the length of the die. Generally two types of dies are used in pultrusion process: (1) the forming die and (2) the heating or curing die. Forming is done immediately after the impregnation process. Forming dies are normally attached to the heating or curing die in order to provide the correct relationship between the forming and the heated curing step. The heating die cures the material into its final shape as it passes through the die. This cured profile is then automatically cut to desired length. Pultrusion is used to manufacture constant cross-section shapes, like box, channels, tubings, etc.

Advantages

a) It is a continuous process.
b) Material scrap rate is low.
c) Labor requirements are low.

Disadvantages

a) Equipment costs are high.
b) Limited to constant or near constant cross-section components.

4.4.4. Fiber-Reinforced Polypropylene Composite

Among lots of PMCs, fiber reinforced polypropylene composites (FRPCs) are discussed in this section. In this chapter, detail discussion on different types of composites is not possible. But little discussion on a particular type of composite is necessary. That's why among lots of PMCs, FRPCs will be discussed in this section. Lot of works has been done using PP matrix and different fibers. PP possesses several useful properties like high heat distortion temperature, transparency, flame resistance, dimensional stability and high impact strength which widen its application. As a matrix material, PP is widely used because it has some excellent characters for composite fabrication. PP is also very suitable for filling, reinforcing and blending. PP with natural fibrous polymers is one of the most promising routes to create natural-synthetic polymer composites.

4.4.4.1. Fabrication of FRPCs

FRPCs are mainly fabricated by extrusion, injection or compression molding machines (Figure 7). These methods have already been discussed. To fabricate FRPCs with unidirectional fibers, extrusion method is not used as this process can only result short fibers reinforced polypropylene composites (few millimeters at most).

In an extruder, a screw pushes mixture of fiber and PP through a heated cylinder, changing PP from solid to liquid and mixing with the fibers as it moves through the barrel followed by the die which will give the mixture a constant cross section area. Extrusion method is widely used to fabricate short fiber-reinforced polypropylene composites. Many researchers fabricated FRPCs by extrusion method for example, Mora´n et al., fabricated flax/PP composite [29], Masudul et al., fabricated betel nut short fiber/PP composite [30], Van der over et al., jute/PP [31] and Fu et al., fabricated glass fiber/PP and carbon fiber/PP [32] composites using extrusion method. For fabrication of flax/PP composite, Mora´n et al., cut flax fiber to an approximate length of 10 mm to make the fiber's incorporation in the extruder easier. He dried fibers (vacuum dry) in an oven at 70°C for 24 hours. Fibers and PP were blended in a twin screw extruder. Fibers and polypropylene were extruded and pelletized five times. Materials extracted after each extrusion step were compression molded to get flax/PP composites [29].

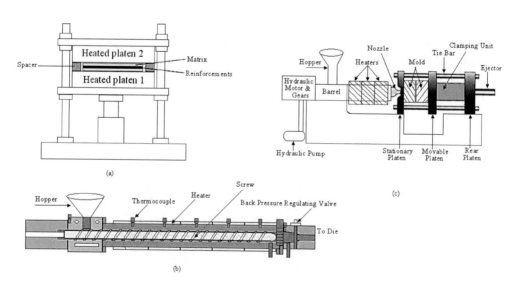

Figure 7. (a) compression, (b) extrusion and (c) injection molding machines.

In an injection molding machine, a large force is used to inject materials into the hollow mold cavity. Only randomly oriented short fibers can be used in this process. The orientation of the fibers in the composite is determined by the flow of the composite melt during filling of the mold cavity. Many researchers fabricated fibers reinforced polypropylene composite by injection molding method for example, abaca/PP, jute/PP and flax/PP composites were fabricated by Bledzki et al., using injection molding method [33]. Similarly, Arzondo et al., fabricated sisal/PP [34], Thi-Thu-Loan et al., jute/PP [35] and Abraham et al., fabricated nylon/PP [36] composites by using injection molding method. For fabrication of abaca/PP composite by injection method, Bledzki et al., dried abaca fibers at 80°C in an oven for 24 hours and mixed with PP by high speed cascade mixer. Then the fiber and PP at different proportions were placed into hot mixer and heated up to the melting temperature of polypropylene (173°C). Hot agglomerate granules were then transferred to the cool mixer where hot agglomerate granules cooled down to room temperature by the cold water. Cold agglomerate granules were then dried again at 80°C (24 hours) before the sample preparation by injection molding process. Test samples were prepared from dried agglomerate by injection molding process (temperature zone 150–180°C) with mold temperature of 80°C and under an injection pressure of 20 kN/mm^2 [33].

Compression molding processes [27] is the process used for the fabrication of unidirectional FRPCs. Shubhra et al., [37] fabricated unidirectional silk/PP composite using compression molding. Similarly, Ruhul et al., fabricated phosphate glass/PP [38] composite and Mubarak et al. fabricated jute/PP [39] composites in the same way using compression process. Shubhra et al., fabricated PP sheets by using granules of PP (about 6 gm) by placing the granules into two steel plates [10]. The steel plates were then placed into the heat press at 185°C and 6 bar pressure for 1 minute. Then it was cooled for 1 minute in a separate press under 6 bar pressure at room temperature. The PP sheets were cut into desired size for composite fabrication. Composites were prepared by sandwiching three layers of silk fibers between four sheets of PP which were placed against two steel plates and heated at 185°C for 1 minute and then cooled in another press to 6 bar pressure for 1 minute [10] to get silk fiber reinforced polypropylene composites.

4.4.4.2. Mechanical Properties of FRPCs

For fiber reinforced composites, mechanical properties are of major concern as reinforcement affects the mechanical properties of FRPCs. Tensile, bending, impact strength and hardness are some mechanical properties significantly considered for a fiber-reinforced PP composite material. The mechanical properties of FRPCs depend on properties of fibers.

The strength of a unidirectional fiber reinforced composite, in the direction of fibers is given by the rule of mixture:

$$\sigma_{cL} = \sigma_f V_f + \sigma_m' (1 - V_f) \tag{1}$$

where V_f is the volume fraction of fibers, σ_f is the tensile strength of fibers and σ_m' is the stress developed in the matrix. For composites with long parallel fibers, the Young's modulus (E_c) in the fiber direction is calculated using the following equation:

$$E_c = E_f V_f + E_m V_m \tag{2}$$

where E_c, E_f, E_m are the Young's moduli of the composite, fiber and matrix respectively and V_f and V_m are the volume fractions of the fiber and matrix, respectively.

Tensile strength [40] is the maximum stress that a material can withstand without tearing apart. It is a measurement of the ability of a material to withstand forces that tend to pull it apart. Tensile strength is measured in unit of force per unit area. Tensile properties of FRPCs can be determined according to ASTM D638 [36, 41] or according to DIN 53455 method [42]. Tensile strength of FRPCs is determined by the following equation:

$$\sigma_{fu} = F_u/A_f \tag{3}$$

where, F_u is the force at failure, A_f is the average filament cross sectional area.

Bending strength [43], also known as flexural strength is defined as a material's ability to resist deformation under load. The bending strength represents the highest stress experienced within the material at its moment of rupture. There are two methods that cover the determination of bending properties of material: three-point loading system and four point loading system. Prior to 1997, one ASTM standard covered both loading modes which is ASTM D 790. However since 1997, only three-point loading is included in ASTM D 790 and a new ASTM standard has been created for four-point loading which is ASTM D 6272. Without ASTM standard, DIN 53452 method [42] can be used also to determine bending properties of FRPCs. For a rectangular sample under a load in a three-point bending setup, the bending strength is calculated by the following formula:

$$\sigma = 3FL/2bd^2 \tag{4}$$

where F is the load (force), L is the length of the support span, b is width and d is thickness. For a rectangular sample under a load in a four-point bending setup where the loading span is one-third of the support span, bending strength is calculated by the following formula:

$$\sigma = FL/bd^2 \tag{5}$$

where F, L, b, d have the same meaning as indicated above. But if the loading span is 1/2 of the support span, bending strength is calculated by the following formula:

$$\sigma = 3FL/4bd^2 \tag{6}$$

where F, L, b, d have the same meaning as indicated above.

Impact strength [44] is the ability of a material to resist suddenly applied loads. In most cases, impact test of FRPCs are carried out by two different types of testing: Charpy test and Izod test. Izod impact strength of FRPCs can be determined according to ASTM D256 [45] or DIN EN ISO 180 and Charpy impact strength can be determined according to ASTM D 6110 or DIN EN ISO 179 [42].

Mechanical properties of different fibers reinforced polypropylene composites are studied by many researchers. Shubhra et al., worked on silk fiber/PP composites [46], Avik et al., worked on calcium alginate fiber/PP [47], Barkoula et al., flax/PP [48], Sushanta et al., bamboo fiber/PP [49], Ajay et al., wood fiber/PP [50], Mominul et al., abaca fiber/PP [51],

Haydaruzzaman et al., coir fiber/PP [52], Mominul et al., palm fiber/PP [53] and Shubhra et al., phosphate glass fiber/PP [46] composites.

From these studies it is clear that strength of FRPCs increase with the strength of fibers used in the composites. For example, the study carried out by Shubhra et al., [10] can be discussed here.

In his study he determined the mechanical properties of jute/PP and silk/PP composites using 20% fibers in both cases. Tensile strength (TS), Young's modulus (YM), bending strength (BS), bending modulus (BM) and impact strength (IS) of the matrix PP used in that study were 22 MPa, 545 MPa, 27 MPa, 2050 MPa, 4.2 kJ/m^2, respectively. Due to fiber reinforcement, these properties increased greatly. Jute composites gained 88, 38, 64, 7.3 and 186% increase in TS, YM, BS, BM and IS, respectively over the corresponding properties of the matrix. On the other hand, silk composites gained 153, 39, 111, 62 and 305% increase in TS, YM, BS, BM and IS, respectively over the corresponding properties of the matrix. TS, TM, BS, BM and IS of the matrix and stated composites are shown in Figure 8, 9, 10, 11, 12, respectively.

These figures give idea on the increase in strength of polymeric PP matrix by the reinforcement of fibers and indicate the reason behind the popularity of fibers as reinforcement. Moreover from these figures, it is clear that the stronger the fiber, the stronger the composites. Silk fibers being stronger than jute, produces high strength composites with PP [10]. It should be noted that, strength of composites varies with the fiber content. In another study, Shubhra et al., found that 10% silk fiber-reinforced PP composites has TS value of 46.2 MPa [46] whereas for 20% fibers it was more than 55 MPa [10].

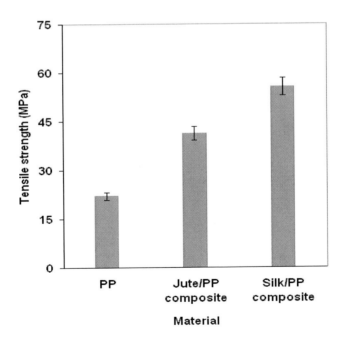

Figure 8. Tensile strength of polypropylene sheet, jute/PP and silk/PP composites. Reprinted from Fibers and Polymers, 11, 2010, 725-731, Characterization of Plant and Animal Based Natural Fibers Reinforced Polypropylene Composites and Their Comparative Study, Quazi Shubhra, T. H. et al., Figure 2, Copyright (2010), with kind permission of Springer Science and Business Media.

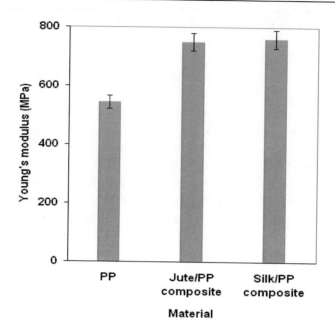

Figure 9. Young's modulus of polypropylene sheet, jute/PP and silk/PP composites. Reprinted from Fibers and Polymers, 11, 2010, 725-731, Characterization of Plant and Animal Based Natural Fibers Reinforced Polypropylene Composites and Their Comparative Study, Quazi Shubhra, T. H. et al., Figure 3, Copyright (2010), with kind permission of Springer Science and Business Media.

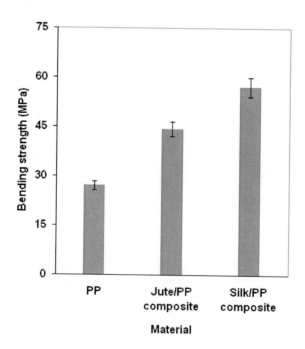

Figure 10. Bending strength of polypropylene sheet, jute/PP and silk/PP composites. Reprinted from Fibers and Polymers, 11, 2010, 725-731, Characterization of Plant and Animal Based Natural Fibers Reinforced Polypropylene Composites and Their Comparative Study, Quazi Shubhra, T. H. et al., Figure 5, Copyright (2010), with kind permission of Springer Science and Business Media.

Fiber-Reinforced Composites

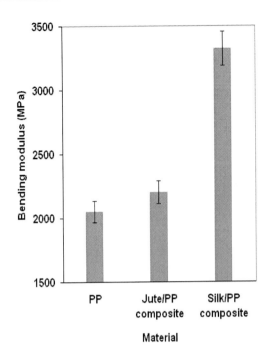

Figure 11. Bending modulus of polypropylene sheet, jute/PP and silk/PP composites. Reprinted from Fibers and Polymers, 11, 2010, 725-731, Characterization of Plant and Animal Based Natural Fibers Reinforced Polypropylene Composites and Their Comparative Study, Quazi Shubhra, T. H. et al., Figure 6, Copyright (2010), with kind permission of Springer Science and Business Media.

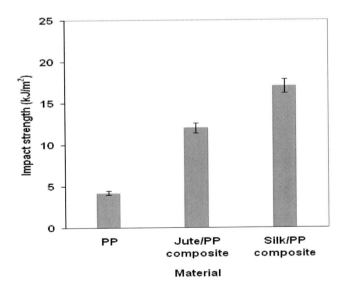

Figure 12. Impact strength of polypropylene sheet, jute/PP and silk/PP composites. Reprinted from Fibers and Polymers, 11, 2010, 725-731, Characterization of Plant and Animal Based Natural Fibers Reinforced Polypropylene Composites and Their Comparative Study, Quazi Shubhra, T. H. et al., Figure 7, Copyright (2010), with kind permission of Springer Science and Business Media.

But it should be noted that with the increase in fiber content, strength of FRPCs increases upto a certain limit and then decreases with further increase in fiber content. Sushanta et al., has worked on bamboo fiber-reinforced PP composites [49]. He observed that the tensile strength increases with the increase in bamboo fiber content up to 30% bamboo fiber in PP with a subsequent decrease in tensile strength at 40% of fiber loading. At lower levels of fiber content, the composites show poor mechanical properties due to poor fiber population and low load transfer capacity to one another. As a result, stress gets accumulated at certain points of the composites and highly localized strains occur in the matrix [52]. At intermediate levels of fiber loading, the population of the fibers is just right for maximum orientation and the fibers actively participate in stress transfer [52]. Decrease in strength of fiber/PP composite at higher fiber content is a direct consequence of poor fiber/matrix adhesion which leads to micro-crack formation at the interface under loading and non-uniform stress transfer due to the fiber agglomeration in the matrix [49]. Higher wt% of fiber content also leads to an increase in fiber–fiber interaction which results difficulties in dispersion of the fibers within the PP matrix [49].

4.4.4.3. Fiber Matrix Adhesion

For FRCs, the stress transfer at the interface between the matrix and fiber phases is determined by the degree of adhesion. Effective transfer of stress and load distribution throughout the interface is possible when strong adhesion exists at the interfaces. The adhesion phenomenon is described in several theories including mechanical interlocking, absorption, primary bonding, interdiffusion, electronic theories, etc.

4.4.4.3.1. Testing of Adhesion in FRPCs: Micromechanical Characterization

Interfacial shear strength (IFSS) is measured by the pullout test. In the first step of this test, one end of the fiber is embedded in a block of a polymer matrix and in the second step a force is applied to the free end to pull it out of the matrix. Applied force is continuously monitored and recorded and the force at which debonding occurs is obtained. The average IFSS can be calculated from using the following equation:

$$\tau = F / \pi \times d \times l \tag{7}$$

Here,

F= Maximum load measured prior to debonding of the fiber,
d = The diameter of used fiber, and
l = The length of embedded portion of fiber.

Single-fiber fragmentation test (SFFT) is also used to calculate the IFSS. A single fiber is embedded in the matrix which is then converted to a dogbone-shaped tensile coupon using appropriate method (e.g. cutting). The specimen is then subjected to a tensile load and tensile forces are transferred from the matrix to the fiber. Depending on the level of fiber-matrix adhesion, build up of tensile stresses occur within the fiber. After certain level of tensile strength, stress concentration reaches to a certain high level at which debonding occurs resulting fracture. This loading process continues until the fiber fragmentation process ceases. The final fiber fragment length is termed as the fiber critical length and denoted by l_c. The critical length (l_c) can be measured using the following formula: $l_c = 4/3\ l_f$, where l_f indicates average fragment length [54]. Critical length is a good indicator of the ability of the interphase to transmit loads between the matrix and fiber. Kelly and Tyson [55] developed the equation by which IFSS is calculated using this length. The equation is as follows:

$$\tau_i = d \times \sigma_f / 2 \times l_c \tag{8}$$

In equation (8), d is the diameter of fiber used, σ_f is the tensile strength of single fiber at the critical length. The ratio (l_c/d) can be used as an indicator of the fiber-matrix bond strength.

Drzal et al. [56] altered the above equation reflecting Weibull statistics and rearranged the Kelly and Tyson's equation, proposing the following modification to calculate the IFSS:

$$\tau_i = \sigma_f \times \Gamma \times [1-1/\alpha] / 2 \beta \tag{9}$$

In equation (9), β is the scale and α is the shape parameters in the Weibull distribution and Γ is the Gamma function. The strength of fiber can be calculated from the extrapolation gauge length using Weibull chart [57]. At the critical fragment length, the strength of fiber is found by the following equation [54]:

$$\sigma_f = \sigma_{f0} (l_c / l_0) \exp{-1}/\rho \tag{10}$$

In equation (10), σ_{f0} is the fiber strength at gauge length l_0 and ρ is the shape parameter of the Weibull distribution for the tensile strength of fiber.

Karlsson et al., [58] used the Wilhelmy plate method to measure the apparent perimeter of regenerated cellulose. He used this parameter to calculate the IFSS. Due to inherent variability of the henequen fibers (*Agave fourcroydes*), equations (8) and (9) were rearranged to use the perimeter P_f instead of the diameter to calculate the IFSS. Following this approach the calculation of IFSS for the pullout test was done using the following equation:

$$\tau_i = F_f / P_f \times l_e \tag{11}$$

In equation (11), F_f is the debonding force, P_f is the measured perimeter of the henequen fiber and l_e is the fiber-embedded length. In the case of the single fiber fragmentation test the IFSS is calculated with the following equation:

$$\tau_i = 2F_f / \beta \times \Gamma \times P_f \times [1+1/\alpha] \tag{12}$$

In equation (12), F_f is the tensile load at break for the henequen fiber, P_f, α, β and Γ have the same meaning as indicated above.

Ruhul et al., [54] studied IFSS of phosphate glass/PP composites. He prepared single fiber composite samples using one single filament of fiber between two sheets of polypropylene. He performed fragmentation test using universal testing machine and he counted the number of fragments by microscope. He found the critical length 3030 μm for the stated system. He measured tensile strength of the fiber at the critical length using the Weibull weakest chart rule. He found the IFSS of phosphate glass fiber/PP system 5.93 MPa. He concluded that, his result suggests excellent fiber matrix adhesion between phosphate glass fiber and PP matrix.

4.4.4.3.2. Improvement of Fiber Matrix Adhesion

Polypropylene is nonpolar in nature. For FRPCs, polar groups affect the fiber–matrix adhesion to a large extent. Most reinforcements used for PP are polar in nature. In absence of polar groups on PP, only physical and/or mechanical forces like friction influences the interface properties. To overcome this problem coupling agents can be used or fibers can be chemically treated. Surface treatments of fibers will make the fiber surface more hydrophilic. On the other hand, coupling agents tends to modify the chemistry of the PP by attaching polar groups onto the molecular backbone, such as acrylic acid (AA) or maleic anhydride (MA).

4.4.4.3.2.1. Improvement of Fiber Matrix Adhesion by Coupling Agent

So far, lots of coupling agents have been used in research of FRPCs. Among them, maleated polypropylene (MAPP) is the most popular as well as the most effective one [59]. Recent works suggest that the use of maleated polypropylene significantly improves the fiber-matrix bonding [60]. So, discussion will be limited to MAPP.

Without MAPP, the tensile and bending strengths of FRPCs are lower than that containing MAPP which suggests that there is comparatively little stress transfer from the matrix to the fibers for FRCPs containing no MAPP than that containing MAPP [60]. The use of MAPP improves interaction and adhesion between the fibers and matrix leading to better stress transfer from matrix to fiber. Incorporation of polar maleic anhydride grafted polypropylene, improves the fiber dispersion and fiber/matrix interfacial interaction through hydrogen bonding between hydroxyl groups of natural fibers and carbonyl groups of the maleic anhydride segment of the MAPP [61] thereby reducing the rate of moisture absorption and increasing the mechanical strength in the composites [61–62].

Xiang et al. studied glass fiber/PP composites. In their study, they used three different compatibilizers including octane–ethylene copolymer, maleic anhydride grafted octane–ethylene copolymer and maleic anhydride. They found that these compatibilizers had different adhesion effects in the order: PP-g-MAH>POE-g-MAH>POE i.e., glass fiber/PP containing MAPP resulted highest mechanical properties as it showed most adhesion [63]. Barkoula et al., studied on flax fiber/PP composites (PP with and without 3 wt% MA-PP). They found a sharp increase in the strength of flax/PP composites containing MAPP over the composites containing only PP [48]. Sanjay et al., found that the tensile strength and modulus of hybrid glass and sisal (15% glass and 15% sisal) fiber-reinforced PP composites compatibilized with 2wt% of MAPP increases to the tune of 17.74 and 14.52%, as compared with uncom-patibilized composites containing same amount of fiber reinforcement [64]. According to them, this is primarily due to covalent bonding between the anhydride groups of MAPP and hydroxyl groups of the sisal fiber, along with chain entanglement between MAPP and PP chains that create a good stress transfer at the interface. They also claimed that, inclusion of 2wt% of MAPP to the hybrid composites prepared at 15:15 wt% of sisal: glass fibers additionally enhances the tensile strength and modulus due to the formation of strong covalent linkage between the –OH groups of sisal fiber and SiO groups of glass fibers with MAPP, thereby resulting in an improved interfacial adhesion between the matrix and both type of fibers [64].

When MAPP is used as coupling agent, bonding between the anhydride groups of MAPP and hydroxyl groups of the cellulosic fibers occurs as shown in Figure 13 [65]. This is the situation for plant based cellulosic fibers reinforced PP composite containing MAPP.

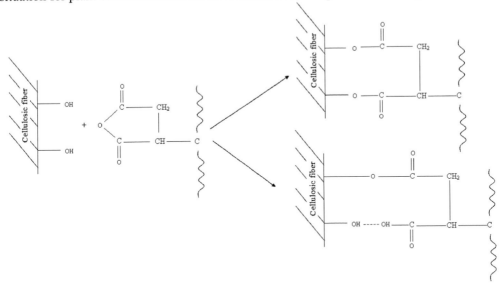

Figure 13. Bond formation between hydroxyl groups of natural fiber and anhydride ring of MAPP.

For proteinous animal based fibers, bonding between amino group of proteinous fibers and MAPP can result increased mechanical properties of FRPCs. In case of synthetic fibers, bonding between MAPP and synthetic fibers occur. Reaction occurs between MAPP and carbon fiber [66], the anhydride groups of MAPP and SiO groups of glass fibers [64] and improved mechanical properties result when MAPP is used with synthetic fibers reinforced PP composites.

Thus MAPP increase fiber-matrix adhesion and increase the mechanical properties of FRPCs. But the content of MAPP should not be high. Because at higher MAPP content, a marginal decrease in mechanical properties occur which may be explained due to self-entanglement of MAPP resulting from the migration of excess MAPP around the fiber surface rather than causing inter-chain entanglement and contributing to the mechanical continuity of the system [67].

4.4.4.3.2.2. Improvement of Fiber Matrix Adhesion by Fiber Treatment

Generally fiber treatment is carried out for natural fibers to improve fiber-matrix adhesion so that their reinforcement increase properties of fiber-reinforced composites. The major drawbacks associated with the use of natural fibers as reinforcements in polypropylene matrix is the poor wettability and weak interfacial bonding between fiber and polypropylene due to the inherently poor compatibility as well as dispersability of the hydrophilic cellulose fibers with the hydrophobic polypropylene. Thus, to improve the fiber-matrix adhesion a pre-treatment of the fiber surface is required. Several studies have been reported based on the influence of various type of chemical modification on the physical and mechanical properties of natural fiber filled polypropylene composites.

Various fiber surface treatments such as alkali/mercerization [68-69], silane [70], combination of alkali and silane [71-72], monomer grafting under UV radiation [73-74] and various other methods such as acetylation, benzyolation, etc. [75-76] have been reported by several authors. Lot of works has been done by many researchers on the treatment of fiber to improve the mechanical properties of fiber/PP composites.

Haydar et al., has studied the effect of surface treatment of jute with $KMNO_4$ and found increased mechanical properties of jute/PP composites due to that treatment [77]. He found better fiber-matrix adhesion and maximum of 20% increase in TS value than that of untreated one for 0.03 % $KMnO_4$ treated jute/PP composite at 2 minutes soaking time. In another study [42], the same author treated jute with starch solutions of different concentrations varying treating times. The maximum TS was found for 5% starch treated composites for 5 minutes soaking time which was 19% higher than the TS, of untreated composites [42]. Mubarak et al., studied on the surface modification of jute with acrylic monomers [78]. He used different formulations using oligomer urethane diacrylate, monomers 2-ethyl hexyl acrylate (EHA), 1-vinyl 2-pyrrolidone (NVP) and photo-initiator in methanol in order to modify the surface of jute. After soaking in the prepared formulations, he cured jute by UV radiation source of varying intensities. He found that, the formulation containing 5% NVP, 5% EHA, 2.5% oligomer and 2% photo-initiator and cured at 15 UV passes showed the highest mechanical properties. For the stated treatment of jute, the author found composites having 20% more TS [78] than that of untreated one which was due to better fiber-matrix adhesion.

Haque et al., chemically treated palm fibers using benzene diazonium salt. According to him, the hydroxyl groups in the raw palm are responsible for high water absorption and weak interfacial bonding between the palm fiber and PP matrix. Diazonium salt converts two hydroxyl groups into diazo group and results azo product [53]. Due to the elimination of most of the hydroxyl groups in the treated palm, the interfacial bonding between the palm fiber and PP matrix increased resulting composites of high strength [53]. Haque et al., also studied the chemical treatment of abaca and coir fiber with benzene diazonium salt. Due to the treatment, the interfacial bonding between the abaca and coir fiber with PP matrix increased resulting composites of high tensile, bending and impact strength [51].

Guillermo et al., treated flax fiber by esterification. He also treated the fiber with silane. He found that treated fibers result composites of higher mechanical properties than untreated one which was due to better fiber-matrix adhesion. Flax fiber was esterified with maleic anhydride (MA) dissolved in boiling acetone (T=50±5°C) with a fiber: solvent ratio of 1:25 (w/v) [79]. For silanization, he dissolved vinyl trimethoxy silane (VTMO) in acidified water (pH=3.5) for 10 minutes to get a better functionalization and then fibers were added and maintained for 1 hour in the solution for obtaining a 2.5 wt.% silane with respect to the weight of flax fiber [79]. Arbelaiz et al., treated flax fiber with alkali and found increased mechanical properties for treated composites. For alkalization, he soaked flax fibers in a 20wt% aqueous solution of sodium hydroxide for 1 hour at room temperature followed by washing. He washed fibers for several times in distilled water followed by neutralization with a few drops of acetic acid. The fibers were then washed again and finally dried in an oven. In his study, the ratio of fiber weight to alkali solution volume was 1:20 [80]. Similarly, Avik et al., treated calcium alginate fibers with vinyl triethoxy silane [H_2=CH-Si-$(OC_2H_5)_3$] (5%, v/v) to reduce the strong hydrophilic nature of the fiber and to improve the mechanical and interfacial properties of the calcium alginate fiber-reinforced PP composites [47]. He found improved fiber-matrix adhesion due to the stated fiber treatment.

Fiber-Reinforced Composites

These studies clearly state that improved fiber-matrix adhesion is obtained for FRPCs by proper treatment of fiber which results increased mechanical properties.

4.4.4.4. Thermal Analysis of FRPCs

Thermal analysis techniques are being used to understand thermal behavior of FRPCs. Differential scanning calorimetry (DSC) is a thermo analytical technique in which the difference in the amount of heat required to increase the temperature of a sample and reference are measured as a function of temperature. The basic principle underlying this technique is that, when the sample undergoes a physical transformation such as phase transitions, more (or less) heat will need to flow it than the reference to maintain both at the same temperature. The flow of heat to the sample depends on whether the process is exothermic or endothermic. For example, when a solid sample melts to a liquid, it will require more heat flowing to the sample to increase its temperature at the same rate as the reference. This is due to the absorption of heat by the sample as it undergoes the endothermic phase transition from solid to liquid. Likewise, as the sample undergoes exothermic processes (such as crystallization) less heat is required to raise the sample temperature. By observing the difference in heat flow between the sample and reference, differential scanning calorimeters are able to measure the amount of heat absorbed or released during such transitions. DSC may also be used to observe more subtle phase changes, such as glass transitions. The result of a DSC experiment is a heating or cooling curve. Differential scanning calorimetry can be used to measure a number of characteristic properties of a sample. Using this technique it is possible to observe fusion and crystallization events as well as glass transition temperatures (Tg). DSC can also be used to study oxidation, as well as other chemical reactions. DSC analysis is carried out in an argon atmosphere with a specific heating rate ($10°C/min$). A static inert gas flow of 50 mL/min and an aluminum sample pan is used. A specimen of approximately 10 mg is scanned over a specific temperature range. The glass transition temperature (Tg), melt temperature (Tm), cold crystallization temperature (Tc) and heat of melting (ΔHm) are determined for matrix material and composite material. Details of cold crystallization and pre-melt crystallization can be found. The percentage crystallinity of each sample can be calculated by using the following relationship:

$$X_C \text{ (\% crystallinity)} = \frac{\Delta H}{\Delta H_m} \times \frac{100}{w} \qquad (13)$$

where ΔH and ΔH_m are the heat of fusion of PP and 100% crystalline PP, respectively (taken as $\Delta H_m^\circ = 138$ J/g for PP [81]) and w is the weight fraction of PP in the composite. To determine the crystallinity index of FRPCs, the heat of cold crystallization and pre-melt crystallization were subtracted from heat of melting.

Thermogravimetric analysis (TGA) is one type of thermal analysis which is used to characterize a wide variety of materials. This technique provides complimentary and supplementary characterization information to DSC analysis. It measures the amount and rate (velocity) of change in the mass of a sample as a function of temperature or time. It determines the thermal and/or oxidative stabilities of materials along with their compositional properties. The technique can analyze materials that exhibit either mass loss or gain due to oxidation, decomposition or loss of volatiles. It is especially useful for the study of fiber reinforced polymeric matrix composites. Information provided by TGA helps to select materials for certain end-use applications, predict product performance and improve product quality. Kinetic parameters for the various stages of thermal degradation were determined from the TGA graphs using the following equation, given by Broido [82]:

$$\ln(\ln\frac{1}{y}) = -\frac{E_a}{RT} + \ln(\frac{RZ}{E_a\beta}T_{max}^2) \qquad (14)$$

where y is the fraction of nonvolatilized material not yet decomposed, T_{max} is the temperature of maximum reaction rate, β is the heating rate, Z is the frequency factor, and E_a is the activation energy.

Initially plots of $\ln\{\ln(1/y)\}$ versus 1/T for various stages of decomposition is generally found to be linear, suggesting good agreement with the Broido equation. The activation energies, E_a, is determined from the slopes of these plots.

Beg et al., studied thermal properties of radiata pine (*Pinus radiata*) fiber reinforced PP composites. Using DSC, he determined the percentage crystallinity of PP matrix and composites. Figure 14 and 15 shows DSC curves of PP and composites, respectively.

From Figure 14 and using equation (13), he calculated the percentage crystallinity of virgin PP which was 38.8. Similarly he calculated percentage crystallinity of eight times reprocessed (injection molded and granulated repeatedly for eight times) PP which was 40.1. In the same way, he determined percentage crystallinity of virgin and eight times reprocessed composites and the values were 43.7 and 45.2, respectively.

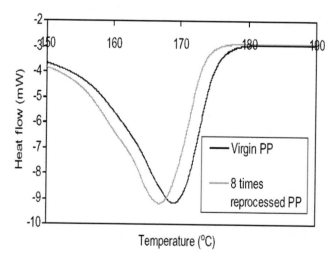

Figure 14. DSC curves of PP. Reprinted from Composites: Part A, M. Beg, M.D.H. et al., Reprocessing of Wood Fibre Reinforced Polypropylene Composites. Part I: Effects on Physical and Mechanical Properties, 1091–1100, Copyright (2008), with permission from Elsevier.

Fiber-Reinforced Composites

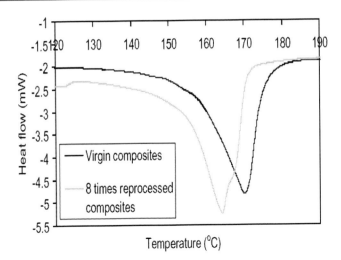

Figure 15. DSC curves of radiata pine (*Pinus radiata*) fiber reinforced PP composites. Reprinted from Composites: Part A, M. Beg, M.D.H. et al., Reprocessing of Wood Fibre Reinforced Polypropylene Composites. Part I: Effects on Physical and Mechanical Properties, 1091–1100, Copyright (2008), with permission from Elsevier.

Using TGA, he determined activation energy. Typical TGA for PP and composites studied by him are shown in Figure 16 and 17 respectively. From the temperature values obtained from TGA curves, he plotted $\ln\{\ln(1/y)\}$ versus $1/T$ for various stages of decomposition (shown in Figure 18) which is almost linear, suggesting good agreement with the Broido equation. The activation energies, Ea can be determined from the slopes of Figure 18.

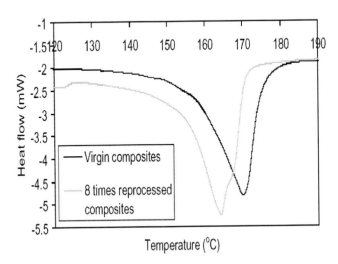

Figure 16. TGA curves of PP. Reprinted from Composites: Part A, M. Beg, M.D.H. et al., Reprocessing of Wood Fibre Reinforced Polypropylene Composites. Part I: Effects on Physical and Mechanical Properties, 1091–1100, Copyright (2008), with permission from Elsevier.

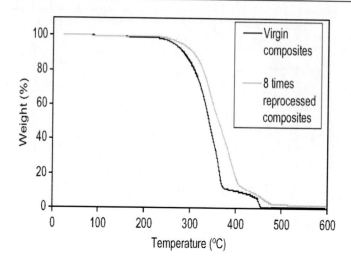

Figure 17. TGA curves of radiata pine (*Pinus radiata*) fiber reinforced PP composites. Reprinted from Composites: Part A, M. Beg, M.D.H. et al., Reprocessing of Wood Fibre Reinforced Polypropylene Composites. Part I: Effects on Physical and Mechanical Properties, 1091–1100, Copyright (2008), with permission from Elsevier.

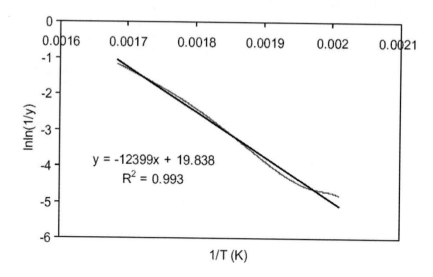

Figure 18. 1/T versus ln{ln(1/y)} for the first decomposition stage of PP. Reprinted from Composites: Part A, M. Beg, M.D.H. et al., Reprocessing of Wood Fibre Reinforced Polypropylene Composites. Part I: Effects on Physical and Mechanical Properties, 1091–1100, Copyright (2008), with permission from Elsevier.

In his study, positions of weight loss on the TGA traces both for PP and composites shifted to higher temperatures with increased number of times the materials were reprocessed (Figure 16 and 17) suggesting increased thermal stability.

CONCLUSION

Among different types of reinforcements, fibers are stronger and stiffer materials than any other form which is the main reason for the overwhelming attraction for fibrous reinforcements. Different types of fibers used in composites have different properties and so affect the properties of the composite in different ways. The mechanical properties of most reinforcing fibers are considerably higher than those of un-reinforced resin systems. The mechanical properties of the fiber/resin composite are therefore dominated by the contribution of the fiber to the composite. Between synthetic and natural fibers, synthetic fibers are usually more uniform in size, more economical to use and behave in a more predictable manner. For engineering applications the most commonly employed significant fibers are synthetic fibers which includes glass fibers, metallic fibers, organically derived synthetic fibers, etc. Most strong and stiff fibers (e.g., ceramic fibers of glass, graphite-carbon, boron carbide and silicone carbide) are usually difficult to use as structural materials in bulk. However, embedding such materials in a ductile matrix (such as a polymer or metal) enables them to behave as a stronger, stiffer and tougher material. On the other hand, natural fiber-reinforced polymer composites have created interest in the automotive and other industries in recent years due to their lower cost than glass and carbon fibers, biodegradability and environment-friendliness. Unlike glass and carbon fibers, natural fibers are CO_2 neutral. The modulus–weight ratio of some natural fibers is greater than that of E-glass fibers indicating that they can be very competitive with E-glass fibers in stiffness-critical designs. Polymer matrix composites are very popular due to their low cost, simple fabrication methods, lightweight and desirable mechanical properties. But the main drawbacks of PMCs are their low thermal resistance and high coefficient of thermal expansion. To overcome this limitation researchers are working to modify the renewable reinforcing materials through different techniques. For evidence the crystallinity index of bleached fiber-reinforced PP composite is higher than the prewashed natural fiber-reinforced PP composites. Composite materials are now well established all over the world and used in almost every type of advanced engineering structures, with their usage ranging from aircraft, helicopters and spacecraft through to ships, boats and offshore platforms and to automobiles, sports goods, chemical processing equipments and civil infrastructures such as bridges and buildings. The development of fiber-reinforced composite materials has not only been a great motivating factor for scientists, but also an important provider of opportunities to improve the standard of living of people around the world. Along with researchers and entrepreneurs, we also expecting the expansion of research for new fiber-reinforced composite materials with high performance at affordable costs.

REFERENCES

[1]   Peply H.E. (1987). *Engineered Materials Handbook*, Reinhart T.J. (Ed), Volume 2, Chapter 2.3. Ohio, USA, ASM International.

[2]   Bullions, T.A. Gillespie, R.A. Price-O'Brien, J. and Loos, AC. (2004). The Effect of Maleic Anhydride Modified Polypropylene on the Mechanical Properties of Feather

Fiber, Kraft Pulp, Polypropylene Composites. *Journal of Applied Polymer Science, 92,* 3771–3783.

[3]   Nick, A. Becker, U. and Thoma, W. (2002). Improved Acoustic Behavior of Interior Parts of Renewable Resources in the Automotive Industry. *Journal of Polymers and the Environment, 10,* 115-118.

[4]   Haydaruzzaman, Ruhul, A. K. Mubarak A. K. and Hossain, M. A. (2009). Effect of Gamma Radiation on the Performance of Jute Fabrics-reinforced Polypropylene Composites. *Radiation Physics and Chemistry, 78,* 986–993.

[5]   Rai, S.K. and Padma Priya, S. (2006). Utilization of Waste Silk Fabric as Reinforcement for Acrylonitrile Butadiene Styrene Toughened Epoxy Matrix. *Journal of Reinforced Plastics and Composites, 25(6),* 565–574.

[6]   Chen, Y. Chiparus, O. Sun, L. Negulescu, I. Parikh, D. V. and Calamari, T. A. (2005). Natural Fibers for Automotive Nonwoven Composites. *Journal of Industrial Textiles, 35,* 47–62.

[7]   Hollaway, L. (1993). Polymer Composites for Civil and Structural Engineering. Chapman and Hall Inc., Glasgow, U.K.

[8]   Sheldon, R.P. (1982). Composite Polymeric Materials. Applied Science Publishers, London, UK.

[9]   Mueller, D.H. and Krobjilowski, A. (2003). New Discovery in the Properties of Composites Reinforced with Natural Fibers. *Journal of Industrial Textiles, 33 (2),* 111-130.

[10] Quazi Shubhra, T. H. Alam, A. K. M. M. Gafur, M.A. Sayed Shamsuddin, M. Mubarak Khan, A. Saha, M. Dipti Saha, Quaiyyum, M. A. Jahangir Khan, A. and Ashaduzzaman, Md. (2010). Characterization of Plant and Animal Based Natural Fibers Reinforced Polypropylene Composites and Their Comparative Study. *Fibers and Polymers, 11(5),* 725-731.

[11] Nabi, D. Saheb and Jog, J.P. (1999). Natural Fiber Polymer Composites: A review, *Advances in Polymer Technology, 18(4),* 351-363.

[12] Mukherjee, P.S. and Satyanarayana, K.G. (1986). Structure and Properties of Some Vegetable Fibres. *Journal of Materials Science, 21,* 51–56.

[13] Schoutens, J.E. (1982). Introduction to Metal Matrix Composite Materials, MMCIAC Tutorial Series, DOD Metal Matrix Composites Information Analysis Center, Santa Barbara, CA.

[14] Soon Hong, H. and Kyung Chung, H. (1995). Effects of Vacuum Hot Processing Parameters on the Tensile Properties and Microstructures of SiC-2124Al Composites. *Material Science and Engineering A, 194,* 165-170.

[15] Jang, J. H. and Han, K. S. (2007). Fabrication of Graphite Nanofibers Reinforced Metal Matrix Composites by Powder Metallurgy and Their Mechanical and Physical Characteristics. *Journal of Composite Materials, 41(12),* 1431-1443.

[16] Laurent, V. Jarry, P. Regazzoni, G. and Apelian, D. (1992). Processing-Microstructure Relationships in Compocast Magnesium/SiC. *Journal of Materials Science, 27 (16),* 4447-4459.

[17] Adam Papworth and Peter Fox. (1996). Oxide Film Casting Defects in Squeeze Cast Metal Matrix Composites. *Materials letters, 29 (4-6),* 209-213.

[18] Amateau, M. F. (1998). Ceramic Composites: Matrix Material. Peters, S. T.(Ed), Handbook of composites: (2nd, 312). Mountain View, California: CHAPMAN and HALL.

[19] Mallick, P.K. (2008). Metal, Ceramic, and Carbon Matrix Composites: Carbon Matrix Composites. Fiber Reinforced Composites: Materials, Manufacturing and Design (3rd, 27). NW, USA. CRC press, Taylor and Francis Group.

[20] Yunfeng Hua, Litong Zhang, Laifei Cheng and Jing Wang (2006). Silicon Carbide Whisker Reinforced Silicon Carbide Composites by Chemical Vapor Infiltration. *Materials Science and Engineering: A, 428 (1-2),* 346-350.

[21] Brydson, J.A. (1999). Plastics Materials (7th, 247-268). Oxford, U.K. Butterworth-Heinemann.

[22] Donald Garlotta (2001). A Literature Review of Poly (Lactic Acid). *Journal of polymers and the environment, 9 (2),* 63-84.

[23] Moura Branco, C. Ferreira, J. M. Fael, P. and Richardson, M. O. W. (1996). A Comparative Study of the Fatigue Behaviour of GRP Hand Lay-up and Pultruded Phenolic Composites. *International Journal of Fatigue, 18 (4),* 255-263.

[24] Lauke, B. and Friedrich, K. (1993). Evaluation of Processing Parameters of Thermoplastic Composites Fabricated by Filament. *Composites Manufacturing, 4(2),* 93-101.

[25] Zeng, X. and Raghavan, J. (2010), Role of Tool-part Interaction in Process-induced Warpage of Autoclave-manufactured Composite Structures. *Composites Part A, 41 (9),* 1174-1183.

[26] Mubarak Khan, A. Johannes Ganster and Hans-Peter Fink (2009). Hybrid Composites of Jute and Man-made Cellulose Fibers with Polypropylene by Injection Molding. *Composites Part A, 40(6-7),* 846-851.

[27] Shubhra, Q. T. H. Saha, M. Alam, A. K. M. M. Beg, M. D. H. and Khan, M. A. (2010). Effect of Matrix Modification by Natural Rubber on the Performance of Silk-reinforced Polypropylene Composites. *Journal of Reinforced Plastics and Composites, 29,* 3338-3344.

[28] Kathleen Van de Velde and Kiekens (2001). Thermoplastic Pultrusion of Natural Fibre Reinforced Composites. *Composite Structures, 54 (2-3),* 355-360.

[29] Mora´n, J. Alvarez, V. Petrucci, R. Kenny, J. and Vazquez. A. (2007). Mechanical Properties of Polypropylene Composites Based on Natural Fibers Subjected to Multiple Extrusion Cycles. *Journal of Applied Polymer Science, 103 (1),* 228-237.

[30] Masudul, H. Manfred, M. Wagner, H. Zaman, H. U. and Mubarak Khan, A. (2010). Physico-Mechanical Performance of Hybrid Betel Nut (Areca catechu) Short Fiber/Seaweed Polypropylene Composite. *Journal of Natural Fibers, 7 (3),* 165-177.

[31] Van den Oever, M. J. A. and Snijder, M. H. B. (2008). Jute Fiber Reinforced Polypropylene Produced by Continuous Extrusion Compounding, part 1: Processing and Ageing Properties. *Journal of Applied Polymer Science, 110 (2),* 1009-1018.

[32] Fu, S.Y. Lauke, B. Ma der, E. Yue, C.-Y. and Hu, X. (2000).Tensile Properties of Short-Glass-Fiber- and Short-Carbon-Fiber-Reinforced Polypropylene Composites. *Composites: Part A, 31,* 1117–1125.

[33] Bledzki, A. K. Mamun, A. A. and Faruk, O. (2007). Abaca Fibre Reinforced PP Composites and Comparison With Jute and Flax Fibre PP Composites. *eXPRESS Polymer Letters, 1,* 755-762.

[34] Arzondo, L.M. Pérez, C.J. and Carella, J.M. (2005). Injection Molding of Long Sisal Fiber–reinforced Polypropylene: Effects of Compatibilizer Concentration and Viscosity on Fiber Adhesion and Thermal Degradation. *Polymer Engineering and Science, 45 (4),* 613-621.

[35] Thi-Thu-Loan Doan, Shang-Lin Gao and Ma der, E. (2006). Jute/Polypropylene Composites, I. Effect of Matrix Modification, *Composites Science and Technology, 66,* 952–963.

[36] Thomas, N. Abraham and George, K.E. (2009). Studies on Recyclable Nylon-reinforced PP Composites: Effect of Fiber Diameter. *Journal of Thermoplastic Composite Materials, 22(1),* 5-20.

[37] Quazi Shubhra, T.H. Alam, A.K.M.M. Mubarak Khan, A. Saha, M. Dipti Saha, and Gafur, M.A. (2010). Study on the Mechanical Properties, Environmental Effect, Degradation Characteristics and Ionizing Radiation Effect on Silk Reinforced Polypropylene/Natural Rubber Composites. *Composites: Part A, 41,* 1587–1596.

[38] Ruhul Khan, A. Parsons, A.J. Jones, I.A. Walker, G.S. and Rudd, C.D. (2010). Preparation and Characterization of Phosphate Glass Fibers and Fabrication of Poly(caprolactone) Matrix Resorbable Composites. *Journal of Reinforced Plastics and Composites, 29,* 1838-1844.

[39] Mubarak Khan, A. Ruhul Khan, A. Haydaruzzaman, Abul Hossain, Khan, A. H. (2009). Effect of Gamma Radiation on the Physico-Mechanical and Electrical Properties of Jute Fiber-Reinforced Polypropylene Composites. *Journal of Reinforced Plastics and Composites, 28 (13),* 1651-1660.

[40] Oksman, K. (2000). Mechanical Properties of Natural Fibre Mat Reinforced Thermoplastic. *Applied Composite Materials, 7,* 403-414.

[41] Rezaur Rahman, Md. Mahbub Hasan, Monimul Huque, Md. and Nazrul Islam, Md. (2010). Physico-Mechanical Properties of Jute Fiber Reinforced Polypropylene Composites. *Journal of Reinforced Plastics and Composites, 29 (3),* 445-455.

[42] Haydaruzzaman, Khan, A. H. Hossain, M. A. Mubarak Khan, A. Ruhul Khan, A. and Hakim, M. A. (2010). Fabrication and Characterization of Jute Reinforced Polypropylene Composite: Effectiveness of Coupling Agents. *Journal of Composite Materials*, published online 5 January, doi: 10.1177/0021998309356604.

[43] van den Oever, M.J.A. Bos, H.L. and van Kemenade, M.J.J.M. (2000). Influence of the Physical Structure of Flax Fibres on the Mechanical Properties of Flax Fibre Reinforced Polypropylene Composites. *Applied Composite Materials, 7,* 387-402.

[44] Garkhail, S.K. Heijenrath, R.W.H. and Peijs, T. (2000). Mechanical Properties of Natural-Fibre-Mat-Reinforced Thermoplastics based on Flax Fibres and Polypropylene. *Applied Composite Materials, 7,* 351-372.

[45] Qiang Yuan, Donglyang Wu, Januar Gotama and Stuart Bateman (2008). Wood Fiber Reinforced Polyethylene and Polypropylene Composites with High Modulus and Impact Strength. *Journal of Thermoplastic Composite Materials, 21,* 195-208.

[46] Quazi Shubhra, T. H. Alam, A. K. M. M. Beg, M.D.H. Mubarak Khan, A.and Gafur, M. A. (2 010). Mechanical and Degradation Characteristics of Natural Silk and Synthetic Phosphate Glass Fiber Reinforced Polypropylene Composites. *Journal of Composite Materials,* doi: 10.1177/0021998310380290.

[47] Avik Khan, Tanzina Huq, Saha, M. Ruhul Khan, A. Mubarak Khan, A.and Gafur, M. A. (2010). Effect of Silane Treatment on the Mechanical and Interfacial Properties of

Calcium Alginate Fiber Reinforced Polypropylene Composite. *Journal of Composite Materials,* doi: 10.1177/0021998310371536.

[48] Barkoula, N.M. Garkhail, S.K.and Peijs, T. (2010). Effect of Compounding and Injection Molding on the Mechanical Properties of Flax Fiber Polypropylene Composites. *Journal of Reinforced Plastics and Composites, 29 (9),* 1366-1385.

[49] Sushanta, K. Samal, Smita Mohanty and Sanjay Nayak, K. (2009). Polypropylene-Bamboo/Glass Fiber Hybrid Composites: Fabrication and Analysis of Mechanical, Morphological, Thermal, and Dynamic Mechanical Behavior, *Journal of Reinforced Plastics and Composites, 28,* 2729.

[50] Ajay Karmarkar, Chauhan, S.S. Jayant Modak, M. and Manas Chanda (2007). Mechanical Properties of Wood–Fiber Reinforced Polypropylene Composites: Effect of a Novel Compatibilizer With Isocyanate Functional Group. *Composites: Part A, 38,* 227–233.

[51] Mominul Haque, Rezaur Rahman, Nazrul Islam, Monimul Huque and Mahbub Hasan (2010). Mechanical Properties of Polypropylene Composites Reinforced with Chemically Treated Coir and Abaca Fiber. *Journal of Reinforced Plastics and Composites, 29 (15),* 2253-2261.

[52] Haydaruzzaman, Khan, A.H. Hossain, M.A. Ruhul Khan, A. Mubarak Khan, A. (2010). Mechanical Properties of the Coir Fiber-reinforced Polypropylene Composites: Effect of the Incorporation of Jute Fiber. *Journal of Composite Materials, 44 (4),* 401-416.

[53] Mominul Haque, Saiful Islam, Sakinul Islam, Nazrul Islam, Monimul Huque and Mahbub Hasan (2010). Physicomechanical Properties of Chemically Treated Palm Fiber Reinforced Polypropylene Composites. *Journal of Reinforced Plastics and Composites, 29,* 1734-1742.

[54] Ruhul Khan, A. Mubarak Khan, A. Sabrina Sultana, Nuruzzaman Khan, M. Quazi Shubhra, T.H. and Farhana Noor, G. (2010). Mechanical, Degradation, and Interfacial Properties of Synthetic Degradable Fiber Reinforced Polypropylene Composite. *Journal of Reinforced Plastics and Composites, 29(3),* 466-476.

[55] Kelly, A. and Tyson, W. R. (1965). Tensile Properties of Fibre Reinforced Metals: Copper/Tungsten, Copper/Molybdenum. *Journal of the Mechanics and Physics of Solids, 13.* 329-350.

[56] Drzal, L. T. Rich, M. J. Koeng, M. F. and Lloyd, P. F. (1983). Adhesion of Graphite Fibres to Epoxy Matrices. *Journal of Adhesion, 16,* 133-152.

[57] Wu, H. F. and Netravali, A. N. (1992). Weibull Analysis of Strength-Length Relationships in Single Nickell on SiC Fibres, *Journal of Material Science, 27,* 3318-3324.

[58] Karlsson, J.O. Blachot, J.F. Peguy, A. and Gatenholm (1996). Improvement of Adhesion Between Polyethylene and Regenerated Cellulose Fiber by Surface Fibrillation. *Polymer Composites, 17,* 300–304.

[59] Lu, J.Z. Wu, Q. and Harold, S.M. (2000). Chemical Coupling in Wood Fibre and Polymer Composites: A review of Coupling Agents and Treatments. *Wood and Fibre Science, 32 (1),* 88 – 104.

[60] Alireza Ashori, Amir Nourbakhsh (2009). Polypropylene Cellulose-Based Composites: The Effect of Bagasse Reinforcement and Polybutadiene Isocyanate Treatment on the Mechanical Properties. *Journal of Applied Polymer Science, 111,* 1684–1689.

[61] Lai, S. M. Yeh, F. C. Wang, Y. Chan, H. C. and Shen, H. F. (2003). Comparative Study of Maleated Polyolefins as Compatibilizers for Polyethylene/Wood Flour Composites, *Journal of Applied Polymer Science, 87,* 487-496.

[62] Maiti, S. N. Subbarao, R. and Ibrahim, M. N. (2004). Effect of Wood Fibers on the Rheological Properties of i-PP/wood Fiber Composites. *Journal of Applied Polymer Science, 91,* 644-650.

[63] Xiang Fu, Bobing He and Xian Chen (2010). Effects of Compatibilizers on Mechanical Properties of Long Glass Fiber-Reinforced Polypropylene. *Journal of Reinforced Plastics and Composites , 29,* 936-949.

[64] Sanjay Nayak, K. and Smita Mohanty (2010). Sisal Glass Fiber Reinforced PP Hybrid Composites: Effect of MAPP on the Dynamic Mechanical and Thermal Properties. *Journal of Reinforced Plastics and Composites, 29,* 1551-1568.

[65] Mohanty, S. Nayak, S. K. Verma, S. K. and Tripathy, S. S.(2004). Effect of MAPP as Coupling Agent on the Performance of Sisal-PP Composites. *Journal of Reinforced Plastics and Composites, 23,* 2047-2063.

[66] Anuar, H. Ahmad, S. H. Rasid, R. and Zakaria, S. (2005). Tensile Behavior and Morphological Studies of TPNR-KF-CF Hybrid Composite. *American Journal of Applied Sciences (Special Issue),* 34-39.

[67] Yuan, X. Zhang, Y. and Zhang, X. (1999). Maleated Polypropylene as a Coupling Agent for PP-waste News Paper Flour Composites. *Journal of Applied Polymer Science, 71,* 333-337.

[68] Ray, D. Sarkar, B. K. Das, S. and Rana, A. K. (2002). Dynamic Mechanical and Thermal Analysis of Vinylester-Resin-Matrix Composites Reinforced with Untreated and Alkali-Treated Jute Fibers. *Composites Science and Technology, 62,* 911–917.

[69] Ray, D. Sarkar, B. K. Basak, R. K. and Rana, A. K. (2002). Study of Thermal Behaviour of Alkali-Treated Jute Fibers, *Journal of Applied Polymer Science, 85,* 2594–2599.

[70] Rong, M. Z. Zhang, M. Q. Liu, Y. Zhang, Z. W. Yang, G. C. and Zeng, H. M. (2002). Mechanical Properties of Sisal Reinforced Composite in Response to Water Absorption. *Polymer and Polymer Composites, 10(6),* 407–426.

[71] Pothan, L. A. and Thomas, S. (2003). Polarity Parameters and Dynamic Mechanical Behaviour of Chemically Modified Banana Fiber Reinforced Polyester Composites. *Composites Science and Technology, 63,* 1231–1240.

[72] Herrera-Franco, P. J. and Valadev-Gonzales, A. (2004). Mechanical Properties of Continuous Natural Fiber-Reinforced Polymer Composites. *Composites: Part A, 35,* 339–345.

[73] Khan, M. A., Rahman, M. M. and Akhunzada, K. S. (2002). Grafting of Different Monomers onto Jute Yarn by In Situ UV-Radiation Method: Effect of Additives. *Polymer Plastic Technology and Engineering, 41(4),* 677–689.

[74] Gassan, J. and Gutowski, V. S. (2000). Effects of Corona Discharge and UV Treatment on the Properties of Jute-Fiber Epoxy Composites. *Composites Science and Technology, 60(15),* 2857–2863.

[75] George, J. Bhagawan, S. S. and Thomas, S. (1998). Improved Interactions in Chemically Modified Pineapple Leaf Fiber Reinforced Polyethylene Composites. *Composite Interfaces, 5(3),* 201–223.

[76] Pothan, L. A. George, J. and Thomas, S. (2002). Effect of Fiber Surface Treatment on the Fiber-Matrix Interaction in Banana Fiber Reinforced Polyester Composites. *Composite Interfaces, 9(4),* 335–353.

[77] Haydar Zaman, U. Mubarak Khan, A. Ruhul Khan, A. Arifur Rahman, M. Lily Das, R. and Al-Mamun Md. (2010). Role of Potassium Permanganate and Urea on the Improvement of the Mechanical Properties of Jute Polypropylene Composites. *Fibers and Polymers 11(3),* 455-463.

[78] Mubarak Khan, A. Ruhul Khan, A. Haydar Zaman, U. Noor-A-Alam, M.and Hoque, M. A. (2010) Effect of Surface Modification of Jute with Acrylic Monomers on the Performance of Polypropylene Composites. *Journal of Reinforced Plastics and Composites, 29 (8),* 1195-1205.

[79] Guillermo Cantero, Arbelaiz, A. Llano-Ponte, R. and Mondragon, I. (2003). Effects of Fiber Treatment on Wettability and Mechanical Behavior of Flax/Polypropylene Composites. *Composites Science and Technology, 63,* 1247–1254.

[80] Arbelaiz, A. Ferna´ndez, B. Cantero, G. Llano-Ponte, R. Valea, A. and Mondragon, I. (2005). Mechanical Properties of Flax Fibre/Polypropylene Composites. Influence of Fibre/Matrix Modification and Glass Fibre Hybridization. *Composites: Part A, 36,* 1637–1644.

[81] Beg, M.D.H. and Pickering, K.L. (2008). Reprocessing of Wood Fibre Reinforced Polypropylene Composites. Part I: Effects on Physical and Mechanical Properties, *Composites: Part A, 39,* 1091–1100.

[82] Broido, A. (1969). A Simple, Sensitive Graphical Method of Treating Thermogravimetric Analysis Data. *Journal of Polymer Science Part A-2, 7,* 1761.

In: Fiber-Reinforced Composites
Editor: Qingzheng Cheng

ISBN: 978-1-61470-303-7
© 2012 Nova Science Publishers, Inc.

Chapter 2

NATURAL FIBER-REINFORCED COMPOSITES

*Di Bella Guido[*1], Fiore Vincenzo[2] and Valenza Antonino[2]*
[1] CNR ITAE; Via Salita Santa Lucia sopra Contesse 5, 98126 Messina, Italy
[2] Department of Civil, Environmental and Aerospace Engineering;
University of Palermo, Viale delle Scienze, 90166 Palermo, Italy

ABSTRACT

Over the last decade, composites of polymers reinforced with natural fibers have received increasing attention, both from the academic world and from several industries. There is a wide range of natural fibers which can be applied as reinforcements or fillers thanks to their properties and availability; i.e. flax, hemp, jute, kenaf and sisal.

Natural fibers are mainly attractive for the following reasons: specific properties, price, health advantages and recyclability.

Particularly, industry is getting more and more interested in environment-friendly products and therefore the research on natural fiber based on composite materials is gaining importance.

Some of the benefits linked to the usage of such natural composites are their low density and good specific properties. Furthermore, they are renewable and have a CO_2-neutral life cycle, in contrast with their synthetic opponents (i.e. glass and carbon).

Despite such good premises, these fibers have also some negative characteristics: they are highly hydrophilic and their properties may vary in time.

The natural fibers have a complex structure of elementary fibers, consisting of cellulose, hemicellulose, pectin, lignin and others and thus they should not be considered as monofilament fibers.

Mechanical, physical and even chemical properties of these fibers are strongly harvest-dependent, influenced by climate, location, soil characteristics, and weather circumstances. They are affected also by fiber processing (i.e. retting, scutching, bleaching, spinning) and by their incorporation into composites: handling, impregnation and consolidation may introduce supplementary changes. Obviously, such variability complicates the prediction and the evaluation of the composite properties.

---

[*] Email: guido.dibella@itae.cnr.it

Another known problem in natural fiber reinforced composites is the poor interface quality between the fibers and the polymer matrix; chemical pre-treatments are often applied in order to enhance the adhesion between these components.

In the last few years several new components based on natural fiber composites have been mainly developed by industries in the automotive field; i.e. in 2003 around 43,000 tonnes of natural fiber have been used by the European automotive industry as composite reinforcement. This interest is due to the excellent tensile mechanical properties of fiber obtained from plants such as flax, hemp and the stinging nettle.

Other main applications of the natural fibers are in the following fields: nautical and biomedical ones.

The aim of this work is to present a review on the main natural fibers investigated by researchers and to report the results of some original studies performed by the Authors.

1. INTRODUCTION

Fibers are hair-like materials that appear as continuous filaments. There are two types of fiber: natural and man-made (or synthetic). Natural fibers are classified according to their origins [1]:

- cellulose (from plants/vegetable): they can be obtained from different parts of plants (i.e. seed, stem, leaves, fruit). They include hairs (cotton, kapok), fiber-sheafs of dicotylic plants or vessel-sheafs of monocotylic plants (flax, hemp, jute, ramie) and hard-fibers (sisal, henequen, coir). They can be divided also in wood (subdivided into soft or wood and hard wood) and non-wood [2, 3] fibers;
- protein (from animals): these fibers are grouped under the categories of hair (wool), fur (angora) and secretions (silk);
- mineral: among these, the most important ones are asbestos, which is hazardous for the health, and basalt.

From this classification, it emerges clearly that the term "natural fibers" covers a wide range of vegetable, animal and mineral fibers. However, in the composite industries, the term usually refers to wood fiber and agrobased bast, leaf, seed, and stem fibers. These fibers often contribute to the structural performance of the plant and, when used in plastic composites, they can produce significant reinforcement [4].

Cellulose fibers, depending on the part of the plant from which they are taken, can be classified as:

- Grasses and reeds: These fibers come from the stems of monocotyledonous plants such as bamboo and sugar cane. Both types of fibers can be used to reinforce plastics.
- Leaf fibers: These fibers run lengthwise through the leaves of most monocotyledonous plants such as sisal, henequen, abaca and esparto. They are also called 'hard fibers' and are the most commonly employed as reinforcing agents in plastics.
- Bast fibers: These fibers (bundles) come from the inner bark (phloem or bast) of the stems of dicotyledonous plants. Common examples are jute, flax, hemp and kenaf.

- Seed and fruit hairs: These fibers come from seed-hairs and flosses, such as cotton and coconut.
- Wood fibers: These fibers come from the xylem of angiosperm (hardwood) and gymnosperm (softwood) trees. Examples include maple, yellow poplar, pine and spruce.

There are also different, but still valid, classification criteria [3]:

- Soft/stem or bast fibers (i.e flax, ramie, hemp and jute): these fibers are typical of many dicotyledonous plants and they are grouped outside the xylem, in the cortex, phloem and pericycle (bark). Typically, they form clusters of several cells that, in some cases, may function as the "fiber" in spinning. Each fiber cell is tenaciously cemented to the adjacent fiber by the pectic middle lamella. The strands are quite strong and durable and they are able to withstand bleaching or other harsh treatments.
- Hard/leaf or structural fibers (i.e. sisal): these ones are strands of small, short cells found in monocotyledonous plants. They form the supportive and conductive strands of leaves (even though few monocots have conspicuous woody stems) and they are known as fibro-vascular bundles. They consist of both xylem and phloem and various sheathing cells that are scattered through a sleek pithy matrix of the leaf or stem. The cells are lignified to a greater or lesser degree and they are hard if compared with the soft fibers found in dicots, in which the cellulose is largely associated with pectic materials. The entire fibro-vascular bundle functions as a single fiber and usually cannot be bleached or chemically treated. They are less durable than soft fibers.
- Surface fibers: these fibers are found on the surface of stems, leaves, fruits, seeds, etc. The most important plant fiber is cotton, others include kapok (used for stuffing).

Despite being used in composites for many years, interest in natural fibers has waned with the development of synthetic fibers such as glass and carbon fibers. However, in recent years, the interest of both the academic world and several industries in the composites of polymers, reinforced with vegetables fibers, has broadened again. There is a wide range of different natural fibers usable as reinforce or filler is wider. Figure 1 presents the classification of the main employed fibers. All these natural fibers are constituted by long cells having relatively thick walls, that gives them stiffness and strength. Both wood and non-wood fibers, are made up of three main cell wall polymers: cellulose, lignin and matrix polysaccharides (such as pectin and hemicellulose) connected to the cellulose and lignin in the cell wall (i.e. flax, hemp, and ramie are the cellulose associated with pectin material, while jute fibers are more lignified). A high percentage of cellulose in fibers has a positive effect on strength and durability and, similarly, low moisture content of the cell wall is usually indicative of physical superiority [5, 6].

Among vegetable fibers, the most used in composite materials are flax, hemp, jute, kenaf and sisal, thanks to their properties and availability. In fact, flax, hemp, jute and kenaf develop in the bast of the plant. They have more or less similar morphologies properties and, consequently, similar functions in the composite.

The following are the reasons that make natural fibers interesting in many different fields [7]:

- specific properties: natural fibers based on cellulose have a relatively low density and they are relatively stiff and strong due to their structure. As a consequence, their specific properties (i.e. the properties divided by the density) are good and actually comparable to those of glass fibers;
- price: natural fibers are in most cases cheaper than glass fibers, depending on their quality. They can subsequently replace glass fibers in composite structures.
- health advantages: natural fibers are thought to have a lower impact on the health of people producing the composites. They do not cause skin irritations and they are not believed of causing lung cancer. This is a strong point in favour of natural fibers as very small glass fibers are indicated as a cause of lung cancer;
- recyclability: natural fibers are recyclable.

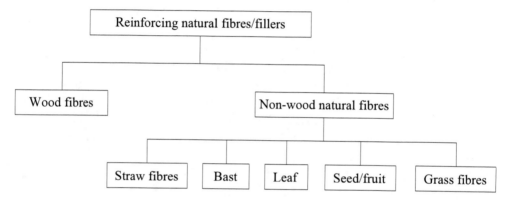

Figure 1. Classification of the vegetable fibers.

Table 1. Consumption of natural fibers tonnes in the European automotive industry [tonnes]

Fiber	1996	1999	2000
Flax	1,800	15,900	20,000
Hemp	-	1,700	3,500
Jute	2,100	3,400	1,700
Sisal	1,100	500	100
Kenaf	400	900	2,000
Coir	-	-	1,000
Total	5,400	22,400	28,300

The growth of natural fiber plastic composites is lead by the automotive industry, particularly in Europe, where they are successfully used because of their low density and the low impact on the environment.

Table 1 shows the increase of consumption of natural fibers by the European automotive in the last years [9]. Increased technical innovations, identification of new applications, continuing political and environmental pressures, and government investment in new methods for fiber harvesting and processing suggest a continuing growth in the use of natural fibers in composites [4].

The use of natural fibers in composite structures has its root in the field of the biomimetic, also known as biomimicry [10], which is considered an abstraction of a good design of nature [11]. This is a new field of science that studies how nature designs, manufactures and assembles/disassembles the molecules in order to produce structures with high performances (i.e. mollusc shells, bone, skin, tendons, insect fly, eggs shells, lotus leaves) creating a fundamental tool to engineer and optimize new materials with specific properties [12, 13]. In more formal terms, biomimetic is "a strategic tool designed for creating an advanced and practical technology or materials whose clues can be obtained from actual biological structures and functions" [12].

The considerations above involve two main issues: how the ideas can be found in nature and how they can possibly apply to materials engineering. Strict contact with nature [14plays then a fundamental role because it allows to detect the feature of the natural world that may lead to innovative applications in material engineering.

As Janine Benyus [14] suggests, to be inspired by nature, four steps have to be undertaken:

1) Quieting (immerse ourselves in nature);
2) Listening (interview the flora and fauna of our own planet);
3) Echoing (encouraging biologists and engineers to collaborate, using nature as a model and measure);
4) Stewarding (preserving life diversity and genius).

Following the indications mentioned, the best approach to biomimetic design of materials seems to be a multidisciplinary one. This project can be put into action by designing new materials from biological tissues, with higher environmental friendliness, and by designing materials which can serve multiple purposes during their lifetime, such as "smart packaging" in eggs or shells and "self cleaning surfaces" in lotus (Nelumbo nucifera) leaves [15].

As these examples state, biomimetics requires knowledge in both the fields of biology and engineering, together with physics, chemistry and science of materials.

This new natural approach has gained importance in the last few years because of the growing interest in sustainable development. The sustainability of a material is becoming an important feature in material technology [16] both in the production phase (because of the immediate impact on man's health) and in end of the life cycle of a product (because of the environmental impact); i.e. in lay-up process the production of composite laminates in yachts manufacturing implies the use of highly toxic materials (polyester resins, glass fibers).

Regrettably, as the studies concerning the recycling of the composite materials are outdated, recycling procedures are expensive and complicated. [16]. The replacement of synthetic fibers with natural or recycled ones can be advantageous because they come from renewable and cheap sources and reduce the problems of incineration-disposal (solid/gaseous residua) [17].

2. Advantages and Limits

Generally, they are mentioned four main reasons which make the application of natural fibers attractive; they are the following:

– specific properties: natural fibers based on cellulose have a relatively low density, and they are relatively stiff and strong. Therefore their specific properties, i.e. the properties divided by the density, or the 'properties per kilo', are rather high, and at present they are comparable to those of glass fibers;
– price: natural fiber reinforced composites are originally aimed at the replacement of glass fiber-reinforced composites. Depending on the exact quality of fiber needed, natural fibers are in most cases cheaper than glass fibers;
– health advantages: natural fibers are also expected to give less health problems for the people producing the composites. Natural fibers do not cause skin irritations and they are not suspected of causing lung cancer. This is especially an issue since the discussion on whether or not very small glass fibers can cause lung cancer, it hasn't still ended.;
– recyclability: natural fiber composites are also often claimed to be recyclable. There is some confusion on this issue and, for example, different sources do not agree on the feasibility of flax fiber composite recycling. For mechanical recycling -in the production process itself- natural fibers have no clear advantage over glass fibers, both fibers will suffer from a second processing step, and for flax fibers there is a chance of additional thermal degradation during this step. A definite advantage of flax fiber composites over glass fiber composites, however, is the fact that they can be burned (euphemistically called thermal recycling) without leaving large amounts of slag. All in all, the use of natural fibers has a definite 'green image'.

Nevertheless, these promising fibers possess also some negative characteristics:

– they are highly hydrophilic;
– they can vary a lot in properties;

which makes accurate predictions of the respective composite properties difficult.???

3. Properties of Natural Fibers

3.1. Production Processes

The major steps in producing natural fibers for use in composites include:

– harvesting of the fiber-bearing plants: there are several methods for harvesting most natural fibers since they are used not only in composites but also in the manufacture of other products. For example, fibers derived from wood are used in paper and forest products industries, flax fiber is used to make linen and cigarette papers, jute fiber is used for rope and burlap [18]. Since many natural fibers come from annual crops, issues such as storage and variability in the growing season have to be considered. Consequently, Europe is largely funding new harvesting and fiber separation technologies for natural fibers such as flax [19];

— extraction of the fibers: fiber extraction procedures depend on the type of plant and the part from which the fibers are derived (i.e. bast, leaves, wood), as well as performances and costs. Fiber bearing plants have very different anatomies (i.e. tree versus dicotyledonous plants) and often fibers are derived from agricultural residues or by-products from industry [19]. Consequently, the needs in processing vary greatly. Wood is primarily composed of hollow, elongated, spindle-shaped cells (called tracheids or fibers) running parallel along the trunk of the tree [20]. These fibers are firmly cemented together and form the structural component of wood tissue. Fibers are extracted from wood with mechanical or chemical treatments during the pulping process. Bast fibers, such as flax or kenaf, have differently structures and, consequently, they are processed quite differently. They are found in the inner bark of the stems of dicotyledons and typically form less than 30% of the stem [19]. Inside the inner bark there is a woody core (called the "shive") with very shorter fibers [18]. Fiber strands are removed from the bast. These fiber strands are very long (i.e. several meters) and are actually fiber bundles of overlapping single ultimate fibers;

— further processing of the raw fiber to meet required purity and performances: bast fibers are processed by various means that include retting, breaking, scutching, hackling and combing [18, 19]. The specific process used depends, to a large extent, on the type of the plant and the fiber source. For example, flax fiber can be obtained from different flax plants or from by-products of linen or flax seed production [18]. As previously stated, useful natural fibers have also been derived from other parts of plants: i.e. leaves (i.e. sisal), seeds (i.e. coir), or grass stems [21]. The production of these fibers changes significantly according to the fiber type. Most natural fiber used in composites is currently made into fiber mats, that are often needled, thermally fixed with small amounts of polymer fibers, or otherwise modified to improve handling and then press moulded [21]. However, this additional step increases cost. The fiber mats can be used also to reinforce thermosetting resins in several processes such as hand lay-up, vacuum bagging or vacuum infusion. The use of short fibers in more conventional processes such as injection moulding is planned to increase in the future [22]. Some progress has been reported on methods for pelletizing fibers to facilitate their introduction into polymer compounding equipment [23].

3.2. Chemical Properties

The structure and chemical make-up of natural fibers changes significantly and depends on the source and a lot of processing variables. Nevertheless, it is possible to identify some common features. Natural fibers are complex, three-dimensional, polymer composites made up primarily of cellulose, hemicellulose, pectins and lignin. Component mean values of some plant-fibers are shown in Table 2 [24]. It is possible to observe that the principal components of the fiber cell walls are cellulose, hemicelluloses and lignin with pectin normally considered to be the main binder (Table 2).

Table 2. Composition of different cellulose based natural fibers

Component	Flax	Jute	Ramie	Sisal
Cellulose	64.1	64.4	68.6	65.8
Hemi-cellulose	16.7	12.0	13.1	12.0
Lignin	2.0	11.8	0.6	9.9
Pektin	1.8	0.2	1.9	0.8
Water	10.0	10.0	10.0	10.0
Water soluble	3.9	1.1	5.5	1.2
Wax	1.5	0.5	0.3	0.3

3.2.1. Cellulose

Cellulose is characterised by the least variation in chemical structure and can be considered the major framework component of the fiber.

Cellulose is a polydisperse linear homopolymer, consisting of regio- and enantioselectively β-1,4-glycosidic linked d-glucopyranose units (so-called anhydroglucose units [AGU]) with a degree of polymerization of around 10000 (Figure 2).

It has been shown by ^{1}HNMR spectroscopy that the β-d-glucopyranose adopts the $^4C_1$ chain conformation, the lowest free energy conformation of the molecule. As a consequence, the hydroxyl groups are positioned in the ring plane (equatorial), while the hydrogen atoms are in the vertical position (axial). The polymer contains free hydroxyl groups at the C-2, C-3 and C-6 atoms. Based on the OH groups and the oxygen atoms of both the pyranose ring and the glycosidic bond, ordered hydrogen bond systems form various types of supramolecular semi-crystalline structures [25].

Cellulose is a strong, linear (crystalline) molecule with no branching. It is the main component providing the strength, stiffness and structural stability. It has good resistance to hydrolysis although all chemical and solution treatments will degrade it to some extent [26].

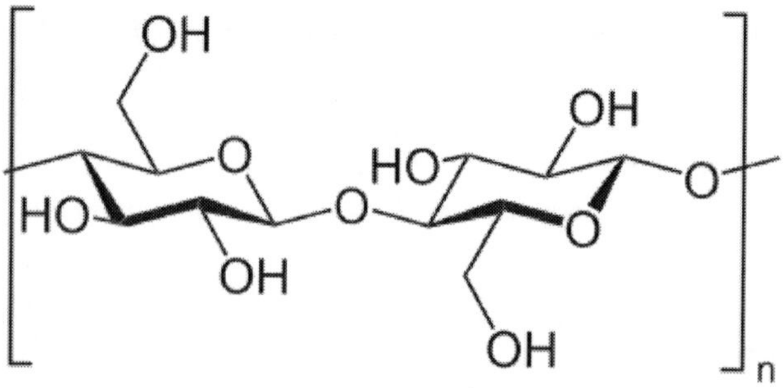

Figure 2. Molecular structure of cellulose.

Vincent [11] states that the elementary fibril of cellulose is about 3.5 nm diameter and contains about 40 molecules. The cellulose elementary fibrils can be arranged into larger fibrils of 20–25 nm diameter. They are formed by adjacent cellulose chains forming hydrogen bonds leading to partially crystalline regions (micelles). These form a strong structural framework in the cell walls [26].

3.2.2. Hemicellulose

Hemicelluloses are lower molecular weight polysaccharides, often copolymers of glucose, glucuronic acid, mannose, arabinose and xylose, that form random, amorphous branched or nonlinear structures with low strength (Figure 3). The hemicellulose differs from cellulose in three important aspects [27]:

1) it contains several different sugar units while cellulose contains only 1,4-b-d-glucopyranose units;
2) it exhibits a considerable degree of chain branching, while cellulose is a strictly linear polymer;
3) the degree of polymerization of native cellulose is ten to one hundred times higher than that of hemicellulose.

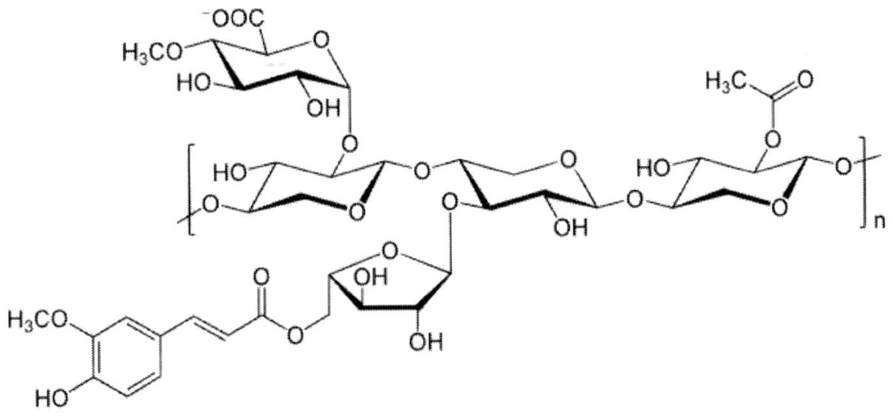

Figure 3. Molecular structure of hemicellulose.

3.2.3. Lignin

Lignin is an amorphous, cross-linked polymer network (Figure 4) consisting of an irregular array of variously bonded hydroxy- and methoxy-substituted phenylpropane units [28]. Its chemical structure varies depending on its source. Lignin is less polar than cellulose and acts as a chemical adhesive both within and between fibers.

It is formed by non-reversible removal of water from sugars (primarily xylose) to create aromatic structures. Lignification progresses as the plant matures thus conferring mechanical stability to the plant. As lignin becomes more rigid, it places away from the lumen surface and porous wall regions to maintain wall strength and permeability and help with the transport of water. Lignin resists attack by most microorganisms as the aromatic rings are resistant to anaerobic processes while aerobic breakdown of lignin is slow. The mechanical properties are lower than those of cellulose [27].

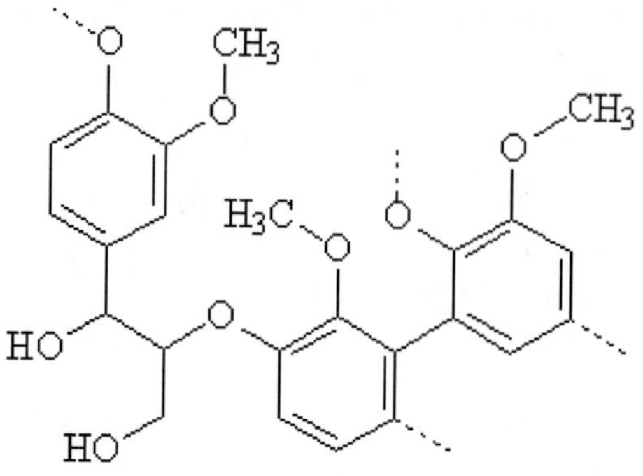

Figure 4. Molecular structure of lignin.

3.2.4. Pectin

Pectins are complex polysaccharides, whose main chains consist of a modified polymer of glucuronic acid and residues of rhamnose. Their side chains are rich in rhamnose, galactose and arabinose sugars. The chains are often cross-linked by calcium ions, improving structural integrity in pectin-rich areas. Pectins are important in non-wood fibers, especially bast fibers. Lignin, hemicelluloses and pectins collectively function as matrix and adhesive, helping to hold together the cellulose framework structure of the natural composite fiber [21]. Pectin is soluble in water only after a partial neutralization with alkali or ammonium hydroxide [27].

3.2.5. Waxes

Natural fibers contain also small amounts of additional extraneous components, including low molecular weight organic components (extractives) and inorganic matter (ash). Though extractives are often small in quantity, they can have large influences on properties such as colour, odour and decay resistance [28]. The high ash content of some natural materials, such as rice hulls, arouses some concern about their abrasive nature.

These waxy materials consist of different types of alcohols, that are insoluble in water as well as in several acids (palmitic acid, oleaginous acid, stearic acid) [27].

3.3. Physical Properties

A single natural fiber consists of several cells. These cells are formed by crystalline microfibrils based on cellulose that are connected to a complete layer, by amorphous lignin and hemicellulose. Multiple of such cellulose–lignin/hemicellulose layers in one primary and three secondary cell walls stick together to a multiple-layer-composites, the cell, as shown in Figure 5 [37].

From Figure 5 it is clear that each elementary fiber normally has a central hollow channel (the lumen). This characteristic introduces additional porosity into the composite (voids within synthetic fibers are comparatively rare) and reduces the cross-section of fiber material available to carry load.

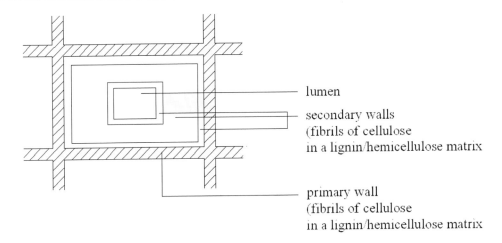

Figure 5. Constitution of a natural fiber cell.

Table 3. Structure parameters of different cellulose base natural fibers

Fiber	Cellulose content [wt %]	Spiral angle [°]
Coir	43	45.0
Flax	71	10.0
Hemp	78	6.2
Jute	61	8.0
Ramie	83	7.5
Sisal	67	20.0

These cell walls differ in composition (ratio between cellulose and lignin/hemicellulose, Table 2) and in orientation (spiral angle, Table 3 [29]) of the cellulose microfibrils. Both the content of cellulose and the spiral angle of the fibrils determine the mechanical properties of the cellulose.

The Krenchel [30] equation allows to calculate the effectiveness of misaligned fiber reinforcement, as a fiber orientation distribution factor, using the proportions of fiber at each angle and the fourth power of the cosine of the angle between the fiber and the reference direction.

3.4. Fiber Dimensions, Density, and Mechanical Performance

The dimensions of natural fiber, as well as their physical and mechanical performance, can vary to a large extent (i.e. different species, natural variability within each species, differences in climates and growing seasons). Methods of producing fibers with more reproducible properties are the main aim of several studies [31]. Most natural fibers have a maximum density of about 1.5 g/cm^3. Though some natural fibers, such as wood, are hollow and have low densities in their native state, they are often densified during processing. Nevertheless, even the maximum density of these fibers is less than that of inorganic fibers such as glass fibers. Consequently, their low density makes them attractive as reinforcements in applications where weight plays an important role (i.e. automotive or naval). Table 4 [20, 32] reports the dimensions of some natural fibers.

Table 4. Dimensions of some natural fibers

Fiber	Length [mm]	Width [μm]
Flax	33	19
Hemp	25	25
Jute	2	20
Kenaf	5	21
Sisal	3	20

Generally, cellulose has been estimated to have a modulus of 140 GPa, using X-ray diffraction to determine the strain [11]. A slightly higher value can be obtained by calculation from the chemical structure of the crystal considering both the straightening of the covalent bonds and the stretching of the interchain hydrogen bonds. If hydrogen bonding is not included in the calculation, the modulus drops by a factor of about 8. Experimentally measured modules will inevitably be lower than theoretical values (due to <100% crystallinity and off-axis fiber orientation).

Vincent [11] has reported values of 100 GPa for dry flax and about 80 GPa for wet flax. As water penetrates the amorphous regions of cellulose, the stiffness can drop by a factor of 2–4 as the contribution of the hydrogen bonding is progressively removed. Table 5 presents some mechanical property data for natural fibers.

Table 5. Mechanical properties of the main natural fibers

Fiber	Density [g/cm^3]	Elongation [%]	Tensile strength [MPa]	E modulus [GPa]
Cotton [33-35]	1.5-1.6	7.0-8.0	287-597	5.5-12.6
Jute [21, 33, 34, 36, 37]	1.3-1.5	1.5-3.0	393-860	20.0-60.0
Flax [21, 34, 38]	1.4-1.5	1.5-4.0	345-1825	27.6-73
Kenaf [39, 40]	1.2	-	240-600	14.0-38.0
Nettle [41]	-	1.3-2.9	954-2234	59.0-115.0
Hemp [21, 37, 42]	1.5	1.6-4	300-920	30.0-70.0
Ramie [34, 35]	-	3.6-3.8	400-938	61.4-128.0
Sisal [33, 35, 43]	1.5	2.0-2.5	511-635	9.4-22.0
Coir [33, 43]	1.2	30.0	175	4.0-6.0
Viscose (cord) [35]	-	11.4	593	11.0
Hardwood [21]	1.4	-	90-180	10.0-70.0
Soft wood kraft [21, 44]	1.4-1.5	-	100-170	10.0-50.0
E-glass [37, 45]	2.5	2.5-3.4	2000-3500	70.0-71.0
S-glass [33, 45]	2.5	2.8	4570	86.0
Aramide [45]	1.4	3.3-3.7	3000-3150	63.0-67.0
Carbon [45]	1.4	1.4-1.8	4000	230.0-240.0

Mechanical performances of fibers are good, but they are not as good as that of synthetic fibers such as glass. Moreover, the variability range is remarkably wider than that of glass-fibers due to the differences in fiber structure. However, their densities are considerably lower. The balance of significant reinforcing potential at low cost and low density is one of the factors that contributes to their success in the industrial field.

3.5. Moisture and Durability

The main chemical constituents of natural fibers contain hydroxyl and other oxygen-containing groups that attract moisture through hydrogen bonding [46]. The moisture content of these fibers can vary greatly depending on the fiber type. The processing of the fiber can also significantly affect moisture sorption. Table 6 [32] shows the wide range of moisture contents for some natural fibers at several values of relative humidity.

Table 6. Equilibrium moisture content at 27 °C of some natural fibers

Fiber	relative humidity		
	30 %	65 %	90 %
Bamboo	4.5	8.9	14.7
Bagasse	4.4	8.8	15.8
Jute	4.6	9.9	16.3

This hygroscopicity creates problems both in composite fabrication and in the performance of the end product. In order to allow the usage of natural fibers, a process that is insensitive to moisture has to be used or the fibers have to be dried before or during processing.

Natural fibers absorb less moisture in the final composites since they are at least partially encapsulated by the polymer matrix. However, even small quantities of absorbed moisture can affect performances. Moisture can plasticize the fiber, altering the composite properties. Moreover, volume changes in the fiber associated with moisture sorption can reduce fiber-matrix adhesion and damage the matrix [46]. Methods of reducing moisture sorption include: adequately dispersing and encapsulating fibers in the matrix, limiting fiber content, improving fiber-matrix bonding, chemically modifying the fiber, treating the fibers surface, or simply protecting the composite from moisture exposure.

Natural fibers undergo photochemical degradation when exposed to UV radiation [47]. They are degraded biologically because organisms recognize the chemical constituents in the cell wall and can hydrolyze them into digestible units using specific enzyme systems [47]. Even though the degradability of natural fibers can be a disadvantage in durable applications where composites are exposed to harsh environments, it can also be an advantage. Due to their low thermal stability, natural fibers are generally processed with plastics which are not required high temperature resistance (less than about 200 °C). Above such temperatures, many of the polymeric constituents in natural fibers begin to decompose.

Since cellulose is more thermally stable than other chemical constituents, highly pulped fibers, made up almost completely by cellulose have been used to extend this processing window [48]. The release of volatile gases can, before, during, and after processing, lead to odour issues in applications where the composite is in an enclosed environment, such as in many automotive applications, especially when moisture is present [49].

4. Methods for Surface Modification of Natural Fibers

The quality of the fiber–matrix interface is significant for the application of natural fibers as reinforcement fibers for plastics. Physical and chemical methods can be used to optimize this interface. These modification methods produce different efficiency for the adhesion between matrix and fiber [50].

4.1. Physical Methods

Physical methods, such as stretching [51], calandering [52, 53], thermotreatment [54], and the production of hybrid yarns do not change the chemical composition of the fibers. Physical treatments change only structural and surface properties of the fiber and thereby influence the mechanical bonding to polymers.

Corona treatment (i.e. electric discharge) is one of the most interesting techniques for surface oxidation activation. This process changes the surface energy of the cellulose fibers [55] and in case of wood surface activation increases the amount of aldehyde groups [56].

The same effects are obtained with cold plasma treatment. Depending on type and nature of the used gases, a variety of surface modifications can occur: introduction of surface crosslinkings, increase or decrease of surface energy, production of reactive free radicals [55] and groups [57].

Electric discharge methods are known [58] to be very effective for "non-active" polymer substrates as polystyrene, polyethylene, polypropylene, etc. They are successfully used for cellulose-fiber modification, to decrease the melt viscosity of cellulose–polyethylene composites [59] and improve mechanical properties of cellulose–polypropylene composites [55].

An old method of cellulose fiber modification is mercerization [42, 60, 61]. It has been widely used on cotton textiles. Mercerization is an alkali treatment of cellulose-fibers, differing on the basis of the type and concentration of the alkaline solution, its temperature, time of treatment, tension of the material as well as on the additives [61, 62]. At the present time there is a tendency to use mercerization on natural fibers as well. Optimal conditions of mercerization grant the improvement of the tensile properties and absorption characteristics [42, 61, 62].

4.2. Chemical Methods

Strongly polarized cellulose fibers [63] are inherently incompatible with hydrophobic polymers [64-66]. When two materials are incompatible, it is often possible to obtain compatibility by introducing a third material that has intermediate properties. There are several mechanisms [67] of coupling in materials:

- Weak boundary layers: coupling agents eliminate weak boundary layers;
- Deformable layers: coupling agents produce a tough, flexible layer;
- Restrained layers: coupling agents develop a highly crosslinked interphase region, with a modulus intermediate between that of substrate and of the polymer;

- Wettability: coupling agents improve the wetting between polymer and substrate (critical surface tension factor);
- Chemical bonding: coupling agents form covalent bonds with both materials;
- Acid-base effect: coupling agents alter acidity of substrate surface.

The development of a definitive theory for the mechanism of bonding by coupling agents in composites is a complex problem. The main chemical bonding theory alone is not satisfactory. So the consideration of other concepts, including the morphology of the interphase, the acid–base reactions at the interface, surface energy and the wetting phenomena, appears to be necessary.

5. SELECTED NATURAL FIBERS

FLAX

Name

Linum usitatissimum; Linum angustifolium

Description

Flax is an annual crop for multipurpose use: fiber and oil production, non-textile applications (i.e. in composites). The parts composing the plant are the roots, the stem, the capsular and the seeds, but the relevant part for fiber production is the stem. The stem has a length which goes from 600 to 1000 mm. The top branches out, developing a rispy inflorescence. The sclerenchyma cells (elementary fiber) are organised in bundles, radial around the cambium and wooden body, surrounding the lumen. The fiber bundle is covered by a bark structure, epidermis and cuticle on the outside, requiring decortication. For textile use, those bundles have to be removed softly in order to keep the maximum bundle fiber length to 70 mm, while the binding between the elementary fiber should be retained. For the non-textile use, the length of 10 to 20 mm is sufficient. The number of bundles per stem varies from 20 to 50 and the number of elementary fiber per bundle varies from 10 to 30. The numbers differ according to the stem length and the fibers are parallel but off-set. For the industrial use, the non-retted stems are dominantly mechanically decorticated [68].

Properties

Flax fiber is an ecologically safe alternative to reinforcing fibers in plastic composites. Flax fibers are less dense than glass fibers, are renewable and combustible, and they are not expensive. Strength and toughness of flax fiber are comparable to those of glass fibers [69].
Density: 1.4-1.5 g/cm^3 / E modulus: 27.6-73 GPa / Tensile Strength: 345-1825 MPa / Elongation: 1.5-4.0 % (Table 5).

Applications

In 1941, flax and hemp (with wheat and spruce pulp) fibers were used in resin matrix composites for the bodywork of a Henry Ford car which was claimed to have an "impact strength 10 times greater than steel" [26]. Flax is the most widely used natural fiber in the European automotive industry, representing 71% of the natural fibers consumed in 2000. Most of this is short-fiber flax obtained as a by-product of the textile industry [70]. Natural fibers are typically combined with polypropylene, polyester, or polyurethane to produce components like door and trunk liners, parcel shelves, seat backs, interior sunroof shields, and headrests [71]. Recently, flax has also been employed to reinforce thermosetting resin (i.e. epoxy) in order to realize structural components for automotive and marine ships with high performances [72].

HEMP

Name

Cannabis sativa

Description

Hemp is an annual plant native of Central Asia and known to have been grown in China over 4500 years ago [73]. It probably reached central Europe during Iron Age (400 BC c.a) and there is evidence of growth in the UK by the Anglo–Saxons (800-1000 AD). It does not require fertiliser, herbicides or pesticides to grow well (hence is potentially of great interest in the context of sustainability). In suitable warm conditions, it can grow to 4 m in 12 weeks. True hemp is a fine, light-coloured, lustrous and strong bast fiber obtained by retting. Colour and cleanliness vary considerably according to the method of preparation of the fiber. The lower grades are dark cream and contain much non-fibrous matter. The main producing areas are Italy, Yugoslavia and Russia [74]. The fiber ranges in length from 1.0 to 2.5 m.

Properties

Although acceptable composite properties have been obtained from mechanically extracted fiber [75], concern arouses over long term stability due to degradation of lignin [76], which mainly acts as an adhesive holding the cellulose fibers together. It is advisable to process the fibers to remove the lignin before including it into composites. Removal of lignin, pectin and hemicellulose by means of alkali treatment improve instantaneous fiber properties due to better packing and increased molecular orientation of cellulose chains [77]. However, over-treatment with alkali reduces fiber properties [78, 79]. Therefore in order to optimise composite strength, it is important to be able to assess the effect of fiber extraction as well as growing conditions on fiber properties, particularly fiber strength.

Density: 1.5 g/cm^3 / E modulus: 30-70 GPa / Tensile Strength: 300-920 MPa / Elongation: 1.6-4.0% (Table 5).

Applications

Hemp, is amongst the natural fibers now finding use in thermoplastic matrix composites for internal structures in similar automotive applications to those for flax fibers.

NETTLE

Name

Urtica dioica; Girardinia diversifolia

Description

Nettle [80-83] is another plant-stem fiber that can find application as reinforcement. It is a common herbaceous plant which counts from 30 to 45 species. It is part of the Urticaceae family such as ramie and belongs to the genus Urtica. The stinging nettle is the most prominent species in Europe. It grows in rich soils and up to 1.20 m high. The yield of nettle fiber ranges from 335 to 411 kg/ha in the second year and from 743 to 1016 kg/ha in the third year [84].

Properties

The fibers are far stronger than cotton but finer than other bast fibers such as hemp. They are a much more environmentally friendly fiber crop than cotton, which requires more irrigation and agrochemical input. As flax and hemp, nettle fibers are biodegradable, they require low energy to produce, and they are extracted from a renewable source.

E modulus: 59-115 GPa / Tensile Strength: 954-2234 MPa / Elongation: 1.3-2.9% (Table 5).

Applications

The entire plant can be used for various purposes such as food, fodder, medicine, cosmetic, biodynamic agriculture and textile production. In the past, nettle fibers have been widely used to make rope, string, and cloth. For example, nettle thread was used in Poland from the 12th to the 17th century when it was replaced by silk [85].

During World War I, Germans used nettle fibers to make tents, rucksacks, undershirts, socks; 85% of their clothes were made out of nettle fibers. The green colour of unbleached nettle fibers was valued by the army for camouflage equipment [86]. Lewington [73] states that, during the Second World War, the British Ministry of Aircraft Production experimented a very strong, high-grade paper made from nettle fiber for reinforcing plastic aircraft panels as well as gear wheels and other machine parts.

In the 1940s, 500 ha of fiber nettle were grown in Germany and Austria, and 70 ha in Great Britain for textile production. Unfortunately, the nettle fiber textile industry was abandoned for technical and cost-effectiveness reasons. Indeed, the fiber extraction could not be entirely mechanized and when the cost of labour increased, farming nettles became unprofitable, and the cultivation ceased [85]. This crisis was emphasized by the ascent of cotton industry which took over the smaller textile industries such as flax, hemp, and nettle.

The use of natural fibers as reinforcing components for composite materials has been the focus of many recent studies.

KENAF

Name

Hibiscus cannabinus

Description

Kenaf is a warm season, short-day, annual herbaceous fiber plant native of central Africa, and common as a wild plant in tropical and subtropical Africa and Asia. It has been cultivated since around 4000 BC for food and fiber extracting purposes. The plant has a unique combination of long bast (about 35% of the stalk dry weight) with short core fibers instead of the hollow core [74]. Kenaf belongs to the Malvaceae, a family noteworthy for both its economic and horticultural importance. Kenaf has a high growth rate, rising to heights of 4–6 m in about 4–5 months. It can then yield a dry weight of 6000-10,000 kg/ha year (new varieties may reach 30,000 kg/ha year).

Properties

Kenaf fiber exhibits low density, non-abrasiveness during processing, high specific mechanical properties and biodegradability [87]. It is similar to jute and can be used either as an alternative to, or in admixture with, jute.

Density: 1.2 g/cm^3 / E modulus: 14-38 GPa / Tensile Strength: 240-600 MPa (Table 5).

Applications

Kenaf has been actively cultivated in recent years for two main reasons: firstly, kenaf absorbs nitrogen and phosphorus included in the soil [88]; secondly, kenaf accumulates carbon dioxide at a significantly high rate [89]. There are numerous advantages in using natural lignocellulosic fibers as reinforcements of the matrix. Kenaf is well known as a cellulosic source with economical and ecological advantages. Recently, kenaf has been used as a raw material, alternative to wood in pulp and paper industries because it avoids the destruction of forests [90]. It has been also used as non-woven mats in the automotive industries [91], textiles [92], fiberboard [93] and civil.

JUTE

Name

Corchorus capsularis; Corchorus olitorius

Description

Corchorus capsularis (white jute) and Corchorus olitorius (dark jute) fiber are obtained from the bast layer of the plants. Each of the above classes is sub-divided into numerous grades denoting quality and other characteristics. Jute is the second most common natural fiber (after cotton) cultivated in the world. It is an annual plant that flourishes in monsoon climates and grows to 2.5-4.5 m [73]. It is primarily grown in Bangladesh, Brazil, China, India and Indonesia.

Properties

The jute fiber shows prominent stiffness and harshness properties [94].
Density: 1.3-1.5 g/cm^3 / E modulus: 20-60 GPa / Tensile Strength: 393-860 MPa / Elongation: 7.0-8.0 % (Table 5).

Applications

It is widely used for the manufacture of flexible packaging fabrics besides its prospective use as carpet backing, decorative fabrics and in some other fields. Some major points in favour of jute are its agro-based, annually renewable and biodegradable nature and availability at a low cost. Reusability of bag/sack is another major advantage of jute [94]. Jute-based thermoplastic matrix composites find a substantial market in the German automotive door-panel industry (growing from 4,000 tonnes in 1996 to over 21,000 tonnes in 1999 and rising) [73]. The fiber is also used in marine application to reinforce polyester resin.

RAMIE

Name

Boehmeria nivea

Description

Ramie is a perennial plant native to China, Japan, and the Malay Peninsula, where it has been used as a textile fiber for centuries due to its excellent fiber. Ramie is readily available since it can be harvested three times per annum and its production is very high. Only the bast fiber in the outer culm of the ramie plant can be used. The typical length of a ramie fiber varies between 60 and 250 mm (the longest can be 500 mm) and its diameter ranges from 20 to 35 $\square$m, so its length-to-radius ratio can be as high as 1900 [95].

Properties

The fibers, obtained from the outer part of the stem, are the longest, and one of the strongest, fine textile fibers. Another advantage of this fiber is the resistance to bacteria, mildew, and insect attack. Its strength slightly increases when wet [96]. It can be washed without any problems and dries quickly and it has a very high tear resistance.
E modulus: 61.4-128 GPa / Tensile Strength: 400-938 MPa / Elongation: 3.6-3.8 % (Table 5).

Applications

Ramie is used mainly in textile industry. Recently, as kenaf, jute, flax, is used to made interior parts in automotive applications.

SISAL

Name

Agave sisalana

Description

Sisal fiber is one of the most widely used natural fibers and it is very easily cultivated. It has short renewal times and grows wildly in the hedges of fields and railway tracks [97]. Nearly 4.5 million tons of sisal fibers are produced every year throughout the world. Tanzania and Brazil are the two main producing countries [98].
Sisal fiber is a hard fiber extracted from the leaves of the sisal plant. Though native to tropical and sub-tropical North and South America, sisal plant is now widely grown in tropical countries of Africa, the West Indies and the Far East [99].
A sisal plant produces about 200-250 leaves and each leaf contains 1000±1200 fiber bundles which are composed of 4% fiber, 0.75% cuticle, 8% dry matter and 87.25% water [97], so a leaf weighing about 600 g will yield about 3% by weight of fiber, with each leaf containing about 1000 fibers.

Sisal leaf contains three types of fibers [99]: mechanical, ribbon and xylem. Mechanical fibers are mostly extracted from the periphery of the leaf. They have a roughly thickened-horseshoe shape and seldom divide during the extraction processes. They are the most commercially useful among sisal fibers. Ribbon fibers occur in association with the conducting tissues in the median line of the leaf.

Properties

Figure (left) shows a cross-section of a sisal leaf and indicates where mechanical and ribbon fibers are obtained [99]. The related conducting tissue structure of the ribbon fiber gives them considerable mechanical strength. They are the longest fibers and can be easily split longitudinally during processing, if compared to mechanical fibers they. Xylem fibers have an irregular shape and occur opposite the ribbon fibers through the connection of vascular bundles. They are composed of thin-walled cells and are therefore easily broken up and lost during the extraction process.

The tensile properties of sisal fiber are not uniform along its length [99]. The root or lower part has low tensile strength and modulus but high fracture strain. The fiber becomes stronger and stiffer at midspan and the tip has moderate properties [100].

Density: 1.5 g/cm^3 / E modulus: 9.4-22 GPa / Tensile Strength: 511-635 MPa / Elongation: 2.0-2.5 % (Table 5).

Applications

At the present time, sisal fiber is mainly used as ropes for the marine and agriculture industry. Other applications of sisal fibers include twines, cords, upholstery, padding and mat making, fishing nets, and fancy articles such as purses, wall hangings, table mats, etc. The use of sisal fiber as a reinforcement in composites has raised interest and expectations amongst materials scientists and engineers [101].

Figure

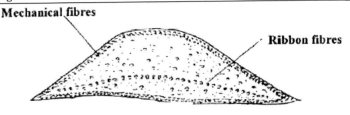

A sketch of sisal plant and the cross-section of a sisal leaf.

COIR

Name
Cocos nucifera

Description

Plantations of coconut are spread all over the world in tropical and sub-tropical regions, and they play an important role in the economy of many of these regions. The annual world production of coconut is about 42 million MT, equivalent to almost 50 billion coconuts [102]. The main use of coconut is for culinary purpose and after extraction of the copra and/or of the liquid endosperm that fills the interior of the fruit, the fruit shell is disregarded. Transformation industries and regions where consumption of coconut is high have, therefore, a large problem to dispose conveniently of this waste, since the fruit shell has a long decay time. The coconut fruit is, in fact, adapted for being dispersed by seawater, and can float for months without rotting.

Growing attention is nowadays being paid to coconut fiber These fibers are extracted from the external layer of the exocarp and from the endocarp of the fruit. The coconut palm, in fact. can be regarded as an integral fiber producer because fibers can be extracted from many parts of the palm, such as from the leaf sheath, the bark of the petiole or from the midribs of leaves [103, 104].

Properties
Coir is inexpensive if compared to the other natural fibers available in world. Furthermore, it has the advantages of a lignocellulosic fiber. It is not brittle as glass fibers, is amenable to chemical modification, is non-toxic and poses no waste disposal problems. Unfortunately, the performance of coir as a reinforcement in polymer composites is unsatisfactory and not comparable with other natural fibers. This inferior performance of coir is due to various factors such as its low cellulose content, high lignin and hemicellulose content, high microfibrillar angle, and large and variable diameter [105]. Density: 1.2 g/cm^3 / E modulus: 4-6 GPa / Tensile Strength: 175 MPa / Elongation: 30% (Table 5).
Applications
Fibers extracted from the husk of the nut, known as coir fiber, are now being commercially used, blended with natural rubber latex in the production of seat cushion parts in automobiles [106].

6. NATURAL FIBER COMPOSITES FOR INDUSTRIAL APPLICATIONS

In the last few years, the depletion of petroleum resources, the increasing awareness towards the environment and a great concern for the greenhouse effect have stimulated the automotive, construction, and nautical packing industries to look for sustainable materials that can replace conventional synthetic polymeric fibers. In reference to this issues, 'green' composites made of renewable agricultural and forestry feedstocks represent a suitable alternative to synthetic fiber reinforced composites.

Moreover, the renewable and biodegradable characteristics of natural fibers help their ultimate disposal by composting or incineration, which are not possible with most industrial fibers.

As discussed in this chapter, if compared with conventional organic fibers (i.e glass and carbon fibers) the natural ones provide many advantages, such as abundance (and therefore low cost), biodegradability, flexibility during processing and less resulting machine wear, minimal health hazards, high sound absorption, fracture resistance and low density.

On the other hand, several disadvantages limit the use of these "green" materials as reinforcement for composites: fiber-matrix interfacial adhesion, poor resistance to moisture absorption, limited processing temperature, lower mechanical properties and low dimensional stability.

Poor fiber–matrix interfacial adhesion affects negatively physical and mechanical properties of the resulting composites due to the surface incompatibility between hydrophilic natural fibers and non-polar polymers (thermoplastics and thermosets). To overcome these problems and to improve the properties of composites, natural fiber are pre-treated chemically (i.e. silanization or mercerization).

To limit the absorption of moisture, during the production of natural fiber reinforcements there is a drying step, during which the natural plant moisture, including that gained during the retting process, is removed. With time, however, the fibers will absorb moisture from the air to return to some equilibrium point, which varies according to the fiber, but is typically in the range of 5 to 15 percent by weight. Encapsulation of the fiber in composite resins delays this pickup significantly, but the overall moisture pickup still exceeds that of fiberglass composites.

The limited processing temperature is due to the thermal degradation of natural fibers that at 120°C, with temperatures above 230°C causing rapid decomposition of cellulose. In order to minimize this degradation lower process temperatures and short thermal exposure are needed.

These limitations preclude the use of natural fiber composites in many applications of industrial field (for instance, structural applications or in high moisture environment), although some breakthroughs and developments are occurring in that area.

6.1. Automotive

During the last years the automotive industry has developed various new components based on natural fiber composites: in 2003 around 43000 tonnes of natural fiber were being used by the European automotive industry as composite reinforcement [107].

For example, only in Germany, the use of natural fibers in automotive industry has increased from 4000 tonnes to 18000 tons per year from 1996 until 2003, with amount of natural fiber press-moulded parts equal to 45000 tons only in the year 2003.

The choice of these materials is principally due to economical and marketing reasons and weight reduction, rather than technical requirements.

For what regards regulations about the automotive end-of-life requirements, the European Union (E.U.) and Asian countries have released stringent guidelines. European Union legislation implemented in 2006 has expedited recent natural-fiber-reinforced plastic automotive insertion; by 2006, 80% of a vehicle must be reused or recycled and by 2015 the threshold value is 85% [108]. Japan requires 88% of a vehicle to be recovered (which includes incineration of some components) by 2005, rising to 95% by 2015.

For these reasons, nowadays most automakers are evaluating the environmental impact of a vehicle's entire lifecycle, and natural fibers are widely used in this industrial field.

As previously discussed, only non-structural or semi-structural components are realised by mainly using polyester resin or thermoplastic polymers and lignocellulosic fibers (i.e. flax, hemp, sisal and jute).

Bast fiber composites are predominantly used in automotive interior panels, such as doors, pillar trim, trunk liners and package or rear-parcel trays. Early composites, replacing wood fiberboard, were a mixture of flax and sisal fibers in an epoxy matrix, first used on the Mercedes E-Class door panels in the early 1990s. The fibers are generally supplied in a needle-punched nonwoven mat format, and the simplest method involves placing the mat into the heated mold, pouring liquid resin on top, and pressing until cured.

A remarkable weight reduction of about 20% was achieved, and the mechanical properties, important for passenger protection in case of an accident, were improved. Furthermore, the flax/sisal material can be moulded in complicated 3D shapes, thus making it more suitable for door trim panels than the previously used materials.

Nowadays the natural fiber mat used in automotive semi-structural components consists of a 50/50 mixture of flax and sisal fibers. Other fibers are usable as well. For example, hemp has been found to have good properties. The fiber products are supplied in rolls with widths adapted to the production process. Areal weight varies between about 700 and 1200 g/m^2, depending on the application.

Nowadays, natural fibers are widely used in the interior of passenger cars and trucks cabins. Despite their use in trim parts such as door panels or cabin linings, natural fibers are broadly used for thermo-acoustic insulation. Such insulating materials, mainly based on cotton fibers recycled from textiles, have relatively high fiber content, more than 80% by weight. Another well established field of application is the use of coconut fibers bonded with natural latex for seat cushions. For this application the capacity of the natural fibers to absorb large amount of humidity leads to an increased comfort that cannot be reached with synthetic materials [109].

At the beginning, initially only thermosetting resins (i.e. polyester or epoxy) were used as matrix for natural fiber composites thanks to their low viscosity, which results in excellent fiber wetting and adhesion. In fact, this kind of matrix, allow the composites to be compression molded in more complex shapes than wood fiber-based materials.

With increasing emphasis on recyclability, in the last few years thermosets have been partially replaced by thermoplastics in natural fiber interior components. On the other hand, thermal degradation of natural fibers limits the selection of the thermoplastics matrix to polymers with lower melt-points, generally polypropylene and polyethylene.

The most important structural applications of natural fiber composites in automotive field are the load floors of several sport utility vehicles (i.e. Volkswagen Touareg, Porsche Cayenne, and Audi Q7). These parts are a sandwich of expanded polypropylene foam covered on each side with 1400 g/m^2 natural fiber/polypropylene composites, and topped with PET carpet. Each load floor is produced in a single molding cycle.

6.2. Biomedical

Definitely, a material suitable for medical application must have very specific characteristics, different from that requested by domestic-used plastic products. The most important requirements are biocompatibility and absence of adverse effect to the host tissues. The most common types of conventional composites (thermosetting resin reinforced by glass, aramid or carbon fibers) lead to the problem of conventional removal after the end of life time, as the components are closely interconnected, relatively stable and thus difficult to separate and recycle. On the other hand, the use of stainless steel and titanium in bone repair, due to their unique biocompatible properties, requires several times of subsequent surgical operations to remove the plates and other fasteners, which may cause unnecessary pain or inconvenience to the patients.

For these reasons in the last years, those traditional structures began to be substituted by bio-engineered composites, consisting of biodegradable polymer as matrix and usually bio-fibers as reinforcing elements which have generally low cost, low density, high toughness, acceptable specific strength properties, good thermal properties, ease of separation, enhanced energy recovery and biodegradability.

Apart from the plant-based fibers, animal-based fibers represent an alternative in the production of biodegradable, biomedical and bio-resorbable composite materials for bioengineering and orthopedic applications. These fibers are mainly made by proteins, like wool, spider and silkworm silk. In particular, silk fiber has been used in biomedical applications such as sutures by which the silk fibroin fiber is usually coated with waxes or silicone to enhance material properties and reduce fraying. However, from 1960s to the early 1980s,the use of virgin silk negatively affected the use of this biomaterial from the surgical practitioner perspective, because of, for examples, the reaction of silk to the host tissue and its inflammatory potential. Recently, silk matrices are being rediscovered and reconsidered as potentially useful biomaterials for a range of applications in clinical repairs and as scaffolds for tissue engineering. Recently, also chicken feather fiber (CFF) has attracted much attention to different product design and engineering industries [110].

6.3. Nautical

Thanks to their total biodegradability, biocomposites represent a valid alternative to glass fibers reinforced thermoset composites also in marine applications. The main disadvantage that limits their use in this field consists in a low durability in a seawater environment due to their hydrophilic nature.

For instance, bamboo is a natural material that offers good potential for composites as a substitute of solid wood for structural uses also in marine applications. In fact it is sufficiently hard, strong and dimensionally stable. Even though the existing Bamboo species are numerous, the one used in technological applications is the Phyllostachys Pubescens, which can reach up to 20 m of height and a diameter of 18 cm. Its cross-section is generally constituted by a near net of micro – openings which make it particularly sensitive to moisture absorption, worsening however the wettability with matrices in composite applications.

In a work research [111] bamboo composites were exposed to the harsh marine environment by recording the variation of weight and of tensile strength through the moisture absorption test – conducted according to UNI EN 2489 with both sea and distillate water. The experimental results (tensile strength decrease only around 20%) showed that bamboo is a valid alternative to synthetic materials for the production of structural component in marine application.

As applications, the same authors realised and tested (axial compression load and for impact) a spinnaker pole and a hull panel. Successively a complete 6 meters boat hull has been realised in order to analyze critical lamination points, vacuum performances and necessity of external treatments.

Flax fibers are widely used as reinforcements in the production of composite components for marine applications too.

A Researchers' group of Palermo's University has realized, by a vacuum bagging process, the hull of a 4,70 meters sailboat named L.E.D. (Linen Epoxy Dinghy), by using unidirectional (areal density equal to 190 g/m^2) and bidirectional (twill balanced weave of 220 g/m^2) flax fabrics.

As Figure 7 shows, the boat hull is a sandwich structure with the core constituted by 8 mm cork of 140 Kg/m^3. The lamination sequence of the structure is [UD 0°/BD ±45°/UD 90°/cork/UD 90°/BD ±45/UD 0°].

As regards greater boats, shipyards Mareehaute, has recently made a 13.30 meters boat racer type Class 40. This boat is realised by an autoclave process using balsa as core and skins reinforced by flax prepreg fabrics.

Furthermore, thanks to their high stiffness, flax fibers can be used as reinforcement to realize nautical components like surf boards, rudders or thin keel.

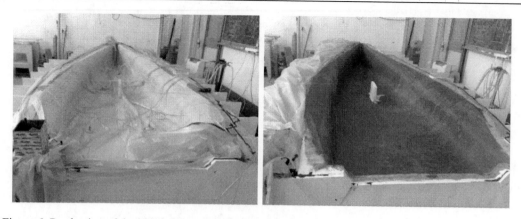

Figure 6. Production of the hull by vacuum bagging process.

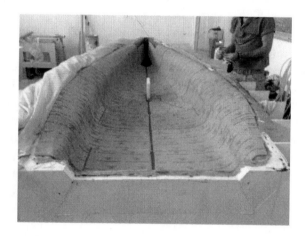

Figure 7. Cork used as core in the sandwich structure of the boat.

6.4. Other Applications

Another possible application of natural fiber is the printed circuit boards (PCBs). Thanks to their total biodegradability, natural fibers represent an optimal alternative to materials currently used in PCBs (fiberglass and brominated flame retardants) as both of them will reduce health risks during manufacturing and at the end of their useful life.

A research group [111] developed composite designs using an epoxy matrix and woven flax fiber for reinforcement in order to improve the environmental performance of PCB manufacturing and disposal.

The experimental characterization of their prototypes gave acceptable results for thermal, mechanical, and electrical properties, with the exception of wet conditioning or water submersion. To improve moisture resistance, flax fibers were treated with sodium hydroxide and octadecyltrichlorosilane.

It has been found that the improved bioepoxy-flax PCB design is a viable alternative to current PCBs; it has potentially lower environmental impacts, it is cheaper, and it has satisfactory thermal, electrical, and mechanical properties. However, additional improvements in moisture absorption properties may be needed for commercial applications.

Another research work of industrial interest [112] is aimed at developing high-performance natural fiber composite systems for structural applications using continuous textile reinforcements like UD-tapes or woven fabrics. Different types of fabrics (i.e., biaxial plain weaves, unidirectional fabrics and non-crimp fabrics) were produced and evaluated as reinforcement in composites manufactured by well manufacturing techniques such as hand lay-up, vacuum infusion, pultrusion and resin transfer moulding. As expected, the developed materials cannot directly compete in terms of strength with glass fiber composites. However, they are clearly able to compete with these materials in terms of stiffness, especially if the low density of flax is taken into account. Furthermore, their properties are very good when compared with non-woven glass composites.

As final applications, they produced a cover lid of diameter 600 mm for siphon type solar panel accumulator tank and a water accumulator tank, by using flax biaxial fabric.

Another possible application of biocomposites regards the packing industries. In fact, most plastic packaging products are discarded after a single use resulting in a redundant supply of waste plastics. For instance, coconut or coir fiber, have a natural resistance to mildew and microbial attack, as well as a natural resistance to burning because they are a lignin-rich natural fiber. Furthermore these fibers are also strong and stiff, which allows to be formed into various shapes for packaging applications.

In addition to their application in biomedical field, chicken feather fibers can be used as reinforcements of composite for electronic application (for/as high-speed data transforming insulator material). It has been recently found [114] that the dielectric properties of chicken feather fibrils using epoxy resin as matrix are dependent on operating frequency and temperature conditions and such composites have potential use as a low dielectric material for typical applications.

The natural fibers are used also in civil applications to realise element for insulating in order to reduce the thermal conductivity of the walls. Exist several panels constituted by only the fibers or by a mixing between the fibers with polymeric or ceramic matrices.

7. ORIGINAL RESEARCHES

One of the original works of the authors in this field regards the influence of areal weight and chemical treatment of bidirectional flax fabrics on the mechanical properties of FFRP structures [72]. In this paper several bidirectional flax fibers, usually used to make curtains, were employed as reinforcement of an epoxy matrix and four different laminates were made by a vacuum bagging process, with varying both the areal weight and the treatment of the fabric. In particular three different kinds of textile were used: BD130 (areal weight of 130 g/m2), BD150 (150 g/m2) and BD170 (170 g/ m2). These ones, chemically treated (i.e. whitening and colouring), were combined in three different structures (in the next "T150", "T170" and "T130") to study the effect of the stacking sequence with varying the external layers, as shown in Table 7. Moreover another laminate ("UT150") was made using only untreated BD150 to evaluate the effect of the chemical treatment, realised in fabric manufacturing.

Table 7. Investigated structures

Layer	T 150	T 170	T 130	UT 150
1	BD150	BD170	BD130	BD150
2	BD150	BD150	BD150	BD150
3	BD150	BD150	BD150	BD150
4	BD150	BD150	BD150	BD150
5	BD150	BD150	BD150	BD150
6	BD150	BD170	BD130	BD150
thickness [mm]	1.57 ± 0.03	1.85 ± 0.03	1.86 ± 0.02	1.89 ± 0.02
P_f [%]	54.0	55.0	55.7	47.9

These structures were mechanically investigated carrying out both tensile and three point flexural tests. In order to evaluate the effect of the identified variables on the mechanical parameters (i.e. Young modulus and maximum stress) a statistical analysis was performed. Finally, by comparing the flax composite with a glass one, the convenience of the replacement of glass fiber was investigated in term of cost, weight and mechanical performances, by evaluating the structural efficiency coefficient of the cost of the material.

Analysing the results, it was possible to draw out the following considerations:

– In flexural, both the parameters (areal weight and treatment) are significant. With varying the areal weight, the properties decrease due to the delamination phenomenon occurs inducing the catastrophic and premature failure of the samples. With varying the treatment, the structure with the treated fibers are more performance because of the layer can be considered as pre-preg.
– In tensile, the areal weight significantly influences only the modulus. In this case the failure interests the fibers and, as a consequence, the strength does not change. The higher amount of resin induces higher strains, with varying the areal weight, and then the modulus decreases. As for the flexural, the treatment gives better properties to the laminate with the treated fabrics.
– Finally, the cost analysis shows that it is advantageous to use the FFRP than the GFRP from a stiffness point of view when the structures are subjected to flexural solicitations.

Another our interesting work regards the replacement of one or more layers of glass mat in GFRP laminates with natural fiber obtained from natural dehydration of cladodes of Opuntia ficus-indica, a plant widespread in dry regions of the world like Sicily, and to evaluate the influence of this replacing on the mechanical performances of hybrid structures [115].

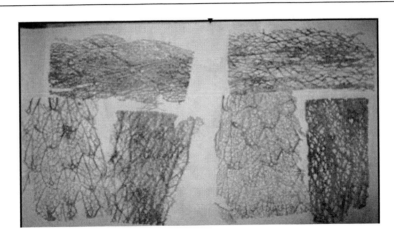

Figure 8. Opuntia ficus- indica mat layers.

All the composite structures have been made with a single lamination using the vacuum bagging technique. This method involves an initial hand lay-up phase and then the polymerization of the matrix in a flexible bag in which negative pressure is reached by a vacuum pump. All the laminates have been cured at room temperature for 24 h and then post-cured at 60 °C for 8 h. The glass composite laminates ("GFRP") have been constituted by six layers of E-glass mat (randomly oriented fibers, with areal weight of 450 g/m2 in a epoxy matrix. The hybrid composites have been produced, starting from GFRP structure, replacing two or four layers of glass mat with similar layers of opuntia ficus-indica (Figure 8).

Three point bending and tensile tests are carried out in order to evaluate the specific properties of the hybrid composite.

The results have shown that, the hybrid structures were characterized, as expected, by lower performances than those of GFRP laminate. Particularly the specific stiffness of hybrid structures was reduced around 20% in tensile and between 25% and 30% in bending. Regarding the specific strength, decreases between 40% and 50% in tensile; 50% and 70% in bending, were obtained.

These results indicate that, changing two or four layer of mat glass with similar layers of opuntia ficus-indica natural fibers it is possible to achieve structures with specific stiffness to be used in dedicated applications.

REFERENCES

[1] Saravana Bavan D, Mohan Kumar GC. Potential use of natural fiber composite materials in India. *Journal of Reinforced Plastics and Composites* 2010: doi:10.1177/0731684410381151.

[2] Mohanty AK, Misra M, Drzal LT, Selke SE, Harte BR, Hinrichsen G. Natural fibers, biopolymers, and biocomposites: an introduction. In: Mohanty AK, Misra M, Drzal LT. Natural fibers, biopolymers and biocomposites. Boca Raton: CRC Press, 2005.

[3] Bledzki AK, Sperber VE, Faruk O. Natural and wood fiber reinforcement in polymers, *Rapra Review Reports* 2002; 3(8).

[4] Craig M, Clemons, Caulfield DF. Natural fibers. In: Marino X. Functional fillers for plastics. Weinheim: Wiley-VCH, 2005.

[5] Fowler PA, Mark Hughes J, Elias RM. Review biocomposites: Technology, environmental credentials and market forces. *Journal of the Science of Food and Agriculture* 2006; 86: 1781-1789.

[6] Alix S, Marais S, Morvan C, Lebrun L. Biocomposite materials from flax plants: preparation and properties. *Composites: Part A* 2008; 39: 1793-1801.

[7] Bos HL. The potential of flax fibers as reinforcement for composite materials. Eindhoven: University Press Facilities, 2004.

[8] Hull D. An introduction to Composite Materials. Cambridge: Cambridge University Press, 1981.

[9] Ellison GC, McNaught R, Eddleston EP. The Use of Natural Fibers in Nonwoven Structures for Applications as Automotive Component Substrates. Research and Development Report, Ministry of Agriculture Fisheries and Food, Agri - Industrial Materials, London, 2000.

[10] Santulli C. Biomimetic interest and possibilities for replacement of glass fibers with plant fibers in composite materials: the case of impact damage. In: International school on advanced material science and technology: smart materials and nanotechnologies, Jesi; 2003.

[11] Vincent JFV. Structural biomaterials. Princeton: Princeton University Press, 1990.

[12] Okamoto H. A dialogue on biomimetic design for natural technology. *Biomimetics* 1994; 2(1):1-11.

[13] Reis RL. Biomimetics. *Current Opinion in Solid State and Materials Science* 2003; 74(5): 263-4.

[14] Benyus JM. Biomimicry: innovation inspired by nature. New York: William Morrow and Company, 1997.

[15] Jeronimidis G, Atkins AG. Mechanics of biological-materials and structures: nature's lessons for the engineer. *Journal of Mechanical Engineering Science* 1995; 209: 221-35.

[16] Marchetti A, Levita G, Forntanelli D, Anguillesi I, Balzelli C. Properties of Spanish broom fibers for low environmental impact composites. In: 9th European conference on advanced materials and processes, Brighto, UK, 2000.

[17] Takashi N, Koichi H, Masaru K, Katsuhiko N, Inagaki H. Kenaf reinforced biodegradable composite. *Composite Science Technology* 2003; 63: 1281-6.

[18] McGovern JN, Coffelt DE, Hurter AM, Ahuja NK, Wiedermann A. Other Fibers. In: Hamilton F, Leopold B, Kocurek MJ. Pulp and Paper Manufacture. Atlanta: Tappi Press, 1987; 3: 110-121.

[19] Young, RA. Processing of Agro-Based Resources into Pulp and Paper. In: Rowell RM, Young RA, Rowell JK Paper and Composites from Agro-Based Resources. Boca Raton : CRC Press, 1997: 135-245.

[20] Miller RB. Structure of Wood. In: Wood Handbook: Wood as an Engineering Material. Madison: USDA Forest Service, Forest Products Laboratory, 1999.

[21] Lilholt H, Lawther JM. Natural Organic Fibers. In: Chou TW. Comprehensive Composite Materials. New York: Elsevier, 2000; 1: 303-325.

[22] Kaup M. Evaluation of a Market Survey 2002: The Use of Natural Fiber in the German and Austrian Automotive Industries: Status 2002, Analysis and Trends". In: EcoComp 2003 Conference; London, 2003.

[23] Jacobson R, Caulfield D, Sears K, Underwood J. Low Temperature Processing of Ultra-Pure Cellulose Fibers into Nylon-6 and Other Thermoplastics". In: 6th International Conference on Woodfiber-Plastic Composites; Madison, 2001: 127-133.

[24] Gassan J, Bledzki AK. Einfluß von Haftvermittlern auf das Feuchteverhalten naturfaserverstärkter. Die Angewandte Makromolekulare Chemie 1996; 236: 129-138.

[25] Klemm D, Schmauder HP, Heinze T. Cellulose. In: Steinbüchel A. Biopolymers. Wiley-VCH, Weinheim, Germany, 2004; 6: 275-287.

[26] Summerscales J, Dissanayake NPJ, Virk AS, Hall W. A review of bast fibers and their composites. Part 1 – Fibers as reinforcement. *Composites: Part A* 2010; 41: 1329-1335.

[27] Bledzki AK, Gassan J. Composites reinforced with cellulose based fibers. Progress in Polymer Science 1999; 24: 221-274.

[28] Pettersen RC. The Chemical Composition of Wood. In: Rowell RM. The Chemistry of Solid Wood Washington: *American Chemical Society*, 1984: 76-81.

[29] Mukherjee PS, Satyanarayana KG. Structure and properties of some vegetable fibers - Part 2: Pineapple fiber. *Journal of Materials Science* 1986; 21: 51-56.

[30] Krenchel H. Fiber reinforcement. Copenhagen: Akademisk Forlag, 1964.

[31] Kenny JM. Natural Fiber Composites in the European Automotive Industry. In: 6th International Conference on Woodfiber-Plastic Composites, Madison, 2001: 9-12.

[32] Rowell RM, Han JS, Rowell JS. Characterization and Factors Affecting Fiber Properties. In: Natural Polymers and Agrofibers Based Composites: Preparation, Properties, and Applications. In: Frollini E, Leão AL, Mattoso LHC. Natural Polymers and Agrofibers Based Composites. São Carlos: Embrapa Instrumentação Agropecuária, 2000: 115-134.

[33] Bisanda ETN, Ansell MP. Properties of sisal-CNSL composites. *Journal of Material Science* 1992; 27: 1690-1700.

[34] Sridhar MK, Basavarajappa G, Kasturi SG, Balasubramanian N. Evaluation of Jute as a Reinforcement in Composites. *Indian Journal of Textile Research* 1982; 7(9): 87-92.

[35] Zeronian SH. Mechanical properties of cotton fibers. *Journal of Applied Polymer Science* 1991; 47: 445-461.

[36] Roe PJ, Ansell MP. Jute-Reinforced Polyester Composites. *Journal of Materials Science* 1985; 20: 4015-4020.

[37] Hancox NL. Fiber composite hybrid materials. Barking: Elsevier Applied Science, 1981.

[38] Baley C. Analysis of the flax fibers tensile behaviour and analysis of the tensile stiffness increase. Composites: Part A 2002; 33(7): 939-948.

[39] Aziz SH, Ansell MP. The effect of alkalization and fiber alignment on the mechanical and thermal properties of kenaf and hemp bast fiber composites: part 1 – polyester resin matrix. *Composites Science and Technology* 2004; 64(9): 1219-1230.

[40] Ochi S. Mechanical properties of kenaf fibers and kenaf/PLA composites. *Mechanics of Materials* 2008; 40(4–5): 446–452.

[41] Bodros E, Baley C. Study of the tensile properties of stinging nettle fibers (Urtica dioica). *Materials Letters* 2008; 62(14): 2143–2145.

[42] Ugbolue SCO. Structure-Property Relationships in Textile Fibers. *Textile Progress* 1990; 20(4): 1-43.

[43] Khazanchi AC, Saxena M, Rao TC. Material science of natural organic fibers reinforced composites in polymer–cement–mud matrix for construction engineering. In: International Conference on Textile Composites in Building Constructions, Primalis, France; 1990: 69-76.

[44] Michell AJ, Willis D. Cellulosic fibers for reinforcement. *Appita* 1978; 31(3): 347-354.

[45] Saechtling H. International plastics handbook. Munchen: Hanser Publications, 1987.

[46] Peyer S, Wolcott M. Engineered Wood Composites for Naval Waterfront Facilities. Yearly Report to Office of Naval Research. Pullman: Washington State University: 2000.

[47] Rowell RM. Penetration and Reactivity of Cell Wall Components. In: Rowell RM. The Chemistry of Solid Wood. Washington: American Chemical Society; 1984.

[48] Sears KD et al. Reinforcement of Engineering Thermoplastics with High Purity Cellulose Fibers. In: 6th International Conference on Woodfiber-Plastic Composites, Madison, 2001: 27-34.

[49] Bledzki AK et al. Odor Measurement of Natural Fiber Filled Composites used for Automotive Parts. In: 9th Annual Global Plastics Environmental Conference: Plastics Impact on the Environment, Detroit, 2003.

[50] Bledzki AK, Reihmane S, Gassan J. Vegetable Fibers for Natural Fiber Composites. *Journal of Applied Polymer Science* 1996; 59: 1329-1336.

[51] Zeronian SH, Kawabata H, Alger KW. Factors Affecting the Tensile Properties of Nonmercerized and Mercerized Cotton Fibers. *Textile Research Journal* 1990; 60(3): 179-183.

[52] Semsarzadeh MA. Fiber matrix interactions in jute reinforced polyester resin. *Polymer Composites* 1986; 7(2): 23-25.

[53] Semsarzadeh MA, Lotfali AR. Mirzadeh H. Jute reinforced polyester structures. *Polymer Composites* 1984; 5(2): 141-142.

[54] Ray PK, Chakravarty AC, Bandyopadhyay SB. Fine structure and mechanical properties of jute differently dried after retting. *Journal of Applied Polymer Science* 1976; 20: 1765-1767.

[55] Belgacem MN, Bataille P, Sapieha S. Effect of corona modification on the mechanical properties of polypropylene/cellulose composites. *Journal of Applied Polymer Science* 1994; 53: 379-385.

[56] Sakata I, Morita M, Tsuruta N, Morita K. Activation of wood surface by corona treatment to improve adhesive bonding. *Journal of Applied Polymer Science* 1993; 49: 1251-1258.

[57] Wang Q, Kaliaguine S, Ait-Kadi A. Catalytic Grafting: A New Technique for Polymer-Fiber Composites. III. Polyethylene-Plasma Treated KevlarTM Fibers Composites: Analysis of the Fiber Surface. *Journal of Applied Polymer Science* 1993; 48: 121-136.

[58] Goa S, Zeng Y. Surface modification of ultrahigh molecular weight polyethylene fibers by plasma treatment. I. Improving surface adhesion. *Journal of Applied Polymer Science* 1993; 47: 2065-2071.

[59] Dong S, Sapieha S, Schreiber HP. Rheological properties of corona modified cellulose/polyethylene composites. *Polymer Engineering and Science* 1992; 32: 1734-1739.

[60] Kroschwitz JI. Polymers: fibers and textiles. New York: Wiley, 1990.

[61] Safonov VV. Treatment of textile materials. Moscow: Legprombitizdat, 1991.

[62] Nevell TP, Zeronian SH. Cellulose chemistry and its applications. New York: Wiley, 1985.

[63] Westerlind BS, Berg JC. Surface energy of untreated and surface-modified cellulose fibers. *Journal of Applied Polymer Science* 1988; 36: 523-534.

[64] Schick MJ. Surface characteristics of fibers and textiles. Part II. New York: Marcel Dekker, 1977.

[65] Felix JM, Gatenholm P. The nature of adhesion in composites of modified cellulose fibers and polypropylene. *Journal of Applied Polymer Science* 1991; 42: 609-620.

[66] Maldas D, Kokta BV, Daneaulf C. Influence of coupling agents and treatments on the mechanical properties of cellulose fiber–polystyrene composites. *Journal of Applied Polymer Science* 1989; 37: 751-775.

[67] Mittal KL. Silanes and other coupling agents: Netherlands: VSP BV, 1992.

[68] Kromer KM. Physical properties of flax fiber for non-textile-use. *Research in Agricultural Engineering* 2009; 55(2): 52-61.

[69] Powell T, Panigrahi S, Ward J, Tabil LG, Crerar WJ, Sokansanj S. Engineering Properties of Flax Fiber and Flax Fiber-Reinforced thermoplastic in Rotational molding. In: ASAE/CSAE North-Central Intersectional Meeting; 2002.

[70] Plackett D. The Natural Fiber-Polymer Composite Industry in Europe - Technology and Markets. In: Progress on Woodfiber-Plastic Composites Conference, Toronto; 2002.

[71] Suddell BC, Evans WJ. The Increasing Use and Application of Natural Fiber Composite Materials within the Automotive Industry. In: 7th International Conference on Woodfiber-Plastic Composites, Madison; 2003.

[72] Di Bella G, Fiore V, Valenza A. Effect of areal weight and chemical treatment on the mechanical properties of bidirectional flax fabrics reinforced composites. *Materials and Design* 2010; 31: 4098-4103.

[73] Lewington A. Plants for people. London: Transworld Publisher; 2003.

[74] Farnfield CA, Alvey PJ. Textile terms and definitions. 7th ed. Manchester: The Textile Institute; 1975 [ISBN 0-900739-17-7]; Denton, Daniels PN. Textile terms and definitions. 11th ed. Manchester: The Textile Institute; 2002.

[75] Hepworth DG, Hobson RN, Bruce DM, Farrent JW. The use of unretted hemp fiber in composite manufacture. *Composites: Part A* 2000; 31: 1279-1283.

[76] Rowell RM. A new generation of composite materials from agrobased fiber. In: 3th international conference on frontiers of polymers and advanced materials, Kuala Lumpar. New York: Plenum Press; 1995.

[77] Gassan J, Bledzki AK. Alkali treatment of jute fibers: relationship between structure and mechanical properties. *Journal of Applied Polymer Science* 1999; 71: 623-629.

[78] Beg MDH, Pickering KL. Effect of fiber pre-treatment on the mechanical properties of wood/polypropylene composites. In: 2nd International Conference on Structure, *Processing and Properties of Materials*, Dhaka, 2004: 240-247.

[79] Placet V. Characterization of the thermo-mechanical behaviour of Hemp fibers intended for the manufacturing of high performance composites. *Composites: Part A* 2009; 40: 1111-1118.

[80] De Leo P, Miceli A, Antonaci C, Vigna G. Characterisation and enzymatic hydrolysis of nettle (Urtica dioica L) deproteinated biomass. *Journal of the Science of Food and Agriculture* 1993; 63(4): 391-395.

[81] Dreyer J, Dreyling G, Feldmann F. Wiederinkulturahme der Fasernessel Urtica dioica L als nachwachsender Rohstoff zur Faser- und Zellstoffproduktion: qualitative und quantitative differenzierung von ehemals genutzten Klonen, Angewandte Botanik. *Journal of Applied Botany* 1996; 70(1–2): 28-39.

[82] Davies GC, Bruce DM. Effect of environmental relative humidity and damage on the tensile properties of flax and nettle fibers. *Textile Research Journal* 1998; 68(9): 623-629.

[83] Singh SC, Shrestha R. Extraction and chemical analysis of Himalayan nettle fiber. *Research and Industry* 1987; 32(4): 259-262.

[84] Hartl A, Vogl CR. Dry matter and fiber yields, and the fiber characteristics of five nettle clones (Urtica dioica L.) organically grown in Austria for potential textile use. *American Journal of Alternative Agriculture* 2002; 17(4): 195-200.

[85] B. Bertrand. Les secrets de l'Ortie. Aspet: Edition du Terran; 2005.

[86] K. Wheeler. A natural history of nettles. Victoria: Trafford Publishing; 2005.

[87] Takashi N, Koichi H, Masaru K, Katsuhiko N, Inagaki H, Kenaf reinforced biodegradable composite, *Composites Science and Technology* 2003; 63: 1281-1286.

[88] Abe K, Ozaki Y. Comparison of useful terrestrial and aquatic plant species for removal of nitrogen and phosphorus from domestic wastewater. *Soil Science and Plant Nutrition* 1998; 44: 599-607.

[89] Amaducci S, Amaducci MT, Benati R, Venturi G. Crop yield and quality parameters of four annual fiber crops in north of Italy. *Industrial Crops Products* 2000; 11: 179-186.

[90] Pande H, RoyDN. Influence of fiber morphology and chemical composition on the papermaking potential of kenaf fibers. *Pulp Paper Can* 1998; 99: 31-34.

[91] Magurno A. Vegetable fibers in automotive interior components. *Die Angewandte Makromolekulare Chemie 1999; 272: 99-107.*

[92] Ramaswamy GN, Craft S, Wartelle L. Uniformityand softness of kenaf fibers for textile products. *Textile Research Journal* 1995; 65: 765-770.

[93] Kawai S, Ohnishi K, Okudaira Y, Zhang M. Manufacture of oriented fiberboard from kenaf bast fibers and its application to the composite panel. In: *International kenaf Symposyum*, 2000:144-148.

[94] Basu G, De SS, Samanta AK. Effect of bio-friendly conditioning agents on jute fiber spinning. *Industrial Crops and Products* 2009; 29: 281-288.

[95] Shihong L, Benlian Z, Qiyun Z, Xianrong B. A new kind of super-hybrid composite material for civil use – ramie fiber/Al. *Composites* 1994; 25: 225-228.

[96] Angelini LG, Lazzeri A, Levita G, Fontanelli, Bozzi C. Ramie (Boehmeria nivea (L.) Gaud.) and Spanish Broom (Spartium junceum L.) fibers for composite materials: agronomical aspects, morphology and mechanical properties. *Industrial Crops and Products* 2000; 11: 145-161.

[97] Murherjee PS, Satyanarayana KG. Structure and properties of some vegetable fibers, part 1. Sisal fiber. *Journal of Materials Science* 1984; 19: 3925-3934.

[98] Chand N, Tiwary RK, Rohatgi PK. Bibliography resource structure properties of natural cellulosic fibers - an annotated bibliography. *Journal of Materials Science* 1988; 23: 381-387.

[99] Bisanda ETN, Ansell MP. Properties of sisal-CNSL composites. *Journal of Materials Science* 1992; 27: 1690-1700.

[100] Li Y., Mai YW, Ye L. Sisal fiber and its composites: a review of recent developments. *Composites Science and Technology* 2000; 60: 2037-2055.

[101] Mukhopadhyay S, Srikanta R. Effect of ageing of sisal fibers on properties of sisal – Polypropylene composites. *Polymer Degradation and Stability* 2008; 93: 2048-2051.

[102] Ohler JG. The coconut palm and its environment. In: Ohler JG. Modern Coconut Management - Palm Cultivation and Products. London: FAO and Intermediate Technology Publications Ltd, 1999: 12-156.

[103] Venkataswamy MA, Pillai CSK, Prasad VS, Satyanarayana KG. Effect of weathering on the mechanical properties of midribs of coconut leaves. *Journal of Materials Science* 1987; 22: 3167-3172.

[104] Satyanarayana KG, Pillai CSK, Sukumaran K, Pillai SGK, Rohatgi PK, Vijayan K. Structure property studies of fibers from various parts of the coconut tree. *Journal of Materials Science* 1982; 17: 2453-2462.

[105] Geethamma VG, Thomas Mathew K, Lakshminarayanan R, Thomas S. Composite of short coir fibers ad natural rubber: effect of chemical modification, loading and orientation of fiber. *Polymer* 1998; 39(6-7): 1483-1491.

[106] Schuh TG, Gayer U. Automotive applications of natural fiber composite. In: Leão AL, Carvalho FX, Frollini E. Lignocellulosic-Plastics Composites. Botucatu: Unesp Publishers, 1997: 181-195.

[107] Liu Q, Stuart T, Hughes M, Sharma HSS, Lyons, G. Structural biocomposites from flax – part II: the use of PEG and PVA as interfacial compatibilising agents. *Composites Part A* 2007; 38: 1403-1413.

[108] European Parliament and Council. End of Life Vehicles. Directive 2000/53/EC. *Official Journal of the European Communities* 2000.

[109] Schuh TG. Renewable Materials for Automotive Applications. Report Daimler-Chrysler AG.

[110] Cheung HY, Ho MP, Lau KT, Cardona F, Hui D. Natural fiber-reinforced composites for bioengineering and environmental engineering applications. *Composites Part B* 2009; 40: 655-663.

[111] Corradi S, Isidori T, Corradi M, Soleri F, Olivari L. Composite boat hulls with bamboo natural fibers. *International Journal of Materials and Product Technology* 2009; 36: 73-89.

[112] Lincoln J, Shapiro A, Earthman J, Saphores J, Ogunseitan A. Design and Evaluation of Bioepoxy-Flax Composites for Printed Circuit Boards. *IEEE Transactions on Electronics Packaging Manufacturing* 2008; 31: 211-231.

[113] Goutianos S, Peijs T, Nystrom B, Skrifvars M. Development of Flax Fiber based Textile Reinforcements foR Composite Applications. *Applied Composite Materials* 2006; 13: 199-215.

[114] Mishra SC, Nayak NB. An investigation of dielectric properties of chicken feather reinforced epoxy matrix composite. *Journal of Reinforced Plastics and Composites* 2009; 29: 2691-2697.

[115] Fiore V, Di Bella G, Valenza A. caratterizzazione di compositi a matrice epossidica rinforzati con fiber di vetro e di opuntia-ficus indica. In: XIX Convegno Italiano di Scienza e Tecnologia delle Macromolecole, Milano, Italy 2009.

In: Fiber-Reinforced Composites
Editor: Qingzheng Cheng

ISBN: 978-1-61470-303-7
© 2012 Nova Science Publishers, Inc.

Chapter 3

POLYVINYL ALCOHOL NANOCOMPOSITES REINFORCED WITH CELLULOSIC NANOFIBERS ISOLATED FROM JUVENILE POPLAR

*Jingxin Wang**, *Qingzheng Cheng**, *Adebola Adebayo and Stephen Difazio*

Division of Forestry and Natural Resources
Biology Department
West Virginia University, Morgantown, WV 26506

ABSTRACT

In this Chapter, cellulosic nanofibers (CNFs) were isolated from juvenile poplar (*P. trichocarpa*) (PT) and juvenile hybrid poplar (HP) using acid hydrolysis. A film casting method was used to fabricate biodegradable nanocomposites from CNFs and a polyvinyl alcohol (PVA) solution. CNFs were characterized using atomic force microscopy and dynamic light scattering. Most nanofibers were bundles of single nanocrystals and the average diameter of the nanofibers was about 21 nm. The mechanical properties and thermal degradation behavior of the nanocomposites were analyzed by tensile test and thermogravimetric analysis, respectively. The elastic modulus of PVA was significantly improved by the CNFs, while the strength of the composites was comparable with that of pure PVA. No significant differences of tensile mechanical properties of PVA nanocomposites reinforced with CNFs isolated from PT and HP were detected. The thermal degradation behavior of PVA was improved by CNFs.

INTRODUCTION

Bio-based nanocomposites are considered as a "green" material for the next generation. The market for these bio-based nanocomposites with a green concept has the potential for

---

* Corresponding authers: jxwang@wvu.edu, q.cheng@mail.wvu.edu

dramatic growth in recent years. The bio-based nanocomposites can be produced from renewable resource-based polymers in combination with nano-filler reinforcement, such as cellulose nanocrystals and nanofibers, for a variety of applications. Nanocomposites of this category are expected to possess improved strength and stiffness with small reduction in toughness, reduced gas/water vapor permeability, a lower coefficient of thermal expansion, and an increased heat deflection temperature (Ray and Okamoto 2003, Samir et al. 2004, Zimmermann et al. 2004). Bio-based nanocomposites have the greatest market growth potential in industrial applications, where biocompatibility and environmentally responsible design and construction are required. The potential applications include biomedical tissue, car interiors, packaging materials, electronics, and housing appliances.

Cellulose is the most abundant natural, renewable, biodegradable polymer in the world. The microfibrils are near defect free crystals with axial physical properties approaching those of perfect crystals. Wood is not only a good building material, but also an abundant cellulose source. Cellulose is a high-molecular-weight linear homopolymer, which consists repeating β-D-glucopyranosyl units joined by (1-4) glycosidic linkages in a variety of arrangements (Figure 1) (Chaplin, 2010). Cellulose chains have a Degree of Polymerization of approximately 10,000 glucopyranose units in wood cellulose (Chaplin, 2010).

Several methods have been developed to obtain cellulose nanocrystals and nanofibers: 1) chemical methods: such as strong acid hydrolysis - cellulose nanocrystals have been obtained from wood fibers (Candanedo et al. 2005) and sugar beet pulp (Dufresne et al. 1997) by this method; and 2) mechanical methods: such as homogenizer (Herrick 1983, Turbak 1983), refining and cryocrushing (Chakraborty et al. 2005), and high-intensity ultrasonication (Cheng et al. 2007, Cheng and Wang 2008, Cheng et al. 2009a, Wang and Cheng 2009, Cheng et al. 2010).

Cellulose nanocrystals and nanofibrils can be used to process nanocomposite materials using a polymer as the matrix. Nanocomposite is a broad concept that includes at least one-dimension of one component in the composite at the nanometer scale (1-100 nm) (Jordan et al. 2005). Because of the nanometric size effect, these composites have some unique outstanding properties with respect to their conventional microcomposite counterparts.

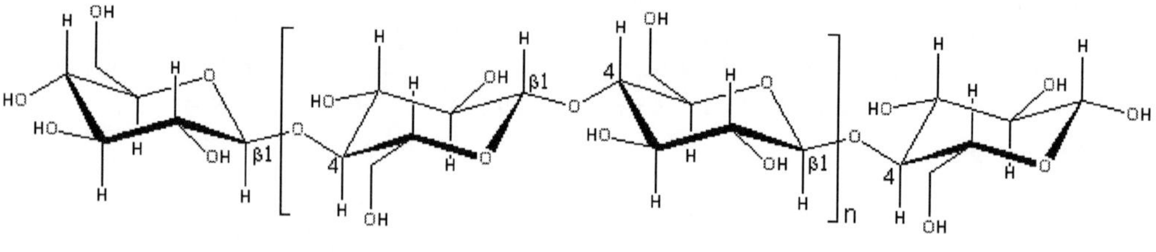

Figure 1. Basic chemical structure of cellulose biopolymer (Chaplin, 2010).

The research on nanocomposites reinforced by cellulose fibrils/crystals has grown very fast in recent years because of its environmentally friendly features and the improvement of stiffness and strength. There are two major fabrications of nanocomposites. One is film casting using water-soluble or solvent-soluble polymers (Favier et al. 1995), and the other is freeze drying followed by classical compression or extrusion processes (Hajji et al. 1996). Pure cellulose and pulp fiber are the common raw materials for the isolation of nanocrystals and nanofibers in current research. In this study, we used poplar wood samples as raw materials to isolate nanofibers and nanocrystals and to fabricate cellulosic nanocomposites.

Many terms have been used to describe different types of cellulose fibrils, such as microfibrillated cellulose (MFC), cellulose nanocrystal, cellulose nanofibril, cellulose whisker, and cellulose nanofibers. This study used cellulosic nanofibers.

Poplar trees (native cottonwood and aspen species from the genus *Populus*) are suitable candidates for genetic improvement for biofuels and bioproducts production due to their modest genome size, fast growth, amenability to genetic engineering and vegetative propagation, and adaptability to a wide range of planting conditions. Thermo-chemical comparisons among juvenile hybrid poplar, *Populus deltoids,* and *Populus trichocarpa* were reported in a separate paper (Adebayo et al. 2010). In order to examine the potential of hybrid poplar and its parents' tree clones for polymer reinforcement to fabricate cellulosic bio-nanocomposites, this were to (1) isolate nanofibers from black cottonwood and a hybrid poplar, (2) fabricate cellulosic nanocompsites using nanofibers and Poly vinyl alcohol, and (3) evaluate mechanical and thermal properties of the nanocomposites.

MATERIALS AND METHODS

Materials

A hybrid poplar plantation (*Populus trichocarpa* x *Populus deltoides*) was established from vegetative cuttings at the West Virginia University Agronomy Farm. The plant material consisted of a single interspecific family produced from a pseudobackcross crossing design. Flowering branches were collected from a female *P. trichocarpa* tree clone (93-968) from Washington State. Pollen was collected from a male *P. deltoides* clone (ILL-101) and used for crosses in a greenhouse setting. This produced an F_1 hybrid, clone 52-225, which has been used extensively in commercial plantations. To dissect the factors contributing to hybrid vigor in this commercial clone, the F1 hybrid was backcrossed to a different *P. deltoides* tree (clone D124), from Minnesota (Riemenschneider et al. 2001). This produced the pseudobackcross family (52124), which was used for establishing the plantation and stool bed at West Virginia University (Figure 2).

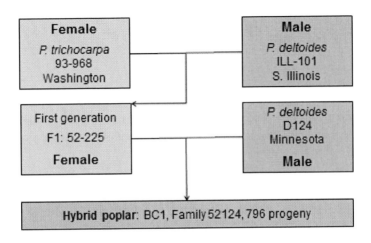

Figure 2. Crossing design for production of the quantitative trait locus (QTL) mapping pedigree.

Two poplar samples, black cottonwood (*Populus trichocarpa* clone 93-968) (PT) and hybrid poplar (HP – offspring from family 52124), were used as raw materials for nanofiber isolation. Microcrystalline cellulose (MCC, Avicel PH-101, supplied by FMC BioPolymer) was used as reference material. Poly vinyl alcohol (PVA) (99-100% hydrolyzed, approx average M_W 86,000, Acros, New Jersey) was used as composite matrix. Sulfuric acid (H_2SO_4, SA170-4, 70% Textile) was purchased from Acros. Sodium chlorite ($NaClO_2$, unstabilized, tech., 80%), sodium hydroxide (NaOH), and acetic acid solution (1N) were purchased from Fisher Scientific.

Cellulosic Nanofiber Isolation and Characterization

Isolation procedure of cellulose nanofibers is shown in Figure 3. The poplar samples were debarked, dried, and milled to particles passing a 60-mesh screen. The lignin in wood particles was removed using acidified sodium chlorite solution (Wise et al. 1946). Briefly, 10 g wood particles were added in 150 ml distilled water in a flasket. The mixture was heated to 75 °C, then 1 ml acetic acid and 3 g $NaClO_2$ were added and the temperature was maintained for 1 h with a stirrer. The same amount of reagents was then added every hour for an additional period of 3 hours, turning the product white. Tap water was used to wash the product 10 times, then washed again using distilled water by centrifuge (>3 times) or filtration until acid free.

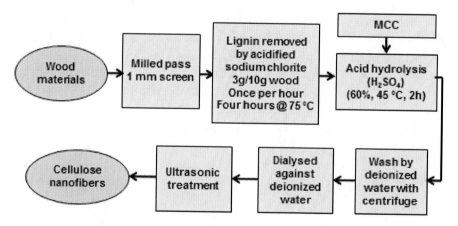

Figure 3. Isolation procedure of cellulose nanofibers.

The lignin removed samples were then placed in a 60% H_2SO_4 (v/v) solution for 2 h at 45°C with medium stirring. The cellulose to acid ratio was 1:10 g/mL. After hydrolysis, the mixture was centrifuged and decanted. A solution of 2% sodium hydroxide was added to the centrifuge solids to neutralize any remaining acid and the centrifugation was repeated several times. Then deionized water was added to the centrifuge solids for rinsing (2-3 times). The solid particles were subjected to ultrasonic irradiation (Sonic Newtown, CT, 20 kHz, Model 750 W) in deionized water for 3 min at 50% power to disperse the nanofibers. The suspension was subjected to ultrafiltration (Dialysis) to remove any remaining ions to decrease the conductivity to <10 µS/cm monitored with a hand-held conductivity instrument (Oakton

Instruments). The cellulose solid percent was measured by fully drying an aliquot of the suspension in an oven at 103°C overnight.

An optical microscope (Fisher Scientific Micromaster I) was used for morphological characterization of cellulose particles before lignin removal and after acid hydrolysis. An atomic force microscopy (AFM) (Veeco Metrology) with non-contact mode after a small drop of fibril suspension was dried on a mica sheet to characterize the morphology of the nanofibers.

Dynamic light scattering (DLS) (Agilent 7030 Nicomp) was used to measure the diameter distributions of the cellulosic nanofibers.

Nanocomposites Preparation and Characterization

PVA water solution (10 wt%) and cellulose nanofiber water suspension or MCC were mixed and stirred manually and then dispersed by ultrasonic treatment (Sonic Newtown, CT, 20 kHz, Model 750 W) for about one minute at a 50% power level. The film casting method was used to make nanocomposite film in propylene dishes. The mixtures were degassed in a desiccator with vacuum and evaporated at room temperature (T: ~20°C) with a relative humidity (RH) of 30% until films were formed, and then the films were heat treated in an oven at 60 °C for more than 4 hours (Cheng et al. 2009b). The samples were kept in a desiccator before the mechanical properties were characterized. The composites with three levels of fibril loading (1%, 2%, and 5%) were manufactured and tested. The nominal thickness of the composites was 150 μm.

The mechanical tests were performed using an Instron testing machine (Model 5869). The crosshead speed was 2.5 mm/min. Crosshead extensions were used as the specimen deformations to calculate the tensile strain. The samples were cut to dogbone shapes with width of 5 mm for the narrow portion and total length of 40 mm (gauge length was 20 mm). Based on ASTM D1708 standard (ASTM 2002), five specimens were tested for each composition. Tensile modulus was tangent modulus from the linear portion in the stress-strain curves while ignoring the initially unstable part.

The maximum tensile stress was used as the tensile strength. The elongation to break was calculated using the extension at break of each sample divided by the original length of the sample. Single parameter test of analysis of variance (ANOVA) was used to examine the influence significance of the cellulose contents on tensile properties of the composites ($\alpha=0.05$).

The thermal degradation behavior of the composites was conducted by a thermogravimetric analyzer (TGA Q50, TA Instruments). The temperature range was from room temperature to 400 °C. Heating rate was 20 °C/min. Nitrogen gas rate was 20 ml/min to avoid oxidation.

The sample weights were between 2-3 mg. TA Instruments Software was used to transform the TG curve (variation of the mass in function of the temperature) to the derivative thermogravimetric (DTG, derivative of loss of mass versus the time) curve. The combination of TG and DTG thermograms gives a clear indication of the number of stages of the thermal degradation (Renneckar et al. 2004).

RESULTS AND DISCUSSION

Characterization of Wood Particles and Nanofibers

The wood particles were changed to white after lignin was removed. The microstructure of wood particles was irregular and most particles were smaller than 200 μm (Figure 4a). After acid hydrolysis, there were still some big particles with length more than 100 μm (Figure 4b) and a turbid suspension presented in all samples, but after ultrasonic treatment, the suspensions were transparent. The average diameter of the nanofibers was about 21 nm measured by dynamic light scattering (Figure 5), which indicated that the nanofiber was still bundles of single nanocrystals since a single wood cellulose crystal is about 3-5 nm in diameter (Beck-Candanedo et al. 2005). The AFM height images were further illustrated that the diameters of most cellulose nanofibers were about 20-30 nm and lengths were in the hundreds nm range (Figure 6). No significant differences were found between the nanofibers from PT and HP poplar juvenile trees.

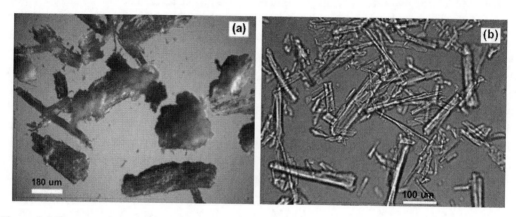

Figure 4. Microscopy images of poplar wood particles before lignin removal (a) and acid hydrolysis (b).

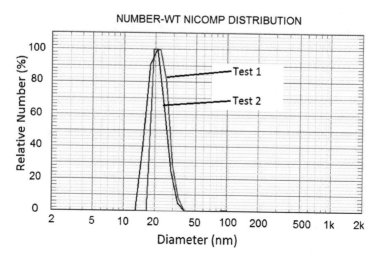

Figure 5. Nanofiber diameter distribution measured by dynamic light scattering with two test runs.

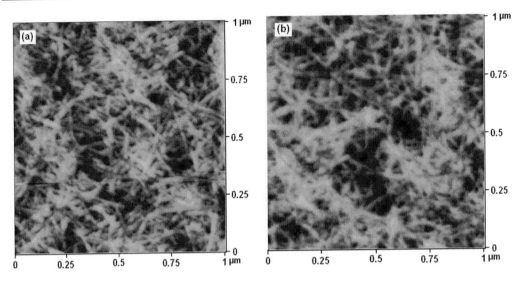

Figure 6. AFM images of nanofibers from *P. trichocarpa* (a) and hybrid poplar (b) show that the diameters of most nanofibers were about 20-30 nm.

Mechanical Properties of PVA and Its Cellulosic Composites

Figure 7 shows typical tensile curves of pure PVA and nanocomposites with 5% cellulose nanofibers, respectively. The elongation to break decreased in different percents of nanofibers and MCC for all composites compared with pure PVA. This indicated that the toughness of PVA increased by adding these cellulose because the elongation could increase the toughness of a material (Choi and Simonsen 2006).

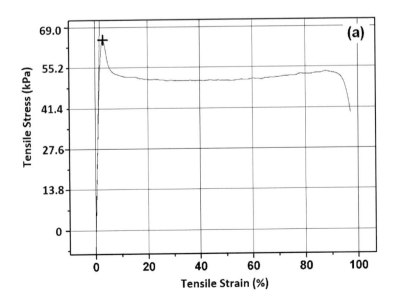

Figure 7. (Continued).

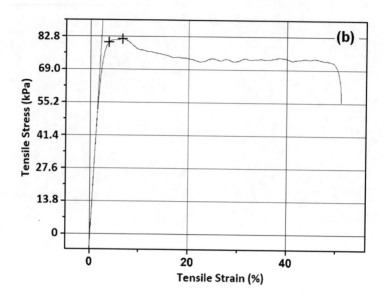

Figure 7. Typical stress strain curves of pure PVA (a) and cellulosic nanocomposites (b) show that the extension of PVA was decreased by adding cellulosic nanofibers.

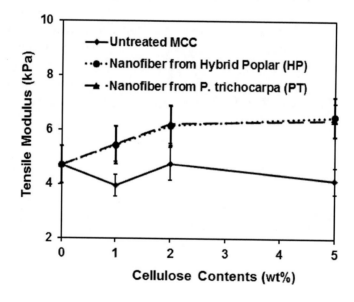

Figure 8. Tensile moduli of the nanocomposites with fillers of untreated MCC, nanofibers from HP (hybrid poplar), and nanofibers from PT (*P. trichocarpa*).

The tensile modulus and strength of neat PVA and its composites reinforced by MCC and nanofibers form MCC, hybrid poplar (HP), and *P. trichocarpa* (PT) with 1, 2, and 5 wt% are shown in Figures 8 and 9, respectively. Generally, the small nanofibers reinforced PVA composites had much higher mechanical properties (both modulus and strength) than those of neat PVA and MCC reinforced composites. This may be because small nanofiber had higher mechanical properties and higher aspect ratios (length/diameter) than those of MCC (Cheng et al. 2009b, Wang and Cheng 2009). The tensile modulus of neat PVA was decreased by untreated MCC, yet not significantly different (P values between 0.07 to 0.90). PVA modulus was increased by all cellulosic nanofibers, which was not significant for 1% nanofibers (P=0.12), but was significantly increased by 2% (P=0.01) and 5% (P=0.002). There was no significant difference between 2% and 5% nanofiber-reinforced composites. And no significant differences were detected between the two nanofibers from HP and PT, which indicated that the nanofibers from HP and PT could have similar structures and dimensions as shown in Figures 5 and 6.

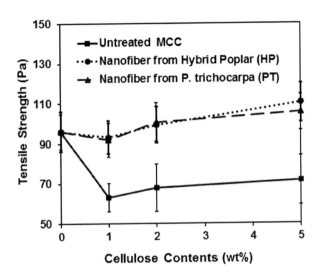

Figure 9. Tensile strength of the nanocomposites with fillers of untreated MCC, nanofibers from HP (hybrid poplar), and nanofibers from PT (*P. trichocarpa*).

The tensile strength of neat PVA was significantly decreased by all three contents of MCC, (P=0.002 for 1%, P=0.003 for 2%, and P=0.007 for 5%). However, PVA strength was not decreased by all cellulosic nanofibers. There was no significant difference among them (P values between 0.09 to 0.64). The maximum modulus improvement of PVA was about 38% by 5% nanofiber (Figure 8), which was not as high as those of microfibers (Wang and Sain 2007). This probably was because the sources of neat PVA, making conditions, the uniform distribution of the fibrils, aspect ratio and orientation of the fibers were different (Cheng et al. 2009b).

Thermal Degradation Behaviors of PVA and Its Cellulosic Composites

The TGA curves of neat PVA, neat HP nanofibers, HP wood, and HP nanofiber-reinforced composites are shown in Figure 10. The thermal degradation behavior of pure PVA film showed the weight loss starting at its melting point (typically 226 °C) and indicated a very narrow temperature range of degradation (Figure 11). Pure nanofibers started to degrade at a lower temperature (~210 °C) than PVA, but showed a very broad degradation temperature range. Similar findings were reported by others (Choi and Simonsen 2006). The degradation behaviors of all the nanocomposite films are generally between those of neat PVA and nanofibers. The degradation temperature of HP wood was different when compared with neat nanofibers because the lignin and hemicelluloses play an important role in the degradation of wood and have different decomposition temperature ranges (Shafizadeh 1985).

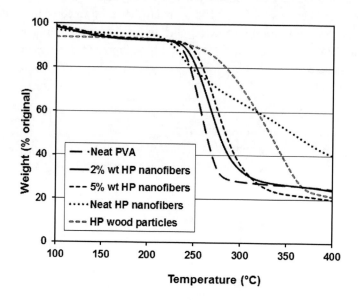

Figure 10. Thermogravimetric analyzer curves of weight losses for PVA, wood particles, nanofiber and its nanocomposites.

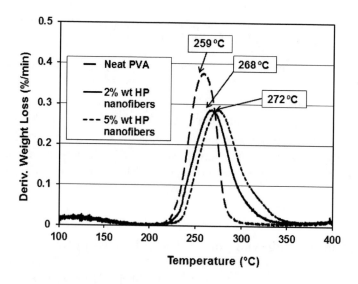

Figure 11. Derivative thermogravimetric thermograms of neat PVA and its nanocomposites with 2% and 5% HP nanofibers.

The addition of cellulosic nanofibers increased the degradation temperature of PVA, showing an apparent increase in the thermal stability of the composite matrix. The decomposition peak temperature of the nanocomposites was clearly increased by cellulosic nanofibers. Figure 11 shows DTG curves of PVA and 2% and 5% HP nanofiber-reinforced composites. The two nanocomposites show thermal events with a shift of 9 °C and 13 °C toward higher temperature compared to the neat PVA. It is postulated that this phenomenon is due to interactions between the PVA and the nanofibers (Magalhaes, 2009). It indicated that only 2% cellulosic nanofibers can significantly improve PVA's thermal stability.

CONCLUSIONS

Cellulose nanofibers can be isolated from hybrid poplar (HP) and *P. trichocarpa* (PT) using acid hydrolysis. Most nanofibers were bundles of single nanocrystals and the average diameter of the nanofibers was about 21 nm measured by DLS and AFM. The nanofibers can be used to reinforce PVA to fabricate bio-nanocomposites. Cellulose nanofibers had much better performance than commercial MCC for PVA reinforcement. No significant differences of tensile mechanical properties were detected for PVA nanocomposites reinforced with cellulosic nanofibers isolated from HP and PT. The decomposition temperatures of the cellulosic nanocomposites were higher than that of neat PVA, indicating that the thermal degradation behavior of PVA was improved by cellulosic nanofibers.

REFERENCES

Abe K, Iwamoto S, Yano H (2007) Obtaining cellulose nanofibers with a uniform width of 15 nm from wood. *Biomacromolecules* 8:3276-3278.

Adebayo A, Wang J, Cheng Q, Difazio S (2010) Bioenergy related properties juvenile hybrid poplar and wild poplar species - *Populus deltoids,* and *Populus trichocarpa. Wood Fiber Sci* (In review).

ASTM D (2002) D 1708-02a, standard test method for tensile properties of plastics by use of microtensile specimens. American Society for Testing and Materials, West Conshohoken, PA.

Beck-Candanedo S, Roman M, Gray, DG (2005) Effect of reaction conditions on the properties and behavior of wood cellulose nanocrystal suspensions. *Biomacromolecules* 6(2):1048-1054.

Candanedo SB, Roman M, Gray DG (2005) Effect of reaction conditions on the properties and behavior of wood cellulose nanocrystal suspensions. *Biomacromolecules* 6:1048-1054.

Chakraborty A, Sain M, Kortschot M (2005) Cellulose micriofibrils: a novel method of preparation using high shear refining and cryocrushing. *Holzforschung* 59:102-107.

Chaplin M. (2010). Water Structure and Science. http://www.lsbu.ac.uk/water/hycel.html, (accessed on January, 2011).

Cheng Q, Wang S (2008) A method for testing the elastic modulus of single cellulose fibrils via atomic force microscopy. *Compos Part A-Appl.* S 39:1838-1843.

Cheng Q, Wang S, Han Q (2010) A novel method to isolate fibrils from cellulose fibers by high intensity ultrasonication. Part II: Fibril characterization. *J. Appl. Polym. Sci.* 115(5):2756–2762.

Cheng Q, Wang S, Harper DP (2009a) Effects of process and source on elastic modulus of single cellulose fibrils evaluated by atomic force microscopy. *Compos Part A-Appl. S* 40:583-588.

Cheng Q, Wang S, Rials TG (2009b) Poly(vinyl alcohol) nanocomposites reinforced with cellulose fibrils isolated by high intensity ultrasonication. *Compos Part A-Appl. S* 40:218-224.

Cheng Q, Wang S, Rials TG, Lee SH (2007) Physical and mechanical properties of polyvinyl alcohol and polypropylene composite materials reinforced with fibril aggregates isolated from regenerated cellulose fibers. *Cellulose* 14(6):593-602.

Choi YJ, Simonsen J (2006) Cellulose nanocrystal-filled carboxymethyl cellulose nanocomposites. *J. Nanosci. Nanotechnol.* 6(3):633-639.

Dufresne A, Cavaille JY, Vigonon MR (1997) Mechanical behavior of sheets prepared from sugar beet cellulose microfibrils. *J. Appl. Polym. Sci.* 64:1185-1194.

Favier V, Canova GR, Cavaille JY, Chanzy H, Dufresne A, Gauthier C (1995) Nanocomposite materials from latex and cellulose whiskers. *Polym. Advan. Technol.* 6(5):351-355.

Hajji P, Cavaille JY., Favier V, Gauthier C, Vigier G (1996) Tensile behavior of nanocomposites from latex and cellulose whiskers. *Polym. Composite* 17(4):612-619.

Herrick FM, Casebier RL, Hamilton JK, Sandberg KR (1983) Microfibrillated cellulose: Morphology and accessibility. *J. Appl. Polym. Sci.: Appl. Polym. Symp.* 37:797-813.

Jordan J, Jacob KI, Tannenbaum R, Sharaf MA, Jasiuk I (2005) Experimental trends in polymer nanocomposites - a review. *Mat. Sci. En.g A-Struct* .393(1-2):1-11.

Magalhaes WLE, Cao XD, Lucia LA (2009) Cellulose Nanocrystals/Cellulose Core-in-Shell Nanocomposite Assemblies. *Langmuir* 25(22):13250-13257.

Ray SS, Okamoto M (2003) Polymer/layered silicate nanocomposites: a review from preparation to processing. *Prog. Polym. Sci.* 28:1539-1641.

Renneckar S, Audrey G, Zink-Sharp A, Thomas C, Ward T, Glasser W (2004) Compositional Analysis of Thermoplastic Wood Composites by TGA. *J. Appl. Polym. Sci.* 93(3):1484-1492.

Riemenschneider DE, Stanton BJ, Vallee G, Perinet P (2001) Poplar breeding strategies. Pages 43-76 *In* DI Dickmann, JG Isebrands, JE Eckenwalder, and J Richardson, eds. Poplar Culture in North America. NRC Research Press, Ottawa, Canada.

Samir M, Alloin F, Sanchez J, Dufresne A (2004) Cross-linked nanocomposite polymer electrolytes reinforced with cellulose whiskers. *Macromolecules* 37:4839-4844.

Shafizadeh F (1985) Pyrolytic reactions and products of biomass. Pages 183–217 *In* RP Overend, TA Milne, and LK Mudge, eds. Fundamentals of Biomass Thermochemical Conversion. Elsevier, London.

Turbak AF, Snyder FW, Sandberg KR (1983) Microfibrillated cellulose, a new cellulose product: Properties, uses, and commercial potential. *J. Appl. Polym. Sci.: Appl. Polym. Symp.* 37:815-827.

Wang B, Sain M (2007) Isolation of nanofibers from soybean source and their reinforcing capability on synthetic polymers. *Compos Sci. Technol.* 67:2521-2527.

Wang S, Cheng Q (2009) A novel method to isolate fibrils from cellulose fibers by high intensity ultrasonication. Part I: Process optimization. *J. Appl. Polym. Sci.* 113:1270–1275.

Wise LE, Murphy M, Addieco AA (1946) Isolation of holocellulose from wood. *Pap. Trade J.* 122:35-43.

Zimmermann T, Pohler E, Geiger T (2004) Cellulose fibrils for polymer reinforcement. *Adv. Eng. Mater.* 6(9):754-761.

In: Fiber-Reinforced Composites
Editor: Qingzheng Cheng

ISBN: 978-1-61470-303-7
© 2012 Nova Science Publishers, Inc.

Chapter 4

PHYSICAL, FLAMMABILITY AND MECHANICAL PROPERTIES OF POLYMER ECO-NANOCOMPOSITES

A. Alhuthali, H. Alamri and I. M. Low[*]

Department of Imaging and Applied Physics,
Curtin University of Technology, Perth, WA 6845, Australia

ABSTRACT

When compared to their synthetic counterparts, natural fibres represent an environmentally friendly alternative by virtue of several attractive attributes that include lower density, lower cost, non-toxicity, ease of processing, renewability and recyclability. In addition, the use of natural fibres in polymer matrix composites has the potential to produce materials with higher specific strength and specific modulus due to their low density. In this chapter, epoxy and vinyl-ester resin matrix composites reinforced with recycled cellulose fibre (RCF) have been fabricated and characterised. Nanocomposites from both polymers reinforced with different concentration (1%, 5% and 10%) of nano-clay platelets (30B) and halloysite nano-tubes (HNTs) have been synthesized. The mechanical properties, such as flexural strength, impact toughness, fracture toughness and the effect of water absorption on impact toughness were evaluated. Porosity and flammability properties have been investigated. Results indicated that flexural strength decreased for the majority of samples due to the poor dispersion of nano-fillers and the formation of micro-voids within the samples. In contrast, impact toughness and fracture toughness were improved for all reinforced samples. Water absorption resulted in enhanced impact toughness as a result of local plasticization at the crack-tip. Addition of nanoclay increased the porosities but improved the flammability of all nanocomposites. The effects of RCF and nanoclay additions on the physical and mechanical properties have been discussed in terms of the observed microstructures.

Keywords: Recycled cellulose fibres (RCF), halloysite nano-tubes (HNTs), nano-clay platelets (30B), epoxy resin, vinyl-ester resin, flexural strength, fracture toughness, impact toughness, flammability.

---

[*] Author to whom correspondence should be addressed; Email: j.low@curtin.edu.au; Tel.: +61-8-9266-7544; Fax: +61-8-9266-2377

1. INTRODUCTION

Composite materials reinforced by natural fibres now constitute a new field of research interest [1]. As opposed to their synthetic counterparts, natural fibres have economic, mechanical, and environmental advantages. Natural fibres were first used historically because of their cheaper acquisition and because of more sophisticated synthetic processing that was not available. One of the contingencies with using natural fibres is that the properties tend to depend on the interfacial properties between matrix and natural fibres [2]. For example, natural fibres such as sisal or flax have a high stiffness, but their variability in diameters and variability in length reduce value in composite applications [3]. Nonetheless, natural fibres tend to exhibit a number of desirable functional characteristics including low weight, low cost, low density, high specific strength to density ratio, stiffness, low abrasiveness and availability from renewable resources [4-6].

Moreover, in an increasingly green conscious society, biodegradability, and being a resource which is renewable are significant draw cards for material scientists to re-visit composites containing such fibres [7-8]. Moreover, natural fibres are non-toxic materials, therefore better for health and safety, and less hazardous during handling [9-11]. Natural fibres were used in the construction industry widely right up until the end of the 19th century. In fact, at the time, hemp fibres crops were the number one agricultural crop globally [12]. Early builders most likely took advantage of the low price of such materials. Nowadays, using natural fibres to create a range of novel materials has fast attracted the interest of a number of research groups, particularly enterprises interested in environmentally friendly products.

One of the most exciting attractions of natural fibres is that wastepaper products such as recycled newspapers, printed paper, cardboard, and other waste paper are able to be recycled [12]. Given the renewed focus on sustainability, and increasing international environmental regulations, the results of increased studies into the use of natural fibres in nanocomposites are important [13]. Via dry processing methods, newspaper fibre reinforced composites have similar properties to wood fibre reinforced composites and have a number of production related advantages such as cost, renewability, and flexibility and minimal wear on processing machinery [14]. In the future, newspaper reinforced materials will be used in a range of construction materials for housing including structural components such as load bearing roof systems, sub-flooring and framing components. The composites could also be used in doors, windows, and furniture [15].

In this chapter, we have investigated the effect of recycled cellulose fibres (RCF), derived from newspapers, as a reinforcement material for vinyl ester resin and epoxy resin and its composites or nanocomposites. The latter were reinforced with different concentrations (1%, 5% and 10%) of halloysite nanotubes (HNTs) and organoclay platelets (30B). Various mechanical properties have been investigated for each group of samples such as impact toughness, fracture toughness and flexural strength. Porosity and flammability have also been evaluated. The effect of recycled cellulose fibres (RCF) and nanoclay addition on the physical and mechanical properties of different types of composites have been discussed. In addition, the effect of water absorption on impact toughness behaviours of all samples has also been investigated and discussed.

2. EXPERIMENTAL

2.1. Materials

The principal material used for this study was RCF sheets. The sheets were derived from newspapers and laboratory prepared to dimensions of 60 mm x 40 mm x 2 mm. The mass of each RCF sheet was 2.3 grams. In addition to the RCF sheets, general purpose vinyl ester resin and epoxy resin, were used in this study. These resins were supplied by Fibreglass and Aesin Sales Pty Ltd, WA, Perth, Australia. The third group of materials used in the study was two different types of nano-clay particles. The first was organoclay Cloisite 30B, supplied by Southern Clay Products in the USA. It is organically modified clay with physical properties shown in Table 1.

Table 1. The physical properties of the nanoclay (Cloisite 30B)

Physical properties of Cloisite 30B	
Color	Off white
Density, g/cc	1.98
d-spacing (d_{001}), Å	18.5
Aspect ratio	200-1000
Surface area, m^2/g	750
Mean particle size, μ	6

The second nanoclay used in the study was halloysite nanotubes (HNTs), supplied by NZCC, New Zealand. The elemental compositions (wt%) of HNTs were 50.4 % SiO_2, 35.5%; Al_2O_3, 0.25; Fe_2O_3, and 0.05; TiO_2.

2.2. Sample Preparation

Four groups of composite samples were prepared for studying their physical and mechanical properties.

(a) Pure Samples

The initial samples which needed to be produced were the pure polymer samples. These samples were needed to provide baseline data of the properties of pure epoxy and pure vinyl ester resin.

(b) Nano-Composites

The following groups of samples were the nano-composite samples. As there were two types of nano-particles under investigation, two sets of nano-composites were prepared. Firstly, samples were prepared with a dispersion of nanoclay (Cloisite 30B) at concentrations of 1%, 5%, and 10%. Secondly, the same procedure was used to prepare samples with dispersions of 1%, 5%, and 10% of HNTs. A mechanical mixer was used to disperse the nanoclay in the resin for five minutes. Then a hardener was added to the mixture and hand-mixed slowly to avoid the creation of air bubbles within the sample. The resultant mixtures were poured into silicon moulds and left at room temperature for 24 hours to cure.

(c) Eco-Composites

The third group samples were the eco-composites. Here, two recycled cellulose-fibre (RCF) sheets were used to make the composites. Each RCF sheet was soaked in the polymer (epoxy or vinyl) for 15 minutes, before they were laid up in silicon mould under small load and left to cure for 24 hours.

(d) Eco-Nanocomposites

The fourth group samples were eco-nanocomposites. Here, nanocomposite matrices with different concentrations of 1%, 5% and 10%, Cloisite 30B and HNTs were prepared. These matrices were then reinforced with two RCF sheets by soaking them in the resin mixture for 15 minutes and pressed together under low pressure and cured at room temperature for 24 hours.

2.3. Porosity Measurements

The apparent porosity P_a of a sample refers to the ratio of open pores in the material to its bulk volume. The Australia Standard 1774.5, 2001 was used to determine the apparent porosity of samples using the following equation:

$$P_a = \frac{m_s - m_d}{m_s - m_i} \times 100 \tag{1}$$

where m_d is dry mass, m_i is mass of sample immersed in water, and m_s is the mass of soaked sample.

2.4. Flexural Strength Test

Rectangular bars of 60 mm x 10 mm x 6 mm were cut from the fully cured samples for three-point bend tests with a span of 40 mm to evaluate the flexural strength (σ_F). This test was performed with a LLOYD Material Testing Machine using a displacement rate of 1.0 mm/min. At least three specimens of each composition were used for the measurements. The value of σ_F was computed using the following equation:

$$\sigma_F = \frac{3}{2} \frac{P_m S}{WD^2} \tag{2}$$

where P_m is the maximum load, S is the span of the sample, D is the specimen thickness and W is the specimen width.

2.5. Charpy Impact Toughness Test

To assess impact toughness (G_{IC}), three specimens in each group with a 40 mm span were evaluated using a Zwick Charpy-impact tester with a 2.0 J pendulum hammer. Each bar specimen was notched with a razor blade to produce a sharp crack. The impact toughness was then calculated using the following equation:

$$G_{IC} = \frac{E}{A}$$

(3)

where E is the impact energy to break a sample with a ligament of area A.

Some samples were immersed in tap water for 30 days in order to evaluate the effect of water absorption on the impact toughness of each group of samples.

2.6. Fracture Toughness Test

For the fracture toughness (K_{IC}) measurement, the ratio of notch length to width of sample (a/w) used was 0.4 and a sharp razor blade was used to initiate a sharp crack. The flexural tests were performed with a LLOYD Material Testing Machine using a displacement rate of 1.0 mm/min. At least three specimens of each composition were used for the measurements. The value of K_{IC} was computed using the following equation [37]:

$$K_{IC} = \frac{P_m S}{WD^{2/3}} f(\frac{a}{w})$$

(4)

where P_m is the load at crack extension, S is the span of the sample, D is the specimen thickness, W is the specimen width, and a is the crack length, and $f(w/a)$ is the polynomial geometrical correction factor given as:

$$f(\frac{a}{W}) = \frac{3(a/W)^{1/2}[1.99 - (a/W)(1 - a/W) \times (2.15 - 3.93a/W + 2.7a^2/W^2)]}{2(1 + 2a/W)(1 - a/W)^{2/3}}$$

(5)

2.7. Flammability Tests

Three separate tests were carried out to determine the flammability characteristics of the samples, namely, ignition time, burn out time, and fire velocity within the sample. To conduct these tests, samples of 60 x 10 mm x 6 mm dimensions for each group were hung with a retort stand. A constant fire source was then bought to the samples and the times were gathered with use of a stop watch.

3. RESULTS AND DISCUSSION

3.1. Sample Porosity

Figures 1-3 show the apparent porosities of all samples. From Figures 1 and 3, it can be seen that the porosity of eco-composites and eco-nanocomposites for both resins is higher than the nanocomposite samples. These results can be attributed to the presence of RCF which created interfacial areas between RCF and matrices. [16-17].

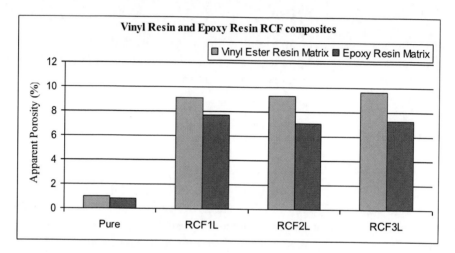

Figure 1. Comparison between porosities of vinyl ester resin and epoxy ester resin reinforced with different number of RCF sheets.

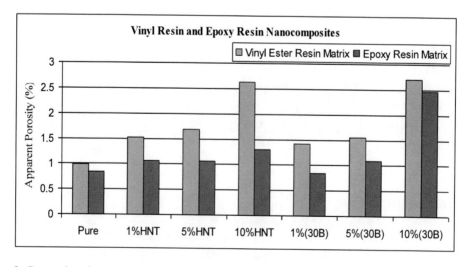

Figure 2. Comparison between porosities of vinyl and epoxy resins nanocomposites which reinforced with different concentrations (1%, 5% and 10%) of halloysite nanotubes (HNTs) and nanoclay platelets (Cloisite 30B).

From Figure 1, we can see that the amount of porosity does not significantly change with the increasing of the number of RCF sheets. Yet, on the other hand, nanocomposites of both polymers (as seen in Figure 2) with a high concentration of nanoclay particles (10%) have high porosities among all nanocomposites due to air-bubbles being trapped within the matrix. This indicates that there are more voids present inside this group of samples [18].

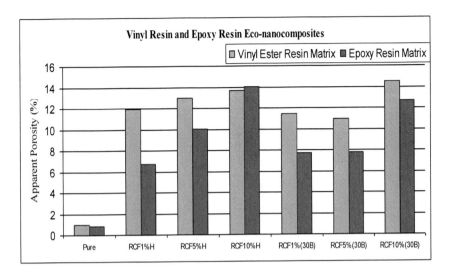

Figure 3. Comparison between porosities of vinyl and epoxy resins eco-nanocomposites reinforced with RCF sheets and different concentrations (1%, 5% and 10%) of halloysite nanotubes (HNTs) and nanoclay platelets (Cloisite 30B).

3.2. Flexural Strength

Figures 4 and 5 show the respective difference in strength behaviors between vinyl ester resin and epoxy resin nanocomposites and eco-nanocomposites respectively. The strengths of the majority of nanocomposites are lower than that of the pure vinyl and pure epoxy (figure 4). As both nanocomposites for both resins were fabricated via direct mechanical mixing technique, it is postulated that this technique created numerous undesirable micro-voids within the samples [19]. Yasmin, et al. [20] defines micro-bubbles as trapped air within the sample due to pouring mixtures into mould or other aspects of the preparation process. Micro-voids are often locations of crack initiation and cause the sample to easily fail, i.e. not withstand loading force [19-21]. Effective dispersion is vital for composites to display the best possible qualities. From Figure 4, it can be seen that the addition of 1% nano-filler resulted in the highest strengths of all of the samples [18]. Put simply, with a lower loading of nano-filler, the potential of the formation of micro-voids is less, and the dispersion is more uniform which both lead to strength improvement.

The results obtained in this study highlight that high concentration of nano-fillers, such as 10%, leads to a mixture which is too viscous for full-wetting with the cellulose fibres. The more nanoclay added, the more viscous of the clay resin mixture [20-21], and this can lead to formation more micro-voids and cause the creation of nanoclay agglomerates. Micro-voids and agglomerates of nano-fillers are considered to be the cause of the strength and strain reduction observed [22-19].

However, porosities results of present study indicated slightly voids percentage within nanocomposites samples, but it displayed a significant voids percentage through eco-nanocomposites which is corresponding well with the results of strength. These consequences of poor dispersion are shown in Figure 6.

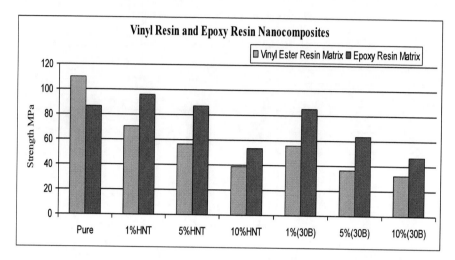

Figure 4. Comparison of strength behaviors of vinyl and epoxy resins nanocomposites reinforced with different concentrations (1%, 5% and 10%) of halloysite nanotubes (HNTs) and nanoclay platelets (Cloisite 30B).

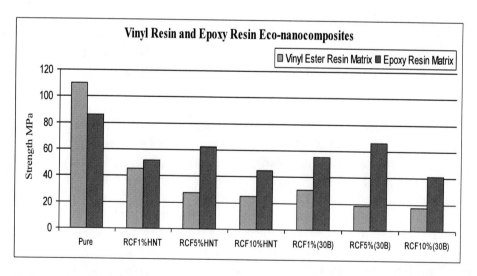

Figure 5. Comparison of strength behaviors of vinyl and epoxy resins eco-nanocomposites reinforced with RCF and with different concentrations (1%, 5% and 10%) of halloysite nanotubes (HNTs) and nanoclay platelets (Cloisite 30B).

Figure 5 shows the respective strengths of the eco-nanocomposites. Overall, the presence of the RCF sheets within the sample led to dramatic reduction in the strength of the sample. The eco-nanocomposites samples with a lower loading of clay nano-fillers, 1% and 5%, displayed better strength reinforcement throughout this group of samples. On the other hand, eco-nanocomposites samples with the high loading of clay nano-fillers, 10%, showed the lowest strength volumes.

This result can be explained by the lack of adhesion between fibres and matrix [23]. In preparation, when the RCF sheets and the matrixes were combined many of the fibres were not completely surrounded by matrix.

It is suggested that this was due to the fabrication method used in the present study, which led to the weak mechanical properties of our study samples due to the poor bonding between RCF and polymer matrixes.

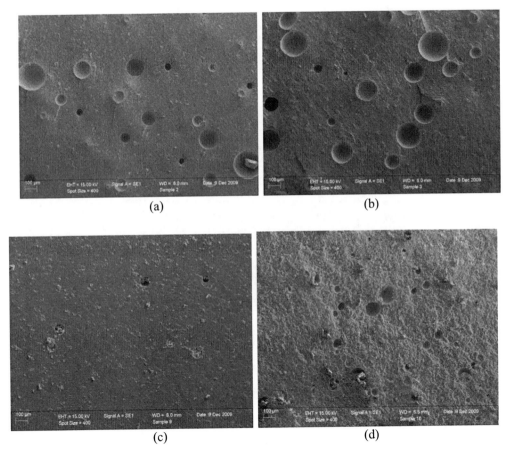

Figure 6. SEM image of fracture surface of vinyl ester/nanoclay composites with (a) 1% nano-clay (b) 10% nano-clay, and epoxy/nanoclay compsites with (c) 1% nano-clay (d) 10% nano-clay, air-bubbles can be seen to get larger in (b) and (d) with increasing nanoclay loading.

The effect of nano-clay addition on adhesion between RCF and polymers can be seen clearly from Figure 7, which displayed fracture surface of eco-nanocomposites reinforced with 10% of nano-clay.

From this figure we can observe that a majority of cellulose fibres are not surrounded by matrixes which can indicate to the low adhesion between the cellulose fiber and the matrix as a result of high loading of nano-clay.

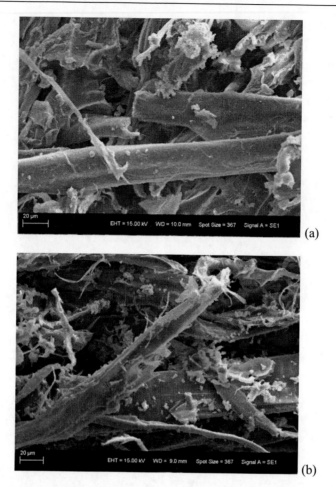

Figure 7. SEM images of fracture surface of (a) vinyl ester/RCF/nanoclay composites with 10% nano-clay (b) fracture surface of epoxy/RCF/nanoclay composites with 10% nano-clay subjected to three points test.

3.3. Impact Toughness

Figures 8 and 9 show an overall improving in toughness in all composites of vinyl ester resin and epoxy resin due to the reinforcement of RCF and clay nano-particles (HNT and 30B).

From both figures, RCF sheets effectively improved the impact toughness of vinyl and epoxy resins. A number of earlier research found that the enhanced fracture resistance of the samples can be attributed to the large amount of energy dissipated in crack deflection at the fibre–matrix interface.

As well as fibre-debonding, fibre bridging, fibre pullout, and fibre fracture all increased the toughness of the RCF-polymers composites [12, 23-25]. Figure 10 displays the typical fracture surface of eco-composites for both polymers, which clearly shows the mechanism of fibre fracture, interfacial debonding and fibre-pull-outs.

Physical, Flammability and Mechanical Properties ... 115

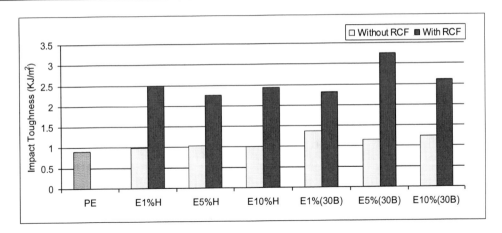

Figure 8. Comparison between epoxy resin nanocomposites with and without RCF.

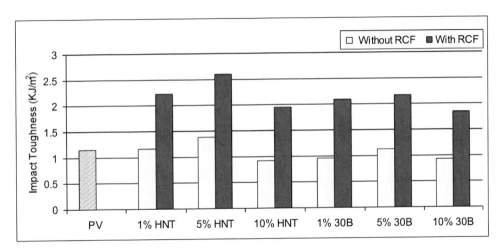

Figure 9. Comparison between vinyl ester resin nanocomposites with and without RCF.

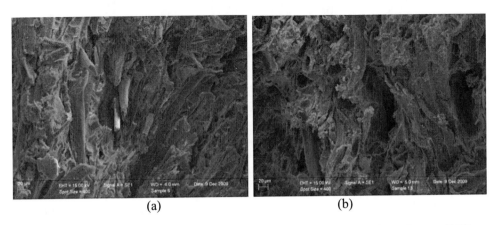

Figure 10. SEM image of fracture surfaces of (a) vinyl ester/RCF eco-composite and epoxy/RCF eco-composites samples subjected to three points test.

From the same data both polymer nanocomposites with the addition of low concentrations of HNT and 30B nano-clays, particularly 1% and 5%, were found to show a better reinforcement than with 10%. This result has been supported by a number of studies [26-27]. These results can be indicated that the dispersion of nano-clay within the matrixes is not fully uniform. This is because toughness improvements upon clay intercalated dispersion have also been reported by a number of studies [18, 19, 28]. That because nano-particles are too small to provide toughening through pining or crack-bridging mechanism. In addition, the direct mechanical mixing method used in this study can lead to produce intercalated dispersion of nano-clay within matrix [20].

From the same figures, nanocomposites of both polymers with 10% of HNT and 30B have the lowest impact toughness. In this case, especially at the high clay loadings, 10%, micro-voids are formed around the large inhomogeneities within the samples. What happens is that these formed microvoids coalesce with formation of larger cracks causing embrittlement, ultimately resulting in reduced toughness [18]. Put simply, the overloading of nanoclay into composites usually leads to reduced mechanical properties. Furthermore, undesirable aggregates of nano-clay particles will emerge in the microstructure of nanocomposites [30]. These clusters or agglomerations of nano-clay particles in the matrix act as additional crack initiation sites. What this means is that there are more locations where the composites may split easily under applied load [20].

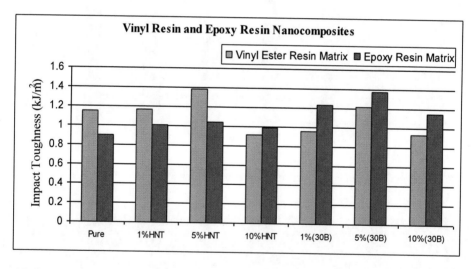

Figure 11. Comparison of impact toughness of vinyl and epoxy resins nanocomposites which reinforced with different concentrations (1%, 5% and 10%) of halloysite nanotubes (HNTs) and nanoclay platelets (Cloisite 30B).

Figures 11 and 12 compare and contrast the impact toughness between vinyl ester resin and epoxy resin as the matrices. There are basically no significant differences between the impact toughness properties for both matrices in this work. Vinyl ester resin and epoxy resin reinforced with 5% of nano-fillers (HNT and 30B) exhibited the highest impact toughness amongst all samples of present study as can be seen in (Figure 11). This same trend is seen in (Figure 12) with the nanocomposites of both resins reinforced with RCF.

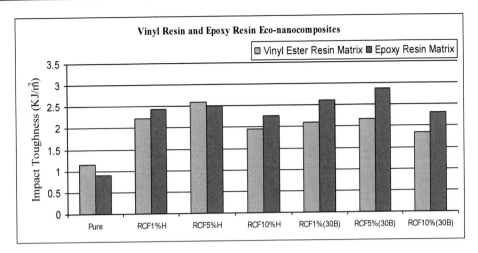

Figure 12. Comparison of impact toughness of vinyl and epoxy resins eco-nanocomposites reinforced with RCF sheets and different concentrations (1%, 5% and 10%) of halloysite nanotubes (HNTs) and nanoclay platelets (Cloisite 30B).

3.4. Effect of Water Absorption on Impact Toughness

To determinate the effect of water absorption on impact toughness three samples from etch group have been immersed in tap water for 30 days. Each bar specimen was notched with a razor blade to produce a sharp crack. Then the impact toughness of each group obtained with Zwick Charpy-impact tester. From Figures 13-16, we can observe the increasing of impact toughness for all samples due to immersion in water. These results may be attributed to the plasticization effect of water and thus a concomitant increase the resistance of composites to impact fracture [10]. Moreover, a number of researchers have reported that molecules of water can act as a plasticizer within polymers composite, which lead to an increase of the maximum strain for the composites after water absorption [16], and this can lead to improve the toughness of the samples.

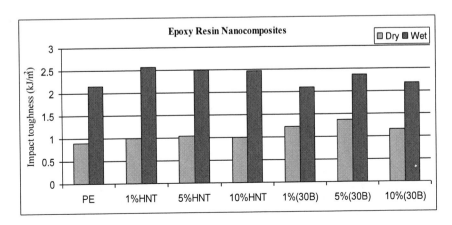

Figure 13. Comparison of impact toughness of Epoxy resin nanocomposites which reinforced with different concentrations (1%, 5% and 10%) of halloysite nanotubes (HNTs) and nanoclay platelets (Cloisite 30B) before and after water absorption.

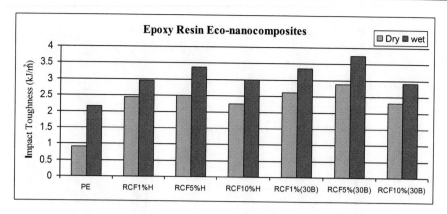

Figure 14. Comparison of impact toughness of Epoxy resin Eco-nanocomposites which reinforced with RCF sheets and different concentrations (1%, 5% and 10%) of halloysite nanotubes (HNTs) and nanoclay platelets (Cloisite 30B) before and after water absorption.

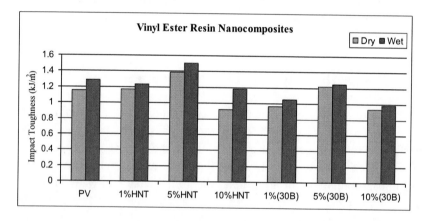

Figure 15. Comparison of impact toughness of vinyl ester resin nanocomposites which reinforced with different concentrations (1%, 5% and 10%) of halloysite nanotubes (HNTs) and nanoclay platelets (Cloisite 30B) before and after water absorption.

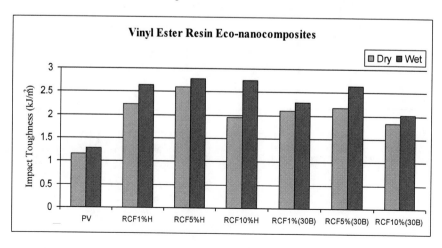

Figure 16. Comparison of impact toughness of vinyl ester resin Eco-nanocomposites which reinforced with RCF sheets and different concentrations (1%, 5% and 10%) of halloysite nanotubes (HNTs) and nanoclay platelets (Cloisite 30B) before and after water absorption.

3.5. Fracture Toughness

Figures 17 and 18 compare and contrast the fracture toughness between vinyl ester resin and epoxy resin as the matrices. The fracture toughness results represented in both figures display an overall improvement. Impact toughness results are similar. From Figure 17, both polymer nanocomposites with addition of low concentrations of HNT and 30B nanoclay (1% and 5%), were found to obtain a better reinforcement than addition of 10% of both types of nanoclay. As discussed earlier high addition of nanoclay particles can lead to more voids more nanoclay agglomerates which result on reduction of toughness properties.

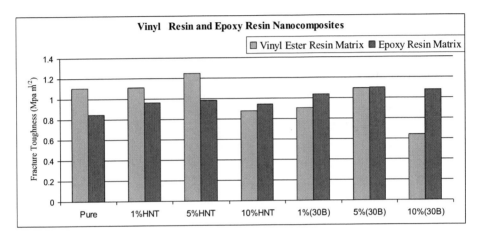

Figure 17. Comparison of fracture toughness of vinyl and epoxy resins nanocomposites which reinforced with different concentrations (1%, 5% and 10%) of halloysite nanotubes (HNTs) and nanoclay platelets (Cloisite 30B).

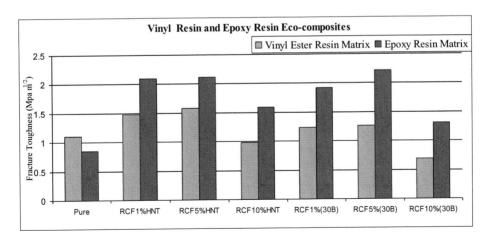

Figure 18. Comparison of fracture toughness of vinyl and epoxy resins eco-nanocomposites reinforced with RCF sheets and different concentrations (1%, 5% and 10%) of halloysite nanotubes (HNTs) and nanoclay platelets (Cloisite 30B).

From Figure 18, eco-nanocomposites of both polymers reinforced with RCF sheets show better toughness due to the role of RCF on toughness mechanism as mentioned previously.

From the same figure, Samples with 10% clay nano-particles had the lowest fracture toughness amongst the eco-nanocomposites. This is due to the high loading of nanoclay particles composites containing more voids and having poor dispersion of nanoclay within the samples Matrices with high loading of nanoclay become highly viscous, which leads to difficult soaking processing the method which applied to fabricate our samples. Cracks within these specimens can initiate at the weak points throughout the samples which can cause the samples to fail at relatively low stress leading to reduced toughness.

3.6. Flammability

Thermal tests have been conducted at ambient conditions to simulate the realistic fire action. During the test, time of burning out and ignition time of each group of sample have been determined, and then fire spreading speed of each group of samples calculated. As can be seen from Tables 2 and 3, pure samples of vinyl ester resin and epoxy resin had the lowest thermal resistance. Samples reinforced with RCF sheets have higher thermal stability than pure samples. It is accepted that this is due to the higher and longer thermal resistance of cellulose fibers present [31]. Interestingly, the addition of clay nano-particles effectively improved thermal stability which can be seen in the improvement of time of burning out and ignition time for all nanocomposites and eco-nanocomposites, This effect of nanoclay particles on thermal stability had been reported by number of studies [32-33-34]. Nanoclay particles can enhance thermal stability for nanocomposites by acting as a high effective insulator and mass transport barrier to the volatile products which is generated during the time of decomposition, nanoclay particles can assist in the formation of char after the burning of the materials [18-35].

Table 2. Thermal properties of vinyl ester resin and its composites

Sample	Burning time (s)	Ignition time (s)	Flame velocity (mm/s)
Pure Vinyl	92	3.2	0.109
V+RCF	103	3.53	0.097
V+ 1%HNT	115	4.74	0.087
V+5%HNT	120	4.81	0.083
V+10%HNT	124	5.23	0.081
V+1% (30B)	98	4.03	0.102
V+5%(30B)	102	4.24	0.098
V+10%(30B)	110	4.6	0.091
V+RCF1%HNT	95	4.56	0.105
V+RCF5%HNT	102	4.87	0.098
V+RCF10%HNT	106	5.81	0.094
V+RCF1%(30B)	97	4.5	0.103
V+RCF5%(30B)	105	5.12	0.095
V+RCF10%(30B)	109	5.31	0.092

From the thermal data of our samples we can find that the addition of HNT improved thermal stability better than nanoclay platelets (30B). This phenomenon can be explained as a result of the hollow tubular structures of the HNTs which can form barriers for heat transport as well as the presence of iron in the chemical structure of HNT [36]. In addition, it is observed from the present study that the pure samples of both polymers burnt out without remains. However, in contrast, all groups of composites, eco-composites, nanocomposite and eco-nanocomposites leave some remaining materials [18].

Table 3. Thermal properties of epoxy resin and its composites

Sample	Burning time (s)	Ignition time (s)	Flame velocity (mm/s)
Pure Epoxy	81	3.6	0.123
E+RCF	98	3.82	0.102
E+1%HNT	123	4.36	0.081
E+5%HNT	126	4.72	0.079
E+10%HNT	135	4.98	0.074
E+1%(30B)	87	4.52	0.115
E+5%(B30)	112	4.96	0.089
E+10%(30B)	121	5.14	0.083
E+RCF1%HNT	98	4.74	0.102
E+RCF5%HNT	111	5.12	0.090
E+RCF10%HNT	115	5.28	0.087
E+RCF1%(30B)	85	5.12	0.118
E+RCF5%(30B)	93	5.26	0.108
E+RCF10%(30)	106	5.38	0.094

CONCLUSION

Epoxy and vinyl-ester composites and nanocomposites reinforced with recycled cellulose fibre (RCF) sheets have been fabricated and evaluated. Flexural strength, impact toughness, fracture toughness and the effect of moisture absorption on impact toughness have been investigated. Porosities and flammability behaviours such as burning out time and ignition have also been studied. The flexural strengths of all composites were found to be lower than pure polymer samples due to the presence of undesirable micro-voids generated during processing. RCF and low addition of nanoclay (1-5 wt%) significantly improved the toughness properties of all composites. Water absorption led to impact toughness improvement due to the water plasticization at the crack-tip. The thermal properties in the terms of burning-out time and ignition-time both increased with the increase of the nanoclay content.

ACKNOWLEDGEMENTS

We are grateful to Mr. Charles Hrubos of NichePlas for providing the organoclay (30B) for this study and we would like to thank Mr. Andreas Viereckl from Mechanical Engineering and Mr. Ross Williams from Applied Physics at Curtin University of Technology for assistance with mechanical tests.

REFERENCES

[1]   Silva, R; Spinellli, D; Filho, B; Neto, C; Chierice, G; Tarpani, J. Fracture toughness of natural fibers/castor oil polyurethane composites. *Composites Science and Technology,* 2006, 66, 1328-1335.

[2]   Mottershead, B; Eichhorn, S. Deformation micromechanics of model regenerated cellulose fibre-epoxy/polyester composites. *Composites Science and Technology,* 2007, 67, 2150-2159.

[3]   Lu, J; Askeland, P; Drzal, L. Surface modification of microfibrillated cellulose for epoxy composite applications. *Polymer,* 2008, 49, 1285-1296.

[4]   Singleton, A; Baillie, P; Beaumont, P; Peijs, T. On the mechanical properties, deformation and fracture of a natural fibre/recycled polymer composite. *Composites: Part B,* 2003, 34, 519 – 526.

[5]   Ghavami, K; Romildo, D; Toledo F; Barbosac, N. Behaviour of composite soil reinforced with natural fibres. *Cement and Concrete. Composites,* 1999, 21, 39-48.

[6]   Manfredi, LB; Rodriguez, ES; Wladyka-Przybylak M; Va´zquez, A. Thermal degradation and fire resistance of unsaturated polyester, modified acrylic resins and their composites with natural fibres. *Polymer Degradation and Stability,* 2006, 91, 255-261.

[7]   Facca, A; Kortschot, M. Predicting the elastic modulus of natural fibre reinforced thermoplastics. *Composites: Part A,* 2006, 37, 1660-1671.

[8]   Juarez, C; Duran, A; Valdez, P; Fajardo, G. Performance of Agave Lecheguilla natural fiber in Portland cement composites exposed to severe environment conditions. *Building and Environment,* 2007, 42, 1151-1157.

[9]   Bessadok, A; Marais, S; Gouanve, F; Colasse, L; Zimmerlin, I; Roudesli, S; Me'tayer, M. Effect of chemical treatments of Alfa (Stipa tenacissima) fibres on water-sorption properties. *Composites Science and Technology,* 2007, 67, 685-697.

[10]   Low, IM; Somers, J.; Kho, HS; Davies, IJ; Latella, BA. Fabrication and properties of recycled cellulose fibre-reinforced epoxy composites. *Composite Interfaces,* 2009, 16, 659-669.

[11]   Satyanarayana, KG; Sukumaran, K; Mukherjee, S; Pavithran, C; Pillai, SGK. Natural fibre-polymer composites. *Cement and Concrete Composites,* 1990, 12, 117-136.

[12]   Low, I; Somers, J; Pang, W. Synthesis and properties of recycled paper nano-clay-reinforced epoxy eco-composites. *Key Engineering Materials,* 2007, 334, 609-612.

[13]   Sanadi, A; Young, R; Rowell, R. Recycled newspaper fibers as reinforcing fillers in thermoplastics: Part I–Analysis of tensile and impact properties in polypropylene. *Journal of Reinforced Plastics and Composites,* 1994, 13, 54-67.

[14] Huda, M; Drzal, L; Mohanty, A; Misra, M. The effect of silane treated- and untreated-talc on the mechanical and physico-mechanical properties of poly(lactic acid)/newspaper fibers/talc hybrid composites. *Composites: Part B*, 2007, 38, 367-379.

[15] Baroulaki, I; Karakasi B; Pappa, G; Tarantili, P; Economides, D; Magoulas, K. Preparation and study of plastic compounds containing polyolefins and post used newspaper fibers. *Composites: Part A*, 2006, 37, 1613-1625.

[16] Kim, HJ; Seo, DW. Effect of water absorption fatigue on mechanical properties of sisal textile-reinforced composites. *International Journal of Fatigue*, 2006, 28, 1307-1314.

[17] Dhakal, HN; Zhang, ZY; Richardson, MOW. Effect of water absorption on the mechanical properties of hemp fibre reinforced unsaturated polyester composites. *Composites Science and Technology*, 2007, 67, 1674-1683.

[18] Pavlidou, S; Papaspyrides, CD. A review on polymer–layered silicate nanocomposites. *Progress in Polymer Science*, 2008, 33, 119-1198.

[19] Qi, B; Zhang, QX; Bannister, M; Mai, YW. Investigation of the mechanical properties of DGEBA-based epoxy resin with nanoclay additives. *composite Structures*, 2006, 75, 514-519.

[20] Yasmin A; Abot, JL; Daniel, IM. Processing of clay/epoxy nanocomposites by shear mixing. *Scripta Materialia*, 2003, 49, 81-86.

[21] Ho, M; Lam, C; Lau, K; Ng, DHL; Hui, D. Mechanical properties of epoxy-based composites using nanoclays. *Composite Structures*, 2006, 75, 415-421.

[22] Deng, S; Zhang, J; Ye, L; Wu, J. Toughening epoxies with halloysite nanotubes. *Polymer*, 2008, 49, 5119-5127.

[23] Benjamin, B; Jorg, M. Review: Impact and tensile properties of PLA/Cordenka and PLA/flax composites. *Composites Science and Technology*, 2008, 68, 1601-1607.

[24] Hetzer, M; De Kee, D. Wood/polymer/nanoclay composites, environmentally friendly sustainable technology: A review. *Chemical Engineering Research and Design*, 2008, 86, 1083-1093.

[25] Monteiro, SN; Lopes, FP; Ferreira, AS; Nascimento, DC. Natural-fiber polymer-matrix composites: Cheaper, tougher, and environmentally friendly. *JOM*, 2009, 61, 17-22.

[26] Liu, X; Wu, Q. Polyamide 66/clay nanocomposites via melt intercalation. *Macromol. Mater. Eng.*, 2002, 287, 180-186.

[27] Ratna, D; Manoj, N; Varley, R; Raman, RS; Simon, G. Clay-reinforced epoxy nanocomposites, *Polym. Int.*, 2003, 52, 1403-1407.

[28] Chen, B; Evans, JRG. Impact and tensile energies of fracture in polymer–clay *Nanocomposites Polymer*, 2008, 49, 5113-5118.

[29] Sun, L; Gibson, RF; Gordaninejad, F; Suhr, J. Energy absorption capability of nanocomposites: A review. *Composites Science and Technology*, 2009, *69*, 2392-2409.

[30] Liu, M; Guo, B; Du, M; Cai, X; Jia, D. Properties of halloysite nanotube–epoxy resin hybrids and the interfacial reactions in the systems. *Nanotechnology*, 2007, 18, 45-52.

[31] Ma, X; Yu, J; Kennedy, JF. Studies on the properties of natural fibers-reinforced thermoplastic starch composites. *Carbohydrate Polymers*, 2005, 62, 19-24.

[32] Paul, MA; Alexandre, M; Degee, P; Henrist, C; Rulmont, A; Dubois, P. New nanocomposite materials based on plasticized poly(L-lactide) and organo-modified montmorillonites: Thermal and morphological study. *Polymer*, 2003, 44, 443-450.

[33] Vyazovkin, S; Dranka, I; Fan, X; Advincula, R. Kinetics of the thermal and thermo-oxidative degradation of a polystyrene–clay nanocomposite. *Macromol. Rapid Commun.,* 2004, 25, 498-503.

[34] Becker, O; Varley, RJ; Simon, GP. Thermal stability and water uptake of high performance epoxy layered silicate nanocomposites. *European Polymer Journal,* 2004, 40, 187-195.

[35] Zeng, QH; Yu, AB; Lu, GQ; Paul, DR. Clay-based polymer nanocomposites: Research and commercial development. *Journal of Nanoscience and Nanotechnology,* 2005, 5, 1574-1592.

[36] Du, M; Guo, B; Jia, D. Thermal stability and flame retardant effects of halloysite nanotubes on poly(propylene). *European Polymer Journal,* 2006, 42, 362-1369.

In: Fiber-Reinforced Composites
Editor: Qingzheng Cheng

ISBN: 978-1-61470-303-7
© 2012 Nova Science Publishers, Inc.

Chapter 5

STATIC AND DYNAMIC BEHAVIOURS OF COMPOSITE LAMINATE

R. Zitoune[1], A. Mir[2], B. Hachemane[1, 2], F. Boyer[1], C. Bouvet[3], F. Collombet[1] and B. Bezzazi[2]*

[1]Institute Clément Ader, "INSA, UPS, Mines Albi, ISAE";
Toulouse University – France
[2]U. R. Materials, Processes and Environment;
Boumerdès University – Algeria
[3]Institute Clément Ader, "INSA, UPS, Mines Albi,
ISAE", ISAE; Toulouse-France

ABSTRACT

The use of composite materials made of natural fibres in buildings and automobiles increases more and more. However, the mechanical properties and the mechanical behaviours of these materials depend on the type of natural fibre used as well as the type of manufacturing process used for obtaining the final part. In this chapter, we are interested in the characterization of the mechanical and thermo mechanical behaviour of composite laminate made of jute/epoxy.

For the mechanical characterization, standard static tests (tensile, shear and bending) are carried out for the identification of the mechanical proprieties. Thanks to the digital image correlation technique, we observe that the mechanical proprieties have important variabilities during the static tests. The impact tests show that these variabilities have little impact on the first force responsible for the first damage during the impact tests. Nevertheless, the maximum forces obtained during impact tests are affected by the variabilities of this material.

---

* Email: Redouane.zitoune@iut-tlse3.fr

I. Introduction

The use of natural fibres in the reinforcement of composites is growing constantly, due to their environmental and economic concerns. Composite materials with natural fibres are considered as potential products to substitute the composites for non renewable synthetic fibres. The most current natural fibres are straw, flax, hemp and jute fibre. The Asian markets have been using natural fibres for many years. For example, jute is a common reinforcement in India. The major market for natural fibres in the year 2000 is in building applications, whereas other applications have been growing at an increasing rate, especially in the automotive industry. For instance, from 1996 till 2003, the use of natural fibres in composites of the German automotive industry increased from 4,000 tons to 18,000 tons per year. In 2006, European companies have taken the lead in natural-fibre composite development, such as Dieffenbacher (Germany), BASF (Germany), and Rieter Automotive (Switzerland). In addition to its mechanical properties, which are comparable to those of the hemp fibre, jute fibre has the advantage of being easy to weave [1-2]. However, even if the mechanical performances of composites with natural fibres offer new perspectives for the structures design (low densities, and acceptable mechanical properties), the disparity of their properties and the variability of their behaviour as well as the misunderstanding of these materials make them difficult to use [1-3]. In addition, being natural materials, the reproducibility of their properties is not easy to assure because it depends on diverse parameters such as the origin of the grains of plants, the type of ground on which they were planted, the climate, etc. Several studies have shown that these materials are sensitive to humidity and to the heat [4]. When these materials are heated up at various temperature levels, at the results, an important reduction of the failure stress in tensile, in bending as well as in chocks. For example, the tensile tests that have been carried out on jute fabric at various temperatures have shown that the failure stress decreases by 43% when the jute fabric is heated to 180°C. For the improvement of the mechanical properties of the jute, especially its behavior in wet medium, several works are initiated to treat the surface of the jute fiber [5, 6]. These treatments, which modify the interphase of the surface of fiber, produce morphological changes [7, 8]. Treatments are carried out using alkaline, silane or alkaline, and silane [7-9] which decrease the absorption of moisture from 4.2% to 3.8%. The failure stress of a fabric treated with silane is higher than that of fabric untreated, on the other hand the breaking stress of a jute fabric untreated is better than a jute treated with alkaline [10-12]. Other treatments also done under UV radiation present an increase of 58% of its flexural strength [6-13].

The uses of jute-reinforced plastic are the current interest of many researchers. The most used thermoplastic resins are polyethylene (EP) [14,15], polypropylene (PP) [16,17], polystyrene (PS) [18,19], vinyl polychloride (PVC) [20,21], and the polyester [22] for economic reasons. The choice of a structural polymeric resin does not pose a problem of provisioning but constitutes a barrier to the recyclability of the unit. However, the ecological or natural resins do not answer the schedule of conditions of the end product because of the weak mechanical properties. These resins are rigid and breakable like the polylactone (PLA) with the jute [23], dissolve in water like the natural polysaccharose (TPS) [24].

Few works have focused on the analysis of the behaviour of jute fibre reinforced laminate subjected to impact loading by falling weight [25-26]. It has conducted an experimental study in order to show the influence of hybridization of glass fibers (jute–glass hybrid composites) on low velocity impact response, damage resistance and damage tolerance capability of woven jute fabric reinforced isothalic polyester composites. The result shows that the jute composites have better energy absorption capacity compared to jute–glass hybrid laminates. However, the hybrid laminate with 16% glass fiber weight is the most optimum combination of jute and glass fiber with minimum deflection, maximum peak load, better damage tolerance, and less costlier than hybrids laminates with 25.2% and 8.2 % glass fiber weight. The acoustic emission and thermoelastic stress analysis, the cyclic post-impact three-point bending tests have been carried out on plain woven jute fabric/polyester plates [26] have shown that damage in natural fiber-reinforced laminates progresses as far as the defects present in the laminate reach a critical energy that allow them to develop.

In this chapter, a study has been carried out to identify the behaviour of the jute/epoxy laminate under static and dynamic loading. At first, tensile, shearing, and three-point bending static on standardized test specimens are investigated. For the tensile tests, the specimens are instrumented on the surface by two techniques. The first concerns extensometric gauges and the second concerns the randomly speckle pattern which is used for the measurement of the strain field by Digital Image Correlation (DIC). The results given by these two techniques of measurements are then compared.

Then, the dynamic behaviour of the jute/epoxy specimens is identified. For the dynamic mechanical analysis, the dynamic proprieties are identified as function of the temperature as well as the time and frequency (storage modulus, loss modulus and damping factor). In the end, impact tests have been carried out with falling weight on jute/epoxy specimens and the influence of impact energy on the mechanical behaviour (force and displacement) as well as the damage size, various impact tests are conducted.

II. DESCRIPTION OF EXPERIMENTAL SETUP AND PROCEDURE

II.1. Materials and Specimens

The jute/epoxy laminate is manufactured by the process "infusion" (Figure 1). The fabric is prepared and cut out with dimensions of 300mm x 300mm. The resin used is an epoxy of type LY 5052 with hardener DY 5052 manufactured by HUNTSMAN company. The unit is polymerized at a temperature of 80°C during 8 h. The stacking sequence studied are $[0]_S$ for the warp direction, $[90]_S$ for the weft direction and $[+45/-45]_S$. The jute/epoxy laminates are cut out with a diamond saw following standard NF IN ISO 527-1.

II.2. Static Tests

For identifying the mechanical properties, in the warp and weft directions, tensile tests and three point bending tests were carried out in accordance with the standard IN ISO 527-5 for the tensile test and NF IN ISO 14125 for the three point bending test. For the three point bending test, the maximum stress is given by Equation (1) with P_{max} (N) the maximum force of rupture, L the distance between supports (mm), b and h are respectively the width and the thickness of the sample.

$$\sigma_{max} = \frac{3P_{max}L}{bh^2} \qquad (1)$$

For the tensile tests, the clamping was done using hydraulic jaws with a pressure of 10 bar and at a speed of 2 mm/min. For measuring deformation, the specimens were mounted on the surface by strain gages of 25 mm length. The gages were connected on a Strainsmart de Vishay bridge.

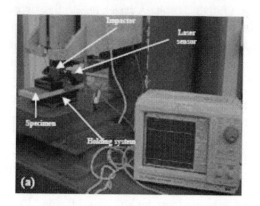 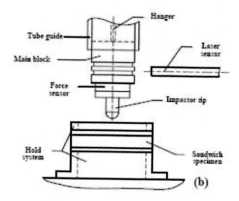

Figure 1. Experimental device for the impact tests.

While conducting tensile tests in warp and weft direction, in addition to strain gages, 3-D digital image correlation technique (3-D DIC) was also used. The 3-D DIC method is based on both digital image correlation (DIC) and stereovision technique and was developed at the end of the last century [27-29]. The technique uses a DIC algorithm to determine point correspondences between two images of a specimen acquired from two rigidly bound cameras. The correlation scores are computed by measuring the similarity of a fixed subset window in the first image to a shifting subset window in the second one. A first order two-dimensional shape function in the subset and a zero normalized sum of square difference (ZNSSD) correlation criterion are used. Sub-pixel correlation is performed using quintic B-spline gray level interpolation. After determining the calibration parameters for each camera as well as the 3-D relative position/orientation of the two cameras (pinhole model and radial distortion of 3rd order), the 3-D specimen shape can be reconstructed from the point correspondences founded by DIC using triangulation. To determine the 3-D displacement field, DIC is also used to determine point correspondences between the stereo pairs acquired before and after deformation. The Green-Lagrange strain field was obtained from the displacement field by numerical differentiation. A complete description of the 3-D DIC technique can be found in literatures, e.g. Luo et al. [29] or Garcia [30].

II.3. Dynamic Tests

II.3.1. Dynamic Mechanical Thermal Analysis

The investigation of the thermo-mechanical behaviour was performed by dynamic-mechanical thermal analysis and DMTA by using a sensor force of 100 N. For the measurements two rectangular specimens were prepared. The first specimen is of 45 mm length, 10 mm width and 2.5 mm thickness, while the second specimen is of 45 mm length, 20 mm width and 2.5 mm thickness. The tests were performed in tensile and compression mode at frequencies varying from 0.002 Hz to 100 Hz with a dynamic strain of 0.05% in a temperature range between 30°C and 150 °C with a heating rate of 5 °C/min.

II.3.2 Impact Tests

The experimental apparatus used for the impact tests is represented in Figure 1. It is mainly composed by a Kistler force sensor equipping an impactor, two laser sensors, an oscilloscope, guide column and a sandwich plate holding device.

The impactor is characterized by:

- Free falling main block (2 Kg appreciatively),
- 10 KN force sensor mounted at main block bottom,
- 12.7 mm diameter hemispherical tip.

The first laser sensor allows to calculate the beginning contact velocity and the second gives the specimen non-impacted side displacement. Both sensors signals are synchronized par an oscilloscope. The sandwich plates are clamped to a rigid table with a 125 x 75 mm² window holding system.

The impact force (F_{impact}) between the impactor and the specimen was determined from the force sensor measured force ($F_{measured}$) by the following relation:

$$F_{impact} = \frac{M_{impactor}}{M_{impactor} - M_{tip}} F_{mesured} \tag{2}$$

with $M_{impactor}$ and M_{tip} are respectively the total mass of the impactor (2.056 Kg) and that of hemispherical tip (0.176 Kg).

These impact tests were carried-out based on the recommendations of ASTM D1736/D1736M-05, which suggests the following expressions for obtaining the main testing results:

$$E_i = \frac{M_{impactor}}{2} V_i^2 \tag{3}$$

$$\delta(t) = \delta_i + V_i\, t + \frac{g}{2} t^2 - \int_0^t (\int_0^t \frac{F_{impact}\,(t)}{M_{impactor}}\, dt)\, dt \tag{4}$$

with: E_i: the impact energy (J), V_i: the impact velocity (m/s), g: the acceleration due to gravity (9.81 m/s²), Vi(t): impactor velocity at time t (m/s), δ_i: the impactor displacement from reference at time t = 0 (m), $\delta(t)$: the impactor displacement from reference at time t (m).

III. RESULTS AND ANALYSIS

III. 1. Static Analysis

The first results of the 3-D DIC obtained during tensile test on jute/epoxy specimens in the warp and weft direction show an important difference between the behaviour in these two directions as well as an important gradient of deformation (Figure 2). It can be explained by the fact that, the variation of the diameter of the wire constituting fabrics well as the jute fibre has microfibril angle in the range of 7°-12° to the fibre axis [31]. Also, we have more wires in the weft direction than warp direction. (Figure 3).

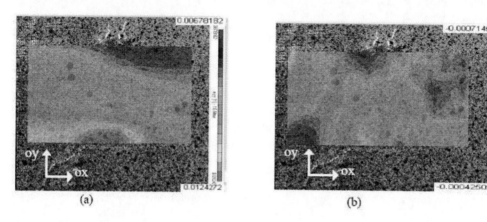

Figure 2. Cartography of strain distribution (a) longitudinal strain ε_{yy} (warp), (b) transversal strain ε_{xx} (weft).

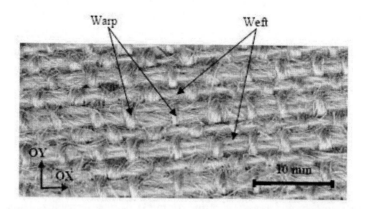

Figure 3. Picture illustrating the form of weaving as well as the wire in the weft direction than warp direction.

These geometrical and dimensional variations on jute fibre lead to the variation of the fibre content in the laminate the presence of a strong gradient primarily dependent on the geometrical variability of fabric and that of the fibre content at any point. From the figure 4 we show a linear elastic behaviour up to 30 MPa and then a non-linear behaviour until the failure stress (43 MPa). However, the weft direction presents a linear elastic behaviour up to

47MPa and finally non-linearity until rupture at 61 MPa (cf. Figure 4). From the Figure 4 and using the curve of the normal strain and the transverse strain (perpendicular to the applied load) during the tensile test in the weft direction we can calculate the Poisson's ratio (v_{yx}). This value is around 0.28.

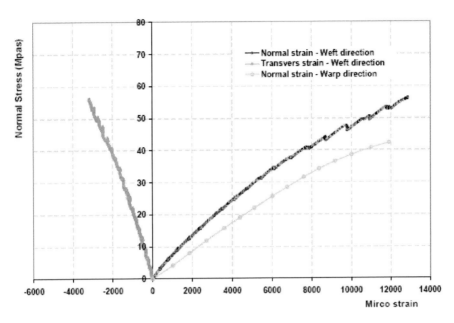

Figure 4. stress versus axial strain for the tensile test on jute/epoxy in weft and warp direction.

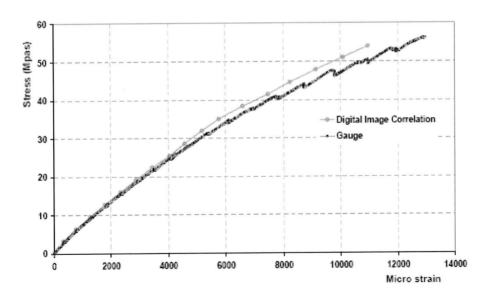

Figure 5. Comparison between 3D-DIC and strain gauge on specimen made in weft direction.

The shear test carried out on jute/epoxy specimen with a staking sequence [+45/-45]$_S$ shown a nonlinear behaviour. This nonlinear behaviour has been studied from tensile test. In this case the shearing modulus is a round 1.5 GPs.

Figure 5 shows a comparison between the axial and transverse deformations obtained by 3-D DIC and by strain gauges. Lesser than 40 MPa, the difference between both the methods is lower than 5%, while approaching the rupture stress the variation reaches to 8%. This maximum variation corresponds to the axial deformation. Figure 6 represents the evolution of the load according to displacement during the three point bending test on laminates jute/epoxy for specimens in the warp and weft direction. It is noted that all the tested specimens present a nonlinear mechanical behaviour. The average value of the failure stress of the specimen is 80 ± 8.78 MPa in warp direction and 84 ± 5.25 MPa in weft direction (see Eq. 1). When the deflections are lower to 1 mm all the specimens tested (in warp and weft direction) present the same behaviour (see Figure 6).When the deflections increase, an important variation of the behaviour was observed between the specimens in the warp and weft direction.

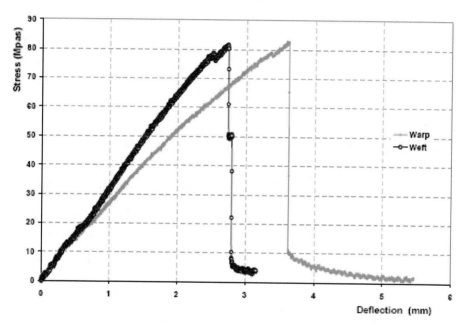

Figure 6. Stress various displacements during three point bending test on jute/epoxy.

Figure 7 (a) shows scanning electron micrograph of the fracture topographies after tensile tests on specimens. We observe a clear rupture of fibre in the warp direction and the presence of some fibrillates shredded in constituent fibre in the weft direction, this mode of rupture was identified to that of a jute/polyester presented by Sabeel et al. [22]. The same mode of rupture was observed after tensile test on specimens made in weft direction, on the other hand in the place of the shredded fibrillates, from the result, the presence of print of wire of warp (Figure 7-b). Figure 7-c represents the fracture topographies in shearing of a laminate jute/epoxy [+45/-45] S, and noted a good cohesion between the various layers of reinforcement (no slip between layers). The modes of rupture are identical to the work of Park et al. [32]. Figure 8 represents the central zone of the jute/epoxy laminate after rupture by three point bending. The part of the specimen observed in tension shows the presence of a package of fibres (constituting warp) stretched (Figure 8-b). This can be explained in the following way, under the effect of loading the fibre of warp elongated by exceeding their yield stress, after rupture, the specimen remains stretched (Figure 8-b).

This phenomenon is not observed on a warp thread tested in compression (Figure 8-c). The analysis with the SEM on the zone of rupture of the specimen in warp and weft direction presented the same observations.

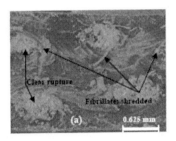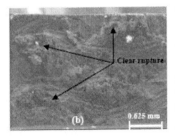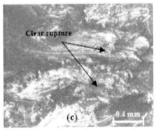

Figure 7. A fracture faces obtained by SEM after tensile tests and shear test: (a) tensile test in the warp direction, (b) tensile test in the weft direction, (c) shear test.

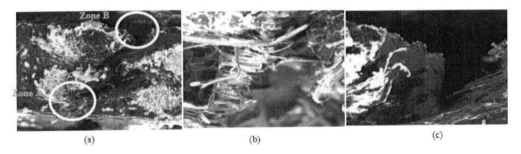

Figure 8. Pictures showed the fracture faces obtained by SEM after flexural tests in the warp direction.

III.2. Dynamic Analysis

III.2.1. Dynamic Mechanical Thermal Analysis

Figure 9 shows the evolution of the storage modulus (E') as well as the damping factor (tan δ) according to the temperature during the loading with 10Hz of frequency. These curves are obtained with a temperature cycle from 30 °C to 150 °C with a heating rate of 5 °C/min. From these results we observe that, the damping factor increases with increasing of the temperature. When the temperature is equal to 135 °C, the damping factor reaches a maximum value of 0.138. The temperature corresponding to the peak for the jute/epoxy composites indicates the glass transition temperature (Tg) for the epoxy matrix. We notice that, the glass transition temperature of the epoxy resin without reinforcement given by HUNTSMAN Company is around 125 °C. This gain in thermostability can be interpreted as a reduction of the mobility of the epoxy matrix around the jute fibres by the interfacial interactions. The same phenomenon is noticed by Florian et al. [33] on epoxy resin compared to the epoxy resin reinforced by carbon nanotubes during the DMTA tests. In addition, the increase of the temperature will have as consequence an increasing in the free volumes and the movements of thermal chains origin.

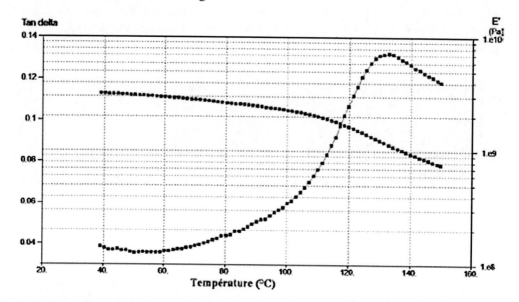

Figure 9. DMA curves represented storage modulus and damping factor as a function of temperature.

Concerning the storage modulus, when the temperature increases from 30°C until 80 °C, a reduction 18 % is noticed. However, an important diminution of the storage modulus is observed (around of 70 %) when the temperature reached the Tg value. Figure 10 shows the evolution of the storage modulus in according to the temperature for various frequencies of loading. We observe that for temperatures lower to 120 °C the frequency variation have small influences on the storage modulus.

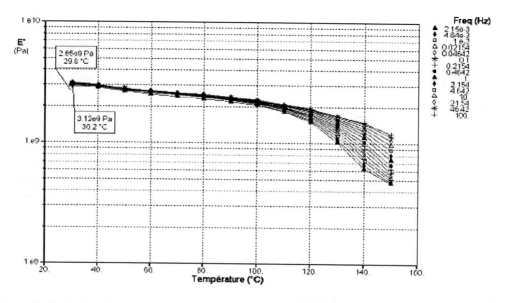

Figure 10. Evolution of storage modulus as function of temperature for various frequencies.

For example, for 110 °C of the heating temperature, the variation of the frequency from 0.002Hz to 100 Hz resulting in increasing of the storage modulus around 10%. However, when the heating temperatures are superior to the Tg, the variation of the frequency from 0.002 Hz to 100 Hz generate an significant increasing of the storage modulus. For 150 °C of the heating temperature the gain of the storage modulus is around 110 %. This gain of the storage modulus when the temperature and the frequency increase, can be likened to the increasing of the damping factors when temperatures are superior to the Tg value.

III.2.2. Impact

The results in Figure 11 shows an example of the evolution of the impact force versus time for various specimens of $[0°]_6$ stacking sequence impacted with the same energy of 7 J. It is noticed that for all impacted specimens, these curves can be regarded as sine waves. From this figure, it can be distinguished two critical values, the first one relates the first damage force, denoted as F_0, the second is the maximum impact force recorded by the force sensor, denoted as F_{max}. From the first contact (t = 0.005 s) until the apparition moment of the first damage (corresponding to the F_0 forces (surrounded zone)), all the specimens tested exactly the same behaviour and we observe a very small differences between these initials forces (F_0). However, the difference between the maximum forces recorded for each specimen is more important. As the tested plates were subjected to the bending solicitation, these differences can be linked to the variability of the thickness plates as well as the intrinsic versatilities of the warp and weft thread. Similar results are obtained for all impact energy tested (from 1.5 J to 15 J).

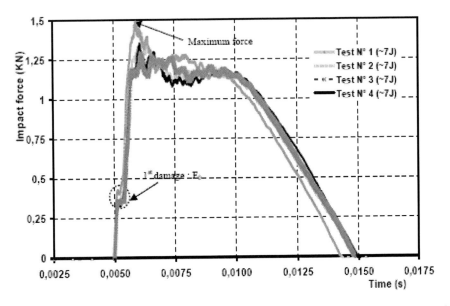

Figure 11. Temporal impact force curves for various jute/epoxy plates with the impact energy of 7 J.

Figure 12 shows the evolution of the average force F_0 vs the impact energy as well as the tendency curve. We note that, on one hand the increases of the impact energy generate an increasing of the first damage force and on the other hand with a polynomial tendency curve of degree can predict these forces with small difference compared to the experimental values.

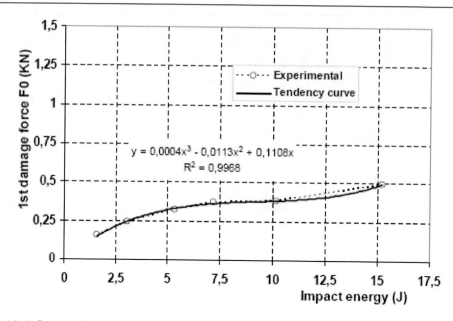

Figure 12. Influence of the impact energy on the first damage force F_0.

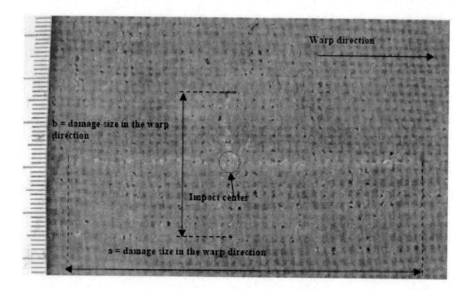

Figure 13. The state of the plate after impact with an energy of 7 J.

The state of the plate was not impacted face after impact is presented on the Figure 13. We observe that, the damages are appeared as break of layer fabrics in warp and weft direction. The lengths of these damages are strongly affected by the impact energy. With regard to the impacted faces, damage forms are similar to those of the not impacted facets with a small wide. We think that, the compression failure stress of the jute/epoxy are more important compared of the tensile failure stress and on the other hand the damage mode is produced as a fracture in mode I. This fracture mode I is produced in weft and warp directions.

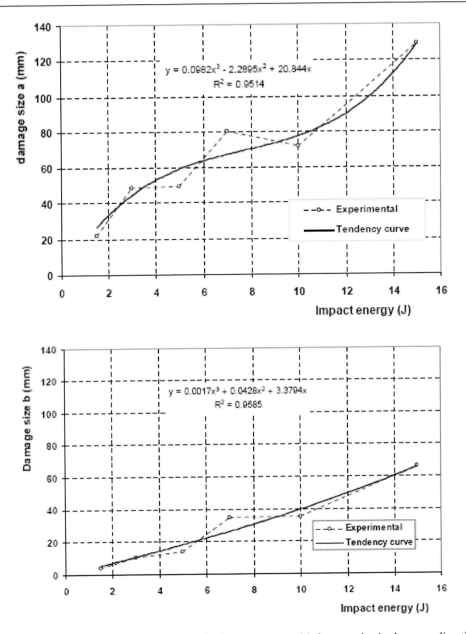

Figure 14. Evolution of the damages sizes vs the impact energy. (a) damage size in the warp direction, (b) damage size in the weft direction.

Evolutions of the length damages in the warp and weft directions various impact energies were presented on the Figure 14. We observe that, the length damage in the warp direction was more important compared to the length damage in the weft direction. This difference can be linked to the dimension specimens tested and to the mechanical properties of this material in the weft direction are better to mechanical properties in the warp direction. From the static tests, the tensile modulus and the tensile stress in the weft direction is about 30 % better compared to the warp direction.

CONCLUSION

This paper presents experimental results of mechanical testes carried –out on composite laminates made from woven fabrics submitted static and dynamic loading. Based on this experimental analysis, the following conclusions were drawn.

1. The static mechanical tests (traction, inflection, and shearing) carried out on the jute/epoxy show certain variability mainly dependent on nature of fabric and its mode of weaving. The dimensional checks of fields by stereo-correlation of digital images highlight this variability. On the cartographies of deformations in warp and weft direction, we observed a microgradient of deformation of about 6000 micro-deformation.
2. The Dynamic Mechanical Thermal Analysis showed that the damping factor increases with the increasing of the temperature prior to 135°C. When the temperature is equal to 135 °C, the damping factor reaches a maximum value of 0.138. The glass transition temperature (Tg) of the resin without jute fibres is lower compared to the glass transition temperature (Tg) of resin reinforced by the jute woven. When the temperatures are lower than 120 °C, the frequency variation of loading has small influences on the storage modulus. However, when the heating temperatures are higher than the Tg, the variation of the frequency from 0.002 Hz to 100 Hz generates a significant increasing of the storage modulus.
3. Contrary to the results of the static tests (who showed variability in the observed mechanical behaviour), the results of the dynamic tests showed that a small variation of the first force responsible of the first damage for all specimens tested in the same impact energy. However, we have observed an important variation on the maximum impact force for all specimens tested in the same impact energy. This variation is around 20 %.

Although the mechanic properties of the jute/epoxy laminate is by far comparable of that of the glass/epoxy or carbon/epoxy, this material can be a good candidate in civil engineering. In addition, due to environmental and economic reasons, jute/epoxy is the best material for habitat applications. The underway development of bio-based resins such as polyhydroxyalkanoate (PHA) biodegradable polyesters can promote in future the increase of the rate using of green materials in the civil engineering field. Moreover, in the future one it is necessary to find a strategy which makes possible to take into account in the numerical modelling variabilities of the mechanical properties observed during the storage of characterization.

REFERENCES

[1]   Bledzki AK, Sperber VE. and Faruk O. 2002. Natural and Wood Fibre Reinforcement in Polymers. Rapra Review Reports. Vol.13, No. 8, Report 152, University of Kassel.

[2] A. Mir, R. Zitoune, F. Collombet, B. Bezzazi. 2009. Study of mechanical and thermo mechanical properties of Jute/epoxy composite laminate. *Journal of Reinforced Plastics and Composites*. Published on October 14, as doi:10.1177/0731684409341672.

[3] J. Gassan, A.K. Bledzki. 1997. Effect of Moisture Content on the Properties of Silanized Jute- Composites. *Polymer Composites*. vol. 18, n°2, p. 179–184.

[4] Mukhopadhyay S., R. Fangueiro. 2009. Physical Modification of Natural Fibers and Thermoplastic Films for Composites – A Review. *J. of Thermoplastic Composite materials, V.22*, Number 2, p. 135-162.

[5] Rao, R. M. V. G. K., Balasubramaniam, N. and Manas Chanda. (1983). Diffusion Model for Permeable Fibre Polymer Composites, *Journal of Reinforced Plastics and Composites*. Vol. 4, pp. 289_299.

[6] Masudul Hassan, M., Islam, M. R. and Khan, M. A. (2003). Improvement of Physicomechanical Properties of Jute Yarn by Photografting with 3 (Trimethoxysilyl) Propylmethacrylate, *Journal of Adhesion Science and Technology*. 17(5), Vol. 737-750.

[7] Garkhail, S. K., Heijenrath, R. W. H. and Peijs, T. (2000). Mechanical Properties of Naturalfibre- mat Reinforced Thermoplastics Based on Flax Fibres and Polypropylene, *Applied Composite Materials*. Vol. 7, pp. 351-372.

[8] Plackett, D. and Va´ zquez, A. (2004). *Green Composites: Polymer Composites and the Environment*, p. 123, Woodhead Publishers, Cambridge.

[9] Herrera-Franco, P. J. and Valadez-Gonzales, A. (2004). Mechanical Properties of Continuous Natural Fibrereinforced Polymer Composites. *Composites Part A*, Vol. 35, pp. 339-345.

[10] Ray, D., Sarkar, B. K., Das, S. and Rana, A. K. (2002). Dynamic Mechanical and Thermal Analysis of Vinylester_Resin_Matrix Composites Reinforced with Untreated and Alkali-treated Jute Fibres, *Composites Science and Technology*. Vol. 62, pp. 911-917.

[11] Khan, M. A., Mina, F. and Drzal, L. T. (2000). Influence of Silane Coupling Agents of Different Functionalities on the Performance of Jute_Polycarbonate Composites, In: 3rd International Wood and Natural Fibre Composite Symposium, Kassel, September.

[12] Park, J.-M., Kim, P. G., Jang, J. H., Wang, Z., Hwang, B. S. and DeVries, K. L. (2008). Interfacial Evolution and Durability of Modified Jute Fibers/Polypropylene (PP) Composites Using Micromechanical Test and Acoustic Emission, Composites Part B. Vol. 39, 1042-1061.

[13] Khan, M. A., Rahman, M. M. and Akhunzada, K. S. (2002). Grafting of Different Monomers onto Jute Yarn by In Situ UV-radiation Method: Effect of Additives, *Polymer-Plastics Technology and Engineering*. Vol. 41(4), pp. 677-689.

[14] Raj, R. G., Kokta, B. V. and Daneault, C. (1990). Wood Flour as a Low-cost Reinforcing Filler for Polyethylene: Studies on Mechanical Properties, *Journal of Materials Science*. Vol. 25, pp. 1851-1855.

[15] Harper, D. and Wolcott, M. (2004). Interaction Between Coupling Agent and Lubricants in Wood_Polypropylene Composites, *Composites Part A: Applied Science and Manufacturing*. Vol. 35(3), pp. 385-394.

[16] Aranberri-Askargorta, I., Lampke, T. and Bismarck, A. (2003). Wetting Behavior of Flax Fibers as Reinforcement for Polypropylene, *Journal of Colloid and Interface Science*. Vol. 263(2), pp. 580-589.

[17] Karmarkar, A., Chauhan, S. S., Modak, J. M. and Chanda, M. (2007). Mechanical Properties of Wood_Fiber Reinforced Polypropylene Composites: Effect of a Novel Compatibilizer with Isocyanate Functional Group, *Composites Part A: Applied Science and Manufacturing.* Vol. 38(2), pp. 227-233.

[18] Naik, J. B. and Mishra, S. (2007). Esterification Effect of Maleic Anhydride on Surface and Volume Resistivity of Natural Fiber/Polystyrene Composites, *Polymer-Plastics Technology and Engineering,* 46(5), pp. 537-540.

[19] Trek Sean, Sy. (2007). Composites from Newsprint Fiber and Polystyrene, *Polymer-Plastics Technology and Engineering.* Vol. 46(4), pp. 421-425.

[20] Keener, T. J., Stuart, R. K. and Brown, T. K. (2004). Maleated Coupling Agents for Natural Fibre Composites, *Composites Part A: Applied Science and Manufacturing.* Vol. 35(3), pp. 357-363.

[21] Jiang, H. and Kamdem, D. P. (2004). Development of Poly(Vinyl Chloride)/Wood Composites: A Literature Review, *Journal of Vinyl and Additive Technology.* Vol. 10(2): pp. 59-69.

[22] Sabeel, K. A. and Vijayarangan, S. (2008). Tensile, Flexural and Interlaminar Shear Properties of Woven Jute and Jute-glass Fabric Reinforced Polyester Composites, *Journal of Materials Processing Technology.* Vol. 207, pp. 330-335.

[23] Placketta, D., Løgstrup Andersenb, T., Batsberg Pedersenc, W. and Nielsenc, L. (2003). *Biodegradable Composites Based on l-Polylactide and Jute Fibres, Composites Science and Technology.* Vol. 63, pp. 1287-1296.

[24] Wollerdorfer, M. and Bader, H. (1998). Influence of Natural Fibres on the Mechanical Properties of Biodegradable Polymers, Industrial Crops and Products. Vol. 8, pp. 105_112.

[25] K.Sabeel Ahmed, S. Vijayarangan and A. KUMAR. (2007). Low Velocity Impact DamageCharacterization of Woven Jute–Glass Fabric Reinforced Isothalic Polyester Hybrid Composites, *Jour. of Reinf. Plast. and Comp*, Vol. 26, No. 10, pp 959-976.

[26] C. Santulli. (2006). Post-impact flexural tests on jute/polyester laminates monitored by acoustic emission, *Journal of Materials Science.* Vol. 41, pp 1255-1259.

[27] Helm J, McNeill S, Sutton M. (1996). Improved three dimensional image correlation for surface displacement measurement. *Opt. Eng.* Vol. 35(7), pp. 1911–20.

[28] Synnergren P, Sjödahl M. (1999). A stereoscopic digital speckle photography system for 3-D displacement field measurements. *Opt. Lasers Eng.* Vol. ;31, pp. 425–43.

[29] Luo P, Chao Y, Sutton M, Peters W. (1993). Accurate measurement of threedimensional deformations in deformable and rigid bodies using computer vision. *Exp. Mech .* Vol. 30(2), pp. 123–32.

[30] Garcia D. (2001). Mesure de formes et de champs de déplacements tridimensionnels par stéréo-corrélation d'images. PhD thesis, INSA de Toulouse, Ecole des Mines d'Albi-Carmaux.

[31] Rials T G and Wolcott M P. 1996. Physical and Mechanical Properties of Agro-Based Fibres in Paper and Composites from Agro-Based Resources. CRC Press, Boca Raton.

[32] J.-M. Park, P. G. Kim, J. H. Jang, Z.Wang, B.S. Hwang, K. L. DeVries. (2008). Interfacial evolution and durability of modified Jutefibers/polypropylene (PP) composites using micromechanical test and acoustic emission. *Composites Part B.* Vol. 39, pp. 1042-1061.

[33] Florian H. Gojny, K. Schulte. (2004). Functionnalisation effect on the thermo-mechanical behaviour of multi-wall carbon nanotube/epoxy-composites, *Composites Science and Technology* 64 2303-2308.

In: Fiber-Reinforced Composites
Editor: Qingzheng Cheng

ISBN: 978-1-61470-303-7
© 2012 Nova Science Publishers, Inc.

Chapter 6

PERFORMANCE PROPERTIES AND MICRO-STRUCTURAL QUANTIFICATION OF CELLULOSE FIBRE-POLYPROPYLENE COMPOSITES

Marianne Lenes[*1], Gary Chinga-Carrasco[1] and Øyvind Gregersen[2]*

[1]Paper and Fibre Research Institute (PFI),
[2]Norwegian University of Science and Technology (NTNU)

ABSTRACT

This study focuses on the effects of cellulose content and cellulose morphology on the structural and mechanical properties of PP/cellulose fibre composites. A detailed structural characterisation was performed, including SEM-BEI analysis of fibre spatial distribution and X-µCT analysis of fibre orientation. The results indicated that the fibres were well-dispersed and oriented mostly along the main direction of the PP/cellulose test specimen.

Based on light microscopy it was found that the cellulose fibres acted as nucleators in the PP matrix. By use of DSC it was shown that addition of cellulose fibres increased both the crystallinity and T_c. The increased crystallinity is mainly a result of the transcrystallinity formation. The tensile properties of the composites were affected by the fibre dimensions and the fibre concentration. The long fibres yielded the highest increase of Young's modulus. Elongation at break was mostly affected by the fibre concentration. The results thus indicate that there is a major potential for improving the mechanical properties of fibre-reinforced composites, provided an adequate concentration of fibres is applied.

Keywords: Short-fibre composites, fibres, Mechanical properties, characterisation, ImageJ.

---

[*] Author to whom correspondence should be addressed; Email: marianne.lenes@pfi.no

INTRODUCTION

Fibre-Reinforced Composites

The interest in composites based on renewable materials has grown since the 1990's. Natural fibres such as wood cellulose fibres are now used as reinforcement for thermosets and thermoplastics due to the environmental advantages of being renewable, biodegradable and recyclable [1]. The potential applications of wood thermoplastic composites include automotive, building industry and furniture [2, 3]. Improvement in the mechanical properties of such composites requires strong adhesion between the wood fibre and polymer matrix. However, the performance and mechanical properties of natural fibre reinforced thermoplastic composite depends on mechanical, chemical and structural properties of the fibre used as well as the structure and sizes of the fibre and composite after the processing [4,5,6,7].

Polypropylene reinforced with cellulose fibres shows an increased fraction of crystallinity, controlled by the nucleation stage and cooling condition of thermoplastic/fibre blend [8]. This crystalline morphology is characterized by a high density of nucleating crystallites, which grow perpendicular to the fibre surface responsible for nucleation [8,9] and differs thereby significantly from the spherulitic crystal growth. Surface energy, surface chemical composition, crystal morphology and moisture content are important to the fibres' nucleation of the polypropylene and has been reported to be essential factors for the formation of this transcrystallisation [10,11]. The orientation of the transcrystalline layer is expected to have several effects on the mechanical properties of composite materials [5]. Especially when processing thermoplastics, changes in matrix morphology in the interfaces must be taken into account [12].

Structural Characterisation

A successful development of fibre-reinforced composites requires a complete understanding of the corresponding structure-properties relationships. Hence, proper characterization is crucial. Structural characterisation has had a remarkable development during the last decade. Advances have been made with respect to sample preparation, image acquisition, visualization and quantification methods. This has given a new perspective in the assessment of multi-scale structures. In addition, 3D assessment is progressing due to the development of 3D image acquisition devices [13,14,15]. Proper image acquisition devices, combined with adequate computerized image analysis, will thus advance our understanding of relevant structure-property relationships.

The purpose of the present study is to assess the effect of cellulose fibres and their morphology on the corresponding structural and mechanical properties. The study emphasizes detailed structural quantification of the fibre-reinforced composite materials based on microscopy and computerized image analysis.

MATERIALS AND METHODS

Composite Materials

Polypropylene (PP) homopolymer (HE 125 MO), produced by Borealis A/S, Norway, was used as a matrix material. The density is 908 kg/m^3 (ISO 1183), melt flow rate 12 g/10 min (ISO 1133), tensile stress at yield (50 mm/min) is 34.5 MPa and the melt processing temperature is between 220 and 260 $^\circ$C [16].

Sulphite softwood fibres produced from Norway Spruce from Borregaard Ind. Ltd., Norway, was blended with the matrix. The chemical pulp properties given by Borregaard Ind. Ltd were: α-cellulose content of 95%, viscosity of 875 mL/g, S18 is 3.5 and S10 is 6.0 and an extractive content of 0.03%. S18 and S10 are alkali solubility in 18 and 10% NaOH respectively (ISO 692:1982) [17]. A portion of sulphite fibres were knife milled and spray dried before use. Such fibres will be termed as knife milled cellulose (KMC). The fibre dimensions are given in Table 1.

Preparation of Composites

Composites were prepared by melt-mixing with a Brabender (5.5 KW, Duisberg Mod GNF1315/2,) and a mixing unit with two counter rotating screws at 10 min. Composites with 5 and 10 wt% cellulose fibres were made and the total amount in each batch was 200 g. In order to get a proper distribution of fibres in the composite, the polymer was melted before the fibres were added. The mixing was carried out at 200 $^\circ$C, 60 rpm and the mixing time was 10 minutes after addition of cellulose.

Table 1. Composites applied in this study. The polypropylene polymer has been reinforced with cellulose fibres and knife milled cellulose (KMC) fibres.

Code	Composites	Original fibre dimensions		Fibre loading (%)
		Length (mm)	Diameter (μm)	
PP	Polypropylene	-	-	0
PC05	Polypropylene + Cellulose fibre	1.716	25	5
PC10	Polypropylene + Cellulose fibre	1.716	25	10
PK05	Polypropylene + KMC fibre	0.804	25	5
PK10	Polypropylene + KMC fibre	0.804	25	10

Properties of the Fibre-Reinforced Composites

The crystallisation was assessed with light microscopy. The composite materials were cut in 20 μm slices with a microtome (Microm Zeiss Stemi SR HM 350, Microm GmbH, Sandhausen, Germany). The samples were prepared by melting and pressing the microtome slices between two microscopy slides before it was put into the hotstage (Leica Microscope

Heating system Stage 350, Leica Microsystems AG, Wetzlar, Germany). The hot-stage unit was heated from 20 to 220°C at 10°C/min and held at 220°C for 5 min to erase the present crystallinity in the PP. The PP was thereafter cooled at about 13°C/min. The isothermal crystallisation was studied between 120 and 110 °C. The times (t) from initiation of transcrystalline layer until completion were about 5 minutes for the studied samples, two replicates. The crystallisation was studied by light transmission microscopy (Leica DMLB, Leica Microsystems AG, Wetzlar, Germany) with cross polarized light. Images were recorded with a Leica DC300 camera (Leica Microsystems AG, Wetzlar, Germany).

Differential scanning calorimetry (DSC) was measured on a Q100 RCS from TA Instruments Ltd., (New Castle DE, USA). The crystallisation and nucleation of PP mixed with different fibre materials were studied. The experiments were carried out with two heating steps (20-225°C at 10°C/min) and one cooling step (225 – 20°C , 10°C/min) with 3 minutes isothermal period after each step. The number of replicates was two or three.

Quantification of Mechanical Properties

Test specimens for the mechanical tests were made by a DSM mini extruder/injection moulder. The melting chamber was adjusted to 220°C during melting and the temperature of the mould was 50°C. The pressure was 7 bar. The length of the test specimens (dog bone) was 75 mm.

The mechanical properties was measured according to ISO 527 [18], by use of a Hounsfield material tester H5KS (Hounsfield Test Equipment Ltd UK) with a built-in extensometer. The measurement range was 0.5-5 kN. For all specimens the gauge length was 30 mm and the speed 5 mm/min. Six or seven samples were measured for each composite. The terminology "tensile strength" will be used for the maximum tensile stress.

Structural Assessment

The fibre-reinforced composites (Figure 1, left) were embedded in epoxy resin (Epofix, Struers) and cured for 48 hrs. The blocks were hand-held ground and automatically polished with polishing cloths of 9 and 1 µm (Figure 1, right).

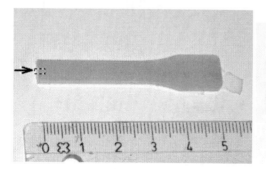 

Figure 1. Preparation of dogbones for structural analysis. Left) A piece of a dogbone after a tensile testing. The arrow indicates the area where the X-µCT and SEM analyses were performed. Right) Eight fibre-PP composite pieces embedded in epoxy resin for SEM-BEI cross-sectional analysis.

From each sample, twenty four SEM cross-sectional images were acquired in backscattered electron imaging (BEI) mode. The applied magnification was 150x. The acceleration voltage was 5 kV and the size of the images was 2560x1920 pixels. The obtained resolution was 0.31 µm. The working distance was 8-10 mm.

3D volume images were acquired from 4 fibre-reinforced composites. The images were acquired with a Skyscan SEM-microCT system. The resolution was approximately 2.4 µm and the field of view was 2 x 2 x 2 mm.

The ImageJ program was applied for image processing and analysis. The SEM-BEI images were thresholded automatically applying an Otsu algorithm. The spatial distributions of the fibres in the polypropylene matrix were quantified as described by Chinga-Carrasco et al. [19]. The quantification of fibre spatial distribution was applied to the 2D SEM BEI and 3D X-µCT images, for comparison purposes.

The Skyscan 3D-µCT images were utilized for estimating the fibre orientation. The quantification is based on applying a gradient analysis to the grey level images. No thresholding was applied. Based on the gradient analysis a global polar plot was estimated in the 3D volume images [20]. The gradient analysis was applied to estimate the orientation (ϕ) and anisotropy (a) based on a mean resultant vector approach, as described by Curray [21].

RESULTS AND DISCUSSION

Crystallisation Behaviour of Polypropylene/Cellulose Composites

The crystallisation behaviour of polypropylene mixed with cellulose fibres was studied by light transmission microscopy with a hotstage unit. As indicated in Figure 2 the untreated cellulose fibres induce a transcrystalline layer during cooling. This is in agreement with other reports, showing that cellulose fibres and microfibres increase the total crystallinity, by enhancing and initiating transcrystallisation growth from the fibre, in a PP matrix [8]. Also spherulites are growing gradually when the PP bulk is cooled. Figure 2 shows the crystals just after initiation, where the transcrystalline layers have been influenced by the growth of the spherulites in the bulk PP matrix.

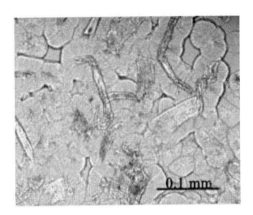

Figure 2. Crystallisation of polypropylene mixed with cellulose fibres (PC).

The elapsing times (t) from initiation of the transcrystalline layer until after completion were in average five minutes with the cooling rate of 10 °C/ min. The faster crystallisation is basically due to the heterogeneous nucleation on the fibres [22].

Differential Scanning Calorimetry (DSC)

It has been reported that pure cellulose fibres act as an efficient nucleation agent for the crystallisation of PP matrix [8,23,24]. In the present work it was observed that the cellulose fibres affect the crystallisation and interfacial morphology of the PP matrix. Addition of unmodified cellulose fibres resulted in a significant increase in the crystallisation temperature (T_c), by 4-6°C, (Table 2). The highest increase in T_c was observed for samples with the highest concentration of cellulose, 10 wt% cellulose and kmc, which is in accordance with the result from Borja et al. [25]. The fibres act as nucleation agent and thereby also the crystallinity increases with increasing concentration of fibres. The fibre length was observed to not affect the crystallisation as the kmc and cellulose fibres have the same influence on the crystallinity.

Table 2. Different scanning calorimetry results for samples 100% polypropylene (PP) and composites of PP with 5 or 10 wt% of cellulose fibres (PC) and knife milled cellulose (PK)

Sample	T_c (°C)	T_m (°C)	ΔH_f (J/g)	X_c (%)
PP	116.2	164.2	112.7 ± 2.7	54 ± 1
PC05	120.1	162.8	110.7 ± 5.5	56 ± 2
PC10	122.4	163.3	107.7 ± 1.1	57 ± 1
PK05	120.5	164.1	111.3 ± 5.2	57 ± 2
PK10	121.9	164.4	105.7 ± 4.0	56 ± 2

The heat of fusion (ΔH_f) was lower for composites with 10 wt%, than with 5 wt% concentration of cellulose fibres. This was expected, since the fibres act as diluents in the PP matrix, as previously reported by Amash and Zugenmaier [8]. Figure 3 shows single endothermic peaks for the composites, which exclude different types of crystalline phases for PP. Addition of cellulose fibres caused a reduction of the melting temperature (T_m) by approximately 2.6°C.

The effect of cellulose fibre on the degree of crystallinity (X_c) of the PP can be calculated from the DSC results. The crystallinity was calculated according to equation I.

Equation I: $X_c = \Delta H_f / \Delta H_m * (100/w)$, [25]

ΔH_f = heat of fusion for matrix,
ΔH_m = heat of fusion for neat PP, 209 J/g
w = mass fraction of polypropylene

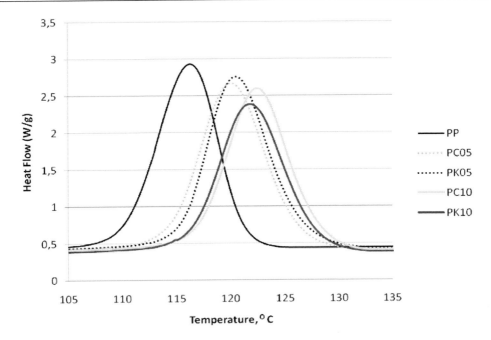

Figure 3. Scanning calorimetry curves of cooled samples of 100% polypropylene (PP) and composites of PP with 5 or 10 wt% cellulose fibres.

The addition of cellulose fiber increases the crystallinity and T_c. The percent crystallinity is lower for pure PP than for most of the composites. Addition of 10 wt% cellulose fibres increased the percent crystallinity from about 54 to 57%, compared to pure PP. This is in agreement with the light transmission microscopy study of the transcrystalline behaviour of the composites (Figure 2), where addition of the cellulose fibres gave the denser transcrystalline structure (Table 2).

Mechanical Properties

The mechanical properties in a composite material depend on the efficiency of stress transfer from the resin matrix to the fibres [5]. The mechanical properties of the studied composites were assessed, including the tensile strength, the modulus and the elongation at break.

Pure PP is a homogenous material, which has a much higher tensile strain at break compared to composites. By adding fibres to the matrix a lot of rupture initiation points may appear. The ruptures are normally initiated either by fibre fracture or by fracture between fibres and matrix. In fibre-reinforced composites, each fibre is a potential fracture initiation point (Figure 8). Stress-strain curves for the different composite materials are shown in Figure 4. The tensile strength (maximum tensile stress) occurs at different strain for the composite materials. The results show maximum tensile stress at about 5% strain for composites added 5 wt% cellulose fibre and about 4% for composites with 10% cellulose fibre. Higher concentration of cellulose fibres gives a maximum stress at a lower strain for the composites.

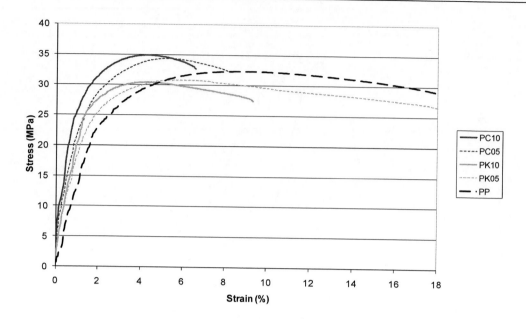

Figure 4. Stress-strain curves for samples of 100% polypropylene (PP) and composites of PP with 5 or 10 wt% of cellulose fibres (PC05, PC10) and knife milled cellulose (PK05, PK10).

Qui et al. [26] concluded that the elongation and tensile strength of fibre-reinforced PP composites decrease significantly while the Young's modulus increases, compared to pure PP. Our results show that the highest tensile strength was obtained in the two composites added cellulose fibres, as shown in Figure 5.

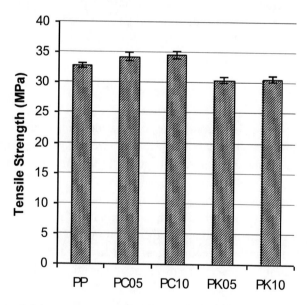

Figure 5. Tensile strength for samples of 100% polypropylene (PP) and composites of PP with 5 or 10 wt% of cellulose fibres (PC05, PC10), knife milled cellulose (PK05, PK10).

The longer cellulose fibres induce an increase in tensile strength while the shorter fibres (kmc) seem to decrease the strength of the composites compared to pure PP. Compared to reported PP/cellulose composite studies, the fibres used in the present study are longer and have a higher aspect ratio. This affects the composite properties due to the reinforcement effect and increased stress transfer of the longer fibres, compared to the short fibres.

The elongation at break dropped markedly for all composites compared to pure PP which have an elongation of more than 100%, as expected. The lowest elongation at break was observed for cellulose fibres, Figure 6. The results also indicate that the elongation depends on the cellulose concentration giving higher elongation for the 5% addition compared to 10% addition. This is in agreement with Bledzki and Gassan [4] and Borja et al.[25], who observed that the elongation is a function of the filler concentration. The elongation thus depends on the fibre content.

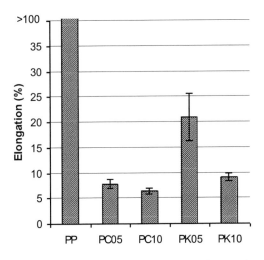

Figure 6. Elongation at break for samples of 100% polypropylene (PP) and composites of PP with 5 or 10 wt% of cellulose fibres (PC05, PC10) and knife milled cellulose (PK05, PK10). The elongation at break for pure PP > 100%.

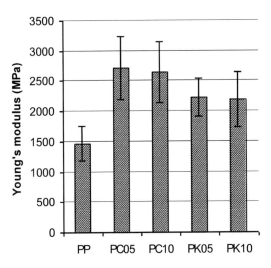

Figure 7. Young's modulus for samples of 100% polypropylene (PP) and composites of PP with 5 or 10 wt% of cellulose fibres (cell), knife milled cellulose (kmc).

As expected, the Young's modulus of the composites (Figure 7) are significantly higher than the measured modulus of pure PP. It is known that an increase in the material stiffness appears when adding cellulose fibres [24,26,27]. Addition of cellulose fibres to the matrix increased the Young's modulus from 1470 MPa for neat PP to 2200 - 2700 MPa for the composites. The highest modulus was obtained for the long cellulose fibres. According to Karimi et al. [28] fillers with high elastic modulus and high aspect ratio can generally cause higher modulus value in composites. In our case we observed a difference between cellulose fibre and kmc, the longer cellulose fibres induce a higher modulus and higher tensile strength than kmc.

Structural Quantification

A SEM cross-sectional analysis was applied to quantify the fibre spatial distribution in the polypropylene matrix. Figure 8A, shows a fracture area where the fibres are clearly observed. Measurements of fibre-fibre inter-distances were quantified in SEM-BEI images of the four assessed samples, see also Chinga-Carrasco et al. [19]. The results indicate that the composites containing 5% cellulose fibres have larger inter-distances (approx. 47 µm) than the fibres containing 10% cellulose fibres (Figure 9).

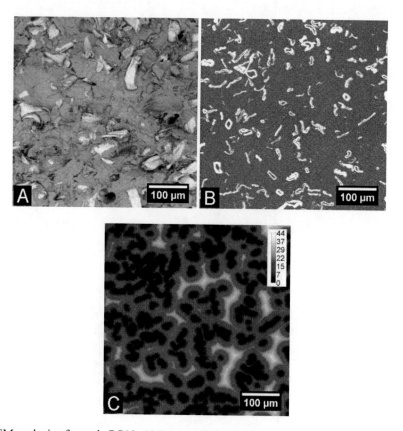

Figure 8. SEM analysis of sample PC10. A) Image of a fracture area. B) SEM-BEI cross-sectional image. C) The corresponding distance map, showing fibre-fibre inter-distances.

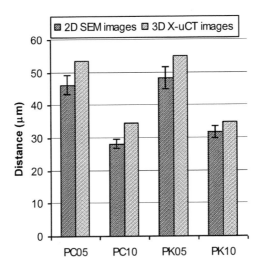

Figure 9. Quantification of fibre-fibre inter-distances.

The scattering of the inter-distance measurements is relatively low, thus indicating a good dispersion of the fibres. The composites containing short fibres (PK05 and PK10) had slightly large inter-distances than the composites containing unmodified cellulose fibres (PC05 and PC10), though the differences are not significant. It is worth to notice that the cross-sectional shapes of the fibres observed in the Figure 8B are mostly tubular, where the lumens are clearly observed. The cross-sectional shapes thus indicate that the fibres are well-oriented along the principal direction of the dogbones. In order to verify this observation an X-µCT analysis was performed.

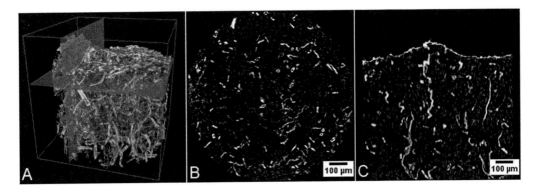

Figure 10. X-ray microtomography of a dog bone structure. A) 3D volume rendering of a top region of a fracture area. B) A cross-sectional slice corresponding to the horizontal plane in A. C) A longitudinal slice corresponding to the vertical plane in A.

One of the major advantages with X-µCT is the availability to 3D data. In fibre-reinforced composite materials, the dispersion and orientation of the fibres may influence the mechanical properties of the material. Quantification of the fibre-fibre inter-distances in the corresponding 3D X-µCT images shows a similar trend as the SEM-BEI analysis. However, the obtained inter-distances are larger than the values obtained with SEM cross-sectional analysis (Figure 9).

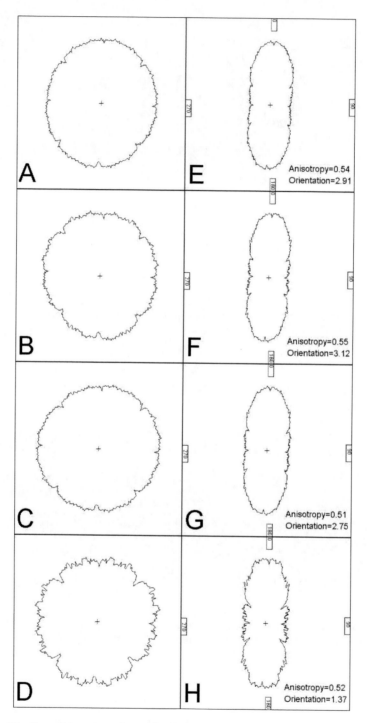

Figure 11. Quantification of the orientation and anisotropy of fibres in the PP-matrix. The orientations have been quantified at horizontal and vertical planes as exemplified in Figure 10. The "0"-degree is towards the fracture area.

The difference may be due to the resolution of the two techniques. The SEM and X-µCT images had resolutions of approximately 0.3 µm and 2.4 µm, respectively. A relatively low resolution may exclude small fibre fragments from the analysis, which will thus increase the quantified inter-distances. It is worth to keep in mind that higher resolution can be obtained with the X-µCT system applied in this study. However the main objective with the X-µCT analysis was to explore the fibre orientation, where sub-micron resolution is not necessary.

Figure 10 exemplifies the quantification of fibre orientations in the two principal directions of a dog bone structure, i.e. cross-sectional and along the dog bone. The orientation distributions given by polar plots show clearly that the fibres are well-oriented along the dog bone direction (Figure 11). The cross-sectional polar plot is mostly circular, indicating low anisotropy and no preferred orientation. The polar plot along the dog bone is elongated, having a relatively high anisotropy and thus a preferred orientation. The quantification of fibre orientation based on gradient analysis is influenced by the noise appearing in the 3D images. The anisotropy values may thus be overestimated. However, the obtained results seem reasonable. Slight differences were detected between the samples with respect to the orientation and orientation anisotropy of the fibres. The orientation and orientation anisotropy of the unmodified cellulose fibres are slightly higher than the kmc fibres. Higher orientation and anisotropy of samples PC05 and PC10, compared to PK05 and PK10, is reasonable. PC05 and PC10 contain longer fibres, which may easily align along the dogbone direction.

Reasonable structural differences were quantified in the assessed composites. The fibres were well-dispersed and oriented mostly along the main direction of the dogbones. The results obtained in this study indicate that the major factors affecting the mechanical properties are the fibre dimensions and the fibre concentration. However, the improvement of the mechanical properties was slight. This is most probably due to 1) the low fibre concentration applied in this study and 2) the incompatibility between the hydrophilic surface of cellulose and the hydrophobic thermoplastics. A number of attempts have been made to improve adhesion between fibres and matrix by different chemical treatments on the fibres or during composite production [4]. There is thus a major potential for improving the mechanical properties of cellulose fibre reinforced composites, provided that efficient processes for improving the compatibility between cellulose fibres and a given polymer matrix are achieved.

CONCLUSION

This study has focused on the effects of cellulose content and cellulose morphology on the mechanical and structural properties of PP/soft wood cellulose fibre composites. A detailed structural characterisation was performed, including SEM-BEI analysis of fibre spatial distribution and X-µCT analysis of fibre orientation. The results indicated that the fibres were well-dispersed and oriented mostly along the main direction of the dogbones. Based on light microscopy it was found that the cellulose fibres acted as nucleators in the PP matrix. The total crystallinity was found to increase when the transcrystallinity appears around the fibres. The tensile properties of the composites were affected by the fibre dimensions and the fibre concentration. The knife milled cellulose fibres induced a negative effect on the mechanical properties. The long fibres yielded the highest increase of Young's modulus and elongation at break was reduced by increasing the fibre concentration. The tensile strength was slightly increased by addition of cellulose fibres, whereas knife milled cellulose fibres significantly reduced the tensile strength. This indicates that the cellulose fibres improve the tensile stiffness and the tensile strength due to their higher elastic modulus and high aspect ratio, although they also act as fracture initiation points. Therefore a proper compatibility between the fibres and the surrounding matrix is important. The results thus indicate that there is a potential for improving the mechanical properties of fibre-reinforced composites, provided an adequate concentration of sufficiently long fibres is applied and a suitable compatibility between the fibres and the polymer matrix is achieved.

ACKNOWLEDGEMENTS

Alexander Sasov (SkyScan) is acknowledged for acquiring the X-μCT images.

REFERENCES

[1] Felix, J. Enhancing interaction between cellulose fibres and synthetic polymers. PhD-thesis, Chalmers University of Technology, *Department of Polymer Technology*, Sweden, 1993.

[2] Bledzki, A.K.; Letman, M; Viksne, A; Rence, L. A comparison of compounding processes and wood type for wood fibre-PP composites. *Composite: Part A,* 2005, 36, 789-797.

[3] Qui, W.; Endo, T.; Hirotsu, T. Structure and properties of composites of highly crystalline cellulose with polypropylene: Effect of polypropylene molecular weight. Eur Polym J, 2006, 42, 1059-1068.

[4] Bledzki, A.K.; Gassan, J. Composites reinforced with cellulose based fibres. *Progr. Polymer Sci.*, 1999, 24, 221-274.

[5] Mukhopadhyay, B.L.; Deopura, B.L.; Alagiruswamy R. Interface behaviour in polypropylene composites. *J. Thermoplastics Comp. Mat.*, 2003, 16, 479-495.

[6] Zafeiropoulos, N.E.; Baillie, C.A.; Matthews, F.L. A study of transcrystallinity and its effect on the interface in flax fibre reinforced composites materials. *Composites, Part A,* 2001, 32, 525-543.

[7] Bledzki, A.K.; Reihmane, S.; Gassan, J. Thermoplastics reinforced with wood fillers: A literature review. *Polym. Plast. Technol. Eng.*, 1998, 37, 451 – 468.

[8] Amash, A.; Zugenmaier P. Morphology and properties of isotropic and oriented samples of cellulose fibre-polypropylene composites. *Polymer*, 2000, 4, 1589-1596.

[9] Harper, D.; Wolcott, M. Interaction between coupling agent and lubricants in wood-polypropylene composites. *Composites: Part A*, 2004, 35, 385-394.

[10] Huson, M.G.; McGill, W.J. Transcrystallinity in polypropylene. *J. Polym. Sci. Polym. Chem.*, 1984, 22, 3571-3580.

[11] Lenes, M.; Gregersen, Ø. Effect of surface chemistry and topography of sulphite fibres on the transcrystallinity of polypropylene, *Cellulose*, 2006, 13, 345-355.

[12] Wu, C-M.,; Chen, M.; Karger-Kocsis, J. Effect of micro morphologic features on the interfacial strength of iPP/Kevlar fiber micro composites. *Polymer,* 2001, 42, 199-208.

[13] Denk, W.; Horstman, H. Serial block-face scanning electron microscopy to reconstruct three-dimensional tissue nanostructure. *Plos Biology,* 2004, 2, 1900-1909.

[14] Mayo, S.C.; Miller, P.R.; Wilkins, S.W.; Davis, T.J.; Gao, D.; Gureyev, T.E., Paganin, D.; Parry, D.J.; Pogany, A.; Stevenson, A.W. Quantitative X-ray projection microscopy: phase-contrast and multi-spectral imaging. *J. Microcopy,* 2002, 207, 79-96.

[15] Wiltsche, M.; Donoser, M.; Bauer, W.; Bischof, H. (2005) A new slice-based concept for 3D paper structure analysis applied to coating layer formation. In S. J. l'Anson, editor, Advances in Paper Science and Technology, Transactions of the 13th Fundamental Research Symposium. The Pulp and Paper Fundamental Research Society, Cambridge (UK), 853-899.

[16] Borealis A/S, Denmark. Datasheet of polypropylene HE 125MO found on www.borealisgroup.com.

[17] International Standard, *Pulps – Determination of alkali solubility,* ISO 692, 1982.

[18] International Standard, *Plastics – Determination of tensile properties,* ISO 527, 1993.

[19] Chinga-Carrasco, G.; Lenes, M.; Johnsen, P.O.; Hult, E.-L. Computer-assisted scanning electron microscopy of wood pulp fibres: dimensions and spatial distribution in a polypropylene composite. *Micron,* 2009, 40, 761-768.

[20] Chinga-Carrasco, G.; Johnsen, P.O.; Øyaas, K. Structural quantification of wood fibre surfaces - morphological effects of pulping and enzymatic treatments, *Micron,* 2010, 41, 648-659.

[21] Curray, J.R. The analysis of two-dimensional orientation data. *J. Geol.,* 1956, 64 117-131.

[22] Biagiotti, J.; Puglia, D.; Torre, L.; Kenny, J.M. A systematic investigation on the influence of the chemical treatment on natural fibres on the properties of their polymer matrix composites. *Polym. Comp.,* 2004, 25, 470-479.

[23] Mi,Y.;Chen, X.; Guo, Q. Bamboo fiber-reinforced polypropylene composites: Crystallization and interfacial morphology. *J. Appl. Polym. Sci.,* 1997, 64, 1267-1273.

[24] Qui, W.; Zhang, F.; Endo, T.; Hirotsu, T. Effect of maleated polypropylene on the performance of polypropylene/cellulose composites. *Polym. Comp.,* 2005, 26, 448-453.

[25] Borja, Y.; Riess, G.; Lederer, K. Synthesis and characterisation of polypropylene reinforced with cellulose I and II fibres. *J. Appl. Polym. Sci.,* 2006, 101, 364-369.

[26] Qui, W.; Zhang, F.; Endo, T.; Hirotsu, T. Isocyanate as a compatibilizing agent on the properties of highly crystalline cellulose/polypropylene composites. *J. Mat. Sci.,* 2005, 40, 3607-3614.

[27] Felix, J.M.; Gatenholm, P. The nature of adhesion in composites of modified cellulose fibers and polypropylene. *J. Appl. Polym. Sci.,* 1991, 42, 609-20.

[28] Karimi, A.; Nazari, S.; Ghasemi, I.; Tajvidi, M.; Ebrahimi, G. Effect of delignification of wood fibres on the mechanical properties of wood fiber-polypropylene composites. *J. Appl. Polym. Sci.,* 2006, 102, 4759-4763.

In: Fiber-Reinforced Composites
Editor: Qingzheng Cheng

ISBN: 978-1-61470-303-7
© 2012 Nova Science Publishers, Inc.

Chapter 7

THE IN-PLANE SHEAR BEHAVIOUR OF A CARBON FABRIC REINFORCED THERMOPLASTIC UNDER QUASI-STATIC AND FATIGUE LOADING CONDITIONS

Ives De Baere[*], *Wim Van Paepegem and Joris Degrieck*

Ghent University, Department of Materials Science and Engineering.
Technologiepark-Zwijnaarde 903, B-9052 Zwijnaarde, Belgium.

ABSTRACT

In this chapter, the nonlinear shear stress-strain relationship of a carbon fabric-reinforced polyphenylene sulphide is investigated, both under quasi-static and fatigue loading conditions. In order to determine the in-plane shear properties of a fibre reinforced composite, there are various possibilities. One of them is the standard three-rail shear test, as described in "ASTM D 4255/D 4255M The standard test method for in-plane shear properties of polymer matrix composite materials by the rail shear method". This set-up, however, requires drilling holes through the specimen. In this study, a new design based on friction and geometrical gripping, without the need of drilling holes through the composite specimen is presented. Next, this setup is used and compared to $[(+45°,-45°)]_{4s}$ tensile tests to study the shear behaviour of the material under consideration.

For the fatigue behaviour, fatigue tests with R = 0 were conducted, to assess the behaviour of the newly designed grips under fatigue loading conditions on one hand and to investigate the in-plane shear behaviour on the other hand. With respect to the grips, excellent results were obtained; the specimen fails under shear loading conditions in the loaded area. With respect to the material, an increase in permanent deformation and a decrease in shear stiffness until a certain point in time is visible, after which a drastic increase in deformation and temperature, higher than the softening temperature of the matrix occurred. Also, the maximum value of the shear stress for fatigue with R = 0 has a large influence on the fatigue lifetime. These findings were also assessed with the $[(+45°,-45°)]_{4s}$ fatigue tensile tests, with very good correspondence.

---

[*] Author to whom correspondence should be addressed; Email: Ives.DeBaere@UGent.be

For the quasi-static behaviour, a previously derived material model with four material constants is considered. Using quasi-static and hysteresis tests, both with the rail shear setup and with $[(+45°,-45°)]_{4s}$ tensile tests, the necessary data for determining the material constants are obtained. Then, the material constants are optimised by comparing finite element simulations with the data derived from the experiments. The conducted experiments are simulated and the results are compared with the experiments, with excellent correspondence.

INTRODUCTION

There are various ways of inducing a state of in-plane shear [1, 2] in a composite. Examples are the Iosipescu test [1, 3, 4, 5], the 10° off-axis test [4, 5, 6, 7], the $[+45°/-45°]_{ns}$ tensile test [7, 8, 9, 10, 11, 12, 13, 14 15, 16], the two-rail shear test [17, 18, 19], the three-rail shear test [20], torsion of a rod [21] and torsion of thin-walled tubes [22, 23, 24, 25]. From all these tests, torsion of a thin-walled tube is practically the only universal method used for determination of both in-plane shear modulus and shear strength [1] and it produces the most desired state of shear stress, free of edge effects [20]. However, this method is rather expensive, since it requires a tension-torsion machine with specialized gripping and it cannot determine the shear characteristics of flat products, fabricated by pressing or contact moulding. Furthermore, such tubes are not easily fabricated. The $[+45°/-45°]_{ns}$ tests do not require any specialised fixtures and as such are a lot easier to perform. On the other hand, they are very sensitive to edge effects due to the $[+45°/-45°]$ lay-up [20]. For the 10° off-axis tests, oblique end tabs are required [4, 5, 6, 7].

The rail shear test positions itself somewhat in the middle. It does not require a sophisticated apparatus like the torsion setup and it induces a stress state which does not differ a lot from pure shear. Furthermore, it requires flat specimens with limited preparation.

If fatigue loading conditions are required, then the rail shear test is only rarely considered [20]. The favourite test setup remains the torsion of thin-walled tubes, sometimes combined with tension or bending in biaxial fatigue [22, 23, 24, 25]. The $[+45°/-45°]_{ns}$ test is also used [9] for fatigue research.

The rail shear test, both two-rail and three-rail, as described in the "ASTM D 4255/D 4255M standard test method for in-plane shear properties of polymer matrix composite materials by the rail shear method" has one large disadvantage: it requires drilling holes through the specimen, so that the clamps can be bolted to the specimen. Drilling in composites should be avoided, since it nearly always causes damage to the composite and it may cause stress concentrations around the holes [18]. Furthermore, the preparation of the specimen takes more time. With this in mind, there has already been a proposal of a new design for the two-rail shear test, described by Hussain and Adams [18, 19]. This design no longer requires holes in the specimen.

In this chapter, a modification for the three-rail shear test is proposed, which no longer requires holes through the specimen, as has been proposed for the two-rail shear test in [18, 19]. Furthermore, this design should allow for fatigue loading conditions, which were not considered by Hussain et al. [18, 19]. The setup used by Lessard et al. [20] for their fatigue research was the standard three-rail setup, which requires the holes. The emphasis of their study was the use of notched specimens, in order to avoid preliminary failure of the specimens. Notches, however, will not be considered for this study.

Finally, the rail shear test is often only considered for unidirectionally reinforced or cross-ply composites, whereas for this study, a carbon fabric reinforcement is considered. The material used for this study will be described in the next paragraph, after which the design of the modified three-rail shear setup is elaborated on. This is followed by a discussion of the performed experiments, under quasi-static, cyclic and fatigue loading conditions. All these loading conditions are also applied for a $[(+45°,-45°)]_{4s}$ tensile test. The purpose of these tests is to verify if the results obtained from the modified three-rail shear setup can also be reproduced by a different kind of experiment, which also induces an in-plane shear stress state. All fatigue tests were performed with a stress ratio equal to zero and the influence of both the loading frequency and maximum shear stress was examined. Creep tests were also performed with both setups, to verify whether the detected permanent deformation was due to creep phenomena or to fatigue damage. The data from the quasi-static tests are then used for a material model which describes the nonlinear quasi-static behaviour.

THE USED COMPOSITE MATERIALS

For this research, a carbon fabric-reinforced polyphenylene sulphide (PPS), called CETEX is taken under consideration. This material is supplied to us by Ten Cate [26].

The Fibres

The fibre type is the carbon fibre T300J 3K which has the properties listed in Table 1.

Table 1. Properties of the carbon fibre

Specific mass [kg/m^3]	1,760
Tensile strength [MPa]	3,600
Ultimate strain [-]	0.014
Young's Modulus [MPa]	231,000

The weaving pattern is a 5-harness satin weave (Figure 1) with a mass per surface unit of 286 g/m^2. The 5-harness satin weave is a fabric with high strength in both directions and excellent bending properties.

Figure 1. Close-up of the 5-harness satin weave pattern [27].

The term '5-harness' means that a fibre bundle in the warp (0°) direction goes over four fibre bundles in the weft (90°) direction before going underneath one. The advantage of this pattern is that the fibre is less bent than in a plain weave (Figure 2) and the fabric is more deformable and allows better draping on curved surfaces.

The strength in the warp direction (0°) is a little higher than the strength in the weft direction (90°). A possible explanation lies in the fact that warp fibres are not bent, contrary to the weft fibres.

'Satin' means that the crossing moves one or more fibre bundles. This increases the uniformity of the strength.

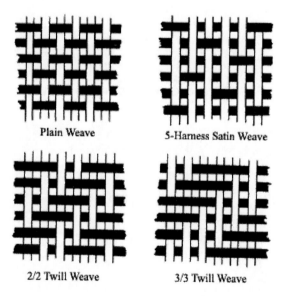

Figure 2. Comparison of different weave patterns [28].

The Matrix

The fibres are embedded in polyphenylene sulphide (PPS). The structural formula is shown in Figure 3 (a) where the hexagon represents benzene (Figure 3 (b)).

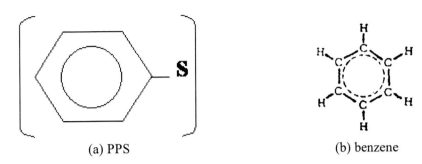

Figure 3. Structural formula of polyphenylene sulphide (PPS) and benzene.

PPS is a thermoplastic polymer matrix with a semi-crystalline structure, a high chemical resistance and good weldability. There are multiple types of this polymer commercially available, but for CETEX the linear variant is used. It has a long chain structure with limited branching resulting in a low viscosity and a faster crystallisation than for the strongly branched polymers.

The linear variant has the properties listed in Table 2.

Table 2. Properties of the PPS matrix [29]

Melting temperature [°C]	280
Glass transition temperature [°C]	90
Optimal crystallisation temperature [°C]	150 – 160
Specific mass	
Amorphous [kg/m^3]	1.30
Crystallised [kg/m^3]	1.35
Tensile strength [MPa]	75
Shear strength [MPa]	130
Ultimate strain [-]	0.08
Young's modulus in tension [MPa]	3,700
Young's modulus in compression [MPa]	3,800

PPS is a semi-crystalline thermoplastic, which means that the amorphous and the crystalline phase coexist. The proportion between both phases is determined by the preceding temperature cycle. In general, the higher the cooling rate, the more amorphous the structure becomes. The temperature cycle of PPS consists of different stages, as is illustrated in Figure 4 and is discussed next.

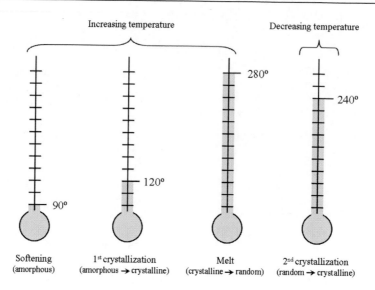

Figure 4. Characteristic temperatures for PPS [29].

Amorphous PPS starts to soften at a temperature of 90°C. This state of slightly altered viscosity lasts until the material starts to crystallise at 120°C. This crystallised material does not soften; the structure is stable and can only be broken by melting.

Melting starts at a temperature of 280°C, the material is completely liquid at 300°C. This material structure is highly disordered and the molecules have a high mobility. Since the molecular forces are very low, the melted material has a very low viscosity.

The cooling of the melt results in a second crystallisation. The number of crystals being formed depends on the cooling rate. Since crystallisation takes time, the lower the cooling rate, the higher the fraction of crystals. If the material is cooled down very fast, the disordered structure is 'frozen' and the resulting material is amorphous.

Production Process

Since PPS weakens at elevated temperatures (see paragraph 1.1.2), it is excellent for hot pressing. This is a fast and flexible method for producing plates from semi-pregs on one hand, and parts from blanks on the other hand (see Figure 5).

A semi-preg consists of two layers: (i) a layer of fabric, for instance the carbon fabric and (ii) a thermoplastic film, for instance a PPS film. The two layers are bonded to each other simply by heating the two, so that the thermoplastic melts and bonds to the fabric. A semi-preg differs from the well known pre-preg, because for the former, the two layers can still be distinguished, whereas for the latter, the fabric is already completely impregnated with the (thermoplastic) matrix.

A plate is then formed by hot pressing a number of semi-pregs together, allowing the films of thermoplastic to fully impregnate the fabric reinforcement, before hardening again. From these plates, blanks can be made, which are no more than flat panels with dimensions fit for the final product. By reheating the blank and pressing it into a mould, such as illustrated in Figure 5, a part can be realised.

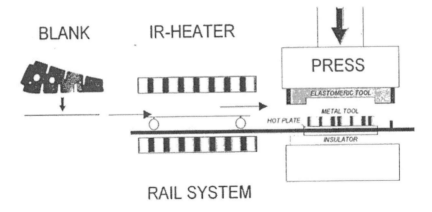

Figure 5. Illustration of hot pressing.

PPS has the advantage that it can be produced in an easy manner, but because of the shrinkage during (re)crystallisation, internal stresses are introduced. When the bonding between carbon fibre and matrix is insufficient, fracture will occur at the fibre-matrix interface and will cause an early failure. However, by varying parameters such as cooling speed and sizing of the fibres, adhesion between fibre and matrix could be optimised.

Material Properties

The following mechanical properties (Table 3) were received for the carbon fabric-reinforced polyphenylene sulphide from the Technical University of Delft. They were determined according to the 'MIL 17 Guidelines for Characterization of Structural Materials' and they apply for a temperature of 23°C and an atmospheric humidity of 50%.

Table 3. Mechanical properties of the Carbon 5H CD0286 T300J 3K film stacked

Tensile strength warp [MPa]	734
Tensile strength weft [MPa]	754
Tensile modulus warp [MPa]	56,000
Tensile modulus weft [MPa]	57,000
Compression strength warp [MPa]	567
Compression strength weft [MPa]	498
Compression modulus warp [MPa]	52,000
Compression modulus weft [MPa]	50,000
In-plane shear strength [MPa]	110
In-plane shear modulus [MPa]	4,175
Open hole tensile strength [MPa]	274
Open hole compressive strength [MPa]	259
Compression after impact [MPa]	238
Yield bearing strength [MPa]	391
Ultimate bearing strength [MPa]	738

Field of Application

Composites are used more and more as surfaces in aeronautical applications since they combine high strength and stiffness with low density. They are also more and more preferred over aluminium for structural applications.

CETEX has a number of advantages when compared to other advanced composite materials, namely its weldability, chemical inertness and the possibility of energy absorption at impact. CETEX is the first thermoplastic composite to be used in structural airline applications, for example the 'J nose' part of the leading edge of an airplane wing (Figure 6 (a)), reinforcement profiles for the wings (Figure 6 (b)) and the 'heavy duty' sandwich floor panels (Figure 6 (c)).

That the confidence in thermoplastic composites for structural applications is growing, may appear from the fact that in the Airbus A340-500/600 there is more than 1000 kg of CETEX and in the new Airbus A380, there is about 6000 kg of CETEX [26].

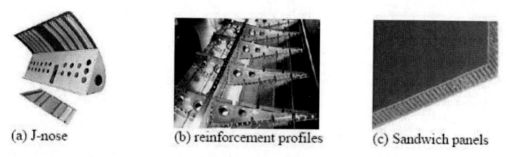

(a) J-nose   (b) reinforcement profiles   (c) Sandwich panels

Figure 6. Structural CETEX parts in aeronautics.

DESIGN OF A MODIFIED THREE-RAIL SHEAR TEST

Principle of the Three-Rail Shear Test

The principle of the three-rail shear test is illustrated in Figure 7 (a) and (b). The specimen is gripped by three rails and during the test, the central rail has a relative vertical motion with respect to the two outer rails. This movement can be either up or down. As a result, a state of shear stress is induced in the specimen.

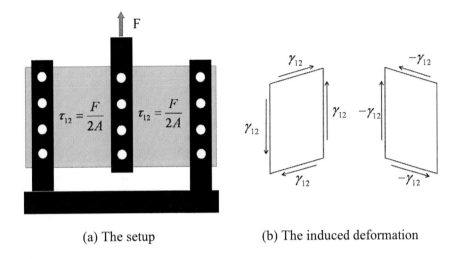

(a) The setup   (b) The induced deformation

Figure 7. Principle of the three-rail shear test.

Figure 7 (b) illustrates the (theoretically) induced deformation state. Near the edges and near the clamps, the stress and deformation state will be slightly different because of the edges and corresponding edge effects.

The shear stress can be calculated by dividing half of the force (each zone carries half of the total force) by the cross section:

$$\tau = \frac{F}{2} \cdot \frac{1}{h.t} \qquad (1)$$

where h is the height of the specimen, t is the thickness and F is the imposed force on the central rail.

To measure the shear strain, the ASTM D 4255/D 4255M standard prescribes to use strain rosettes, but if the loading is symmetrical and no bending of the specimen occurs, even one simple strain gauge will suffice. This can be visualised by presenting the deformation state, given in Figure 7 (b) on Mohr's circle, illustrated in Figure 8.

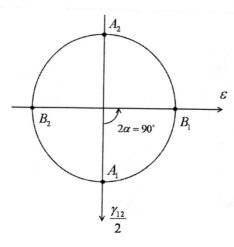

Figure 8. Representation of the deformation state on Mohr's circle.

Points A_1 and A_2 correspond with the occurring deformation, a state of pure shear. Rotating over $2\alpha = 90°$ on Mohr's circle to points B_1 and B_2 yields the principal in-plane strains, which can be measured with strain gauges. This corresponds with a rotation of $\alpha = 45°$ on the surface of the specimen for B_1 and of $\alpha = 135°$ for B_2, meaning that the strain gauges should be mounted under an angle of +45° and -45° with respect to the fibre orientation. The shear strain is then calculated as:

$$\gamma = |\varepsilon_{+45} - \varepsilon_{-45}| \qquad (2)$$

If only one strain gauge is mounted, then the shear strain can even be calculated as:

$$\gamma = 2|\varepsilon_{+45}| \qquad (3)$$

The latter is also mentioned in [18], but this assumes symmetry of the loading conditions. The instrumentation of the specimens used here is discussed in paragraph 3.1. Next, the new design is commented on.

Design of the Setup

Since the setup is designed for fatigue loading conditions, some modifications should be made so that the setup itself does not fail under fatigue loading. Lessard et al. had also made some modifications to the standard three-rail shear setup [20]. However, for the design presented here, the modifications are far more drastically, since there are no more bolts through the clamp, holding it together. The same principle as in [17, 18] is used, meaning that the specimen is gripped by pressing a plate against the specimen. This pressure is applied by bolts which go through only one side of the clamp. However, the force required to press this load transfer plate against the specimen has a similar but opposite reaction force that pushes the two sides of the grip outwards. The latter is clarified in Figure 9, where the different application of the bolts is illustrated.

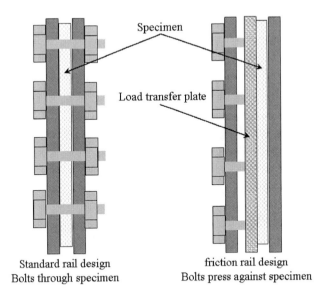

Figure 9. Difference between the use of bolts in the standard and the new design.

It is obvious that because of this outward force, the clamps will need to be more massive because they should withstand the fatigue loading conditions. As a starting point, the grips are designed in one piece, whereas the standard grips are two separate pieces, bolted together. In the grip, a rectangular cavity is milled away for the specimen. Since sharp corners produce unwanted stress concentrations, circular holes are drilled at the ends of this rectangular cavity, to soften the stress concentrations. This results in the grip, of which a cross section is illustrated in Figure 10; some general dimensions are added.

The grips should of course be able to withstand the same dynamic load range as the servo-hydraulic tensile machine it is mounted on, meaning that the grips should withstand a dynamic longitudinal force of 100 kN. Since the gripping is based on friction, a value of the friction coefficient is estimated. The assumption was made (for design purposes) that a friction coefficient of 0.5 should be feasible, with the use of additional rubber films or layers that increase the friction, should the friction between steel and composite be insufficient.

After a few preliminary tests, it became obvious that for some materials, such as the carbon fabric-reinforced PPS used here, the friction coefficient of 0.5 could not be reached, even if extra layers of high frictional materials were added. In some cases the rubber film was pushed out of the grips, in other cases the film failed under the shear loads. This, however, means that the first gripping design, depicted in Figure 11 (a) and based on pure friction, will not suffice. Therefore, the geometrical gripping was added, which is illustrated in Figure 11 (b). The load transfer plate is now supported by flattened cylinders, so that the load transfer to the grip is not only achieved by friction, but also by these cylinders.

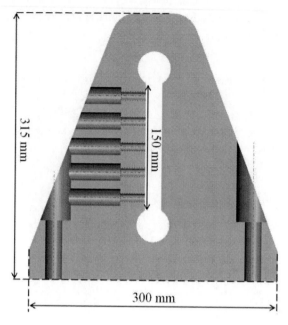

Figure 10. Vertical cross section of the grip, indicating some general dimensions.

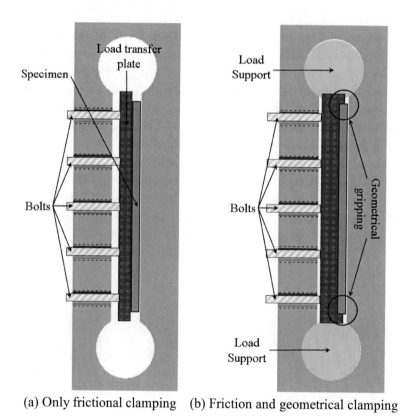

(a) Only frictional clamping   (b) Friction and geometrical clamping

Figure 11. Illustration of the used gripping principles for the proposed design.

This final design, implementing both frictional and geometrical clamping, is illustrated in Figure 12, as an exploded view. This design was used for all rail shear experiments conducted in this study.

Figure 12. Exploded view of one clamp for the three-rail shear test.

For the design, the CAD/CAE package 'Solidworks 2005' was used. In this package, there is also a finite element module, 'COSMOS Express' which was used to determine the stress distribution in the clamp.

In order to ensure an infinite fatigue life under loading of 100 kN, a safety factor of 3 with respect to the yield stress was taken into account when designing the clamps. Because of the high loads, a high-strength steel was chosen. For the simulation, done with the COSMOS express package, the following material constants for the clamps were used:

In this setup the central clamp carries the highest load; due to symmetry of the three-rail shear setup, each of the outer grips carries half the load of the central one. Therefore, the central grip is considered for the simulations. Furthermore, it is assumed that the vertical force of 100 kN is evenly distributed over the two vertical faces, so that each face carries 50 kN. Using a friction coefficient of 0.5, this results in a necessary horizontal load of 100 kN by the load transfer plate on each face.

Table 4. The used material constants in COSMOS express

Elastic modulus [GPa]	210
Poisson's ratio [-]	0.28
Yield strength [MPa]	620
Mass density [kg/m^3]	7,700

These loading conditions are the worst case scenario for the grips and assume the use of only frictional gripping (Figure 11 (a)). If geometrical gripping is added (Figure 11 (b)), then the horizontal forces will be less, since part of the vertical force is transported via the cylinders to the grips. As such, the vertical load remains the same, the horizontal force will decrease.

The results of the simulation for the worst case scenario yields the stress distribution depicted in Figure 13, where the values of the Von Mises criterion are illustrated; the deformation is scaled with a factor of 773.2.

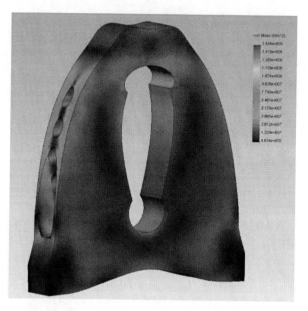

Figure 13. Von Mises stress distribution in the clamp under the given load conditions. Deformation factor is 773.2; maximum stress level is 154.8 MPa.

Since the maximum stress level is 154.4MPa, the minimum factor of safety is 3.86 with respect to the estimated yielding stress of 620 MPa. This should ensure infinite fatigue life.

The clamps were made out of THYROPLAST 2344 EFS steel, which is a quenched and tempered mould steel; the chemical composition is given in Table 5. After quenching, the material has a hardness of 54 HRC and a tensile strength of 1910 MPa. The material has no typical yield behaviour, it is linear till failure.

As a result, the safety factor under fatigue loading conditions is a lot higher than 3, ensuring infinite fatigue life. This material was recommended for the highest demands and fitted the request for high strength, high yield stress and high hardness. The latter was necessary to avoid damage to the surfaces from glass or carbon fibres.

The DIN material number is 1.2344 and is referred to as X 40 CrMoV 5 1.

Table 5. Chemical composition for the THYROPLAST 2344 EFS, typical analysis in %

C	Si	Cr	Mo	V
0.40	1.0	5.3	1.4	1.0

After production, the clamps were nitrated for a higher surface hardness, since thermal hardening would result in large, intolerable deformations. The other parts were coated with a nitrate layer with a 'niblox' treatment to avoid micro-welding of the surfaces.

Finally, two extra plates were designed in order to be able to mount this setup on a standard servo-hydraulic testing machine. The final setup, mounted on the tensile machine, is shown in Figure 14.

Figure 14. The final design, mounted on the tensile machine.

During fatigue tests, the plunger of the tensile machine may start rotating if no precautions are taken. However, this possible rotation of the plunger and therefore the bottom clamp is prevented, using a special guiding system which is mounted on the plunger. Therefore, once alignment of the clamps is achieved, this alignment is guaranteed throughout the fatigue test.

The desired stress state for this setup is to have pure and equal shear load for both of the loaded zones. However, due to misalignment of the clamps, the zones on the left and the right of the central clamp may not experience the same load. Furthermore, buckling or bending of the specimen may occur, which means that the front and the back of one loaded zone (left or right) may have a different stress state. Hence, a few quasi-static hysteresis tests were done to assess the occurrence of buckling and whether the setup induces a symmetrical load. By performing these cyclic loadings, possible problems regarding symmetry or buckling will be more easily detected than in a quasi-static test with monotonic loading until saturation or debonding of the gauges, because differences in the strains will probably increase with each successive loading, due to permanent deformation of the specimen.

The results of these hysteresis experiments are documented extensively in paragraph 5, but it could be concluded that the setup does not induce bending and both sides are loaded symmetrically.

SAMPLE GEOMETRY AND INSTRUMENTATION

The Three-Rail Shear Test

The dimensions of the used specimen are given in Figure 15.

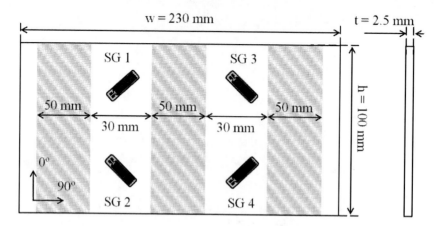

Figure 15. Dimensions of the used [(0°,90°)]$_{4s}$ three-rail shear specimen. The locations of the strain gauges, as well as the clamps, are also illustrated.

For all quasi-static and hysteresis tests, the force F, the strains ε_i, i = 1,2,3,4 and the temperature were recorded. With these values, the shear stress and strain can be calculated as

$$\tau_{12} = \frac{1}{2}\frac{F}{t.h}$$
$$\gamma_{12} = |\varepsilon_i - \varepsilon_j| \qquad (4)$$

where h is the height of the specimen and t the thickness; (i,j) is either (1,2) or (3,4) with respect to the numbering of the strain gauges in Figure 15. For the fatigue experiments, no strain gauges were mounted since they would de-bond after a few dozen cycles due to the large deformations.

[(+45°,-45°)]$_{4s}$ Tensile Test

For these experiments, the longitudinal strain ε_{xx} is measured with the extensometer and the transverse strain ε_{yy} is measured with a transverse strain gauge. When shear strains are expected, a third strain measurement can be performed on the coupon, under an angle θ (see Figure 16, α equal to 45°).

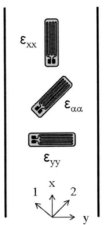

Figure 16. Positioning of the strain gauges ε_{yy} and $\varepsilon_{\alpha\alpha}$; ε_{xx} can be measured with either the extensometer or a strain gauge.

With the three strains ε_{xx}, ε_{yy} and $\varepsilon_{\alpha\alpha}$ the shear strain γ_{xy} can be calculated using a coordinate transformation:

$$\varepsilon_{\alpha\alpha} = \varepsilon_{xx} \cos^2 \alpha + \varepsilon_{yy} \sin^2 \alpha + 2\varepsilon_{xy} \sin\alpha \cos\alpha \tag{5}$$

with α = 45°, as illustrated in Figure 16, this becomes:

$$2\varepsilon_{xy} = \gamma_{xy} = 2\varepsilon_{\alpha\alpha} - \varepsilon_{xx} - \varepsilon_{yy} \tag{6}$$

However, the strain gauge under 45° is not always necessary. If the strains in the xy-coordinate system are transformed to the 1,2-coordinate system for the [(+45°,-45°)]$_{4s}$ stacking sequence, the following expression is found for the shear strain γ_{12} [30]:

$$\gamma_{12} = \varepsilon_{xx} - \varepsilon_{yy} \tag{7}$$

For the longitudinal and transverse strains, the following formulas can be derived for the [(+45°,-45°)]$_{4s}$ stacking sequence, using coordinate system transformation formulas:

$$\varepsilon_{11} = \frac{1}{2}(\varepsilon_{xx} + \varepsilon_{yy} + \gamma_{xy})$$
$$\varepsilon_{22} = \frac{1}{2}(\varepsilon_{xx} + \varepsilon_{yy} - \gamma_{xy}) \tag{8}$$

ε_{11} being the strain along the fibre in the warp direction and ε_{22} the strain along the fibre in the weft direction.

However, for a tensile test, the following stress state is applied:

$$\sigma_{xx} = \sigma \qquad \sigma_{yy} = \sigma_{zz} = \sigma_{xy} = \sigma_{yz} = \sigma_{zx} = 0 \tag{9}$$

σ being the stress applied to the specimen.

$$
\begin{pmatrix} \varepsilon_{xx} \\ \varepsilon_{yy} \\ \varepsilon_{zz} \\ \gamma_{yz} \\ \gamma_{xz} \\ \gamma_{xy} \end{pmatrix} =
\begin{pmatrix}
\dfrac{1}{E_{xx}} & -\dfrac{\nu_{xy}}{E_{xx}} & -\dfrac{\nu_{xz}}{E_{xx}} & 0 & 0 & 0 \\[2mm]
-\dfrac{\nu_{yx}}{E_{yy}} & \dfrac{1}{E_{yy}} & -\dfrac{\nu_{yz}}{E_{yy}} & 0 & 0 & 0 \\[2mm]
-\dfrac{\nu_{zx}}{E_{zz}} & -\dfrac{\nu_{zy}}{E_{zz}} & \dfrac{1}{E_{zz}} & 0 & 0 & 0 \\[2mm]
0 & 0 & 0 & \dfrac{1}{E_{yz}} & 0 & 0 \\[2mm]
0 & 0 & 0 & 0 & \dfrac{1}{G_{xz}} & 0 \\[2mm]
0 & 0 & 0 & 0 & 0 & \dfrac{1}{G_{xy}}
\end{pmatrix}
\begin{pmatrix} \sigma_{xx} \\ \sigma_{yy} \\ \sigma_{zz} \\ \tau_{yz} \\ \tau_{xz} \\ \tau_{xy} \end{pmatrix} \tag{10}
$$

Using Hooke's law for an orthotropic material (Equation 10) [31], this results in $\tau_{xy} = 0$ which combined with Equation 8 yields:

$$\varepsilon_{11} = \varepsilon_{22} = \frac{\varepsilon_{xx} + \varepsilon_{yy}}{2} \tag{11}$$

As such, for the $[(+45°,-45°]_{4s}$ stacking sequence, all strains in the 1,2-coordinate system can be calculated using only the longitudinal strain ε_{xx} and the transverse strain ε_{yy}. The third strain gauge, however, can be mounted to control the alignment of the test specimen in the tensile machine, since it is known that γ_{xy}, calculated using Equation 6 should be zero.

Finally, the dimensions of the used coupons, according to the "ASTM D3518/D3518M-94 (2007) standard test method for in-plane shear response of polymer matrix composite materials by tensile test of a $\pm 45°$ laminate", are shown in Figure 17.

For every quasi-static and hysteresis experiment the force F, the longitudinal and transverse strains ε_{xx} and ε_{yy} and the temperature were recorded. With these values, the shear stress τ_{12} and shear strain γ_{12} can be calculated as

$$\tau_{12} = \frac{1}{2} \frac{F}{w.t} \tag{12}$$

$$\gamma_{12} = \varepsilon_{xx} - \varepsilon_{yy}$$

where w is the width of the specimen and t is the thickness. The transverse strain was measured using a strain gauge and the longitudinal strain was measured using the extensometer. For the fatigue experiments, no strain gauges were mounted since they would de-bond after a few dozen cycles due to the large deformations.

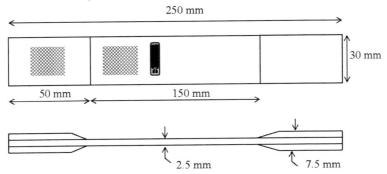

Figure 17. Dimensions of the used [(+45°,-45°)]$_{4s}$ tensile coupon, equipped with chamfered tabs of [(+45°,-45°)]$_{4s}$ carbon-PPS.

The following three sections will describe the quasi-static, hysteresis and fatigue experiments conducted with both the three-rail shear setup and the [(+45°,-45°)]$_{4s}$ tensile test. For each of these sections, first the experiments performed with the three-rail shear setup are discussed, after which the [(+45°,-45°)]$_{4s}$ tensile experiments are commented on and the correspondence between the results is assessed.

QUASI-STATIC TESTS

Three-Rail Shear Tests

All tests were done in a displacement-controlled manner with a displacement speed of 1 mm/min, during which the force F, the strains ε_i, i = 1,2,3,4 and the temperature were recorded and shear stress and strain were calculated using Equation 4.

A few shear stress-strain results are given in Figure 18. Since from each rail shear test, two curves can be derived, one from each instrumented loaded zone, there should be six curves for three specimens. However, in Figure 18, only the curves from the strain gauges which lasted the longest are given; the tests were stopped once all strain gauges de-bonded or saturated which means that the maximum value of the shear stress and strain in Figure 18 do not correspond with failure. After saturation or failure of the strain gauges, the test was stopped so the possibility for non-destructive evaluation of the specimens remained an option.

It should be noticed that the results from these quasi-static tests are very reproducible and that the calculated stiffness corresponds very well with the value given in Table 3 (paragraph 1.3, G_{12} = 4,175 MPa). During the quasi-static rail shear tests, no temperature increase was measured.

Finally, Figure 19 illustrates the evolution of the measured strains and the derived shear strains for the quasi-static test on specimen N5 until saturation of the strain gauges; the numbering of the gauges corresponds with Figure 15. As such, it can already be remarked that both sides of the specimen are loaded symmetrically.

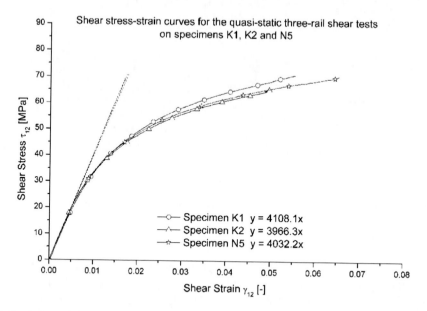

Figure 18. Evolution of the shear stress as a function of the shear strain for the quasi-static three-rail shear experiments.

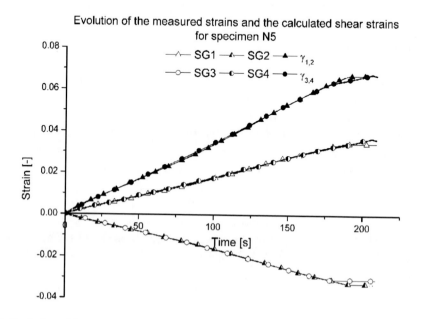

Figure 19. Evolution of the measured strains and calculated shear strains for specimen N5.

[(+45°,-45°)]₄ₛ Experiments

In this paragraph, the quasi-static [(+45°,-45°)]₄ₛ experiments will be discussed and the results will be compared with those obtained in the previous paragraph. All tensile tests were done in a displacement-controlled manner with a displacement speed of 2 mm/min, in correspondence with the ASTM D3518/D3518M-94, during which the force F, the longitudinal and transverse strains and the temperature were recorded and shear stress and strain were calculated using Equation 12.

Figure 20 illustrates the highly nonlinear shear stress-strain evolution for two quasi-static experiments, M1 and M3. The curve is only depicted until the transverse strain gauge either saturated which was the case for M3, or de-bonded which happened for M1. The failure stresses were 105.4 MPa and 105.3 MPa for M1 and M3 respectively. These values show good correspondence with the value given in Table 3 (S = 110 MPa). The stiffness could also be calculated from these results, as is shown in Figure 20. Although some scatter is present on these values, they still correspond quite well with the values given in Table 3 (G_{12} = 4,175 MPa). It should be noted that these results correspond excellently with the shear stress-strain curves from the three-rail shear tests.

During these tests, no increase in temperature was recorded.

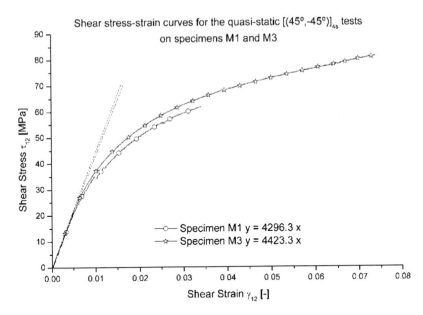

Figure 20. Evolution of the shear stress as a function of the shear strain for the quasi-static tensile experiments.

Finally, Figure 21 illustrates the evolution of the shear strain ε_{12} and the strains along and perpendicular to the fibre, ε_{11} and ε_{22}, both calculated using Equation 11 as a function of time until saturation of the strain gauges for specimen M3. It should be noted that the strains ε_{11} and ε_{22} are very low, so they will have only a very limited influence on the damage behaviour, thus the imposed stress state should correspond very well with the pure shear state in the three-rail shear test.

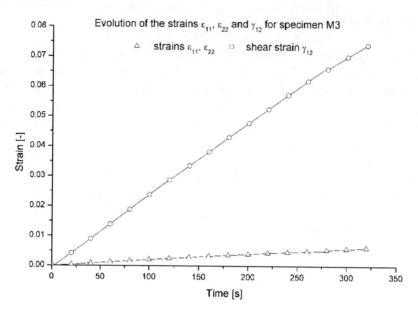

Figure 21. Evolution of ε_{11}, ε_{22} and γ_{12} for specimen M3.

QUASI-STATIC HYSTERESIS TESTS

Three-Rail Shear Tests

As was already mentioned in the design section of the chapter, the desired stress state for this setup is to have pure and equal shear load for both of the loaded zones. However, due to misalignment of the clamps, the zones on the left and the right of the central clamp may not experience the same load. Furthermore, buckling or bending of the specimen may occur, which means that the front and the back of one loaded zone (left or right) may have a different stress state. Hence, quasi-static cyclic loading (hysteresis) was applied, meaning that the specimens were loaded until a maximum shear stress of 10 MPa was reached and then completely unloaded. For each of the next cycles, the maximum shear stress was increased with 10 MPa with respect to the previous cycle. This was repeated until the strain gauge failed or de-bonded. The low starting value of 10 MPa was chosen to verify if damage already occurs for these low load levels. All experiments were done in a displacement controlled manner with a displacement speed of 1 mm/min; the signals were sampled at 20 Hz.

By performing these cyclic loadings possible problems regarding symmetry or buckling will be more easily detected than in a quasi-static test with monotonic loading until saturation or de-bonding of the gauges, because differences in the strains will probably increase with each successive loading, due to permanent deformation of the specimen.

To verify the symmetry, four strain gauges were placed on the same side of the specimen, but distributed over the two loaded zones, as illustrated on Figure 15. The time evolution of the different strain gauges for such a test is given in Figure 22. If the setup is symmetrical, then strain gauge 1 and 3 and strain gauge 2 and 4 respectively should give the same strain. It can clearly be seen that the signals coincide almost perfectly, meaning symmetry is achieved.

The de-bonding or failing of the strain gauges can also be seen: after about 350 seconds, strain gauge 3 de-bonded; after 450 s, gauge 1 failed and after 500 s, gauge 2 failed. Gauge 4 failed soon after gauge 2. As a result; the corresponding curves no longer coincide from that point on.

To assess whether buckling occurs, the four strain gauges are placed on the same loaded zone, but on front and rear surfaces. Strain gauges 1 and 2 are placed as depicted in Figure 15 and strain gauges 3 and 4 are placed on the same position on the opposite side of the specimen in such a way that the strains from strain gauge 1 and 2 should correspond with the signals from strain gauges 3 and 4 respectively. The result from such a test is depicted in Figure 23. Again, the signals coincide almost perfectly. Similar to the previous test, the curves are only shown until the strain gauges de-bonded. This happened at 470 s for gauge 3 and at 510 s for gauge 4. The other gauges failed soon after 600 s.

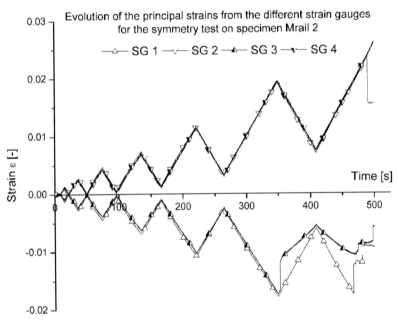

Figure 22. Evolution of the strain as a function of time for all four strain gauges in the symmetry test.

Similar experiments have been conducted, all with corresponding results. As such, it may be concluded that the stress state induced by the clamps is symmetrical and that no buckling or bending of the specimen occurs. As a result, there is no longer need for four strain gauges. Two strain gauges, mounted on the same surface of one loaded zone under +45° and -45° suffice for the measurement.

The shear stress-strain results from these, as well as two other experiments are given in Figure 24 and again, only the curves from the strain gauges which lasted the longest, are plotted. The omitted curves, however, showed very good correspondence with the corresponding depicted ones. Again it should be noted that the results are very much alike and that the initial stiffness corresponds very well with the value given in Table 3 (G_{12} = 4,175 MPa).

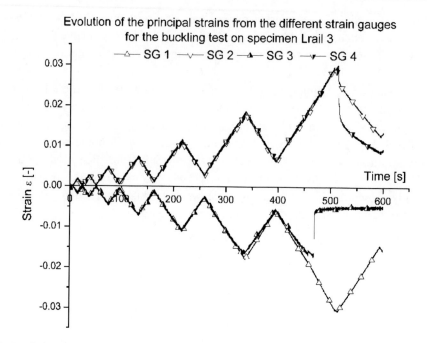

Figure 23. Evolution of the strains from al four strain gauges as a function of time for the buckling test.

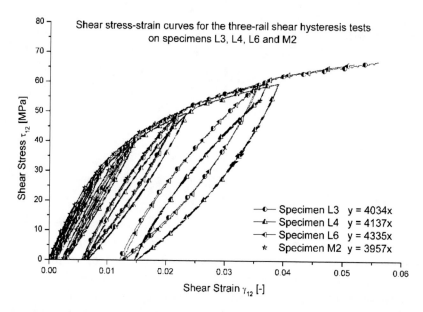

Figure 24. Evolution of the shear stress as a function of the shear strain for the rail shear hysteresis experiments.

For all measurements mentioned above, the temperature was also monitored, using a thermocouple. Because of the shear loads, a temperature increase is expected. However, an increase of only 1 °C was measured, which was considered negligible.

[(+45°,-45°)]₄ₛ Experiments

For these experiments, the specimens were loaded until a maximum shear stress of 20 MPa was reached and then completely unloaded. For each of the next cycles, the maximum shear stress was increased with 10 MPa with respect to the previous cycle, as was the case for the three-rail shear experiments. This was repeated until the strain gauge failed or de-bonded. The value of 20 MPa was chosen because only very limited nonlinear behaviour is visible in Figure 20 before this value is reached and for the three-rail shear tests, very limited damage occurred at 10 MPa.

Figure 25 illustrates the shear stress-strain evolution for three specimens M2, M4 and M5. For M4, the cycle after 60 MPa was reversed just before the strain gauge was expected to saturate. For M2 and M5, the strain gauge de-bonded after the cycle of 60 MPa. Failure stresses were 109.8 MPa for M2, 116.3 MPa for M4 and 103.8 MPa for M5. Again these values correspond very well with the value given in Table 3 (S = 110 MPa). The initial stiffness is also calculated and is shown in Figure 25. Again, there is some scatter on the results, but the values correspond quite well with those found in Figure 20 and with the value determined at the Technical University of Delft (Table 3, G_{12} = 4,175 MPa). It should be noted that the reproducibility of M2 and M4 is very high. Specimen M5 tends to behave stiffer during the entire experiment, without any apparent reason. These results show very good agreement with the cyclic shear stress-strain behaviour seen for the three-rail shear tests. This shape of the τ_{12}- γ_{12} curve was also reported in [7, 10] for a glass fibre-reinforced epoxy and in [11, 12, 13, 14] for a carbon fibre-reinforced epoxy. The loops are the result of friction between the fibre and the matrix [12, 13].

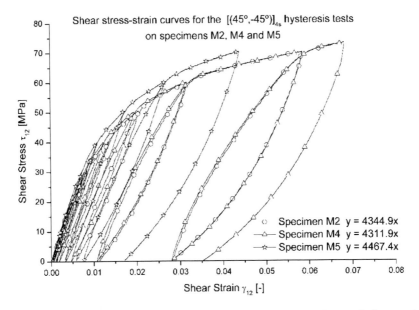

Figure 25. Evolution of the shear stress as a function of the shear strain for the tensile hysteresis experiments.

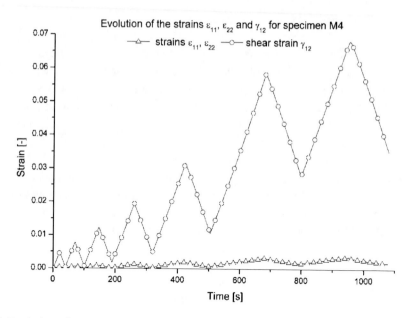

Figure 26. Evolution of ε_{11}, ε_{22} and γ_{12} for specimen M3.

Figure 26 illustrates the evolution of the shear strain γ_{12} and the strains along and perpendicular to the fibre, ε_{11} and ε_{22}, both calculated using Equation 11, as a function of time until saturation of the strain gauges for specimen M4. Again, the strains ε_{11} and ε_{22} are very low, so the behaviour should correspond well with the pure shear state from the three-rail shear test.

Finally, Figure 27 illustrates the occurring fracture for specimen M2, but this type of failure was seen for all the $[(+45°,-45°)]_{4s}$ tensile tests. The local narrowing before final failure can clearly be distinguished.

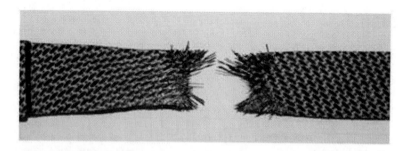

Figure 27. Image of the fracture of a $[(+45°,-45°)]_{4s}$ specimen.

During these experiments, a very slight increase in temperature was noted of about 1 °C, which again was considered negligible.

Shear Strain-Displacement Correlation

Because of the combined frictional and geometrical clamping of the design, the modified three-rail shear setup has another interesting advantage. Given the definition of the shear strain γ (Figure 28), it can be calculated as:

$$\tan \gamma = \frac{D}{W} \tag{13}$$

If small displacements are assumed, the shear strain will be small and can be calculated as:

$$\tan \gamma \approx \gamma \approx \frac{D}{W} \tag{14}$$

This means that there should be a linear correlation between the shear strain γ and the imposed displacement from the central rail D. For the given value of W = 30 mm (see Figure 15), this relation becomes

$$\gamma = \frac{1}{30}D = 0.033D \tag{15}$$

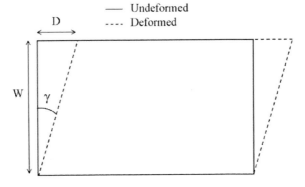

Figure 28. Definition of the shear strain γ.

During the quasi-static and hysteresis experiments it was noted that there is a correlation between the imposed displacement and the resulting shear strain in the specimen, but it is rather parabolic than linear. This is illustrated in Figure 29; the different experiments are given an offset al.ong the x-axis for a clear image. It can be noted that for both the hysteresis tests, the curves are very reproducible.

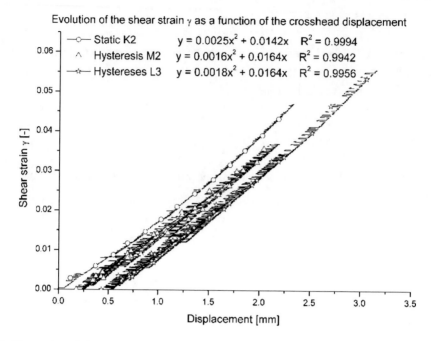

Figure 29. Illustration of the correlation between shear strain and displacement.

The reason for the difference between the experimental and theoretical correlation can be explained as follows. Theoretically, a displacement of 2 mm on the side edge is exactly 2 mm, whereas a displacement of 2 mm of the hydraulic actuator will be less than 2 mm displacement of the side edge of the loaded zone, because of the following reasons:

i. The height of the specimen is not exactly equal to the height of the geometrical gripping, since then, mounting the specimen would be practically impossible. As a result, some slippage may occur.
ii. The load transfer plates do not fit exactly in the space provided, the height is a little less since otherwise, the plates could not be mounted. Again, some slippage may occur.
iii. The mounting mechanism on the tensile machine, as well as the tensile machine deforms under the occurring loads. Hence, the displacement of the side of the specimen will always be a little smaller than the imposed displacement by the plunger. This is the same reason why the calculated strain from the displacement in a uni-axial tensile test differs from the strain, measured with extensometer or strain gauges.

Of course, after some time, enough sliding will have occurred and all free space will be occupied, so for higher displacements, this effect should no longer be present. For instance, if the slope is calculated for specimen L3 for a displacement of 3 mm, this yields:

$$\left.\frac{df}{dx}\right|_{x=3} = 0.0036x + 0.0164\big|_{x=3} = 0.0272 \qquad (16)$$

which approximates the expected 0.033. The remaining difference is due to the elastic deformation of the entire setup. This is also the reason why there is a larger difference between the curve for the quasi-static test and the hysteresis tests than between both hysteresis tests respectively. After one loading, there will still be some free space that allows sliding, whereas after a few successive loadings, this free space will no longer exist.

It is expected that a similar phenomenon is present when using the standard three-rail shear test setup. However, nothing on this matter is documented by Lessard et al. [20].

The finite element modelling of quasi-static nonlinear behaviour will be given at the end of this chapter, first the fatigue experiments are discussed.

FATIGUE LOADING CONDITIONS

Three-Rail Shear Tests

The fatigue experiments were done according to the "ASTM D4255/D4255M standard test method for in –plane shear properties of polymer matrix composite materials by the rail shear method" but under fatigue loading conditions. They were done in a load-controlled manner and force, displacement and temperature were registered. By using load-control, it is assured that the specimen is always loaded between the desired τ_{min} and τ_{max}, whereas with displacement-control, no two experiments would be the same since the little amount of sliding in the grips would never be equal for different experiments. Because of the large deformations, strain gauges were not considered, since they de-bond after a few dozen cycles. As such, the displacement is chosen as parameter for information about the deformed state, since there is a relationship between the shear strain and the displacement, because of the geometrical clamping of the grips.

From the quasi-static experiments, described in paragraph 4, it can be derived that a maximum stress level between 30 MPa and 50 MPa for the fatigue tests should give interesting results. Lower than 30 MPa there is only very limited nonlinear behaviour and beyond 50 MPa it is expected that the specimen would fail after a few dozen cycles.

Since the objective is to compare the results of the three-rail shear test with the [(+45°,-45°)]$_{4s}$ tensile tests, all fatigue experiments were done with a shear stress ratio R equal to 0, meaning that the shear stress varies between 0 and a certain maximum value. For the [(+45°,-45°)]$_{4s}$ tensile test, it would be rather difficult to have negative shear stresses, since this would mean that compressive tests should be performed, with the corresponding difficulties related to buckling.

During the fatigue experiments, every five minutes five fatigue cycles were registered and from these signals, the maximum, minimum and average value was calculated and stored.

A first experiment was done at 1 Hz with a maximum load of 40 MPa, which is halfway the range suggested earlier. Since the test was load-controlled, an increase in the displacement amplitude corresponds with shear stiffness degradation. While monitoring the experiment, it was noticed that the permanent deformation and stiffness degradation was less than expected and no temperature increase occurred, so after 250,000 cycles, the frequency was increased to 2 Hz. The results of this experiment are shown in Figure 30. After 770,000 cycles, the test was stopped.

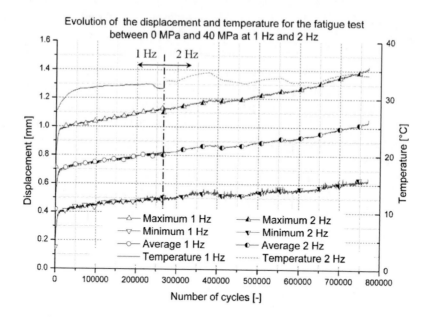

Figure 30. Maximum, minimum and mean value of the displacement as a function of the number of cycles for a 0-40 MPa fatigue test at a combined frequency of 1 Hz and 2 Hz.

The temperature tends to be slightly higher for 2 Hz than for 1 Hz. Therefore, a next test was done between 0 MPa and 40 MPa at 2 Hz until failure. Figure 31 illustrates the run-in of the experiment and Figure 32 shows the fatigue data. It should be noted that already some permanent deformation occurs during the run-in, followed by a steady state regime. That is also the reason why the minimum of the displacement during the fatigue test does not start at 0 in Figure 32. It can also be noticed that the evolution of the displacement does not have the expected sinusoidal evolution with increasing amplitude, as expected. The test was load-controlled with a sinusoidal evolution and due to stick-slip in the grips, the corresponding displacement differs a little.

The test was stopped after 1,400,000 cycles, because the temperature had already exceeded the glass transition temperature of the matrix, which is 90°C. From Figure 32 it can be seen that after the run-in there is a gradual increase in permanent deformation and stiffness degradation, until about 1,200,000 cycles. Until that time, the temperature remains constant. Then, there is a sudden increase in both temperature and permanent deformation. This can be explained as follows. During the first 1,200,000 cycles, damage grows in the material, with little effect on the stiffness. Probably the fibre-matrix interface deteriorates. Once there is sufficient damage to this interface, the fibres start to slide inside the matrix under influence of the shear load, causing heating of the matrix, with softening of the matrix as result. This, however, causes more deteriorating of the fibre-matrix interface, so that even more sliding and heating occurs. This process ends with final failure. The heating due to frictional sliding of the fibre in the matrix has also been reported in [32, 33] for fibre-reinforced ceramics and in [34] for fibre-reinforced polymers.

The following experiment was done between 0 MPa and 45 MPa at 2 Hz, to assess the influence of maximum load level. The run-in gave similar results as for the 0-40 MPa test, so it is not shown. Figure 33 depicts two of these experiments.

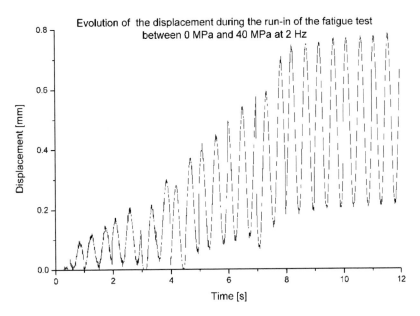

Figure 31. Displacement as a function of time during the run-in of the 0-40 MPa, 2 Hz fatigue test.

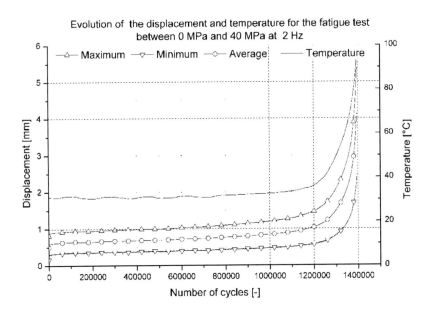

Figure 32. Maximum, minimum and mean value of the displacement as a function of the number of cycles for a 0-40 MPa fatigue test at 2 Hz.

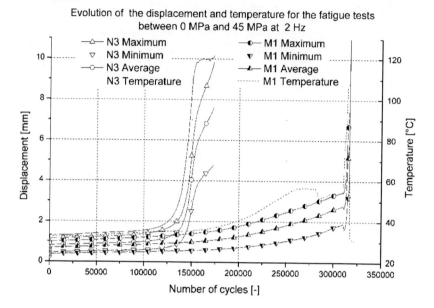

Figure 33. Maximum, minimum and mean value of the displacement as a function of the number of cycles for a 0-45 MPa fatigue test at 2 Hz.

Again, a similar evolution as for the 0-40 MPa test can be seen, although the initiation of the large stiffness decrease and sudden temperature increase happens a lot earlier. For these tests, the sudden jump happens around 140,000 cycles for N3 and around 310,000 cycles for M1. This is five to ten times lower than for the 0-40 MPa at the same frequency. It should, however, be noted that the increase in permanent deformation is much more sudden for N3 than for M1. For the latter, the sudden jump is preceded by a gradual increase in permanent deformation and stiffness degradation, which both start around 150,000 cycles. The latter is also the point in time where the temperature starts increasing. The sudden drop in temperature around 300,000 cycles for M1 was due to a loss of contact of the thermocouple, because of the large deformations. The thermocouple was re-attached soon afterwards.

In Figure 34, the fracture of this specimen is shown. Although the failure started near the clamped edges, where stress concentrations exist, final failure occurred in the centre of the specimen, under shear loading conditions.

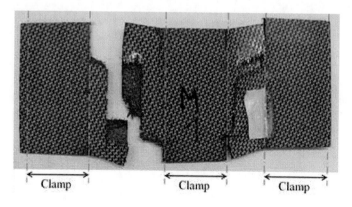

Figure 34. Illustration of the failure of specimen M1.

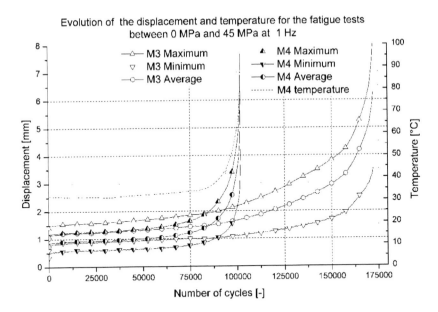

Figure 35. Maximum, minimum and mean value of the displacement as a function of the number of cycles for a 0-45 MPa fatigue test at 1 Hz.

Because of this difference in lifetime, the following experiments were conducted in the same stress range, 0-45 MPa, but at 1 Hz, to assess the influence of the frequency. The results of these experiments are shown in Figure 35. The temperature measurement was unavailable for specimen M3, because the thermocouple de-bonded very early. Again, similar evolutions are found for the evolution of the displacement. Specimen M4 has a sudden increase in temperature and displacement after about 80,000 cycles, whereas specimen M3 has a more gradual increase, starting around 100,000, until there is a sudden increase at 160,000 cycles. Apparently, lowering the frequency tends to decrease the lifetime.

To verify this behaviour, fatigue experiments were done at a load range of 0-50 MPa and again at 1 Hz and 2 Hz. Figure 36 shows the run-in for one of these experiments, but a similar evolution as for the 0-40 MPa@2 Hz test is found (see Figure 31), namely a relatively large increase, followed by a steady state regime.

Figure 37 shows the results for the test at 2 Hz. The sudden increase for the displacement occurs around 55,000 cycles for N1 and around 45,000 cycles for L5. It should be noted that due to the large deformations, the thermocouple de-bonded after 59,000 cycles for N1, hence the sudden drop in temperature. Furthermore, the temperatures lie lower than in the previous experiments, because the thermocouple lost contact with the specimen's surface, due to the large deformations, but still remained quite close to the surface. As such, the actual temperatures would be in the same range as for the previous tests, which means higher than the softening temperature of PPS (90°C).

Considering the lifetime, a reduction factor of about three with respect to the 0-45 MPa@2 Hz measurements is found.

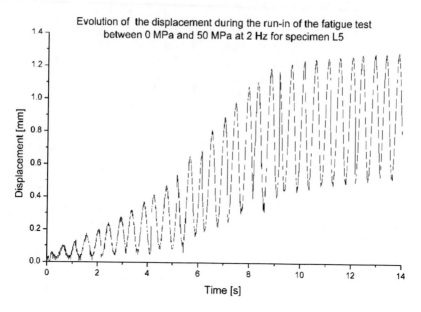

Figure 36. Displacement as a function of time during the run-in of a 0-50 MPa, 2 Hz fatigue test.

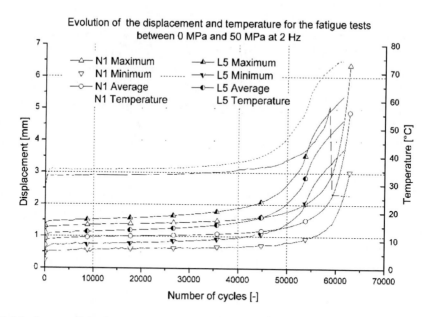

Figure 37. Maximum, minimum and mean value of the displacement as a function of the number of cycles for a 0-50 MPa fatigue test at 2 Hz.

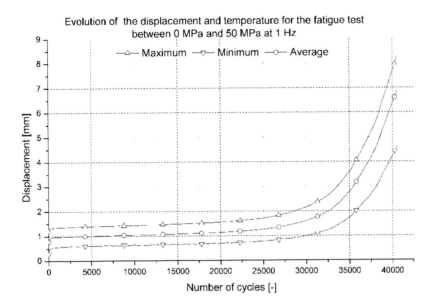

Figure 38. Maximum, minimum and mean value of the displacement as a function of the number of cycles for a 0-50 MPa fatigue test at 1 Hz.

Finally, Figure 38 illustrates the results from the 0-50 MPa@1 Hz experiment. Unfortunately, the temperature measurement was not successful, because of the de-bonding of the thermocouple. The sudden increase in displacement occurs after about 27,500 cycles, which indeed is lower than for the corresponding experiment at 2 Hz.

In conclusion, increasing the maximum shear stress for fatigue tests with R equal to zero decreases the fatigue life, while increasing the loading frequency for a given load range seems to increase fatigue life. The fatigue life is very sensitive to the maximum shear stress at lower maximum stress amplitudes; an increase from 40 MPa to 45 MPa decreased fatigue life with a factor ten. For higher stresses, the sensitivity is lower; an increase from 45 MPa to 50 MPa only reduces fatigue life with a factor of about three. During run-in, there is first a relatively large increase in displacement. Then, there is some sort of steady state with a very gradual increase and finally, a sudden large increase in both temperature and displacement occurs, before failure.

[(+45°,-45°)]$_{4s}$ Experiments

These experiments were done according the "ASTM D3479/D3479M-96 (2007) standard test method tension-tension fatigue of polymer matrix composite materials" and the "ASTM D3518-76 standard practice for in-plane shear stress-strain response of unidirectional reinforced plastics". The fatigue experiments were done in a load-controlled manner; because of the large deformations, strain gauges were not considered since they would de-bond after a few dozen of cycles. The longitudinal strain was measured with the extensometer, but since there is no data of the transverse strain, the shear strain cannot be calculated. Of course, load, displacement and temperature were also registered. As previously mentioned, the same experiments as were done with the three-rail shear test are repeated, with exception of the test between 0 MPa and 40 MPa, because of long duration of this experiment. For each experiment it was verified that no slipping in the grips occurred. For comparison with the three-rail shear tests, the evolution of both temperature and displacement is only considered until there is a significant temperature increase and stiffness degradation. Afterwards, a new regime reinstates itself, which is discussed at the end of this paragraph. Also, the specimens tend to deform into a 'dog-bone' like shape. This deformation was most apparent for the highest stress range and frequency, which is illustrated in Figure 39, where the specimen N2, on which a 0-50 MPa, 2 Hz fatigue experiment was done, can be compared with another specimen which is still to be tested, but with the same dimensions as specimen N2 had before testing.

Figure 39. Un-deformed (top) and deformed (bottom) $[(+45°,-45°)]_{4s}$ specimen.

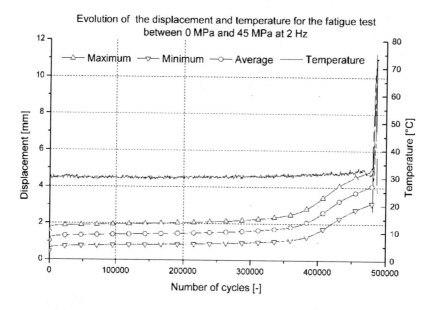

Figure 40. Maximum, minimum and mean value of the displacement as a function of the number of cycles for a 0-45 MPa tensile fatigue test at 2 Hz.

For the discussion of the fatigue experiments, the same order with respect to test frequency and load range as used for the three-rail shear tests is followed here. Figure 40 shows the results from a fatigue test between 0 MPa and 45 MPa at 2 Hz. The initiation of the increase in temperature and deformation is situated around 400,000 cycles, which is higher than the 140,000 and 310,000 cycles found for the corresponding rail shear test.

Figure 41 shows the evolution of the displacement of a fatigue test between 0 MPa and 45 MPa at 1 Hz. The initiation of both the increase in temperature and deformation growth occurs around 280,000 cycles. For the corresponding rail shear test, this happened at 80,000 and 160,000 cycles.

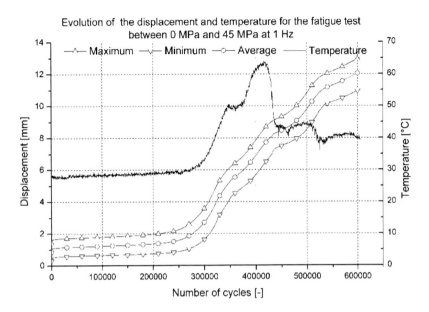

Figure 41. Maximum, minimum and mean value of the displacement as a function of the number of cycles for a 0-45 MPa tensile fatigue test at 1 Hz.

For the same experiment, Figure 42 shows the evolution of the longitudinal strain. It must be noted that the sudden increase in longitudinal strain only occurs around 400,000 cycles and stops after 510,000 cycles, whereas the temperature and displacement increase (see Figure 41) already started at 280,000 cycles and carried on until the test was stopped at 600,000 cycles. This illustrates that the narrowing of the specimen to a 'dog-bone' like shape, as illustrated in Figure 39, does not happen in a uniform manner over the entire specimen, but starts near the clamped ends and then gradually grows along the entire specimen length. Since the extensometer is mounted near the centre of the specimen, it only measures the narrowing and the resulting elongation when it occurs between the blades of the extensometer. This growing of the localization of the shear strain could clearly be seen in the experiments.

A similar effect occurs with the temperature measurement; the temperature is the highest in the zone of localization, but since the carbon fibres are highly conductive, this effect is not so distinct.

Figure 43 shows the results from a 0-50 MPa@1 Hz experiment. It can again be remarked that the behaviour is very similar to the three-rail shear tests. The initiation of the increase in temperature and deformation is situated at 45,000 cycles, which is again higher than the 27,500 cycles found for the corresponding rail shear test. However, an increase in frequency tends to yield an increase in lifetime, which corresponds with the rail shear test.

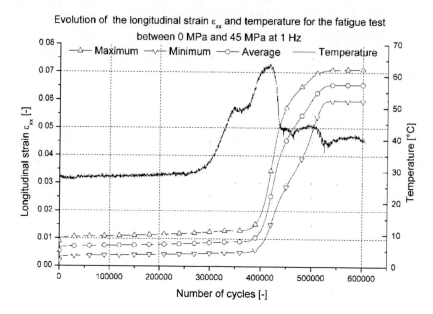

Figure 42. Maximum, minimum and mean value of the longitudinal strain as a function of the number of cycles for a 0-45 MPa tensile fatigue test at 1 Hz.

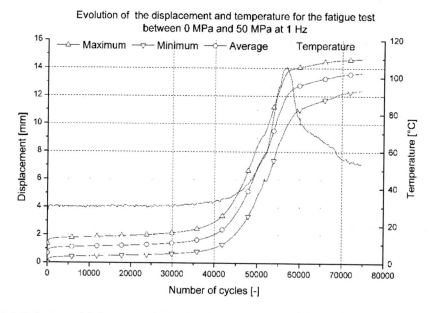

Figure 43. Maximum, minimum and mean value of the displacement as a function of the number of cycles for a 0-50 MPa tensile fatigue test at 1 Hz.

Figure 44 shows the results from two of the conducted 0-50 MPa@2 Hz experiments. It can be remarked that the behaviour is very similar to the three-rail shear tests. The sudden jump occurs at 70,000 cycles for N2 and at 125,000 cycles for N5, which is higher than the corresponding rail shear tests, where the jump occurred at 45,000 and 50,000 cycles.

However, an increase in frequency again tends to yield an increase in lifetime, corresponding with the rail shear test.

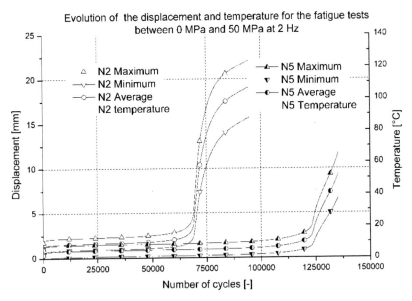

Figure 44. Maximum, minimum and mean value of the displacement as a function of the number of cycles for a 0-50 MPa tensile fatigue test at 2 Hz.

Finally, the entire measured evolution of the displacement for the 0-50 MPa@2 Hz from specimen N2 (see Figure 44) is given in Figure 45. As can be seen, a different behaviour manifests itself after the rise in temperature and decrease in stiffness.

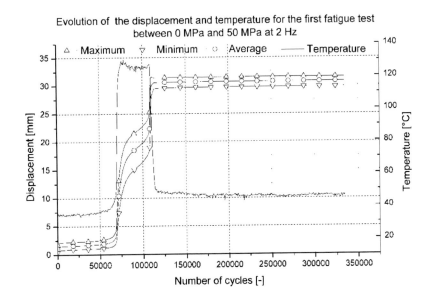

Figure 45. Maximum, minimum and mean value of the displacement as a function of the number of cycles for the preliminary 0-50 MPa tensile fatigue test at 2 Hz.

The evolution of both displacement and temperature is very similar to the three rail shear test until the point of the sudden rise in temperature and deformation, which occurs here at about 60,000 cycles. For the three-rail shear test, the specimen failed soon after the rise in temperature, but for the [(+45°,-45°)]$_{4s}$ tensile test, a new regime initiates. After the sudden rise in temperature, which again exceeds the softening temperature of the matrix, the specimen cools down and no further increase in displacement or stiffness degradation manifests itself. This can be explained by re-observing the images of the specimen after the test (Figure 39).

This 'dog-bone' like shape arose during the heating of the specimen. Because of the softening of the matrix, the fibres were able to realign considerably with respect to the loading direction. This can be seen in a detailed image of specimen N2 (see Figure 46).

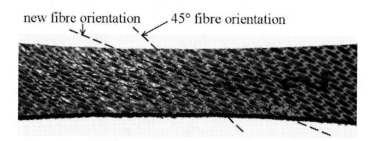

Figure 46. Illustration of the change in fibre orientation due to the softening of the matrix.

As such, a new load distribution occurs, with less shear load and more load along the fibres and Equation 12 for the shear stress and strain is no longer valid.

Therefore, after the sudden rise in temperature and displacement, no more increase in stiffness degradation or deformation occurs, because the specimen is now loaded more along the fibres.

During the increase in temperature, the force-displacement hysteresis loops become much wider, as illustrated in Figure 47, indicating that a lot of energy is dissipated, both in damage and of course in the heating of the specimen.

In conclusion, the same remarks as with the three-rail shear test could be made. Increasing the maximum load and decreasing the frequency both decrease the lifetime of the specimen. The point in time where both the sudden temperature increase as the stiffness decrease occur, is situated a little later than in the three-rail shear test, which can been seen in Figure 48. Also, after this increase, the specimen does not fail, but a new regime re-instates itself, due to a re-alignment of the fibres along the loading direction, because of the softening of the matrix.

CREEP LOADING CONDITIONS

Three-Rail Shear Setup

Since all fatigue experiments were done with a shear stress between 0 MPa and a maximum value τ_{max}, the specimen is subjected to an average load of half of τ_{max}.

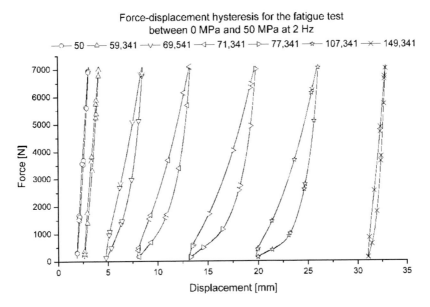

Figure 47. Force as a function of the displacement for a few measurements of the tensile fatigue test on specimen N2.

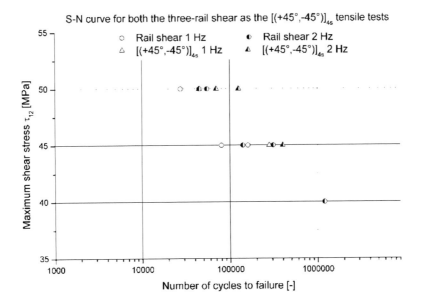

Figure 48. S-N curve for all fatigue experiments discussed in this chapter.

To verify whether the detected deformation during the fatigue loading originates from fatigue damage, a creep test was performed. The specimen was instrumented with two strain gauges and a constant shear stress of 25 MPa was applied. The latter corresponds with the fatigue tests with a maximum shear stress of 50 MPa, which are the highest fatigue loading conditions for this study. The load was applied for about 700,000 seconds, which equals

700,000 cycles at 1 Hz or 1,400,000 cycles at 2 Hz. The evolution of the shear strain γ_{12} is depicted in Figure 49. As can be seen, the absolute increase in shear strain is less than 0.0008.

Since the shear strain could not be registered during the fatigue experiments, only information about the displacement is available. For the latter, an increase of only 0.05 mm over 700,000 seconds occurred, so it may be concluded that the increase in displacement during the fatigue test is mainly caused by fatigue damage, since the measured increase is larger.

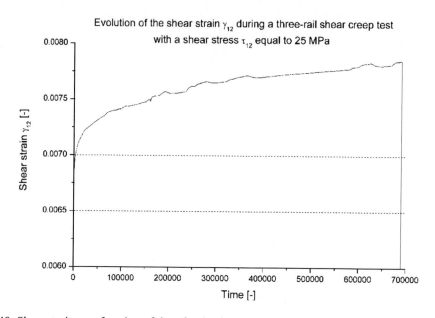

Figure 49. Shear strain as a function of time for the three-rail shear creep test.

[(+45°,-45°)]$_{4s}$ Experiments

Again to verify if the occurring growth in deformation is due to fatigue damage, a creep test was performed. The same constant shear stress as with the rail shear test was applied, namely 25 MPa, in order to compare the results; the test was stopped after 600,000 seconds, corresponding with 600,000 cycles for a 1 Hz test and 1,200,000 cycles for a 2 Hz test. Figure 50 depicts the shear strain during this experiment and again, very limited growth is measured, only 0.0008 absolute increase in shear strain, which corresponds excellently with the rail shear test.

To be able to assess the creep in the fatigue results, where the displacement was illustrated, the evolution of the occurring displacement during the creep test was also registered and an increase of 0.05 mm was measured over the 600,000 seconds, meaning that the occurring deformation during the fatigue is dominated by fatigue damage.

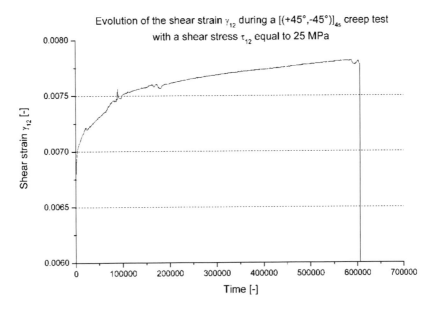

Figure 50. Shear strain as a function of time for the [(+45°,-45°)]₄ₛ tensile test.

FINITE ELEMENT MODELLING OF THE QUASI-STATIC NONLINEAR BEHAVIOUR

In this paragraph, the mechanical behaviour of the carbon fabric-reinforced polyphenylene sulphide under in-plane shear loading is modelled for the quasi-static loading conditions. The authors have already modelled the nonlinear shear stress-strain behaviour of a glass fibre-reinforced epoxy, by performing [+45°/-45°]$_{ns}$ tensile tests and 10° off-axis tests [7, 35]. The carbon fabric-reinforced PPS, which is considered here is of a totally different nature than the one used in the previous study [7, 35]: (i) carbon fabric versus unidirectional glass fibre-reinforcement and (ii) thermoplastic matrix versus thermosetting epoxy. However, rather than developing an entirely new model, the same model as given in [35] is used to prove that the developed approach may be considered applicable to a wide range of materials. Once the material constants are determined by optimising the model, the quasi-static experiments, both the three-rail shear test and the [(+45°,-45°)]₄ₛ tensile test, are simulated to verify whether the model can predict the nonlinear behaviour. All data necessary for the material model are derived from the experiments described in previous paragraphs. The finite element software ABAQUS™/standard was chosen for all finite element calculations in this research. This software also has the ability to implement a user-defined material model, using the UMAT routine, which requires the FORTRAN programming language.

Modelling the Quasi-Static Experiments

The Material Model

As mentioned in the introduction of this section, a previously derived material model [7, 35] will be used in this manuscript, to prove the wide applicability of the model. The shear stress-strain relationship is given by:

$$\tau_{12} = G_{12}^0 (1 - D_{12})(\gamma_{12}^{total} - \gamma_{12}^{perm})$$
$$D_{12} = 1 - \frac{G_{12}^*}{G_{12}^0} \quad (17)$$

In this equation:

- G_{12}^0 is the initial shear stiffness
- G_{12}^* is the shear stiffness of the damaged material
- D_{12} is the damage parameter, which indicates the stiffness degradation
- γ_{12}^{total} is the total shear strain, given by the sum of the elastic and the permanent shear strain
- γ_{12}^{perm} is the permanent shear strain

Modelling the shear behaviour using stiffness degradation and permanent shear strains was also done in [12, 13, 14]. Figure 51 shows how the values of G_{12}^* and γ_{12}^{perm} can be derived from the experimental data. A similar approach was used in [8, 36, 37].

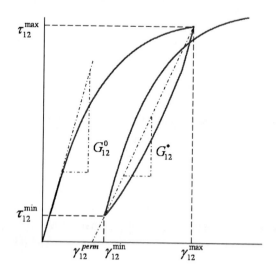

Figure 51. Definition of the shear modulus G_{12}^* and the permanent shear strain γ_{12}^{perm}.

Using these definitions, γ_{12}^{perm} and D_{12} (Figure 53) show the following evolutions (Figure 52 and Figure 53 respectively) for the cyclic rail shear experiments, described in paragraph 5.1.

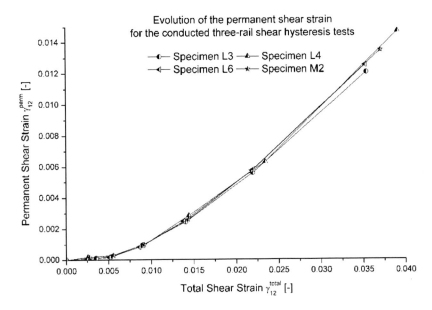

Figure 52. Evolution of the permanent shear strain during the three-rail shear hysteresis experiment.

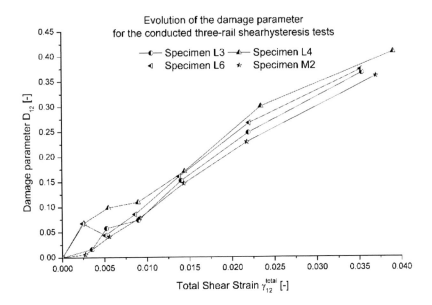

Figure 53. Evolution of the damage parameter during the three-rail shear hysteresis experiment.

It should be noted that there is a significant scatter on the damage parameter D_{12} for low total shear strains. This is due to the fact that the strains are very low and therefore, the determination of the slope of the hysteresis loop, as described in [7], is a lot more sensitive to noise and scatter on the strain measurement. These data will be used for determining the material constants in section 8.1.2.

For the $[(+45°,-45°)]_{4s}$ tensile tests, the following evolutions for permanent deformation (Figure 54) and stiffness degradation (Figure 55) are obtained from the cyclic experiments described in paragraph 5.2.

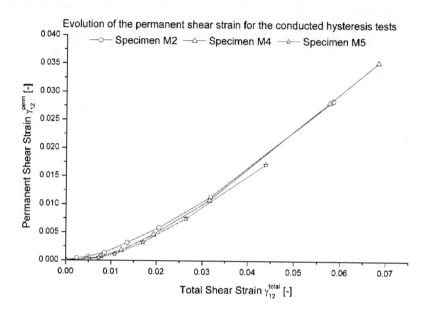

Figure 54. Evolution of the permanent shear strain during the tensile hysteresis experiment.

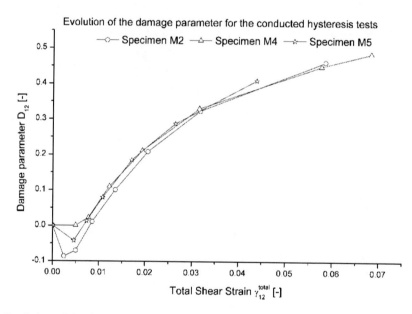

Figure 55. Evolution of the damage parameter during the tensile hysteresis experiment.

The results are very reproducible for both damage parameters. Again, there is some scatter on the damage parameter D_{12} for low values of the total shear strain, but as mentioned before, this is due to the fact that the strains are very low and therefore, the determination of the slope of the hysteresis loop, as described in [7] is more sensitive to scatter and noise on the measured strains. It should be noted that these evolutions show very good correspondence with the damage parameters obtained from the three-rail shear tests. These data will also be used for determining the material constants in the following section.

Implementing the Material Model

Several assumptions have been made to establish the models for the shear damage D_{12} and permanent shear strain γ_{12}^{perm}:

- both damage variables show an exponential behaviour,
- the value of both variables must be monotonically increasing,
- the shear modulus degradation should depend on the elastic part $\gamma_{12}^{e} = \gamma_{12} - \gamma_{12}^{p}$ of the total shear strain γ_{12},
- the accumulation of permanent shear strain γ_{12}^{p} is assumed to depend on the total shear strain γ_{12}.

On the basis of these equations, the following phenomenological model has been put forward [35]:

$$\frac{d\gamma_{12}^{perm}}{d\gamma_{12}^{total}} = C_{1} \cdot \left|\gamma_{12}^{total}\right| \cdot \exp\left(C_{2} \cdot \left|\gamma_{12}^{perm}\right|\right)$$

$$\frac{dD_{12}}{d\gamma_{12}^{elast}} = C_{3} \cdot \exp\left(C_{4} \cdot D_{12}\right) \tag{18}$$

where C_1 till C_4 are material constants and still need to be determined. The permanent shear strain is assumed to depend on the total shear strain, and the increase shows an exponential behaviour. The factor $\left|\gamma_{12}^{total}\right|$ is added so that the increase in γ_{12}^{perm} is lower for low values of the total shear strain, but higher for higher values.

The damage parameter D_{12} is assumed to depend on the elastic part of the strain and the increase also shows an exponential behaviour. Since the growth depends on the elastic part of the strain, first the permanent deformation must be determined, after which D_{12} can be calculated.

In order to implement the material model in UMAT, it must first be written using finite differences. This yields for Equation 18, with $\gamma_{12}^{elast} = \gamma_{12}^{total} - \gamma_{12}^{perm}$:

$$
\begin{aligned}
\gamma_{12}^{perm,new} &= \gamma_{12}^{perm,old} \\
&+ C_{1} \cdot \left|\gamma_{12}^{total,new}\right| \cdot \exp\left(C_{2} \cdot \left|\gamma_{12}^{perm,old}\right|\right) \cdot \left(\gamma_{12}^{total,new} - \gamma_{12}^{total,old}\right) \\
D_{12}^{new} &= D_{12}^{old} \\
&+ C_{3} \cdot \exp\left(C_{4} \cdot D_{12}^{old}\right) \cdot \left[\left(\left|\gamma_{12}^{total,new}\right| - \left|\gamma_{12}^{perm,new}\right|\right)\right. \\
&\left. - \left(\left|\gamma_{12}^{total,old}\right| - \left|\gamma_{12}^{perm,old}\right|\right)\right]
\end{aligned}
\tag{19}
$$

where the superscript 'old' refers to the values obtained from the previous time step and the superscript 'new' refers to the value necessary for the current time step.

Next, Equation 17 can be implemented. However, after some test runs with UMAT, it became clear that a converging solution could only be found if strain increments are used rather than the total strains, meaning that the total stress is calculated using stress increments. As such, Equation 17 must be rewritten with finite differences. This yields:

$$\tau_{12}^{new} = G_{12}^0.(1-D_{12}^{new}).(\gamma_{12}^{total,new} - \gamma_{12}^{perm,new})$$
$$\tau_{12}^{old} = G_{12}^0.(1-D_{12}^{old}).(\gamma_{12}^{total,old} - \gamma_{12}^{perm,old}) \quad (20)$$

Hence, the following equation is derived for the stress:

$$\tau_{12}^{new} = \tau_{12}^{old}$$
$$+ G_{12}^0.(1-D_{12}^{new}).\left[(\gamma_{12}^{total,new} - \gamma_{12}^{total,old})\right.$$
$$\left. -(\gamma_{12}^{perm,new} - \gamma_{12}^{perm,old})\right] \quad (21)$$
$$- G_{12}^0.(D_{12}^{new} - D_{12}^{old}).(\gamma_{12}^{total,old} - \gamma_{12}^{perm,old})$$

The sum in Equation 21 consists of two terms: the first is the expected one, giving a stress increment using the increment in elastic strain. The second term is the result of the use of stress increments and must not be forgotten.

Next, the modelling in ABAQUS™ of the two test geometries is discussed.

Modelling the Three-Rail Shear Test

For these simulations, only the loaded part of the specimen was modelled and because of symmetry, only half of the specimen was drawn, Figure 56 depicts the used mesh and boundary conditions.

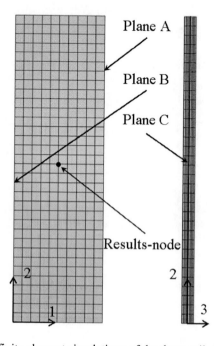

Figure 56. Used mesh for the finite element simulations of the three-rail shear test.

The specimen was meshed using a 3D quadratic brick element with reduced integration; the size of the mesh was 3 mm and eight elements through the thickness were used. An extra simulation was performed with a mesh size of 1 mm, also to assess the influence of the mesh size on the output of the material model. The width of the modelled specimen was 30 mm and the height 100 mm, as was the case in the experiments. The following boundary conditions were applied:

- Plane A, the right side of the loaded zone, was fixed along the 1- and 2-axis.
- Plane B, the left side of the loaded zone, was fixed along the 1-axis and given a displacement of 3 mm along the 2-axis. This value corresponds with the value at which point the strain gauges in the experiments de-bonded or saturated.
- Plane C, the central plane of the loaded zone, was fixed along the 3-axis.

Since it is a 3D analysis, there are no rotational degrees of freedom. Because of the large deformations, a geometrically nonlinear simulation was performed.

Modelling the [(+45°,-45°)]4s Test

For these simulations, only the loaded part of the specimen was modelled, resulting in a rectangular specimen of 150 mm by 30 mm. Symmetry was not used, since this would yield images of stress distributions which are a little more difficult to interpret than if the entire specimen is modelled. Figure 57 depicts the used mesh and boundary conditions.

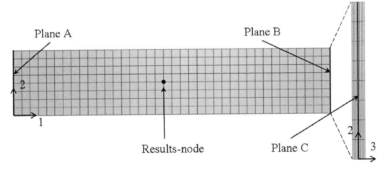

Figure 57. Used mesh for the finite element simulations of the [(+45°,-45°)]$_{4s}$ tensile test.

The specimen was meshed using a 3D quadratic brick element with reduced integration; the size of the mesh was 4 mm and eight elements through the thickness were used. An extra simulation was performed with a mesh size of 2 mm, to assess whether the mesh size has significant influence on the model. The following boundary conditions were applied:

- Plane A, the left side of the loaded zone, was fixed along the 1- and 2-axis, simulating the rigid grip.
- Plane B, the right side of the loaded zone, was fixed along the 1-axis and given a displacement of 10 mm along the 2-axis. This value corresponds with the value at which point the strain gauges in the experiments de-bonded or saturated.
- Plane C, the central plane of the loaded zone, was fixed along the 3-axis.

Since it is a 3D analysis, there are no rotational degrees of freedom. Because of the large deformations, a geometrically nonlinear simulation was performed.

Discussion of the Simulations

Of course, first the material constants C_1 till C_4 in Equation 18 need to be determined. These values have been determined by the combined optimisation of the evolutions of the $\tau_{12} - \gamma_{12}$ curve, γ_{12}^{perm} and D_{12} on the experimental results, given in Figure 52 till Figure 55.

This yielded the following combination of material constants:

$$C_1 = 29 \quad C_2 = -35 \quad C_3 = 16 \quad C_4 = -0.1 \tag{22}$$

For both the three-rail shear and the $[(+45°,-45°)]_{4s}$ tensile test simulation, the evolution of the shear stress, shear strain, permanent shear strain γ_{12}^{perm} and damage parameter D_{12} was plotted for a node in the centre of the specimen (Results-node in Figure 56 and Figure 57), so that the stress concentrations near the clamps did not have any influence. These results are given in Figure 58, Figure 59 and Figure 60 respectively.

In the legend, the parameter M represents the mesh size used in the finite element simulation, for instance M = 3 means a mesh size of 3 mm is used. It should be remarked that there is an excellent correspondence for all three evolutions and that refining the mesh, for both the three-rail shear as the $[(+45°,-45°)]_{4s}$ simulations does not have a significant effect.

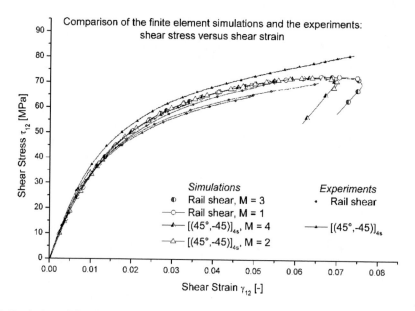

Figure 58. Evolution of the shear stress as a function of the shear strain for both the simulations and the experiments.

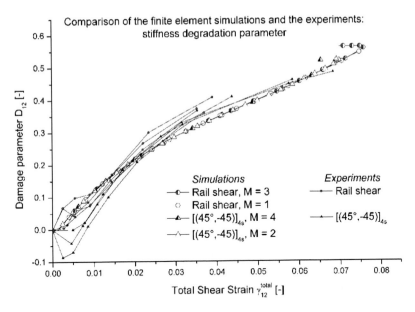

Figure 59. Evolution of the permanent shear strain for both the simulations and the experiments.

In the shear stress-strain evolution, depicted in Figure 58, it can be seen that after a certain strain level is reached, the total strain tends to decrease, as well as the corresponding shear stress. The corresponding damage and permanent shear strain remain constant during this decrease.

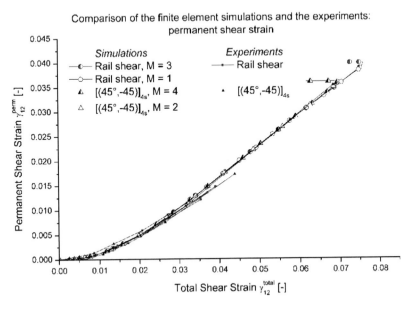

Figure 60. Evolution of the damage parameter for both the simulations and the experiments.

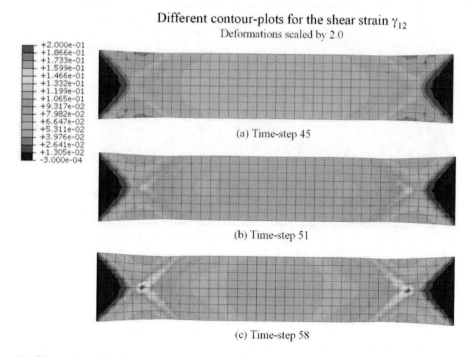

Figure 61. Illustration of the decrease in the shear strain in the centre of the specimen.

This happens for both types of simulations and is the result of a stress relief in the centre of the specimen, due to localization of the total shear strain elsewhere in the specimen. This can be easily shown in the [(+45°,-45°)]$_{4s}$ simulations. Figure 61 depicts the total shear strain at three well chosen time-steps. Time-step 45 represents the end of the increase in shear strain, Time step 58 is the last time-step available and time-step 51 is chosen for an intermediate distribution plot. The deformation is scaled by 2.0 to visualise the reason more clearly.

Figure 62. Illustration of the experimental localization of the total shear strain.

From these different stages in time, it can clearly be seen that the fast narrowing of the specimen near the clamped ends results in a high increase in shear strain. Since the simulation is done displacement-controlled, corresponding with the experiment, the shear strain in the centre of the specimen is allowed to relax. Given the fact that the permanent shear strain is not allowed to decrease, this renders a decrease in elastic shear strain and as a result, a decrease in shear stress. Figure 62 shows this localization in three stages as it could be seen in the experiment.

This decrease in shear stress and strain was not seen in the experiments, but the localization of the total shear strain occurred after the strain gauge saturated and the strain measurement was stopped, meaning that the material model predicts this effect too early.

Finally, in Figure 63 and Figure 64, the distributions of the shear stress and the damage model parameters are given for the time step at which maximum stress in the central area is reached. The shear stress distribution is very uniform, except near the clamped ends, where a cross-like shape can be distinguished. However, this uniform distribution is still the main reason why the $[(+45°,-45°)]_{4s}$ is chosen to characterise the shear behaviour of a composite. The permanent shear strain is also relatively uniform, although higher values are reached where the specimen is most likely to narrow. The damage parameter D_{12} also indicates where failure is most likely to occur.

For the three-rail shear test, also a very uniform shear stress distribution can be seen. The distribution of both damage and permanent deformation clearly indicate where failure will start, namely in the vicinity of the clamps. This was also derived by Lessard et al. [20]. A possibility for future research could be to investigate whether the use of notches, as was done in [20] would have the same effects on a fabric-reinforced composite.

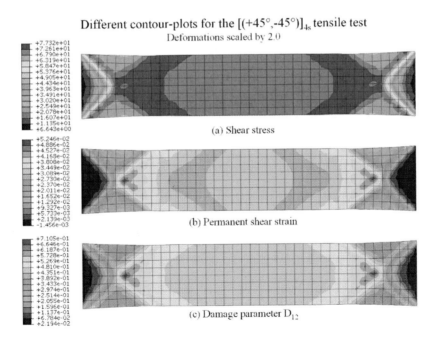

Figure 63. Distribution of the (a) shear stress, (b) permanent shear strain and (c) damage parameter D_{12} for the $[(+45°,-45°)]_{4s}$ simulation.

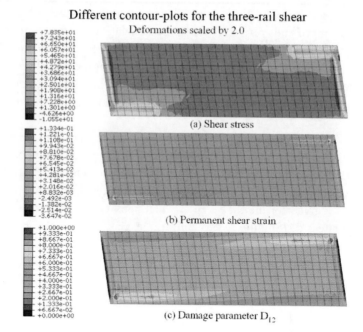

Figure 64. Distribution of the (a) shear stress, (b) permanent shear strain and (c) damage parameter D_{12} for the three-rail shear simulation.

Finally, in Figure 65, the calculated relation between shear strain and displacement is depicted for the three-rail shear test, as a verification for the theoretical deduction given in paragraph 5.3. It can be seen that there is a slight difference, possibly because in the theoretical deduction, where a factor of 0.033 was derived, small displacements were assumed, whereas this simulation has taken geometrical nonlinearities into account.

To assess the effect of the boundary condition on this factor, a second simulation was performed, using the same boundary conditions as in [20], which mean that for plane A, only zero displacement along the 2-axis was prescribed (BC2). This, however, does not seem to have any influence on the shear strain-displacement relationship.

CONCLUSION

A modified design of the standard three-rail shear setup, as described in the "ASTM D 4255/D 4255M standard test method for in-plane shear properties of polymer matrix composite materials by the rail shear method" has been presented. This new design uses friction and geometrical gripping, without the need of drilling holes through the composite specimen. From the performed quasi-static tests, it could be concluded that both sides of the specimen are loaded symmetrically and no buckling occurs.

Fatigue tests have been performed to assess the behaviour of the grips and to investigate the material behaviour under fatigue loading conditions. With respect to the grips, excellent results were achieved; the specimen fails under shear loading conditions in the loaded area.

The material behaviour was investigated by means of quasi-static, hysteresis, fatigue and creep tests.

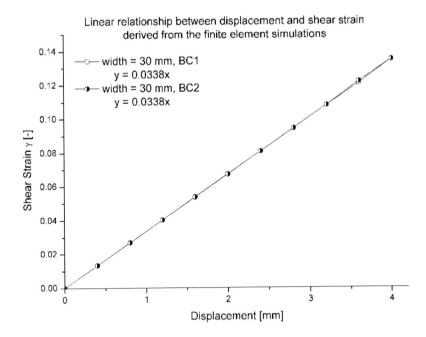

Figure 65. Numerical evolution of the shear strain as function of the displacement.

For the quasi-static tests a highly nonlinear shear stress-strain relationship was found. The stress-strain data were obtained by performing both the $[(+45°,-45°)]_{4s}$ tensile test, as described in the ASTM D3518/D3518M-94 (2007) Standard and the three-rail shear test, as described in the ASTM D 4255/D 4255M Standard. Both types of experiments yielded very similar results. Quasi-static cyclic loadings were applied, to derive the evolution of the permanent deformation and the stiffness degradation, which both lead to the highly nonlinear behaviour. Again, the results from both the three-rail shear test and the $[(+45°,-45°)]_{4s}$ tensile test corresponded very well.

Using the obtained stress-strain data, the nonlinear shear stress-strain behaviour was modelled using a previously derived material model, which has two damage parameters: (i) D_{12} which represents the shear stiffness degradation and (ii) the permanent shear strain γ_{12}^{perm}, which represents the permanent shear deformation. This material model was then implemented in the finite element software ABAQUS™/Standard with the UMAT routine and both test setups were modelled. The four material constants of the material model were optimised based on the experimentally derived evolutions of shear stress-strain, D_{12} and γ_{12}^{perm}. Afterwards, the results from the simulations were compared with the experimental curves and a very good agreement was achieved.

All fatigue experiments were done with a stress ratio equal to 0, meaning that the shear stress varied between 0 MPa and a certain maximum. For both the three-rail shear test and the $[(+45°,-45°)]_{4s}$ tensile test, the same conclusions could be drawn with respect to the frequency and the maximum shear stress: (i) increasing the maximum shear stress decreases fatigue lifetime; (ii) increasing the frequency seems to increase the fatigue lifetime. To explain this frequency effect, a more profound study of the micro-mechanical behaviour of this specific material will be necessary.

The results from both types of tests show good correspondence, although the lifetime was always higher for the $[(+45°,-45°)]_{4s}$ tensile test. The main advantage of the new design of the three-rail shear test over the $[(+45°,-45°)]_{4s}$ tensile test is the absence of free-edge effects. The material under study was not very sensitive to delaminations, but for most fibre-reinforced epoxies, the damage behaviour in the $[(+45°,-45°)]_{4s}$ tensile test will be dominated by delaminations, caused by these free-edges. Due to the absence of these edge effects in the three-rail shear setup, the damage behaviour of these reinforced epoxies will be dominated by in-plane shear, also because there are no longer holes from which the delaminations can grow.

The material behaviour itself can be described in three stages: (i) run-in of the fatigue test, where a certain amount of permanent deformation occurs without an increase in temperature; (ii) a steady-state phase, where there is a gradual increase in permanent deformation, without an increase in temperature and (iii) the end-of-life phase, where there is a sudden increase in temperature, above the softening temperature of the matrix and a sudden growth in permanent deformation. It can of course be expected that the increase in temperature will accelerate the damage occurring under fatigue loading conditions, but it would be very difficult to separate both effects, since keeping the specimen at a constant temperature is more easily said than done.

One remark should be made: for the three-rail shear test, the end-of-life phase ended with failure of the specimen, whereas for the $[(+45°,-45°)]_{4s}$ test, a 'dog-bone' like shape was formed, with re-alignment of the fibres along the loading direction.

Finally, creep tests were performed to verify whether the constant growth of permanent deformation during the 'steady-state' phase in the fatigue life was due to fatigue damage. It could be concluded that the material did creep, but the main damage occurred due to the fatigue loading conditions. Yet again, the creep results corresponded excellently for both types of testing. Over 700,000 seconds, an absolute increase in shear strain of only 0.0008 at a constant load of 25 MPa was measured.

REFERENCES

[1]   Tarnopol'skii YM, Arnautov AK, Kulakov VL, Methods of determination of shear properties of textile composites. *COMPOSITES PART A-APPLIED SCIENCE AND MANUFACTURING* 30 (7): 879-885 1999.

[2]   Whitney, J.M., Daniel, I.M. and Pipes, R.B. (1984). Experimental mechanics of fiber reinforced composite materials. Chapter 4: Composite characterization. Connecticut, The Society for Experimental Mechanics, pp. 160-202.

[3]   Odom E.M., Blackketter D.M. and Suratno B.R., Experimental and analytical investigation of the modified Wyoming shear-test fixture. *EXPERIMENTAL MECHANICS* 34 (1): 10-15 MAR 1994.

[4]   Odegard G. and Kumosa M. Determination of shear strength of unidirectional composite materials with the Iosipescu and 10° off-axis shear tests. *COMPOSITES SCIENCE AND TECHNOLOGY* 60 (2000) 2917-2943.

[5]   Pierron, F and Vautrin A. New ideas on the measurement of the in-plane shear strength of unidirectional composites. *JOURNAL OF COMPOSITE MATERIALS* 31(9) pp. 889-895, 1997.

[6] Pierron, F and Vautrin A. The 10° off-axis tensile test: a critical approach. *COMPOSITES SCIENCE AND TECHNOLOGY* 56 (1996) 483-488.

[7] W. Van Paepegem, I. De Baere and J. Degrieck, Modelling the nonlinear shear stress-strain response of glass fibre-reinforced composites. Part I: Experimental results, *COMPOSITES SCIENCE AND TECHNOLOGY* 66 (10): 1455-1464 AUG 2006.

[8] Echtermeyer, A.T. (1994). Evaluation of the [±45]s inplane shear test method for composites reinforced by multiaxial fabrics. In : Hogg, P.J., Schulte, K. and Wittich, H. (eds.). ECCM-CTS 2 : Composites testing and standardisation. European Conference on composites testing and standardisation, 13-15 September 1994, Hamburg, Germany, Woodhead Publishing Limited, pp. 305-313.

[9] Shalom S, Harel H, Marom G, Fatigue behaviour of flat filament-wound polyethylene composites. *COMPOSITES SCIENCE AND TECHNOLOGY* 57 (9-10): 1423-1427 1997.

[10] Thollon, Y. Hochard, C. Herman, M. And Charles J.P. Mechanical behaviour of unbalanced woven fabric composite up to first failure. 12th European Conference on Composite Materials (ECCM 12) 29 August – 2 September 2006, Biarritz, France.

[11] Payan J, Hochard C, Damage modelling of laminated carbon/epoxy composites under static and fatigue loadings. *INTERNATIONAL JOURNAL OF FATIGUE* 24 (2-4): 299-306 FEB-APR 2002.

[12] Hochard C., Lahellec N. and Bordreuil C. A ply scale non-local fibre rupture criterion for CFRP woven ply laminated structures. *COMPOSITE STRUCTURES* 80 (3): 321-326 OCT 2007.

[13] Hochard C., Aubourg P.A. and Charles J.P. Modelling of the mechanical behaviour of woven-fabric CFRP laminates up to failure. *COMPOSITES SCIENCE AND TECHNOLOGY* 61 (2): 221-230 2001.

[14] Thollon, Y. Hochard, C. Herman, M. And Charles JP. Mechanical behaviour of unbalanced woven fabric composite up to first failure. 12th European Conference on Composite Materials (ECCM 12) 29 August – 2 September 2006, Biarritz, France.

[15] Khashaba UA, In-plane shear properties of cross-ply composite laminates with different off-axis angles. *COMPOSITE STRUCTURES* 65 (2): 167-177 AUG 2004.

[16] Maeda T, Baburaj V, Koga T, Evaluation of in-plane shear modulus of composite laminates using holographic interferometry; *OPTICAL ENGINEERING* 36 (7): 1942-1946 JUL 1997.

[17] Hussain AK, Adams DF, The Wyoming-modified two-rail shear test fixture for composite materials. *JOURNAL OF COMPOSITES TECHNOLOGY and RESEARCH* 21 (4): 215-223 OCT 1999.

[18] Hussain AK, Adams DF , Experimental evaluation of the Wyoming-modified two-rail shear test method for composite materials. *EXPERIMENTAL MECHANICS* 44 (4): 354-364 AUG 2004.

[19] Hussain AK, Adams DF, Analytical evaluation of the two-rail shear test method for composite materials. *COMPOSITES SCIENCE AND TECHNOLOGY* 64 (2): 221-238 FEB 2004.

[20] Lessard, L.B., Eilers O.P., Shokrieh M.M., Testing of in plane shear properties under fatigue loading. *JOURNAL OF REINFORCED PLASTICS AND COMPOSITES* 14 (9): 965-987 SEP 1995.

[21] Ferry L, Perreux D, Varchon D, Sicot N, Fatigue behaviour of composite bars subjected to bending and torsion. *COMPOSITES SCIENCE AND TECHNOLOGY* 59 (4): 575-582 1999.

[22] El-Assal, Ahmed M., Khashaba, U. A., Fatigue analysis of unidirectional GFRP composites under combined bending and torsional loads. *COMPOSITE STRUCTURES* 79 (4): 599-605 AUG 2007.

[23] Qi, DT, Cheng, GX, Fatigue behavior of filament-wound glass fiber reinforced epoxy composite tension/torsion biaxial tubes under loading. *POLYMER COMPOSITES* 28 (1): 116-123 FEB 2007.

[24] Kawakami H, Fujii TJ, Morita Y, Fatigue degradation and life prediction of glass fabric polymer composite under tension torsion biaxial loadings. *JOURNAL OF REINFORCED PLASTICS AND COMPOSITES* 15 (2): 183-195 FEB 1996.

[25] Fujii T and Lin F, Fatigue behaviour of a plain-woven glass fabric laminate under tension-torsion biaxial loading. *JOURNAL OF COMPOSITE MATERIALS* 29 (5): 573-590 1995.

[26] http://www.TenCate.com, 01/10/2010.

[27] http://www.tech.plym.ac.uk/sme/MATS324/MATS324C2%20fabrics.htm, 01/10/2010.

[28] http://www.fibermaterialsinc.com/2Dws.htm, 01/10/2010.

[29] CETEX – PPS Guide Lines, Ten Cate Advanced Composites.

[30] ASTM D3518-76, standard practice for in-plain shear stress-strain response of unidirectional reinforced plastics, 1982.

[31] Van Paepegem W., Mechanics of Materials, Faculty of Engineering, Ghent University, 2005.

[32] Cho C.D., Holmes J.W., Barber J.R., Estimation of interfacial shear in ceramic composites from frictional heating measurements, *JOURNAL OF THE AMERICAN CERAMIC SOCIETY* 74 (11): 2802-2808 NOV 1991.

[33] Jacobsen TK, Sorensen BF, Brondsted P, Measurement of uniform and localized heat dissipation induced by cyclic loading, *EXPERIMENTAL MECHANICS* 38 (4): 289-294 DEC 1998.

[34] Gamstedt EK, Redon O, Brondsted P, Fatigue dissipation and failure in unidirectional and angle-ply glass fibre/carbon fibre hybrid laminates, *EXPERIMENTAL TECHNIQUES AND DESIGN IN COMPOSITE MATERIALS 5 KEY ENGINEERING MATERIALS* 221-2: 35-47 2002.

[35] Van Paepegem W., De Baere I. and Degrieck J., Modelling the nonlinear shear stress-strain response of glass fibre-reinforced composites. Part II: Model development and finite element simulations, *COMPOSITES SCIENCE AND TECHNOLOGY* 66 (10): 1465-1478 AUG 2006.

[36] Lafarie-Frenot, M.C. and Touchard, F. Comparative in-plane shear behaviour of long-carbon-fibre composites with thermoset of thermoplastic matrix. *COMPOSITES SCIENCE AND TECHNOLOGY* 52 (1994) 417-425.

[37] Ladeveze, P. and Le Dantec, E. Damage modelling of the elementary ply for laminated composites. *COMPOSITES SCIENCE AND TECHNOLOGY* 43 (1992) 257-267.

[38] Van Paepegem W. Development and finite element implementation of a damage model for fatigue of fibre-reinforced polymers. *Doctoral thesis at Ghent University*, 2002. ISBN 90-76714-13-4.

In: Fiber-Reinforced Composites
Editor: Qingzheng Cheng

ISBN: 978-1-61470-303-7
© 2012 Nova Science Publishers, Inc.

Chapter 8

ON TOUGHENING OF EPOXIES AND THEIR CARBON FIBRE-REINFORCED COMPOSITES

Jin Zhang[*], Tong Lin and Xungai Wang
Centre for Material and Fibre Innovation, Deakin University,
Geelong, VIC 3217, Australia

ABSTRACT

Epoxy resins were introduced commercially in the late 1940s and have been used widely in industrial applications, including fibre-reinforced polymer matrix composites, adhesives and coatings. While epoxies have numerous advantages, they suffer from brittleness as a result of their high crosslink polymer structure. The present work is a brief review of the toughening methods used previously to improve the crack growth resistance of epoxies and their fibre reinforced composites. Toughening agents discussed include rubbers, inorganic fillers, thermoplastics and hyperbranched polymers. The chapter provides the state-of-the-art knowledge in this field as well as indicates the future directions towards improving the fracture toughness of epoxies.

1. INTRODUCTION

Epoxies are predominately used as matrices in aerospace polymer matrix composites, because of their excellent adhesion to carbon fibres, low shrinkage on curing, high chemical and corrosion resistances and good mechanical and thermal properties [1, 2]. However, their brittle nature resulted from the highly crosslinked structure limits their applications in high performance composites where superior impact resistance is critical. Modification of thermosetting epoxy matrices is crucial in material development aiming at reducing the delamination tendency of composite structures [3].

Many types of filler from inorganic materials, rubbers and thermoplastics have been used previously for toughening epoxy resins. Among them, thermoplastic shows great advantages,

---

[*] Author to whom correspondence should be addressed; Email: jin.zhang@deakin.edu.au

since they can significantly toughen epoxies without compromising the thermal-mechanical properties, elastic modulus and yield stress, which is usually associated with rubber toughening [4]. For that reason, high performance thermoplastics have served as toughening agents for modifying brittle epoxies for over 20 years. Thermoplastic toughening is generally achieved through phase separation of an initially homogeneous thermoplastic/epoxy blend, which leads to the formation of different phase structures such as particulate (discrete thermoplastic particles dispersed in the epoxy matrix), co-continuous and phase inverted morphologies, depending on factors such as the composition of the initial blend, the cure temperature, the competition between the rate of cure and the rate of phase separation, and the diffusion rate [5, 6].

In this review, previous research on toughening of carbon/epoxy composites using thermoplastics, hyperbranched polymer and thermoplastic nanofibres is summarized. The phase morphology, thermal-mechanical property, toughening effect and mechanism are discussed for the toughening of multifunctional epoxies with different modifiers.

2. RUBBER AND INORGANIC TOUGHENING FILLERS

As early as the 1970's, Sultan et al. [7] first reported their work on using a soft rubbery particulate phase to improve the crack resistance of epoxy resins. Two main methods have been used to apply rubber to epoxy, such as using reactive oligomers and preformed elastomeric particles. In the first approach, toughening is achieved by the generation of a second particulate phase during curing of an epoxy that contains initially dissolved reactive oligomer. Liquid rubber copolymers of butadiene-acrylonitrile with various functional end groups, including carboxyl-terminated group (CTBN) [8, 9], amine terminated group (ATBN) [10], epoxy terminated group (ETBN) [11], hydroxyl-terminated group (HTBN) [12] and vinyl-terminated group (VTBN) [13], have been intensely studied as modification agents. In the second approach, preformed elastomeric particles such as core-shell particles with a rubbery core and a glassy shell have been employed, where the particle size, volume fraction and handling characteristics can be better controlled [14]. Recent progress has been made when block copolymers are incorporated into rubber modifiers which can self-assemble into hierarchical nanostructures such as vesicles, spherical micelles and worm-like micelles [13, 15, 16]. This field of research requires further study to examine their applicability in toughening of epoxies.

A number of theories have been proposed to explain the toughening mechanism of rubber modification. Particle cavitations and shear yielding is the most accepted theory currently. The cavitations of rubber particles releases the hydrostatic tension in resin matrix, with the stress state between voids being converted from a triaxial to a more uniaxial tensile state, which is favourable for the initiation of shear bands [17]. Improved toughness is obtained through effective energy dissipation at the crack tip as a result of shear yielding.

Rubber modifiers are able to increase the fracture toughness of epoxies effectively. In contrast to unmodified epoxies, however, they also cause deterioration in other important properties, such as modulus and glass transition temperature of the modified epoxies. As an alternative, engineering thermoplastics demonstrates great advantages by enhancing the fracture toughness as well as maintaining the desired properties, which is favourable when high thermal stability and toughness are both needed for industry applications.

Besides rubber, several inorganic toughening agents, such as glass beads, alumina, silica, barium titanate, dolomite, aluminium hydroxide have also been considered for toughening epoxies [18, 19]. To enhance the modulus, strength and toughness of rubber modified epoxies, inorganic fillers have been added to prepare hybrid systems. Recently, inorganic nanofillers have been widely investigated as the modifiers of epoxies. Polymer nanocomposites, as a new family of engineering materials, have attracted intense attention in both industry and research field. Depending on their nanostructures, nanofillers can be divided to three types. For example, montmorillonite clay falls into the two dimensional category [20], carbon nanotubes [21], carbon nanofibres and cellulose whiskers belong to the one dimensional nanofillers, and spherical silica nanoparticles are typical zero dimensional nanofillers. Apart from increased toughness in epoxies, significantly enhanced strength and modulus can be achieved, and new functionality may also be added to the system. Nanofillers constrain the matrix deformation less than the micro particles, because their smaller size allows them to integrate better into the polymer microstructure. They may influence the deformation of matrices on nano or micro scales, based on their more or less strong interaction with the matrices [22].

3. THERMOPLASTIC TOUGHENING AGENTS

Thermoplastic toughening has been widely used in aerospace industry because it enhances toughness without significantly compromising other desirable properties of epoxies such as high modulus and high glass transition temperature. The toughening result is achieved by phase separation of an initially homogeneous blend of thermoplastics and epoxies. The phase separation process of thermoplastic/epoxy blends has been studied extensively. Nucleation and growth (NG), spinodal decomposition (SD), or a mixed mode of both can occur during the separation process [5, 23]. When its concentration is low, the thermoplastic modifier tends to segregate from the matrix by the NG mechanism. For blends containing higher thermoplastic content, phase separation proceeds by more complex SD mechanism and leads to co-continuous or phase inverted morphology [24]. Different types of thermoplastics have been used as toughening agents in epoxies, including polyethersulphones [25, 26], polyetherimides [27], polyester [28], polyacryletherketone [29], and polyhydration [30].

3.1. The Influence of Processing Rates

The end mechanical properties are directly related to the phase behaviour of epoxy/thermoplastic blends. By controlling various parameters such as thermoplastic concentration, molar mass, curing temperature and reaction rate, it is possible to generate various phase morphologies, which result in different mechanical properties. Several studies have been conducted to investigate the influence of cure rate on the phase separation mechanism of thermoplastic modified epoxy systems. Jo et al. [31] applied a simulation approach to study the effect of reactivity on the phase separation behaviour. Girard-Reydet et al. [32] selected cure temperature as a parameter to change the cure rate. Peng et al. [33] changed the cure temperature and the amount of cure agent to modify the cure rate. Zhang et al. [34-36] used heating rate as a variable to investigate the effect of cure rate on the cure kinetics and phase morphology of polyether sulphone (PES) modified multifunctional epoxies, triglycidyl aminophenol (TGAP) and tetraglycidyl diamino diphenylmethane (TGDDM), cured with 4,4'-diamino diphenylsulfone (DDS). Thermoplastic/epoxy blends were prepared by solution casting from methylene chloride at room temperature. The blends were dried in a vacuum oven at 50 °C for 15 hours to remove the trace solvent. Polymer blends were cured with two heating rates, i.e. a low heating rate of 1.5 °C/min and a high heating rate of 10 °C/min. The cure temperature was 175 °C. The cure kinetics was investigated by performing real-time Fourier transform infrared spectroscopy (FTIR). The oxirane ring stretching mode at 908 cm^{-1} was used for measuring the conversion of epoxy groups.

For a TGAP/PES/DDS (100/30/30 w/w) blend, the cure kinetics following with different cure schedules can be seen in Figure 1. The slow temperature rise (1.5 °C/min) resulted in a very slow epoxy conversion which can almost be seen as zero in the first 80 min. The epoxy conversion of the blend heated at 10 °C/min reached a plateau when it was heated for 40 min. The final conversion after the blend was heated at 175 °C for 60 minutes was 0.865 for the cure with the heating rate of 10 °C/min, but was 0.807 for the cure with lower heating rate.

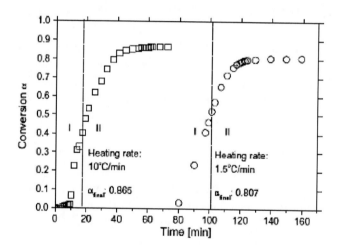

Figure 1. Cure kinetics followed with different cure schedules of the TGAP/PES/DDS (100/30/30) blend: I. heating stage (from 20 °C to 175 °C); II. isothermal curing stage (at 175°C) [34].

The Tg vs conversion curve was followed by DSC which is shown in Figure 2. The Tg vs conversion relationship shows increase in the concentration of branching points and high-functional crosslink as a function of conversion [37]. The data points were simulated by using an MMF function within the sigmoidal regression models' family which is shown in equation 1:

$$Y = \frac{ab + cx^d}{b + x^d} \quad (1)$$

where a, b, c and d are fitting parameters. The initial formulation exhibits a shift in the baseline at T_{g0} = -24.9 °C. T_g increases with cure time at the constant temperature (175 °C). After the blend is cured for 320 minutes, the T_g value increases to 194.8 °C. The conversion of the specimens is defined by:

$$\infty = 1 - \frac{-\Delta H_R}{-\Delta H} \qquad (2)$$

where ($-\Delta H$) is the reaction heat for the initial formulation, and ($-\Delta H_R$) is the residual reaction heat measured from the partially cured specimens.

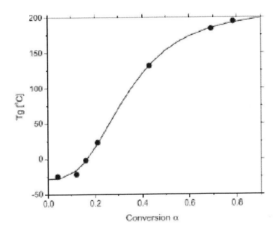

Figure 2. T_g as a function of conversion for the TGAP/PES/DDS (100/30/30) blend cured at 175°C for different times. The solid line is the curve simulated using equation 1 (The values of the fitting parameters are: a = -28.45, b = 0.046, c = 213.77, and d = 2.84) [34].

A phase contrast optical microscope was employed to investigate the morphology evolution. To slow down the phase separation process for observation, a specimen was chosen to be monitored at 160 °C (a heating rate of 10 °C/min was used to heat up the specimen). As shown in Figure 3, the blend remained homogeneous after 56 minute heating at 160 °C. Phase separation started at 70 minutes. A co-continuous phase structure was observed at 72 minutes, indicating that the phase separation was likely to take place via a spinodal decomposition mechanism. As the curing reaction proceeded, the periodic distance of the co-continuous structure increased. The growing macrophases connected to each other to form larger phases. In the mean while, fine particles started to appear and dispersed in both the dispersed macrophase and the matrix macrophase. These visible particles imply that second and possibly third stage of microphase separation continued to proceed after the primary macrophase separation. At 79 minutes, the morphology was finally fixed in which the irregular macrophases dispersed in a continuous matrix macrophase.

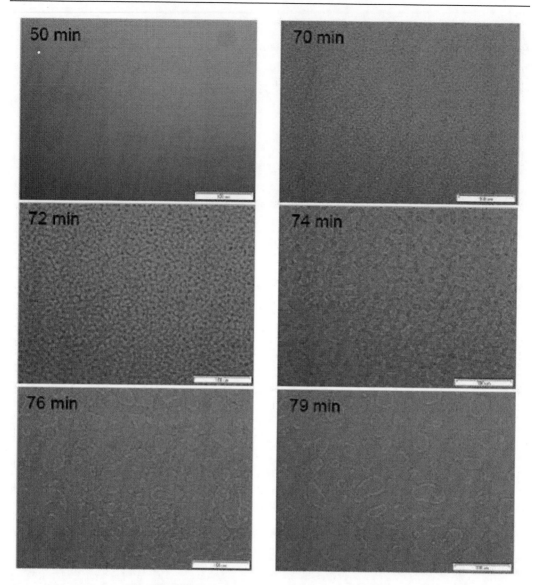

Figure 3. The process of phase separation of the TGAP/PES/DDS (100/30/30) blend monitored by a phase contrast optical microscope at 160 °C [34].

Figure 4 shows the optical and SEM images for this blend cured with two different heating rates. Optical images provide a two dimensional view of the phase structure. For comparison, an image processing software (ImagePro+ 4.5.1) was used to measure the domain size and the percent area occupied by the dispersed macrophases.

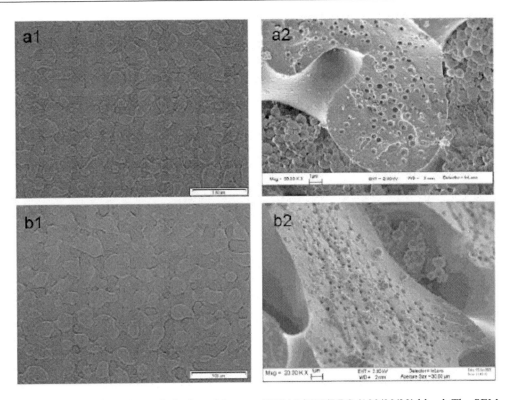

Figure 4. Optical and SEM morphologies of the cured TGAP/PES/DDS (100/30/30) blend. The SEM images were taken on the etched fracture surface: (a1 and a2) heating rate: 1.5 °C/min; (b1 and b2) heating rate: 10 °C/min [34].

Table 1. Comparison of macrophase size and distribution in the cured TGAP/PES/DDS (100/30/30) blend [34]

		D_{mean} (ηm)	$Area_{mean}$ (μm^2)	Percentage area (%)	No. of dispersed macrophases	Size in length (μm)	Size in width (μm)
Cure with a heating rate of 1.5°C/min	Average	9.67	107.80	14.94	202	12.30	8.08
	Standard deviation	0.64	11.22	1.33	15	0.76	0.64
Cure with a heating rate of 10°C/min	Average	11.41	162.75	18.74	167	14.60	9.35
	Standard deviation	0.62	11.87	1.98	17	0.60	0.66

The results are shown in Table 1. With an increase in the heating rate, the number of the dispersed macrophases became smaller. However, the domain size becomes larger and the percent area occupied by the dispersed macrophases became larger as well. The reason is that when the system is cured under a linear rise of temperature, a slow temperature rise leads to longer curing time at lower temperatures [12]. The SEM images show a three dimensional view of the phase-separated structure. Since the thermoplastic phase was etched away, it can be seen that there existed two different types of macrophases. The epoxy micro-particles within the PES-rich macrophase and the cavities in the epoxy-rich macrophase left by the PES-rich particles can be seen in these structures.

Figure 5 demonstrates the possible toughening mechanism in the TGAP/PES/DDS (100/30/30) blend. The material can be improved in its toughness by increasing the plastic deformation ability of the material. The idea is to initiate a very large number of small local yield events [38]. In this case, PES-rich particles were plastically stretched in the area of the crack tip to prevent premature crack propagation, hence increasing the toughness of epoxy. SEM images show the actual examples of the plastically deformed PES-rich particles which contribute to the toughening effect.

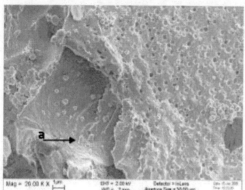
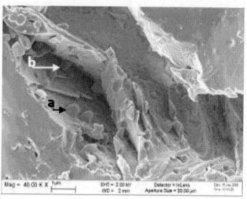

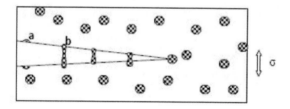

Figure 5. Toughening mechanism as observed in the TGAP/PES/DDS (100/30/30) blend cured with the heating rate of 10 °C/min. The illustrative graph was redrawn from [38]. A and b in the SEM images correspond with a and b in the illustrative graph [34].

SEM observation reveals that more and larger cavities were left in the blend when heating at a higher ramp rate. It suggests a more complete and higher degree of phase separation occurring in the rapidly heated blend. It has been found that phase separation is inhibited when the blend is of high viscosity which results in none or incomplete phase separation [24]. Significant decrease in the minimum viscosity has been demonstrated by applying a higher heating rate [39], which may assist with achieving a more complete phase separation.

For a TGAP/TGDDM/PES/DDS (90/10/30/30) blend, the cure kinetics at different heating rates is shown in Figure 6. The heating rate had a great influence on the cure process. At the low heating rate of 1.5 °C/ min, the cure reaction took place by two stages as evidenced by the two-step conversion rise monitored by FTIR. This is caused by the presence of more reactive tetrafunctional resin - tetraglycidyldiaminodiphenylmethane (TGDDM). Slower temperature ramp drove the possibility to separate particularly curing processes of

TGDDM and TGAP. The final epoxy conversion after the blend was heated at 175 °C for 60 minutes is 0.875 for curing with the heating rate of 10 °C/min, which was higher than 0.853 for curing with the heating rate of 1.5 °C/min. The Tg vs conversion curve is displayed in Figure 7. Tg increased from 20.7 °C to 178.8 °C with the increase in curing time from 0 minute to 320 minutes at 175 °C. A heating rate of 10 °C/min was used for the cures.

The cured TGAP/TGDDM/PES/DDS (90/10/30/30) blend showed a finer phase separation structure compared to the TGAP/PES/DDS (100/30/30) blend. The morphologies in Figure 8 display a co-continuous phase structure. The periodic distance of the macrophases for the blend heated at 10 °C/min was longer than that heated at 1.5 °C/min.

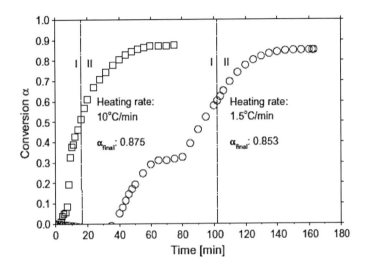

Figure 6. Cure kinetics following with different cure schedules of the TGAP/TGDDM/PES/DDS (90/10//30/30) blend: I. heating stage (from 20 °C to 175 °C); II. isothermal curing stage (at 175 °C) [34].

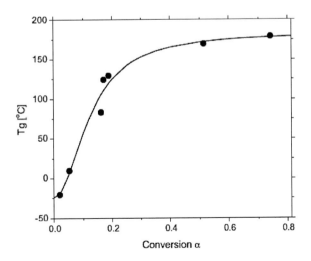

Figure 7. Tg as a function of conversion for the TGAP/TGDDM/PES/DDS (90/10/30/30) blend cured at 175 °C for different times. The solid line is the curve simulated using Equation 1 (the values of the fitting parameters are: a = -23.93, b = 0.015, c = 183.52, d = 2.02) [34].

The epoxy micro-particles within the PES-rich macrophase and the PES-rich particles within the epoxy-rich macrophase indicate further phase separation after the primary phase separation. The SEM images of the un-fractured surface for the fast heated specimen (after etching) show more cavities left due to the removal of PES-rich particles.

Similar phenomena were found on the aerospace composite system HexPly914. This commercial composite prepreg consists of carbon fibres (Torayca T300H) embedded within a blend of TGDDM (MY720) and TGAP (ERL0510) hardened with dicyandiamide (DICY) and/or DDS, to which is added a small percentage of PES. Due to the existence of residual carbon fibres, there were interferences in the conversion versus time curves shown in Figure 8. As for the sample heated at 1.5 °C/min, the epoxy conversion started to reach a plateau when the temperature increased to 175 °C. However, the epoxy conversion of the sample material heated at 10 °C/min seemed to keep increasing even after 60 minutes of curing at 175 °C. The final epoxy conversion after the sample was cured at 175 °C (for 60 minutes) was 0.849 for the rapidly heated material and was 0.595 for the slowly heated one. To ensure the complete cure of the HexPly914 composite, a post heat treatment at 190 °C for 4 hours is required.

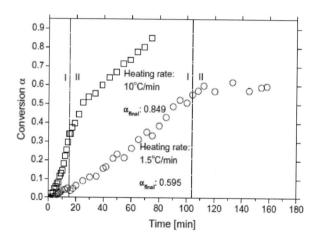

Figure 8. Cure kinetics following with different cure schedules of the T300/914 prepreg: I. heating stage (from 20 °C to 175 °C); II. isothermal curing stage (at 175°C) [34].

Composite laminates laid up with the HexPly914 prepregs were cured at heating rates of 1.5 °C/min or 10 °C/min, followed by a post-curing treatment at 190 °C for 4 hours. SEM observation on the delaminated fracture surface reveals a connected-globule phase-separated structure (Figure 9).

The domain sizes of these microphases, i.e. epoxy globules, are quite uniform, around 0.88 μm for the fast heated specimen, which is slightly larger than those of the slowly heated specimen (0.80 μm).

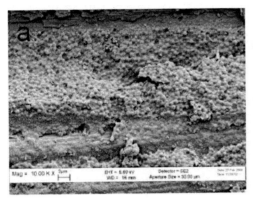 

Figure 9. SEM morphology of delamination fracture surface after etching treatment. Composite specimen HexPly914, cured with heating rates of (a) 1.5 °C/min and (b) 10 °C/min.

3.2. Interlayer Toughening by Using Thermoplastic Nanofibres

Two main strategies to toughen thermosetting matrix composites have been developed: blending a toughener with the matrix resin [40-42] or incorporating a discrete toughening interlayer between composite plies [43-45]. The addition of tougheners to the entire matrix resin could lead to the decrease in in-plane compression and shear strengths. The increased viscosity arising from the addition of tougheners could also reduce the processing ability of matrix resin. However, interlayer toughening is less likely to cause those issues, and it is thus used increasingly in the composite industry. Indeed, several interlayer toughening methods have been developed, such as using thermoplastic or thermosetting films [46, 47], porous membranes [48], and rubber and/or thermoplastic particles [49, 50]. Zhang et al. [51] used electrospun polyetherketone cardo (PEK-C) nanofibres to improve the interlaminar fracture toughness of carbon/epoxy composites. In this work, PEK-C nanofibres were directly deposited onto carbon fabrics. The effects of nanofibre diameter and interlayer thickness on the mode I delamination fracture toughness, flexural property and thermal mechanical properties of the resultant composites were examined.

Electrospun PEK-C nanofibres were obtained from PEK-C solutions with various PEK-C concentrations. The nanofibre morphologies are shown in Figure 10. It has been established that electrospinning a polymer solution could result in individual beads, beads-on-structure and bead-free fibres [52].

Figure 10. SEM images of PEK-C nanofibres electrospun from (a) 20 wt%, (b) 23 wt%, (c) 25 wt% and (d) 30 wt% PEK-C solutions [51].

In the case of PEK-C solution, a similar trend on the fibre morphology was also observed. For example, electrospinning of 10 wt% PEK-C solution in N, N-dimethylformamide led to beads-on-string fibres. When the PEK-C concentration was larger than 20 wt%, bead-free uniform nanofibres were produced. With an increase in the PEK-C concentration from 20 wt% to 30 wt%, the average fibre diameter increased from 275 nm to 950 nm, respectively.

PEK-C nanofibres were directly electrospun onto dry carbon fabrics to form a randomly orientated nanofibre nonwoven membrane, and they served as an interlayer toughening agent for preparing carbon/epoxy composite laminates. To investigate the influence of fibre diameter on the mechanical performance of interlayer toughened composites, PEK-C nanofibres with average diameters of 450 nm, 750 nm and 950 nm were used. With an equal weight loading of PEK-C nanofibres on the carbon fabric, the effects of fibre diameter on the Mode I delamination fracture toughness and flexural properties were examined. To achieve approximately 0.44% weight fraction of PEK-C nanofibres in the composite laminates, the electrospinning time for different polymer concentrations varied. The average interlayer thickness of the cured composite laminates was 58 μm, 65 μm and 70 μm for the membranes with fibre diameters of 450 nm, 750 nm and 950 nm, respectively.

Figure 11a reveals the delamination resistance curves for both the control and the composite laminates modified by nanofibres with different diameters. For the control specimen, there existed a considerable scatter in the calculated critical strain energy release rate G_{IC} from mode I double cantilever beam (DCB) tests, which was attributed to the $[0/90]_4$ carbon fibre orientation and the bonding stitches pre-existed in the carbon fabrics [53, 54]. Compared with the control, the PEK-C nanofibre modified composite specimens showed increased G_{IC}. The delamination resistance curves showed unstable crack growth and the stick/slip fracture behaviour. The fluctuation of the delamination resistance curves became more significant as the fibre diameter increased.

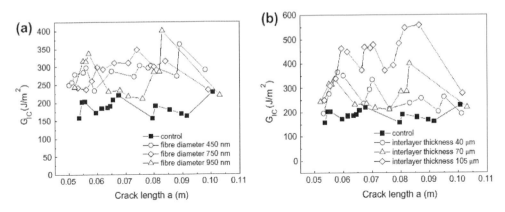

Figure 11. Delamination fracture toughness of the composite specimens [51].

Visual detection was applied as the criteria to define the initiation of crack propagation. The strain energy release rates for crack initiation (G_{IC-INI}) and for crack propagation ($G_{IC-PROP}$) influenced by nanofibre diameter were compared and included in Figure 12a. The average G_{IC-INI} for the control specimen was 151 J/m^2, whereas the G_{IC-INI} value for the nanofibre-modified specimens with average fibre diameters of 450 nm, 750 nm and 950 nm was 249 J/m^2, 228 J/m^2 and 241 J/m^2, respectively, indicating increased G_{IC-INI} value for the nanofibre-modified composite specimens. The presence of nanofibre interlayer also led to increased average $G_{IC-PROP}$. Here, since the weight loading of the nanofibres in the nanofibre-modified specimens was less than 1 wt% (around 0.4 wt%), the improvement in G_{IC-INI} and $G_{IC-PROP}$ due to such a low interlayer weight loading is quite significant.

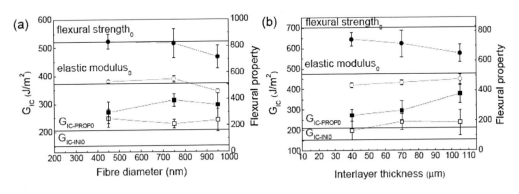

Figure 12. Mode I delamination fracture toughness and flexural properties of nanofibre modified composites as a function of (a) nanofibre diameter and (b) interlayer thickness. –□– G_{IC-INI}; –■– G_{ICPROP}; –○– elastic modulus (E^{-1}·GPa); –●– flexural strength (MPa). The dotted lines underneath different properties indicate the corresponding average data for the control composite specimen [51].

The effect of fibre diameter on the flexural strength and elastic modulus is also shown in Figure 12a. By comparison with the control specimen, the flexural strength of the nanofibre-modified specimens decreased slightly when the average nanofibre diameter increased from 450 nm to 750 nm. However, a noticeable decrease in the flexural strength was observed in the specimen with a larger average fibre diameter (950 nm). Similarly, a considerable reduction in the elastic modulus was observed when coarse nanofibres (e.g. 950 nm) were employed as interlayers. This suggested that composites modified with finer nanofibres had greater improvement in the interlaminar property without compromising the in-plane performance of the toughened composites.

The interlayer thickness was adjusted through the electrospinning time. Here nanofibres with the average diameter of 950 nm were chosen. With an increase in the electrospinning time from 23 min, to 46 min and 68 min, the average interlayer thickness of the cured composite laminates changed from 40 μm, to 70 μm and 105 μm, which led to changes of the G_{IC-INI} from 198 J/m^2, to 241 J/m^2 and 236 J/m^2 and the $G_{IC-PROP}$ from 272 J/m^2, to 296 J/m^2 and 376 J/m^2. In comparison with the control specimens, whose G_{IC-INI} and $G_{IC-PROP}$ values are 151 J/m^2 and 207 J/m^2, respectively, the presence of nanofibre interlayer resulted in higher delamination fracture toughness (Figures 11b and 12b). It was also noted that the weight fraction of PEK-C in the cured composites was approximately 0.22 %, 0.44 % and 0.65 % for the laminates with an average interlayer thickness of 40 μm, 70 μm and 105 μm, respectively. The influence of interlayer thickness on the flexural properties is also shown in Figure 12b. Both the flexural strength and the elastic modulus showed a decreasing trend as the nanofibre interlayer thickness increased.

Figure 13a shows the morphology of the delaminated control specimen, and the phase morphologies of delaminated composite specimens toughened by PEK-C nanofibres with average diameters of 450 nm, 750 nm and 950 nm are shown in Figure 13b, 13c and 13e, respectively. In contrast to the control specimen, PEK-C-rich particulate phases with various sizes were shown on the delamination fracture surface of modified composites. Despite the equal weight loading of the PEK-C nanofibres, the size of PEK-C-rich phases and their distance increased with the increase in the fibre diameter.

It should be noted that the penetration of epoxy at room temperature did not lead to dissolution or "deterioration" of nanofibres and PEK-C nanofibres started dissolving into the epoxy resin as the curing temperature increased to 70 °C. However, the high viscosity of PEK-C retained the diffusion of dissolved PEK-C into the epoxy resin, which led to unique phase structure after curing. Coarser nanofibres caused higher local concentration of PEK-C in epoxy resin, resulting in larger PEK-C-rich particulate phases as revealed on the delamination fracture surface. The PEK-C-rich phases create stress concentrations at their equators and also act as sites for initiating shear bands. When the shear bands created by one toughener-enriched phase interact with another, they may stop propagating and keep the matrix yielding localised [18, 55]. Therefore the cracks were hindered intermittently by the PEK-C-rich phases and forced to propagate between two unstable jumps. The membrane composed of finer fibre diameters gave rise to reduced distance between the ductile PEK-C-rich phases, which assisted with maintaining a more stable crack growth during delamination. Analogous findings were reported where poly(acrylonitrile-butadiene-styrene) particles (ABS) particles were used for interlayer toughening of glass fibre/vinyl-ester resin composites [56]. By using particles with a similar size, the stability of crack propagation was improved by increasing the particle concentration in the interlayer. Figures 13d, 13e and 13f show the morphology of delamination fracture surface influenced by nanofibre interlayer thickness. Larger PEK-C-rich phases were observed to scatter on the fracture surface of thicker nanofibre interlayer.

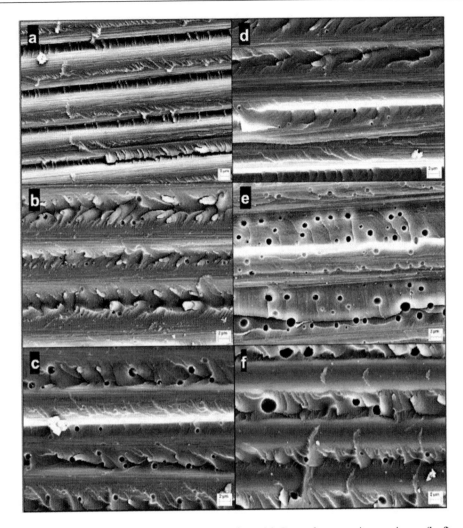

Figure 13. SEM images of delamination fracture surface. (a) Control composite specimen, (b–f) nanofibre-modified composites; (b) fibre diameter: 450 nm, interlayer thickness: 70 μm; (c) fibre diameter: 750 nm, interlayer thickness: 70 μm; (d) diameter: 950 nm, interlayer thickness: 40 μm; (e) fibre diameter: 950 nm, interlayer thickness: 70 μm; (f) fibre diameter: 950 nm, interlayer thickness: 105 μm. The PEK-C-rich particulate phases have been removed by solvent DMF and are shown as dark holes in the images [51].

Figure 14 shows the tan δ curves for the control and PEK-C nanofibre-modified composite specimens. The control sample displayed a well-defined relaxation peak at 281 °C, which corresponds to the glass transition temperature of the epoxy resin. All the PEK-C nanofibre modified composites showed two separate glass transitions. The glass transition temperature for PEK-C powder was measured by differential scanning calorimetry (DSC), where T_g was determined from the midpoint of the slope change of the heat capacity plot of the second scan. A glass transition at 219 °C was obtained from the DSC curve. Since the T_g values measured by DSC are generally 20-30 °C lower than those measured by DMA at the same compositions, the relaxation peak at lower temperature shown in the tan δ curves of modified composites was attributed to the PEK-C-rich phase, which is T_{gPEK-C}. The relaxation peak at higher temperature displayed in the spectrum is ascribed to the glass transition of the cured epoxy resin, which is T_{gER}. Figure 14a displays the tan δ curves influenced by the nanofibre diameter.

It is noted that there was almost no reduction or slight increase in T_{gER} with the addition of PEK-C nanofibres. The peaks centred at 283 °C, 280 °C and 285 °C correspond to the T_{gER} for the composites modified by PEK-C nanofibres with average fibre diameter of 450 nm, 750 nm and 950 nm respectively. Similar shifts of relaxation peaks to higher temperature have been reported in the literature on the thermal mechanical properties of nanocomposites [57]. Nano-reinforcements have been shown to affect the segmental motions of polymer matrices when they are well dispersed [58].

The large surface areas of nanofibre membranes provide complete interaction with the epoxy matrix, which may prohibit the epoxy chains from moving freely as in the neat resin [59]. No noticeable shifts occurred in the relaxation peaks for the T_{gPEK-C} as the average fibre diameter changed. It is also noticed from figure 15b that both the T_{gER} and T_{gPEK-C} reduced to lower temperatures as the thickness of PEK-C interlayers increased to 105 μm. This result indicates the level of interaction and compatibility between the PEK-C nanofibres and the epoxy resin deteriorated after the inclusion of nanofibre membranes increased to a certain extent.

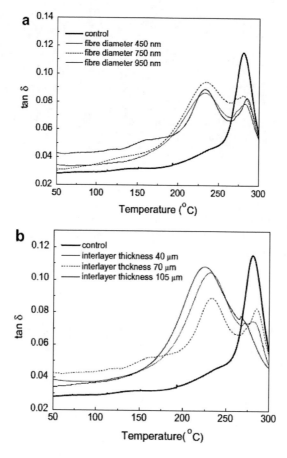

Figure 14. Tan δ traces of both control and nanofibre-modified composites [51].

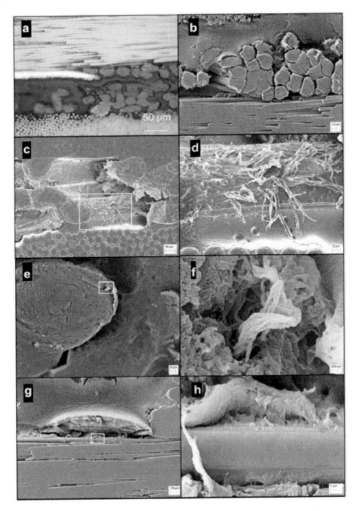

Figure 15. Phase structure of interlayer region of a nanofibre-modified composite specimen (average nanofibre diameter: 950 nm, average interlayer thickness: 105 μm). Optical micrographic (a) and SEM (b) images of interlayer region; (c–f) phase structures of the PEK-C-rich macrophases after etched with DMF; (d and f) are magnified images of the selected areas in (c and e); (g and h) phase structure of the epoxy-rich macrophase [51].

Figure 15 presents the phase structure of the interlayer region from PEK-C nanofibre modified carbon fibre reinforced TGAP/DDS epoxy matrix composites. Figure 15a is an optical micrograph for the cross section of a modified composite laminate. The bright part is a pre-inserted aluminium film, which was located in the midplane of composite laminate. A dark interlayer region composed of both an epoxy-rich macrophase and PEK-C- rich macrophases was found to be sited between the top [0°] and the bottom [90°] carbon fibre layers.

As shown in Figures 15 b~e, cylinder-shaped macrophase structures were observed to be embedded in the continuous epoxy resin. Figures 15 d~f are enlarged views of the PEK-C-rich macrophases showing a fibrous structure inside of the cylinder-shaped macrophase. Since the specimens have been etched with N, N-dimethylformamide, the PEK-C should be removed.

These fibrous structures could come from undissolved PEK-C nanofibres which were covered by cured epoxy resin. Figures 15 g and h show the morphology of a detached epoxy-rich macrophase from carbon fibre. It clearly indicated that the PEK-C-rich microphases (particulate phases) dispersed in the continuous epoxy-rich matrix.

4. Hyperbranched Polymers as Toughening Agents

4.1. Introduction

Since being introduced theoretically by Flory in 1952 [60], hyperbranched polymers (HBPs) reawakened the interests of many researchers in the early 1990's when Kim and Webster reported the reduced melt viscosity of polystyrene with the addition of a small amount of hyperbranched polyphenylene [61]. The unique three-dimensional dentritic architecture of hyperbranched polymers leads to significant differences in properties from linear polymers, which result in unique physical and chemical properties. The potential applications of hyperbranched polymers cover versatile areas including conjugated functional materials, polymer electrolytes, supramolecular chemistry, coatings, nanomaterials, modifiers and additives [62, 63]. One of the most important properties of hyperbranched polymers is their low melt viscosities even for high molecular weights, resulting from a much lower degree of chain entanglement relative to linear polymers [64]. This characteristic has particular significance when HBPs are used as additives to modify inherently brittle thermosets. Conventional modifiers such as rubber [10, 65] and thermoplastic [66-69] toughening agents are able to enhance the fracture toughness of unmodified thermosets, however, they tend to cause significant increase in the viscosity of the blends with the addition of only a relatively small amount. It is well known that liquid moulding technologies are economical alternatives to expensive autoclave processing of polymer matrix composites. Viscosity and resin flow are critical issues for obtaining good quality composite components. As a result, HBPs are promising candidates for modifying thermosets with minimal decrease in processing-ability of the modified resins. A number of works reported the modification effect of hydroxyl and epoxy functionalised HBPs on different epoxy resin formulations.

4.2. HBPs as Modifiers for Difunctional and Trifunctional Epoxies

Both hydroxyl and epoxy functionalised HBPs have been shown to significantly increase the toughness of less-reactive and lower crosslink-density difunctional epoxy resin systems such as the diglycidyl ether of bisphenol A (DGEBA) resin. Two hydroxyl functionalized hyperbranched polyester, Boltorn H30 and H40, which have theoretically 32 primary hydroxyl end groups and 64 primary hydroxyl end groups per molecule respectively, were studied by Cicala et al. [70] for their modification result on a DGEBA epoxy (Epon 828) cured with 3, 3′ diamino diphenyl sulphone (DDS). The stress intensity factor K_C for fracture toughness tests increased from 0.93 $MPa \sqrt{m}$ to 1.56 $MPa \sqrt{m}$ with 30 wt% H30 and to 1.67 $MPa \sqrt{m}$ with 30 wt% H40. All the blends studied, except at low percentages of H30, showed particulate morphologies and two relaxation peaks in the tan δ spectra. Two other works also showed similar trends when a fifth generation hydroxyl functionalised HBP, Boltorn G5, and an epoxidised HBP Boltorn E1 were used as additives in DGEBA epoxy resins. The Boltorn G5, Ratna and Simon [71] used for their study, was a fifth-generation Perstorp hyperbranched polyols which were developed on polyester chemistry and have an average of 128 –OH end groups per molecule and a theoretical molecular weight and polydispersity index of 14,000 g mol^{-1} and 2.0, respectively. It was noted that the HBP strongly enhanced the curing rate due to the catalytic effect of hydroxyl groups. The impact strength of the cured blends showed the most significant increase when the content of the HBP accounted for 20% of the blends, which was almost double the value of the unmodified epoxy system. The impact behaviour was explained by the phase separated morphology of the cured blends. Up to 10% of HBP, the globular HBP particles were dispersed in the continuous epoxy matrix and as the content of HBP increases, the connectivity between the dispersed HBP particles increased and the phase morphology became nearly co-continuous at higher concentrations. Varley [72] evaluated the effect of epoxy-terminated aliphatic polyester HBP as a low viscosity toughener for modifying the less-reactive DGEBA epoxy resin. The main advantage of using Boltorn E1 is that it has minimal influence on processing parameters such as viscosity and gel time, while enhancing the fracture toughness by around 54 % with 15 wt% addition of modifier. Similarly, particulate phase structure was formed from phase separation.

For the more reactive trifunctional epoxies, HBPs have also displayed good suitability as toughening agents. Ratna, Simon and Varley reported that the HBP Boltorn G5 [73] and E1 [74, 75] modified triglycidyl p-amino phenol (TGAP) (MY0510, Ciba Speciality Chemicals) epoxy resins had 50 % higher impact strength than unmodified epoxy resin with 20 wt% Boltorn G5 and 100 % higher impact strength than unmodified epoxy resin with 20 wt% Boltorn E1. The smaller improvement of Boltorn G5, in comparison with an 80% increase in K_{IC} of DGEBA using the same HBP, was believed to be caused by the single phase morphology due to the compatibility of Boltorn G5 with TGAP epoxy resin. In contrast, the trifunctional epoxy resin modified with the epoxy-functionalised HBP showed a two-phase morphology, where globular HBP-rich particles dispersed uniformly in the continuous TGAP-rich matrix. The addition of Boltorn E1 demonstrated little influence on gel time and vitrification time up to 10% of HBP content.

4.3. HBPs as Modifiers for Tetrafunctional Epoxies

Nevertheless, the modification effect of HBPs on the highly reactive and highly crosslinkable tetrafunctional epoxy resins have not been shown to be desirable. Varley tried to add HBP Boltorn E1 into an epoxy resin based on tetraglycidyl methylene dianiline (TGDDM) (Araldite MY721, Ciba) and a cycloaliphatic diamine hardener [76]. Not only the E1 modified TGDDM exhibited decreased K_{IC} , but also the TGDDM modified by other low viscosity additives such as carboxy-terminated polybutadiene, CTBN rubber (Goodyear, OH, USA) and the aminopropyl-terminated polydimethylsiloxane, PS510 (United Chemical Technologies, PA, USA), showed little toughening result. High service temperature and good chemical and water resistance provide tetrafunctional epoxies with great advantages in the application of aerospace polymer matrix composites. In fact, they have been used as matrix materials for > 90% composites [77]. Modification of the brittleness of these highly crosslinked epoxies is crucial in material development aimed at reducing the delamination tendency of composite structures. Zhang et al. [78] modified a tetrafunctional epoxy resin tetraglycidyl methylene dianiline (TGDDM) (Araldite MY720, Huntsman) with a fourth generation hydroxyl functionalised hyperbranched polyester (Boltorn H40, Perstop Speciality Chemicals, Sweden). The cure agent used was 3, 3'-diaminodiphenyl sulfone (DDS).

The DDS-cured epoxy/HBP blends displayed a decreased transparency and gradually became opaque with the increase of the HBP content, indicating the occurrence of phase separation. The DSC curves obtained for the DDS-cured neat epoxy resin and epoxy/HBP blends during the second heating scan are shown in Figure 16. The glass transition of the epoxy-rich phase, T_{gER} can be seen in all DSC curves. There was a reduction in T_{gER} with increasing HBP content, which may be caused by the residual miscibility between HBP and the crosslinked epoxy resin. The dilution effect of HBP might also produce an incomplete curing reaction, which resulted in a lower T_g for the crosslinked epoxy resin [79]. The glass transition of the HBP-rich phase T_{gHBP} was not detected by DSC for the blends with HBP content up to 20 part per hundred (phr). However, the cured blend with 30 phr HBP exhibited two separate glass transitions. The glass transition at higher temperature in the blend containing 30 phr HBP was associated with the cured epoxy phase (T_{gER}). The other at lower temperature was attributed to the HBP-rich phase (T_{gHBP}). The values for T_{gER} and T_{gHBP} measured by DSC are included in Table 2 along with the measurements carried out by DMA.

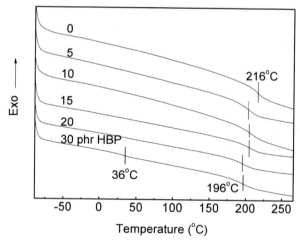

Figure 16. DSC thermograms of the second run of the DDS-cured neat epoxy resin and epoxy/HBP blends [78].

Table 2. T_{gER} and T_{gHBP} for DDS-cured neat epoxy resin and epoxy/HBP blends measured by DMA and DSC [78]

HBP (p.h.r)	T_{gER} (DMA) (°C)	T_{gER} (DSC) (°C)	T_{gHBP} (DMA) (°C)	T_{gHBP} (DSC) (°C)

0	245	216		
5	237	204		
10	235	205		
15	230	205	63	
20	230	195	59	
30	230	196	64	36

Figure 17 shows the dynamic mechanical spectra for the DDS-cured neat epoxy resin and epoxy/HBP blends with the HBP content of up to 30 phr. The DMA measurements of the cured epoxy/HBP blends with HBP content up to 20 phr exhibited storage modulus G' close to the cured neat epoxy resin. G' was not strongly influenced by the HBP content. A reduction in G' at ~40°C was observed in the blend with 30 phr HBP content, which corresponds to the low-temperature thermal transition of HBP. The tan δ versus temperature curve for the cured neat epoxy resin displayed a well-defined relaxation peak at 245 °C, attributed to its glass transition.

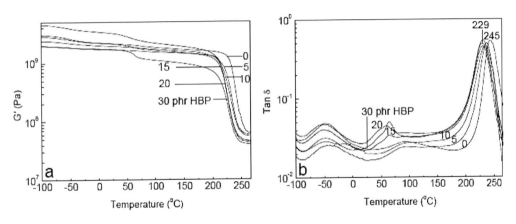

Figure 17. Temperature dependence of a) storage modulus (G') and b) tan δ of the DDS-cured neat epoxy resin and epoxy/HBP blends [78].

This relaxation peak shifted slightly to a lower temperature in the cured blends with increasing HBP content, a reduction from 237°C to 230°C for T_{gER}, corresponding to the increase of HBP content from 5 to 30 phr. It is noted from Figure 18b that a well-defined lower-temperature relaxation peak started to appear in the cured blend with HBP content of 15 phr. This indicates the existence of a HBP-rich phase in the cured blends. The relaxation peaks at around -50 °C are the β relaxation peaks and those at around 60 °C are the α relaxation peaks of the HBP-rich phase. The values for T_{gER} and T_{gHBP} measured by DMA also can be found in Table 2. As expected, the T_g values measured by DMA are generally 20 - 30 °C higher than those measured by DSC at the same composition. DMA is more sensitive to the detection of the HBP-rich phase; the two separate relaxation peaks in the tan δ spectra for cured blends with 15-30 phr showed the occurrence of phase separation. It was interesting to see that the glass transition temperature of the HBP-rich phase (T_{gHBP}) was decreased in the cured blends with addition of HBP. This result implies that the epoxy resin dissolved in the HBP-rich phase was not sufficiently cured. The HBP-rich phase composed of HBP and some uncured and/or insufficiently cured epoxy resin thus had a reduced T_g [64].

Analogous results have been shown in other researcher's work, such as toughening TGDDM (Araldite MY721, Ciba) using a hyperbranched epoxy-functional HBP Boltorn E1 [76]. Increasing additive content produced a gradual decrease in T_{gER}, resulting in a 24 °C reduction from the cured neat epoxy resin when 15 wt% HBP is used. In contrast, the T_{gER} was almost unaffected when HBP was added into a difunctional epoxy DGEBA (DER 331, Dow) [76]. Previous work also supports unchanged T_{gER} in the DDM-cured diglycidyl 1, 2 – cyclohexanedicarboxylate (DGCHD) / Boltorn H40 blends with HBP content up to 30 wt% [64].

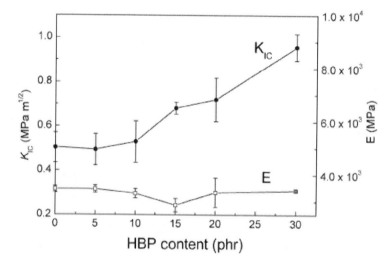

Figure 18. Effect of HBP content on the fracture toughness and Young's modulus of DDS-cured neat epoxy resin and epoxy/HBP blends [78].

The fracture toughness and phase structure were studied as a function of the modifier content. The K_{IC} values as a function of the HBP content for the cured neat epoxy resin and epoxy/HBP blends are shown in Figure 18. The addition of HBP led to a 90% increase in the K_{IC} values, from 0.503 $MPa\sqrt{m}$ for the cured neat epoxy resin to 0.956 $MPa\sqrt{m}$ for the cured epoxy/HBP blend with 30 phr HBP. Meanwhile, the Young's modulus of the cured epoxy/HBP blend showed limited decrease over that of the unmodified epoxy. The HBP modified tetrafunctional epoxy systems in this work exhibited significantly higher fracture toughness than thermoplastic modified tetrafunctional epoxies [80-82]. Fernandez et al. [80], added polyethersulfone (PES) into a DDM-cured TGDDM epoxy resin. However, no significant improvement was observed in fracture toughness after modification. The deficiency of modification was believed to be caused by the nanoscale phase separation. The high crosslink density of the tetrafunctional epoxy resin hindered the occurrence of higher enhancement in fracture toughness. Song et al. [83] used phenolphthalein poly(ether ether ketone) (PEK-C) to modify a DDM-cured TGDDM resin. The fracture toughness of epoxy/thermoplastic blends showed a slight decreasing tendency, which was attributed to the reduced crosslink density of the epoxy net work and the homogenous phase structure of the cured blend.

The morphology of the cured epoxy/HBP blends was investigated by SEM. Figure 19 presents the phase structures of the cured epoxy/HBP blends by cryogenically breaking the samples and then immersing in tetrahydrofuran (THF). Heterogeneous morphologies were observed for the cured blends with 10, 15, 20 and 30 phr HBP contents. In comparison with the DMA and DSC results, the cured blend with HBP content of 10 phr was also revealed to have a two-phase morphology. For the cured blend containing 5 phr HBP, a homogenous morphology was observed (Figure 19a). One of the most important criteria for obtaining increased fracture toughness in rubber or thermoplastic modified epoxy resins is a two-phase morphology after phase separation [84]. Similarly, with 5 phr HBP, no toughening result was shown. With the increase in the HBP content, a particulate morphology in which the discrete HBP-rich phases were dispersed in a continuous matrix of epoxy resin was obtained [85]. The discrete HBP-rich particles with an average size of less than 1 μ m in diameter appear in the cured blend containing 10 phr HBP (Figures 19 bandc).

Figure 19c is an enlargement to reveal the small HBP-rich particles which cannot be seen clearly in Figure 19b using the same scale as the other SEM micrographs. The toughening effect was not obvious for the cured blend with 10 phr HBP content despite the occurrence of phase separation. Inclusion of 15 phr HBP produced much more and larger HBP-rich phases with a bimodal particle size distribution, resulting in a 35% enhancement of K_{IC}. When the HBP content increased to 20 phr, the particle size ratio in the bimodal phase structure became higher (Figure 19e), leading to higher fracture toughness in the cured blend. Okamoto et al. [86] have studied the relationship between the Izod impact strength and the blend ratios of two different monomodal high impact polystyrene (HIPS) for two bimodal systems. They observed that the particle size ratio in the bimodal system affected the extent of the toughening effect. The greater the difference between the particle size of the constituent bimodal HIPS, the higher was the impact strength. The maximum toughening effect was demonstrated for the cured blend containing 30 phr HBP, when a uniform distribution of monomodal HBP-rich particles was achieved.

It has been noted that all the epoxy/HBP blends with HBP content from 10 to 30 phr experienced phase separation through the nucleation and growth (NG) mechanism [5, 24]. An important factor which may influence the phase behaviour of the epoxy/HBP blends needs to be considered: as a low viscosity modifier, HBP does not have the semi permeable characteristics required for mass transfer (usually, when the modifier is a viscous thermoplastic, the dispersed primary phase will continuously receive material from the continuous primary phase but not deliver species back [23]), which can be associated with significantly different thermodynamic simulations from thermoplastic modified epoxies.

For the cured blends containing 15 and 20 phr HBP, both phase structures displayed bimodal particle-size distributions (Figure 19 d and e). The explanation could be based on the following: when the average distance between growing particles is high and the conversion rate is significantly high, concentration profiles of the modifiers become very sharp. The high supersaturation at these points (concentration excess with respect to equilibrium at the particular conversion level) causes a new nucleation process to occur. The different concentration profiles are repeated as the cure conversion proceeds, resulting in the bimodal particle size distributions. However, the inclusion of higher HBP content (30 phr) gave rise to a reduced cure conversion rate when phase separation occurred, which may not be sufficient enough to initiate the second NG process. Under these conditions, the cured blend containing 30 phr HBP exhibited a monomodal particle size distribution (Figure 19f).

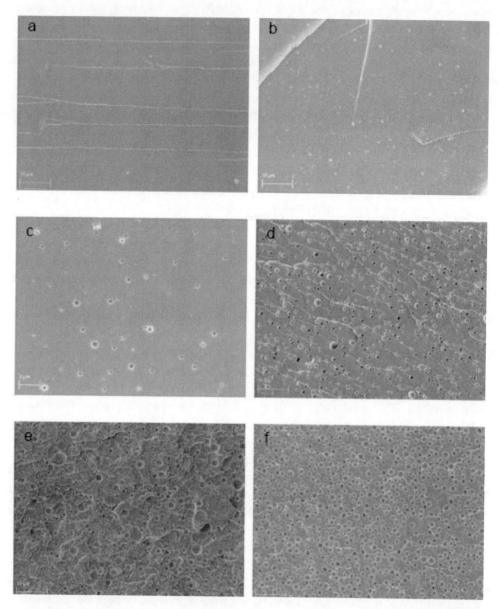

Figure 19. SEM micrographs of cryogenically fractured surfaces of DDS-cured epoxies with (a) 5, (b), (c) 10, (d) 15, (e) 20, (f) 30 phr HBP modifiers. The scale bars are 10μm except for the micrograph (c) which is 2μm [78].

Figure 20 presents the SEM micrographs on the fracture surfaces of the cured neat epoxy resin and the epoxy/HBP blends after single-edge-notch bending tests. For the neat epoxy resin (Figure 20a), the regularly oriented cracks were developed freely, indicating typical brittle characteristics of the fracture surface. The inclusion of 5 phr HBP did not change the brittle nature of the fracture surface (Figure 20b), accounting for its unimproved fracture toughness. As it has been known earlier, phase separation occurred when the HBP content increased to 10 phr. Although the dispersed HBP-rich particles cannot be distinctly observed in Figure 20c, a great amount of tortuous and fine cracks appeared on the fracture surface, resulting from the formation of shear bands near the plastic HBP-rich particles. The plastic characteristic was further developed on the surface of the cured blend containing 15 phr HBP (Figure 20d). The dispersed HBP-rich particles created stress concentrations at their equators and also acted as sites for initiating shear bands. When the shear bands created by one particle interact with another, they may stop propagating and keep the yielding localised [18, 55]. More matrix yielding was induced locally surrounding the particles due to a larger number of dispersed particles.

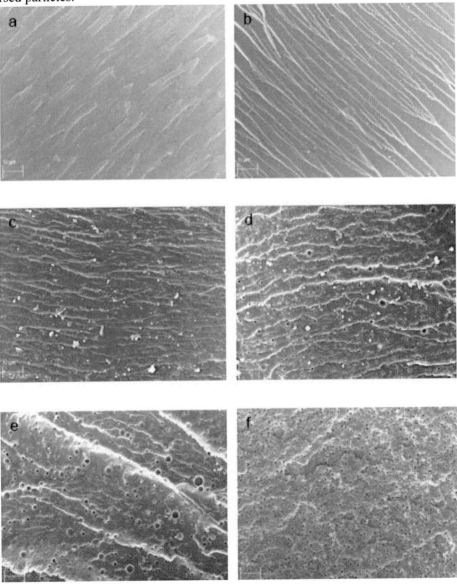

Figure 20. SEM micrographs on K_{IC} fracture surfaces of DDS-cured epoxies with (a) 0, (b) 5, (c) 10, (d) 15, (e) 20, (f) 30 phr HBP modifiers. The scale bars are 10μm [78].

Some cracks deviated from their original planes, leading to significantly increased surface area. All these factors contributed to more absorbed surface energy thereby increasing fracture toughness. For the cured blend containing 20 phr HBP (Figure 20e), the crack tips became obscured and large plastically deformed planes were aligned regularly on the surface. As the particle size ratio in the bimodal phase structure increased, the surface relief became more pronounced, which absorbed even more energy and led to a higher K_{IC} value. The SEM micrograph for the 30 phr HBP blends (Figure 20f) showed distinctly different characteristics from the others. No more lengthy cracks could be detected. Extensive plastic deformation took place in the continuous matrix embedded with uniformly distributed HBP-rich particles with considerably high concentration. Apparently, the most surface energy was absorbed through this way, contributing to the highest K_{IC} achieved among all the HBP modified epoxies in this study. In addition to the above observations, particle cavitations were present in all the investigated epoxy/HBP blends which experienced phase separation. The cavitated HBP-rich particles induced large stress concentrations, leading to extensive shear deformation. This high energy absorbing mechanism assisted with toughening the tetrafunctional epoxy resin more effectively [87].

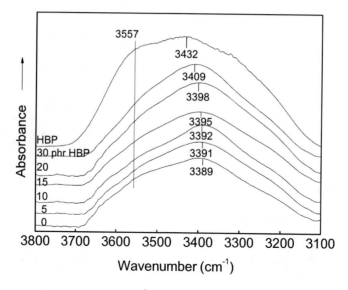

Figure 21. FTIR spectra in the 3100-3800 cm^{-1} region of the HBP (Boltorn H40) and the DDS-cured epoxy/HBP blends [78].

Figure 21 shows the FTIR spectra of the HBP and the DDS-cured epoxy/HBP blends in the stretching region from 3100 to 3800 cm^{-1}. Two components can be seen from both the spectrum for the HBP and for the DDS-cured neat epoxy resin. For the cured epoxy resin, a broad band centred at 3389 cm^{-1} is attributed to the self-associated hydroxyl groups (i.e., hydrogen bonded hydroxyl groups) and a shoulder centred at 3557 cm^{-1} is assigned to non-associated, free hydroxyl groups [88]. For the HBP, a broad band centred at 3432 cm^{-1} is attributed to the hydrogen bonded hydroxyl group and the shoulder centred at 3557 cm^{-1} is assigned to free hydroxyl groups. The relative intensity of the free hydroxyl band to the hydrogen bonded hydroxyl band is the highest for the HBP, indicating the existence of a higher fraction of free hydroxyl groups in the HBP than both the cured neat epoxy resin and the epoxy/HBP blends. The spectrum for the cured epoxy resin shows a reduced relative intensity of the free hydroxyl band to the hydrogen bonded hydroxyl band. The inclusion of HBP to the epoxy resin caused gradually decreased relative intensity of the free hydroxyl band to the hydrogen bonded hydroxyl band, which indicates the formation of hydrogen bonded hydroxyls in the cured epoxy/HBP blends. It can also be seen from Figure 21 that with the increase in HBP content, the associated hydroxyl band shifted to higher frequencies. However, the non-associated hydroxyl band did not move. The frequency difference between the non-associated hydroxyl absorption and the associated hydroxyl absorption ($\Delta\nu$) is a measure of the average strength of the intermolecular interactions [89]. This indicates that the average strength of the hydrogen bond in the cured blends ($\Delta\nu = 148$ cm^{-1} for the cured epoxy/HBP blend containing 30 phr HBP) is lower than the hydrogen bond in the cured neat epoxy resin ($\Delta\nu = 168$ cm^{-1}).

CONCLUSION

In reviewing the research endeavour contributed to improving the fracture toughness of epoxies and their carbon fibre reinforced composites, this chapter has briefly summarised the main toughening agents used in the past. Rubbers were firstly accepted as effective modifiers to toughen epoxies and commercial adhesives and composite matrix resins based on these systems have been widely used. However, rubber toughening has its main disadvantages of lowering the modulus and glass transition temperatures of the host epoxy resin. Thermoplastic toughening was proposed to overcome this barrier by replacing rubbers by high performance thermoplastics, which have similar values of elastic modulus and yield stress as those of the epoxy network. This method has been successful to modify epoxies with higher crosslink-density; as a result, many aerospace composite systems nowadays are based on the thermoplastic toughened epoxy matrix resin systems. A key factor influencing the application of laminating resins is the processing-ability of the resin, i.e. the initial viscosity of the resin system. Both rubbers and high performance thermoplastics lead to significantly increased viscosity of the modified epoxy resins, resulting from their linear polymer structure which involves a high degree of polymer chain entanglement. As promising candidates for modifying epoxies with minimal decrease in processing-ability of the modified resins, hyperbranched polymers with unique three-dimensional dentritic polymer structures have attracted great attention. The low viscosity, high solubility and high degree of functionality of these hyperbranched polymers have been taken advantage of for the modification of epoxies.

Future research in toughening of epoxies should focus on improving the fracture toughness as well as maintaining the initial properties of epoxies. In the mean time, important parameters such as the size and distribution of the ductile secondary phases, the interfacial adhesion between modifier particulate phases and matrix, and the viscosity and resin flow should also be carefully controlled. Newly emerged areas include using rubbers incorporating block copolymers that are able to self-assemble into hierarchical substructures at nanometre scale. The work of utilizing nano-sized rubber modifiers is currently under investigation.

REFERENCES

[1] Peters, S.T., ed. Handbook of Composites. 2nd ed. 1998, ChapmanandHall. 5.

[2] May, C.A., ed. Epoxy Resin Chemistry and Technology. 2nd edition ed. 1988, Marcel Dekker: New York.

[3] Kim, J.K. and Mai, Y.W., *Engineered interfaces in fibre reinforced composites*. 1998, New York: Elsevier.

[4] Hodgkin, J.H., Simon, G.P. and R.J.Varley, Thermoplastic toughening of epoxy resins: a critical review. *Polymers for Advanced Technologies*, 1998. 9: p. 3-10.

[5] Inoue, T., Reaction-induced phase decomposition in polymer blends. *Progress in Polymer Science*, 1995. 20: p. 119-153.

[6] Swier, S. and Van Mele, B., Reaction-induced phase separation in polyethersulfone-modified epoxy-amine systems studied by temperature modulated differential scanning calorimetry. *Thermochimica Acta*, 1999. 330(1-2): p. 175-187.

[7] Sultan, J.N., Liable, R.C. and McGarry, F.J., Microstructure of two-phase polymers. *Applied Polymer Symposia*, 1971. 16: p. 127-176.

[8] Wise, C.W., Cook, W.D. and Goodwin, A.A., CTBN rubber phase precipitation in model epoxy resins. *Polymer*, 2000. 41(12): p. 4625-4633.

[9] Tripathi, G. and Srivastava, D., Effect of carboxyl-terminated poly(butadiene-co-acrylonitrile) (CTBN) concentration on thermal and mechanical properties of binary blends of diglycidyl ether of bisphenol-A (DGEBA) epoxy resin. *Materials Science and Engineering: A*, 2007. 443(1-2): p. 262-269.

[10] Chikhi, N., Fellahi, S. and Bakar, M., Modification of epoxy resin using reactive liquid (ATBN) rubber. *European Polymer Journal*, 2002. 38(2): p. 251-264.

[11] Hwang, J.F., Manson, J.A., Hertzberg, R.W., Miller, G.A. and Sperling, L.H., Structure-property relationships in rubber-toughened epoxies. *Polymer Engineering and Science*, 1989. 29(20): p. 1466-1476.

[12] Sankaran, S. and Chanda, M., Chemical toughening of epoxies. II. Mechanical, thermal, and microscopic studies of epoxies toughened with hydroxyl-terminated poly(butadiene-co-acrylonitrile). *Journal of Applied Polymer Science*, 1990. 39(8): p. 1635-1647.

[13] Bagheri, R., Marouf, B.T. and Pearson, R.A., Rubber-toughened epoxies: a critical review. *Polymer Reviews*, 2009. 49(3): p. 201-225.

[14] Maazouz, A., Sautereau, H. and Gérard, J.F., Toughening of epoxy networks using pre-formed core-shell particles or reactive rubbers. *Polymer Bulletin*, 1994. 33(1): p. 67-74.

[15] Guo, Q., Thomann, R., Gronski, W. and Thurn-Albrecht, T., Phase Behavior, Crystallization, and Hierarchical Nanostructures in Self-Organized Thermoset Blends of Epoxy Resin and Amphiphilic Poly(ethylene oxide)-block-poly(propylene oxide)-block-poly(ethylene oxide) Triblock Copolymers. *Macromolecules*, 2002. 35(8): p. 3133-3144.

[16] Hammeed, N., Guo, Q., Xu, Z., Hanley, T.L. and Mai, Y.-W., Reactive block copolymer modified thermosets: highly ordered nanostructures and improved properties. *Soft Matter*, 2010.

[17] Ratna, D., Epoxy composites: impact resistance and flame retardancy in Rapra Review Reports. 2005. p. 11.

[18] Garg, A.C. and Mai, Y.-W., Failure mechanisms in toughened epoxy resins - a review. *Composites Science and Technology*, 1988. 31: p. 179-223.

[19] Kawaguchi, T. and Pearson, R.A., The effect of particle-matrix adhesion on the mechanical behavior of glass filled epoxies. Part 2. A study on fracture toughness. *Polymer*, 2003. 44(15): p. 4239-4247.

[20] Wang, K., Chen, L., Wu, J., Toh, M.L., He, C. and Yee, A.F., Epoxy Nanocomposites with Highly Exfoliated Clay: Mechanical Properties and Fracture Mechanisms. *Macromolecules*, 2005. 38(3): p. 788-800.

[21] Gojny, F.H., Wichmann, M.H.G., Köpke, U., Fiedler, B. and Schulte, K., Carbon nanotube-reinforced epoxy-composites: enhanced stiffness and fracture toughness at low nanotube content. *Composites Science and Technology*, 2004. 64(15): p. 2363-2371.

[22] Sue, H.J., Gam, K.T., Bestaoui, N., Clearfield, A., Miyamoto, M. and Miyatake, N., Fracture behavior of [alpha]-zirconium phosphate-based epoxy nanocomposites. *Acta Materialia*, 2004. 52(8): p. 2239-2250.

[23] Williams, R.J.J., Rozenberg, B.A. and Pascault, J.P., Reaction-induced Phase Separation in Modified Thermosetting Polymers, in Polymer Analysis Polymer Physics. 1997. p. 96-156.

[24] Chen, J.L. and Chang, F.C., Temperature-dependent phase behaviour in poly (e-caprolactone) -epoxy blends. *Polymer*, 2001. 42: p. 2193-2199.

[25] Kim, B.S., Chiba, T. and Inoue, T., Morphology development via reaction-induced phase separation in epoxy/poly(ether sulfone) blends: morphology control using poly(ether sulfone) with functional end-groups. *Polymer*, 1995. 36(1): p. 43-47.

[26] Kim, B.S., Chiba, T. and Inoue, T., A new time-temperature-transformation cure diagram for thermoset/thermoplastic blend: tetrafunctional epoxy/poly(ether sulfone). *Polymer*, 1993. 34(13): p. 2809-2815.

[27] Hourston, D.J. and Lane, J.M., The toughening of epoxy resins with thermoplastics: 1. Trifunctional epoxy resin-polyetherimide blends. *Polymer*, 1992. 33(7): p. 1379-1383.

[28] Sala, G., Impact behaviour of heat-resistant toughened composites. *Composites: Part B*, 2000. 31: p. 161-173.

[29] Bennett, G.S., Farris, R.J. and Thompson, S.A., Amine-terminated poly(aryl ether ketone)-epoxy/amine resin systems as tough high performance materials. *Polymer*, 1991. 32(9): p. 1633-1641.

[30] Stenzenberger, H.D., Recent developments of thermosetting polymers for advanced composites. *Composite Structures*, 1993. 24: p. 219-231.

[31] Jo, W.H. and Ko, M.B., Effect of reactivity on cure and phase separation behaviour in epoxy resin modified with thermoplastic polymer: a Monte Carlo simulation approach. *Macromolecules*, 1994. 27(26): p. 7815-7824.

[32] Girard-Reydet, E., Vicard, V., Pascault, J.P. and Sautereau, H., Polyetherimide-modified epoxy networks: influence of cure conditions on morphology and mechanical properties. *Journal of Applied Polymer Science*, 1997. 65: p. 2433-2445.

[33] Peng, L., Cui, J. and Li, S., Studies on the phase separation of a polyetherimide-modified epoxy resin, 4-Kinetic effect on the phase separation mechanism of a blend at different cure rates. *Macromolecular Chemistry and Physics*, 2000. 201: p. 699-704.

[34] Zhang, J., Guo, Q. and Fox, B., Study on thermoplastic-modified multifunctional epoxies: influence of heating rate on cure behaviour and phase separation. *Composites Sicence and Technology*, 2009. 69: p. 1172-1179.

[35] Zhang, J., Guo, Q. and Fox, B., Structual and material property of rapidly cured thermoplastic toughened epoxy systems. *Journal of Applied Polymer science*, 2009. 113: p. 485-491.

[36] Zhang, J., Guo, Q., Huson, M., Slota, I. and Fox, B., Interphase study of thermoplastic modified epoxy matrix composites: Phase bahviour around a single fibre influenced by heating rate and surface treatment. *Composites Part A,* 2010. 41: p. 787-794.

[37] Pascault, J.-P., Sautereau, H., Verdu, J. and Williams, R.J.J., Thermosetting Polymers. 2002: CRC Press.

[38] Michler, G.H., Micromechanics of polymers. *J. Macromol. Sci- Phys.*, 1999. B38(5and6): p. 787-802.

[39] Davies, L.W., Day, R.J., Bond, D., Nesbitt, A., Ellis, J. and Gardon, E., Effect of cure cycle heat transfer rates on the physical and mechanical properties of an epoxy matrix composite. *Composites Science and Technology*, 2007. 67: p. 1892-99.

[40] Tsotsis, T.K., Interlayer toughening of composite materials. *Polymer Composites*, 2009. 30(1): p. 70-86.

[41] Quaresimin, M. and Varley, R.J., Understanding the effect of nano-modifier addition upon the properties of fibre reinforced laminates. *Composites Science and Technology*, 2008. 68(3-4): p. 718-726.

[42] Lowe, A., Matrix dominated tensile behaviour of unidirectional T300/914 and structural modeling of the material. *Journal of Materials Science*, 1996. 31: p. 983-993.

[43] Arai, M., Noro, Y., Sugimoto, K.-i. and Endo, M., Mode I and mode II interlaminar fracture toughness of CFRP laminates toughened by carbon nanofiber interlayer. *Composites Science and Technology*, 2008. 68(2): p. 516-525.

[44] Hojo, M., Matsuda, S., Tanaka, M., Ochiai, S. and Murakami, A., Mode I delamination fatigue properties of interlayer-toughened CF/epoxy laminates. *Composites Science and Technology*, 2006. 66(5): p. 665-675.

[45] Li, G., Li, P., Zhang, C., Yu, Y., Liu, H., Zhang, S., Jia, X., Yang, X., Xue, Z. and Ryu, S., Inhomogeneous toughening of carbon fiber/epoxy composite using electrospun polysulfone nanofibrous membrances by in situ phase separation. *Composites Science and Technology*, 2008. 68: p. 987-994.

[46] Masters, J.E., Improved impact and delamiantion resistance through interleaving Key *Engineering Materials*, 1989. 37: p. 317.

[47] Aksoy, A. and Carlsson, L.A., *Composites Science and Technology*, 1992. 43(1): p. 55.

[48] Tsotsis, T.K., Highly porous interlayers to toughen liquid-molded fabric-based composites. 2004: US.

[49] Odagiri, N., Kishi, H. and Nakae, T. T800H/3900-2 toughened epoxy prepreg system: toughening concept and mechanism. in American Society for Composites Sixth Technical Conference: Composite Materials, Mechanics, and Processing. 1991. Albany, New York.

[50] Hillermeier, R.W. and Seferis, J.C., Interlayer toughening of resin transfer molding composites. *Composites Part A: Applied Science and Manufacturing*, 2001. 32(5): p. 721-729.

[51] Zhang, J., Lin, T. and Wang, X., Electrospun nanofibre toughened carbon/epoxy composites: Effects of polystherketone cardo (PEK-C) nanofibre diameter and interlayer thickness. *Composites Science and Technology*, 2010. 70: p. 1660-1666.

[52] Lin, T. and Wang, X., Controlling the morphologies of electrospun nanofibres, in Nanofibre and nanotechnology in textiles, Brown, P. and Stevens, K., Editors. 2007, Woodhead publishing Ltd.: Cambridge, UK. p. 90-110.

[53] de Morais, A.B., de Moura, M.F., Marques, A.T. and de Castro, P.T., Mode-I interlaminar fracture of carbon/epoxy cross-ply composites. *Composites Science and Technology*, 2002. 62(5): p. 679-686.

[54] Rhee, K.Y., Koh, S.K. and Lee, J.H., Fracture resistance characteristics of graphite/epoxy laminated composites. *Polymer Composites*, 2000. 21: p. 155-64.

[55] Atkins, A.G. and Mai, Y.W., Elastic and plastic fracture - metals, polymers, ceramics, composites, biological materials. second ed. 1988: Ellis Horwood Limited.

[56] Stevanovic, D., Kalyanasundaram, S., Lowe, A. and Jar, P.Y.B., Mode I and mode II delamination properties of glass/vinyl-ester composite toughened by particulate modified interlayers. *Composites Science and Technology*, 2003. 63(13): p. 1949-1964.

[57] Petersson, L. and Oksman, K., Biopolymer based nanocomposites: comparing layered silicates and microcrystalline cellulose as nanoreinforcement. *Composites Science and Technology*, 2006. 66: p. 2187-2196.

[58] Lin, S., Cai, Q., Ji, J., Sui, G., Yu, Y., Yang, X., Ma, Q., Wei, Y. and Deng, X., Electrospun nanofiber reinforced and toughened composites through in situ nano-interface formation. *Composites Science and Technology*, 2008. 68(15-16): p. 3322-3329.

[59] Alexandra, M. and Dubois, P., Polymer-layered silicate nanocomposites: preparation, properties and uses of a new class of materials. *Material Science and Engineering R: Reports*, 2000. 28: p. 1-63.

[60] Flory, P.J., Molecular size distribution in three-dimensional polymers. VI. Branched polymer containing A-R-Bf-1-type units. *Journal of the American Chemical Society*, 1952. 74: p. 2718-23.

[61] Kim, Y.H., Webster, O. W. "Water soluable hyperbranched polyphenylene: a unimolecular micelle? " *Journal of American Chemical Society*, 1990, 112: p. 4592-3.

[62] Gao, C. and Yan, D., Hyperbranched polymers: from synthesis to applications. *Progress in Polymer Science*, 2004. 29: p. 183-275.

[63] Mezzenga, R., Boogh, L. and Manson, J.A.E., A review of dendritic hyperbranched polymer as modifiers in epoxy composites. *Composites Science and Technology*, 2001. 61(5): p. 787-795.

[64] Guo, Q., Habrard, A., Park, Y., Halley, P.J. and Simon, G.P., Phase separation, porous structure, and cure kinetics in aliphatic epoxy resin containing hyperbranched polyester. *Journal of Polymer Science: Part B: Polymer Physics*, 2006. 44: p. 889-899.

[65] Ratna, D. and Simon, G.P., Mechanical characterization and morphology of carboxyl randomized poly(2-ethyl hexyl acrylate) liquid rubber toughened epoxy resins. *Polymer*, 2001. 42(18): p. 7739-7747.

[66] Francis, B., Thomas, S., Jose, J., Ramaswamy, R. and Lakshmana Rao, V., Hydroxyl terminated poly(ether ether ketone) with pendent methyl group toughened epoxy resin: miscibility, morphology and mechanical properties. *Polymer*, 2005. 46(26): p. 12372-12385.

[67] Mimura, K., Ito, H. and Fujioka, H., Improvement of thermal and mechanical properties by control of morphologies in PES-modified epoxy resins. *Polymer*, 2000. 41(12): p. 4451-4459.

[68] Guo, Q., Phase behaviour in epoxy resin containing phenolphthalein poly (ether ether sulphone). *Polymer*, 1993. 34(1): p. 70-76.

[69] Zhong, Z., Zheng, S., Huang, J., Cheng, X., Guo, Q. and Wei, J., Phase behaviour and mechanical properties of epoxy resin containing phenolphthalein poly (ether ether ketone). *Polymer*, 1998. 39(5): p. 1075-1080.

[70] Cicala, G., Recca, A. and Restuccia, C., Influence of hydroxyl functionalized hyperbranched polymers on the thermomechanical and morphological properties of epoxy resins. *Polymer Engineering and Science*, 2005. 45: p. 225-237.

[71] Ratna, D. and Simon, G.P., Thermomechanical properties and morphology of blends of a hydroxyl-functionalized hyperbranched polymer and epoxy resin. *Polymer*, 2001. 42: p. 8833-8839.

[72] Varley, R.J. and Tian, W., Toughening of an epoxy anhydride resin system using an epoxidized hyperbranched polymer. *Polymer International*, 2004. 53: p. 69-77.

[73] Ratna, D. and Simon, G.P., Thermal and mechanical properties of a hydroxyl-functional dendritic hyperbranched polymer and trifunctional epoxy resin blends. *Polymer Engineering and Science*, 2001. 41: p. 1815-1822.

[74] Ratna, D., Varley, R. and Simon, G.P., Toughening of trifucntional epoxy using an epoxy-functionalized hyperbranched polymer *Journal of Applied Polymer Science*, 2003. 89: p. 2339-2345.

[75] Ratna, D., Varley, R. and Simon, G.P., Processing and chemorheology of epoxy resins and their blends with dendritic hyperbranched polymers. *Journal of Applied Polymer Science*, 2004. 92: p. 1604-1610.

[76] Varley, R.J., Toughening of epoxy resin systems using low-viscosity additives. *Polymer International*, 2004. 53: p. 78-84.

[77] Lin, K.F. and Chen, J.C., Curing, compatibility, and fracture toughness for blends of bismaleimide and a tetrafunctional epoxy resin. *Polymer Engineering and Science*, 1996. 36(2): p. 211-217.

[78] Zhang, J., Guo, Q. and Fox, B., Thermal and mechanical properties of a dendritic hydroxyl-functional hyperbranched polymer and tetrafunctional epoxy resin blends. *Journal of Polymer Science: Part B: Polymer Physics*, 2010. 48: p. 417-424.

[79] Zheng, S., Wang, J., Guo, Q., Wei, J. and Li, J., Miscibility, morphology and fracture toughness of epoxy resin / poly(styrene-co-acrylonitrile) blends. *Polymer*, 1996. 37(21): p. 4667-4673.

[80] Fernandez, B., Arbelaiz, A., Diaz, E. and Mondragon, I., Influence of polyethersulfone modification of a tetrafunctional epoxy matrix on the fracture behaviour of composite laminates based on woven carbon fibers. *Polymer Composites*, 2004. 25(5): p. 480-488.

[81] Blanco, I., Cicala, G., Costa, M. and Recca, A., Development of an epoxy system characterized by low water absorption and high thermomechanical performances. *Journal of Applied Polymer Science*, 2006. 100(6): p. 4880-4887.

[82] Hourston, D.J., Lane, J.M. and Macbeath, N.A., Toughening of epoxy resins with thermoplastics. II. Tetrafunctional epoxy resin polyethermide blends. *Polymer Inernational*, 1991. 26(1): p. 17-21.

[83] Song, X., Zheng, S., Huang, J., Zhu, P. and Guo, Q., Miscibility and mechanical properties of tetrafunctional epoxy resin/phenolphthalein poly(ether ether ketone). *Journal of Applied Polymer Science*, 2001. 79: p. 598-607.

[84] Bucknall, C.B. and Patridge, I.K., Phase separation in epoxy resins containing polyethersulphone. *Polymer,* 1983. 24: p. 639-644.

[85] Guo, Q., Huang, J., Ge, L. and Feng, Z., Phase separation in anhydride-cured epoxy resin containing phenolphthalein poly(ether ether ketone). *European Polymer Journal,* 1992. 28(4): p. 405-409.

[86] Okamoto, Y., Miyagi, H., Kihara, H. and Mitsui, S., Impact toughening mechanisms in glassy polymers, in Polymer blends and alloys, Shonaike, G.O. and Simon, G.p., Editors. 1999, CRC Press.

[87] Boogh, L., Pettersson, B. and Manson, J.-A.E., Dendritic hyperbranched polymers as tougheners for epoxy resins. *Polymer*, 1999. 40: p. 2249-2261.

[88] Guo, Q., Harrats, C., Groeninckx, G., Reynaers, H. and Koch, M.H.J., Miscibility, crystallization and real-time small-angle X-ray scattering investigation of the semicrystalline morphology in thermosetting polymer blends. *Polymer*, 2001. 42(14): p. 6031-6041.

[89] Purcell, K. and Drago, R., *Journal of the American Chemical Society*, 1968. 89: p. 2874.

In: Fiber-Reinforced Composites
Editor: Qingzheng Cheng

ISBN: 978-1-61470-303-7
© 2012 Nova Science Publishers, Inc.

Chapter 9

PLASMA SURFACE MODIFICATION OF GLASSY CARBON PLATES, CARBON FIBRES AND UHMWPE FIBRES FOR ADHESION IMPROVEMENT

Yukihiro Kusano[1], Henrik Mortensen, Steluta Teodoru, Joanna M. Drews, Tom L. Andersen, Bent F. Sørensen, Noemi Rozlosnik[1], Stergios Goutianos, Peter Kingshott, Susanta Mitra, Frank Leipold, Kristoffer Almdal[*], Poul K. Michelsen and Henrik Bindslev

Risø National Laboratory for Sustainable Energy,
Technical University of Denmark, DK-4000 Roskilde, Denmark
[1] Department of Micro- and Nanotechnology,
Technical University of Denmark, DK-2800 Kongens Lyngby, Denmark

ABSTRACT

Carbon fibres and ultra-high-molecular-weight polyethylene (UHMWPE) fibres are extensively used for improving mechanical properties of fibre reinforced polymer (FRP) composites. Strong adhesion between the fibre surfaces and the polymer matrix is one of the key issues for improving the longitudinal tensile strength of FRP composites. However, adhesion to these fibres is often difficult or weak because of their low surface energy and poor reactivity. Therefore, adhesion improvement of the fibre surfaces is of considerable industrial importance, which is enhanced by adding chemical functional groups and micro- or nano-scale roughness at the surfaces. Plasma treatment is attractive for this application because of its environmental compatibility and high treatment efficiencies without affecting the textural characteristics of the bulk material. Such a plasma is generally obtained at low pressures, but can also be generated at atmospheric pressure.

---

[*] Author to whom correspondence should be addressed; Email: yuki@risoe.dtu.dk

In the present chapter, low and atmospheric pressure plasmas are used to modify the surfaces of glassy carbon plates, carbon fibres and UHMWPE fibres for adhesion improvement. Here, a glassy carbon plate, consisting of amorphous-like sp^2 hybridized carbon material, can be an ideal model specimen for fundamental studies of adhesive properties of carbon fibres due to the structural similarity and easier handling than carbon fibres. The plasma treatment typically improved wettability of the surface, increased the oxygen containing polar functional groups at the surface, and enhanced the surface roughness. The results of the surface characterization were compared with adhesion properties with epoxy resin.

1. INTRODUCTION

Carbon fibres and ultra-high-molecular-weight polyethylene (UHMWPE) fibres are extensively used for improving the mechanical property and reducing the weight of fibre reinforced polymer (FRP) composites. Carbon fibres themselves exhibit high strength, high toughness and light weight, to which glassy carbon plates are structurally and chemically similar [1], but easier to handle. Glassy carbon plates are thus thought to be ideal model specimens for fundamental studies on the adhesion phenomena resulting from various surface modifications aimed at increasing the mechanical strength of carbon-FRP composite materials. Here, glassy carbon is a disordered polymer-like, non-graphitizing dense material, consisting of randomly interwoven ribbons of sp^2-hybridized elemental carbon [2-5]. UHMWPE has also been widely used because of its high chemical stability, high impact strength, flexibility, lightweight and low cost.

In order to achieve high mechanical strength of the FRP composites, fibre distribution, alignment, fibre damage, and interface between fibre surfaces and a polymer matrix need to be considered. In particular strong adhesion between the fibre surfaces and the polymer matrix is one of the key issues for improving the longitudinal tensile strength of FRPs [6]. However, due to the non-polar nature of those fibres they are difficult to wet and almost impossible to chemically bond to general polymer matrices. Proper surface modification should be chosen so that fibre surfaces can be wettable by the polymer matrix and bond to it tightly. Adhesion can be improved by surface treatment of the fibres, mainly by oxidation of the surfaces, introducing reactive polar functional groups onto the fibre surfaces. For this application, low temperature plasma surface modification is attractive due to its environmental friendliness, high treatment efficiencies without affecting the textural characteristics of the bulk material. Plasma surface modification can usually be divided into two categories with opposite effects, depending mainly on the process-gas(es) used. The first one mainly ablates the surfaces, and is usually called "plasma treatment", "plasma surface modification", "plasma ablation", or "non-polymer-forming plasma". The second one is usually called "plasma polymerization", "polymer-forming plasma", "plasma assisted chemical vapour deposition (PACVD)" or "plasma enhanced chemical vapour deposition (PECVD)". In the following "plasma surface modification" is meant to cover both types while "plasma treatment" is used for the first one. If the used gas(es) has high proportions of carbon and hydrogen atoms, double- or triple- bonds in its composition such as methane, ethylene, acetylene and ethanol, or if they are precursors such as metal-organic (organo-metallic) gas(es), the plasma often results in plasma polymerization or PECVD. Here, metal-organic gases are those which contain a metal, and particularly compounds in which the metal atom has a direct bond with a carbon atom. Otherwise, the plasma will have a tendency of ablation (plasma treatment).

Extensive research has been devoted to the surface modification of carbon fibres in order to improve their bonding to the polymer matrix. An epoxy resin is often preferred for the host matrix due to excellent electrical properties, high mechanical strength, high resistance against aging/hydrolysis, and high bond strength to many other polymer materials [7]. It is noted that for adhesion improvement the chemical effect of oxygen containing polar functional groups such as -OH, =O, and -COOH at the carbon fibre surfaces is known to be more important than the mechanical effect of rough surfaces [8]. The methods of surface modification of carbon fibres and glassy carbon plates include dry or wet oxidation, vacuum heat treatment [9], high pressure high temperature treatment [10], chemical oxidation [11,12], electrochemical oxidation [13], ultraviolet/ozone treatment [14], laser ablation [15], laser induced surface modification [16-19], reactive ion etching [20], thermal plasma [21], low temperature plasma treatment, polymer coatings, plasma polymerization, and PECVD [6,22,23]. The wet oxidation and electrochemical methods use nitric acid, $KMnO_4$, H_2SO_4, sodium hypochlorite, chromic acid, and electrolytic $NaOH$, while the dry methods use oxygen, ozone, and catalysis. However, these kinds of chemical methods may be least preferable. For example, when the carbon fibres are oxidized in concentrated nitric acid, the equipment used must have good corrosion resistance and the acid absorbed on the fibre surfaces must be properly removed by subsequent washing, which is time-consuming and inevitably damages and tangles the carbon fibres [6.23]. These methods can also produce environmental pollution, and thus plasma surface modification techniques are attractive for this application.

Low pressure radio-frequency (RF) O_2 or Ar plasma treatment is proved to be a rapid, reproducible and contaminant-free method for the introduction and enhancement of oxidic functional groups on the glassy carbon surfaces [24], while the glassy carbon surface can be hydrophobic by RF SF_6 plasma treatment [25]. RF O_2 plasma can reveal selective etching effects at the glassy carbon surfaces due to different etching rates of various domain types in the glassy carbon [26]. It is also reported that extended use of plasma treatment modified only the outermost surface without interfering with the bulk properties [27]. The surface damage by RF Ar plasma is investigated referring to the width of the Raman disorder (D) band at 1360 cm^{-1} on the glassy carbon surface [28]. It is found that microwave O_2 plasma treatment mainly produces constrictions at the entrance of the micro-pores [29]. Electron cyclotron resonance SO_2 plasma can induce etching of a glassy carbon surface without increasing the surface roughness [30]. The surface coverage with sulphur was shown to be dependent on the power and the pressure in the plasma chamber and also on the material that was used.

Reports of plasma polymerization on glassy carbon plates are limited, but plasma polymerization on different substrates for improvement of adhesion or mechanical properties is reported. Alexander *et al.* [31-34] studied plasma polymerized coatings of acrylic acid deposited on aluminium substrates, and the interface between the coatings and epoxy resin. Friedrich *et al.* [35,36] reported on the adhesion of metal layers to polymer surfaces using a pulsed plasma system. Their work showed a dependence of the Al peel strength on the polymer surface with different functional groups of carboxylic acid, hydroxyl and amino groups. They found that the carboxylic acid groups yield the strongest interfacial bonds [31-36]. Regarding the increase of the interfacial adhesion, [37] have shown increase in tensile strength, elongation, impact strength, and melt index of an interfacial system using maleic anhydride (MAH) grafted on high-density polyethylene-kaolin composites. Chiang and Ku

[38] reported on the grafting of MAH on Ar plasma pretreated low density polyethylene (PE) for flame retardation using ultraviolet (UV) light. They observed an increase in the mechanical strength as a function of the amount of grafted MAH. The use of low-power plasma surface modification [39] was first explored to improve the glassy carbon/matrix adhesion. In the past the monomer MAH has often been studied using low pressure RF plasma polymerization technique [40,46] and a pulsed or continuous plasma processing [47-53]. Conversely, the monomer 1,2-methylenedioxybenzene (MDOB) has rarely been studied [54].

Attempts to improve the interfacial adhesion between UHMWPE fibres and polymeric materials have been reported using plasma treatment [55-58]. Bond breakage induced by energetic ions, metastables or photons (UV), and reaction of free radicals with surfaces can induce etching and chemical modification. The removal of the monomer unit from the polymer chain, formation of free radicals and chain reactions can occur during plasma treatment [59].

These plasmas are often generated at low pressures. Plasma surface modification at low pressures, however, suffer from the drawbacks that they require expensive vacuum systems, and methods are only well-developed for batch or semi-batch treatments. To overcome these drawbacks an atmospheric pressure plasma treatment system can be used, which not only avoids the need for vacuum equipment but also permits the treatment of large objects, and continuous treatment on production lines [60].

In the present study glassy carbon surfaces were modified by plasma polymerization at low pressure [61], a cold plasma torch [62] and a dielectric barrier discharge (DBD) [63,64] at atmospheric pressure for adhesion improvement. In addition, carbon fibres [60,65] and UHMWPE fibres [66] were continuously treated by the DBD. The modified surfaces were characterized and the adhesive properties were evaluated, comparing with conventional methods.

2. Experimental Methods

2.1. Materials

Polished glassy carbon plates (Sigradur G) were purchased from HTW Hochtemperatur-Werkstoffe GmbH, Germany in sizes of $65 \times 5 \times 1$ mm^3 in most cases and $20 \times 20 \times 1$ mm^3 for some of the plasma-polymerized coatings. The glassy carbon plates for the cold plasma torch and the DBD were generally cleaned and degreased by wiping with acetone. Samples for x-ray photoelectron spectroscopic (XPS) analysis and for plasma polymerized coatings were cleaned ultrasonically for 5 minutes in dichloromethane followed by twice 5 minutes in acetone and finally 5 minutes in methanol, and then dried with an Ar jet.

Poly(acrylonitrile) (PAN) based unsized electrochemically-treated carbon fibres (TOHO TENAX HTA5001, 800tex) were used for the DBD treatment at atmospheric pressure without pre-cleaning. UHMWPE fibres (1300 filament yarn with a tex number of 145, 12 µm diameter of a multifilament) were purchased from Goodfellow, UK. They were ultrasonically cleaned in acetone for twice 5 minutes and in methanol for 5 minutes for removing the silicon containing contaminants from the surfaces.

Polyethylene (PE) films were used as substrates for measuring the Attenuated Total Reflectance Fourier transform Infrared (ATR-FTIR) spectra of plasma polymerized coatings.

MAH was obtained from Sigma-Aldrich as briquettes. These were ground into fine powder in a hood and stored in a metal container under vacuum before use. MDOB was purchased from Fluka (purity 99.0%) and used without further purification. Ultra-pure water was obtained using a MilliQ system from Millipore. Poly-(D, L-lactide) was received from Sigma-Aldrich in MW 90 000– 120 000 and was stored in a freezer.

2.2. Plasma Polymerization

Plasma polymerization was carried out using MAH, MDOB or their mixtures, driven by 50 Hz AC voltage. This plasma setup [54,67-68] can be used with a power input no higher than 1 W/L in most cases. The glassy carbon plates were mounted with double-sided tape on a glass tube where the plasma was expanded. A special benefit of this technique is that the power introduced to the system is enough to run the polymerization in a way comparable to liquid based polymerization of the monomer but without plasma induced destruction of the monomer. Several studies already put special emphasis on preserving the chemical structure of the monomers—either the ring structure [69] or the functional groups [47-50,70]. The composition of the mixture was controlled by adjusting the gas pressure of each component incrementally from a base pressure below 5 Pa, starting with MDOB (usually adding 2–3 Pa) and then adding MAH until the desired final pressure is obtained, typically an additional 2–3 Pa. The total pressure was about 10 Pa. In the case of pure monomer synthesis, the base pressure is below 5 Pa and the addition of the monomer was up to 9–10 Pa. The polymerization procedure consisted of two steps: surface activation followed by the plasma polymerization. The surface activation step is a sputtering process to remove oxygen from the outer layer of the surface. The polymerized layer cannot properly attach to the surface without activation. The activation time was varied in order to determine a suitable value for use in the experiment (data not shown here). For the activation step Ar or H_2 was fed through the chamber for 1 minute at the Ar flow rate of 5 sccm, followed by H_2 for 10 minutes at 50 sccm, and finally Ar for 20 s. During this time a plasma was maintained at a current of 30 mA $(1.21\pm0.16$ W/L). After the surface activation, the monomer was introduced to initiate plasma polymerization. The initial gas pressure of Ar was set prior to the activation. The plasma polymerization proceeded for 10 minutes, and in separate experiments the current was varied from 10 to 30 mA $(0.32\pm0.01-1.21 \pm0.05$ W/ L).

2.3. Plasma Treatment by Cold Plasma Torch and DBD at Atmospheric Pressure

The plasma torch is based on the design by Koinuma *et al.* [71] and is operated in a ventilated box at atmospheric pressure. Figure 1 shows a schematic of the setup. The powered electrode is a 0.5-mm diameter tungsten wire placed inside a glass pipette, which serves as the dielectric barrier. The inner and outer diameters of the pipette are 5.5 mm and 7 mm, respectively, tapering to approximately 1.5 mm and 2.2 mm at the plasma outlet. Wrapped around the glass tube a strip of Al foil constitutes the ground electrode. It is connected to ground through a 50-Ω current viewing resistor to facilitate measurement of the DBD current. The voltage and current were recorded by means of a digital oscilloscope which was also used to calculate the corresponding power dissipation. The total power dissipated in the torch cannot be measured directly, but it is found to be fair to assume that it is within the same order of magnitude as the measured power. The plasmas were generated by applying an AC voltage (typically 4 kV$_{p-p}$ at 14 kHz) between the electrodes in a flow of 3 L/min (~28 m/s) of pure He or He premixed with 5 mL/min of NH_3. The measured power was around 1 W. During plasma treatment, the samples were electrically isolated. A stable glow was observed between the pipette wall and the powered electrode as well as in the 5-mm gap between the torch orifice and the substrate. The light emitted by the effluent plasma was focussed by a quartz lens onto a spectrometer grating with 3600 grooves/mm (for wavelengths $\lambda < 400$ nm) or 1200 grooves/mm ($\lambda > 400$ nm), and OES were captured by means of a CCD camera.

An atmospheric pressure DBD has been used to treat the glassy carbon plates. Figure 2 shows a schematic diagram of the experimental setup of the DBD with parallel plate water cooled metal electrodes (50 × 50 mm^2) covered with alumina plates (100 × 100 × 3 mm^3). The DBD was generated by an AC (approximately 40 kHz, V$_{pp}$ = 23 ~ 32 kV) power supply (Generator 6030. SOFTAL Electronic GmbH). The average power input was obtained by measuring voltage and current with a high voltage probe and a 50 Ω current viewing resistor, respectively. He, Ar, a mixture of He with O_2, CO_2 or NH_3, or a mixture of Ar with NH_3 was fed into the DBD. O_2 and CO_2 gases were used for enhanced surface oxidation [72], while NH_3 was attempted for introducing amino groups onto the glassy carbon surfaces. The flow rate of He and Ar was 1 L/min, while that of O_2, CO_2 and NH_3 was 0.1 L/min. The applied power to the discharge was 80-100 W and the treating time was 600 s.

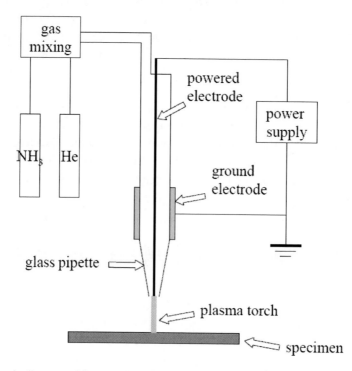

Figure 1. Schematic diagram of the experimental setup for atmospheric pressure cold plasma torch treatment of a glassy carbon plate [62].

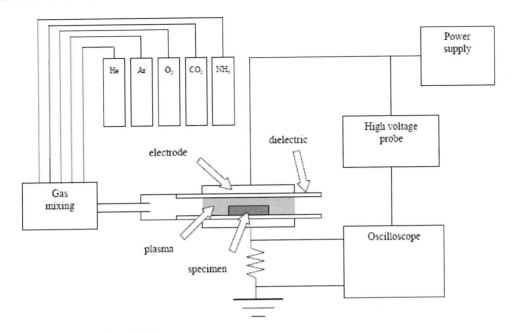

Figure 2. Diagram of DBD [63].

The carbon fibres were continuously treated with the atmospheric pressure DBD. The average power input was approximately 100 W corresponding to a power density of 8 W/cm. He, Ar, a mixture of He with O_2, N_2 or NH_3, or a mixture of Ar with NH_3 was fed into the DBD. The flow rates of He, Ar, O_2, N_2 or NH_3 are 1, 1, 0.1, 0.1, and 0.001 L/min, respectively. The carbon fibres were also treated with the DBD in ambient air [65].

The UHMWPE fibres were continuously treated with the atmospheric pressure DBD [60,65] with a feed speed of 100 mm/min. Ar, He, O_2, N_2 or a mixture of He with 9 vol. % O_2 was fed into the DBD, with a total flow rate varying between 1 and 30 L/min. However, contaminations and the leakage of N_2, O_2, and H_2O from ambient air to the discharge is often inevitable, resulting in introduction of oxygen and/or nitrogen at the exposed surfaces in a discharge, even if no oxygen or nitrogen containing gas is apparently supplied [60,65,73]. In fact such a leakage sometimes plays an important role in oxidizing and nitriding the surface. The reproducibility of the treatment associated with the level of the leakage is indirectly ensured by confirming that the measured voltage and power are reproducible in the same conditions. The average power input was obtained by multiplying measured voltage and current numerically and integrating over an AC period. The input power into the DBD was adjusted by varying the frequency. The change of the frequency from 40 to 35 kHz corresponds to an input power from approximately 50 to 150 W.

2.4. Water Stability Test

Some specimens with plasma polymerized coatings were dipped in water for 24 h in order to check the stability of the film (water treatment). In all cases the surfaces were dried in a fume hood in clean Petri dishes.

2.5. Surface Analysis

Contact angles with deionized water were measured in air at room temperature both before and after the plasma treatment for evaluation of the wettability of the surfaces (CAM100. Crelab Instruments AB). Plasma torch treated samples for the contact angle measurement were prepared for each gas composition with total exposure times of 1 and 5 minutes during which the sample was manually moved continuously to achieve a quasi-homogeneous exposure of the surface in the ~100-mm^2 zone between ~32 mm and 52 mm from one end of the sample.

FTIR was performed using a PerkinElmer Spectrum B instrument that was equipped with a Germanium crystal. All spectra were collected within the range from 700 to 4000 cm^{-1} with a resolution of 4 cm^{-1} and averaging over 32 scans. The measurements were obtained at normal atmospheric pressure. Measurements were performed before and after water stability test.

X-ray photoelectron spectroscopy (XPS) was performed for the surface characterization of the glassy carbon plates and the carbon fibres using a SPECS Sage 100 instrument with a nonmonochromatic Mg $K\alpha$ x-ray source at a power of 275 W (11keV and 25 mA), resulting in a maximum probe depth of ≈ 10 nm. The pressure in the chamber was always at a pressure of $< 10^{-5}$ Pa. Two kinds of spectra were obtained: survey and high resolution. The survey spectra were acquired at 100 eV pass energy in the range from 0 to 1100 eV. Atomic concentrations were calculated from the intensities of the peaks in this spectrum. The high resolution spectra were acquired at 23 eV pass energy. The spectra were deconvoluted through curve-fitting. The technique employs 100% Gaussian components and linear background subtraction that relies on a least squares minimization routine and the full width at half maximum of the peaks was constrained to 1.8 or 2 eV. Accurate binding energies were obtained by referencing the $\underline{C}$–C/$\underline{C}$–H component in the C1s spectrum to 285.0 eV [74].

For the UHMUPE fibres XPS were collected using monochromated photons (1486.6 eV) with a lateral resolution of 30 μm (K-Alpha, ThermoFischer Scientific, UK). The binding energies (BEs) were referred to the hydrocarbon component ($\underline{C}$-C, $\underline{C}$-H) at 285 eV. The valence band (VB) spectra were collected over 100 scans and a pass energy 25 eV. The collection time was less than 20 min for a survey spectrum, and less than 5 hours for a VB spectrum. For deconvolution of VB spectra a Savitsky-Golay smoothing technique with linear background was applied, using a fourth order polynomial and repeating 100 times.

Raman microscopy was used to characterize the bonding states of sp^2 hybridized carbon atoms of the glassy carbon plates. Raman scattering was produced by 785-nm diode laser excitation at a laser power of ≈ 17 mW. Spectra were recorded using a Renishaw NIR 780TF series, which was equipped with a charged couple detector and a microscope (Leica).

Topographical and thickness measurements of the plasma polymerized coatings were performed with a PSIA XE-150 atomic force microscopy (AFM) using BS-Tap 300 tips (17-μm long SiO$_2$ symmetric probe) (Budget Sensors) in tapping mode and intermittent contact mode. The masking method of [75] was used for thickness measurements. In brief, a droplet of a 5% poly-(D,L-lactide)(PDLLA) solution in acetone was pipetted onto the clean glassy carbon surface and dried. This substrate was then subjected to plasma polymerization. Afterwards, the sample was placed in water for 24 h to allow for the PDLLA to undergo hydrolysis. This was then removed with sharp tweezers. As a result of this procedure there was a region on the substrate which is free from the coating. The scan size was set up for 20 μm and the scan rate to 0.5 Hz for determination of the cross-section picture. For topographical measurements the scan size was set up for either 5 or 1 μm with the scan rate 1 Hz. All data obtained were analyzed with the PSIA, XEI software (version 1.5).

Plasma torch treated samples for AFM measurements were prepared for each gas composition with total exposure times of 1 and 5 minutes during which the sample was manually moved continuously to achieve a quasi-homogeneous exposure of the surface in the $\sim$100-mm^2 zone between $\sim$32 mm and 52 mm from one end of the sample. The induced microscopic roughness of the surfaces was evaluated by the intermittent contact AFM (XE-150, PSIA). For estimating the etch rate we used a stylus profilometer (Dektak V200-Si, Veeco Instruments) with a 700-nm diameter stylus to record height profiles across treated spots on the surface.

The UHMWPE fibre surfaces were also observed using the AFM (XE-150, PSIA) with 2 μm scan width in the fibre direction in the non-contact mode. The fibres were fixed with a double-side tape on a silicon wafer.

2.6. Adhesion Tests

A simple adhesion test was applied for the preliminary evaluation of the effect of the plasma torch treatment on adhesion with epoxy. A $\sim$1.5-mm thick layer of epoxy (Strong Epoxy Rapid 2806, Casco) was applied to one side of the samples and cured at 100 °C for 60 minutes. Delamination was then attempted by trying to insert a blade at the interface between the glassy carbon surface and the cured epoxy.

Another simple qualitative mechanical test has been performed for the preliminary evaluation of the plasma treatment effect on adhesion of UHMWPE fibres with epoxy for six conditions: untreated, and treated in Ar (1 L/min), He (1 L/min), He/O$_2$ (1 L/min), N$_2$ (20 L/min) and O$_2$ (10 L/min). The input power was 140 W. The 200 mm long fibres were aligned in parallel on a metal plate. Two mixtures of an uncured epoxy resin and amine (Strong Epoxy Rapid 2806, Casco) were prepared. The first mixture contains 50 vol. % of the epoxy and 50 vol. % of the amine, showing optimized mechanical strength after curing. The second mixture contains much a smaller amount of amine with lower mechanical strength than the first one. Approximately 3-mm thick mixtures were applied onto the plate so that they covered the fibre surfaces completely. They were cured at room temperature for a day, and then at approximately 80 °C in air for one hour. 90° peeling test was performed to evaluate the adhesive property of the fibres with the epoxy at room temperature by gently pulling the fibres from the epoxy composites manually.

For quantitative measurements of the strength of adhesion of epoxy to the treated surfaces, double cantilever beam (DCB) sandwich specimens produced with the glassy carbon plates (65 × 5 × 1 mm^3) were tested. The DCB sandwich specimens were prepared as follows: the glassy carbon plates were placed in a silicone mould. 0.5 mm thick Teflon inserts were placed at each end of the zone to be bonded, ensuring a uniform adhesive thickness. The uncured epoxy resin (Prime 20, SP Systems) fluid was mixed with the hardener (diethylenetriamine/polyoxypropylamine) at a ratio 4:1 and degassed under vacuum for 2 × 15 minutes or more to remove air bubbles before being introduced between the inserts. The specimens were left in the mould for at least a day at room temperature and for 16 h at 50 °C for curing the epoxy resin. Steel beams were glued to the outside of the sandwich structure with an epoxy adhesive (Scotch Weld) to support the glassy carbon plates. The curing time was 24 h at room temperature. With the 6-mm thick steel beams attached, the test specimens fit the testing fixture, which was placed under an optical microscope for observation of crack growth.

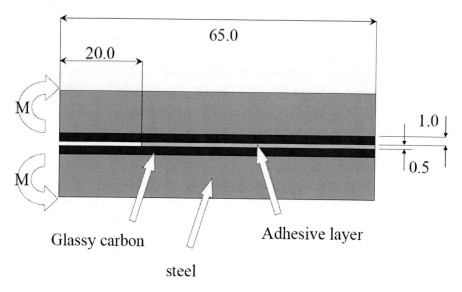

Figure 3. The schematic diagram of the DCB specimen loaded with pure bending moments M (All dimensions are in mm) [63].

A schematic diagram of the prepared specimens is shown in Figure 3. Pure bending moments were applied to the DCB sandwich specimens by a loading arrangement with wires and special grips [76]. During the test, the moment was recorded along with the opening of the cracked end of the specimen. An acoustic emission sensor taped to the specimen aided our visual detection of crack growth events and thus allowed recording for each specimen several load/unload cycles each associated with crack growth with a loading rate between 3.6×10^{-3} and 2.9×10^{-2} Nm/s [77]. The corresponding fracture energy was calculated as the critical energy release [78] using the highest moments applied during the last few cycles before failure. For this calculation the effect of resins is ignored, since their stiffness is much smaller compared to the glassy carbon plates and the steel beams. Additionally this is one of the few testing methods where the energy release rate is independent of the crack length (the energy release rate is calculated from a J integral solution and is a function of the moment, geometry, and elastic properties of the materials [76].

Carbon fibre reinforced epoxy (CFRE) plates for DCB specimens were prepared using both untreated, 1- and 2-s He plasma-treated carbon fibres with the epoxy resin. Steel beams were glued to the outside of each composite plate with the epoxy adhesive (Scotch Weld). Adhesive property was evaluated using fracture resistance values at the crack growth initiation and at the steady state as shown in Figure 4. Data are taken when the stable crack growth is seen in the mid plane of the CFRE plates so that the adhesive property between the carbon fibre surfaces and the epoxy matrix can be properly evaluated.

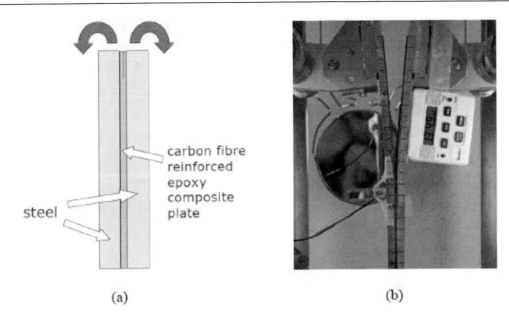

Figure 4. The carbon fibre reinforced epoxy composite plate sandwiched by steel plates. A schematic view (a) and a photo image during the fracture test (b).

For measurements of the etch rate, one sample per gas composition was prepared by stationary exposure to the plasma torch in three separate areas, which were exposed for 3 s, 30 s, and 300 s, respectively. The spots with a 3-s treatment time were characterised by AFM prior to the profilometry. Both types of treatment procedures were employed for XPS samples.

3. RESULTS AND DISCUSSION

3.1. Glassy Carbon Plate

3.1.1. Plasma Polymerization
Extensive surface characterization was carried out for the different plasma polymerized coatings made from MAH, MDOB, and MDOB/MAH. The aim was to elucidate the properties that influence the adhesion between the glassy carbon plates and the epoxy. The techniques used included ATR-FTIR and XPS for investigating the surface chemistry, and AFM for topography and thickness measurements. Finally, fracture mechanics results for the different plasma treatments are given.

All ATR-FTIR spectra were performed for plasma polymerized coatings deposited on PE films.

Representative ATR-FTIR spectra of the ppMAH coatings freshly deposited onto PE films at a plasma current of 20 mA (0.75±0.03 W/L) are shown in Figure 5 (a) (bold line). The saturated five-membered ring of anhydride groups absorbs near 1860 and 1780 cm^{-1} region. There is also significant absorption at 1730 and 1620 cm^{-1}. These belong to the carbonyl and esters [47]. Moreover, absorption bands attributed to non-cyclic anhydrides and five-member ring anhydrides, respectively, are observed at 1070 and at 930 cm^{-1} [79].

Additionally, in the spectra of ppMAH coatings an absorption band is found at 1230 cm^{-1}. This originates from ring stretching. The spectra in Figure 5 (a) show the effects of water treatment (dashed curve) on the chemistry of the ppMAH coatings. As expected the anhydride groups are hydrolyzed into carboxylic acid groups by the water treatment. This conversion is clearly observed in the difference spectrum between the hydrolyzed and polymerized samples (Figure 5 (b)). The absorption bands of the anhydride groups (at 1860, 1780, 1230, 1070 as well as 930 cm^{-1}) diminish, while the intensity of the absorption band at 1730 cm^{-1} increases [48].

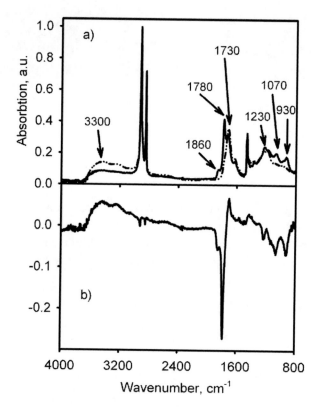

Figure 5. ATR-FTIR spectra of (a) a ppMAH surface (20 mA) before (solid line) and after (dashed line) water saturation; (b) the difference spectrum for ppMAH surfaces before and after water treatment for 20 mA [61].

The ATR-FTIR spectrum of the ppMDOB coating (Figure 6) shows a number of peaks including the C(O)O stretching (1710 cm^{-1}), stretching in C=C (1600 cm^{-1}) and in C–H (1350 cm^{-1}) as well as in the aromatic structure (1100 and 1240 cm^{-1}), paradisubstituted benzene, and (1040 cm^{-1})- orthodisubstituted benzene. The stretch from the acids in the range 3000–3600 cm^{-1} is also observed here. There is also a stretch at 930 cm^{-1} which comes from C–O–C and the ring stretching in agreement with the monomer structure.

The ATR-FTIR spectra for the plasma polymerized coatings synthesized with mixtures of MAH and MDOB are shown in Figure 7. The spectra show clear signals from both of the monomers.

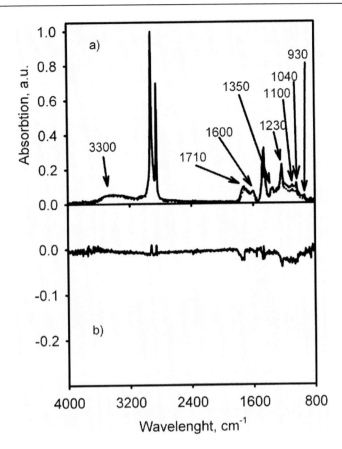

Figure 6. ATR-FTIR spectra of (a) a ppMDOB surface (20 mA) before (solid line) and after (dashed line) water saturation; (b) the difference spectrum for ppMDOB surfaces before and after water treatment for 20 mA [61].

The signals from ppMAH originate mainly from the anhydride groups (absorption bands at 1860 and 1780, 1230, 1070 as well as 930 cm^{-1}), which disappear after water treatment. On the other hand, carboxylic acid groups (1730 cm^{-1}) are created during hydrolyses of anhydride groups as also observed from the ATR-FTIR spectra, where the intensity of this absorption band increases. The carboxylic acid groups originate not only from the hydrolyzed anhydride groups but also from ppMDOB as it was seen in Figure 6. The difference between the freshly deposited and hydrolyzed films is particularly clear in the difference spectrum (Figure 7 (b)). The significant increase of signal at 1730 cm^{-1} is clear to see on the difference spectrum as well as the increase in the range of 3300 cm^{-1}, which is in response to the acid group stretching. The carbonyl (C=O) stretch is also present at 1620 cm^{-1}. Peaks from ppMDOB are also present at 1350 cm^{-1} from C–H as well as 1240 and 1040 cm^{-1} from the ring structure.

The atomic composition of untreated glassy carbon surface has 95% and 5% carbon and oxygen contents, respectively (molar O/C ratio 0.05, Table 1), measured by XPS. This residual oxygen is most likely from surface oxidation that occurs during or after processing of the substrates.

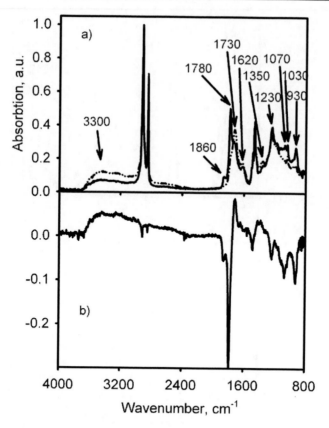

Figure 7. ATR-FTIR spectra of (a) a ppMAH/MDOB surface (20 mA) before (solid line) and after (dashed line) water saturation; (b) the difference spectrum for ppMAH/MDOB surfaces before and after water treatment for 20 mA [61].

The C1s spectrum in Figure 8 (a) consists of peaks assigned to C–C/C–H/C=C (peak A) at 285.0 eV and a small peak assigned to C–O–C/C–OH at 286.5 eV (peak B). All the other peaks are π–π* shake-up peaks from unsaturated sp^2 bonding of carbons (peaks C–G).

XPS measurements were also performed on the ppMAH deposited on glassy carbon plates. Only carbon and oxygen should be present on the surface, based on the monomer structure. The theoretical molar O/C ratio of the monomer is 0.75. However, there is significant loss of oxygen during the plasma polymerization since the molar O/C ratio for all plasma conditions (10 – 30 mA: 0.36±0.01 – 1.21±0.05 W/L) is approximately 0.5 (Table 1). The molar O/C ratio does not change before and after the water treatment.

Figure 8 (b) shows the high resolution C1s spectrum with peak fitted. All the C1s spectra were fitted with four components (C–C, C–O, C=O, C(O)O) corresponding to the bonds expected on the modified substrate from the MAH monomer. Table 1 gives an overview of the curve fitting results which contribute to the C1s spectra for all plasma conditions. There are no significant differences between the treatments with different power as well as before or after water treatment. The peak with the highest binding energy appears at 289.3 eV, and it can be assumed that the registered spectrum comes only from the plasma modified layer. If not, then there should be registered counts of higher bonding energies from the substrate itself (from π*–π shake up). Since the probe depth of XPS is ~10 nm, the thickness of the plasma polymerized layer was at least 10 nm.

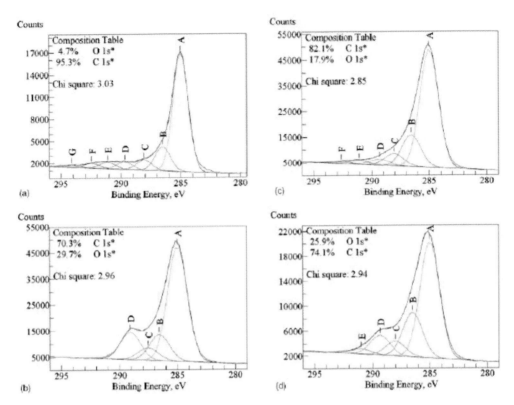

Figure 8. Representative XPS high resolution C1s spectra for (a) clean glassy carbon surface; (b) a ppMAH 20 mA surface; (c) a ppMDOB 20 mA surface; and (d) a ppMAH/MDOB 20 mA surface [61].

Table 1. Summary of the XPS results for a ppMAH surface before and after water treatment samples. Due to insignificant changes before and after water treatment the data was averaged. The ratio O/C from the monomer is 0.75 [61].

Power density [W/L]	O/C	C1s [%] 285.0 eV CH/CC (A)	~286.5 eV CO/COH (B)	~288 eV C=O/OCO (C)	~289.3 eV C(O)O (D)
Glassy carbon	0.05	68.0	13.9	6.0	
0.35±0.01	0.47±0.02	61.1±3.4	16.6±1.7	7.4±0.8	16.6±3.1
0.55±0.01	0.48±0.02	58.1±1.9	18.0±1.6	6.5±1.9	18.7±1.5
0.77±0.03	0.44±0.04	60.9±3.5	17.1±2.2	7.0±1.0	17.2±2.3
0.99±0.02	0.45±0.04	57.5±2.1	19.8±1.6	7.2±2.9	16.7±1.5
1.23±0.05	0.43±0.01	58.7±1.2	19.8±2.0	7.7±1.4	15.3±2.3

[a] Summary of the curve fits for the glassy carbon not included. There are five peaks that are assigned to π–π* satellites. Refer to Figure 8 (a).

The ppMDOB coating on glassy carbon was also examined by XPS before and after water treatment. The coatings tested were generated by two different plasma conditions (10 mA 0.35±0.01 W/L and 20 mA 0.78±0.02 W/L). The theoretical molar O/C ratio of monomer is 0.286, and during the plasma deposition process only a marginal loss of oxygen (the molar

O/C ratio is 0.23–0.24, Table 2) was observed compared to the ppMAH coating. Figure 8 (c) shows the curve fitted C1s spectrum for a representative sample. The ppMDOB is fitted with five different peaks assigned to C–H/C–C at 285.0 eV, C–O/C–OH at 286.5 eV, C=O/O–C–O at 288.0 eV, C(O)O at 289.3 eV, and the last one corresponding to the unsaturated sp^2 bonding ($\pi*-\pi$ shake up) at approximately 291 eV. The peaks are in agreement with the monomer structure. Winther-Jensen *et al.* [54] have previously argued that during such plasma polymerization it is possible to get carboxylic acid groups from opening of the dioxy-ring.

Table 2. Summary of the XPS results for a ppMDOB surface before and after water treatment. Due to insignificant changes before and after water treatment the data was averaged. The ratio O/C from the monomer is 0.286 [61].

Power density [W/L]	O/C	C1s [%]				
		285.0 eV CH/CC (A)	~286.5 eV CO/COH (B)	~288 eV C=O/OCO (C)	~289.3 eV C(O)O (D)	~291 eV $\pi*-\pi$ (E)
0.32±0.01	0.24±0.04	65.0±4.4	20.6±2.8	8.2±1.7	3.6±1.2	2.5±0.5
0.73±0.02	0.23±0.03	68.1±1.8	18.2±1.0	7.8±1.0	3.3±0.5	2.2±0.3

The ppMDOB/MAH samples were examined with XPS before and after the water treatment. In the C1s spectra, there are not only peaks from C–C/CH, C–O, C=O/O–C–O, and C(O)O but also components with higher bonding energies; cf. Figure 8 (d). The last of the fitted peaks is due to $\pi*-\pi$ shake up from the benzene ring in the MDOB monomer structure preserved in the polymerized film. The resulting fitted distributions, positions, and fracture percentages are shown in Table 3. There are no significant differences between the treatments. The molar O/C ratio is 0.37.

Table 3. Summary of the XPS results for a ppMA/MDOB surface before and after water treatment samples. Due to insignificant changes before and after water treatment the data was averaged. The ratio O/C from the monomers is 0.52 [61].

Power density [W/L]	O/C	C1s [%]				
		285.0 eV CH/CC (A)	~286.5 eV CO/COH (B)	~288 eV C=O/OCO (C)	~289.3 eV C(O)O (D)	~291 eV $\pi*-\pi$ (E)
0.35±0.01	0.37±0.02	55.9±1.6	22.4±1.0	9.0±2.9	10.3±3.1	2.7±0.4
0.79±0.02	0.37±0.03	54.7±4.6	23.5±3.8	9.7±3.1	10.0±1.6	2.2±0.3

The thicknesses of the plasma polymerized films were measured by cross sectioning the middle of the specimen after removal of the PDLLA mask. Figure 9 shows the film thickness as a function of the power used during plasma polymerization for all three coatings tested. From the data it can be seen that the ppMAH coating thickness increases from 25 nm at 0.4 W/L to 100 nm at 1.2 W/L during the 10-minute plasma polymerization time. With the ppMDOB coating the same tendency is observed, although the deposition rate is higher than that of the ppMAH coating.

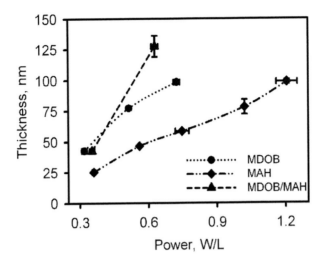

Figure 9. Thickness of the polymerized films vs. power. All measured samples were plasma polymerized for 10 minutes as well as water treated for ~24 h and dried in the fume hood [61].

The thickness of the ppMDOB coating increases from 43 nm at 0.3 W/L to about 100 nm at 0.7 W/L. Interestingly, the ppMDOB/MAH coating has the highest deposition rate and increases up to 125 nm at 0.8 W/L. Since the pressure during all three types of depositions is constant at 10 Pa the different deposition rates and thicknesses observed are a function of the plasma power and flow rate. The effect of increasing the deposition rate with increasing plasma power has previously been shown [79]. This can be explained by the increase in the number of energetic electrons generating a higher population of energetic ions hitting the surface, particularly at low powers. Yasuda [81] has shown that the polymer deposition rate (R) increases as the parameter, defined by the power (P) divided by the flow rate (F) and the molecular weight of the monomer (M), by $R \propto (P/FM)$. Figure 9 shows that the deposition rate is smaller for "lighter" MAH than for MDOB at constant power and flow rate (i.e., the pressure is unchanged). This is not surprising as the Yasuda equation is only valid for homologous monomers. In the present study monomers of very different chemistry natures are compared. The dependence of single monomer polymerization follows the Yasuda parameters. The results of roughness measurements of the plasma modified layers by AFM after water treatment are shown in Table 4. The roughnesses of the plasma modified layers are seen to generally increase with the polymerization power. Plasma polymerization is a random process. The roughness changes are due to polymer thickness. Even though the plasma polymerization was performed for consistent period of time, the speed of polymerization has been changed.

Figure 10 shows the fracture energy for crack initiation (intrinsic interface fracture energy) for the various conditions of plasma polymerization tested. Despite the significant scatter in the results, an increase in fracture energy for all the plasma modified substrates can be seen when it is compared to the fracture energy of untreated glassy carbons (the fracture energy was between 5 and 7 J /m^2).

Table 4. Roughness data for plasma polymerized coatings deposited on glassy carbon plates [61].

Power density [W/L]	Scan size [μm]	Roughness R_a [nm] ppMAH	ppMDOB	ppMDOB/MAH
0.4	1	0.74±0.49	0.79±0.42	1.98±1.90
	5	0.97±0.53	0.98±0.48	2.15±2.18
0.5	1	0.78±0.11		
	5	1.37±0.19		
0.7	1	0.83±0.35	2.28±0.75	21.18±12.78
	5	4.53±5.45	2.54±0.55	22.76±9.94
1.0	1	1.11±0.24		
	5	1.08±0.31		
1.2	1	3.70±1.40		
	5	4.39±1.80		

The degree of improvement in the fracture energy depends on the monomer used and on the time and power of the plasma as well. Specimens modified with ppMDOB show an increasing trend of fracture energy with increase of plasma power for the two different powers examined. The stronger interface regardless of the power is obtained with the ppMAH coating. The fracture energy of ppMAH has an average maximum of 78.7±2.4 J /m^2 at 0.7 W/L, while the weakest (on average) of the interfaces has a fracture energy of around 20.4±7.0 J /m^2 at 0.3 W/L.

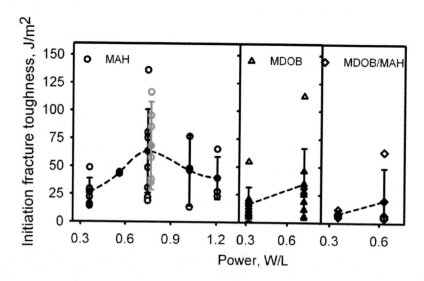

Figure 10. Fracture energy values for crack initiation as a function of power during polymerization (time: 10 minutes (black) and 3.2 minutes (grey)). Open points represent experimental data whereas closed symbols represent the average value and the uncertainty for each treatment [61].

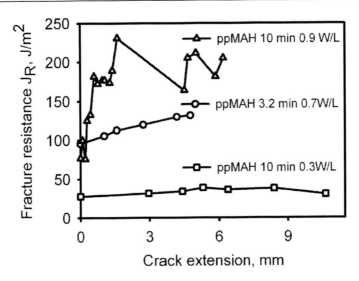

Figure 11. Fracture resistance curves for glassy carbon modified with ppMAH at different powers and different plasma treatment time [61].

In Figure 11 some representative fracture resistance curves are shown for a weak interface (ppMAH: 10 minutes, 0.3 W/L) and two strong interfaces (ppMAH: 10 minutes, 0.9 W/L and 3.2 minutes, 0.7 W/L). As expected for the strong interfaces, an increase in fracture energy with crack extension was observed. In contrast, for the weak interface the fracture energy remained almost constant.

A study of the dependence of polymerization time on fracture energy was also made. In this case, a number of experiments were carried out using ppMAH deposited at 0.7 W/L for only 3.2 min. The interface was found to be as strong as the ppMAH deposited with the same power for 10 minutes. The following discussion concerns the ppMAH coatings since more data presently exist for this treatment. The pp-MAH coatings were found to be sensitive to the power applied during plasma polymerization, whereas surface characterization did not reveal any significant difference in chemistry for the various deposition conditions. The explanation for the difference in fracture energy for samples of similar surface chemistry is unknown at present. However, it is speculated that there could be a subtle difference in cross-link density at the outer surface of the plasma polymerized layer generated from the different treatments. This is not depicted in the XPS data. This could result in a different degree of diffusion of the epoxy resin during the curing process and a subsequent variation in the number of chemical links at the interface between the plasma polymer and the matrix. In addition, a different degree of cross linking results in films that have differences in cohesive strength. Perhaps both lateral bonds in the layer and maximum penetration of the resin into the film are desirable for a tougher interface. Another point that needs to be mentioned is the significant scatter in fracture energy for crack initiation and fracture resistance curves among specimens of the same treatment. The different notch geometry from specimen to specimen could contribute to the scatter observed in the fracture energy for crack initiation but this cannot explain by itself the large differences experimentally observed. Surface contamination during specimen handling should also have an effect on the measured values. Furthermore, almost all experiments were terminated by the onset of unstable crack growth. In some cases the interface crack kinked into the glassy carbon plate. The instability is characterized by a very high crack velocity and it is possible that there is a viscoelastic or viscoplastic effect of the epoxy resin. Pure epoxy specimens were also tested and the crack growth became unstable usually immediately after crack initiation. Similar results were reported by Du *et al.* [81], who related this behaviour to stick-slip fracture instabilities commonly observed in polymers. Although it was mentioned above that the experimental setup used here provided stable crack growth, this is not necessarily the case for rate dependent materials. However, despite the significant scatter it has been shown that adhesion between glassy carbon plates and an epoxy resin was altered by changing the parameters during plasma polymerization.

3.1.2. Cold Plasma Torch

Here, glassy carbon plates were treated by the cold plasma torch at atmospheric pressure. If the plasma etches or erodes the glassy carbon surface, the plasma above the surface can be expected to contain a substantial amount of electronically excited carbonaceous species. Yamagata *et al.* [83,84] used OES to detect such species in the plasma plumes resulting from laser ablation from various other elemental carbon surfaces.

Here OES of the torch plasmas covering the range of wavelengths between 187 and 800 nm was recorded. The dominant features in the spectra are due to $N_2^*(C^3\Pi_u-B^3\Pi_g)$, $N_2^+(B^2\Sigma_u^+-X^2\square\Sigma_g^+)$, and He I. No signatures of carbonaceous species could be identified in the spectra, which did not depend on the distance from the substrate surface to the probed volume.

As can be seen from Figure 12 (a), a comparison of the emission from the NH_3/He discharge with that from the He one revealed that the NH_3 resulted in a substantially more intense $NH(c^1\Pi-a^1\Delta)(0, 0)$ band at 324 nm and additional bands around 336-337 nm that may partly be attributed to the $NH(A^3\Pi-X^3\Sigma)(0, 0)$ and $(1, 1)$ transitions. By contrast, O I lines and the OH $(A-X)(0,0)$ band at 306.4 nm (see Figure 12 (b)) were much more intense in the pure He discharge spectra, which also displayed a higher ratio of the intensity from N_2^+ ions to that from N_2^* neutrals (see Figure 12 (c)).

The quasi-uniformly treated samples, the spots created by 3-s stationary treatment, and a cleaned, but untreated sample were imaged by AFM. The He plasma treated samples could be readily imaged in intermittent contact mode, whereas neither intermittent contact nor contact mode enabled stable imaging of the samples treated by the NH_3/He plasma. This behaviour is tentatively attributed to patches of electrical charge on the surface possibly due to deposition from the plasma, but further studies are required in order to determine the true cause of the unstable imaging. Examples of images obtained on the untreated surface vs. a He plasma treated surface are shown in Figure 13.

The difference is immediately apparent. All images were flattened (offset) line by line to allow a meaningful evaluation of the roughness. The average root mean square roughness (R_q) measured on the untreated surface was 1.3 nm, whereas on the He plasma treated samples it was similar for the 1-minute quasi-uniform treatment, 3.3 nm for the 5-minute, and 2.2 nm for the 3-s spot treatments.

As Figure 13 suggests, the increase of the R_q-values is composed of a small, relatively uniform roughening and an additional appearance of randomly, but less uniformly and rather sparsely distributed protrusions ("bumps") with an apparent height of up to 50 nm. Both will naturally increase the effective surface area, but the effect of the "bumps" on the strength of an adhesive joint is uncertain, as their chemical composition as well as their cohesive and adhesive properties are undetermined thus far.

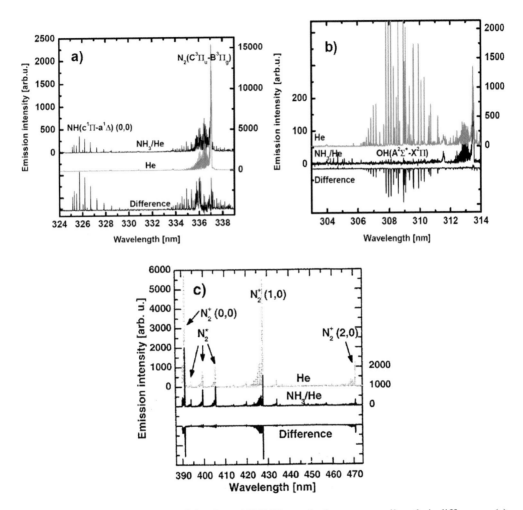

Figure 12. Features from the OES of the He and NH$_3$/He torch plasmas as well as their difference: (a) the NH bands around 324 nm and 336 nm, (b) the 306.4 nm system of OH, and (c) bands of the second positive and first negative systems of N$_2$*, respectively. In the difference spectrum, negative peaks represent species dominant in the He/NH$_3$ plasma. All spectra have been rescaled to equalize the N$_2$* bands [62].

Several adjacent traces were measured with the profilometer across each area exposed in the stationary mode. The volume of material removed from each area was estimated by selecting from each set of such measured profiles the one displaying the deepest depression and integrating it over 180 degrees around its centre axis, thus assuming an axisymmetric hole. The results are summarized in Table 5.

The data are consistent with a constant etch rate for each plasma throughout the 300-s maximum treatment. Visually, the NH$_3$/He treated samples appeared somewhat more discoloured, and ammonia is found to remarkably promote etching.

The NH₃/He mixture etched approximately 2×10^4 µm³/s compared to the 1.2×10^3 µm³/s of the pure He plasma. It is interesting to compare the above data to standard low-pressure reactive ion etching (RIE) techniques applied for microstructuring surfaces. Kuhnke *et al.* have recently applied low-pressure RIE to the micromachining of glassy carbon electrodes for micro fuel cells with potential applications in portable electronics [85,86].

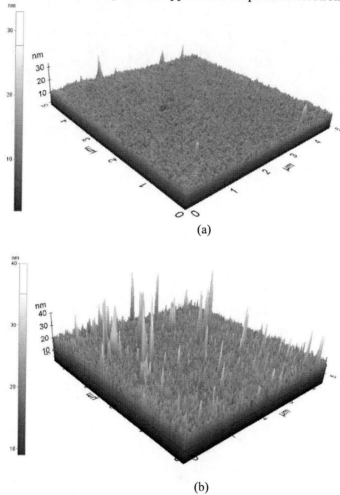

Figure 13. 5×5µm² intermittent contact AFM micrographs of the glassy carbon surface (a) untreated, and (b) after exposure to He plasma for approximately 2 s (5 minutes. total treatment time of 100 mm² with a torch of 1-mm diameter) [62].

Table 5. Estimated plasma etched volumes (*V*) in µm³ and maximum depths (*d*) in µm [62].

Exposure time [s]	He		NH₃/He	
	V	d	V	d
3			8.5×10^4	0.18
30	3.9×10^4	0.27		
300	3.7×10^5	2.3	5.5×10^6	9.2

Employing an RF oxygen plasma, they reported a maximum etch rate of 40 µm/h and an efficiency of ~10^6 µm³/s per W of RF power with a substrate bias of 800 V at 13 Pa (100 mTorr) pressure. In the case of the present atmospheric pressure NH_3/He plasma torch the corresponding parameters are 110 µm/h and 2×10^4 µm³/Ws, respectively, for an unbiased substrate. The plasma torch thus etches slightly faster perpendicular to the surface, but acts very locally and (assuming that no other power than that supplied at RF to maintain the plasma is acting in the low-pressure setup) with a relatively poor energy efficiency compared to low-pressure RIE systems.

The water contact angle (θ) on untreated glassy carbon is ~97°. On the treated samples, the angles were first measured approximately 30 minutes after treatment. All of the four samples (2 gas compositions, 1 and 5 minutes total treatment) were highly hydrophilic with θ too small to be measured. The same was true 100 minutes after treatment. After 170 minutes, the He plasma treated samples displayed θ ~5-10°, whereas the NH_3/He plasma treated ones had θ ~12-24°, still significantly lower than the untreated sample. The aging mechanism has not yet been studied, but our previous experiments with plasma treated glassy carbon show that a residual effect remains even after 3 months of storage under ambient conditions.

Table 6. Elemental composition [%] of the glassy carbon surfaces as derived from XPS [62].

treatment	C	O	N	Si
untreated	96.8	3.2	-	-
He 300 s	77.9	18.1	2.6	1.4
NH_3/He 300 s	77.5	14.7	5.4	2.3
He 300-s spot	72.4	23.3	0.9	3.4
NH_3/He 300-s spot	91.4	6.7	1.0	1.0

With the above aging effect in mind, care was taken to perform XPS analysis as soon as possible after plasma treatment. All samples were transferred to high vacuum within 95 minutes after treatment, and XPS measurements were initiated immediately thereafter. In Table 6 the elemental compositions derived from XPS of two uniformly treated, two spot treated, and one untreated sample are given. It is clear that the plasma treatment strongly increases the surface contents of oxygen and nitrogen, as expected. A slight predominance of oxygen or of nitrogen is observed with He or NH_3/He plasma treatment, respectively. In addition, the treated samples contain some surface silicon, most likely eroded from the glass pipette.

Additional information on the oxygen and nitrogen containing functional groups is obtained from the peak shape analysis of the 1s peaks of C, N, and O [74]. The results of this analysis are summarized in Table 7. In addition to the main hydrocarbon peak at 285.0 eV, components at higher binding energies up to 296.7 eV appear in the C1s signal. All treatments give rise to an increase in the components corresponding to $\underline{C}$-OH, $\underline{C}$-O-C, or $\underline{C}$-N (~286.4 eV) and $\underline{C}$=O or O-$\underline{C}$-O (~287.5 eV), whereas the component that could be due to acid or ester groups (~289.0 eV) is enhanced mainly by the pure He plasma. The O1s peaks after plasma treatment primarily consist of components with binding energies in the range of

531.9-533.8 eV, which can be assigned to $\underline{O}$=C-O, O=C$\underline{O}$ [74], or C-$\underline{O}$ groups [87]. The lowest energy (~530.8 eV) components could be due to C=$\underline{O}$ [88], whereas the highest energy components might be attributed to adsorbed H_2O or O_2 [89]. There are, however, several possible peak-fits to the O1s data, so the binding energies and weights stated in the table should only be regarded as a relatively rough indication of the peak shape. Similarly, due to the low N content, any assignment to functional groups will be rather uncertain. The N1s peak shapes observed are consistent with a mixture of C-C=O($\underline{N}$), C=O($\underline{N}$)C=O, or amide (399.8-400.6 eV) along with possibly NH_3^+ (401.5 eV), $\underline{N}$-H (400.2 eV), and $\underline{N}$-H$\cdots$O (399.3 eV) [90]. In conclusion, the predominant modifications of the surface are likely to be the production of carboxylic acid, ester, amide, and some NH_3^+ groups, with the nitrogen containing groups predominantly being produced by the NH_3/He plasma.

Table 7. Binding energies (BE) and weights (%) of the components of the XPS peaks from C1s (A-H), O1s (I-L), and N1s (M-O). The assignment of the C1s peaks is as follows; A: $\underline{C}$-H, $\underline{C}$-C or $\underline{C}$=C; B: $\underline{C}$-OH, $\underline{C}$-O-C, or $\underline{C}$-N; C: $\underline{C}$=O or O-$\underline{C}$-O; D: acid or ester; E-H: aromatic carbon (π–π* satellite peaks). For the assignment of O1s and N1s, the reader is referred to the text [62].

treatment	C1s [%]				
	285.0 eV CH/CC (A)	~286.4 eV CO/COH (B)	~287.6 eV C=O/OCO (C)	~289 eV C(O)O (D)	>290 eV π–π* (E-H)
untreated	70.0	13.2	5.7	3.1	7.9
He 300 s	59.9	16.6	9.8	7.0	6.8
NH_3/He 300 s	62.7	18.3	7.9	4.9	6.1
He 300-s spot	58.6	16.1	11.2	6.8	7.2
NH_3/He 300-s spot	67.4	13.4	6.1	3.9	9.4

treatment	O1s [%]				N1s [%]		
	~531eV	~533eV	~534eV	~536eV	~400eV	~401.5eV	~403eV
untreated	45.5	54.5	-	-	-	-	-
He 300 s	23.5	57.8	14.4	4.3	41	59	-
NH_3/He 300 s	-	30.7	63.3	6.0	71	18	10
He 300-s spot	-	70.2	29.8	-	100	-	-
NH_3/He 300-s spot	11.0	43.4	36.8	8.7	61	39	-

The preliminary, qualitative adhesion tests showed that the epoxy resin very readily delaminates from the untreated glassy carbon surface, whereas on the samples plasma treated quasi-uniformly for 5 minutes with either gas mixture, it was impossible to remove the epoxy cleanly using the blade.

Three DCB sandwich specimens were made for each of the three conditions applied in the above qualitative test. The specimens containing one untreated glassy carbon plate were very fragile, and two specimens, as well as a spare one, accidentally delaminated before they could be tested. Testing the one remaining, a critical energy release rate of 8 J/m^2 was obtained. Regarding the plasma treated samples, the DCB test confirmed the result of the qualitative test. In all but one case for each plasma composition, the adhesive bond was so strong that fracture of one or the other glassy carbon plate occurred when the crack reached the treated region. The critical energy release rate almost invariably increased monotonically as the crack propagated along the interface being tested. Although this apparent increase in fracture resistance may be due to toughening due to plasticity [91], it is thought that this behaviour is more likely to be an effect of the plasma treatment not being completely localized to the area directly below the discharge plasma. This property of the treatment was confirmed by contact angle measurements. For the two specimens that ultimately failed at the interface, the applied moment required for crack propagation in the last 5 cycles before failure was within 17% of that of the last cycle. Also, at 5 cycles before failure, when observed in the optical microscope, the crack appeared to cross the boundary of the directly plasma treated zone at 32 mm from the sample end. The average of the corresponding energy release rates was adopted as the fracture energy of the respective adhesive bond. For the He treated surface this value was 100 ± 8 J/m^2, whereas for NH$_3$/He treatment it was 152 ± 24 J/m^2. The trend of the NH$_3$/He treatment being more effective was observed in the cohesively failing specimens as well: the energy release rates equivalent to the maximum moments resisted by these were 108 J/m^2 for He treatment and 210 J/m^2 for NH$_3$/He treatment.

3.1.3. DBD

In this section, glassy carbon plates were treated by the air DBD at atmospheric pressure. The contact angle of an untreated glassy carbon surface is 98°. Each treated glassy carbon surface showed improved wettability, and the contact angle became less than 5° which is too low to be measured. It is indicated that polar functional groups could be introduced on the surfaces, and that surface roughness could be increased. These effects potentially cause adhesion improvement with epoxy resin, as is seen for carbon fibres [92].

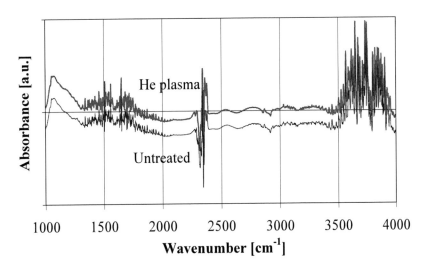

Figure 14. FTIR spectra of the untreated and He plasma-treated glassy carbon plates [63].

Figure 14 compares the FTIR spectra of the glassy carbon plates with and without He plasma treatment. Complex absorption bands around 3500 – 4000 cm^{-1} and 1300 – 2000 cm^{-1} were observed. Since these bands were not detected for general polymers such as polyethylene and polyethylene terephthalate, they are unlikely to be associated with moisture in the FTIR system. Instead, they can be attributed to the presence of water in the specimens. On the other hand, the absorption at 2920 cm^{-1} can be attributed to sp^3 hybridized carbon [93,94]. No clear differences can be seen between the FTIR spectra of the two samples. The FTIR spectra of the glassy carbon plates treated by different plasma conditions showed the similar results. It is therefore suggested that this technique is not sensitive enough for detecting chemical changes of the glassy carbon surfaces by the plasma treatment, and/or that glassy carbons are rather inert to the oxidation [15]. However, the chemical changes of glassy carbon surfaces by low pressure plasma treatment has been detected by FTIR analysis [12,27,29]. There are some differences of experimental conditions. Firstly, the untreated glassy carbon specimens in the references [12,27,29] show significant infrared absorption bands for example at 750, 810, 880 (γ(C-H)), 1210-1430 (δ(CH$_2$), δ(OH), ν(C-O)), 1600 (ν(C=C)), 1700 (ν(C=O)), and 3410 cm^{-1} (ν(O-H)), indicating that the structure of the glassy carbon specimens is different from that of the present chapter, and that the influence of plasma treatment could be also different. Secondly, the glassy carbon specimens were treated until the desired mass loss is achieved in the references, suggesting that the surfaces could be treated severely.

XPS was carried out to analyze the elemental composition of the glassy carbon surfaces before and after the plasma treatments. The results are summarized in Table 8. Silicon was detected in each specimen. When the glassy carbon plates were cleaned and degreased with acetone only, the observed silicon content on the surface before and after the plasma treatment was higher (typically up to 5 at. %) than that after the ultrasonic cleaning. It is therefore indicated that the specimen surfaces were contaminated possibly with organic silicon. After each treatment, the oxygen content increased drastically up to 10 - 20 at. %, while the nitrogen content increased slightly. The introduction of nitrogen at the surface is most likely due to the air leakage into the discharge, supported by the observed optical emission of N$_2$ lines (not shown here). Nitrogen is more effectively introduced onto the surfaces by He, He/NH$_3$ and Ar/NH$_3$ plasma treatments than by the other treatments. The difference of nitrogen introduction by He and Ar plasma treatments can be due to different excited energy levels. Representative XPS C1s core level spectra for the untreated and plasma treated glassy carbon plates are shown in Figure 15.

Table 8. Elemental composition of the glassy carbon surfaces by XPS [%] [63].

	C	O	N	Si
Untreated	91.9	7.3	0.0	0.9
He plasma	78.0	18.3	3.1	0.7
He/O$_2$ plasma	80.4	17.0	1.8	0.9
He/CO$_2$ plasma	82.1	15.5	1.5	1.0
He/NH$_3$ plasma	76.3	19.0	3.8	1.0
Ar plasma	82.0	16.0	1.2	0.9
Ar/NH$_3$ plasma	82.9	12.7	3.8	0.7

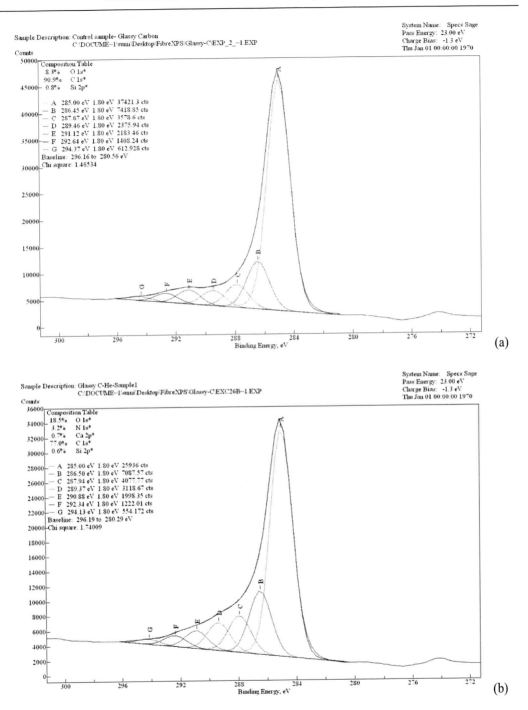

Figure 15. XPS C1s region for the (a) untreated and (b) He plasma treated glassy carbon plates [63].

Table 9 summarizes curve fittings of C1s and N1s. Peaks A, B, C, and D are assigned to C-H/C-C/C=C, C-OH/C-O-C/C-N, C=O/O-C-O and acid or ester, while peaks E, F, and G are π-π* satellite peaks from the aromatic carbons. The peaks B, C and D increased similarly after each treatment. The N1s spectra in XPS consist of a peak at approximately 400 eV and in some cases one at 407 eV. The lower binding-energy peak, which is seen in each treated spectrum, can be assigned to C-C=O(N), C=O(N)C=O [74] or C=N-OH [95]. The higher binding-energy peak, which is observed after He, He/NH$_3$, Ar and Ar/NH$_3$ plasma treatments, could be assigned to ONO$_2$ nitrogen [74]. Introduction of these oxygen containing polar functional groups on the glassy carbon surfaces increases the polar component of surface energy, indicating that the adhesive properties can be improved [92]. On the other hand, the dispersive component of the surface free energy can be unaffected by the increases of polar functional groups [92], assuming the similarity between glassy carbon and carbon fibre surfaces. However, further investigation is necessary to discuss the modified functional groups on the plasma treated glassy carbon surfaces.

Table 9. Summary of the C1s and N1s curve fittings of XPS for the treated and untreated glassy carbon surfaces (BE: binding energy in eV) [63].

Peak	C1s [%]					N1s [%]	
	285.0 eV CH/CC (A)	~286.5 eV CO/COH (B)	~288 eV C=O/OCO (C)	~289.5 eV C(O)O (D)	>290 eV π–π* (E-G)	~401 eV	~407 eV
Untreated	67.6	13.7	6.6	4.4	8.1		
He plasma	58.8	16.0	9.2	7.3	8.9	62	38
He/O$_2$ plasma	60.9	14.7	9.1	7.0	8.5	100	
He/CO$_2$ plasma	59.8	16.1	9.5	6.9	8.0	100	
He/NH$_3$ plasma	62.3	15.0	8.1	6.5	8.3	45	55
Ar plasma	62.4	14.4	8.3	6.6	8.5	63	37
Ar/NH$_3$ plasma	64.2	14.1	7.7	5.9	8.2	72	28

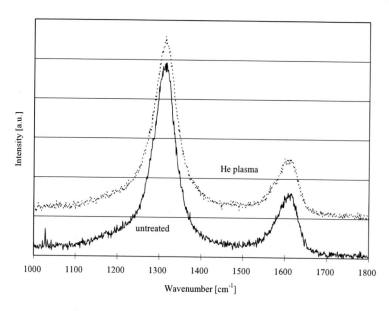

Figure 16. Raman spectra of the untreated and He plasma treated glassy carbon plates [63].

Characterization of microstructure of glassy carbon plates provides information not only on the bulk properties but also on the surface structure. The microstructures of carbon materials can be characterized with Raman spectroscopy [95]. In particular the presence of the disorder (D) band at 1360 cm^{-1} is attributed to any decrease in long range order from an ideal graphite crystal, the existence of nano-crystallites and the presence of edges of graphite nano-crystallites. Therefore the intensity [96] and bandwidth [28] of the D band can be used to characterize defects on the glassy carbon surface. In addition broadening of bandwidth of the graphitic (G) band around 1580 – 1600 cm^{-1} [97] indicates an increase in hardness and density of amorphous carbon surfaces [98]. Raman spectra of the untreated and He plasma treated glassy carbon plates are compared in Figure 16. The first order region up to 1800 cm^{-1} features dominant broad bands at approximately 1320 cm^{-1} and 1605 cm^{-1} attributed to the D and G bands, respectively. The full bandwidth at half maximum of the D band increased from 64 cm^{-1} to 70 cm^{-1} with He plasma treatment. This increase indicates that the defect density at the glassy carbon surface increased. However, it should be noted that the sampling depth of conventional Raman spectroscopy is about 29 nm for glassy carbon [99], and thus increase of the D band does not necessarily mean that the bulk glassy carbon plates are damaged by the plasma treatment. On the other hand the full bandwidth at half maximum of the G band slightly increased from 50 cm^{-1} to 52 cm^{-1} with He plasma treatment, indicating that the hardness and density increased by the plasma treatment. It is suggested that the plasma treatment etched and removed low density, soft domains from the glassy carbon surface. It is also noted that the observed peak frequency at approximately 1320 cm^{-1} is slightly lower and that at 1605 cm^{-1} is slightly higher than the reported values, but that the peak frequencies of the D and G bands can shift, for example, with different excitation wavelengths for Raman spectroscopy [100].

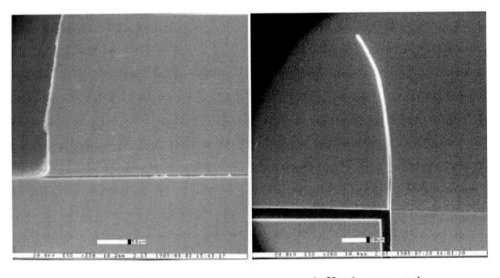

a. untreated   b. He plasma treated

Figure 17. Scanning electron micrographs of crack propagation in the DCB specimens; (a) untreated, and (b) He plasma treated [63].

Adhesion tests were carried out for the untreated and treated glassy carbon plates. Typical crack propagation images of treated and untreated DCB specimens are shown in Figure 17. The interface strength for the untreated specimen was ca. 8 J/m^2, while those of the treated ones were not measurable, since they showed fracture in the glassy carbon plate rather than along the interface between the adhesive and the glassy carbon surface. It is therefore indicated that strong bonding is achieved by the treatments.

3.2. Carbon fibres

The major effect of the conventional electrochemical treatments on carbon fibres for fibre/resin adhesion is believed to be the removal of weakly bound crystallites at the fibre surfaces, while the number of functional groups introduced onto the fibre surfaces by those treatments is too small to have a significant effect [101]. They are less effective when they are applied to highly graphitised surfaces [102]. It is therefore interesting to know whether the plasma treatment can effectively add oxidative functional groups onto the carbon fibre surfaces.

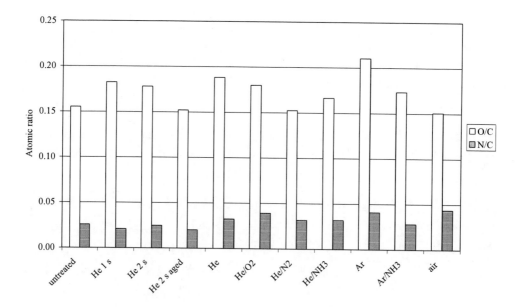

Figure 18. O/C and N/C ratios at the carbon fibre surfaces before and after plasma treatments as determined by XPS. The treatment time is 100 s unless it is stated. Aging conditions: ambient air, room temperature, one month [60].

The elemental composition of the carbon fibre surfaces before and after the plasma treatments was analyzed using XPS. The surfaces are dominated by carbon, oxygen and nitrogen. Figure 18 shows ratios of O/C and N/C for each specimen. It is found that oxygen can be effectively introduced by He, He/O$_2$, and Ar plasmas, while the nitrogen content increased slightly after He/O$_2$, Ar and air plasma treatments. Additionally the He plasma treatment for 1 s resulted in significant oxidation, while the longer treatments did not enhance oxidation.

Table 10. O/C ratios (%) at carbon fibre surfaces characterized using XPS (HM: high modulus fibre (type I), HS: high strength fibre (type II)) [60].

Pressure [Pa]	Gas	Plasma Frequency [Hz]	Power [W]	Carbon fibre producer	Type	Treatment time [s] 0	1	2	20	30	60	100	180	Ref.
100	O_2	2.45G	150	Sigri Carbon	PAN, HS	6.4							8.0	[105]
			75	Clemson University	Pitch,	3.3							12.2	
			150	Amoco	HM	2.8							8.7	
			100										9.7	
			75										10.1	
2	O_2		75	Amoco		1.7							7.4	[102]
26.7	Ar	35M	?	Hercules		12.3							12.6	[107]
	O_2												18.1	
						5.8							15.0	
?	Air	13.56M	50	Courtaulds	PAN	3.0				19.0				[108]
133	Air	13.56M	50	Hercules	PAN, HS	11.0				27.0	27.0			[109]
66.7	O_2	13.56M	30	Taekwang	PAN	12.0				17.2	21.1		21.4	[110]
53.3	O_2	13.56M	50	Hercules	PAN, HS	11.7					25.3		25.4	[111]
2	O_2	13.56M	100	Hercules	PAN	5.7			12.5					[103]
						11.0			21.8					
						11.0			28.0					[112]
	Ar								21.0					
10^5	O_2	?	<300	Courtaulds	PAN, HM	2.6				8.9				[105]
	He/O_2	5k		Hercules	PAN, HS	4.0					8.9		10.8	[104]
	Ar	40k	100	Toho Tenax	PAN, HS	15.5						21.0		This
	He		100				18.2	17.8					18.8	work

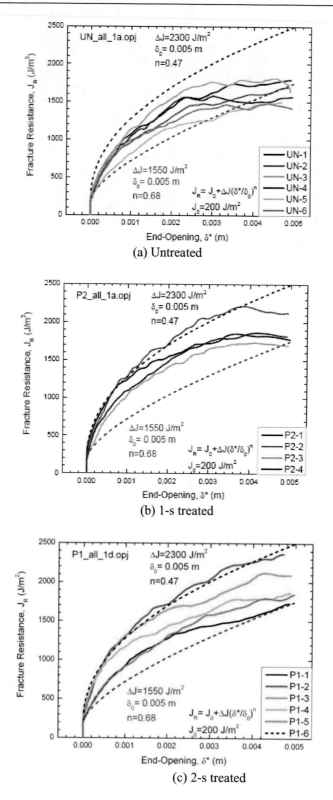

Figure 19. Fracture energy of the CFRE plates.

A regional analysis on C1s spectra shows decrease in the binding energy peak at 285 eV and increase in that at 286.4 eV after each plasma treatment, indicating that a surface density of the C-O single bond increases after the plasma treatments and that the surfaces can be potentially further oxidized to generate carboxyl groups.

After one month exposure in an ambient air at room temperature, the O/C ratio at the He plasma treated carbon fibre surfaces decreased to 0.152, which is close to the O/C ratio of the untreated carbon fibre surfaces. However, the regional analysis on C1s spectra shows that the binding energy peak at 285 eV remains almost unchanged after the aging, indicating that the chemical structure at the carbon fibre surfaces before the treatment and after the treatment with the subsequent aging is different. Unsized carbon fibres are known to be almost insensitive to aging in air due to their rigid graphitic structures [103]. It is therefore suggested that the decrease in the O/C ratio by aging in air could be due to the adsorption of hydrocarbon contamination at the carbon fibre surfaces.

Table 10 compares O/C ratios (in %) at carbon fibre surfaces before and after plasma treatments as determined by XPS reported in the literature and the present chapter. The O/C ratios of carbon fibre surfaces before plasma treatments range between 0.017 and 0.334, depending on surface pre-treatment and manufactures. It can be seen that carbon fibres produced by Hercules tend to be more easily oxidised than the others, and that oxidation by oxygen microwave plasma treatments is less effective than the other plasmas. If carbon fibres produced by Hercules and those treated with microwave plasmas are excluded for further discussion, the typical O/C ratio after plasma treatments can be up to ca. 0.2, which is significantly higher than those by the atmospheric pressure plasma treatments in refs.[104,105], while comparable to those observed in the present study.

Untreated, 1-s and 2-s He plasma treated carbon fibres were used to prepare carbon fibre reinforced epoxy (CFRE) plates for the adhesion tests. The fracture energy at the mechanical test is shown in Figure 19. The fracture energy of the CFRE with untreated (Figure 19 (a)) and 1-s treated (Figure 19 (b)) carbon fibres first increased as the end-opening increased, and approached to saturating values. The CFRE with 1-s treated one shows higher saturation values, indicating that the mechanical strength was improved even by 1-s treatment. The fracture energy of the 2-s treated specimens steadily increased as the end-opening increased. It is therefore concluded that the plasma treatment improved the mechanical properties of the composites, and that the longer treatment further improved them. Since the O/C ratios at the surface after 1 and 2 s treatment were almost same, possible explanation of this improvement is increased roughness at the carbon fibre surfaces after longer treatment [10].

3.3. UHMWPE Fibres

FTIR spectroscopy is a sensitive method to investigate plasma treated UHMWPE fibres due to the large area/volume ratio. The FTIR spectrum of the untreated surface contains the doublet symmetric and asymmetric stretching modes $v(CH_3)$ at 2851 cm^{-1} and 2919 cm^{-1}, the doublet bending mode $\delta(CH_2)$ at 1463 cm^{-1} and 1473 cm^{-1}, and the doublet rocking mode $v(CH_2)$ at 721 cm^{-1} and 734 cm^{-1} [113] as shown in Figure 20. After plasma treatment, the most pronounced changes appear in the ranges between 830 - 1300 cm^{-1} and 1500 - 1730 cm^{-1}. The new absorption bands can be assigned as follows: epoxide and acetal at 830 cm^{-1}, $\delta(OH)$ at 1076 cm^{-1}, $v(C-O)$ between 1100 and 1282 cm^{-1}, $v(C=C)$ at 1638 cm^{-1}, ketone, quinone and lactone at 1732 cm^{-1}, COOH at 2600 cm^{-1} and ν(O-H) at 3250 cm^{-1}. It is noted that C=C and C=O contents markedly increased after Ar, He, He/O$_2$ and N$_2$ plasma treatments. Broad absorption corresponding to the hydroxyl group is seen after Ar, He or He/O$_2$ plasma treatment. Adsorbed water was detected after the N$_2$ and O$_2$ plasma treatments. However, optimization of the treatment conditions was not attempted using FTIR, as it is a qualitative characterization.

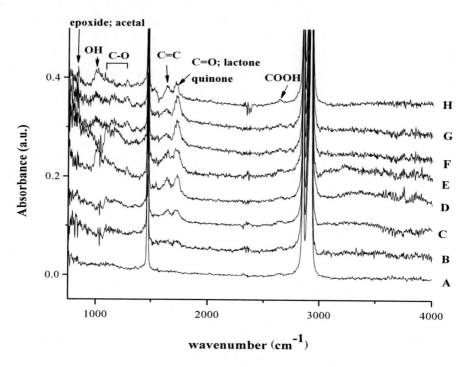

Figure 20. FT-IR absorption spectra of the untreated and plasma treated UHMWPE fibres. A) Untreated; B) O$_2$ plasma 10 L/min; C) O$_2$ plasma 30 L/min; D) He plasma 1 L/min; E) N$_2$ plasma 30 L/min; F) He/O$_2$ plasma 1 L/min; G) He/O$_2$ plasma 2 L/min; H) Ar plasma 1L/min [66]. Copyright Wiley-VCH Verlag GmbH and Co. KGaA. Reproduced with permission.

XPS was used for quantitative analysis of chemical composition at the fibre surfaces. Figure 21 (a) shows the O/C ratio obtained from the XPS survey spectra versus the applied power. Oxygen content steadily increases as the power increases from 50 to 150 W, when Ar (1 L/min), N$_2$ (20 L/min), He/O$_2$ (1 L/min, 2 L/min), He (1 L/min) or O$_2$ (10 L/min) was fed, showing good agreement with corona treated polyethylene (PE) in He/O$_2$ [114]. However, the O/C ratio of the He (2 L/min) plasma treated surfaces seems to approach to maximum as the power increases. The highest amount of oxygen content was observed after Ar (1 L/min) plasma treatment with 105 W power. The flow rate can also influence the O/C ratio at the surfaces as shown in Figure 21 (b).

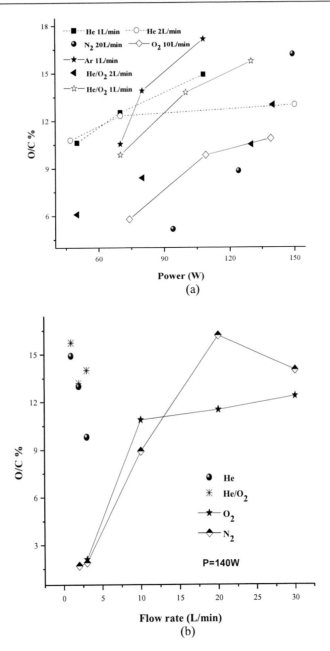

Figure 21. (a) The O/C ratio from XPS Survey analysis versus the applied power. A) He plasma 1 L/min; B) He plasma 2 L/min; C) N_2 plasma 20 L/min; D) O_2 plasma 10 L/min; E) Ar plasma 1 L/min; F) He/O_2 plasma 2 L/min; G) He/O_2 plasma 1 L/min. (b) The O/C ratio from XPS against the flow rate for He, He/O_2, N_2 and O_2 plasmas [66]. Copyright Wiley-VCH Verlag GmbH and Co. KGaA. Reproduced with permission.

The oxygen content at the surface tends to increase with the increase of the flow rates of N_2 and O_2. However, the O/C ratio seems insensitive to the change of the flow rates of He and He/O_2 gases (Figure 21 (b)). One of the possible explanations of these differences is that at higher flow rates filamentary plasmas in N_2 and O_2 can diffuse and extend more, and the surfaces can be treated more uniformly, resulting in higher oxidation than at lower flow rates.

On the other hand, He and He/O₂ plasmas were glow discharges and the feature of the discharge remained almost unchanged for these flow rates. However, further investigation is necessary for better understandings of these differences.

He plasma is efficient in exciting all the impurities present in the glow discharge, implying a more uniform treatment of the surface than filamentary discharges. Detection of the optical emission of He, N_2, CO_2, N_2^+ (Penning effect), CO^+ and OH in an atmospheric pressure He DBD is reported [115], and similar optical emission spectra were observed with the same DBD setup of the present chapter. CO_2 and CO^+ are related to polymer etching. N_2^+ ions might be useful for introducing N atoms at the surface, as the chemical bonding at the polymer surface can be broken by the excess energy of N_2^+ ions [115]. He plasma treatment induces significant cross-linking at a PE surface [116], whose reaction steps are reported [117].

O_2 plasma is a filamentary discharge. Using mass spectroscopy CO, CO_2, H_2O, H_2 and hydrocarbon fragments in O_2 plasma near a PE surface were detected [118] as well as attachment of oxygen atoms bonding to carbon atoms. Possible mechanisms of oxidation by O_2 plasma are chain scission and cross-linking [117] as well as introduction of oxygen at the surface.

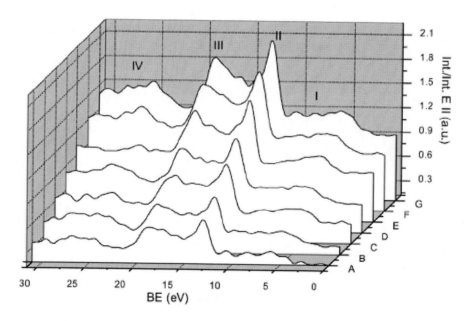

Figure 22. The VB spectra for A) O_2 plasma 30 L/min; B) N_2 plasma 30 L/min; C) O_2 plasma 20 L/min; D) N_2 plasma 20 L/min; E) Untreated; F) N_2 plasma 10 L/min; G) O_2 plasma 10 L/min. The intensities of all the VB spectra were normalized to the intensity of the peak corresponding to BE = 13 eV of the untreated case (EII) [66]. Copyright Wiley-VCH Verlag GmbH and Co. KGaA. Reproduced with permission.

Table 11 shows a summary of a deconvolution of the region I. Four peaks are observed at approximately 5.3, 6.4, 7.4 and 9 eV [119] both before and after plasma treatment. The peaks at 4 and 10 eV appear after the plasma treatment.

The regions II and III can be deconvoluted to four peaks in total as also shown in Table 11. The peaks having the BE of approximately 13 eV and 18.8 eV are attributed to the antibonding and bonding of the C2s molecular orbital, respectively [120]. Two peaks at around 14 and 16 eV might be partly the result of an x-ray exposure of the surfaces during XPS analysis. The same effect for UHMWPE after 5 hours x-ray exposure was observed [121]. However, as the intensities of the peaks differ with different plasma treatments, they can be partly attributed to plasma treatment. A peak at 14eV in a VB spectrum of an N_2 plasma treated linear-low-density PE surface was assigned to cross-linking [120], while a new peak at 16 eV after UV laser irradiation of PE was observed, attributed to the creation of a lateral chain, reticulation or cross-linking [122]. Two additional peaks at around 15.5 and 17 eV are observed in these regions after plasma treatment, possibly corresponding to cross-linking or structure folding modification at the surfaces. The appearance of a new peak in the regions II and III of the PE VB spectrum is also reported when the ordered linear zig-zag fold changes to a nonlinear one [123]. The region IV is deconvoluted in several peaks, and can be the most informative for adhesion improvement among all regions. The number of the peaks varies with the plasma treatment conditions. It is difficult to assign each peak in this region without proper $X\alpha$ calculation. It is theoretically predicted that the O2s-C2s in the VB spectrum can be separated into the peaks of C-O-C, C=O and C-OH [124] or epoxide, ether, hydrogen bonding, and hydroxide [125]. If the surface contains the carboxyl functional group, O2s can form bonding and non-bonding interactions with the C2s orbital [126]. The lower (25 eV) and higher (30 eV) BEs in this region can be assigned to hydroxide and epoxide groups, respectively [124,125]. The other peaks could be the combination of the other oxygen containing functional groups mentioned above.

It is indicated that VB spectra analysis can be useful for identification of specific functional groups such as hydroxide and epoxide that is impossible with general core level XPS. However, further investigation is necessary for more reliable assignment of the VB spectra.

Table 11. Summary of curve fitting of VB spectra in regions I, II, and III. Possible assignments; antibonding of C1s at approximately 13 eV, crosslinking, structure folding, lateral chain, or between 14 and 18 eV, bonding of the C2s at approximately 19 eV [66]. Copyright Wiley-VCH Verlag GmbH and Co. KGaA. Reproduced with permission.

BE [eV]	Region I [%]						Region II			Region III		
	~4	~5	~6.5	~8	~9	~11	~13	~14.3	~15.7	~16.5	~17.5	~19
Untreated	-	6.2	2.3	3.7	3.2	-	17.6	16.3	-	17.6	-	28.7
N_2 10 L/min	-	4.7	4.8	1.5	5.0	-	22.1	7.5	-	19.8	-	21.8
N_2 20 L/min	-	5.1	4.6	-	4.5	3.2	19.5	9.3	1.6	12.4	2	23.4
N_2 30 L/min	1.5	4.7	2.7	3.0	4.9	4.0	12.8	11.8	4.5	6.6	5.2	17.9
O_2 10 L/min	-	4.5	3.6	2.5	4.2	2.4	17.5	13.0	-	15.8	-	20.1
O_2 20 L/min	0.7	6.0	3.6	1.0	3.9	-	19.2	12.1	-	16.3	-	26.6
O_2 30 L/min	1.2	1.7	5.8	4.1	1.3	3.0	14.8	12.9	3.6	3.4	11.8	16.0

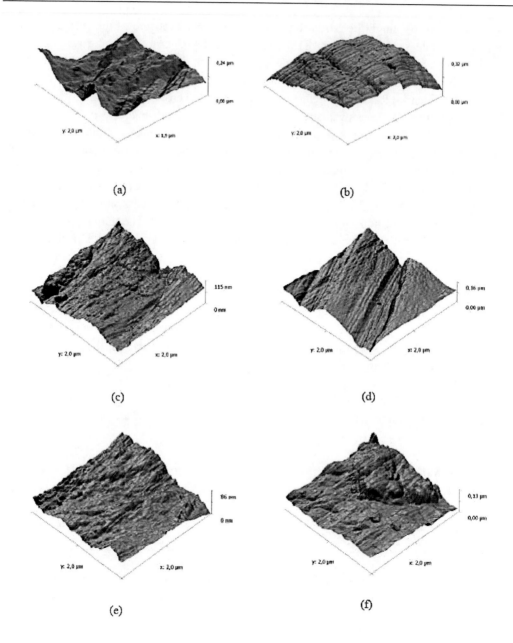

Figure 23. The AFM images of UHMWPE fibres. a) untreated, b) He plasma 2 L/min, c) He/O$_2$ plasma 1 L/min, d) Ar plasma 1 L/min, e) N$_2$ plasma 20 L/min and f) O$_2$ plasma 10 L/min. [66] Copyright Wiley-VCH Verlag GmbH and Co. KGaA. Reproduced with permission.

Figure 23 shows the AFM images, revealing a smoother surface after He plasma treatment, where a lamellar structure can be recognized (Figure 23 (b)). This might be an effect of trans-cristallinity after a β transition of UHMWPE [127]. He/O$_2$ plasma introduced micro-pits at the fibre surface, one of which is seen near the left corner of the AFM image in Figure 23 (c), possibly due to the cross-linking and the chain scission [55]. Figure 23 (d) indicates a cleaner and more uniformly etched surface by Ar plasma. N$_2$ plasma introduced micro-pits at the surface and the formation of ridges as shown near the left corner of the AFM image in Figure 23 (e) [114]. It can be seen that after O_2 plasma treatment the surface was restructured to oriented rows (Figure 23 (f)), corresponding to a preferential etching of the amorphous regions compared to the crystalline ones [128, 129].

The adhesion test indicated that the untreated fibres were easily pulled off from both epoxy mixtures, showing poor adhesion. The plasma treated fibres showed high interaction with the epoxy resins and were difficult to pull off. He and He/O_2 plasma treated fibres could be pulled off from the second mixture completely, and $20 - 30$ mm of the first mixture. It can thus be said that the interfacial strength of the He or He/O_2 plasma treated surface to the first mixture is comparable with the mechanical strength of the first mixture. N_2 and O_2 plasma treated fibres could also be pulled off from the second mixture, but not from the first one at all. It is therefore indicated that the N_2 and O_2 plasma treated fibres showed higher adhesive strength to the epoxy resins than untreated, He and He/O_2 plasma treated fibres. It is suggested that the interaction of the second mixture with the surfaces can be weaker than that of the first one. The Ar plasma treated fibres showed the highest adhesive strength, and could be pulled off from neither of the mixtures.

CONCLUSION

Using the low pressure AC plasma polymerization system it was possible to create thin polymer films with MAH and MDOB on glassy carbon plates and PE films with powers less than 1 W/L. The deposition rate of the plasma polymer changes with the power used in the system as shown by thickness changes of the plasma polymer layer. Despite differences in the deposition rate, the surface chemistry is somewhat seen to remain the same. This is confirmed by XPS measurements that showed insignificant changes during plasma polymerization using different rates of power consumption. The XPS results were largely confirmed by ATR-FTIR measurements, although this technique penetrates up to the first 1 µm of the surface. These observations could perhaps be attributed to the low power $(0.3 - 1.2$ W/L) of the AC plasma system which was found to preserve the ring structures and functional groups of the monomers used. Anhydride groups were still detected by ATR-FTIR.

The effect of treating glassy carbon surfaces with cold atmospheric pressure He and 0.17% NH_3/He plasmas from a torch source was studied. The NH_3/He plasma etched the surface approximately one order of magnitude faster than He plasma alone and at a rate comparable to low-pressure reactive ion etching. A substantial surface roughening resulted from the treatment. In addition, the water contact angle was dramatically reduced, as oxygen and nitrogen containing functional groups were introduced on the surface. The nature of these groups is somewhat uncertain, but likely to include carboxylic acid, ester, amide, and NH_3^+ groups. A larger quantity of nitrogen containing groups was introduced when NH_3 was present in the discharge feed gas, whereas, when it was absent, oxygen was more abundant. This variation correlates well with the predominant observation of NH in the NH_3/He plasma and of atomic oxygen and OH in the He plasma. The modification of the surface led to at least a 12-fold increase in the adhesion energy of the glassy carbon to epoxy with pure He plasma treatment and at least a 19-fold increase with the NH_3/He plasma treatment.

Carbon fibre surfaces were continuously treated by atmospheric pressure DBD plasma for adhesion improvement with epoxy resin. He, He/O_2 and Ar plasma treatments increased oxygen containing polar functional groups on to the carbon fibre surfaces. He plasma treatment of carbon fibres significantly increased mechanical strength of the CFRE composite plates.

UHMWPE fibres were continuously treated using atmospheric pressure DBD, and chemical and morphological modification was characterized using FTIR, XPS and AFM. FTIR and XPS analysis indicates that after plasma treatment oxygen containing functional groups are introduced at the surfaces. VB spectra in XPS are turned out to be useful for identifying epoxide and hydroxide at the surfaces. The AFM observation shows different surface roughening with different plasma treatments. The simple adhesion test with the epoxy resin shows that Ar plasma treatment is the most efficient to improve adhesive property of the UHMWPE fibres.

ACKNOWLEDGEMENTS

This work was carried out as under the Research Project "Atmospheric pressure plasma treatment of polymeric material surfaces for adhesion improvement" and the Framework Program "Interface Design of Composites Materials" funded by the Danish Technical Research Council (STVF Fund Nos. 26-04-0251 and 26-03-0160). The Graduate School of Polymer Science and the Technical University of Denmark are gratefully acknowledged for financial support. Lene Hubert is gratefully acknowledged for helping with the XPS measurements. Søren Nimb and Martin Jessen are gratefully acknowledged for design and construction of plasma sources.

REFERENCES

[1]   Launay; H.; Hansen, C.M.; Almdal; K. *Carbon* 2007, *45*, 2859-2865.
[2]   Jenkins, G.M.; Kawamura, K. *Polymeric Carbons;* Cambridge University Press: Cambridge, UK, 1976.
[3]   Modestov, A.D.; Gun, J.; Lev, O. *Surf. Sci.* 1998, *417*, 311-322.
[4]   O´Malley, B.; Snook, I.; McCulloch, D. *Phys. Rev. B* 1998, *57(22)*, 14148-14157.
[5]   Dasgupta, K.; Sathiyamoorthy, D. *Mater. Sci. Technol.* 2003, *19*, 995-1002.
[6]   Dilsiz, N. *J. Adhesion Sci. Technol.*2000, *14(7)*, 975-987.
[7]   Jones, C. *Comp. Sci. Technol.* 1991, *42*, 275-298.
[8]   Fukunaga, A.; Ueda, S. *Comp. Sci. Technol.* 2000, *60(2)*, 249-254.
[9]   Fagan, D.T.; Hu, I.F.; Kuwana, T. *Anal. Chem.* 1985, 57, 2759-2763.
[10] Jin, C.Q.; Wang, X.; Liu, Z.X.; Zhang, Y.L.; Li, F.Y.; Yu, R.C. *Brazilian J. Phys.* 2003, *33(4)*, 723-728.
[11] Taylor, R.J.; Humffray, A.A. *J. Electroanal. Chem.* 1973, *42*, 347-354.
[12] López-Garzón, F.J.; Domingo-García, M.; Pérez-Mendoza, M.; Alvarez, P.M.; Gómez-Serrano, V. *Langmuir* 2003, *19*, 2838-2844.
[13] Bowling, R.; Packard, R.T.; McCreery, R.L. *Langmuir* 1989, *5*, 683-688.
[14] Zhou, J.; Wipf, D.O. *J. Electroanal. Chem.* 2001, *499*, 121-128.
[15] Srinivasan, R.; Braren, B. *Chem. Rev.* 1989, *89*, 1303-1316.

[16] Pontikos, N.M.; McCreery, R.L. *J. Electroanal. Chem.* 1992, *324*, 229-242.

[17] Vitali, G.; Rossi, M.; Terranova, M.L.; Sessa, V. *J. Appl. Phys.* 1995, *77(9)*, 4307-4311.

[18] Saitoh, H.; Shinada, T.; Ohkawara, Y.; Ohshio, S.; Hiraga, H.; Inoue, T. *Jpn. J. Appl. Phys.* 2002, *41 Pt1 (6A)*, 3890-3895.

[19] Saitoh, H.; Shinada, T.; Ohkawara, Y.; Ohshio, S.; Hiraga, H.; Inoue, T. *Jpn. J. Appl. Phys.* 2002, *41 Pt1 (8)*, 5359-5366.

[20] Kuhnke, M.; Lippert, Th.; Ortelli, E.; Scherer, G.G.; Wokaun, A. *Thin Solid Films* 2004, *453-454*, 36-41.

[21] Kurihara, M.; Maruyama, S.; Oh'e, K.; Ishigaki, T. *Chem. Let.* 1998, *8*, 715-716.

[22] Fitzer, E.; Weiß, R. *Processing and uses of carbon fibre reinforced plastics;* VDI-Verlag GmbH; Düsseldorf, DE, 1981; pp 45.

[23] Hughes, J.D.H. *Comp. Sci. Technol.* 1991, *41(1)*, 13-45.

[24] Schreurs, J.; van den Berg, J.; Wonders, A.; Barendrecht, E. *Recl. Trav. Chim. Pays-Bas.* 1984, *103*, 251-259.

[25] Katoh, M.; Izumi, Y.; Kimura, H.; Ohte, T.; Kojima, A.; Ohtani, S. *Appl. Surf. Sci.* 1996, *100/101*, 226-231.

[26] Brown, N.M.D.; Cui, N.; McKinley, A. *Appl. Surf. Sci.* 1998, *133*, 157-165.

[27] Domingo-García, M.; López-Garzón, F.J.; Pérez-Mendoza, M. *Carbon* 2000, *38*, 555-563.

[28] Izumi, Y.; Ohte, T.; Kojima, A. *Jpn. J. Appl. Phys.* 1997, *36 Pt1(4A)*, 2372-2378.

[29] Pérez-Mendoza, M; Domingo-García, M.; López-Garzón, F.J. *Carbon* 1999, *37*, 1463-1474.

[30] Coen, M.C.; Keller, B.; Groening, P.; Schlapbach, L. *J. Appl. Phys.* 2002, *92(9)*, 5077-5083.

[31] Alexander, M.R.; Duc, T.M. *J. Mater. Chem.* 1998, *8*, 937-943.

[32] Alexander, M.R.; Zhou, X.; Thompson, G.E.; Duc, T.M.; McAlpine, E.; Tielsch, B.J. *Surf. Interface Anal.* 2000, *30*, 16-20.

[33] Alexander, M.R.; Payan, S.; Duc, T.M. *Surf. Interface Anal.* 1998, *26*, 961-973.

[34] Alexander, M.R.; Beamson, G.; Blomfield, C.J.; Leggett, G.; Duc, T.M. *J. Electron Spectrosc. Relat. Phenom.* 2001, *121*, 19-32.

[35] Friedrich, J.; Kuhn, G.; Mix, R.; Unger, W. *Plasma Proc. Polym.* 2004, *1*, 28-50.

[36] Friedrich, J.F.; Mix, R.; Kuhn, G. *Surf. Coat. Technol.* 2005, *200*, 565-568.

[37] Hindryckx, F.; Dubois, P.H. ; Patin, M.; Jérôme, R. ; Teyssié, P.H.; Marti, M.G. *J. Appl. Polym. Sci.* 15, *56*, 1093-1105.

[38] Chiang, W.Y.; Ku, Y.A. *Polym. Degrad. Stab.* 2002, *76*, 281-293.

[39] Friedrich, J.; Mix, R.; Kuhn, G.; Retzko, I.; Schonhals, A.; Unger, W. *Compos. Interf.* 2003, 10, 173-223.

[40] Fujimori, K.; Schiller, W.S.; Craven, I.E. *Makromol. Chem.* 1991, *192*, 959-966.

[41] De Roover, B.; Sclavons, M.; Carlier, V.; Devaux, J.; Legras, R.; Momtaz, A. *J. Polym. Sci., PtA: Polym. Chem.* 1995, *33*, 829-842.

[42] Kuo, P.L. ; Ni, S.C. ; Lai, C.C. *J. Appl. Polym. Sci.* 1992, *45*, 611-617.

[43] Samay, G.; Nagy, T.; White, J.L. *J. Appl. Polym. Sci.* 1995, 56, 1423-1433.

[44] Rengarajan, R.; Parameswaran, V.R.; Lee, S.G.; Vicic, M.; Rinaldi, P.L. *Polym.* 1990, *31*, 1703-1706.

[45] Ho, R.M.; Su, A.C.; Wu, C.H.; Chen, S.I. *Polym.* 1993, *34*, 3264-3269.

[46] Oostenbrink, A.J.; Gaymans, R.*J. Polym.* 1992, *33(4)*, 3086-3088.

[47] Ryan, M.E.; Hynes, A.M.; Badyal, J.P.S. *Chem. Mater.* 1996, *8(1)*, 37-42.

[48] Schiller, S.; Hu, J.; Jenkins, A.T.A.; Timmons, R.B.; Sanchez-Estrada, F.S.; Knoll, W.; Forch, R. *Chem. Mater.* 2002, *14(1)*, 235-242.

[49] Hu, J.; Yin, C.; Mao, H.Q.; Tamada, K.; Knoll, W. *Adv. Funct. Mater.* 2003, *13*, 692-697.

[50] Siffer, F.; Ponche, A.; Fioux, P.; Schultz, J.; Roucoules, V. *Anal. Chim. Acta* 2005, *539*, 289-299.

[51] Gaboury, S.R.; Urban, M.W. *Langmuir* 1993, *9*, 3225-3233.

[52] Jenkins, A.T.A.; Hu, J.; Wang, Y.Z.; Schiller, S.; Foerch, R.; Knoll, W. *Langmuir* 2000, *16*, 6381-6384.

[53] Evenson, S.A.; Fail, C.A.; Badyal, J.P.S. *Chem. Mater.* 2000, *12*, 3038-3043.

[54] Winther-Jensen, B.; Norrman, K.; Kingshott, P.; West, K. *Plasm. Proc. Polym.* 2005, *2*, 319-327.

[55] Qiu, Y.; Hwang, Y.J.; Zhang, C.; Bures, B.L.; McCord, M. *J. Adh. Sci.Technol.* 2002, *16(4)*, 449-457.

[56] Moon, S.I.; Jang, J. *J. Mat. Sci.* 1999, *34*, 4219-4224.

[57] Zheng, Z.; Tang, X.; Shi, M. Zhou, G. *J. Polym. Sci. PtB*, 2004, *42*, 463-472.

[58] Borcia, G.; Anderson, C.A.; Brown, N.M.D. *Plasm. Sources Sci. Technol.* 2003, *12*, 335-344.

[59] Guddeti, R.R.; Knight, R.; Grossmann, E.D. *Plasm. Chem. Plasm. Proc.* 2000, *20(1)*, 37-64.

[60] Kusano, Y.; Andersen, T.L.; Michelsen, P.K. "Atmospheric pressure plasma surface modification of carbon fibres" *J. Phys. Conf. Ser.* 2008, *100*, 012002-1-012002-4.

[61] Drew, J.; Goutianos, S.; Kingshott, P.; Hvilsted, S.; Rozlosnik, N.; Almdal, K.; Sørensen, B.F. "Plasma polymerized thin films of maleic anhydride and 1,2-methylenedioxybenzene for improving adhesion to carbon surfaces" *J. Vac. Sci. Technol. A* 2007, *25(4)*, 1108-1117.

[62] Mortensen, H.; Kusano, Y.; Leipold, F.; Rozlosnik, N.; Kingshott, P.; Sørensen, B.F.; Stenum, B.; Bindslev, H. "Modification of glassy carbon surfaces by atmospheric pressure cold plasma torch" *Jpn. J. Appl. Phys.* 2006, *45(10B)*, 8506-8511.

[63] Kusano, Y.; Mortensen, H.; Stenum, B.; Goutianos, S.; Mitra, S.; Ghanbari-Siahkali, A.; Kingshott, P.; Sørensen B.F.; Bindslev, H. "Átmospheric pressure plasma treatment of glassy carbon for adhesion improvement" *Int. J. Adhes. Adhes.* 2007, *27*, 402-408.

[64] Kusano, Y.; Mortensen, H.; Stenum, B.; Rozlosnik, N.; Strange, M.; Kingshott, P.; Goutianos, M.; Sørensen, B.F.; Bindslev, H. In *Proc. 27th Risø Int. Symp. Mater. Sci.* Risø National Laboratory: Roskilde, DK, 2006; pp 191-196.

[65] Kusano, Y.; Andersen, T.L.; Sørensen, B.F.; Michelsen, P.K. In *Proc. 28th Risø Int. Symp. Mater. Sci.* Risø National Laboratory: Roskilde, DK, 2007; pp 207-212.

[66] Teodoru, S.; Kusano, Y.; Rozlosnik, N.; Michelsen, P.K. "Continuous plasma treatment of ultra high molecular weight polyethylene (UHMWPE) fibres for adhesion improvement" *Plasm. Proc. Polym.* 2009, *6*, S375-S381.

[67] Winther-Jensen, B.; Christensen, S.F.; Petersen, S.G. *U.S. Patent* No. 0086660 A1, 2004.

[68] Ademovic, Z.; Wei, J.; Winther-Jensen, B.; Hou, X.; Kingshott, P. *Plasm. Proc. Polym.* 2005, *2*, 53-63.

[69] Han, L.M.; Timmons, R.B.; Bogdal, D.; Pielichowski, J. *Chem. Mater.* 1998, *10*, 1422-1429.

[70] Rinsch, C.L.; Chen, X.; Panchalingam, V.; Eberhart, R.C.; Wang, J.H.; Timmons, R.B. *Langmuir* 1996, *12*, 2995-3002.

[71] Koinuma, H.; Ohkubo, H.; Hashimoto, T.; Inomata, K.; Shiraishi, T.; Miyanaga, A.; Hayashi, S. *Appl. Phys. Let.* 1992, *60*, 816-817.

[72] Allred, R.E.; Schimpf, W.C. *J. Adhes. Sci. Technol.* 1994, *8(4)*, 383-394.

[73] Kusano, Y.; Mortensen, H.; Stenum, B.; Kingshott, P.; Andersen, T.L.; Brøndsted, P.; Bilde-Sørensen, J.B.; Sørensen, B.F.; Bindslev, H. "Atmospheric pressure plasma treatment of glass fibre composite for adhesion improvement" *Plasm. Proc. Polym.* 2007, *4*, S455-S459.

[74] Beamson G, Briggs D, "High resolution XPS of organic polymers" The Scienta ESCA300 database, John Wiley and Sons, Chichester, UK, 1992.

[75] Hartley, P.G.; Thissen, H.; Vaithianathan, T.; Griesser, H.J. *Plasmas Polym.* 2000, *5*, 47-60.

[76] Sørensen, B.F.; Horsewell, A.; Jørgensen, O.Kumar, A.N. *J. Am. Ceram. Soc.* 1998, *81*, 661-69.

[77] Sørensen, B.F.; Brethe, P.; Skov-Hansen, P. *J. Eur. Ceram. Soc.* 1996, *16*, 1021-1025.

[78] Bao, G.; Ho, S.; Suo, Z.; Fan, B. *Int. J. Solids Structures*, 1992, *29*, 1105-1116.

[79] Lin-Vien, D.; Colthup, N.B.; Fateley, W.G.; Grasselli, J.G. *Handbook of Infrared and Raman Characteristic Frequencies of Organic Molecules;* Academic: New York, US, 1991; pp 143.

[80] Biederman, H. *Vacuum* 1987, *37*, 387-394.

[81] Yasuda, H. *J. Polym. Sci. Macromol. Rev.* 1981, *16*, 199-293.

[82] Du, J.; Thouless, M.D.; Yee, A.F. *Int. J. Fract.* 1998, *92*, 271-286.

[83] Yamagata, Y.; Sharma, A.; Narayan, J.; Mayo, R.M.; Newman, J.W.; Ebihara, K. *J. Appl. Phys.* 1999, *86*, 4154-4159.

[84] Yamagata, Y.; Sharma, A.; Narayan, J.; Mayo, R.M.; Newman, J.W.; Ebihara, K. *J. Appl. Phys.* 2000, *88*, 6861-6867.

[85] Kuhnke, M.; Lippert, T.; Ortelli, E.; Scherer, G.G.; Wokaun, A. *Thin Solid Films*, 2004, *453-454*, 36-41.

[86] Kuhnke, M.; Lippert, T.; Scherer, G.G.; Wokaun, A. *Surf. Coat. Technol.* 2005, *200*, 730-733.

[87] Cruz-Barba, L.E.; Manolache, S.; Denes, F. *Langmuir* 2002, *18*, 9393-9400.

[88] Biniak, S.; Szymanski, G.; Siedlewski, J.; Swiatkowski, A. *Carbon* 1997, *35*, 1799-1810.

[89] Xie, Y.M.; Sherwood, P.M.A. *Appl. Spectrosc.* 1989, *43*, 1153-1158.

[90] Flamia, R.; Lanza, G.; Salvi, A.M.; Castle, J.E.; Tamburro, A.M. *Biomacromolecules* 2005, *6*, 1299-1309.

[91] Tvergaard, V.; Hutchinson, J.W. *Phil. Mag.* 1994, *A 70*, 641-656.

[92] Lee, J.; Drzal, L.T. *Int J Adhes. Adhes.* 2005, *25*, 389-294.

[93] Tanabe, Y.; Burkhard, G.; Ishikura, T.; Tsunoda, K.; Hasuo, H.; Tamaru, H.; Tamura, H.; Sawaoka, A.; Uematsu, K. *Jpn. J. Appl. Phys. Pt1* 1994, *33(12A)*, 6684-6690.

[94] Mukhopadhyay, K.; Sharon, M. *Mater. Chem. Phys.* 1997, *49*, 105-109.

[95] Losito, I.; Malitesta, C.; Bari, I.D.; Calvano, C.D. *Thin Solid Films* 2005, *473*, 104-113.

[96] Alsmeyer, Y.W.; McCreery, R.L. *Langmuir* 1991, *7*, 2370-2375.

[97] Ferrari, A.C.; Robertson, J. Phys. Rev. B 2000, 61, 14095-14107.

[98] Tamor, M.A.; Vassell, W.C. J. Appl. Phys. 1994, 76, 3823-3830.

[99] Alsmeyer, Y.W.; McCreery, R.L. Anal. Chem. 1991, 63, 1289-1295.

[100] Yoshikawa, M.; Nagai, N.; Matsuki, M.; Fukuda, H.; Katagiri, G.; Ishida, H.; Ishitani, A. Phys. Rev. B 1992, 46(11), 7169-7174.

[101] Farrow, G.J.; Jones, C. J Adhesion 1994, 45, 29-42.

[102] Montes-Morán, M.A.; Martínez-Alonso, A.; Tascón, J.M.D.; Young, R.J. Composites A 2001, 32, 361-371.

[103] Morra, M.; Occhiello, E.; Garbassi, F.; Nicolais, L. Composites Sci. Technol. 1991, 42(4), 361-372.

[104] Qiu, Y.; Shao, X.; Jensen, C.; Hwang, Y.J.; Zhang, C.; Mccord, M.G. In Polymer surface modification: Relevance to adhesion; Mittal, K.L.; Ed.; VSP: Zeist, NL, 2004; Vol. 3, pp 33-24.

[105] Desimoni, E.; Salvi, A.M.; Biader Ceipidor, U.; Casella I.G. J. Electron. Spectrosc. Relat. Phenom. 1994, 70, 1-9.

[106] Pavia, M.C.; Bernardo, C.A.; Nardin, M. Carbon 2000, 38, 1323-1337.

[107] Bascom, W.D.; Chen, W.J. J. Adhes. 1991, 34, 99-109.

[108] Atkinson, K.E.; Farrow, G.J.; Jones, C. Composites A 1996, 27A, 799-804.

[109] Heisey, C.L.; Wood, P.A.; McGrath, J.E.; Wightman, J.P. J. Adhes. 1995, 53, 117-147.

[110] Jin, B.S.; Lee, K.H. Choe, C.R. Polym. Int. 1994, 34, 181-185.

[111] Chang, T.C. J Ind. Technol. 1998, 15(1), 2-7.

[112] Bian, X.S.; Ambrosio, L.; Kenny, J.M.; Nicolais, L.; Occhiello, E.; Morra M, Garbassi, F. Dibenedetto, A.T. J Adhes. Sci. Technol. 1991, 5(5), 377-388.

[113] [113] Lien-Vien, D.; Colthup, N.B.; Fateley, W.G.; Grasselli, J.G. In The Handbook of Infrared and Raman Characteristic Frequencies of Organic Molecules; Acad. Press, San Diego, US, 1991.

[114] Lynch, J.B.; Spence, P.D.; Baker, D.E.; Postlethwaite, T.A. J. Appl. Polym. Sci., 1999, 71, 319-331.

[115] Massines, F.; Gouda, G. J. Phys. D Appl. Phys. 1998, 31, 3411-3420.

[116] Shonhorn, H.; Hansen, R.H. J. Appl. Polym. Sci. 1997, 11, 1461-1474.

[117] Hall, J.R.; Westerdahl, C.A.L.; Devine, A.T.; Bodnar, M.J. J. Appl. Polym. Sci. 1969, 13, 2085-2096.

[118] Wheale, S.H.; Barker, C.P.;Badyal, J.P.S. Langmuir 1998, 14, 6699-6704.

[119] Delhalle, J.; Andre, J.M.; Delhalle, S.; Pireaux, J.J.; Caudano, R.; Verbist, J.J. J. Chem. Phys. 1974, 60(2), 595-600.

[120] Foerch, R.; Beamson, G.; Briggs, D. Surf. Interf. Anal. 1991, 17(12), 842-846.

[121] Buncick, M.C.; Thomas, D.E.; McKinny, K.S.; Jahan, M.S. Appl. Surf. Sci. 2000, 156(1-4), 97-109.

[122] Pireaux, JJ.; De Meulemeester, R.; Roberfroid, E.M.; Gregoire, C.; Chtaib, M.; Novis, Y.; Riga, J.; Caudano, R. Nucl. Instrum. Methods Phys. Res. B, 1995, 105, 186-191.

[123] Delhalle, J.; Delhalle, S.; Riga, J. J. Chem. Soc. Faraday Trans. 1987, 83(3), 503-511.

[124] Xie, Y.; Sherwood, P.M.A. Chem. Mater., 1990, 2, 293-299.

[125] Xie, Y.; Sherwood, P.M.A. Chem. Mater., 1991, 3, 164-168.

[126] Sherwood, P.M.A. J. Elect. Spec. Rel. Phen., 1996, 81, 319-342.

[127] Alon, Y.; Marom, G. Macromolecular Rapid Commun., 2004, 25, 1387-1391.

[128] Intrater, R.; Hoffman, A.; Lempert, G.; Gouzman, I.; Cohen, Y.; Grossman, E. *J. Mater. Sci.*, 2006, *41(5)*, 1653-1657.

[129] Banik, I.; Kim, K.S.; Yun, Y.I.; Kim, D.H.; Ryu, C.M.; Park, C.S.; Sur, G.S.; Park, C.E. *Polym.*, 2003, *44*, 1163-1170.

In: Fiber–Reinforced Composites
Editor: Qingzheng Cheng

ISBN: 978-1-61470-303-7
© 2012 Nova Science Publishers, Inc.

Chapter 10

ADHESION IMPROVEMENT OF GLASS FIBRE-REINFORCED POLYESTER COMPOSITE BY ATMOSPHERIC PRESSURE PLASMA TREATMENT

Yukihiro Kusano[*1], *Tom L. Andersen*[1], *Bent F. Sørensen*[1], *Noemi Rozlosnik*[2], *Henrik Mortensen*[1], *Steluta Teodoru*[1], *Peter Kingshott*[1], *Frank Leipold*[1], *Shailendra Vikram Singh*[1], *Jørgen Bilde-Sørensen*[1], *Povl Brøndsted*[1], *Poul K. Michelsen*[1], *Henrik Bindslev*[1], *Alexander Bardenshtein*[1,3] *and Niels Krebs*[3]

[1] Risø National Laboratory for Sustainable Energy,
Technical University of Denmark, DK-4000 Roskilde, Denmark
[2] Department of Micro- and Nanotechnology,
Technical University of Denmark, DK-2800 Kongens Lyngby, Denmark
[3] FORCE Technology, DK-2605 Brøndby, Denmark.

ABSTRACT

Glass fibre reinforced polyester (GFRP) composites exhibit high strength-weight ratio and corrosion resistance, and are therefore used for a variety of applications in civil engineering, aerospace and automobile industry as well as medical and dental applications. Adhesives are often required for joining these materials to components fabricated from similar GFRPs or other types of materials. GFRPs usually have smooth surfaces composed mainly of the polymer matrix materials with low surface energies. In addition, they often have a wide range of contaminants on the surfaces, which can form a weak boundary layer in a bond. Therefore, an adhesive joint usually requires careful surface preparation. Among several surface modification techniques for adhesion improvement, atmospheric pressure plasma treatment is promising, since it shows high treatment efficiency, and is applicable to a variety of materials without affecting the bulk properties and easily up-scalable with low equipment cost. Its example includes dielectric

---

[*] Author to whom correspondence should be addressed; Email: yuki@risoe.dtu.dk

barrier discharge (DBD) generated between electrodes covered with dielectrics, and gliding arc generated between divergent electrodes and extended by a high speed gas flow. In addition it is reported that atmospheric pressure plasma treatment can be highly enhanced by high-power ultrasonic irradiation onto the surface during the treatment.

In the present chapter, GFRP plates are treated with a DBD in air, helium and gliding arc in air for adhesion improvement. The effect of ultrasonic irradiation on surface modification is investigated during helium DBD treatment. The treated surfaces are characterized and the adhesive properties with vinylester resin are tested. It is found that the treatment increased the density of oxygen-containing polar functional groups and polar components of surface energy. The ultrasonic irradiation during the treatment enhanced the treatment efficiency.

1. INTRODUCTION

Fibre-reinforced polymer (FRP) materials exhibit high strength-to-weight ratios and corrosion resistance, and are therefore used for a variety of applications in civil engineering (e.g. reinforcing bars in concrete, concrete column wraps for seismic rehabilitation, bridge deck construction, pilings, sheet pile retaining walls, offshore structures, pipelines, and beams, walls and slabs in existing buildings and bridges), aerospace and automobile industry [1-5], as well as medical and dental applications [6,7]. Adhesives are often required for joining these materials to components fabricated from similar FRPs or other types of materials. FRPs usually have smooth surfaces composed mainly of the polymer matrix materials with low surface energies. In addition, they often have a wide range of contaminants on the surfaces, which can form a weak boundary layer in a bond [8]. Therefore, an adhesive joint usually requires careful surface preparation. In the case of conventional FRPs based on thermoplastic polymeric matrices such as poly(ether-ether ketone), polyamide copolymer, poly(ether imide), polyimide and poly(phenylene sulphide), a treatment employing abrasion followed by solvent cleaning is inadequate to obtain high strength. Instead, various surface treatment techniques have been investigated, including flame treatment, corona treatment, plasma treatment, excimer laser, acid etching, and primer treatment [9-14]. On the other hand in the case of thermosetting composites such as unsaturated polyester, vinylester and epoxy resins, mechanical roughening has been applied for the surface preparation [15,16]. However, it needs laborious abrasion followed by solvent cleaning prior to adhesive bonding for achieving high joint strength.

Plasma treatment is attractive for this application due to its environmental compatibility and high treatment efficiency without affecting the textural characteristics of the bulk material [17]. It proceeds mostly by the interaction of radicals and ions with polymer surfaces and ultraviolet irradiation to the surfaces [18]. Plasma processing is often performed at low pressures. The plasma treatments at low pressures, however, suffer from the drawbacks that they require expensive vacuum systems, and methods are only well-developed for batch or semi-batch treatments. To overcome these drawbacks, an atmospheric pressure plasma treatment system can be used, which not only avoids the need for vacuum equipment but also permits both the treatment of large objects, and continuous treatment on production lines [19,20]. Among atmospheric pressure plasmas, a dielectric barrier discharge (DBD) is generated between electrodes covered with dielectrics by applying a time varying voltage. Advantages of a DBD include its compactness, low-cost, and stable operation [21]. In order to ensure stable DBD operation, however, the gap between the electrodes is limited to a few millimetres, which restrict the size of the specimens to be treated.

Gliding arc discharges can also be generated at atmospheric pressure. They can provide a high degree of non-equilibrium, high electron temperature and high electron density simultaneously, and thus potentially enable selective chemical process with high productivity [22]. They are generated between divergent electrodes, and are extended by a high speed gas flow. A fast transition into a non-equilibrium state occurs once heat losses from the discharge column exceed the supplied energy. Up to 80 % of the electrical energy may be directly absorbed by endothermic chemical reactions [23]. Although a similar effect is expected for microwave plasmas, a gliding arc system is less expensive than microwave plasma devices [24]. It is indicated that a gliding arc discharge can be preferably applied for large scale exhaust gas cleaning, pollution control, fuel conversion, hydrogen production and surface modification [23]. In particular, the gliding arc combines a number of industrially attractive features of plasma based surface treatments [25]: it is environmentally much cleaner than mechanical or wet chemical processes, it operates well in air at atmospheric pressure with low costs; being a torch-like plasma source, it can treat surfaces of bulky objects, and it allows fast processing. Therefore it is useful in many industrial applications that involve coating, painting, printing, dying, and adhesion. In spite of the advantages of this technique, however, the application for adhesion improvement of structural materials has been rarely reported.

Atmospheric pressure plasma surface modification is generally performed by feeding a process gas into the plasma. However, even when the motion of gas is fully turbulent, a boundary gas layer sticks at a material surface. Reactive species such as ions, electrons, high-energy neutrals and radicals generated in the plasma are diffused through the boundary gas layer and activate, interact and/or react with the surface. Due to the short lifetime of these species, only a small fraction can reach the surface. It is reported that powerful ultrasonic waves with a sound power level (SPL) above approximately 140 dB can reduce the thickness of the boundary gas layer [26], and that the treatment efficiency of atmospheric pressure plasma can be highly enhanced by simultaneous high-power ultrasonic irradiation onto the treating surface [27,28]. It is because the acoustic energy can be delivered efficiently at high gas pressures, which reduces the thickness of the boundary gas layer. As a result, more reactive species generated in the plasma can reach the surface before becoming inactivated, and can be utilized for surface modification efficiently.

Combinations of a plasma and ultrasonic waves have been investigated for understanding the interaction between plasma and ultrasound [29-34], electrical discharge machining [35], plasma etching [36], ozone generation [37,38], decomposition of volatile organic compounds [39], improvement of charging performance of corona chargers [40], and surface modification [27,41]. Here, electrical discharge machining is a manufacturing process with which a wanted shape of an object is obtained using electrical discharge plasmas. Ultrasonic waves at a material surface are often generated by vibrating the material using a piezoelectric or other solid state electro-acoustic transducer. However, due to the significant difference in acoustic impedances between a solid and a gas (acoustic impedance mismatch), most of the generated acoustic power by such a transducer cannot be emitted into the surrounding gas, and is converted into thermal energy instead. Therefore, ultrasound generated by such a transducer cannot efficiently reduce the thickness of the boundary gas layer at a material surface. On the other hand, a high-power gas-jet ultrasonic generator generates acoustic waves directly in a gas without acoustic impedance mismatch. Moreover, it can generate at least one order of magnitude higher acoustic power than any electro-acoustic transducer, and the powerful acoustic waves are capable of eliminating the boundary gas layer [26].

In the present chapter, glass fibre reinforced polyester (GFRP) plates are treated with a DBD [42,43] and gliding arcs [43,44] in air at atmospheric pressure for adhesion improvement. Optical emission spectra (OES) were measured for plasma diagnostics. They are also treated using an atmospheric pressure DBD in helium with and without ultrasonic irradiation [45]. The treated surfaces are characterized using contact angle measurement, x-ray photoelectron spectroscopy (XPS), attenuated total reflectance Fourier transform infrared (ATR-FTIR) spectroscopy, scanning electron microscopy (SEM), energy dispersive x-ray (EDX) analysis, and atomic force microscopy (AFM). The adhesive properties with a vinylester resin are tested.

2. EXPERIMENTAL METHODS

Three different kinds of GFRP plates were used; commercial G-Etronax PM 953 (GFRP-I, Elektro-Isola, Vejle, Denmark http://www.elektro-isola.com/Products/sheets.htm) for the DBD treatment in air, commercial G-Etronax PM material which contains 35 – 40 wt. % glass fibres (GFRP-II, Elektro-Isola, Vejle, Denmark) for surface characterization after gliding arc and ultrasound enhanced DBD treatment, and industrially used vacuum-infused orthophalic GFRP (GFRP-III) for XPS and adhesion test after the gliding arc treatment. The samples were cleaned and degreased with methanol and acetone. However, for XPS and AFM analyses they were ultrasonically cleaned in acetone for 2×5 minutes and in methanol for 5 minutes before plasma treatment.

An atmospheric pressure DBD has been used to treat the GFRP-I plates as shown in Figure 1. It was generated between parallel plate water cooled metal electrodes (50 mm × 50 mm) covered with alumina plates (100 mm × 100 mm × 3 mm) by an AC (approximately 36 kHz, $V_{pp} \sim 43$ kV) power supply (Generator 6030. SOFTAL Electronic GmbH, Germany). The average power input was 200 W corresponding to a power density of 8 W/cm, obtained by measuring voltage and current with a high voltage probe and a 50 Ω current viewing resistor. A synthetic air of an N_2/O_2 gas mixture was fed into the DBD. The flow rates of N_2 and O_2 were 2.0 L/min and 0.5L/min, respectively. Each plate was exposed to the filamentary plasma for 2, 5, 15, 30, 45 or 60 s.

Figure 2 shows two atmospheric pressure gliding arc sources used to treat the GFRP-II and -III plates. The first gliding arc (GA-I) was generated between two 2-mm diameter stainless steel wire electrodes with a diverging configuration [9] (Figure 2 (a)). The second one (GA-II) was built with two 1-mm thick stainless steel blade-electrodes with a similar diverging configuration (Figure 2 (b)). In both cases the high speed air flow was fed between the electrodes to extend the discharges. They were driven by an AC power supply (Generator 6030. SOFTAL Electronic GmbH, Germany). The average power input was obtained by measuring voltage and current with a high-voltage probe and a current viewing resistor, respectively. The input power was adjusted by varying the frequency. A change of the frequency from 33 to 40 kHz corresponds to an input power from approximately 2.2 to 1.6 kW. In order to treat a GFRP plate surface, the plate was fixed on a holder which moved forward and back at a speed of 180 mm/s. The distance between the specimen surface and the position of the arc ignition was 55 mm during the treatment. The specimen surfaces were exposed to the discharge 8 times in 40 s unless mentioned otherwise.

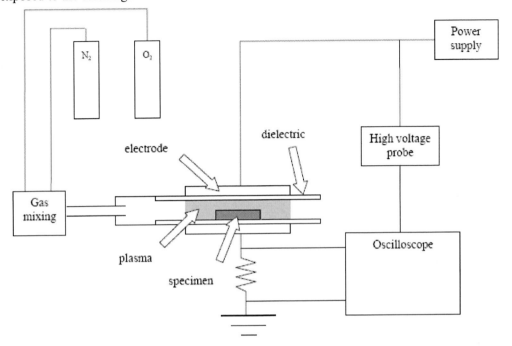

Figure 1. Diagram of DBD.

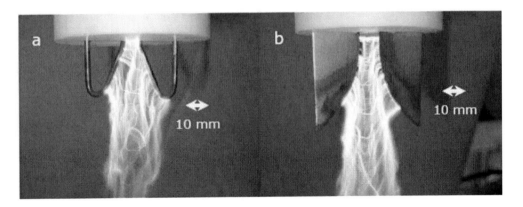

Figure 2. Gliding arc sources used for the treatment of GFRP plates; GA-I operated at a flow rate of 25 L/min with 1.6 kW (a) and GA-II operated at a flow rate of 30 L/min with 2.2 kW (b) [44].

Another atmospheric pressure DBD was used to treat the GFRP-II plates with/without ultrasonic irradiation. The plasma was generated between two parallel plate electrodes and driven by an AC (approximately 40 kHz, 100 W) power supply (Generator 6030. SOFTAL Electronic GmbH, Germany). A schematic diagram of the setup is shown in Figure 3.

The voltage and current were measured with a high voltage probe (PPE20kV, LeCroy) and a 50 Ω current-viewing resistor, respectively. The ground aluminium electrode (50 mm × 50 mm) was water cooled and covered with an alumina plate (100 mm × 100 mm × 3 mm), while the powered aluminium electrode (50 mm × 50 mm) had a 42 mm diameter perforated hole covered with a stainless steel mesh.

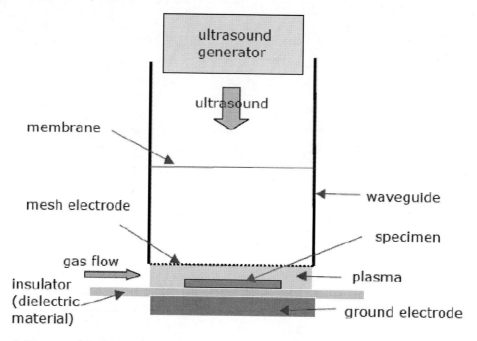

Figure 3. Diagram of the DBD setup with a high-power gas-jet ultrasonic generator [45].

A 40 mm inner diameter poly(methyl methacrylate) (PMMA) waveguide was attached above the powered electrode for the introduction of ultrasound. A high-power gas-jet ultrasonic generator (SonoSteam®, FORCE Technology, Denmark) was placed near the top of the waveguide. The frequency diapason of the ultrasound ranged between 20 and 40 kHz, and the SPL was approximately 150 dB at the GFRP surface. The plasma gas and ambient air was separated using a thin polyethylene film clamped in the waveguide. Helium gas was fed into the plasma at a flow rate of 3 L/min. Each GFRP plate was exposed to the DBD for 0, 5, 10, 30, 60, 120 or 300 s.

OES measurements of the GA-I and GA-II were performed without the GFRP plate using an optical fibre and a 0.75 m spectrometer equipped with a grating with 3600 grooves/mm and a charge-coupled device (CCD) camera (PI-MAX 1024, Princeton Instruments) for the identification of excited species and estimation of the rotational temperature T_{rot} in the discharge. The optical fibre directs the position at the discharge distancing approximately 55 mm from that of the arc ignition. The three nearest points' approximation method was applied for smoothing the OES. T_{rot}, which is assumed to be close to the gas temperature, was evaluated by the relative intensities of two groups of rotational lines corresponding to the R and P branches of the OH A-X (0,0) vibrational band [46].

Contact angles with deionized water and glycerol were measured in air at room temperature both before and after the plasma treatment for evaluation of the surface energy using a contact angle measurement system (CAM100. Crelab Instruments AB, Sweden). Dispersive (Lifshitz-van der Walls interaction γ_S^d) and polar (acid-base or electron-acceptor/electron-donor interaction γ_S^p) components of the solid surface energy of the GFRP plates were obtained by the two-liquid geometric method [47-50], namely by solving the simultaneous equations [51,52],

$$\left(1+\cos\theta_i\right)\cdot\left(\gamma_{Li}^d+\gamma_{Li}^p\right)=2\left(\gamma_{Li}^d\cdot\gamma_S^d\right)^{\frac{1}{2}}+2\left(\gamma_{Li}^p\cdot\gamma_S^p\right)^{\frac{1}{2}} \quad (i=1,2)$$

where θ is a measured contact angle, γ_L^d and γ_L^p are dispersive and polar components of the surface energy of test liquids, and 1 and 2 denote deionized water and glycerol. γ_L^d and γ_L^p are 21.8 and 51 mJ/m^2 for deionized water [17], and 34 and 30 mJ/m^2 for glycerol [53].

The GFRP-I surfaces were observed with scanning electron microscopy (SEM. JEOL 5310LV). Energy dispersive X-ray (EDX) analysis (Quest) was carried out for detection of chemical components of the surfaces.

ATR-FTIR spectroscopy and XPS were employed to study the changes of the functional groups on the GFRP-I surfaces. The FTIR measurements were performed using a Perkin Elmer instrument (Spectrum One) with 32 scans and a resolution of 4 cm^{-1}.

XPS (Sage 100, SPECS, Berlin, Germany) was employed to study the changes of the elemental composition at the GFRP surfaces treated by the air DBD and the gliding arc. Atomic concentrations of each element were calculated by determining the relevant integral peak intensities using a linear background [54]. XPS data were also collected using a micro-focused, monochromatic Al Kα X-ray source (1486.6 eV) with a lateral resolution of 30 μm (K-alpha, ThermoFischer Scientific, UK) to study the changes in elemental compositions and the functional groups on the GFRP-II surfaces treated by He DBD with/without ultrasonic irradiation. Atomic concentrations of all elements were calculated by determining the relevant integral peak intensities using the Shirley background. A high resolution analysis was performed on the carbon 1s (C1s) and oxygen 1s spectra (pass energy 25eV) acquired over 30 scans. The binding energies were referred to the hydrocarbon component (C-C, C-H) at 285 eV. The spectra were de-convoluted through curve fitting, taking purely Gaussian components with linear background subtraction.

The average roughness of the GFRP surfaces was evaluated by AFM (XE-150, PSIA, USA) in Milli-Q water (Millipore Corp. USA) in the contact mode. It is defined as the arithmetic mean deviation of the surface heights from the mean line through the profile [55].

Double cantilever beam (DCB) specimens were prepared for the adhesion test of GFRP-I plates. For the preparation of the DCB specimen a 2 mm thick chamfered spacer was placed between a pair of GFRP-I plates, ensuring a uniform glue thickness. Crack initiation was established in the interface between the glue and one of the plates using a 0.05 mm thin PTFE slip foil (Teflease MG2, Airtech, USA) in a position 40 mm from the end of the plate and thereby leaving a bond area of 40 × 40 mm^2. The plates were bonded using vinylester paste (740-0112 Oldopal® VE Bonding paste, DSM) [42]. The paste was blended 5 minutes by hand stirring and the GFRP plates were bonded within 20 minutes from the blending. The specimens were left for a day at room temperature for curing. Steel hinges were glued to the outside of the DCB structure with an epoxy adhesive (Scotch Weld 460, 3M). The curing time of the epoxy adhesive was 24 h at room temperature. The adhesion test was performed by pulling the hinges of the DCB specimens using an Instron 8532 test machine. The load was measured using a 5-kN static load cell and the test was run at a constant displacement speed of 5 mm/min. Time and load data were sampled in data files at 5 Hz.

The adhesion test for GFRP-III was performed by preparing similar DCB specimens and applying transverse forces to them using an Instron 8532 test machine. A special loading device with a wire was used [56]. The load was measured using a 5-kN static load cell.

3. RESULTS AND DISCUSSION

3.1. Air DBD

The exposure to plasma results in significant changes of the GFRP surface morphology. The SEM images of treated and untreated GFRP-I plates are shown in Figure 4. The density of 1 – 10 μm particles at the surface increased as the treatment time increased. EDX analysis indicates that a major component of these particles is aluminium, and thus they are thought to be alumina particles used as fillers for reducing flammability and enhancing the electrical characteristics of the polymer composites [57]. During the treatment weak polymeric domains would be selectively ablated while alumina particles would remain on the surface. It is well-known that plasma treatment leads to ablation of polymer surface layer and that the ablation rate depends on the chemical structure of the polymeric chain [58]. Since oxygen containing polymers tend to be ablated more easily than C-H polymers [18], ablation of the GFRPs in a short time is understandable.

The water contact angle of the GFRP plate was ~84° before plasma treatment, dropped markedly to 22° after a 2-s treatment, decreased further after longer treatments, and became ~0° when the GFRP plates were treated for more than 30 s. The dispersive and polar components of the surface energies before and after the plasma treatment are shown in Figure 5. The polar component increased markedly after the 2-s treatment, and levelled off to a saturation state. There is a much smaller change of the dispersive component, corresponding to minor cross-linking effect by the treatment [59,60]. The slight decrease of the polar component after longer treatments appears to be due to an increase in surface density of alumina particles and surface roughness. The polar and dispersive components of surface energy for alumina are 9.8 and 75.7 mJ/m, respectively [61]. The increase in the surface particle density leads to the decrease in the polar component and the increase in the dispersive component of the surface energy.

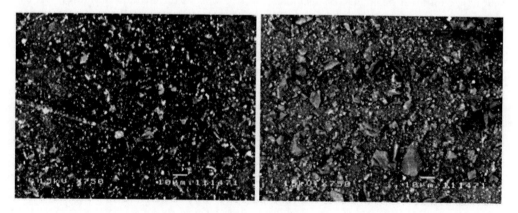

Figure 4. SEM images of the untreated (left), and 30-s treated (right) GFRP plates [42]. Copyright Wiley-VCH Verlag GmbH and Co. KGaA. Reproduced with permission.

In addition the increase in surface roughness results in systematic errors with the two-liquid geometric method that assumes smooth surfaces [62]. The result indicates that polar functional groups could be introduced on the surfaces in a short time treatment, and that the surface roughness could increase as the treatment time increases. These effects potentially improve the adhesion to vinylester resin due to the improvement of wetting with vinylester and enhancement of interlocking.

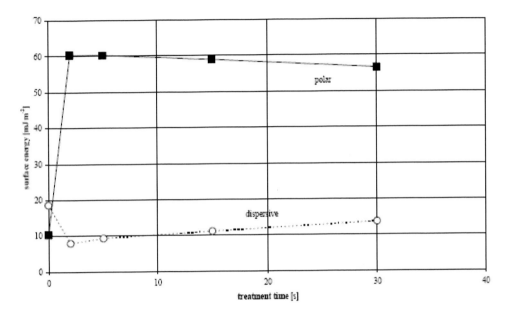

Figure 5. Dispersive and polar components of surface energy of the GFRP-I plates [42]. Copyright Wiley-VCH Verlag GmbH and Co. KGaA. Reproduced with permission.

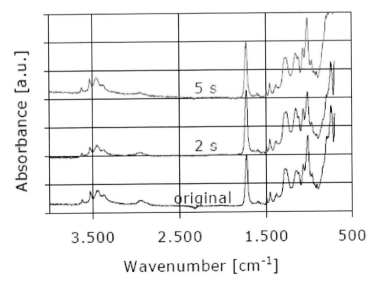

Figure 6. ATR-FTIR spectra of the untreated, 2-s and 5-s treated GFRP plates [42]. Copyright Wiley-VCH Verlag GmbH and Co. KGaA. Reproduced with permission.

Figure 6 compares the ATR-FTIR spectra of the GFRP plates before and after the plasma treatment. Absorption bands were observed between $1000 - 1500$ cm^{-1} (δ(CH$_2$) [63], δ(OH), ν(C-O)), at 1380 and 1454 cm^{-1} [64], at 1494 and 1600 cm^{-1} (aromatic ring vibration) [65], at 1580 cm^{-1} (asymmetric COO- stretching) [64], at 1728 cm^{-1} (ν(C=O) stretching) [65], between 2800 - 3000 cm^{-1} (ν(CH) stretching), and between $3300 - 3700$ cm^{-1} (ν(O-H) stretching) [63-69]. The broad absorption between $2800 - 3000$ cm^{-1} is associated with the dominant anti-symmetric sp^2 CH$_2$ and sp^3 CH$_2$ vibrational modes at 2960 and 2920 cm^{-1} respectively [70,71], slightly decreasing after the treatment. Similar change in the absorption band between $2800 - 3000$ cm^{-1} was reported for other polymer surfaces after plasma treatments [72-74]. The increase in the ratio of the OH peak at 3430 cm^{-1} to the CH stretching peak at 2960 cm^{-1} is often used as an indicator of the formation of oxygen containing functional groups [75] under the assumption that the CH peak remains unchanged. The present result, however, correlates the previous indications [72-74], and suggests that the change of the CH peak should also be considered.

XPS analysis was carried out to analyze the elemental composition of the GFRP surfaces before and after the plasma treatments. The results are summarized in Table 1. A small amount of bromine and zirconium was detected, which can be components of inorganic fire retardants of the FRP [76,77]. The oxygen content increased approximately 10 % after the 2-s treatment, indicating the addition of oxygen containing polar functional groups on the GFRP surface. XPS analysis also indicated that the contents of aluminium as well as oxygen at the surface increased with the longer plasma treatments, and that the further increase in oxygen content can be primarily attributed to the increase in the density of alumina particles at the surface.

Table 1. Elemental composition of the GFRP surfaces by XPS [%] [42]. Copyright Wiley-VCH Verlag GmbH and Co. KGaA. Reproduced with permission.

Treatment time [s]	C	O	N	Al	Br	Zr	Ti	Na
0	76.8	21.3	0.0	1.0	0.1	0.3	0.0	0.4
2	66.4	32.1	0.0	0.9	0.0	0.3	0.0	0.6
5	64.9	31.6	0.0	2.2	0.1	0.2	0.3	0.8
15	59.0	35.2	0.4	4.4	0.0	0.1	0.0	0.9
30	51.9	38.7	1.1	6.7	0.0	0.1	0.4	1.0

Table 2 summarizes curve fittings of C1s spectra. Peaks A, B and C are assigned to C-H/C-C/C=C, C-OH/C-O-C/C-N, and acid or ester, while a peak D is the π^*-π satellite peak from the aromatic carbons. No peaks associated with $\underline{C}$=O/O-$\underline{C}$-O were seen. The peaks B and D moderately increased as the treatment time increased, while peak C increased after the 2-s treatment and remained almost unchanged with the longer treatments. It is noted that the XPS results show good agreement with the evaluation of the surface energies, as introduction of these oxygen containing polar functional groups on the GFRP-I surfaces corresponds to the increase in the polar component of surface energy.

Adhesion tests were performed and the adhesive strength was evaluated with the measured maximum load. The results are shown in Figure 7. The adhesive strength could be expected to increase with the treatment time, since the surface energy increased in a short time treatment while the surface roughness increased monotonically. However, the adhesive strength of the 2-s treated specimen was (145 ± 10) N which is comparable to or higher than that achieved by conventional sand blasting ((127 ± 10) N) which has been regarded as the sufficient treatment for ensuring good adhesion [9,10,14-16].

Table 2. Summary of C1s curve fitting of XPS for the treated and untreated GFRP surfaces [42]. Copyright Wiley-VCH Verlag GmbH and Co. KGaA. Reproduced with permission.

	C_{1s} [eV]			
	285 eV	286.7 eV	289 eV	> 290 eV
Treatment time [s]	A (C-H/C-C)	B (C-O)	C (C(O)O)	D ($\pi^*-\pi$)
0	69.3	18.7	9.8	2.3
2	57.3	22.7	16.9	3.1
5	56.3	24.2	15.5	4.0
15	55.1	24.4	16.4	4.1
30	49.5	29.0	15.8	5.8

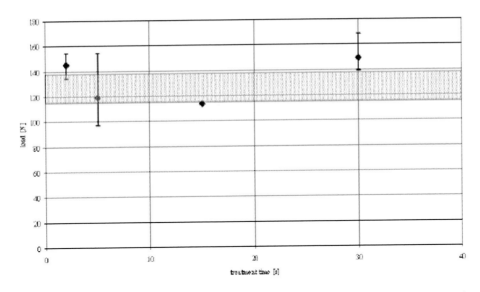

Figure 7. Adhesive strength of the DCB specimens. The grey zone indicates the adhesive strength achieved by sand blasting [42]. Copyright Wiley-VCH Verlag GmbH and Co. KGaA. Reproduced with permission.

It decreased when the surfaces were treated for 5 and 15 s, and recovered and achieved the highest value of 149 N for the 30-s treatment. The 2-s and 30-s treated specimens showed local crack propagated just inside the GFRPs. On the other hand, the 5-s and 15-s treated specimens showed no crack propagation in to the GFRPs, while white, fine alumina particles were seen on the vinylester resin after the adhesion tests. These adhesion results do not correlate with the change of surface energy and morphology, highlighting that understanding adhesion is a complex task [78]. However, it can be suggested that the short treatment increased the polar component of the surface energy and roughness, that longer treatments for

5 – 15 s would form weak boundary layers, and that the 30-s treatment would remove these layers so that the highest adhesive strength was demonstrated.

3.2. Gliding Arc

The input power and the flow rate are important system parameters to determine the properties of the gliding arc discharges [56]. With the input power of more than 1.7 kW, the temperature of the wire electrodes of the GA-I increased within 10 s at the flow rate of 25 L/min, recognized by incandescence of the electrodes. Subsequently the plasma column became highly fluctuated with the occasional generation of unwanted discharges between the powered electrode and the grounded workbench due to the increase in the electrical resistance of the electrodes. In order to stabilize a discharge using the GA-I, the input power has to be less than 1.7 kW. On the other hand, GA-II can be operated at higher input powers possibly due to larger surface areas of the electrodes and easier heat release.

OES was measured to identify excited species in the discharge and to evaluate the rotational temperature. Figure 8 (a) shows a typical OES from the GA-II, including the N_2, N_2^+, NO, OH [79], and CN bands. The OH A-X (0,0) vibrational band is detailed in Figure 8 (b). The OH comes from ambient air humidity. At a flow rate of 30 L/min with an input power between 1.6 and 2.2 kW, T_{rot} was estimated to be between 3200 and 3900 K, while at a flow rate of 40 L/min with an input power of 2.2 kW, T_{rot} was approximately 2200 K. The dispersive and polar components of the surface energy of the untreated GFRP-II plates are 18.0 and 12.1 mJ/ m^2, respectively. Figure 9 shows the surface energy of the GFRP plates treated using the GA-I and GA-II as a function of the air flow rate. In each case, as the flow rate decreased the polar component increased markedly while the dispersive component decreased moderately. However, as was discussed, if the flow rate was lower than 20 L/min, the discharge became unstable due to the increase in the temperature of the electrodes. On the other hand at a flow rate of 40 L/min the extension of the discharge became shorter due to excessive cooling of the discharge, which was estimated by the OES, resulting in lower treatment effect as shown in Figure 9.

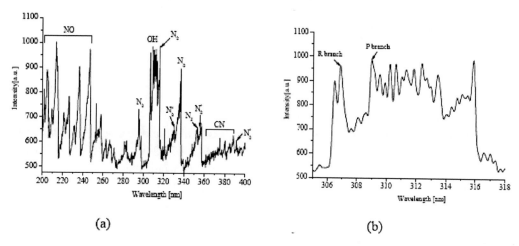

Figure 8. OES of the GA-II at a flow rate of 30 l min[-1] with an input power of 2.2 kW (a). The OH A-X (0,0) vibrational band (b) [44].

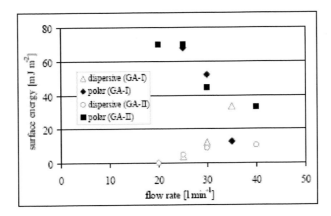

Figure 9. Dispersive and polar components of surface energy of the GFRP-II plates treated by the GA-I and the GA-II with an input power of 1.6 kW [44].

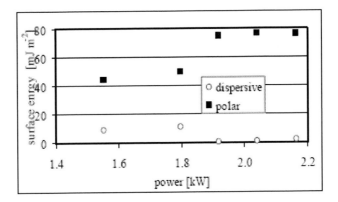

Figure 10. Dispersive and polar components of surface energy of the GFRP-II plates treated by the GA-II as a function of input power at a flow rate of 30 L/min [44].

The surface energy of the GFRP-II plates treated with the GA-II as a function of input power is shown in Figure 10. It is seen that the polar component of the surface energy approaches to 80 mJ/m^2 at the power of 1.9 kW or more.

XPS analysis was carried out to compare the elemental composition of the GFRP-II and -III surfaces before and after the plasma treatment. The results are summarized in Table 3. Silicon was detected on the surfaces in some cases, indicating that the specimen surfaces were contaminated. The oxygen content increased after GA-I treatment, while GA-II treatment further oxidized the surfaces with addition of nitrogen, indicating that adhesive properties can be improved.

AFM was used to measure the surface roughness of the GFRP-II plates, taking a fixed area of 2 × 2 μm^2. The average roughness (r_a) of the untreated surface was approximately 0.7 – 1.1 nm. The GFRP-II plates were treated using the GA-II at a flow rate of 30 L/min with 2.2 kW, exposed 16 times in 80 s, and 32 times in 160 s. The AFM micrographs of these surfaces are shown in Figure 11. The roughness slightly increased to approximately 1.4 – 3.5 nm after the 16 time-exposure (Figure 11 (b)), and decreased to 0.5 – 0.8 nm after the 32 time-exposure (Figure 11 (c)). A possible explanation for this change by the treatment is that the

GFRP-II surfaces could be etched and roughened by the 16 time exposure, and that further exposure could result in micro-scale melting at the surfaces. However, further investigation is necessary to discuss the morphological changes by the treatment, including surface contamination.

Table 3. Elemental composition of the GFRP-I and GFRP-II surfaces by XPS (at %) [44].

GFRP	Treatment	C	O	N	Si
II	Original	75.0	25.0	0.0	
	GA-I	70.7	28.7	0.0	0.6
	GA-II	67.6	31.7	0.8	0.0
III	Original	79.1	20.1	0.0	0.9
	GA-II	63.5	33.2	1.2	2.1

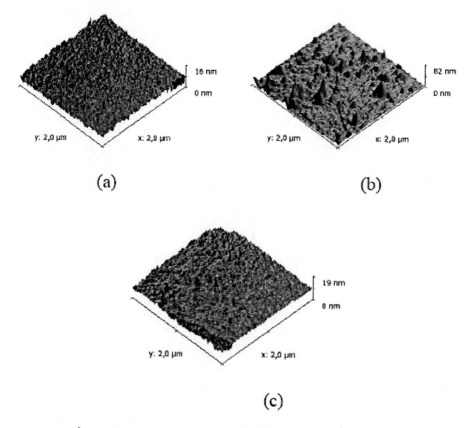

Figure 11. 2 × 2 µm² intermittent contact AFM micrographs of the GFRP-II surfaces; untreated (a), after exposure to the GA-II for 16 times (b) and 32 times (c) at a flow rate of 30 L/min with 2.2 kW [44].

Adhesion tests were performed and the fracture energy was measured as shown in Table 4. It is indicated that adhesive strength of GA-I treated specimens is higher than that by conventional mechanical roughening, and that GA-II treatment further improved adhesion.

This result shows good agreement with the evaluation of surface energy and the XPS analysis. Since major bond mechanism with vinylester resin is non-specific interaction [80], effect of the gliding arc treatment can be primarily improved wetting with the uncured vinylester resin.

Table 4. Fracture energy of the DCB specimens. The GFRP plate surfaces were exposed to the GA-I for 4 and 8 times, and to the GA-II for 8 and 16 times [44].

	Mechanical roughening	GA-I 4 times	GA-I 8 times	GA-II 8 times	GA-II 16 times
Fracture energy [J/m^2]	130 – 170	180 – 240	140 – 240	160 – 320	160 - 360

3.3. He DBD with/without Ultrasonic Irradiation

The waveforms of voltage and current in the helium DBD without ultrasonic irradiation are shown in Figure 12 (a). The current waveform shows a narrow spike in each excitation, indicating that a glow discharge was generated [81]. When the ultrasound was irradiated to the DBD, a higher voltage was required to sustain the plasma at the same power than the plasma without ultrasonic irradiation. In addition, formation of a filamentary discharge was observed with ultrasonic irradiation. It can be identified with complex spiky current waveforms of micro-discharges as shown in Figure 12 (b).

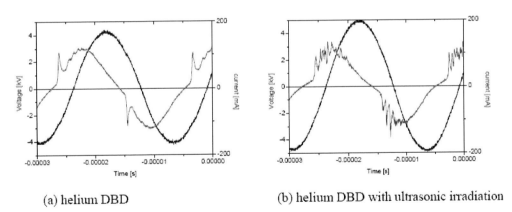

(a) helium DBD (b) helium DBD with ultrasonic irradiation

Figure 12. Voltage (black) and current (red) waveforms for helium DBD without (a) and with (b) ultrasonic irradiation Input power: 100 W.

The treatment drastically changed the GFRP-II surface wettability. The water contact angle on the GFRP-II plate was ~84° before the treatment, dropped markedly to around 30° after 5-s plasma treatment without ultrasonic irradiation, and tended to decrease further after longer treatments as shown in Figure 13. Ultrasonic irradiation during plasma treatment consistently improved the wettability.

The treatment also changed the polar component of the surface energy. It was 12.1 mJ/m^2 before the treatment, between 58 and 67 mJ/m^2 after plasma treatment without ultrasonic irradiation, and between 68 and 73 mJ/m^2 after plasma treatment with ultrasound irradiation as shown in Figure 14. The polar component of the surface energy increased with ultrasonic irradiation, but it was insensitive to the treatment time. Similar result was observed for GFRP composites treated with atmospheric pressure dielectric barrier discharge (Section 3.1 [42]).

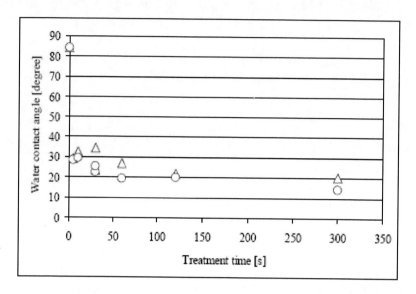

Figure 13. Water contact angle before and after plasma treatment with (circle) and without (triangle) ultrasonic irradiation [45].

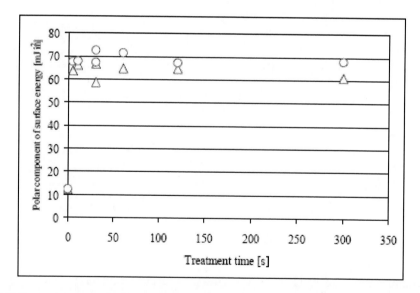

Figure 14. Polar component of surface energy before and after plasma treatment with (circle) and without (triangle) ultrasonic irradiation [45].

XPS survey analysis was carried out to analyze the elemental composition of the GFRP-II surfaces before and after the 30-s treatments with and without ultrasonic irradiation. The results are summarized in Table 5. After the treatment without ultrasonic irradiation, the oxygen content increased approximately 5%, indicating the introduction of oxygen containing polar functional groups on the surfaces. Ultrasonic irradiation further enhanced oxidation at the surface. Although the mesh electrode was exposed to the plasma during treatment, contamination from the electrode is not apparent. Since Si contents at the treated and untreated surfaces were negligible, the surfaces of the GFRPs were almost covered with the polyester matrix and thus most of the glass fibre surfaces were not exposed to the plasma. It is therefore unlikely that the glass fibre surfaces were modified during the treatment.

Table 5. Elemental composition [at. %] and O/C ratio at the GFRP surfaces characterized by XPS [45].

	Elemental composition [at. %]					O/C ratio
Treatment	C	O	N	Si	Ca, Cu, S, Na, or Zn	
Untreated	76.5	22.6	0.6	0.1	0.3	0.295
Plasma, 30 s	71.5	27.5	0.5	0.1	0.7	0.385
Plasma + ultrasound, 30 s	68.3	30.5	0.2	0.1	0.2	0.447

Table 6 summarizes curve fitting of the C1s spectra of the GFRP surfaces before and after 30-s helium DBD treatment with and without ultrasonic irradiation. Peaks at approximately 285, 286.5, 289 and > 290 eV can be assigned to $\underline{C}$-H/$\underline{C}$-C, $\underline{C}$-O-C/$\underline{C}$-OH, $\underline{C}$OO (carboxyl), and plasmon (π^*-π shake-up), respectively [82]. The C1s component peaks at approximately 286.5 eV ($\underline{C}$-OH/O-$\underline{C}$-O) and 289 eV (carboxyl) increased after plasma treatment without ultrasonic irradiation. Ultrasonic irradiation further increased the peak at 289 eV (carboxyl). It indicates an enhanced oxidation at the surfaces during the plasma treatment due to ultrasonic irradiation. It is noted that introduction of these oxygen containing polar functional groups on the surfaces increases the polar component of the surface energy, and that the adhesion properties can be improved [45]. Similar results were obtained for the GFRP plates treated by the air DBD and the gliding arc with and without ultrasonic irradiation [83-85].

Table 6. Summary of C1s curve fitting for the GFRP-II surfaces characterized by XPS [45].

	C_{1s} [eV]			
	285 eV	286.5 eV	289 eV	> 290 eV
Treatment	C-H/C-C	C-O	C(O)O	$\pi^*-\pi$
Untreated	66.6	19.1	12.9	1.4
Helium DBD	59.6	22.5	16.9	1.0
Helium DBD with ultrasound	54.2	20.3	24.0	1.3

AFM was used to measure the surface roughness of the GFRP-II plates, taking a fixed area of 2×2 μm^2. The average roughness (r_a) of the untreated surface was approximately 1.7 – 4 nm. The GFRP plates were treated with the helium DBD with and without ultrasonic irradiation for 30 s. The roughness slightly increased to approximately $2 - 7$ nm and $2 - 10$ nm after the plasma treatment with and without ultrasonic irradiation, respectively. It is therefore unlikely that ultrasonic irradiation further enhances surface roughening during the treatment.

CONCLUSION

GFRP-I plates were treated by atmospheric pressure DBD plasma in air for adhesion improvement. The plasma treatment increased the polar component of the surface energy, and enhanced the surface roughness. However, these results do not correlate with the adhesion test results, suggesting that formation of weak boundary layers could influence the strength of adhesion to vinylester resin.

GFRP plates were treated by gliding arc plasmas. The treatment increased the polar component of the surface energy, oxygen containing polar functional groups at the surfaces, and bond strength with the vinylester resin. Major effect of the adhesion improvement can be attributed to improved wetting with uncured resin. There is an optimum condition of flow rates and input power for adhesion improvement, associated with the temperatures of the electrodes and the discharge.

An atmospheric pressure helium glow DBD changed to a filamentary discharge by the ultrasonic irradiation. Plasma treatment of the GFRP-II surfaces improved the wettability, increased the polar component of the surface energy, and increased the roughness. The ultrasonic irradiation during the plasma treatment further improved the wettability and the polar component of the surface energy. The principal effect of ultrasonic irradiation can be attributed to enhanced surface oxidation during plasma treatment.

ACKNOWLEDGEMENTS

This work is supported by grants from the Danish Research Council (Statens Teknisk-Videnskabelige Forskningsråd (STVF). Grant number 26-04-0251) and the Ministry of Science, Technology and Innovation through the National Danish Proof of Concept funding Scheme (07-017766). Søren Nimb and Lene Hubert are gratefully acknowledged for design and construction of the plasma sources, and XPS measurements, respectively.

REFERENCES

[1] Davis, P.; Cantwell, W.J.; Jar, P.Y.; Bourban, P.E.; Zysman, V.; Kausch, H.H *Composites*, 1991, *22(6)*, 425-431.

[2] Wade, G.A.; Cantwell, W.J.; Pond, R.C. *Interf. Sci.*, 2000, *8*, 363-373.

[3] Dagher, H.J.; Iqbal, A.; Bogner, B. *Polym. Polym. Composites*, 2004, *12(3)*, 169-182.

[4] Kaiser, H.; Karbhari, V.M. *Int. J. Mater. Product Technol.*, 2004, *21(5)*, 349-384.

[5] Bakis, C.E.; Bank, L.C.; Brown, V.L.; Cosenza, E.; Davalos, J.F.; Lesko, J.J.; Machida, A.; Rizkalla, S.H.; Triantafillou, T.C. *J. Composites for Construction*, 2002, *6(2)*, 73-87.

[6] Huttunen, M.; Ashammakhi, N.; Törmälä, P.; Kellomäki, M. *Acta Biomaterialia*, 2006, *2*, 575-587.

[7] Jokstad, A.; Gökçe, M.; Hjortsjö, C. *Int. J. Prosthodont.* 2005, *18(6)*, 489-496.

[8]   Shenton, J.; Lovell-Hoare, M.C.; Stevens, G.C. *J. Phys. D Appl. Phys.*, 2001, *34*, 2761-2768.

[9]   Blackman, B.R.K.; Kinloch, A.J.; Watts, J.F. *Composites* 1994, *25(5)*, 332-341.

[10] Wingfield, J.R.J. *Int. J. Adhes. Adhes.* 1993, *13(3)*, 151-156.

[11] Kinloch, A.J.; Taig, C.M. *J. Adhes*, 1987, *21*, 291-302.

[12] Kinloch, A.J.; Kodokian, G.K.A.; Watts, J.F. *Phil. Trans. R. Soc. Lond. A* 1992, *338*, 83-112.

[13] Kodokian, G.K.A.; Kinloch, A.J. *J. Mater. Sci. Let.* 1988, *7*, 625-627.

[14] Kodokian, G.K.A.; Kinloch, A.J. *J. Adhes.* 1989, *29*, 193-218.

[15] Adams, R.D.; Atkins, R.W.; Harris, A.; Kinloch, A.J. *J. Adhes.* 1986, *20*, 29-53.

[16] Pocius, A.V.; Wenz, R.P. *SAMPE J.* 1985, *20(5)*, 50-58.

[17] Cognard, J. *C.R. Chimie* 2006, *9*, 13-24.

[18] Hegemann, D.; Brunner, H.; Oehr, C. *Nucl. Insum. Methods Phys Res B* 2003, *208*, 281-286.

[19] Kanazawa, S.; Kogoma, M.; Moriwaki, T.; Okazaki, S. *J. Phys. D Appl. Phys.* 1988, *21*, 838-840.

[20] Tendero, C.; Tixier, C.; Tristant, P.; Desmaison, J.; Leprince, P. *Spectrochimica Acta B* 2006, *61*, 2-30.

[21] Kogelschatz, U. *Plasm. Chem. Plasm. Proc.* 2003, *23*, 1-46.

[22] Fridman, A.; Chirokov, A.; Gutsol, A. *J. Phys. D Appl. Phys.* 2005, *38*, R1-R24.

[23] Czernichowski, *A. Pure Appl. Chem.* 1994, *66(6)*, 1301-1310.

[24] Fridman, A.; Nester, S.; Kennedy, L.A.; Saveliev, A.; Mutaf-Yardimci, O. *Progress in Energy and Combustion Sci.* 1999, *25*, 211-231.

[25] Janča, J.; Czernichowski, A. *Surf. Coat. Technol.* 1998, *98*, 1112-1115.

[26] Krebs, N. *Int. Patent* No. 2006 015,604, 2006.

[27] Krebs, N.; Bardenshtein, A., Kusano, Y.; Bindslev, H.; Mortensen, H.J. *Int. Patent* No. 2008 138,901 A1, 2008.

[28] Kusano, Y. *Surf. Eng.* 2009, *25*, 415-416.

[29] Nakane, T.; Hirata, T.; Seya, K. *Jpn. J. Appl. Phys.* 1987, *26 Suppl.* 203-205.

[30] Hirata, T.; Nakane, T.; Seya, K. *Jpn. J. Appl. Phys.* 1991, *30 Suppl.* 54-56.

[31] Kagiwada, T.; Hirata, T.; Nakane, T.; Otsuka T.; Seya, K. *Proc. IEEE Ultrasonic Symp.* 1991, 1015-1018.

[32] Otsuka, T.; Teranishi K.; Nakane, T. *Jpn. J. Appl. Phys.* 2001, *40 Pt1*, 3797-3800.

[33] Nakane, T.; Miyajima T.; Otsuka, T. *Jpn. J. Appl. Phys.* 2004, *43*, 2852-2856.

[34] Bálek, R.; Pekárek, S.; Bartáková, Z. *Ultrasonics* 2006, *44*, e549-e553.

[35] Zhang, Q.H.; Du, R.; Zhang, J.H.; Zhang, Q.B. *Int. J. Machine Tools Manufacture* 2006, *46*, 1582-1588.

[36] Hatsuzawa, T.; Hayase, M.; Oguchi, T. *Precision Eng.* 2002, *26(4)*, 442-447.

[37] Pekárek, S.; Bálek, R. *J. Phys. D Appl. Phys.* 2004, *37*, 1214-1220.

[38] Bálek, R.; Pekárek, S.; Bartáková, Z. *Ultrasonics* 2007, *46*, 227-234.

[39] Pekárek, S.; Bálek, R.; Khun, J.; Pospíšil, M. *Czech. J. Phys. Suppl. B* 2006, *56*, B982-B989.

[40] Choi, K.-S.; Nakamura, S.; Murata, Y. *Jpn. J. Appl. Phys.* 2005, *44(5A)*, 3248-3252.

[41] Koulik, P.; Samsonov, M.; Cherepanov, A.; Petrov, E. *US Patent* No. 2003 165,636 A1, 2003.

[42] Kusano, Y.; Mortensen, H.; Stenum, B.; Kingshott, P.; Andersen, T.L.; Brøndsted, P.; Bilde-Sørensen, J.B.; Sørensen, B.F.; Bindslev, H. "Atmospheric Pressure Plasma Treatment of Glass Fibre Composite for Adhesion Improvement" *Plasm. Proc. Polym.* 2007, *4(S1)*, S455-S459.

[43] Kusano, Y.; Teodoru, S.; Leipold, F.; Rozlosnik, N.; Sørensen, B.F.; Andersen, T.L.; Goutianos, S.; Michelsen, P.K. In *Proc. 29th Risø Int. Symp. Mater. Sci.* Risø National Laboratory: Roskilde, DK, 2008; pp 305-312.

[44] Kusano, Y.; Teodoru, S.; Leipold, F.; Andersen, T.L.; Sørensen, B.F.; Rozlosnik, N.; Michelsen, P.K. "Gliding arc discharge — Application for adhesion improvement of fibre reinforced polyester composites" *Surf. Coat. Technol.* 2008, *202*, 5579-5582.

[45] Kusano, Y.; Singh, S.V.; Bardenshtein, A.; Krebs, N.; Rozlosnik, N. "Plasma Surface Modification of Glass-Fibre-Reinforced Polyester Enhanced by Ultrasonic Irradiation" *J. Adhes. Sci. Technol.* 2010, *24,* 1831-1839.

[46] Laux, C.O.; Spence, T.G.; Kruger, C.H.; Zare, R.N. *Plasm. Source Sci. Technol.* 2003, *12*, 125-138.

[47] Ma, K.; Chung, T.S.; Good, R.J. *J. Polym. Sci. B Polym. Phys.* 1998, *36*, 2327-2337.

[48] Owens, D.K.; Wendt, R.C. *J. Appl. Polym. Sci.* 1969, *13*, 1741-1749.

[49] Kaelble, D.H. *J. Adhes.* 1970, *2*, 66-81.

[50] Schwarcz, A. *J. Polym. Sci.*, 1974, *12*, 1195-1205.

[51] Choi, D.M.; Park, C.K.; Cho, K.; Park, C.E. *Polym.* 1997, *38(25)*, 6243-6249.

[52] Park, S.J.; Kim, M.H.; Lee, J.R.; Choi, S. *J. Colloid Interf. Sci.* 2000, *228*, 287-291.

[53] Dilsiz, N.; Wightman, J.P. *Colloids Surf.*, 2000, *164*, 325-336.

[54] Mortensen, H.; Kusano, Y.; Leipold, F.; Rozlosnik, N.; Kingshott, P.; Sørensen, B.F.; Stenum, B.; Bindslev, H. *Jpn. J. Appl. Phys.* 2006, *45(10B)*, 8506-8511.

[55] Hutchings, I.M. *Tribology;* Arnold, London, UK, 1992.

[56] Mutaf-Yardimci, O.; Saveliev, A.V.; Fridman, A.A.; Kennedy, L.A. *J. Appl. Phys.* 2000, *87(4)*, 1632-1641.

[57] Mansour, S.H. *J. Elastomers Plastics*, 2000, *32*, 248-264.

[58] Švočík, V.; Kolářová, K.; Slepička, P.; Macková, A.; Novotná, M.; Hnatowicz, V. *Polym. Degradation Stability*, 2006, *91(7)*, 1219-1225.

[59] Liston, E.M.; Martinu, L.; Wertheimer, M.R. *J. Adhes. Sci. Technol.*, 1993, *7*, 1091-1127.

[60] Arpagaus, C.; Sonnenfeld, A.; von Rohr, P.R. *Chem. Eng. Technol.* 2005, *28*, 87-94.

[61] Eliades, T.; Eliades, G.; Brantley, W.A. *Am. J. Orthodonics Dentofacial Orthopedics*, 1995, *108(4)*, 351-360.

[62] Chang, T.C. *J. Ind. Technol.* 1998, *15(1)*, 2-7.

[63] Honeybone, P.J.R.; Newport, R.J.; Walters, J.K.; Howells, W.S.; Thomkinson, J. *Phys. Rev. B* 1994, *50(2)*, 839-845.

[64] Raghavan, D.; Egwim, K. *J. Appl. Polym. Sci.* 2000, *78*, 2454-2463.

[65] Gan, B.K.; Bilek, M.M.M.; Kondyurin, A.; Mizuno, K.; McKenzie, D.R. *Nucl. Instr. Methods Phys. Res. B*, 2006, *247*, 254-260.

[66] Fraga, A.N.; Alvarez, V.A.; Vázquez, A.; De la Osa, O. *J. Comp. Mater.* 2003, *37(17)*, 1553-1574.

[67] Patel, M.R.; Patel, J.V.; Sinha, V.K. *Polym. Degrad. Stability*, 2005, *90(1)*, 111-115.

[68] Louh, S.P.; Leu, I.C. ; Hon, M.H. *Diam. Relat. Mater.* 2005, *14*, 1005-1009.

[69] El-Hendawy, A-N. A. *J. Anal. Appl. Pyrolysis* 2006, *75*, 159-166.

[70] Heitz, T.; Drévillon, B.; Godet, C. ; Bourée, J.E. *Carbon*, 1999, *37(5)*, 771-775.

[71] Veres, M.; Koós, M.; Pócsik, I. *Diam. Relat. Mater.* 2002, *11*, 1110-1114.

[72] Pastor-Blas, M.M.; Martín-Martínez, J.M.; Dillard, J.G. *Surf. Interf. Anal.* 1998, *26*, 385-399.

[73] Romero-Sánchez, M.D.; Pastor-Blas, M.M.; Martín-Martínez, J.M.; Zhdan, P.A.; Watts, J.F. *J. Mater. Sci.* 2001, *36*, 5789-5799.

[74] Raacke, J.; Giza, M.; Grundmeier, G. *Surf. Coat. Technol.* 2005, *200*, 280-283.

[75] Gupta, B.; Hilborn, J.; Hollenstein, C.H.; Plummer, C.J.G.; Houriet, R.; Xanthopoulos, N. *J. Appl. Polym. Sci.* 2000, *78*, 1083-1091.

[76] Kicko-Walczak, E. *Polymery*, 1999, *44(11-12)*, 724-729.

[77] Levchik, S.V.; Weil, E.D. *Polym. Int.* 2005, *54*, 11-35.

[78] Shenton, M.J.; Lovell-Hoare, M.C.; Stevens, G.C. *J. Phys. D Appl. Phys.* 2001, *34*, 2754-2760.

[79] Czernichowski, A.; Nassar, H.; Ranaivosoloarimanana, A.; Fridman, A.A.; Simek, M.; Musiol, K.; Pawelec, E.; Dittrichova, L. *ACTA Phys. Polonica A* 1996, *89*, 595-603.

[80] Brown, J.R.; Mathys, Z. *J. Mater. Sci.* 1997, *32*, 2599-2604.

[81] Yokoyama, T.; Kogoma, M.; Moriwaki, T.; Okazaki, S. *J. Phys. D Appl. Phys.* 1990, *23*, 1125-1128.

[82] Beamson G, Briggs D, "High resolution XPS of organic polymers" The Scienta ESCA300 database, John Wiley and Sons, Chichester, UK, 1992.

[83] Kusano, Y.; Singh, S.V.; Norrman, K.; Leipold, F.; Drews, J.; Morgen, P.; Bardenshtein, A.; Krebs, N. *J. Adhes.* Accepted for publication.

[84] Kusano, Y.; Norrman, K.; Leipold, F.; Drews, J.; Singh, S.V.; Morgen, P.; Bardenshtein, A.; Krebs, N. *Surf. Coat. Technol.* in press.

[85] Kusano, Y.; Singh, S.V.; Norrman, K.; Drews, J.; Leipold, F.; Bardenshtein, A.; Krebs, N. *Surf. Eng.* submitted.

In: Fiber–Reinforced Composites
Editor: Qingzheng Cheng

ISBN: 978-1-61470-303-7
© 2012 Nova Science Publishers, Inc.

Chapter 11

ULTRA-HIGH TEMPERATURE CERAMICS REINFORCED WITH SiC SHORT FIBER OR WHISKER

Laura Silvestroni[1] and Diletta Sciti

CNR-ISTEC, Institute of Science and Technology for Ceramics-National Research Council of Italy, Via Granarolo 64, I-48018 Faenza (RA), Italy

ABSTRACT

The IV and V group transition metals borides, carbides and nitrides are widely known as *ultra-high temperature ceramics* (UHTCs), thanks to their high melting point above 2,500°C. These ceramics possess outstanding physical and engineeristic properties, such as high hardness and strength, low electrical resistivity and good chemical inertness.

The materials more deeply investigated are the ZrB_2-based ones, in view of the lower density, compared to Hf and Ta composites, and the better oxidation resistance of the borides, compared to the carbides and nitrides.

So far, the research activity has mainly focused on the achievement of high strength UHTCs, usually based on ZrB_2 and HfB_2 composites containing SiC particles in amounts of 20-30vol%. These ceramics can in fact reach strength up to 1 GPa and even higher, through careful tailoring of processing parameters. However, these strong ceramics still have very low fracture toughness, with values ranging from 2.5 to 4 $MPam^{1/2}$ for ZrB_2-based materials. The demand to design and realize reinforced UHTC materials comes from the need to overcome this brittleness, which makes the scale-up of big components a technological issue. The basic concept is the improvement of the fracture toughness through the introduction of elongated reinforcing phases, like SiC fibers (macro-reinforcement), or whiskers (micro-reinforcement) into UHTC matrices.

A number of critical issues rise when elongated secondary phases are added to the matrix: it is difficult to obtain a homogeneous dispersion, fracture of the reinforcing phase occurs during milling and it is problematic to achieve dense ceramics without deteriorate the reinforcing phase.

For example, whiskers tend to degenerate into particles if the sintering temperature is too high and then lose their reinforcing action. For carbon or silicon carbide fibers there is a strong interface reaction with the ZrB_2 matrix, which should be instead avoided to promote significant pull out.

---

[1] Corresponding author. Tel: +39 546 699723; e.mail: laura.silvestroni@istec.cnr.it

This chapter presents the latest results on ZrB_2-based materials reinforced with SiC fibers or whiskers. Several sintering additives are experienced in order to preserve the integrity of the reinforcing element, thus to exploit at best their toughening action.

The microstructure of the dense materials is studied by SEM and TEM techniques to investigate the effect of the sintering additive and the interface between matrix and reinforcing phase.

The mechanical properties are compared to those of the reference material to assess the real variation obtained by the introduction of these reinforcing agents. Following the inspection of the crack/microstructure interaction, the experimental fracture toughness is quantitatively compared to theoretical models. The addition of such reinforcements can improve the fracture toughness, from 3.7 to 5.5-6.3 $MPam^{1/2}$ upon a proper choice of sintering additive and right amount of reinforcing phase. The main mechanisms invoked for explaining such improvements are crack deflection and crack pinning. The values of high temperature strength are also presented and discussed.

Mention is also made to the effect of the introduction of SiC short fibers to other UHTCs, namely to HfB_2 and TaC.

Keywords: UHTC, whiskers, fibers, microstructure, fracture toughness.

INTRODUCTION

1. Foreword

Ceramic materials seem to be created for the extremes: their melting points are high and their density and thermal expansion coefficients are generally low, they possess high hardness, strength and corrosion resistance, but they are extremely brittle. They can offer thousands of possibilities for modern technology, but they need to be properly tailored.

Brittleness is one of the most critical aspect, which leads ceramics to be highly susceptible to catastrophic failure. While this shortcoming has been tolerated in conventional ceramics for many years, the warning of brittle failure has limited the use of ceramics to very circumscribed areas. Technological measures are mandatory to stem the capability of ceramics to oppose to crack propagation.

In contrast with monolithic ceramics, where the appearance of crack will directly destroy the structure, the cracks within a composite ceramic enable a controlled stress release giving to the material a better damage tolerance. The use of long fibers in composites would allow higher reliability and high temperature performances. However, it has to be underlined that dedicated processing routes and technologies are required to produce long fiber reinforced composites, i.e. polymer infiltration pyrolisis (PIP), chemical vapour deposition (CVD), melt infiltration of porous preform and others, which all necessitate particular care in the optimization of a variety of parameters and are time consuming, without the guarantee of homogeneity and repeatability.

The great advantage of using short fibers over long fibers relies in the process. Short fibers and whisker reinforced ceramic composites have the added attractive features of being fabricated by conventional powder processing techniques and of having the potential for isotropic mechanical properties when properly fabricated. In addition, slurry mixtures of ceramic matrix powders and short fibers or whiskers can be manipulated to behave like other ceramic powder suspensions. Thus, green composite components can often be formed to near net shape by methods such as slip casting, extrusion, and injection molding [1].

The discussions in this chapter attempt to summarize the latest results achieved in the field of a special class of reinforced composites, Ultra-High Temperature Ceramics, so labeled in view of their extremely high melting points. A comprehensive overview of the manufacturing issues, microstructure evolution upon sintering and mechanical properties, whith particular emphasis on the fracture toughness, will be provided. This study tries to furnish an explanation to the complex mechanisms occurring during densification and under mechanical solicitations of ceramic composites containing discontinuous reinforcements. The approach is to implement the existing composites, i.e. ZrB_2-SiC, HfB_2-SiC, with addition of short fibers or whiskers, with the ultimate goal of improving the fracture toughness and creating a new promising class of composites which can be produced through conventional ceramic technology.

1.1 Basic Requirements for High Temperature Materials

High performance materials for structural applications should embody several properties such as sufficient strength and toughness at room temperature, good mechanical properties at high temperature and excellent oxidation resistance. Current high temperature applications require operating conditions up to 2,000°C or even higher. Furthermore, high temperature materials operate in environments with a wide spectrum of mechanical and chemical conditions. Resistance to oxidation is an essential characteristic of high-temperature materials. However many applications involve significantly more complex corrosive environments than simple oxidation.

Ceramics have long been known for their refractoriness, or ability to bear load at elevated temperatures. For a while, oxide ceramics, such as zirconia, alumina, mullite were preferably used for refractory applications, as they are characterized by outstanding chemical resistance over a wide range of temperature, relatively low density, moderately high melting point, brittleness at low temperature and low to moderate cost. Their fracture toughness compares favourably with that of the high-temperature metals and they are unexcelled in resistance to oxidation. However, these materials all possess some characteristic that precludes their use for very high stress and high temperature applications. At temperature above 1,500°C, indeed, all of the oxides have low mechanical properties [2].

The requisite properties for severe application conditions have been found in non-oxide materials, such as silicon nitride and silicon carbide, because they exhibit high strength even at high temperatures, low thermal expansion, excellent thermal shock resistance and oxidation resistance by virtue of protective silica-based glass oxidation layers.

Recently, another class of high temperature materials has attained considerable interest, i.e. ultra-refractory materials based on intermetallic diboride, carbides and nitrides. The limiting characteristics of the carbides include high cost, instability in oxidizing atmosphere and poor thermal shock resistance. With the exception of silicon carbide, none of the carbides can be used at temperature above 1,000°C for long periods of time when oxygen is present, but in reducing environment they can withstand temperatures up to 2,500°C. As for the nitrides, while a number of them have high melting points, their usefulness is generally limited because of their relatively high dissociation pressures [2].

It was soon determined that intermetallic diboride compounds offer the highest degree of oxidation resistance over the other two classes and HfB_2 and ZrB_2 were chosen as the most promising candidates for use in high temperature applications [3-5]. The technical importance of the diborides emanates from their unique engineeristic property combination such as high refractoriness, high electrical and thermal conductivity, chemical inertness against molten metals or non-basic slugs and good oxidation resistance [3-5]. All these properties make them attractive candidates for high-temperature applications where corrosion, wear and oxidation resistance are demanded.

During the cold war, through the mid 1950s to the 1970, many of these compounds were extensively studied in the U.S. and U.S.S.R. for potential aerospace applications in the frame of the Space Race. After a period of relative inactivity, research on UHTCs has experienced a resurgence in the early 1990s pushed by NASA and the US. Air Force, with significant efforts in countries including the United States, China, Japan, Italy and Ukraine.

Properties of these non-oxide structural ceramics have been continually upgraded over these last years through improvements of the starting powders, of the processing techniques and through an increasing understanding of the processing-microstructure-property relationship.

Since 2000, ISTEC group (Italy) has been continuously working on ZrB_2 and other UHTCs in order to improve the processing technique and final properties. The use of ceramic additives like Si_3N_4 [6] or AlN [7] was found to improve the sinterability, the microstructure and the mechanical properties of UHTCs. At the same time other research groups in the US and in China have intensified the research activity on this class of materials focussing on the process, microstructure tailor, on the improvement of the mechanical properties and on the understanding of the thermal behavior under extreme environments [3-5,8-11].

The most investigated additive for ZrB_2 was silicon carbide [12-16], which was found to be beneficial for the densification, for the obtainment of a refined microstructure and for a better oxidation resistance. Molybdenum disilicide has been also found to be a promising additive for both boride and carbides, acting as protective barrier against high temperature oxidation and improving the mechanical properties [17,18].

1.2   Toughening of UHTC Materials trough Discontinuous Elongated Phases

A major concern in using UHTCs is their inherent brittleness which actually greatly restricts their reliability in service. Within the last 3 years a great effort has been devoted to improve the toughness of ceramics.

The incorporation of discontinuous elongated reinforcement, into the UHTC matrix provides the possibility of improving toughness and strength. The most important toughening mechanisms are briefly summarized in the following.

- Microcracking: a misfit between the coefficient of thermal expansion introduces a stress field which can produce spontaneous microcracks if the residual stresses are sufficiently large. The ideal situation is to create a stress field that will not produce cracks without the application of a superimposed extrinsic stress field. In this way, microcracks are present only at the vicinity of the tip of a macrocrack, such that the stress intensity at the crack tip is reduced and the toughness is increased [19,20].
- Crack deflection and branching: the direction of the extension of a crack may change when the crack meets on obstacle, such as a second phase. Briefly, the deviation in direction means that the crack travels a longer path and the stress intensity a the

crack tip is reduced. Crack branching results when the crack bifurcates into more than one direction. The modelling of crack deflection by secondary phase reinforcement has been theoretically developed and experimentally verified by Faber and Evans [21,22]. It was concluded that a rod-shaped second phase with high aspect ratio is more effective in enhancing toughness than particulate reinforcement. The crack deflection mechanisms theorized by Faber-Evans was further developed, including microstructural parameters such as volume fraction, shape, size, orientation and distribution of the second phase particles. It was then demonstrated that in the case of platelets-reinforcement, the direct contribution of toughening by crack deflection may be less than predicted by earlier theory [23].

- Crack bridging: the front of the crack passes beyond the reinforcing phase but this phase remains intact and bridges the fracture surfaces in the wake of the crack. The open displacement of the crack is then limited and this makes further propagation of the crack difficult. This toughening mechanism is especially observed when the reinforcing phases are fibers. It can also be activated by large grains in the matrix. Theoretical analyses of the bridging process by rigid and ductile inclusions were proposed by several authors [24,25].
- Crack pinning: unlike crack bridging, where a crack tip passes beyond a reinforcement fiber, crack propagation may be stopped and the crack pinned at the fibers [26].
- Residual stress toughening: the model of Faber and Evans ignores the local stress field at and near the interface between the matrix and the filler, which generates another toughening mechanisms, known as residual stress toughening. When a second phase is combined into a matrix there is usually a mismatch between them in terms of the structural, mechanical and physical properties. Two mismatches pertinent to the toughening (and strengthening) of ceramics are the differences in thermal expansion coefficients and the elastic *moduli*. The different thermal contractions shown by the second phase and the matrix phase give rise to residual stresses when the composite is cooled down from the fabrication temperature. The toughening contribution due to thermal residual stress was treated by Taya et al. [27] only considering the case of particles. If the particles have a higher thermal expansion coefficient compared to the matrix, the local stress field is compressive and the model predicts a decrease in the stress intensity factor and hence an increase in the fracture toughness.

Having said that, it should be noted that in real materials, more than one mechanism usually operates simultaneously. Individual mechanisms may interact or interfere with each other; for instance, crack deflection may be attracted by microcracking.

1.3 State of the Art of Toughened UHTCs

At the state of the art, only few literature data are available on long fiber reinforced UHTC and very little is reported on short fiber-reinforced UHTCs as well. Concerning the first series of materials, carbon fibers were selected as reinforcement to improve the thermal shock resistance of UHTCs and to reinforce the UHTC matrix [28,29]. A summary of the literature pertinent to UHTCs toughened through discontinuous phases explored so far is presented in Table I.

Table I. Summary of the state of the art on the activity aiming to the toughening of UHTC matrices. Note: all the materials are considered fully dense. SiC is in form of particles unless other wording, w: whiskers, n: nanosized, f: fiber, CNT: carbon nanotubes.

Ref.	Composition vol%	Sintering °C, min, MPa	K_{Ic} MPam$^{1/2}$	Method	σ MPa	Method	Toughening phase
12	ZrB$_2$	HP:1900,45,32	3.5±0.3	SEPB	565±53	4-pt	-
	ZrB$_2$+10SiC	HP:1900,45,32	4.1±0.3	SEPB	713±48	4-pt	SiC particles
	ZrB$_2$+20SiC	HP:1900,45,32	4.4±0.2	SEPB	1003±94	4-pt	SiC particles
	ZrB$_2$+30SiC	HP:1900,45,32	5.3±0.5	SEPB	1089±152	4-pt	SiC particles
13	ZrB$_2$+20SiCn	HP:1900,30,30	6.4±0.3	SENB	925±28	3-pt	nanosized SiC
30	ZrB$_2$+20SiC	HP:1950,60,30	4.8±0.5	SENB	362±39	3-pt	SiC particles
	ZrB$_2$+30SiC	HP:1950,60,30	5.6±0.5	SENB	517±63	3-pt	SiC particles
	laminated 20/30SiC	HP:1950,60,30	8.8±0.3	SENB	960±84	3-pt	laminated
31	ZrB$_2$+20SiC	HP:2000,60,30	4.9±0.2	SENB	502±45	3-pt	SiC particles
	ZrB$_2$+20SiC+ 10Cf	HP:2000,60,30	6.6±0.1	SENB	445±36	3-pt	C fiber
32	ZrB$_2$+20SiC	HP:1900,60,30	4.0±0.3	DCM	582±102	4-pt	SiC particles
	ZrB$_2$+20SiC+2 (wt)CNT	HP:1900,60,30	4.6±0.6	DCM	616+97	4-pt	CNT
33	ZrB$_2$+20SiCw+ 3YAG	SPS: 1600,5,30	6.02±0.22	SENB	708±11	3-pt	SiCw
	ZrB$_2$+20SiCw+ 3YAG	HP: 1800,60,30	6.60±0.14	SENB	753±16	3-pt	SiCw
34	ZrB$_2$+20SiCw	HP:1800,60,30	6.2	SENB	640	3-pt	SiCw
	ZrB$_2$+20SiCw+3 YAG	HP:1800,60,30	6.7	SENB	775	3-pt	SiCw +interface
	ZrB$_2$+20SiCw+ 6YAG	HP:1800,60,30	6.2	SENB	725	3-pt	SiCw +interface
	ZrB$_2$+20SiCw+ 10YAG	HP:1800,60,30	6.25	SENB	725	3-pt	SiCw +interface
37	ZrB$_2$+3YAG	HP:1800,60,30	5.4±0.2	SENB	629±31	3-pt	-
	ZrB$_2$+3YAG+ 10SiCw	HP:1800,60,30	6.5±0.1	SENB	786±24	3-pt	SiCw
	ZrB$_2$+3YAG+ 20SiCw	HP:1800,60,30	6.7±0.1	SENB	783±28	3-pt	SiCw
	ZrB$_2$+3YAG+ 30SiCw	HP:1800,60,30	7.1±0.2	SENB	761±16	3-pt	SiCw
36	ZrB$_2$+20SiCw	HP:1800,60,40	5.97±0.3	SENB	-	-	SiCw
37	ZrB$_2$+10SiCw	SPS:1550,5,40	6.21	SENB*	550	3-pt	SiCw
	ZrB$_2$+10SiCw+ 3-5AlN	SPS:1550,5,40	8.15-8.5	SENB*	553-566	3-pt	SiCw +interface
	ZrB$_2$+10SiCw+ 3-5Si$_3$N$_4$	SPS:1550,5,40	6.97-7.15	SENB*	457-508	3-pt	SiCw +interface

Ultra-High Temperature Ceramics Reinforced with SiC Short Fiber or Whisker 325

Ref.	Composition vol%	Sintering °C, min, MPa	K_{Ic} MPam$^{1/2}$	Method	σ MPa	Method	Toughening phase
38	ZrB_2+20SiCw	HP:1800,60,30	6.2	SENB	648	3-pt	SiCw
	ZrB_2+20SiCw+2SiC	HP:1800,60,30	6.15	SENB	640	3-pt	SiCw+ SiC particles
	ZrB_2+20SiCw+5SiC	HP:1800,60,30	6.2	SENB	733	3-pt	SiCw+ SiC particles
	ZrB_2+20SiCw+7SiC	HP:1800,60,30	6.9	SENB	673	3-pt	SiCw+ SiC particles
	ZrB_2+20SiCw+10SiC	HP:1800,60,30	6.1	SENB	592	3-pt	SiCw+ SiC particles
39	ZrB_2+20SiCw	HP:2000,60,30	4.05±0.2	SENB	510±25	3-pt	SiCw
	ZrB_2fine+20SiCw	HP:1800,60,30	6.03±0.3	SENB	651±31	3 pt	SiCw+fine powder
40	ZrB_2+20SiC	HP:1800,60,30	4.1±0.05	SENB	795±105	3-pt	SiC particles
	ZrB_2+20SiC+10SiCw+10Cflakes	HP:1800,60,30	6.4±0.6	SENB	600±75	3-pt	SiCw+Cflakes
41	ZrB_2+20SiC+10WC	PLS:2200,120,Ar	6.5±0.2	SENB*	530	3-pt	ZrB_2 platelets
42	ZrB_2+20MoSi$_2$	RHP:1800,30,20	4.9±0.2	DCM	457±27	3-pt	ZrB_2 platelets
43	ZrB_2+5MoSi$_2$	PLS: 1850,60,-	2.9±0.1	CNB	570±50	4-pt	MoSi$_2$
	ZrB_2+5MoSi$_2$+20α-SiC	PLS: 2100,120,-	4.7±0.1	CNB	350±80	4-pt	SiC platelets
	ZrB_2+5MoSi$_2$+20ß-SiC	PLS: 2100,120,-	5.0±0.1	CNB	410±40	4-pt	SiC platelets
	ZrB_2+5Si$_3$N$_4$+20ß-SiC	PLS: 2100,120,-	3.8±0.1	CNB	300±40	4-pt	SiC platelets
44	ZrB_2+10SiC+10ZrO$_2$	HP:1750,60,30	6.4±0.4	SENB	771±36	3-pt	ZrO_2
	ZrB_2+10SiC+10ZrO$_2$	HP:1850,60,30	6.0±0.2	SENB	788±78	3-pt	ZrO_2
	ZrB_2+10SiC+10ZrO$_2$	HP:1950,60,30	4.5±0.2	SENB	532±45	3-pt	ZrO_2
45	ZrB_2+10SiC+10 ZrO$_2$	HP:1850,60,30	6.0±0.3	SENB	758±79	3-pt	ZrO_2
	ZrB_2+10SiC+20 ZrO$_2$	HP:1850,60,30	6.1±0.3	SENB	803±61	3-pt	ZrO_2
	ZrB_2+10SiC+30 ZrO$_2$	HP:1850,60,30	6.4±0.3	SENB	829±171	3-pt	ZrO_2
46	ZrB_2+20MoSi$_2$	HP:1850,60,30	4.25	SENB	490	3-pt	MoSi$_2$
	ZrB_2+20MoSi$_2$+5ZrO$_2$	HP:1850,60,30	5.6	SENB	820	3-pt	ZrO_2
	ZrB_2+20MoSi$_2$+10ZrO$_2$	HP:1850,60,30	6.1	SENB	890	3-pt	ZrO_2
	ZrB_2+20MoSi$_2$+20ZrO$_2$	HP:1850,60,30	6.25	SENB	950	3-pt	ZrO_2
47	ZrB_2+15ZrO$_2$	HP:1850,60,30	5.6	SENB	667	3-pt	ZrO_2
	ZrB_2+20ZrO$_2$	HP:1850,60,30	5.8	SENB	740	3-pt	ZrO_2
	ZrB_2+25ZrO$_2$	HP:1850,60,30	6	SENB	790	3-pt	ZrO_2
	ZrB_2+30ZrO$_2$	HP:1850,60,30	6.5	SENB	803	3-pt	ZrO_2

Sintering: HP=hot pressing, SPS= spark plasma sintering, PLS= pressureless sintering, RHP= reactive hot pressing. Method for K_{Ic} evaluation: SEPB= Single edge pre-cracked beam, SENB= Single edge notched beam, DCM= direct crack measurement, CNB= Chevron notched beam.

Values reported for fracture toughness are quite scattered due to the use of different reinforcing elements, as well as different measuring techniques, that are mainly direct crack measurements (DCM) and Single Edge Notched Beam (SENB). For SiC particles reinforced ZrB_2, values range from 4.1 to 5.6 $MPa \cdot m^{1/2}$ were measured. Values up to 8.8 $MPa \cdot m^{1/2}$ were reported by single edge notched beam (SENB) on laminated $ZrB_2+20/30$ SiC [30].

As for the reinforcing action of discontinuous elongated phases, it has been reported that the addition of short fibers, nanotubes, whiskers or flakes gives promising results for the increase of the fracture toughness [31-40]. For ZrB_2-based composites, fracture toughness values up to 6.6-8.5 $MPa \cdot m^{1/2}$ were reported by single edge notched beam (SENB) [36].

Generally speaking, these works indicate that the addition of such reinforcements can improve the fracture toughness and the mechanisms invoked for explaining such improvements are crack deflection and, to a lower extent, crack bridging. Although these first attempts open a window on the concrete possibility of increasing the fracture toughness of UHTCs, they also pointed out that some critical parameters, like dimension and purity of the starting phases, mixing procedure and sintering temperatures, can severely compromise the success of the composites.

One of the most critical issues is indeed the reaction or degeneration of the reinforcement during sintering. Zhang et al. [39] found that whiskers degenerate into particles if the sintering temperature is higher than 1,900°C, thus losing their reinforcing action. For carbon fibers there is a strong interface reaction with the matrix, which should be instead avoided to promote significant bridging [31].

Hence, to achieve high toughness values using fiber or whisker additions, three major obstacles have to be overcome: a difficult dispersion in the matrix due to the high aspect ratio, the degradation of the reinforcing elements at high temperature and the development of a strong interface after reaction with the matrix.

Other attempts concern the toughening of ZrB_2 matrix through the in-situ elongation of ZrB_2 grains [41,42], or the ß→α transition of SiC above 1,900°C [43], or the monoclinic to tetragonal transition of ZrO_2 [44-47].

In this chapter the microstructure and mechanical properties of ultra high temperature ceramics reinforced with SiC fibers or whiskers are presented. Particular attention is devoted to the effect of the sintering additive, namely Si_3N_4, $MoSi_2$, $ZrSi_2$ and $TaSi_2$, on the microstructure evolution upon sintering, the interface between matrix and reinforcing phase and the fracture toughness change with respect to the unreinforced baseline. ZrB_2-based composites will be mainly discussed and mention to other two matrices, HfB_2 and TaC, will be done.

Finally, a note for the fracture toughness measurement: a direct comparison among the fracture toughness values given in the literature is pointless, because these values are obtained with different methods, i.e. indentation crack direct measurement (DCM), single-edge notched beam (SENB), chevron notched beam (CNB), etc.. Often, no details on the notching technique and size are provided and, considering that the results obtained strongly depend on the notch–root radius [48], it could happen that the values of fracture toughness reported are affected by experimental artefacts. To this purpose, the composites here presented are compared to the baseline material produced with the same starting powder, sintering additive type and content, and tested with the same technique. This allows a systematic determination of the real fracture toughness variation obtained by the introduction of fibers or whiskers.

1.4 Reinforcing Phases: SiC Chopped Fibers and SiC Whiskers

1.4.1 SiC Fibers

Silicon carbide fibers are commercially processed through many different routes [49]. The most widely studied, developed and commercialized SiC fiber is derived from the polymer precursor process first introduced by Yajima [50]. The fibers are commercially available under the trade name of Nicalon™ (Nippon Carbon Co., Tokyo, Japan), Tyranno™ (Ube Industries, Ltd., Ube, Japan), and Sylramic™ (COI Ceramics, Inc., Salt Lake City, UT). A comparison of the properties of the various generations and brand SiC-based fibers is given in Table II [51-53].

Table II. Advanced SiC fibres – Main characteristics.

Generation	Commercial name	Composition wt%	Diameter μm	N° of filaments	Tensile strength GPa	Tensile modulus GPa	Density g/cm^3	Thermal conductivity W/m-K
I	Nicalon NL-200	SiC$_{1.31}$ 12% O	14	500	3.0	220	2.55	3 (RT) 2.2 (500°C)
	Tyranno LoxM	SiC$_{1.36}$ 11% O, 2% Ti	11	800	3.3	190	2.48	1.4 (RT)
II	Hi-Nicalon	SiC$_{1.39}$ 0.5% O	14	500	2.8	270	2.74	7.8 (RT) 10 (500°C)
III	Hi-Nicalon Type S	SiC$_{1.05}$ 0.2% O	11	500	2.6	420	3.10	18 (RT) 16 (500°C)
	Tyranno SA3	SiC$_{1.07}$ 2% Al	7.5	1600	2.8	380	3.10	65 (RT)
	Sylramic	SiC TiB$_2$, B	10	800	3.2	380	3.10	46 (RT)

The first stage of the fiber process involves the low-temperature melt-spinning of the polycarbosilane (PCS) or polytitanocarbosilane (PTCS) polymer. These spun fibers, which are in the green state, are then stabilized by elevated temperature exposure to oxygen and successively ceramized in a inert atmosphere to a final temperature of ~1300°C. It is important to note that due to the presence of excess oxygen and carbon, these fibers are more correctly classified as SiC-based fibers, rather than SiC fibers. Nicalon™ NL-200 and Tyranno™ LoxM belong to this fiber category, or Generation-I SiC fibers, comprising a dispersion of beta-SiC crystallites of a few nanometers in size embedded in a continuum glassy silicon oxycarbide matrix (SiO_xC_y, where x+y is approximately 4).

The Nicalon™ fiber's thermomechanical properties have been improved by altering the method of cross-linking the spun polymer. Rather than curing the PCS in air, the polymer is subjected to ionizing radiation in an inert environment. This process reduces the atomic oxygen content from more than 15% to less than 1% and is the process with which Hi-Nicalon™, or Generation-II SiC fiber, is made [54]. The average SiC crystallite size for this product increases by more than a factor of two over the ceramic grade fiber, and the elastic modulus undergoes a large increase while the strength decreases slightly. The density of the Hi-Nicalon fiber is also increased from 2.55 g/cm^3 to 2.74 g/cm^3, which is approximately 85% of the theoretical SiC density. The stability of the Generation-II SiC fibers has been improved over that of the previous generation fibers, however, it exhibits serious instability in air due to the presence of large amount of excess carbon.

Further improvement in the SiC fibers has been achieved. The Hi-Nicalon™ process has been taken a step further by incorporating a decarburization process in an hydrogen environment. This resulted in the Hi-Nicalon™ Type-S SiC fiber with a near theoretical density and very low excess of carbon and oxygen [55]. At the same time, the Tyranno™ SA3 fiber was developed through a very different route, i.e. sintering of the original polymer precursor with aluminium addition, which resulted in similarly very low contents of excess carbon and oxygen [56]. Moreover, the Sylramic™ SiC fiber was developed also by applying sintering technique, but with an addition of boron to the polytitanocarbosilane polymer precursor [57]. All three SiC fibers, categorized as the Generation-III SiC fibers, are characterized by near-stoichiometry, high crystallinity (crystallite sizes 50 to 200 nm), and low excess carbon and oxygen contents.

Among these types of fibers, the only one available on the market in form of chopped fibers, 1 mm length, are the Hi-Nicalon and the Tyranno. The advantage of using chopped fibers instead of continuous fiber is a potential isotropic reinforcement and moreover the possibility to process them as particles, that means no need to set special equipment for the incorporation within the matrix.

1.4.2 SiC Whiskers

In the past, SiC whiskers have been widely used for the development of high-temperature ceramic matrix composites with enhanced toughness. Whiskers chemistry and morphology can widely vary among different manufacturers and among different batches from the same producer. [58] While the difference may be minor, such differences can have a significant influence on the fabricability of a dense composite and on the toughening behavior of the whiskers in a composite material. The differences arises due to different process method used by the manufacturers, such as the raw materials, catalysts, reaction atmosphere, furnace configuration and process temperature. The toughening behaviour of whisker-reinforced composites depends both on shape factors, such as diameter and aspect ratio, and on the ability of the whiskers to debond at the whisker-matrix interface with consequent inhibition of crack propagation by crack bridging, crack deflection and whisker pullout [59].

2. Theoretical Background

In the design of a component, to predict the overall mechanical properties of the material is very important for the final application. In this section the theoretical models and the input values used to foresee the properties of the fiber and whiskers reinforced composites are briefly summarized.

a)   Crack deflection

The increase in strain energy release, G_c, by rod-shaped particles is given by [21]

$$G_C = (G^m / \langle G \rangle_{rod}) G_c^m \tag{1}$$

where G^m is the strain energy release for the undeflected crack and $\langle G \rangle_{rod}$ is the average strain energy release across the deflected crack front which accounts for the tilted (Figure 1) and twisted (Figure 2) part of the crack.

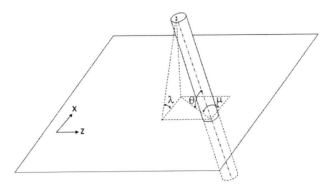

Figure 1. Schematic of a rod-shaped particle tilted at angle Θ with the x-z plane and rotated by angle μ to the direction of crack propagation parallel to the x-axis. The angle λ defines the tilt angle.

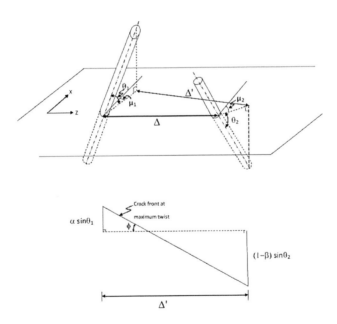

Figure 2. Schematic of crack twist between two adjacent rod-shaped particles whose centers in the x-z plane are separated by a distance Δ.

The effective twist angle λ of the crack front is described by

$$\lambda = \tan^{-1}(\tan \theta / \tan \mu) \qquad (2)$$

whose average can be expressed as

$$\langle \lambda \rangle = \frac{(\alpha/2)\,\lambda_1\,\sin\theta_1 + [(1-\beta)/2]\,\lambda_2\,\sin\theta_2}{\alpha\,\sin\theta_1 + (1-\beta)\,\sin\theta_2} \tag{3}$$

where α and β are the relative positions at which the crack plane intercepts the rods. The twist angle, Φ, is denoted by

$$\phi = \tan^{-1}\left\{\frac{\alpha\,\sin\theta_1 + (1-\beta)\,\sin\theta_2}{\Delta'}\right\} \tag{4}$$

with

$$\Delta' = \left\{\left[\frac{\Delta}{H} - \alpha\,\cos\theta_1\,\sin\mu_1 + (1-\beta)\,\cos\theta_2\,\sin\mu_2\right]^2 + [\alpha\,\cos\theta_1\,\sin\mu_1 - (1-\beta)\,\cos\theta_2\,\sin\mu_2]^2\right\}^{1/2} \tag{5}$$

where H is the aspect ratio of the rod-shaped particles and the interparticle spacing Δ is given by, for a volumetric fraction of the rod-shaped particle V_f by

$$\Delta \approx r\,\frac{e^{4V_f}}{v_f^{1/2}}\int_{4V_f}^{\infty} x^{1/2}\,e^{-x}dx \tag{6}$$

whith r as the radius of the rod-shaped reinforcement. The strain energy release rate due to the twist of the crack front can be therefore written, for $\Phi < \Pi/2$, as

$$\frac{\langle G\rangle^T}{G^\infty} = \frac{4}{\pi^4}\int_{-\pi/2}^{\pi/2}\int_{-\pi/2}^{\pi/2}\int_0^1\int_0^1\int_{-\pi/2}^0\int_0^{\pi/2}\left[\left|\cos^4\left(\frac{\langle\lambda\rangle}{2}\right)\right.\right.\left\{2v\sin^2\phi + \cos^2\phi\cos^2\left(\frac{\langle\lambda\rangle}{2}\right)\left[1+2\sin^2\left(\frac{\langle\lambda\rangle}{2}\right)\right]\right\}^2 +$$

$$\left\{\sin\phi\cos\phi\cos^2\left(\frac{\langle\lambda\rangle}{2}\right)\left[\cos^2\left(\frac{\langle\lambda\rangle}{2}\right)-2v\right]+\sin^2\left(\frac{\langle\lambda\rangle}{2}\right)\left[3\cos^2\left(\frac{\langle\lambda\rangle}{2}\right)-2v\right]\right\}^2\right]d\theta_1\,d\theta_2\,d\alpha\,d\beta\,d\mu_1\,d\mu_2 \tag{7}$$

For θ_1 and θ_2 of like-sign, the strain energy release for the tilted crack front can be written as

$$\frac{\langle G\rangle^t}{G^\infty} = 4/\pi^4\int_{-\pi/2}^{\pi/2}\int_{-\pi/2}^{\pi/2}\int_0^1\int_0^1\int_0^{\pi/2}\int_0^{\pi/2}\cos^4(\bar{\lambda}/2)\,d\theta_1\,d\theta_2\,d\alpha\,d\beta\,d\mu_1\,d\mu_2 \tag{8}$$

where

$$\bar{\lambda} = 1/2\,[\tan^{-1}(\tan\theta_1/\cos\mu_1) + \tan^{-1}(\tan\theta_2/\tan\mu_2)] \tag{9}$$

The normalized total strain energy release is then

$$\langle G\rangle_{rod} = \eta/2\,\langle G\rangle^T + \xi/2\,\langle G\rangle^t \tag{10}$$

where η and ζ are the ratios of the undeflected and deflected crack front lengths which can be expressed as

$$\eta = \frac{[(\Delta/H)-\alpha\cos\theta_1\sin\theta_1+(1-\beta)\cos\theta_2\sin\mu_2]}{[(\Delta\prime)^2+[\alpha\sin\theta_1+(1-\beta)\sin\theta_2]^2]^{1/2}} \tag{11}$$

$$\xi = \frac{[(\Delta/H)-\alpha\cos\theta_1\sin\theta_1+(1-\beta)\cos\theta_2\sin\mu_2]}{[(\Delta\prime)^2+[\alpha\sin\theta_1-(1-\beta)\sin\theta_2]^2]^{1/2}} \tag{12}$$

b)   Thermal residual stresses
According to Taya et al. [27], the toughness variation due to thermal residual stresses arising from the mismatch of reinforcement particles and matrix is

$$\Delta K^{res} = 2q\sqrt{\frac{2(\Delta-d)}{\pi}} \tag{13}$$

where Δ is the interparticle distance (Eq. 6), d the mean size and q the matrix residual stress given by:

$$q = \frac{-V_f\bar{\sigma}}{1-V_f} \tag{14}$$

whith V_f as the reinforcement volumetric fraction and $\bar{\sigma}$, in the case of fiber-shaped inclusion, is given by

$$\bar{\sigma} = \frac{2\sigma_{11}+\sigma_{33}}{3} \tag{15}$$

For fiber-shaped inclusion, the principal thermal residual stresses are: σ_{33} acting along the axis of the fiber-shaped inclusion, and σ_{11} and σ_{22} on the plane perpendicular to this axis. Their expression reads as follows [60]:

$$\sigma_{11} = \sigma_{22} = A\left\{\left[\frac{1}{E_p}+\frac{V_f}{(1-V_f)E_m}\right](\alpha_m-\alpha_1)+\left[\frac{v_p}{E_p}+\frac{V_f v_m}{(1-V_f)E_m}\right](\alpha_m-\alpha_3)\right\}\Delta T \tag{16}$$

$$\sigma_{33} = A\left\{2\left[\frac{v_p}{E_p}+\frac{V_f v_m}{(1-V_f)E_m}\right](\alpha_m-\alpha_1)+\left[\frac{1-v_p}{E_p}+\frac{1+V_f+(1-V_f)v_m}{(1-V_f)E_m}\right](\alpha_m-\alpha_3)\right\}\Delta T \tag{17}$$

with

$$A = \left[\frac{(1+v_p)(1-2v_p)}{E_p^2}+\frac{V_f(2-v_p-v_m-4v_p v_m)+1+v_m}{(1-V_f)E_p E_m}+\frac{V_f(1+v_p)(1+f_p-2f_p v_m)}{(1-V_f)^2 E_m^2}\right]^{-1} \tag{18}$$

E_p, E_m, v_p and v_m are the Young's modulus and the Poisson ratio of the inclusion and matrix, respectively; α_1, α_2 and α_3 are the coefficient of linear expansion (CTE) of the fiber-shaped inclusion in the three directions and α_m is the CTE of the matrix.

For the rod-shaped particles, the interparticle distance was taken as (see eq. 6)

$$\Delta \approx r \, \frac{e^{4V_f}}{v_f^{3/2}} \int_{4V_f}^{\infty} x^{1/2} \, e^{-x} dx \tag{19}$$

c)   Crack bowing

The toughening increment, ΔK^{bowing} due to crack bowing is a monotonic function of the reinforcement fraction as follows [26]:

$$\Delta K = 2 \sqrt{\frac{4r}{\pi}} V_f \sigma_f \tag{20}$$

where r is the reinforcement radius and σ_f its fracture strength.

EXPERIMENTAL PROCEDURE

3.   Composites Production

3.1   Raw Materials

As mentioned above, the choice of the short fibers and whiskers as reinforcing phase is related to the concept of obtaining a 3D toughening effect without basically modifying the powders processing. The compositions investigated are indicated in Tables III, IV.

Table III. Composition, sintering parameters and density of the hot-pressed unreinforced material and of the samples containing SiC chopped fibers. T_{ON}= temperature at which the shrinkage started, T_{MAX}= maximum temperature achieved and held for the dwell time.

Sample	Composition vol%	T_{ON} °C	T_{MAX} °C	Dwell time min	Pressure MPa	Th. density g/cm³	Fin. density g/cm³	Rel. density %
ZS	ZrB_2+5 Si_3N_4	1580	1700	15	30	5.95	5.89	99.0
ZS10f	ZS+10 SiCf	1625	1730	11	35-50	5.53	5.42	97.8
ZS20f	ZS+20 SiCf	1600	1700	10	35-50	5.22	5.07	97.0
ZS30f	ZS+30 SiCf	1680	1700	7	30-40	4.95	4.49	91.0
ZZ	ZrB_2+10 $ZrSi_2$	1300	1600	10	30	5.98	5.73	95.8
ZZ20f	ZZ+20 SiCf	1320	1650	8	40-50	5.34	5.30	99.0
ZM	ZrB_2+15 $MoSi_2$	1500	1750	20	30	6.14	5.98	97.4
ZM20f	ZM+20 SiCf	1530	1900	12	30	5.46	5.45	99.8

Ultra-High Temperature Ceramics Reinforced with SiC Short Fiber or Whisker   333

Table IV. Composition, sintering parameters and density of the hot-pressed samples containing SiC whiskers. T_{ON}= temperature at which the shrinkage started, T_{MAX}= maximum temperature achieved and held for the dwell time.

Sample	Composition vol%	T_{ON} °C	T_{MAX} °C	Dwell time min	Pressure MPa	Th. density g/cm³	Final density g/cm³	Rel. density %
ZS10w	ZS+10 SiCw	1590	1730	11	35-50	5.57	5.45	97.9
ZS20w	ZS+20 SiCw	1545	1650	10	35-50	5.28	4.89	94.0
ZS30w	ZS+30 SiCw	1530	1700	10	30-40	5.09	4.23	83.0
ZZ20w	ZZ+20 SiCw	1300	1600	9	30-40	5.43	4.93	91

As reinforcing phases, SiC fibers or whiskers were used:

- SiC chopped fibers (HI Nicalon, COI Ceramics Inc., Magna, UT), 1-5 wt% vinyl alcohol polymer with vinyl acetate wt% Si:C:O= 62:37:0.5, diameter: 14 µm, length: 1 mm plus debris, aspect ratio ~70.
- SiC whiskers (Tokai Carbon Co., LtD, Grade 3, Japan) at% Si:C=55.3:44.7 C, high level of surface oxygen in the form of SiO_2 and Si-O-C glass, dimensions calculated by image analysis: average diameter 1 µm and average length 30 µm, aspect ratio 30.

The starting chopped fibers and whiskers utilized for the composites are shown in Figure 3a,b, respectively.

Commercial powders were used to prepare the ceramic composites:

- hexagonal ZrB_2 Grade B (H.C. Starck, Germany), specific surface area 1.0 m²/g, impurity max content: C: 0.25%, O: 2%, N: 0.25%, Fe: 0.1%, Hf: 0.2%, particle size range 0.1-8 µm;
- hexagonal HfB_2 (Cerac Incorporated, Milwaukee, USA), particle size range 0.5-5 µm, impurities: Al:0.07%, Fe: 0.01%, Zr: 0.47%;
- cubic TaC (Cerac Inc, Milwaukee, WI), particle size range 0.2-1.5 µm.

As sintering additive, the following commercial powders were added in amount from 5 to 15 vol%, as reported in Tables III, IV.

- hexagonal α-Si_3N_4 Baysind (Bayer, Germany), specific surface area 12.2 m²/g, impurity max content: O: 1.5%;
- orthorhombic $ZrSi_2$-F (Japan New Metals Co., LTD, Osaka, Japan) particle size 2-5 µm, C≤0.15%, Fe≤0.30, O≤1.00.
- hexagonal $TaSi_2$ (ABCR, GmbH and Co, Karlsruhe, Germany), particle size <45 µ;
- tetragonal $MoSi_2$ (<2 µm, Aldrich, Steinbeim, Germany), particle size range 0.3-5 µm and oxygen content ~1%.

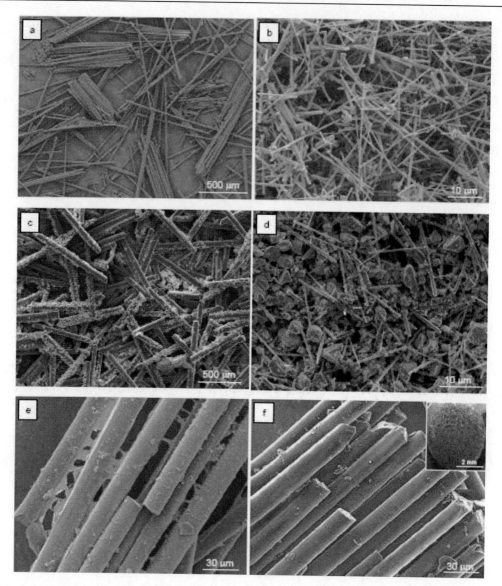

Figure 3. SEM images of the starting (a) SiC chopped fibers and (b) SiC whiskers used in this study. In (c) SiC chopped fibers and (d) SiC whiskers after mixture with ZrB$_2$. In (e) a magnified image of the raw SiC fibers and in (f) after several washing cycles in acetone. The inset in (f) shows agglomeration after the removal of the sizing from the fiber surface finishing occurred upon powder mixing.

3.2 Handling Hazards

When we deal with elongated phases, such as whiskers or fibers, particular attention has to be paid during handling.

Health effects and hazards of SiC whiskers are not well known, however, because of their similarity in size and shape to asbestos, they should be handled with extreme care, especially if there is a possibility of air-borne contamination.

Concerning the SiC chopped fibers used in the present study, they can be considered less dangerous, owing to the bigger size, around 15 μm in diameter and 1 mm in length, and one

Ultra-High Temperature Ceramics Reinforced with SiC Short Fiber or Whisker 335

single fiber can be clearly seen by necked eye and seized with common microscopy tweezers. Despite the size of the fibers is outside the respirable range, in operations where respirable dusts and particles may be generated, the use of respiratory and skin protection is recommended to minimize any inhalation exposure and mechanical irritation.

3.3 Mixture Preparation

The powder mixtures were gently ball milled for 24 hours in absolute ethanol using silicon carbide media. Subsequently, the slurries were dried in a rotary evaporator. After mixing, SEM analysis showed that the fibers length was reduced to about 300 μm Figure 3c, whilst the whiskers remained unaltered (Figure 3d).

Stated the fiber breaking and interphase reaction with the matrices, that will be discussed later in the text, several attempts were performed in order to minimize these two negative consequences, which diminishes the efficacy of the fiber. First the mixing time was reduced from 24 to 10 hours of ball milling, but problems were encountered in the following steps, as the fibers did not pass through the grid funnel and agglomerated inside the matrix. To remove the sizing which covers the fibers, these were washed in acetone and filtered several times (compare Figure 3e-f). However, once added to the matrix, the fibers packed in balls, as depicted in the inset of Figure 3f.

Once the mixture prepared in the conventional way was dried, the powder was combed and carefully disagglomerated manually with a spatula.

3.4 Green Shaping

Cylinders 44- mm in diameter were obtained by the application of 10 MPa for whisker containing composites and 15 MPa for fiber containing composites. In the green shaping it is important that the applied pressure is lower than the pressure that will be applied during sintering. Moreover, dealing with elongated and particle-shaped powders, it is of paramount importance to leave more than 50% of porosity for particles rearrangement during sintering. To note that, in the case of high amount of whiskers in the composite, the application of pressure higher than 15 MPa leads to a sponge-like pellet which is very hard to handle.

3.5 Sintering

The sintering was preceded by a debonding cycle in a graphite furnace at 500 °C for 1 hour holding time, with a heating rate of 50°C/h under flowing argon. This treatment enabled the removal of organic species covering the fibers.

Hot-pressing cycles were conducted in low vacuum (~100 Pa) using an induction-heated graphite die with an uniaxial pressure of 30 MPa during the heating and generally increased up to 40-50 MPa at the maximum temperature. This was set on the basis of the shrinkage curve. Free cooling followed. The schedule of each sintering runs is reported in Tables III, IV.

4. Materials Characterization

4.1 Microstructure

The bulk densities were measured by Archimedes' method.

Crystalline phases were identified by X-ray diffraction (Siemens D500, Germany).

The microstructural features were analyzed on the fractured and polished surfaces using scanning electron microscopy (SEM, Cambridge S360, Cambridge, UK) and energy dispersive spectroscopy (EDS, INCA Energy 300, Oxford instruments, UK).

TEM specimen preparation was quite difficult, because whiskers and fibers tend to fall during ion milling procedure due to the high aspect ratio. TEM samples were prepared by cutting 3 mm discs from the sintered pellets. These were mechanically ground down to about 15 μm and then further ion beam thinned until small perforations were observed by optical microscopy. Local phase analysis was performed using transmission electron microscopy (TEM) equipped with an energy-dispersive X-ray system (FEI, CM12, Eindhoven, The Netherlands; EDS, EDAX Genesis 2000, Ametek GmbH; Wiesbaden, Germany) operating at a nominal voltage of 120 keV. High-resolution investigations were performed using a FEI CM20 STEM operating at a nominal voltage of 200 keV.

Mean grain sizes, amount of porosity, and amount of secondary phases were determined through image analysis on SEM micrographs of polished surfaces using a commercial software program (Image Pro-Plus 4.5.1, Media Cybernetics, Silver Spring, MD). Thermodynamical calculations were carried out by a commercial software HSC Chemistry (Roine A. HSC Chemistry for Windows 5, Outukumpu Research Oy Pori, Finland).

4.2 Thermo-Mechanical Characterization

Vickers microhardness (HV) was measured on polished surfaces, with a load of 9.81 N, using a Zwick 3212 tester (10 indentations for each material).

Young's modulus (E) was measured by resonant frequency on specimens $28 \times 8 \times 0.8$ mm^3 using a Hewlett and Packard gain-phase analyzer.

Fracture toughness (K_{Ic}) was evaluated using the chevron-notched beam (CNB) in flexure. The bars, $25 \times 2 \times 2.5$ mm^3 (length × width × thickness, respectively), were notched with a 0.08 mm diamond saw. The flexural tests were performed on a semi-articulated alumina four-pt jig with a lower span of 20 mm and an upper span of 10 mm on an universal screw-type testing machine Instron mod. 6025. The specimens were deformed with a crosshead speed of 0.05 mm/min. The "slice model" equation of Munz et al. [61] was used for the calculation of K_{Ic}. The fracture toughness is given by:

$$K_{Ic} = \frac{P_{max}}{BW^{1/2}} y *$$

(21)

where P_{max} is the maximum load at fracture, B the width of the bar, W the thickness of the bar. According to the "slice model", $y*$ is defined as:

$$y* = [3.08 + 5a_0/W + 8.33(a_0/W)^2] \frac{(S_1 - S_2)}{W} \left(1 + 0.007\sqrt{\frac{S_1 S_2}{W^2}}\right) \left(\frac{a_1 - a_0}{W - a_0}\right)$$

(22)

with a_0 and a_1 are the notch parameters defined in Figure 4 ($a_0 = 0.12W$ and $a_1 = 0.08\ W$) and S_1 and S_2 the outer and inner span.

The interaction of the crack front with the microstructure was analyzed by introducing cracks onto polished surfaces with a 98.1 N-Vickers indentation.

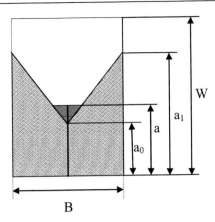

Figure 4. Notch plane section.

Flexural strength (σ_c), up to 1200°C in air, was measured on chamfered bars 25 × 2.5 × 2 mm³ (length × width × thickness, respectively), using a crosshead speed of 0.5 mm/min. For the high temperature tests, a soaking time of 18 min was set to reach thermal equilibrium. Five specimens were used for each temperature point.

RESULTS AND DISCUSSION

In this section the microstructural features of the fiber and whisker reinforced materials are illustrated with special emphasis on the effect of different sintering aids on densification and microstructure, i.e. Si_3N_4, $ZrSi_2$ and $MoSi_2$. Subsequently, the mechanical properties are investigated for ZrB_2-based materials: the experimental results are compared to the theoretical models according to the main toughening mechanisms identified.

Finally, two other matrices are considered HfB_2 and TaC sintered with addition of Si_3N_4 or $TaSi_2$ and reinforced with SiC fibers. Also for these composites the microstructure, the densification behavior and the mechanical properties are discussed.

5. ZrB₂-Based Materials: Microstructure and Mechanical Properties

5.1 Materials Sintered with Si₃N₄

5.1.1 Sintering Behavior of ZrB₂-Si₃N₄ Composites

Since the reinforcing phases, both fibers and whiskers, tend to degenerate during thermal treatment, the sintering temperature must be kept as low as possible. To this purpose, Si_3N_4 was first selected as sintering aid, as it promotes the full densification of ZrB_2 at 1,700°C [6]. The reinforced ceramics were sintered at temperatures from 1,650 to 1,730°C, as shown in Tables III,IV.

With the addition of Si_3N_4, the densification of ZrB_2 started at 1,580°C and was completed at 1,700°C (sample ZS, Table III). The maximum densification rate occurred at

1,620°C, d($\Delta l/lo$)dT= $6.1 \cdot 10^{-3}$ °C^{-1}. During the sintering process the twofold effect of sintering aid was the removal of surface oxides from boride particles and the formation of a liquid Zr-Si-B-N-O phase, that promoted matter transfer mechanisms.

As the introduction of elongated reinforcement is thought to hinder the shrinkage, the applied pressure was increased from 30 to 40-50 MPa when the maximum temperature was achieved. From the densification curves recorded during the hot pressing cycle (not shown), the temperature at which the composites started to densify was 1,580°C for the unreinforced material, around 1540°C for the material containing whiskers, due to the presence of a higher amount of SiO_2 on the surface of SiC, which is known to help the formation of liquid phases and above 1,600°C for the composites containing fibers.

Final densities in the range between 94-98% of the theoretical values were reached at 1,650-1,730°C for the composites containing up to 20 vol% of reinforcing phase, whilst for those containing 30 vol% the density did not overcome the 90%.

However, generally the sudden increase of the densification rate, which occurred at 1,545-1,630°C, indicated the activation of the mass transfer mechanisms through the formation of a liquid phase.

It has to be put in evidence that the final relative densities were expressed as the ratio of experimental and theoretical values calculated with the rule of mixture considering the starting compositions and taking 5.95 g/cm^3 as the density of baseline material, 2.73 g/cm^3 and 3.20 g/cm^3 for the fibers and whiskers density, respectively. As can be seen in Tables III, IV, with the exception of ZS30f, ZS20w and ZS30w, the relative densities were higher than 97%. Furthermore, these values can be still underestimated due to significant amounts of glassy phase formed by reaction of the matrix with Si_3N_4 and the reinforcement, as illustrated later.

5.1.2 Microstructure of Si_3N_4-Containing Materials

Baseline unreinforced material. The baseline ZrB_2 material doped with 5 vol% of Si_3N_4 was fully dense with a mean grain size around 2.7 μm. The secondary phases observed in the sintered microstructure were mainly concentrated at triple points and were identified as ZrO_2, BN, Zr-Si phases, discrete pockets of SiO_2 and a borosilicatic glass containing Zr-Si-B-N-O. An example of the microstructure is given in Figure 5. The origin of the formation of these secondary phases will be discussed in Section 5.1.3.

Fiber-reinforced composites. The fractured and polished sections of the fibers-reinforced composites are reported in Figure 6. The dispersion of the fibers into the matrix was homogeneous, since no agglomeration was observed, except for the material containing 30 vol% of fibers (Figure 6e).

Porosity was nearly absent in ZS10f and ZS20f, however in the ZS30f composite residual porosity can be observed in Figure 6f, indicating that for such high amount of fibers the powder mixing process needs further optimization. As expected, the fibers showed a strong tendency to align their long axis perpendicular to the direction of applied pressure. The length of the fibers was further reduced from the starting dimensions, from 1 mm to 200-300 μm, due to both the action of the milling media and the applied pressure during sintering, the aspect ratio hence decreased too from around 70 to 15-20.

Ultra-High Temperature Ceramics Reinforced with SiC Short Fiber or Whisker 339

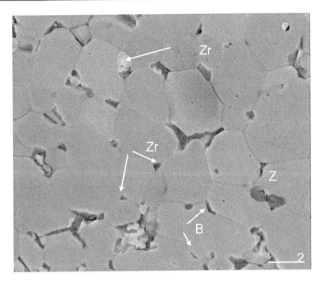

Figure 5. SEM image of the polished section of the ZS sample showing the overall microstructure.

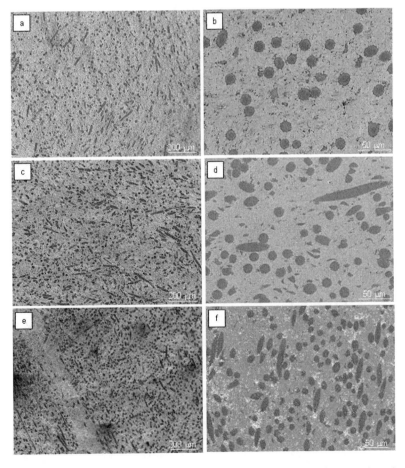

Figure 6. SEM images of the fractured (left) and polished (right) section of the ZrB$_2$-based composites with Si$_3$N$_4$ as sintering additive and containing a-b) 10, c-d) 20, e-f) 30 vol% of SiC short fibers.

On the fracture surfaces, no trace of fiber pullout was observed. The fibers themselves showed a multilayered core-shell morphology (Figure 7a): the inner part was constituted by stoichiometric SiC, the surrounding shell was partial amorphous Si-C-O and the outermost jagged layer was a Si-C phase with low C content. In fact, Hi Nicalon fibers are constituted by a polycarbosilane-derived amorphous/microcrystalline Si-C-O that suffers a degradation as it transforms to equilibrium products during exposure to high temperatures [62].

During densification, a Zr-Si-C-O-based interface formed between the matrix and the SiC fibers due to the SiC/ZrB$_2$ interaction, Figure 7a. By TEM analysis the interface was disclosed to be constituted by ZrC nanocrystals embedded into a partially amorphous Si-C-O phase (Figure 7b,c). BN, ZrO$_2$, ZrN and ZrSi$_2$ were identified as secondary phases (Figure 8). The high resolution TEM mode showed non-wetted grain boundaries, when BN was present at the interface (Figure 9a,b), and wetted interface in presence of a Zr-Si-B-N-O glassy phase.

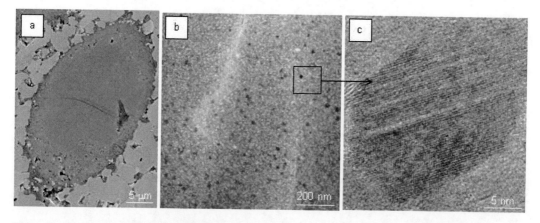

Figure 7. (a) Example of SEM images of the ZrB$_2$-SiCf material sintered with Si$_3$N$_4$ showing the formation of an interface reaction between matrix and fiber. (b) TEM image of the interface layer constituted by amorphous Si-C-O and nanocrystals (c).

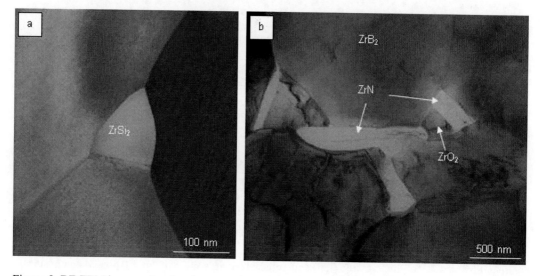

Figure 8. BF-TEM images showing the secondary phases in the ZrB$_2$-SiCf-Si$_3$N$_4$ composites.

Ultra-High Temperature Ceramics Reinforced with SiC Short Fiber or Whisker   341

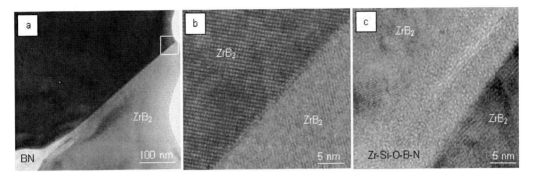

Figure 9. TEM images of the ZS20f composite showing (a,b) non wetting tendency of BN towards ZrB_2 grains and (c) grain boundaries wetted by a Zr-Si-O-B-N glassy phase.

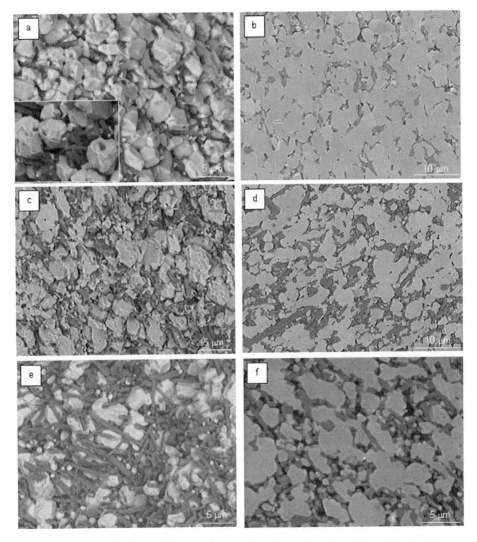

Figure 10. SEM images of the fractured (left) and polished (right) section of the ZrB_2-based composites with Si_3N_4 as sintering additive and containing a-b) 10, c-d) 20, e-f) 30 vol% of SiC whiskers.

Whisker-reinforced composites. The fractured and polished sections of the whiskers-reinforced composites are reported in Figure 10.

As it can be seen, porosity is not so apparent in the fractured surfaces; only some small pores trapped into the ZrB_2 grains can be observed, especially in the sample containing 20 and 30 vol% of whiskers. In the fracture surfaces, Figure 10a,c,e, some whisker pullout is evident, but in the case of addition of 30 vol% the pullout can be ascribed to a not complete densification.

The whiskers were generally well dispersed into the matrix, still recognizable as elongated dark structure in back-scattered SEM images. After sintering, the whiskers assumed a squatter aspect with irregular edges.

On the polished surface (Figure 10b,d,f), diffused microcracking was observed, as reported for the baseline material [6], due to the mismatch of thermal coefficients and elastic constants among ZrB_2, SiC and the newly formed glassy phase. At higher magnification, the presence of ZrO_2 agglomerates on the SiC and Zr-Si phases was confirmed, together with a Zr-Si-B-N-O glassy phase which surrounded the whiskers (see Figure 11).

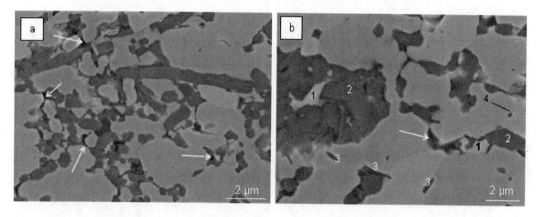

Figure 11. Examples of microstructure of the whiskers-reinforced composites. Note the glassy phase indicated by arrows and the secondary phases: 1-ZrO_2, 2-SiC, 3-BN, 4-$ZrSi_2$.

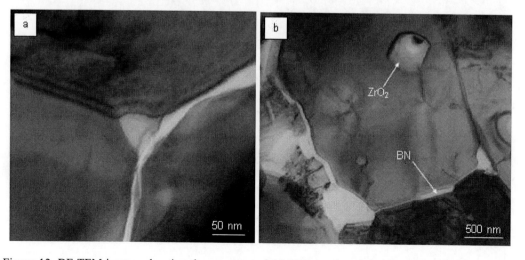

Figure 12. BF-TEM images showing the presence of (a) $ZrSi_2$ at the triple junction with wetted grain boundaries constituted by Zr-Si-B-N-O glassy phase and (b) the BN and ZrO_2 phases.

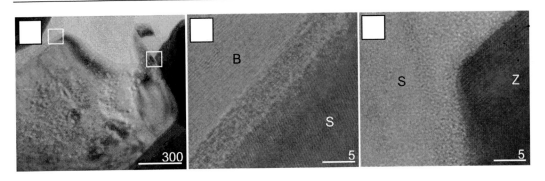

Figure 13. (a) BF-TEM images showing the wetted interfaces between SiC whiskers and (b) BN or (c) ZrB$_2$.

Transmission electron microscopy (TEM) evidenced a Zr-Si-B-N-O amorphous intergranular layer containing calcium impurities, crystalline ZrSi$_2$ at the triple point junctions and BN as intergranular pockets (Figure 12). The interface between matrix and whiskers was wet by the same Zr-Si-B-N-O intergranular phase, as depicted in Figure 13.

5.1.3 Densification Mechanisms in Presence of Si$_3$N$_4$

On the basis of thermodynamic calculations for the system Zr–B–Si–O–N, ZrB$_2$ and Si$_3$N$_4$ phases cannot coexist under the process conditions. In fact, the occurring reactions lead to the complete disappearance of Si$_3$N$_4$ and produced various compounds (BN, ZrN, t-ZrO$_2$, ZrSi$_2$, amorphous Zr-Si-B-N-O). The steep increase in the densification rate at around 1560°C is due to the formation of a eutectic liquid by reactions between the present phases. During the heating up, the sintering additive, Si$_3$N$_4$, could first react with B$_2$O$_3$, which covers the matrix powder, according to the following reaction:

$$2\ B_2O_3 + Si_3N_4 \rightarrow 4\ BN + 3\ SiO_2 \tag{23}$$

This reaction, characterized by a negative Gibbs free energy (-304 kJ/mol, at 1500°C), describes the removal of the boron oxide present on the surface of the ZrB$_2$ particles. However, the liquid phase formed at about 1560°C is probably due to the reaction among Si$_3$N$_4$, SiO$_2$ and ZrB$_2$. The formation of liquid Si is also favored but, in the final microstructure, only a residual borosilicate amorphous phase was found. Si-containing phases, i.e. ZrSi$_2$, were found at the triple junctions in the sintered materials, while borosilicate pockets were frequently detected at the triple points.

The addition of SiC whiskers and fibers introduced further sources of SiO$_2$ which in turn increased the amount of the liquid phase. Considering the overall densification process, on one hand it should be improved by the higher volume of liquid phase, which favor grain rearrangement and mass transport mechanisms, but on the other hand it could be negatively affected by the physical obstruction of the rod-shaped reinforcement, which tend to form a rigid skeleton. Hence, densification, as well as grain growth of the matrix grains, was not significantly altered compared to the baseline material due to the just mentioned counterbalancing effects.

The morphology of the fibers/matrix interface upon sintering induces to hypothesize that, as the temperature increases the external part of the fibers, rich in oxygen, softens and partially incorporates ZrB_2 powder particles which are still free to move. The ZrB_2 particles may react with silica and the Si-C-O phase according to reaction (24):

$$ZrB_2 + Si\text{-}C\text{-}O + SiO_2 \rightarrow ZrC + B_2O_{3(l)} + 2\ Si_{(l)} \tag{24}$$

The liquid silicon formed is then squeezed out and reduced to SiC_{1-x} by the C-rich environment. The irregular morphology of this crest attests the solidification from a liquid phase (see Figure 7a). At the same time, once reached the melting temperature of the liquid phase in the matrix, constituted by the reaction products among Si_3N_4 and B_2O_3, the interaction and mass transfer between the two liquid phases are active and a strong chemical bonding are created upon cooling. The nitrogen peak detected by EDS in the darkest region surrounding the core of the fibers is the proof of an intimate bonding between the liquid phase of the matrix and the fibers.

5.2 Materials Sintered with $ZrSi_2$

5.2.1 Sintering Behavior of ZrB_2-$ZrSi_2$ Composites

In the materials containing $ZrSi_2$, the densification started at around 1300°C and was fully completed at 1600-1650°C, with the maximum densification rate of $4.8 \cdot 10^{-3}°C^{-1}$ at around 1480°C. Like above, when the reinforcing phase was added, the pressure applied was increased from 30 to 40-50 MPa. According to the theoretical density calculated with the rule of the mixture as a function of the nominal composition, Tables III, IV show that the relative density of the composite containing fibers is almost full, on the contrary, the density of the composite containing whiskers does not exceed the 91% of the theoretical density. Also in this case, the final theoretical density could be underestimated due to the presence of glassy phase.

5.2.2 Microstructure of $ZrSi_2$-Containing Materials

Baseline unreinforced material. The baseline material densified with 10 vol% of $ZrSi_2$ (ZZ), showed ZrB_2 rounded grains, with dimensions around 3.5 μm, bright ZrO_2 particles, SiC particles and $ZrSi_2$, which appears as irregular grey phase in Figure 14. Moreover, Zr-Si phases with various stoichiometries were concentrated at triple points and along the grain boundaries.

Fiber-reinforced composite. The material containing 20 vol% of fibers, ZZ20f, was sintered at 50°C higher than the corresponding containing whiskers, and resulted in a fully dense microstructure, as it can be observed in Figure 15. The fracture is mainly transgranular, the fibers are crossed by the crack and there is no pullout (Figure 15a). The polished surface at low magnification of Figure 15b shows that the fibers are homogeneously dispersed along the microstructure and there are no agglomerates.

In the fracture surface and in the Figures 15c-e, it can be noticed that significant fiber degeneration occurred, their surface is rough with little porosities and the interface region is jagged. It can also be noticed, especially in Figure 15c, which shows the longitudinal section of a fiber, that squared ZrB$_2$ grains are incorporated in the fiber, indicating that the fiber itself lost its original structure. This aspect was not observed for the analogous composite sintered with addition of Si$_3$N$_4$ (ZS20f). The fiber reaction interface is constituted by Si-C-O phases, Zr-Si with various stoichiometry and traces of B, C, O.

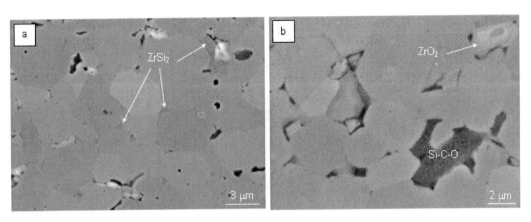

Figure 14. SEM images of the polished section of the ZZ sample showing the overall microstructure and the secondary phases.

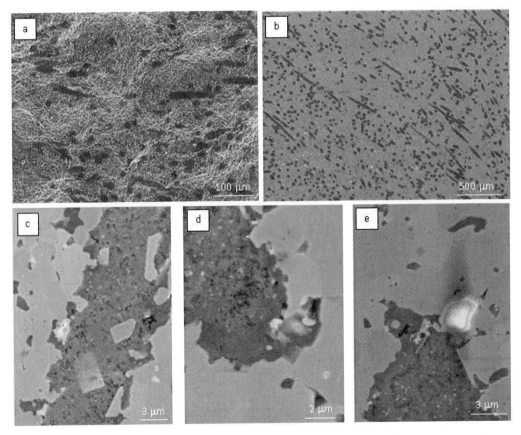

Figure 15. SEM images of the ZZ20f composite: (a) fracture and (b) polished sections. (c-d) Fiber morphology showing strong reaction with matrix and sintering additive.

The bright field TEM image in Figure 16a shows a similar feature of the SEM image in Figure 15d: a ZrB$_2$ squared grain penetrating the fiber, and the SiC crystallites which are coarser compared to the original ones, around 500 nm versus to 5-20 nm [62], owing to grain growth during the hot pressing cycle. High resolution imaging revealed that the interface between SiC/ZrB$_2$ was clean, as Figures 16b,c show.

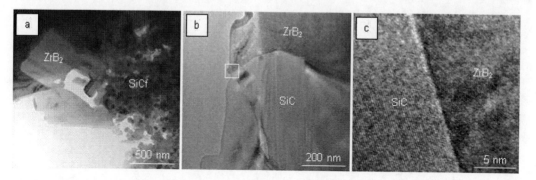

Figure 16. (a) Bright field TEM image showing a ZrB$_2$ grain penetrating the SiC fiber. Interface between ZrB$_2$ and SiC: (b) conventional TEM imaging and (c) high resolution of the box in (b).

Whisker-reinforced composite. The material containing 20 vol% of whiskers, ZZ20w, achieved a density of 4.93 g/cm^3, which corresponds to a relative density of 91%, considering the starting nominal composition.

However from the microstructures in Figure 17 it can be noticed that a high amount of glassy phase with dark contrast is present, but still around 3% of residual porosity can be seen. ZrB$_2$ mean grain size is around 3 μm, the whiskers, observed in the surface perpendicular and parallel to the direction of the applied pressure (Figure 17b,c) are homogeneously dispersed in the microstructure and generally surrounded by a dark phase with irregular shape constituted by Zr-Si-O containing Ca impurities, coming from whisker impurities (Figure 17d). The dimensions of the whiskers are around 10 μm in length and around 500 nm in diameter, indicating that they broke during milling procedure, as the initial length was around 30 μm. It can be affirmed that 1600°C seems to be the maximum temperature the whisker can withstand without degeneration, in fact the whiskers in the ZZ system look completely different from the ZS system which achieved a maximum temperature of 1700°C (Tables III, IV, compare Figures 10 and 17).

5.2.3 Densification Mechanism in Presence of ZrSi$_2$

As already said, for this system the densification started at a lower temperature compared to the composites processed with silicon nitride. In the reference material, the presence of wetting Zr-Si phases at triple points suggests formation of a liquid phase. Pure ZrSi$_2$ is indeed a relatively low melting compound (~1600°C) and, in presence of Si sources, Zr-Si-based liquids may start to form at even lower temperatures, see the phase diagram in Figure 18 [63]. The following reaction can be hypothesized in agreement with experimental evidence:

$$ZrSi_2 + B_2O_3 = Zr\text{-}Si\text{-}B\text{-}O_{(l)} \qquad (25)$$

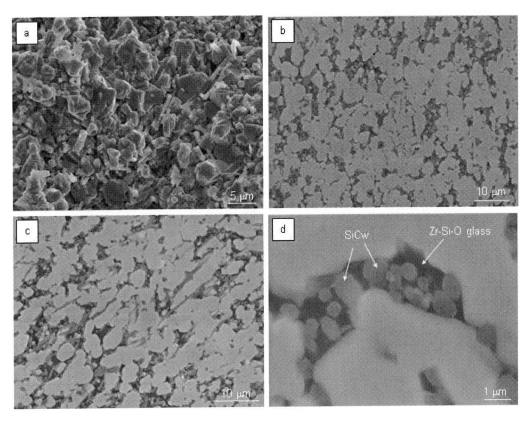

Figure 17. SEM images of the ZZ20w sample showing (a) not complete densification and little whiskers pullout in the fracture surface, (b) the overall microstructure in the polished surface perpendicular and (c) parallel to the direction of the applied pressure and in (d) a magnification of the glassy phase which inglobes the whiskers.

As for other sintering aids, the presence of $ZrSi_2$ very likely helped the removal of surface boron oxide from the diboride particles and the liquid phases enhanced mass transfer mechanisms. The addition of SiC whiskers and fibers introduced a further source of SiO_2 which in turn increased the amount of the liquid phase. The presence of Zr-Si-O-C phases suggests a strong interaction between the sintering agent and the reinforcing phases. When $ZrSi_2$ came in contact with SiC fibers, $ZrSi_2$ reacted with the Si-C-O intergranular phase, according to the following possible reaction:

$$3\ ZrSi_2 + 2\ SiOC = 8\ Si(l) + ZrO_2 + 2\ ZrC \qquad (26)$$

The partial degeneration of the fiber intergranular phase into a Si liquid led to initial degeneration and opening of the fiber, as well as partial incorporation of ZrB_2 grains inside the fiber, as can be observed in Figure 15c.

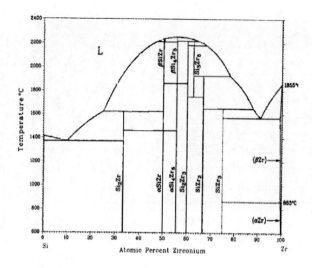

Figure 18. The Zr-Si phase diagram according to [63].

5.3 MoSi$_2$ –Containing Materials

5.3.1 Sintering Behavior

The baseline ZM material achieved the full density at 1,750°C, on the contrary the ZM20f composite required 1,900°C (Table III). This temperature was too high for the preservation of the fibers integrity, as demonstrated by the following microstructural analysis. However from Table III it can be noticed that reinforced and unreinforced materials both started to shrink at the same temperature, around 1,500°C, due to the formation of the same liquid phases.

5.3.2 Microstructure

Baseline unreinforced material. Figure 19a shows the polished surface of the fully dense baseline ZrB$_2$ +20 vol% MoSi$_2$ material, ZM.

ZrB$_2$ and MoSi$_2$ generate very similar contrast, the globular grains are ZrB$_2$, with mean grain size around 2.5 μm, whilst the MoSi$_2$ phase is brighter and characterized by irregular shape. The microstructure analysis evidenced the presence of oxide phases, i.e. SiO$_2$ and ZrO$_2$. Silica pockets, in amount around 4%, were easily recognizable as dark contrasting phases [64].

Fiber-reinforced composite. The microstructure of the material containing ZrB$_2$-MoSi$_2$ and fibers, ZM20f, is reported in Figure 19b-c. The fibers were well distributed along the material, but they completely lost their original aspect, owing to the high processing temperature. In the inset of Figure 19b, the degeneration of the fiber is clearly visible in the fracture surface, EDS analyses revealed a high amount of oxygen among the crystallites of the fibers, which are around 500 nm big. Similarly to the composites sintered with ZrSi$_2$, also in this case several squared ZrB$_2$ particles collapsed on the fibers, as visible in Figure 19c, indicating that these were already very friable at temperatures at which the ZrB$_2$ particle were still able to move.

To note that all the oxide phases, identified in the baseline, i.e. SiO$_2$ and ZrO$_2$, almost disappeared in the composite. This phenomenon can be explained in terms of high temperature, which induces carbo-reduction, and in terms of tendency of the SiC crystallites of the fibers to absorb the oxygen-bearing species, which is confirmed by the O-peak revealed by EDS in the fiber.

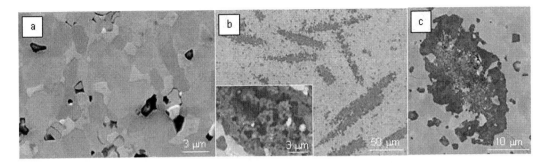

Figure 19. (a) SEM image of the baseline material ZrB$_2$ + 20 vol% MoSi$_2$ (ZM). Polished section of the material ZrB$_2$ + MoSi$_2$ + 20vol% fibers (ZM20f) showing (b) a panoramic view with a fracture surface of a fiber in the inset and (c) thinning and progressive degeneration of the fiber.

5.3.3 Densification Mechanism in Presence of MoSi$_2$

MoSi$_2$ was reported to help the densification in different ways [65]. Removal of the surface oxide present on boride starting particle occurs by reactions such as:

$$MoSi_2 + B\text{-}O \text{ species} = MoB + Mo\text{-}Si\text{-}B\text{-}O \text{ species} + SiO_2 \qquad (27)$$

MoSi$_2$–MoB–Si–B species can locally form liquid phases at 1,350°C [66]. In the same system another eutectic temperature takes place at 1,802°C, among MoB-Mo$_5$Si$_3$-MoSi$_2$, and another one at 1,885°C, among Mo-MoB-Mo$_5$Si$_3$-Mo$_5$SiB$_2$ [66]. Therefore, it is assumed that small amounts of liquid form in this system, which promotes the matter transfer mechanisms.

MoSi$_2$ did not seem to interact with the SiC fibers, as no intermediate reaction phases containing Mo were detected in the proximity of the matrix/fiber interface. The degeneration of fibers is rather ascribed to the gradual volatilization of the intergranular SiCO phase into SiO(g) and CO(g), occurring at temperatures higher than 1,800°C according to:

$$2SiOC = SiO(g) + CO(g) + SiC \qquad (28)$$

and progressive grain growth and coalescence of SiC nanograins.

5.4 Mechanical Properties

5.4.1 Fiber-Reinforced ZrB_2

The mechanical properties of the starting matrix and fiber-reinforced composites are reported in Table V. Among the different systems, the maximum increase of fracture toughness was achieved for the composition containing $ZrSi_2$ as sintering additive, exhibiting a value of 6.2 MPa·m$^{1/2}$. For Si_3N_4 addition, the toughness reached 5.7 MPa m$^{1/2}$ when the fiber content was 20 vol%. It can be noticed that for these two systems the toughness increase was about 50% of the reference material. No toughness increase was observed for $MoSi_2$ doped materials. The fibers degeneration occurring during sintering at 1900°C rendered the reinforcements completely ineffective.

On the other hand, the toughness increase was counterbalanced by a strength decrease at room temperature compared to the reference matrices, for both systems containing Si_3N_4 (-35-60%) and $ZrSi_2$ (-50%). No clear fracture origin could be identified on the fracture surfaces. Using the Griffith equation, $\sigma = 1.3\ K_{Ic}/C^{1/2}$, the size of the mean critical defect (C) was estimated to be about 100 μm, which is of the same order of magnitude of the fibers, indicating that the fibers themselves could represent critical flaws.

Table V. Mechanical properties of the baseline materials and of the fiber-reinforced composites. HV1.0=Vickers hardness, K_{Ic}=fracture toughness, σ_{RT}=room-temperature flexural strength, σ_{1200}=flexural strength at 1200°C. The Δvalues are referred to the baseline composites. The values are expressed as mean±1 standard deviation.

Sample	HV1.0* GPa	K_{Ic} MPam$^{1/2}$	ΔK_{Ic} %	σ_{RT} MPa	$\Delta\sigma$ %	σ_{1200} MPa	$\Delta\sigma_{1200}$ %
ZS	13.4±0.6	3.75±0.10	-	600±90	-	240±30	-
ZS10f	14.1±0.6	5.32±0.33	+42	389±37	-35	409±9	70
ZS20f	14.6±0.3	5.65±0.30	+52	413±17	-31	335±50	40
ZS30f	12.4±1.0	5.18±0.22	+38	240±30	-60	-	-
ZZ	17.8±0.9	4.25±0.04	-	808±31	-	-	-
ZZ20f	18.5±1.0	6.24±0.35	+47	385±13	-52	379±15	-
ZSM	14.9±0.5	3.5±0.6	-	704±98	-	-	-
ZSM20f	-	3.67±0.26	+5	-	-	-	-

*measured on the surface normal to the direction of the applied pressure.

The flexural strength was also tested at 1,200°C in air. It can be noticed that all the composites tend to maintain the same value displayed at room temperature. This value is always around 400 MPa, irrespective of the type of sintering aid ($ZrSi_2$ or Si_3N_4) or amount of fibers (10, 20 vol%). The comparison with baseline materials suggests that the SiC chopped fibers effectively strengthened the composites at this temperature, notwithstanding the higher amount of intergranular glassy phase.

The hardness, did not vary significantly between the baseline material and the reinforced ones, for systems with the same sintering additive. It can be noted that in the case of $ZrSi_2$-doped materials, the hardness was higher than for Si_3N_4 doped materials (~18 GPa, compared to ~14 GPa, respectively).

5.4.2 Toughening Mechanisms

The toughness of the composites can be interpreted as the sum of the unreinforced matrix toughness plus the increment ΔK_{Ic}^{fiber} due to toughening mechanisms exerted by the fibers:

$$K_{Ic}^{reinforced} = K_{Ic}^{matrix} + \Delta K_{Ic}^{fiber} \qquad (29)$$

In order to understand the toughening mechanisms giving rise to ΔK_{Ic}^{fiber}, the interaction of the crack front was analyzed in the baseline and in the reinforced materials, introducing cracks onto polished surfaces by 98.1 N Vickers indentations. In Figure 20a and 21a the baseline materials (ZS, ZZ) are shown: the crack propagates mainly intergranularly, suggesting a weak grain boundary interface. Examples of crack paths in the fiber-reinforced materials are shown in Figure 20b, sintered with Si_3N_4, and Figure 21b, sintered with $ZrSi_2$.

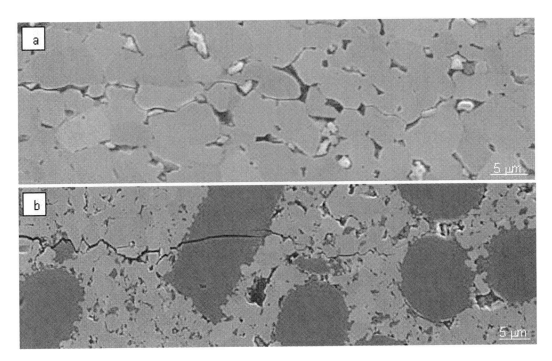

Figure 20. Crack path generated by a 10-kg indentation in (a) ZS, (b) ZS20f. The cracks propagate from left to right.

Irrespective of the type of sintering aid, no fiber pullout was observed, Figure 6a,c,d and Figure 15a. In most cases, the crack propagated straight through the fibers with little deflection.

This is a consequence of the matrix/fiber interface which does not allow the debonding of the fiber and results in no pullout. Considering the high fracture strength of the fibers, crack bowing was considered the dominant toughening mechanisms for this composite rather than crack deflection or bridging (Figure 22).

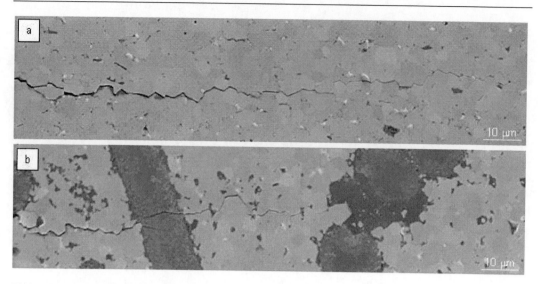

Figure 21. Crack path generated by a 10-kg indentation in (a) ZZ and (b) ZZ20f. The cracks propagate from left to right.

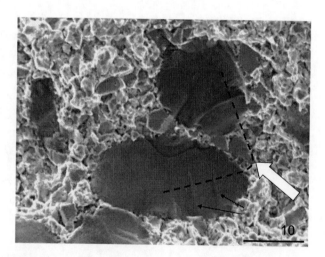

Figure 22. Crack bowing in the fracture surface. The white arrow indicates the crack propagation direction, black arrows the formation of steps and the dotted lines a possible profile of the crack front.

Furthermore, the thermal residual stress contribution was considered in the analysis of crack/microstructure interaction. Thermal residual stresses are a consequence of the mismatch between the thermal expansion coefficients and elastic constants of matrix and fibers [27].

Equation (29) can be thus rewritten as:

$$K_{Ic}^{reinforced} = K_{Ic}^{matrix} + \Delta K_{Ic}^{bowing} + \Delta K_{Ic}^{residual\ stress} \tag{30}$$

Due to the fact that $\alpha_{SiC} < \alpha_{matrix}$, thermal residual stresses give a negative contribution to fracture toughness [27]. In order to compare experimental results and theoretical expectations, these two contributions, crack bowing and thermal residual stresses, were modeled. It was assumed that the values of K_{Ic}^{matrix} were those of unreinforced matrices, ZS, ZZ, ZM (Table

V). The numerical integrations were performed using a commercial software (MATHEMATICA 7, Wolfram Research Inc., Champaign, IL). The toughening increment, ΔK_{Ic}^{bowing} was calculated according to the model proposed by Rouxel [67] taking the mean fiber diameter as 14 μm and the fiber fracture strength as 2.7 GPa. The toughening increment due to thermal residual stresses, $\Delta K^{residual\ stress}$, was instead evaluated according to model of Taya et al. [27]. This model was originally developed for a particle-reinforced composite. We adapted this model to a fiber-reinforced composite considering the appropriate interparticle distance and, for coherence, considering the particle size as the reinforcement diameter. All the details of equations used for the calculations can be found in Section 2. The input values for E_{fiber}, ν_{fiber}, and E_{matrix}, ν_{matrix}, were 300 GPa [62], 0.17 [68] and 500 GPa, 0.11 [3], respectively. Since the thermal expansion of the fibers is isotropic [60], a common value of $4.45 \cdot 10^{-6}$ C^{o-1} was taken for the α_{fiber}, the corresponding value for the matrix, α_{matrix}, was $6.20 \cdot 10^{-6}$ C^{o-1}. Irrespective of the type of reinforcement, for ΔT a stress-free temperature of -1500°C was considered.

The results of this computational modeling are shown in Figure 23a,b where the experimental points for the systems with Si_3N_4 and $ZrSi_2$ are superimposed to the theoretical curves showing the individual contributions of crack bowing, of residual stress and the overall toughening increment.

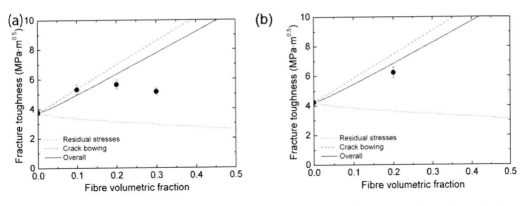

Figure 23. Fracture toughness experimental values and comparison with single and overall toughening contributions. Composites containing fibers and processed with (a) Si_3N_4 and (b) $ZrSi_2$.

The plots indicate that this modeling can properly describe the toughness behavior only for amounts of fibers up to 20 vol%. According to these models, the fracture toughness should increase with increase of the fiber content, in contrast with the experimental results that show that when the fiber content was increased to 30 vol%, the fracture toughness did not vary significantly. One reason for this unexpected behavior is that increasing the reinforcement content it becomes increasingly difficult to properly process the powder mixtures, achieving a dense matrix and a good fiber dispersion. Thus, in the case of the composite with 30 vol % fibers, it can be hypothesized that due to residual porosity in the matrix, the matrix toughness is lower than the reference value, 3.8 MPa·m$^{1/2}$. Furthermore, for composites containing $ZrSi_2$ or Si_3N_4 sintering aid and 20 vol% of fibers, there is approximately the same offset between unreinforced and reinforced materials, Table V. This implies that the higher value found for ZZ20f is only due to the higher toughness of its respective matrix (4.3 vs 3.8 MPa·m$^{1/2}$), whilst the toughness increment ΔK_{Ic}^{fiber} is the same as for ZS20f.

5.4.3 Comparison with Whisker-Reinforced ZrB$_2$

In this section, the mechanical properties of the ZrB$_2$-based composites reinforced by fibers are compared to those reinforced by whiskers, whose values are summarized in Table VI.

The addition of whiskers allowed toughening of the starting reference material without losing strength. The fracture toughness increased from 3.75 to 5.0-5.3 MPa·m$^{1/2}$ for the composites sintered with Si$_3$N$_4$ and from 4.3 to 5 MPa·m$^{1/2}$ for the composites sintered with ZrSi$_2$.

The room-temperature flexural strength was stable or slightly increased, from 600 to about 700 MPa for the ZS system.

The hardness did not vary significantly between the baseline and the reinforced materials for the composite containing 10 % whiskers. It should be observed that the apparent decrease of properties for the composites with 20 vol% of whiskers or more should be related to the presence of higher fractions of residual porosity and glassy phase.

Table VI. Mechanical properties of the baseline materials and of the whisker-reinforced composites. HV1.0=Vickers hardness, K$_{Ic}$=fracture toughness, σ_{RT}=room-temperature flexural strength, σ_{1200}=flexural strength at 1,200°C. The Δvalues are referred to the baseline composites. The values are expressed as mean±1 standard deviation.

Sample	HV1.0* GPa	K$_{Ic}$ MPam$^{1/2}$	ΔK_{Ic} %	σ_{RT} MPa	$\Delta\sigma$ %	σ_{1200} MPa	$\Delta\sigma_{1200}$ %
ZS	13.4±0.6	3.75±0.10	-	600±90	-	240±30	-
ZS10w	13.3±0.5	5.00±0.09	+33	708±35	+18	356±37	48
ZS20w	9.5±0.8	5.28±0.33	+41	614±75	=	352±22	47
ZS30w	8.4±0.7	4.63±0.2	+23	-	-	-	-
ZZ	17.8±0.9	4.25±0.04	-	808±31	-	-	-
ZZ20w	11.1±1.0	5.01±0.05	+18	-	-	-	-

*measured on the surface normal to the direction of the applied pressure.

The interaction of the crack front was analyzed also in the whisker-reinforced materials, introducing cracks onto polished surfaces by 98.1 N Vickers indentations (Figure 24). As can be seen in Figure 24, in the whisker-reinforced composites, very few whiskers were crossed by the crack front and the crack propagated mainly along the matrix grain boundaries or whisker/matrix interface. The comparison with the crack propagation in the baseline material evidences that the tortuosity, calculated as the ratio between the actual path length and the shortest distance between the start and end points of the crack, changed from 1.2 in the unreinforced material to 1.5-1.8 in the whisker-doped materials.

The extent of whisker pullout was instead very low as also observed on the fracture surfaces, see Figures 10a,c,e and Figure 18a. Although some sporadic bridging whiskers were noticed, their frequency was so low that the main toughening mechanism was identified as crack deflection. Beside crack deflection, residual stress toughening must be considered, as for the fiber-reinforced materials.

Also in this case, residual stress results in a negative contribution for the overall fracture toughness. Equation (29) can be thus rewritten as:

$$K_{Ic}^{reinforced} = K_{Ic}^{matrix} + \Delta K_{Ic}^{deflection} + \Delta K_{Ic}^{residual\ stress} \qquad (31)$$

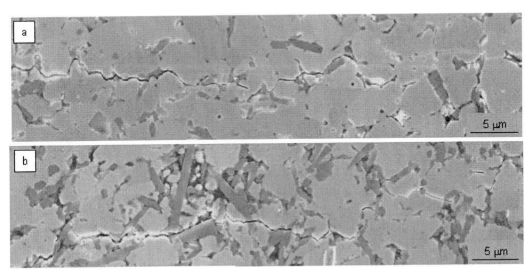

Figure 24. Crack path generated by a 10-kg indentation in (a) ZS10w, (b) ZS20w. The cracks propagate from left to right.

As for the fiber-reinforced materials, toughening contributions from crack deflection and thermal residual stresses, were modeled (Figure 25a,b). The contribution of crack deflection, $\Delta K^{deflection}$, was evaluated according to the model of Faber and Evans [21] for rod-shaped particles, considering an average aspect ratio of 4.6, as estimated by image analysis on SEM micrographs of composites containing both 10 and 20 vol% whisker. For the calculations, the corrected version of the equations as presented in Section 2 [69] was adopted. The mean reinforcement size, d, was evaluated by image analysis and was found to be 0.6 μm for the whisker-reinforced composites. The input values for E_w, E_m, v_w, e v_m, α_w, α_m, were the same as previously reported.

As can be seen in Figure 25a,b, considering all the simplifying assumptions, the agreement between the theoretical model and the experimental values can be considered fairly good, at least up to additions of 20 vol% whiskers. A significant deviation was instead found for the composite containing 30 vol% of whiskers. As for the fibers, increasing the reinforcement content it becomes increasingly difficult to properly process the powder mixtures and achieve a dense composite. The deviation is thus due to the presence of residual porosity that affected all the properties, including hardness and strength (see Table VI).

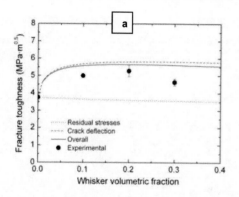 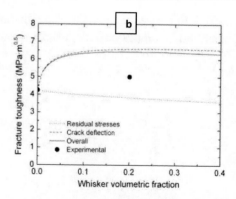

Figure 25. Fracture toughness experimental values and comparison with single and overall toughening contributions. Composites containing whiskers and processed with (a) Si_3N_4 and (b) $ZrSi_2$.

5.4.4 Overall Considerations on the Toughening Effects

The analysis of the interaction between the advancing crack and the microstructure confirmed that the intergranular wetting phases or interface reactions prevented any possibility of reinforcement pullout, independently of the sintering additive or reinforcement type.

The theoretical and experimental results presented in the previous sections clearly indicate that with a 20 vol% of reinforcement of the type used in this work the fracture toughness cannot be improved by more than 50% with respect to the baseline material. In order to overcome the limits shown by this and similar studies, it is our opinion that future developments for these composites should have a two-fold approach:

- maximization of the toughening contributions by: the minimization of the negative contribution of thermal residual stress which could be accomplished either using a different kind of reinforcement with CTE similar to those of ZrB_2 or tailoring shape/dimensions of the reinforcement;
- avoid the fiber or whisker degeneration through the addition of the proper sintering additives and the use of the most suitable sintering technology (for example spark plasma sintering), in order to limit the degeneration reaction of the reinforcing phase.
- careful control of the reinforcement/matrix interface in order to prevent the formation of a strong interface and introduce bridging as ulterior toughening mechanism. Indeed, assuming that crack deflection in the whisker reinforced materials can be replaced by crack bridging, an increase of fracture toughness up to 7-8 $MPa \cdot m^{1/2}$ can be estimated using the model of Becher et al. [70].

6. Other UHTCs Reinforced with SiC Fibers

In this section, attempts to reinforce other UHTCs such as HfB_2 and TaC by introduction of SiC chopped fibers is described. Microstructure and properties are illustrated and compared. No available data can be found in the literature, hence these results can be considered the starting point towards the obtainment of tougher ultra high temperature ceramics reinforced by discontinuous phase.

HfB_2 was selected because, after ZrB_2, is the most investigated system for applications in oxidative environment and thus needs optimization of the performances as well. The choice of TaC relies in its intrinsic higher fracture toughness compared to the other borides and carbides and with a proper reinforcement it should have the highest value attainable among the UHTCs.

6.1 HfB₂-Based Materials: Microstructure and Mechanical Properties

6.1.1 Sintering Behavior of HfB₂-Si₃N₄ Composites

The baseline material, containing 5 vol% of Si_3N_4, HS, was sintered by hot pressing with an applied pressure of 30 MPa (Table VII). The shrinking started around 1530°C, the maximum temperature achieved was 1800°C and resulted in a final relative density around 96% of the theoretical value [71]. The addition of SiC fibers in amount 10-20 vol% increased the shrinkage temperature above 1600°C and the applied pressure was increased to 50 MPa, the maximum temperature, held for 8-10 minutes was 1750-1800°C. However the final density remained below 90%.

6.1.2 Microstructure

Baseline unreinforced material. The microstructure of the unreinforced HfB_2-based material is shown in Figure 26a. HfB_2 grains are rounded, with dimensions of about 2.5 µm. It can be noticed that little amount of residual porosity is present, in agreement with Archimedes' results. In addition to HfB_2, secondary phases were detected by EDS microanalysis, i.e. HfO_2, $HfSi_2$, BN and spurious phases in the systems Hf-N-O and Hf-N-C. These secondary phases were mainly concentrated at the triple points.

Table VII. Composition, sintering parameters, density and mechanical properties of the hot-pressed samples based on HfB₂ and containing SiC fibres. T_ON= temperature at which the shrinkage started, T_MAX= maximum temperature achieved and held for the dwell time, HV1.0=Vickers hardness, K_Ic=fracture toughness, σ_RT=room-temperature flexural strength. σ₁₂₀₀=flexural strength at 1,200°C. The Δvalues are referred to the HS composite. The values are expressed as mean±1 standard deviation.

Sample	Composition vol%	T_{ON} °C	T_{MAX} °C	Dwel time min	Pressure MPa	Th. density g/cm³	Final density g/cm³	Rel. density %	HV1.0 GPa	K_{Ic} MPam$^{1/2}$	ΔK_{Ic} %	σ_{RT} MPa	$\Delta\sigma$ %
HS	HfB_2+8 Si_3N_4	1530	1800	30	30	10.54	10.13	96.1	12.6±0.8	3.75±0.10		478±45	
HS10f	HB+10 SiCf	1600	1800	8	30-50	9.9	8.27	84.0	9.6±0.7	-	-	-	-
HS20f	HB+20 SiCf	1630	1750	10	30-40	9.1	8.13	89.3	12.8±0.3	4.15±0.17	11	326±52	-32

*measured on the surface normal to the direction of the applied pressure.

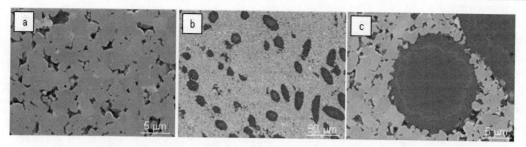

Figure 26. Polished section of (a) HS unreinforced material showing the secondary phases, (b) the overall fibers dispersion in the matrix in HS20 and (c) a magnification of the fiber interface.

Fiber-reinforced composites. The materials containing fibers resulted not fully dense, with porosity mainly concentrated in proximity of the fibers. These were homogeneously dispersed along the matrix, as visible in Figure 26b. HfB_2 grains was slightly reduced compared to the unreinforced materials from about 2.5 to 1.5 µm, thanks to the presence of SiC fibers which inhibited the grain growth even at higher temperature. The fibers broke transgranularly, as no pullout was observed. Despite the high sintering temperature, the interface reaction fiber/matrix was notably reduced compared to the ZrB_2-based composites, confirming that the higher chemical stability of HfB_2 phase compared to ZrB_2, Figure 26c. The core of the fiber was SiC and in the rim Al traces were also detected by EDS analysis. Finally, both core and rim showed little oxygen amounts.

6.1.3 Densification Mechanism

Due to the similarity between ZrB_2 and HfB_2, we can expect that analogous densification mechanisms occur during sintering. However, from the microstructural analysis it is evident that some important differences are present in these two system. HfB_2 is more refractory than ZrB_2, as sintering temperature and amount of sintering aid had to be increased in order to obtain a good degree of densification.

It is assessed that during powder processing, oxygen contamination is unavoidable and induces an enrichment in oxygen, by amorphizing a Hf-B-O coating on the diboride surface particles [72]. The sintering additive, Si_3N_4, undergoes a similar phenomenon and introduces contaminants in the form of SiO_2. The formation of reaction products during hot pressing can be explored on the basis of the chemical interaction between Si_3N_4, HfB_2 and the oxygen bearing species. From thermodynamical calculations it resulted that HfB_2 does not react with Si_3N_4, on the contrary the sintering additive first reacts with boron oxide and then with hafnium oxide. At about 1600°C Si_3N_4, SiO_2, HfO_2 interact consistently and give rise to a fluid Si-based reaction product.

Generally, liquid phase sintering results in the rearrangements of grains by capillary action during the first stage and solution-reprecipitation on the undissolved grains during the second stage. The different sintering behavior compared to ZrB_2, is probably due to HfB_2 higher stability and scarce solubility in the liquid, which did not allow the achievement of the full density at 1700°C. The addition of fibers, which are covered by SiO_2 themselves, could in principle favor the formation of a liquid phase at lower temperature, but the elongated shape dragged the particles movement and rearrangement.

Ultra-High Temperature Ceramics Reinforced with SiC Short Fiber or Whisker   359

It is presumed that this system would achieve a higher density if sintered at 1850°C, as demonstrated by Monteverde and Bellosi [72]. However this temperature would be certainly deleterious for the fibers integrity.

6.1.4 Mechanical Properties

The mechanical properties for HfB_2-based materials are reported in Table VII.

The hardness of the three HfB_2-based materials resulted very low compared to typical values reported for HfB_2 matrix, ranging from 9.5 to 13 GPa [3-5,17,72]. This property was mainly affected by the residual porosity and the formation of glassy phases.

As far as the HfB_2 material containing 20 vol% of fibers is concerned, fracture toughness increase of around 10% occurred, leading to 4.15 $MPa \cdot m^{1/2}$, which is still in the range of unreinforced HfB_2-based composites [3-5,17,72]. It is probable that the reinforcing effect of the fibers was partially annulled by the residual porosity of the composite.

The room temperature flexural strength decreased from 480 to 330 MPa, around 30%, as for the ZrB_2-based composites. The reason for such decrease is the same invoked for ZrB_2-Si_3N_4-SiCf ceramics (see Section 5.4.1).

6.2   TaC-Based Materials: Microstructure and Mechanical Properties

6.2.1 Sintering Behavior of TaC-TaSi$_2$ Composites

The baseline TaC-TaSi$_2$ composite (TCT) started shrinking at 1400°C and required a maximum temperature of 1,750°C to achieve a density of 13.3 g/cm^3 (Table VIII). The relative density, determined as the ratio between bulk and theoretical densities calculated on the basis of starting compositions, was >97. However, since post-sintering SEM analyses ascertained the presence of extra phases (SiC and SiO_2) having much lower density than the starting ones, the relative density indicated in Table VIII can be significantly underestimated, as discussed in the following section.

A similar densification behavior was observed for the material containing TaSi$_2$ as sintering additive and 20 vol% of SiC fibers, TCT20f. A final density of about 98% was achieved after a dwell of 9 minutes at 1,750°C.

The densification curve showed that the addition of TaSi$_2$ was very effective for sintering this carbide, as the shrinkage of composition containing TaSi$_2$ started at a much lower temperature than the composites containing Si_3N_4 or $MoSi_2$.

Table VIII. Composition, sintering parameters, density and mechanical properties of the hot-pressed samples based on TaC and containing SiC fibres. T_{ON}= temperature at which the shrinkage started, T_{MAX}= maximum temperature achieved and held for the dwell time, HV1.0=Vickers hardness, K_{Ic}=fracture toughness, σ_{RT}=room-temperature flexural strength. σ_{1200}=flexural strength at 1,200°C. The Δvalues are referred to the TCT composite. The values are expressed as mean±1 standard deviation.

Sample	Composition vol%	T_{ON} °C	T_{MAX} °C	Dwell time min	P MPa	Th. density g/cm^3	Final density g/cm^3	Rel. density %	HV1.0 GPa	K_{Ic} MPam$^{1/2}$	ΔK_{Ic} %	σ_{RT} MPa	$\Delta\sigma$ %
TCT	TaC+15 TaSi$_2$	1400	1750	9	30	13.69	13.3	97.2	14.6±0.4	4.7±0.1	-	679±18	-
TCT20f	TCT+20 SiCf	1430	1750	9	30	11.60	11.38	98.1		5.26±0.39	+12	548±64	-19

*measured on the surface normal to the direction of the applied pressure.

6.2.2 Microstructure

Baseline unreinforced material. In the TCT material, TaC and TaSi$_2$ were the only crystalline phases detected after sintering by x-ray diffraction. The polished surface, displayed in Figure 27a, put in evidence that little or no porosity was present in the microstructure. The fracture surface (not shown) gave evidence that larger grains (5–7 μm) were transgranularly fractured, while smaller grains (1–3 μm) were intergranularly fractured. The mean grain size of the carbide phase was 2.5 μm, but coalescence of grains led to the formation of larger grains, up to 6–7 μm. The TaSi$_2$ grains were recognizable as grey phase in a brighter matrix, and tended to form large pockets as wide as 3-8 μm (Figure 27a). Moreover, it can be noticed that TaSi$_2$ phase has very low dihedral angles at the interface with the matrix, showing very ductile behaviour. TaSi$_2$ is known to a have a brittle to ductile transition, similar to MoSi$_2$ [64], however, the high wettability displayed by the silicide could also be due to local liquid phase formation. Dark contrasting features were identified as SiC, Si-C-O and SiO$_2$ (Figure 27a). These dark phases were observed to be adjacent to the TaSi$_2$ phase, indicating that they formed consuming TaSi$_2$. According to image analysis, the volumetric amount of TaSi$_2$ in the final microstructure was around 10 vol%, i.e. slightly lower than the initial composition, 15%. Considering the volumetric amount of these phases (2 vol% SiC, 0.5 vol% SiO$_2$) in the calculation of the theoretical density, the final theoretical density is ~13 g/cm^3 and hence the relative density is 99%. This value of relative density is more consistent with the low quantity of porosity ascertained by SEM observations.

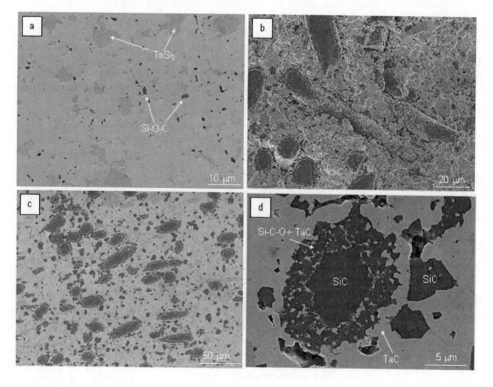

Figure 27. Polished section of (a) TCT unreinforced material showing the overall microstructure. (b) Fracture surface of the TCT20f composite, (c) overview of the TCT20f microstructure and (d) a magnification of the fiber interface in the polished surface.

By TEM inspections, a denticulate substructure was observed in the matrix: the inner and the external part were epitaxial and the corrugate feature of the sub-grain boundary denotes a lattice misfit [73]. However, the chemical analysis in the two regions is very difficult, because from EDS the stoichiometry of the TaC constituting the core and the shell could not be identified, being carbon strongly dependent on the specimen thickness. The only assured composition difference between the two areas, was the presence of oxygen in the outer part, suggesting it might derive from re-precipitation from a Ta-Si-C-O liquid phase. At the triple point junctions Ta_5Si_3 and $Ta_{4.8}Si_3C_{0.3}$ were detected.

Fiber-reinforced composite. The microstructure of the composite containing 20 vol% of fibers is displayed in Figures 27b-d. In the fracture surface of Figure 27b two main features have to be noticed: little fibers pullout and a continuous rough layer which surrounds the fibers. It seems that this wide interface reaction phase is weakly boded to the matrix and helped the fibers pullout, but at the same time a diffused fibers degeneration occurred. In the polished section of Figures 27c,d it can be observed that the fibers are well dispersed in the matrix, their diameter is notably decreased, as the SiC core is around 7 μm compared to the initial diameter of 15 μm; the same occurred on the longitudinal direction, which was reduced to 50 μm-length. According to Figure 27d, around the SiC core an irregular jagged layer of Si-C-O was present with TaC bright grains as small as 300 nm. Beyond this layer, another 2 μm bright stratum of coarser TaC grains was present, which was surrounded itself by coarse SiC particles with squared shape and grain size from 2 to 4 μm. A schematic of the fiber morphology in this composite is depicted in Figure 28.

It is evident that remarkable reactions took place during sintering in presence of $TaSi_2$, even if the maximum temperature was the same as the one used for densification of ZrB_2- and HfB_2-based composites. $TaSi_2$ seems to be responsible for the fibers degeneration. One possible reaction is:

$$2\ SiOC + 12\ TaSi_2 = 19\ Si_{(l)} + 2TaC + SiO_2 + 2\ Ta_5Si_3 \qquad (32)$$

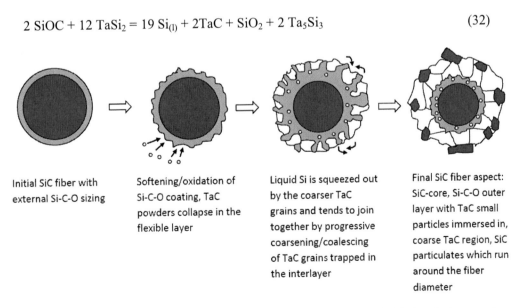

Figure 28. Schematics of the fiber evolution in the composite $TaC+TaSi_2+SiCf$ during the hot pressing cycle at 1,750°C.

6.2.3 Densification Mechanism

Densification results show that the addition of $TaSi_2$ strongly improves the sintering of TaC. The sintering temperatures is 1750 °C, that is much lower than those reported in the literature for similar composites (>2200°C) [74-77]. The shrinkage curves of TCT displays a slope change at T~1450 °C, that is, an acceleration in the sintering rate, which is usually attributed to the formation of liquid phase, which activates matter transfer mechanisms. Microstructural features of the composites confirmed the formation of Si- and SiO_2-based phases, which were presumed to be liquid at the sintering temperatures. All these observations lead to the conclusion that densification of TaC was promoted by liquid phases.

In a previous study concerning boride-$TaSi_2$ composites [78], it was found that $TaSi_2$ undergoes decomposition in presence of CO, which is typically present in graphite-based furnaces. The following reaction was hypothesized to occur:

$$TaSi_2 + CO(g) = TaC + Si_{(l)} + SiO_{(g)} \tag{33}$$

favorable T > 1100°C. The formation of silicon, which is liquid at 1415 °C, is very close to the onset of densification for this composite. The decrease of $TaSi_2$ content after sintering and the presence of Si-based phases in the final microstructure are consistent with the occurrence of reaction (33). Finally, the formation of SiC phases could derive from the carburization of silicon (34).

$$Si(l) + C = SiC \tag{34}$$

Other sources of liquid can be related to the Ta-Si phases: a eutectic point is reported at 1400°C between $TaSi_2$ and Si [79], in addition, $TaSi_2$ and Ta_5Si_3 form a eutectic at 1970°C [79], and the Ta_2O_5-SiO_2 phase diagram [80] foresees a eutectic at 1887°C, however these last two temperatures are out of the sintering temperature range for the composite (1750°C).

Hence, in reducing environment (C/CO-rich), $TaSi_2$ partially dissociated into Ta and $Si/SiO/SiO_2$ phases, which favoured the formation of liquid phases where the matrices were partially soluble in. The liquid was mainly constituted by Ta-Si-C-O and, upon cooling, it crystallized at the triple points leaving clean grain boundaries of the matrix. The presence of traces of oxygen at the triple junctions phases confirmed that the liquid was effective in removing the oxide layer covering the starting powder, which is well known to hinder the densification of non-oxide ceramics.

Considering the fibers morphology upon sintering (Figure 27d), it is very probable that the free Ta and the Si/Si-O compounds strongly interacted with the fibers in a catastrophic way. The oxygen bearing species, reacted with the SiC crystallites of the fibers which coarsened and provoked the fibers deformation and aperture. It is presumable that the fibers degeneration occurred at the beginning of the sintering, because in the inner layer of the fibers, TaC small particles, of the dimension of the order of the initial powder, were immersed in the Si-C-O phase, indicating that they were excluded by the overall densification and coarsening processes. The coarse SiC particles which form a sort of crown surrounding the first TaC-ring might have origin from the Si released by $TaSi_2$ or by liquid Si squeezed out from the fiber, which was slowly carbo-reduced to SiC by the reducing atmosphere. This process is sketched in Figure 28.

6.2.4 Mechanical Properties

The mechanical properties of TaC-based composites are reported in Table VII.

The fracture toughness of the baseline TCT resulted 4.7 MPa·m$^{1/2}$ and the addition of the fibers increased this value up to 5.3 MPa·m$^{1/2}$, about 10% higher. TaC is known to have a more metallic nature, compared to the other ultra-high temperature ceramic based on Zr and Hf, owing to the d-shell containing one electron more [81]. This aspect has as a consequence a generally higher fracture toughness (around 5 MPa·m$^{1/2}$) compared to other covalent compounds (3-4 MPa·m$^{1/2}$ for Zr and Hf-borides and carbides). The fibers contribution to the toughness increase was not so effective, certainly due to their degeneration.

As for the room temperature flexural strength, it passed from 680 MPa for the unreinforced material, to 550 MPa for the fibers-containing composites, that is a decrease of less than 20%. In the ZrB$_2$-composites containing 20 vol% of fibers and sintered with addition of either Si$_3$N$_4$ or ZrSi$_2$, the strength decrease was of the order of 30-50%. For TaC-based composite, the lower strength reduction could be related to the multiple effect of a coarser matrix, 3-7 μm, to a fibers thinning from 200 to 50 μm length and to the formation of SiC particulates of dimension around 10 μm, which give an overall mean grain size around 20 μm, which is not extremely different from the dimension of TaC grains, fibers and the newly formed SiC particles. It follows that the fibers did not result so extremely bigger than the matrix and hence they did not constitute a new defect population like it happened for the ZrB$_2$- and HfB$_2$-based composites.

CONCLUSION

7. Conclusions and Future Perspectives

Group IV-V transition metal borides and carbides are commonly referred to as ultra-high temperature ceramics. These materials possess very high melting point, above 3,000°C and are candidates for applications in aerospace engineering due to their extraordinary chemical-physical properties. Among the main obstacles to their production and use the *sintering* process and the low *fracture toughness* are the two most critical issues.

To obtain a fully dense material, pressure assisted techniques and temperature exceeding 2,000°C are usually required to obtain fully dense materials, but with addition of proper additives, it has been demonstrated that it is possible the reach a complete densification below 1,900°C, without negatively affecting the mechanical properties from room temperature to 1,500°C and the oxidation resistance. With a proper tuning of the secondary phase and sintering additive, fully dense materials can now be obtained, possessing excellent mechanical properties, such as hardness above 20 GPa and strength around 1 GPa. However, the fracture toughness remains the major concern for these composites, as only values around 2-4 MPa·m$^{1/2}$ can be achieved.

The new challenge was here to toughen some UHTCs by addiction of reinforcing phases, such as SiC short fibers (Hi-Nicalon) or SiC whiskers, in order to improve the resistance to crack propagation.

Of paramount importance is the processing temperature: the reinforcing phase can catastrophically react with the matrix or the sintering additive at high temperature and make pointless its beneficial effect on the mechanical properties.

ZrB$_2$-based materials reinforced with 10-30 vol% SiC short fibers or SiC whiskers were hot-pressed at temperature between 1,650 and 1,900°C. The addition of Si$_3$N$_4$ or ZrSi$_2$ as sintering additive allowed achieving densities higher than 95% of the theoretical values at temperatures in the range of 1,650-1,730°C. When MoSi$_2$ was added, it was necessary to raise the sintering temperature up to 1,900°C, with unavoidable fibers degeneration.

Fibers were more stable than whiskers, but a layered strong interface formed upon sintering. The core of the fibers was constituted by SiC, but the Si-C-O intergranular phase acted as a grab for the outermost jagged layer composed of SiC$_{1-x}$ and Si-C-N or Zr-Si-C phases. This graded structure of the fibers formed a very strong interface with the matrix and hindered any fiber pullout.

Whiskers were observed to be very sensitive to the temperature and generally appeared squatter after sintering.

The fracture toughness increased of the 30% to 50% when 10-30 vol% of SiC fibers or whiskers were added. The highest value obtained was for the ZrB$_2$-based composite containing 20 vol% of SiC fibers and ZrSi$_2$ as sintering additive, 6.3 MPa·m$^{1/2}$.

The supposed active toughening mechanisms were crack bowing, in the fibers-reinforced composites, and crack deflection, in the case of whiskers-reinforced composites. The negative contribution of residual stress was considered for bot the two classes of composites. The comparison between the experimental values and the expected toughness increment due to these mechanisms resulted fairly good.

The room-temperature strength increased or remained unaltered for whiskers addition, but decreased of about 30-50% when fibers were added, due to a change of the defects population. On the other hand, both whiskers and fibers strengthened the baseline material at high temperature.

Other two matrices were explored in combination with SiC fibers: HfB$_2$ and TaC.

In the first case, Si$_3$N$_4$ was used as sintering additive. Despite the high temperature, the fibers maintained their original shape, indicating a higher stability of HfB$_2$ phase, compared to ZrB$_2$. However, the residual porosity, around 10-15%, negatively affected the fracture toughness resulting in no appreciable toughening of the material.

In the case of TaC-composites, TaSi$_2$ was used as sintering additive, which allowed the obtainment of a completely dense material at 1,750°C. TaSi$_2$ was identified to play a key role in the interaction with the fiber, upon its decomposition into Ta and SiO/SiO$_2$ species. The final microstructure showed almost no porosity, but the fibers were severely damaged by the interaction of the SiO/SiO$_2$ phases with the SiC crystallites of the fibers, which coarsened and induced the fiber opening. Traces of liquid Si, subsequently reduced to SiC, were also detected running around the fiber core, whose diameter was reduced from 15 to 7 μm.

The study presented in this chapter pointed out that effective toughening and strengthening effects can be exercised by SiC short fibers or SiC whiskers. However, to fully exploit their potential, several conditions should be fulfilled.

In the case of the fibers-reinforced composites, a careful control of the reinforcement/matrix interface should be obtained in order to prevent the formation of a strong interface and introduce bridging and pull out as ulterior toughening mechanism. The matrix or the sintering additive should not interact with them or avoid their oxidation, which was noticed to produce a coarsening of the fibers' SiC crystallites and induce fibers degeneration.

In the case of the whiskers-reinforced composites, the processing temperature must be kept lower than 1,650°C to avoid whiskers degradation. For both fiber and whisker, the mixture homogeneization has to be very accurate when the amount of reinforcing phase exceeds the 20 vol%. Furthermore, the composites should have a density higher than 95%, to maximize the toughening effect, irrespective of the type of reinforcement used.

Among the sintering additives used, two classes can be identified: Si_3N_4 and the transition metals silicides ($ZrSi_2$, $MoSi_2$, $TaSi_2$). The two groups seems to behave in different ways. Si_3N_4 is completely converted into a liquid phase in the final microstructure that forms strong bonds between matrix and fibers/whiskers, but does not promote the reinforcement degeneration. The silicides, especially $TaSi_2$, notably react with the fibers, even at temperature as low as 1,650°C.

The plot of Figure 29 summarizes the averaged fracture toughness values of ZrB_2-based materials obtained in the last 10 years of ISTEC activity, showing that there has been a continuous improvement of this property from unreinforced materials or SiC particle-reinforced to SiC chopped fiber reinforced composites.

Although further efforts directed to fracture toughness improvement are still mandatory, the addition of elongated reinforcement, especially for the chopped fibers, has shown to be a very promising strategy.

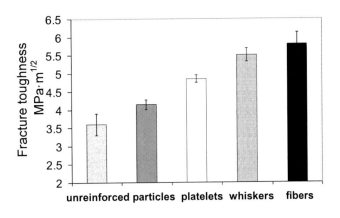

Figure 29. Plot of the fracture toughness values collected at ISTEC (Italy) in the last 10 years of experience on UHTCs. The values refer to pure ZrB_2 and with addition of SiC in form of particles, platelets, whiskers and fibers. Measurement technique: chevron-notched beam.

ACKNOWLEDGEMENTS

The authors gratefully acknowledge the financial support of The Air Force Research Laboratory through the research grant FA8655-09-M-4002, in particular the contract monitor, Dr. Joan Fuller. D. Dalle Fabbriche is acknowledged for hot pressing cycles, S. Guicciardi and C. Melandri for mechanical testing and Prof. H-J. Kleebe's group for the precious help in TEM specimens' preparation.

REFERENCES

[1] Becher, P. F.; Tiegs, T.N.; Angelini, P. In *Fiber Reinforced Ceramic Composites;* Mazdiyasni, K.S.; Ed.; William Andrew Publishing/Noyes: Maryland Heights, MO, 1990, Chap. 11, p. 311.

[2] Boland, P.; Walton, J. D. *Aerospace ceramics-Characteristics and Design principles*; 1965, Final Technical Rept. Jan 64-Apr 65.

[3] Gasch, M. J.; Ellerby, D. T.; Johnson S. M. In:*Handbook of Ceramic Composites*; Bansal, N. P.; Ed.; Kluwer Academic Publishers, Boston, MA, 2005, Part II, 197-224.

[4] Wuchina, E.; Opila, E.; Opeka, M.; Fahrenholtz, W.; Talmy, I. *Interf.* 2007, *16*, 30-36.

[5] Fahrenholtz, W.G.; Hilmas, G.E.; Talmy, I.G.; Zaykoski, J.A. *J Am Ceram Soc.* 2007, *90*, 1347-1364.

[6] Monteverde, F.; Bellosi, A. *Scripta Mater.* 2002, *46*, 223-228.

[7] Monteverde, F.; Bellosi, A. *Adv Eng Mater.* 2003, *5*, 508–512.

[8] Zhu, S.; Fahrenholtz, W.G.; Hilmas, G.E.; Zhang, S.C. *J Am Ceram Soc.* 2007, *90*, 3660-3663.

[9] S. N.; Karlsdottir, J. W.; Halloran, C. E.; Henderson, *J Am Ceram Soc.* 2007, *90*, 2863–2867.

[10] Marschall, J.; Pejaković, D.A.; Fahrenholtz, W.G.; Hilmas, G.E.; Zhu, S.; Ridge, J.; Fletcher, D.G.; Asma, C.O.; Thömel, J. *J Thermoph Heat Transf.* 2009, *23*, 267-278.

[11] Zhou, X.J.; Zhang, G.J.; Li, Y.G.; Kan, Y.M.; Wang, P.L. *Mater. Lett.* 2007, *61*, 960–963.

[12] Chamberlain, A. L.; Fahrenholtz, W. G.; Hilmas, G. E.; Ellerby, D. T. *J Am Ceram Soc.* 2004, *87*, 1170-1172.

[13] Liu, Q.; Han, W; Zhang, X.; Wang, S.; Han, J. *Mater Lett.* 2009, *63*, 1323-1325.

[14] Monteverde, F.; Guicciardi, S.; Bellosi, A. *Mat Sci Eng A.* 2003, *346*, 310-319.

[15] Fenter, J.R. *SAMPE Quarter.* 1971, *2*, 1-15.

[16] Monteverde, F. *Appl Phys. A.* 2006, *82*, 329-337.

[17] Silvestroni, L.; Sciti, D. *Scripta Mater.* 2007, *57*, 165–168.

[18] Silvestroni, L.; Sciti, D. *Adv Mater Sci Eng.* 2011, *10*, 1-11. doi:10.1155/2010/835018.

[19] Buljan, S.T.; Pasto, A.E.; Kim, H.J. *Ceram Bull.* 1989, *68*, 387-394.

[20] Evans, A.G.; Faber, K.T. *J Am Ceram Soc.* 1984, *67*, 255-260.

[21] K.T.; Faber, A.G.; Evans. *Acta Metall.* 1983, *31*, 565-576.

[22] K.T.; Faber, A.G.; Evans. *Acta Metall.* 1983, *31*, 577-584.

[23] Pezzotti, G. *Acta Metall Mater.* 1993, *41*, 1825-1839.

[24] Budiansky, B.; Amazigo, J. C.; Evans, A.G. *J Mech Phys Solids.* 1988, *36*, 167-187.

[25] Becher, P. F. *J Am Ceram Soc.* 1991, *74*, 255-269.

[26] Lange, F.F. *Phil. Mag.* 1970, *22*, 983-992.

[27] Taya, M.; Hayashi, S.; Kobayashi, A.S.; Yoon. H.S. *J Am Ceram Soc.* 1990, 73, 1382-1391.

[28] Tang, S.; Deng, J.; Wang, S.; Liu, W.; Yang, K. Mater Sci Eng A. 2007. 465, 1–7.

[29] Wang, J.; Hu, H.; Zhang, Y.; Wang, Q.; He, X. In *Ceramic Materials and Components for Energy and Environmental Applications*; Ceramic Transactions, 2010. Jiang, D.; Zeng, Y.; Singh, M.; Heinrich, J. Ed.; John Wiley andSons Inc., pp. 467-472.

[30] Zhou, P.; Hu, P.; Zhang, X.; Han W. *Scripta Mater.* 2011, *64*, 276-279.

[31] Yang, F.; Zhang, X.; Han, J.; Du, S. *J Alloy Comp*. 2009, *472*, 395-399.

[32] Tian, W. B.; Kan, Y. M.; Zhang, G. J.; Wang, P. L. *Mater Sci Eng A*. 2008, *487*, 568-573.

[33] Zhang, X.; Xu, L.; Du, S.; Liu, C. Han, J.; Han W. *J Alloy Comp*. 2008, *466*, 241-245.

[34] Zhu, T.; Xu, L.; Zhang, X.; Han, W. Hu, P.; Weng. L. J Europ Ceram Soc. 2009, 29, 2893-2901.

[35] Zhang, X. Xu, L.; Han, W; Weng, L. Han, J.; Du, S. *Solid State Sci*. 2009, *11*, 156-161.

[36] Zhang, X.; Xu L.; Du, S.; Han; J., Hu, P.; Han,W. Mater Lett. 2008, 62, 1058-1060.

[37] Wang, H.; Wang, C.A.; Yao, X.; Fang, D. *J Amer Ceram Soc*. 2007, *90*, 1992-1997.

[38] Chen, D.; Xu, L.; Zhang, X.; Ma, B.; Hu, P. *Int J Refract Metals Hard Mater*. 2009, *27*, 792-795.

[39] Zhang, P.; Hu, P.; Zhang, X.; Han J.; Meng, S. *J Alloy Comp*. 2009, *472*, 358-362.

[40] Zhang, X.H.; Wang, Z.; Hu, P.; Han, W.B.; Hong C.Q. *Scripta Mater*. 2009, *61*, 809-812.

[41] Zou, J; Zhang, G. J.; Kan,Y. M. *J Mater Res*. 2009, *24*, 2428-2434.

[42] Wu, W. W.; Wang, Z.; Zhang, G. J.; Kan, Y. M.; Wang, P. L. *Scripta Mater*. 2009, *61*, 316-319.

[43] Sciti, D.; Silvestroni, L.; Medri, V.; Guicciardi S. submitted to *J Europ Ceram Soc*.

[44] Zhang, X.; Li, W.; Hong, C.; Han, W.; Han, *J. Mater Lett*. 2008, *62*, 2404-2406.styg.

[45] Zhang, X.; Li, W.; Hong, C.; Han, W. *Int. J. Appl. Ceram. Technol*. 2008, *5*, 499–504.

[46] Li, W.; Zhang, X.; Hong, C.; Han, W.; Han, J. *Scripta Mater*. 2009, *60*, 100-103.

[47] Li, W.; Zhang, X.; Hong, C.; Han, W.; Han, J. *J Europ Ceram Soc*. 2009, *29*, 779-786.

[48] Damani, R.; Gstrein, R.; Danzer, R. *J Europ Ceram Soc*. 1996, *16*, 695–702.

[49] Katoh, Y.; Wilson, D.F.; Forsberg, C.W. Technical report ORNL/TM-2007/168, Oak Ridge National Laboratory, September 2007.

[50] Yajima, S.; Hayashi, J. *Nature*, 1976, *261*, 683-685.

[51] Ichikawa, H. *J Ceram Soc Jap*. 2006, *114*, 455-460.

[52] Nakayasu, T.; Sato, M.; Yamamura, T.; Okamura, K.; Katoh, Y.; Kohyama, A. *Ceram Eng Sci Proc*. 1999, *20*, 301-308.

[53] H.M.; Yun, J.A. Di Carlo, *Ceram Eng Sci Proc*. 1999, *20*, 259-272.

[54] Ishikawa, T. *Composite Sci Tech*. 1994, *51*, 135-144.

[55] Takeda, M.; Saeki, A.; Sakamoto, J.; Imai, Y.; Ichikawa, H. *J Amer Ceram Soc*. 2000, *83*, 1063-1069.

[56] Ishikawa, T.; Kohtoku, Y.; Kumagawa, K.; Yamamura, T.; Nagasawa, T. *Nature*, 1998, *391*, 773-775.

[57] Morscher, G.N.; Martines-Fernandez, J. *J Amer Ceram Soc*. 1999, *82*, 145-155.

[58] Karasek, K. R.; Bradley, S. A.; Donner, J. T.; Yeh, H. C.; Schienle, J. L. *J Mater Sci*. 1991, *26*, 103-111.

[59] Tiegs, T. N.; Weaver, S. C. In *Carbide, Nitride and Boride Materials Synthesis and Processing*, Weimer, A. W. Ed. Chapman and Hall, London, UK, 1997, Chap. 16, pp. 411.

[60] Hsueh, C. H.; Becher, P. F. *Mater Sci Eng*. 1996, *A212*, 29–35.

[61] Munz, D. G.; Shannon, J. L. Jr; Bubsey, R. T. *Int J Fract*. 1980, *16*, R137–41.

[62] Berger, M. H.; Hochet, N.; Bunsell, A. R. *J Microsc*. 1995, *177*, 230–241.

[63] Okamoto, H. *Bull Alloy Phase Diagr*. 1990, *11*, 513-519.

[64] Jeng, Y.L.; Lavernia, E.J. *J Mater Sci*. 1994, *29*, 2557-2571.

[65] Silvestroni, L.; Kleebe, H.J.; Lauterbach, S.; Müller, M.; Sciti. D. *J Mater Res.* 2010, *25*, 828-834.

[66] Katrych, S.; Grytsiv, A.; Bondar, A.; Rogl, P.; Velikanova, T.; Bohn, M. *J Alloys Compd.* 2002, *347*, 94-100.

[67] Rouxel, T.; Laurent Y. *Int J Fract.* 1998, *91*. 83-101.

[68] Guo, S. Q.; Kagawa, Y. *Acta Mater.* 1997, *45*, 2257-2270.

[69] Liu, H.; Weisskopf, K.L.; Petzow, G. *J Am Ceram Soc.* 1989, *72*, 559-563.

[70] Becher, P.F.; Hsueh, C.H.; Angelini, P.; Tiegs, T.N. *J Am Ceram Soc.* 1988, *71*, 1050-1061.

[71] Medri, V. submitted to *Adv Eng Mat A*.

[72] Monteverde, F.; Bellosi, A. *Adv Eng Mat A*. 2004, *6*, 331-336.

[73] Silvestroni, L.; Sciti, D. submitted to *J Eur Ceram Soc*.

[74] Samonov, G.V.; Petrikina, R.Y. *Phys Sinter.* 1970, *2*, 1-20.

[75] Opeka, M.M.; Talmy, I.G.; Wuchina, E.J.; Zaykoski, J.A.; Causey, S.J. *J Eur Ceram Soc.* 1999, *19*, 2405-2414.

[76] Ramqvist, L. *Powder Metall.* 1966, *9*, 26-46.

[77] Wuchina, E.J.; Opeka, M.M.; Causey, S.J., Buesking, K.; Spain, J.; Cull, A.; Routbort, J.; Guitierrez-Mora, F. *J Mater Sci.* 2004, *39*, 5939-5949.

[78] Sciti, D.; Silvestroni, L.; Guicciardi, S.; Dalle Fabbriche, D.; Bellosi, A. *J Mater Res.* 2009, *24*, 2056–2065.

[79] Schlesinger, M. E. *J Phase Equil.* 1994, *15*, 90-95.

[80] Roth, R. S.; Waring, J. L. *Phase Diagr Ceram*, 1970, Diagr. 4448.

[81] Dashora, A.; Ahuja, B.L. *Radiat Phys Chem.* 2010, *79*, 1103-1110.

In: Fiber–Reinforced Composites
Editor: Qingzheng Cheng

ISBN: 978-1-61470-303-7
© 2012 Nova Science Publishers, Inc.

Chapter 12

CARPET WASTE FIBERS APPLICATIONS IN CIVIL ENGINEERING

Reza Jamshidi Chenari[*1] *and Hamed Shahbazi Fashtali*[2]

[1]Assistant Professor, Faculty of Engineering, University of Guilan, Rasht, Guilan, Iran
[2]M.Sc Student, Faculty of Engineering, University of Guilan, Rasht, Giulan, Iran

ABSTRACT

Soil reinforced with fiber acts as a composite material and fiber inclusion has a significant effect on soil permeability. With regard to the higher void ratio of carpet fibers, at the first glance it is expected that an increase in the void ratio of the reinforced soil would result in an increase in permeability of the mixture. However, the present chapter indicates that the more the carpet fiber content, the lower the permeability of sand-fiber composite in contrary.

This chapter investigates the effect of fiber inclusion on the permeability and hydraulic properties of coarse grained soils. For this purpose a series of constant head permeability tests have been carried out in order to show the effects.

A new system of phase relationships was introduced to calculate the degree of compaction in sand portion of the composite. It was also tried to maintain a constant relative density so as to isolate the effect of fiber inclusion on the model. It was found that carpet fiber inclusion in particular media will render a reduction of hydraulic conductivity of the reinforced soil and this was shown to be due to the clogging of the porous spaces inherent within the carpet fibers with sand fine grains.

1. INTRODUCTION

As the crisis of municipal waste disposal continued to receive considerable public attention in recent years, some policy makers and environmentalists have advocated the recycling of wastes, particularly non-decaying materials, into useful products as a possible solution. One promising reuse of these materials lies in soil reinforcement as reported in the

---

[*] Author to whom correspondence should be addressed; Email: jamshidi_reza@yahoo.com

literature; Shredded tires (Foose et al.1996 [1], Bosscher et al. 1997 [2], Tatlisoz et al. 1997 [3], Lee et al. 1999 [4]), plastic bottle strips (Khaled and Mashnad et al. 2003 [5], Consoli et al. 2003 [6]), shredded plastic bags (Muntohar, 2009 [7]) and carpet strips and fibers (Wang 1997 [8], Frost et al. 2000 [9], Ghiassian et al. 2004 [10] and Miraftab and Lickfold 2008 [11]) are good examples of the application among all.

The use of natural and synthetic fibers in soil reinforcement can enhance shear strength and volume change characteristics of soil, and improve its load-bearing capacity and durability. (McGown et al. 1987 [12], Gray and Ohashi 1983 [13], Gray and Al-Refeai 1986 [14], Maher and Gray 1990 [15], Al-Refeai 1991 [16], Maher and Ho 1994 [17], Ranjan et al. 1996 [18], Andersland and Khattak 1997 [19], Consoli and Prietto 1998 [20], Michalowski and Cermak 2003 [21], Yetimoglu and Salbas 2003 [22]).

Dynamic characteristics of fiber reinforced soils have also been studied to some extent, indicating that some major improvement can be developed due to the influence of fiber inclusions (Noorani and Ozdavinez 1989 [23], Maher and Woods 1990 [24], Shewbridge and Sousa 1991 [25], Feng and Sutter 2000 [26], Li and Ding 2002 [27], Boominathan and Hari 2002 [28]).

In line with previous studies on the utilization of carpet waste in soil reinforcement, and in order to advance the knowledge about the dynamic behavior of fiber reinforced soils, some research works were carried out to investigate the response of these soils under lateral and vertical cyclic excitations. In the first part of this three-part study, the influence of carpet strip parameters (i.e. aspect ratio and fiber content) were examined on the simulated earthquake response of retaining sheet pile wall models that were backfilled with fine reinforced sand and tested in shaking table apparatus (Jamshidi et al. 2010 [29]). Large cyclic triaxial test apparatus was also utilized to show the effects more clearly (Ghiassian et al., 2008 [30]). The result indicated that some important improvements were achieved in the load and deformation characteristics of the models under cyclic and harmonic sinusoidal horizontal and vertical motions. The second part of this study was attempted to further scrutinize these improvements in small cyclic triaxial testing on saturated fine sand specimens reinforced with carpet strips. Findings of this second part about dynamic properties such as shear modulus and damping ratio have been presented elsewhere (Shahnazari et al. 2009 [31], Ghiassian et al. 2010 [32]). The materials presented in this chapter cover the results regarding the characterization of hydraulic properties of the reinforced samples. For this aim, constant head permeability test scheme was adopted to investigate the effect of fiber inclusion on hydraulic conductivity of sand with different relative densities. Parameters under study consist of fiber weight fraction, fiber aspect ratio, skeletal relative stiffness of sand and the effective size of sand particles.

In order to isolate the coupled effect of soil skeletal density and carpet fiber inclusion on permeability of reinforced sand, a new phase relationship has been introduced for sand-fiber mixture taking different components into account. For isolation of the effect of carpet fiber inclusion and density of sand portion it is needed to keep the skeletal relative density of sand constant when using different fiber contents.

Carpet and geotextile fibers are porous materials which are subject to void ratio reduction depending on the level of effective stress. Furthermore, fine grained sand particles will be entrapped within entangled hairline nature of the porous fibers. The new phase relationship will allow for the sand absorption adoption, and the fiber void ratio reduction adjustment. The permeability and hydraulic conductivity of plain fibers which is referred to zero sand entrapment is at least one order of magnitude larger than that of plain sand. However fibers mixed with sands especially fine grained particles show a dramatic drop in hydraulic conductivity which generally becomes even less permeable than plain sand. The outcome of this effect is that the equivalent permeability of sand-fiber mixtures degrades with any increase in the fiber weight fraction.

2. MATERIALS

2.1. Soil

The soil samples used in this study were obtained from the region of Anzali located in the north of Iran. The soil is fine sand classified as poorly graded sand (*SP*) according to Unified Soil Classification System (USCS) with the specific gravity (G_s) of 2.65 (ASTM D854). The grain size distribution is 96% sand and 4% silt size material, with a Uniformity Coefficient (C_u) of 1.44 and Coefficient of Curvature (C_c) of 0.96. The soil has plasticity Index (I_p) of about zero and is essentially non-plastic. The gradation curve for the sand used in this study is provided in Figure 1.

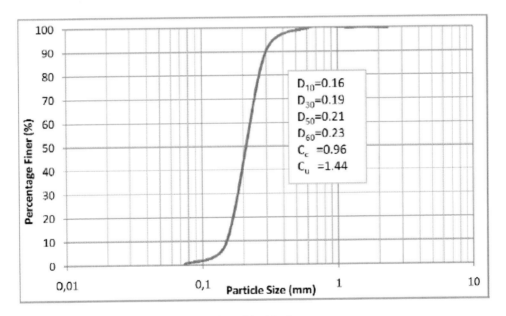

Figure 1. Grain size distribution of sandy soil used in this chapter.

2.2. Reinforcement

The reinforcing fibrous strips were prepared by cutting 5mm thick carpet waste fabrics into short elements with 5×5 mm^2 square cross section and lengths of 15, 25, 35 and 45*mm* corresponding to the aspect ratios (A_R, length/width) of 3, 5, 7 and 9 respectively. The specific gravity of strips (G_{sf}) is 1 as reported by the carpet company. Tensile strength and elastic modulus are 800 *kPa* and 730 *kPa* respectively, with an almost linear strain at failure of 105% (ASTM-D-4595). Various amounts of strips were added to the soil to get strip content or weight fraction (w_f), defined as the weight of strips to the dry weight of sand, as 0%, 0.25%, 0.5%, 0.75% and 1% respectively. The latter is an upper limit for compacting the sand mix into the mold (Poorebrahim 2004 [33]).

3. SAMPLE PREPARATION

The specimens were separated by mixing soil and water, with the addition of appropriate amounts of strip to achieve specified fiber contents. Samples were compacted in 7 layers into a 100 *mm* (4 inch) diameter and 220 *mm* (8.8 inch) high cylindrical mold, with a target skeletal relative density of 50%. This density was calculated for the sand part only based on the new phase relationships and the weight of fibers was not considered. Calculations provided later on this chapter will show that the difference when the fibers are also included in calculation would actually be crucial and nontrivial.

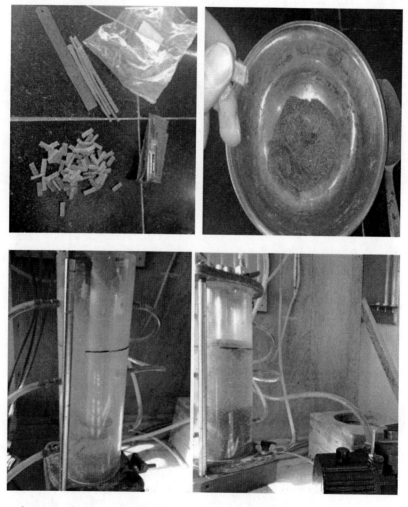

Figure 2. Sample preparation procedure, a) fiber preparation, b) fiber and sand mixing, c) molding by dry tamping and d) saturation.

The dry tamping method was carried out as part of sample preparation. The amount of sand for each layer was first weighted and mixed thoroughly with the required sand portion prior to compaction into the mold to the predetermined height.

After molding and compaction of the fiber-sand mixture into the desired compaction level, vacuum pressure was applied to the top of the mold for 5 minutes, so as to make sure all the entrapped air bubbles inside the mixture were extracted. Then deaired and distilled water was soaked into the sample from the bottom valve provided to the test apparatus while maintaining the vacuum facility at the top of the sample during the saturation procedure.

4. PHASE RELATIONSHIPS

Sand skeletal density and fiber inclusion have coupled effects on mechanical and hydraulic properties of reinforced soil. In order to isolate the mix effects the relative density of free sand portion should be maintained constant while changing the fiber contents. Noorany (1989) [23] proposed a simple method to do so, by estimating the solid volume of fibers and subtracting it from the total mold volume so as to obtain the net volume occupied by sand particles.

The above mentioned simple calculation would make it possible to mix proper amounts of sand particles to maintain constant relative density; however the aforesaid procedure is applicable solely for fibers which do not experience compression and squeezing during mixing or application of the compaction efforts. Hairline fiber used by Noorany (1989) [23] meets the above criterion, however it is not the case for the porous carpet fibers which are subject to void ratio changes due to the applied stress.

Volume change characteristics of the carpet fibers on the one hand and sand particle entrapment within porous spaces on the other hand make the skeletal density estimation in reinforced sand complicated and approximated.

The sensitivity analysis employing new phase definitions and diagram provided in Figure 3 paves the way toward a clear understanding of the mentioned effect and the way it is encompassed by different affecting parameters. Hence, the sand dry solid weight was assumed to be partially entrapped within entangled nature of porous fibers and this portion, W_{ss2} was found to be a factor of the weight of fibers.

Different values of zero, 1% and 2% sand entrapment factors, α, were adopted to see the impact on weight calculations. It was experimentally found that fibers thoroughly mixed with sand particles used in this study are almost doubled in terms of weight. In this study, a value of 1% was selected for α; however it is expected to vary depending on the sand effective and average size.

Another terminology introduced in this study is the fiber void ratio, e_f which is subject to variation due to the stress level changes imposed during compaction.

Figure 4 shows the experimental results of void ratio variation with stress level changes for the fiber used in this chapter. It is evident that the zero-stress void ratio for fiber under study is near 5 reflecting high porosity of the kind of fiber; however it shows dramatic reduction with stress level increasing from zero to 4 kPa.

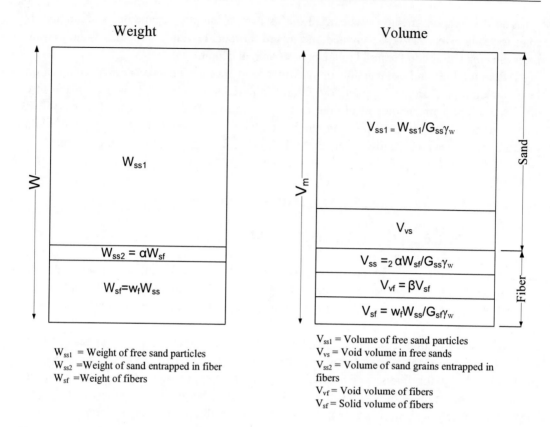

Figure 3. Phase diagram for reinforced sand.

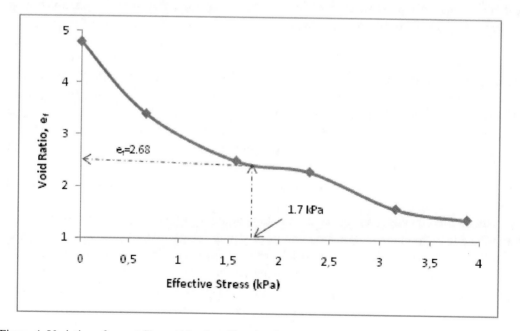

Figure 4. Variation of carpet fiber void ratio with stress level.

With regard to the changes in stress level through the height of the test apparatus, 22 cm, the relative density of 50% for sand portion was employed to calculate the average stress level in constant head permeability test apparatus. This average value was found to be 1.7 kPa which brings the final void ratio of the fibers to 2.68 in average sense. This is indicated in Figure 4.

In the following, the effect of skeletal density changes due to the assumption that a constant amount of sand to is used throughout the test, irrespective to the fiber content values, sand entrapment and the stress-induced void ratio reduction will be shown.

The skeletal void ratio for plain (unreinforced) model will be evaluated in Equation 1.

$$e_{unreinforced} = \frac{V_{vs}}{V_{ss}} = \frac{V_m}{V_{ss}} - 1 = \frac{V_m}{W_{ss}} \times G_{ss}\gamma_w - 1 \qquad (1)$$

where v_m is the volume of mold denote in Figure 3.

The corresponding void ratio in reinforced model is however calculated by adopting different terms and definition, illustrated in Figure 3. Equation 2 provides the basic definition for the above mentioned void ratio.

$$e_{reinforced} = \frac{V_{vs}}{V_{ss1}} = \frac{V_m - (V_{ss1}+V_{ss2}+V_{vf}+V_{sf})}{V_{ss1}} = \frac{V_m - (V_{ss}+V_{vf}+V_{sf})}{V_{ss1}} \qquad (2)$$

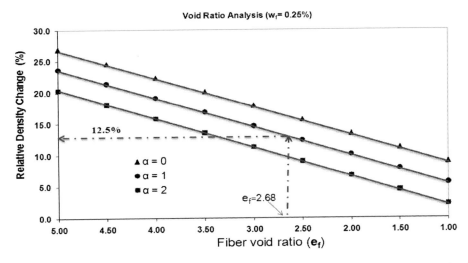

Figure 5. Effect of different reinforcement parameters on relative density of sand, $w_f = 0.25\%$, $G_{sf}=1$ and $G_{ss}=2.65$.

Manipulation of Equation 2 by substitution of the respective weight-volume relationship and definitions of basic phase parameters leads to the following equation for the skeletal void ratio (Equation 3).

$$e_{reinforced} = \frac{V_m - (\frac{W_{ss}}{G_{ss}\gamma_w} + \frac{(1+e_f)w_f W_{ss}}{G_{sf}\gamma_w})}{W_{ss}(\frac{1}{G_{ss}\gamma_w} - \frac{\alpha w_f}{G_{sf}\gamma_w})} \qquad (3)$$

where:
α = Sand entrapment factor

e_f = Fiber void ratio
G_{ss} = Specific gravity of sand grains
G_{sf} = Specific gravity of fiber solids.

The other parameters used in Equation (3) have been previously defined in Figure 3.

Adopting Equations 1 and 3 for the void ratio of plain and reinforced models respectively, it is then possible to calculate the relative density for both methods.

For reinforced model, adopting an amount of sand grain calculated for plain sand in relative density of 50%, will lead to a change in skeletal relative density which is drawn in Figures 5 to 8.

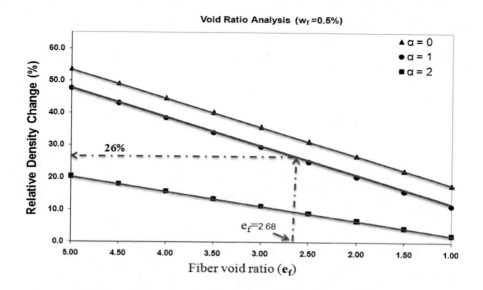

Figure 6. Effect of different reinforcement parameters on relative density of sand, w_f = 0.5%, G_{sf} = 1 and G_{ss} = 2.65.

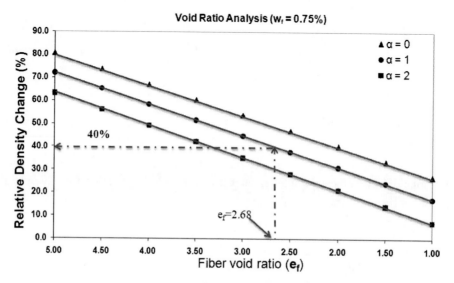

Figure 7. Effect of different reinforcement parameters on relative density of sand, w_f = 0.75%, G_{sf} = 1 and G_{ss} = 2.65.

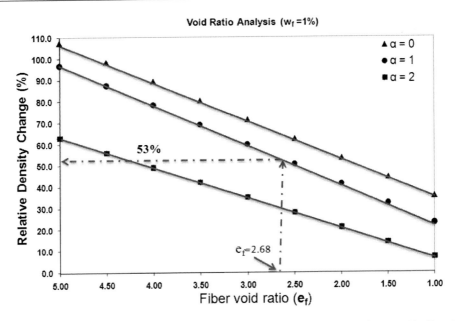

Figure 8. Effect of different reinforcement parameters on relative density of sand, $w_f = 1\%$, $G_{sf} = 1$ and $G_{ss} = 2.65$.

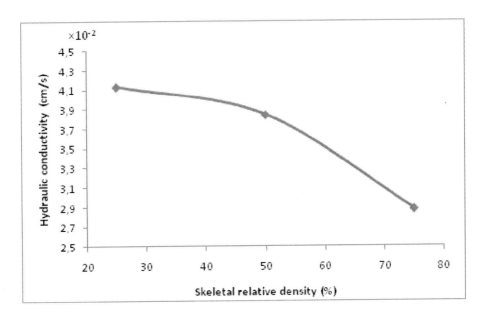

Figure 9. Variation of hydraulic conductivity of sand with the relative density.

It is concluded from the aforesaid figures that introducing reinforcement fibers into the sand model will cause a notable increase in skeletal relative density depending on the fiber content. Generally, it can be seen that fiber inclusion would increase the skeletal relative density when the overall solid sand weight is unchanged. For 0.25% to 1% fiber weight content it is observed that skeletal relative density will increase 12.5% to 52% assuming

e_f=2.68 which is expected to have significant impact on mechanical and also hydraulic properties of mixture.

The effect of skeletal relative density on the hydraulic behavior of sand is confirmed by noting to the hydraulic conductivity of plain sand models with different relative densities. Figure 9 illustrate how the permeability of sand changes with the degree of compaction or the relative density in other words. This will highlight the importance and the role of skeletal relative density and thus it becomes crucially important to isolate the effect of the relative density and the fiber content.

5. WEIGHT ADJUSTMENT

Due to the calculated inherent error related to the skeletal relative increase, it seems that instead of assuming a constant solid weight for dry sand between different tests, the efforts should concentrate on maintaining a constant skeletal relative density in different experiments. For this purpose, according to the new phase relationships, the solid weight for sand grains was calculated according to Equation 4.

$$W_{ss} = \frac{V_m}{[e_{50\%}\left(\frac{1}{G_{ss}\gamma_w} - \frac{\alpha w_f}{G_{sf}\gamma_w}\right) + \left(\frac{1}{G_{ss}\gamma_w} + \frac{(1+e_f)w_f}{G_{sf}\gamma_w}\right)]} \tag{4}$$

where:

$$e_{50\%} = e_{max} - D_r(e_{max} - e_{min}) \tag{5}$$

and other parameters are listed in Table 1.

The above equation will render the sand solid weight for different fiber contents assuming to maintain the skeletal relative density at 50% and sand entrapment factor 1, which is confirmed by weighing porous carpet fiber after mixing thoroughly with sand grains.

Preparation of laboratory samples was carried out by dry tamping in seven layers where effective stress will be different between layers. Therefore there should be different weight adjustment, depending on the layer number. To accumulate this and for clarification of the issue, two different procedure were adopted.

Table 1. Input parameters for new phase relationships

Relative density	$D_r = 50\%$
Maximum void ratio	$e_{max} = 0.815$
Minimum void ratio	$e_{min} = 0.667$
Specific gravity of sand grain	$G_{ss} = 2.65$
Specific gravity of fiber solids	$G_{sf} = 1$
Sand entrapment factor	$\alpha = 1$
Fiber void ratio	$e_f = 2.68$
Mold volume	$V_m = 1400\ cm^3$
Water unit weight	$\gamma_w = 1\ gr/cm^3$

In the first procedure, the fiber void ratio, e_f in different layers was calculated from Figure 4 according to the effective stress level at that specific layer and then the dry weight of sand for each layer was calculated according to the estimated fiber void ratios and the total volume of the layers which were presumed equal.

In the second method, an average fiber void ratio was adopted for the whole layers accounting for average effective stress level and the total solid weight of dry sand was calculated according to the average representative fiber void ratio and the mold volume, adopting Equation 4.

Table 2 presents the results of total sand solid weight calculation, for different fiber content and two procedures.

Table 2. Comparison of two different calculation procedures for solid weight of sand

Fiber Content (%)	Layered Calculation (1st procedure)	Average Sense (2nd procedure)
0.25	2109.8	2109.7
0.5	2087.1	2087
0.75	2064.9	2064
1	2043.2	2042

Looking at Table 2 reveals that the difference in weight estimations between two different procedures is insignificant and always less than 1%. Therefore, the weight calculation was pursued in average sense and the calculated solid weight of dry sand was divided into seven portions and mixed with the respective fiber contents and molded sequentially.

6. CONSTANT HEAD PERMEABILITY TESTS

Permeability (or hydraulic conductivity) refers to the case with which water can flow through a soil. This property is necessary for the calculation of seepage through earth dams or under sheet pile walls, the calculation of the seepage rate from waste storage facilities (landfills, ponds, etc), and the calculation of the rate of settlement of clayey soil deposit.

There are two general types of permeability test methods that are routinely performed in the laboratory: (1) the constant head test method, and (2) the falling head test method. The constant head test method is used for permeable soil ($k > 10^{-4}$ cm/s) and the falling head method is mainly used for less permeable soil ($k < 10^{-4}$ cm/s).

Weight adjustment scheme presented earlier highlights a clearance to be made between two distinctive effects, namely skeletal relative density and the porous fiber inclusion on the hydraulic conductivity of fiber-reinforced sand. This chapter has only considered the effect of carpet fiber inclusion on permeability of reinforced sand by maintaining the skeletal relative density at 50%. For this reason laboratory permeability test apparatus was adopted.

As the permeability of plain sand was measured in the order of 10^{-2}, it was decided to use constant head permeability test.

The purpose of this test is to determine the permeability (hydraulic conductivity) of the reinforced soil by constant head test method.

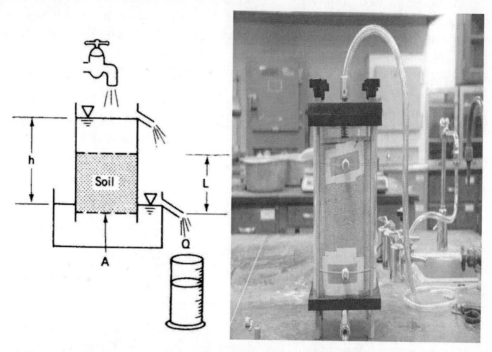

Figure 10. Constant head permeability test, a) working principle, b) assembled permeameter.

ASTM D2434 provides the standard test method for permeability of granular soils. Figure (10-a) schematically illustrates the working principle of constant head test and Figure (10-b) shows the assembled constant head permeameter.

A device called a permeaeter is used in the laboratory (Figure 10(a) and (b)) and the volume of water Q collected in time is:

$$Q = A\upsilon t \quad (6)$$

where υ is the discharge velocity, calculated from Darcy's law and provided in Equation 7.

$$\upsilon = Ki = K\,h/L \quad (7)$$

where K is the hydraulic gradient of the porous medium calculted from Equation 8.

$$K = \frac{QL}{Aht} \quad (8)$$

where:
Q = total discharge volume, m^3, in time t, s and
A = cross-sectional area of soil sample, m^2

7. EXPERIMENTAL RESULTS

Reinforced sand models of different fiber contents and constant skeletal relative density of 50% were tested upon constant head permeability test apparatus as explained before. Carpet fibers of different weight fractions varying from 0.2 to 1% were added to the uniformly graded sand of average size, $D_{50}=0.21$, effective size, $D_{10}=0.16$ and mixed thoroughly. It was then molded in a layered structure as explained earlier. Another investigated parameter under study is the fiber aspect ratio, A_R which was defined as the length to width ratio. Different aspect ratios of 3, 5, 7 and 9 were adopted at each weight fraction.

Figure 11 illustrate how the hydraulic conductivity (permeability) of reinforced sand is affected by the fiber content. It was unexpectedly observed that the equivalent permeability of the reinforced sand decreases as the fiber weight content increases. This is in contrary to the normal expectation as the hydraulic conductivity of plain sand was measured to be 3.84×10^{-2} cm/s while it was measured to be 2.72×10^{-1} cm/s for pure fibers model.

The extraordinary hydraulic behavior of reinforced sand can be explained by the fact that fine grained sands are entered into the porous spaces within fibers and entrapped there. This would lead to a significant reduction in hydraulic conductivity of the fiber which itself induces a decrease in hydraulic conductivity of composite. Figure 12 shows the reduction level in hydraulic conductivity (permeability) of reinforced sand with different fiber contents. Based on the experimental studies it was found evident that the permeability reduction rate increases by increasing the fiber weight fraction. To be more specific, for reinforced sand with 0.2% carpet fiber content, the reduction is 5-17.5% and it shows a dramatic change when the fiber content is 1% (32-48%).

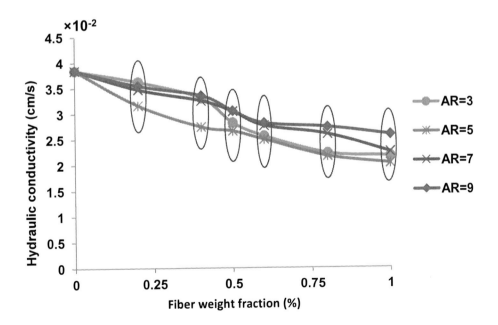

Figure 11. Hydraulic conductivity of reinforced sand.

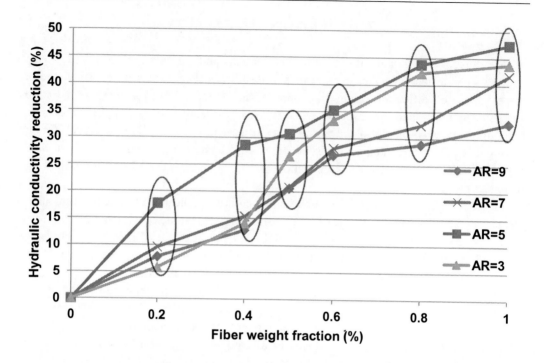

Figure 12. Reduction in hydraulic conductivity of reinforced sand.

Table 3 provides the range which the hydraulic conductivity of reinforced sand decreases depending on the fiber weight fraction. It is observed that max 48% reduction in permeability of sand-fiber composite, w_f, may be expected due to fiber inclusion.

Fiber aspect ratio, A_R is another affecting parameter which was proven to influence the hydraulic conductivity (permeability) of sand-fiber composite. The effect was isolated and depicted in Figure 13 for different fiber contents. It is shown that, there is an optimum value for the fiber aspect ratio when the inclusion shows the least hydraulic conductivity (permeability). The optimum aspect ratio for different fiber weight fraction appears to remain between 3 and 5 depending on the fiber content.

The presence of optimum aspect ratio is explained by soliciting to the fact that by increasing the fiber aspect ratio, the sand entrapment potential of a single fiber increases. Through this phenomenon the overall fiber weight fraction remains constant; however the number of less pervious fibers decreases in composite.

Table3. Ranges of reduction in permeability of reinforced sand

Fiber Content (%)	Reduction Range (%)
0	0
0.2	5-17.5
0.4	12.5-28
0.5	20-31
0.6	27-35
0.8	28-44
1	33-48

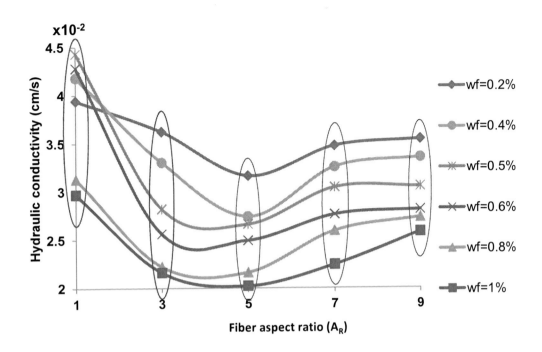

Figure 13. Variation of hydraulic conductivity with fiber aspect ratio.

This means that the hydraulic behavior of fiber reinforced sand with specific weight fraction is the mutual and interactive effect of the sand entrapment potential and the number of fibers in composite and it should be noted that the sand entrapment potential decreases the permeability while the impact of the number of fibers in composite is vice versa.

CONCLUSION

This chapter investigated the effect of carpet fiber inclusion on hydraulic conductivity (permeability) of clean fine sands. For this reason, the coupled effects of skeletal relative density and porous fiber inclusion were first pointed out and a new system of phase relationships was introduced. Employing the new phase diagram enabled isolation of the aforesaid effects and dry sand solid weight was then calculated to focus on the effect of fiber content on hydraulic properties of reinforced sand while maintaining the skeletal relative density fixed at 50%.

Constant head permeability test was conducted to investigate the effect of different reinforcement parameters namely fiber content and aspect ratio on permeability of the sand-fiber composite.

Following results were obtained:

1) Due to the fine sand particles entrapment within porous fibers, the composite permeability decreases despite the fact that plain porous fibers are even more permeable than plain sand.

2) There is an optimum aspect ratio of 3 to 5, which renders the least hydraulic conductivity for different fiber contents.

This result is explained by emphasizing the dependence of hydraulic properties of composite on both the number of less permeable elements and also the sand entrapment potential for single fibers which vary in reverse order with fiber aspect ratio.

Further investigation may be needed to accommodate the effects of sand grains effective size, skeletal relative density and overburden pressure. Current chapter has only considered one specific size of sand which is conceived to be uniformly graded fine sand. However, sand entrapment potential for porous fibers will largely depend on the sand grains effective size, D_{10}. Effective size controls the hydraulic properties of skeletal sand. On the other hand it determines the entrapment potential for porous fibers.

Overburden pressure in practice is far larger than assumed in this chapter; expected to decrease the void ratio of fibers which which it would also induce more reduction in permeability of composite .

REFERENCES

[1] Foose, G.J., Benson, C.H. and Bosscher, P.J. 1996. Sand reinforced with shredded waste tires. *J. of Geot. Eng.* 122 (9): 760-767.

[2] Bosscher, P. J., Edill, T. B. and Kuraoka, S. 1997. Design of highway embankments using tire chips. *Journal of Geotechnical and Geoenvironmental Engineering, ASCE, Vol. 123,* No. 4, pp. 295–304.

[3] Tatlisoz, N., Benson, C., and Edil, T., 1997. Effect of fines on mechanical properties of soil tire chip mixtures. *Testing Soil Mixed with Waste or Recycled Materials*, ASTM STP 1275, ISBN 0-8031-2470-8, pp. 93–108.

[4] Lee, J.H., Salgado, R., Bernal, A., Lovell, C.W., 1999. Shredded tires and rubber-sand as lightweight backfill. *ASCE Journal of Geotechnical and Geoenvironmental Engineering* 125,132–141.

[5] Sobhan K. and Mashnad M. 2003, Mechanical stabilization of cement soil-fly ash mixtures with recycled plastic strips. *Journal of Environmental Engineering*, ASCE, Vol.129 No. 10, *pp.* 943-947.

[6] Consoli, N., Casagrande, M. D., Prietto, D. M. and Thome, A. 2003. Plate load test on fiber reinforced soil. *Journal of Geotechnical and Geoenvironmental Engineering*, ASCE, Vol.129, No. 10, pp. 951–955.

[7] Muntohar, A. S, 2009. Influence of plastic waste fibers on the strength of lime-rice husk ash stabilized clay soil.

[8] Wang, Y. 1997. Carpet waste for soil stabilization. Proceedings of Second Conference on recycling of fibrous textile and Carpet Waste, Atlanta, GA.

[9] Murray, J.J., Frost, J.D. and Wang, Y. 2000. Behaviour of a sandy silt reinforced with discontinuous recycled fiber Inclusions, *Transportation Research Record*, 1714, 9–17.

[10] Ghiassian, H., Poorebrahim, G. R., and Gray D. H. 2004. Soil reinforcement with recycled carpet wastes. *Journal of Waste Management and Research (WMR), International Solid Waste Association* (ISWA), 22(2), 108-114.

[11] Miraftab, M. and Lickfold, A. 2008 Utilization of carpet waste in reinforcement of substandard soils. *Journal of industrial textiles*, 38, 2, 167-174.

[12] McGownA, Andrawes KZ, Al-HasaniMM 1987. Effect of inclusion properties on the behavior of sand. *Geotechnique;* 28 (No.3):327–46.13.

[13] Gray, D. H. and Ohashi, H. 1983. Mechanics of fiber reinforced in sand. *Journal of Geotechnical and Geoenvironmental Engineering*, ASCE, Vol. 109, No. 3, pp. 335-353.

[14] Gray, D. and Rafeai, A. 1986. Behavior of fabric-versus fiber reinforced sand. *Journal of Geotechnical and Geoenvironmental Engineering*, ASCE, Vol. 12, No. 8, pp. 804-820.

[15] Maher, M. H. and Gray, D. H. 1990. Static response of sand reinforced with distributedfibers. *Journal of Geotechnical Engineering*, ASCE, Vol. 116, No. 11, pp. 1661–1677.

[16] AlRefeai 1991.TO.Behavior of granular soils reinforced with discrete randomly oriented inclusions. *Geotextile and Geomembranes*;10(No.3):319–33.

[17] Maher, M. H, and Ho, Y. 1994. Mechanical properties of Kaolinite/fiber soil composition . *Journal of Geotechnical Engineering* (ASCE), Vol. 120, No. 8. pp. 1380 - 1393.

[18] Ranjan, G., Vassan, R. M. and Charan, H. D. 1996. Probabilistic analysis of randomly distributed fiber reinforced soil. *Journal of Geotechnical and Geoenvironmental Engineering*, ASCE, Vol. 122, No. 6, pp. 419-426.

[19] Andersland, O. B., and Khattak, A. S. 1997. Shear strength of kaolinite/fiber soil mixtures. *Proc. Int. Conf. on Soil Reinforcement*, Paris, France.

[20] Consoli, N., and Prietto, D. M. 1998. Influence of fiber and cement addition in sandy soils. *Journal of Geotechnical Engineering* (ASCE), Vol. 116, No -, pp 805-821.

[21] Michalowski, R. L. and Cermak, J. 2003. Triaxial compression of sand reinforced withfibers. *Journal of Geotechnical and Geoenvironmental Engineering*, ASCE, Vol. 129, No.2,pp. 125–136.21.

[22] Yetimoglu, T. and , O. 2003. A study on shear strength of sands reinforced with randomly distributed discrete fibers. *Journal of Geotextiles and Geomembranes*, Vol. 21, No.21, pp. 103–110.

[23] Noorany, I. and Uzdavines, M. 1989. Dynamic behavior of saturated sand with geosynthetic fibers. Geosynthetics Conference, Vol. 2, pp. 385–396, San Diego, USA.

[24] Maher, M. H. and Woods, R. D. 1990. Dynamic response of sand reinforced with randomly distributed fibers. *Journal of Geotechnical Engineering*, ASCE, Vol. 116, No. 7, pp. 1116–1131.

[25] Shewbridge, S. E. and J. Sousa, 1991, Dynamic Properties of a Reinforced Sand, *Jour. Of Geotechnical Eng.*, ASCE, Vol.117, No. 9, pp.1402-1422.

[26] Feng, Z. Y. and Sutter, K. G. 2000. Dynamic properties of granulated rubber sand mixtures. *Geotechnical Testing Journal*, GTJODJ, VOL. 23, No. 3, pp. 338-344.

[27] Li, J. and Ding, D. W. 2002. Nonlinear elastic behavior of fiber reinforced soil under cyclic loading. *Journal of Soil Dynamics and Earthquake Engineering*, Vol. 22, No. 22, pp. 977-983.

[28] Boominathan, A. and Hari, S. 2002. Liquefaction strength of fly ash reinforced with randomly distributed fibers. *Journal of Soil Dynamics and Earthquake Engineering*, Vol. 22, No.22,pp. 1027-1033.

[29] Jamshidi, C. R., Towhata, I., Ghiassian, H., and, Tabarsa, A. 2010. Experimental evaluation of dynamic deformation characteristics of sheet pile retaining walls with fiber reinforcement backfill. *Journal of Soil Dynamics and Earthquake Engineering*. Volume 30, Issue 6, June, Pages 438-446.

[30] Ghiassian, H., Jamshidi C, R., and Tabarsa, A. R. 2008. Dynamic Performance of Toyoura Sand Reinforced with Randomly Distributed Carpet Waste Strips. 4th Decennial 20 Geotechnical Earthquake Engineering and Soil Dynamics Conference: Sacramento, California, USA.

[31] Shahnazari, H., Ghiassian, H., Noorzad, A., Shafiee, A., Tabarsa, A.R., and Jamshidi, R. 2009. Shear modulus of silty sand reinforced by carpet waste strips. *Journal of Seismology and Earthquake Engineering*, Tehran, Iran, Volume 11, No. 3.

[32] Ghiassian, H., Shahnazari, H., Tabarsa, A., and Jamshidi, C. R. (2010). "Damping ratio of silty sand reinforced with carpet fibers" *Amirkabir Journal of Science and Technology*, Tehran, Iran, Vol. 42, No. 1.

[33] Poorebrahim G. 2003. Soil reinforcement with carpet wastes. PhD dissertation, Iran University of Science and Technology, Tehran, Iran.

In: Fiber–Reinforced Composites
Editor: Qingzheng Cheng

ISBN: 978-1-61470-303-7
© 2012 Nova Science Publishers, Inc.

Chapter 13

MACHINING OF FIBRE-REINFORCED COMPOSITES

Luís Miguel Durão

Dept. Engenharia Mecânica ISEPGab. F355. Email: lmd@eu.ipp.pt

ABSTRACT

Although composites components are produced to near-net shape, finishing operations like drilling, in order to fulfil requirements related with tolerances or assembly needs, are usually required. It is known that machining operations in composites can be carried out with conventional tools and machinery with proper adaptations. However, due to their inhomogeneity, this operation can lead to different damages, like delamination, intralaminar cracks, fibre pull out or thermal damage. The most frequent and noticeable evidence of these damages is the existence of a border around the machined hole in consequence of the drilling process. From these damages, delamination is considered the most serious as it can affect the mechanical properties of the composites parts, hence, lower reliability.

Production of higher quality holes, with dimensional precision, acceptable surface quality and damage minimization, is a challenge to everyone related with composites industry. However, the knowledge on this subject is yet limited. So, it means that a better understanding of the damage mechanisms is needed until composites machining reach the same level of confidence that can be found about metal machining. Another difficulty in getting that level of knowledge may be related with the composites nature itself, as their inhomogeneity results in properties dependence on fibre and matrix materials, fibre volume fraction, ply lay-up sequence, orientation and other factors. Hence, conclusions that are valid for a certain type of fibre–reinforced composite with some fibre orientation may not be valid if changed from unidirectional to cross-ply or other, just to quote an example.

The purpose of this chapter is to be organised as follows. The first part is dedicated to a review of the principles of conventional composites machining, including orthogonal cut and fibre orientation influence. It will include a state-of-the-art of recent advances on dedicated tools and equipment. Materials for tools will also be evaluated. In the end of this part, a brief review of non-conventional machining will be presented. The second part will be dedicated to machining related damage, focusing on delamination. The concept of critical thrust force for delamination onset and existent analytical models for its determination will be presented. As delamination involves fracture related mechanisms, fracture modes will be reviewed and relevance of interlaminar fracture toughness discussed. The third part will deal with damage evaluation criteria. Different numerical ratios are normally referred to when damage extension is to be compared. Those criteria are based on the availability of non-destructive inspection methods, from tool maker's microscope to computerized tomography. These and other techniques will be presented. Results of damage extension will be related with main drilling parameters - cutting speed or feed rate. Finally, a brief reference to the mechanical tests normally used for damage assessment, like bearing test, and their correlation with damage extension will be the object of the fourth part.

2. OVERVIEW OF MATERIAL REMOVAL PROCESSES

Composites machining began with the use of machines, tools and processes already existent and well established in the machining of metals and metallic alloys. In the last decades, due to the rising interest in composites, research on its machining has increased, with the purpose of optimizing processes developed to metal machining.

Machining is a manufacturing process in which a cutting tool is used to remove excess material from a workpiece. This process is classified as orthogonal or bi-dimensional, or as oblique or tri-dimensional. The latter is a more generic model that considers an inclination angle not equal to 90° between the cutting edge and the direction of cutting speed. This geometric model, closer to real machining operation, is somewhat complex. In orthogonal cutting model, the cutting edge is perpendicular to the direction of the cutting speed. This enables the process to be represented in a bi-dimensional way, which reduces geometric complexity and allows for simple geometric relations, force and speed vectorial analysis (Figure 1). The orthogonal cutting model, by its simplicity, has been used for the development of theoretical and practical work.

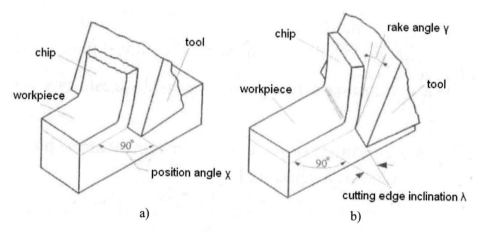

Figure 1. Orthogonal –a) -and oblique –b) - cutting models.

Drilling is usually performed with a rotating cylindrical tool that has two cutting edges on its working end. The tool is called a *drill* or *drill bit*. The rotating drill feeds, parallel to machine axis, into the stationary work part to form a hole whose diameter is determined by the drill diameter. Vertical movement rate is called feed f and is specified in mm/rev. Feed can be converted to feed rate, f_r, expressed in mm/min, using equation

$$f_r = f \times N \qquad (1).$$

Rotational speed of the drill N is given in rpm (revolutions per minute). Knowing this speed and the drill diameter D, it is possible to determine the cutting speed v_c, which is the surface speed at the outside diameter of the drill in m/s by equation

$$v_c = \frac{\pi * D * N}{1000} \qquad (2).$$

It should be remembered that in drilling, as the centre of the drill is approached, cutting speed reduces linearly until zero. In drilling, depth of cut is equal to drill radius.

The standard twist drill geometry is illustrated in Figure 2. The body of the drill has two spiral flutes, whose angle is called the helix angle, typically around 30° or 35°. One of its functions is to evacuate the chips. The web provides support for the drill over its length. The point of the drill has the general shape of a cone, which most common design is the *chisel edge*, as the one in figure 2 and a typical value of point angle is 118°. This angle has a decisive influence in the way drill penetrates into the material until maximum diameter is reached. Connected to the chisel edge are two cutting edges – lips – that lead into the flutes. The portion of each flute adjacent to the cutting edge acts as the rake face of the drill.

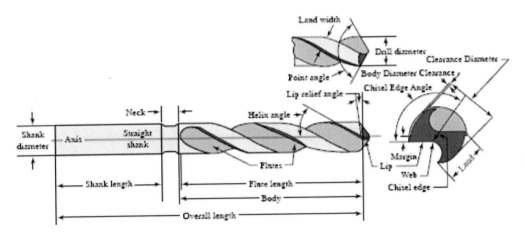

Figure 2. Standard geometry of a drill.

The main tool system angles of a twist drill – rake, clearance and edge angles – are shown in Figure 3. During work an increase in feed causes a variation in tool angles, increasing rake angle and diminishing clearance angle.

There are two forces to be considered during drilling: thrust force and torque. Thrust force is the necessary force to maintain a given feed rate into the material. Torque is the amount of tangential force necessary to maintain drill rotational speed. These two forces are affected by cutting parameters like speed, feed, drill geometry or work piece material.

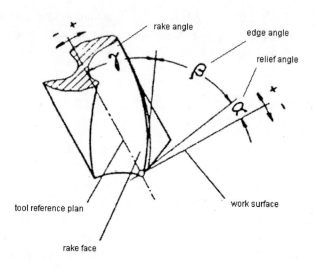

Figure 3. Tool angles in a twist drill.

From the torque value M_t, it is possible to derive the specific cutting pressure by equation

$$k_c = \frac{4000 \times M_t}{\pi \times f_n \times D^2} \qquad (3).$$

When feed increases, the chip thickness also increases, the specific cutting pressure is lower and tangential force and chip section are greater, causing a torque increase. Chip thickness can also be increased by the use of greater point angles, reducing torque and specific cutting pressure.

The cutting action of a twist drill is complex. As the cutting speed varies as function of the distance from its axis, so does the cutting efficiency, being less efficient in the centre. Chip removal mechanism is similar to oblique cutting, but more complicated as the cutting speed and rake angle change continuously. The relative velocity of the drill point is zero and so no cutting takes place, except on the external edges. The chisel edge of the drill pushes aside the material as it penetrates into the hole, requiring a large thrust force to drive the twist drill forward [1]. On the other side, contribution to torque is very small. When drilling, chisel edge action is more similar to an extrusion.

In summary, it is possible to say that drilling is a very complex process that depends on a large diversity of factors.

2.1. Composites Machining

Although their physical properties make composites an attractive group of materials, machining poses a certain number of specific issues, when compared with metallic materials. Generally, machined parts have poor surface appearance and tool wear is higher. One of the problems of machining composites is related with the fibre reinforcement that is usually very abrasive and causes rapid tool wear and deterioration of the machined surfaces [2]. Machining can cause several damages such as delamination, fibre pull-out, thermal damages [3] as well as other damages. When considering the drilling of a composite part, good results are mainly fibre related and less dependent of the matrix material [4].

Koplev et al. [5] examined the cutting process of unidirectional carbon fibre reinforced plastics in directions perpendicular, as well as parallel to fibre orientation. A series of quick-stop experiments were carried out to examine the area near the tool tip. The author stated that machining of CFRP consists in a series of fractures, each creating a chip. When this material is machined, a large number of chips are created and the examination of these chips revealed that they are not subjected to large plastic deformation as is normally found in metal chips. Machining parallel to the fibres gives smoother surfaces than machining perpendicular to fibres, and matrix cracking is also smaller.

2.2. Cutting Mechanisms and Machinability

Composite materials are characterized for being non-homogeneous and, in most cases, anisotropic. Fibre orientation towards cutting direction is an influencing factor in the development of forces and stresses during machining. Definition of fibre orientation angle – θ – is shown in Figure 4.

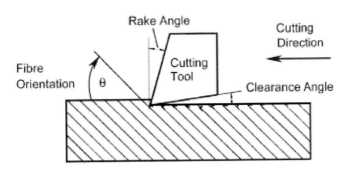

Figure 4. Definition of fibre orientation angle and other cutting variables.

In an experimental research performed with a carbon/epoxy composite with unidirectional fibres, Wang and Chang [6] verified that fibre orientation has a determining influence on the forces that developed during the cut, no matter what tool rake angle – γ. Depth of cut – DOC – can have some influence, function of fibre orientation angle. Their conclusions are summarized in Table 1.

Table 1. Effect of fibre orientation angle on cutting forces

Fibre orientation θ	Cutting force	Notes
0 to 60°	increase	------
60 to 120°	decrease	increase if DOC great
> 120°	increase	decrease if DOC great

Table 2. Deformation and cutting mechanisms with fibre orientation

Fibre orientation θ	Fibre stress	Fibre breakage	Notes
0°	compression	delamination	
0°<θ<90°	bending and tensile	breaking	45° best value
90°	bending	shear off	
>90°	bending and compression	breaking	least favourable

Fibre fracture plays an important role in the cutting mechanism which differs according to the machining angle to the fibre orientation θ.

Klocke et al. [7] had drawn some conclusions regarding fracture in orthogonal cutting, which are summarized in Table 2. When the angle to the fibre orientation is 0°, the laminate is subjected to compression. Material failure occurs by delamination or interfacial fibre failure. When machining at an angle of 90°, fibres are subjected to bending and are sheared off. As that angle increases to 135°, fibres are compressed and bent in the direction opposite to fibre orientation. Fibre breaking is a result of bending and compression. Angles in the interval between 120 and 150° are the least favourable, resulting in poor surface quality. An orientation of 45° is more favourable to cut. Fibres are subjected to bending and tensile stresses and break in bundles. Sometimes, individual fibres are ripped off the matrix due to insufficient adhesion.

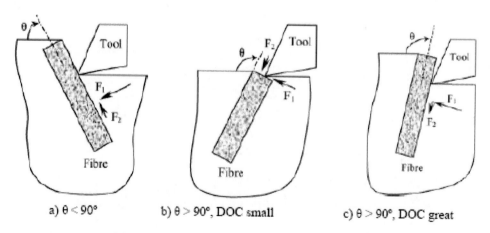

Figure 5. Schematic cutting mechanisms [6].

Wang and Chang [6] had related fibre deformation mechanisms in the cutting zone with fibre orientation angle θ. When this angle is less than 90° (Figure 5a) the fibre is pushed by the tool in a direction perpendicular to its axis and towards the workpiece subsurface. Fibre is supported by uncut material behind and the bending of the fibre is small. A tensile stress is created along the fibre axis and carbon fibres, brittle, fracture easily. Surface roughness and subsurface damage is small. For angles greater than 90°, they divided the mechanism according to the depth of cut. For small depths of cut (Figure 5b), fibre is subjected to an axial compression. As a result of that force, matrix material around the fibre is fractured. Thus a machined surface has protruded fibres resulting in greater surface roughness. When the depth of cut is great (Figure 5c), fibres are pushed towards the external of the workpiece. As the support from surrounding material is weaker, the result is fibre bending and fibre-matrix debonding. This causes a rougher surface finish and deeper subsurface damage. Figure 5 helps on the explanation of the above mentioned mechanisms.

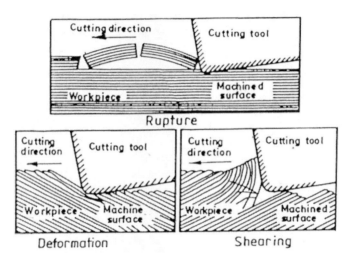

Figure 6. Rupture, deformation and shearing of fibre reinforced composites during machining [9].

Krishnamurthy et al. [8] conducted a study on the machinability of three types of composites with different reinforcement fibres – glass, carbon and aramid – using K20 carbide tools. Coated carbides, TiC and TiN, did not perform well owing to coating deformation and subsequent chipping and spalling. Tool have a rake angle of -6°, a clearance angle of 6° and an edge radius of 0.8 mm. Cutting speeds were in a large range of 12.5 to 200 m/min and feed between 0.025 and 0.15 mm/rev. They have observed that reinforcement fibres were a dominant factor in machinability. Depending on fibre nature, its orientation and angle to cutting edge removal mechanisms may be associated with rupture, deformation or shearing, as illustrated in Figure 6.

A different view can be given by published studies on machinability, relating cutting parameters with cutting force and torque. El-Sonbaty et al. [10] investigated the influence of some parameters on the thrust force, torque and surface roughness in drilling process of glass fibre-reinforced epoxy. The authors were able to demonstrate the relationship between feed rate and peak values of the thrust force and torque – Figure 7-a) and b). The sudden jump of torque at the end of the cycle was due to the drill movement to emerge out of the laminate. The effect of increasing cutting speed on thrust force was a slight increase while torque had decreased.

The effect on both cutting speed and feed on surface roughness was insignificant. Mohan et al. [11] used Taguchi optimization technology and analysis of variance (ANOVA) to study the effect of process parameters on machining process. The lower cutting thrust force result from a combination of the highest speed with the lowest feed rate. Sardinas et al. [12] implemented a micro-genetic algorithm to carry out the optimization process and to obtain a set of optimal solutions. As a result of a multi-objective optimization is was possible to choose the most adequate solution for each particular operation.

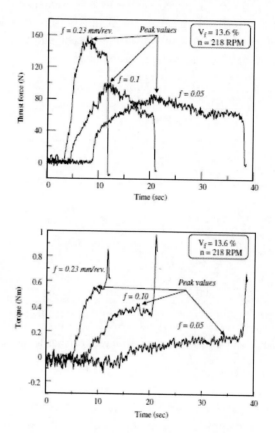

Figure 7. a) Thrust force over a complete cycle; b) Torque over a complete cycle [10].

Gaitonde et al. [13] studied the effects of cutting speed and feed rate on machining force, cutting power and specific cutting pressure during turning on reinforced and unreinforced polyamides. From their experimental work, it was possible to conclude that machining force and cutting power both increase with feed rate, while specific cutting pressure decreased, for a given value of cutting speed. The cutting power was more sensitive to variations when high cutting speed and feed rates were used.

In an analysis about the effect of machining parameters on delamination, surface roughness and bearing strength [14] as well as in tool wear [15], Khashaba et al. confirmed the importance of reducing the thrust force during drilling in order to minimize delamination. In order to achieve that result feed rate should be kept low. This will also reduce tool wear which in turn also avoids machined surface deterioration and keeps thrust force below critical values for delamination onset.

2.3. Composites Drilling

Composites are produced to near-net shape, minimizing machining needs of manufactured parts. However, machining operations like trimming to final shape and drilling to enable assembly of parts in complex structures, like aircrafts, are usual operations in industries that deal with these kinds of materials.

Drilling is a complex process which is characterized by the existence of extrusion and cut mechanisms, the former performed by drill chisel edge that has null or very small linear speed and the latter by the existence of rotating cutting lips at a certain speed. The most common drill is the conventional conical point drill. The cutting process is unique and can be divided in two distinct regions: chisel edge and cutting lips. In a common drill, there is a small region around the centre of the chisel edge where the tool does not cut but extrudes the material, called the indentation zone. At the region outside the indentation zone, called secondary cutting edges area, the rake angle is highly negative. As fibre reinforced plastics are more brittle than metals, it is unlikely that extrusion really takes place and orthogonal cutting could be assumed for the entire chisel edge. However, model predictions based in this assumption do not agree with experimental data. Along the cutting lips, cutting action of a drill is a three-dimensional oblique cutting process. The cutting speed, rake angles and other varies along the cutting lips with the radial distance from the centre. The cutting action is more efficient at the outer regions of the cutting lips than near drill centre [16, 17].

The drilling process can be understood by the observation of thrust force and torque evolution. As drilling progresses, thrust force increases until a nearly constant value corresponding to steady drilling through the thickness of the laminate is reached. Then it drops sharply as the tool bit exits through the opposite side [2]. Torque increases rapidly until the cutting edges of the tool are completely engaged. Then, it increases linearly until a maximum value is reached and is followed by a slight drop after hole completion [3]. Fernandes and Cook [18] followed a similar procedure in order to divide force and torque into stages and discuss common problems and damages associated with each stage. Stage I corresponds to entrance and the possible problems identified were skidding, wandering or deflection of the drill bit. Stage II is drilling and the main problems associated are delamination and tool wear. The risk of delamination is high at the end of this stage as the last plies of the laminate are pushed by the chisel edge [18]. Stage III is drilling and reaming corresponding to the moment after drill tip reaches the bottom surface of the workpiece. Drilling action is replaced by the reaming action and maximum torque is reached. Stage IV is reaming and stage V is backing out where drill movement is reversed for withdrawal.

As composites are non-homogeneous and anisotropic, drilling raises specific problems that can affect parts strength and fatigue life [19 - 21]. Typical damages after drilling are push-out delamination at the exit side, peel-up delamination at entrance side, fibre pullout, burrs, splintering and swelling [22]. Some thermal damage can also occur, due to the low thermal conductivity of polymer matrix composites. Due to their abrasiveness, composites drilling cause high tool wear, leading to the need of frequent tool changes that affect the production cycle. From all these problems, delamination is the major concern as it reduces severely the load carrying capacity of laminated composite structures and must be avoided [2]. Delamination is a damage that occurs in interlaminar regions, that is to say, in the contact plan between adjacent layers, so it depends not only on fibre nature but also on resin type and respective adhesive properties. Delamination mechanisms are divided according to the laminate region where it occurs, exit or entrance, respectively called push-out and peel-up.

Peel-up is caused by the cutting force pushing the abraded and cut materials to the flute surface. Initially, the cutting edge of the drill will abrade the laminate. As the drill moves forward it tends to pull the abraded material along the flute and the material spirals up before being effectively cut. This action creates a peeling force upwards that tends to separate the upper laminas of the plate (Figure 8). This peeling force is a function of tool geometry and friction between tool and workpiece [23].

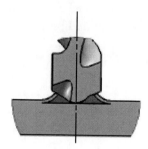

Figure 8. Peel-up delamination at entrance.

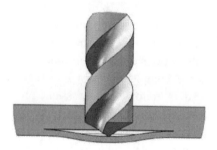

Figure 9. Push-out delamination at exit.

Push-out is a consequence of the compressive thrust force that the drill always exerts on the workpiece. The laminate under the drill tends to be drawn away from the upper plies, breaking the interlaminar bond in the region around the hole. As the drill approaches the end of the laminate, the uncut thickness becomes smaller and the resistance to deformation decreases. At some point, the loading exceeds the interlaminar bond strength and delamination occurs, before the laminate is totally penetrated by the drill (Figure 9). A different tool geometry that lowers thrust force can reduce delamination [23]. Thermal stress that develops during drilling can also facilitate delamination by matrix softening.

Fibre/matrix debonding is observed along the machined hole walls and is characterized by the existence of fibres torn away by the action of drill cutting edges. Intralaminar cracks are also reported in literature as a typical damage mode occurring in a composite plate after drilling [24].

Tool wear is a consequence of composite abrasiveness and low thermal conductivity that causes tool heating superior to what is normally observed in the drilling of metallic materials. In carbon-epoxy drilling 50% of the energy is absorbed by the tool and the remainder is absorbed almost equally by the workpiece and chips. When drilling metals 75% of the thermal energy is eliminated with chips, 18% by the tool and only 7% by the workpiece [2]. Effect of tool wear on the thrust force and torque throughout the drilling process was also investigated in [18]. Tool wear increases both with thrust force and torque. The higher the feed the bigger the effect that the tool wear will have on the thrust force.

After drilling is completed, circularity shall be observed, as there is a bouncing back tendency of the material that causes hole deformation. The return to its initial position causes tightening around the drill, and the drilled diameter is less than the tool diameter. This roundness error is due to the anisotropy of the material [25].

The above mentioned damages are responsible for the rejection of producing parts and contribute to the rise of the fabrication costs.

In [26] a survey on the published literature about drilling of polymeric composite materials had focused on aspects such as tool materials and geometry, machining parameters and their influence on the thrust force and torque. Main conclusions referred the need to develop tools with special geometry in order to achieve best performance. Another point of agreement among the authors is the need to develop devices and/or procedures in order to allow for the reduction of the damaged area without using backup material, which makes the drilling operation longer and dearer.

A recent advance on machining strategy was given by Schulze et al. [27] minimizing damage by directing the process forces towards the centre of the workpiece. This is achieved through a combined process of circular and spiral milling on a three-axial machining centre. The advantages of this process still require further research.

2.3.1. Drilling of Glass/Epoxy Composites

Tagliaferri et al. [28] carried out drilling tests on a glass/epoxy panel obtained from prepreg in a quasi-isotropic stacking sequence, high speed steel (HSS) drills, without backup plate or cutting fluid. By analyzing the variation of cutting speed and feed, they concluded that when feed remains constant, damage reduction is accomplished by an increase in cutting speed, and when speed remains constant, lower feeds show better results in terms of damage reduction. They also said that the tensile strength of a GFRP containing a hole is not dependent on damage extent and that bearing strength only correlates with damage extent when this damage is quite large. Finally, they suggested that an optimal ratio between speed and feed seems to exist for maximum bearing strength, adopting lower drilling speeds.

Bongiorno et al. [29] drilled several glass/epoxy plates in different process conditions, generating different kinds and levels of damage, using HSS drills with 5 mm diameter. Plates were subjected to fatigue bearing tests. Results showed that the presence of damages like intralaminar cracks along the hole section deeply affects fatigue behaviour as these cracks propagate quickly into the material, while delamination played a minor role. To avoid these types of cracks a low feed should be adopted. They also verified that the presence of a backup support plate reduces push-out delamination only, but does not affect internal hole damage.

Khashaba [30] conducted a study to determine the effect of various values of fibre volume fractions (V_f) and drill sizes in notched and point bearing strengths of randomly oriented GFRP composites. Results showed that fibre volume fraction plays a significant role on bearing tests and stress intensity factor. He also concluded that the ratio of width to diameter must be greater than five for the development of full bearing strength.

More recently, the same author [31] carried out an experimental investigation about the effect of cutting speed and feed on thrust force, torque and delamination of GFRP composites. He concluded that push-out delamination is more severe than peel-up delamination. Other conclusions were the increase of delamination size with increasing feed and decreasing speed. A change in material matrix or ply orientation can affect delamination as well as thrust forces and torque. Composites with polyester matrix had higher torque and greater delamination than those with epoxy matrix. Using of variable feed and cross-plying led to delamination free holes. A drilling analysis of chopped composites by the same author [32] had investigated the effect of the drilling parameters like speed and feed on cutting forces and torque. The authors stated that delamination could be decreased with decreasing the feed, while no clear effect of cutting speed on the delamination size was observed.

Davim et al. [33] studied the effect of cutting parameters on specific cutting pressure, thrust force, damage and surface roughness using analysis of variance (ANOVA) during drilling with carbide tools. They concluded that feed rate is the cutting parameter which has greater influence on the specific cutting pressure and thrust force. On the other hand, the cutting speed had higher influence on damage and surface roughness. Surface roughness increase with feed rate and decrease with cutting speed in GFRP.

Velayudham et al. [34] evaluated the drilling characteristics of high volume fraction fibre glass reinforced composite. The results indicate that thrust force and torque increase with increasing feed rate, but also with tool wear.

Ramkumar et al. [35] studied the effect of workpiece vibration on drilling of GFRP unidirectional laminates. In their work, the part was subjected to a vibration using a variable frequency generator close to the hole being drilled. Frequency used was 220 Hz and amplitude was 10-15 µm. Drill performance had increased, delamination reduced, and tool life was longer.

Aoyama et al. [36] studied the drilling of glass/epoxy, plain woven cloth, with 1 mm diameter drills, typically used for printed wiring board hole drilling. Cutting speed was 15.7 mm/min and feed in the range of 5 to 63 µm/rev. An aluminium plate as top stiffener and a bakelite plate as bottom stiffener were used. It was clearly found that damage caused by drilling was larger at a fibre angle of 45° and that the damage increased when feed increases.

Palanikumar et al. [37] presented a mathematical model to predict delamination based on a study of the effects of feed speed, cutting speed and drill point angle. An empirical model was developed to correlate the drilling parameters to the delamination factor. Feed speed was the factor with greatest influence on delamination. Drill point angle influence was found to be limited.

A different approach can be given by the use of high speed machining, as demonstrated by Lin and Shen [38] studying the effects of increasing cutting speed ranging from 9550 to 38650 rpm, or 210 to 850 m/min, on average thrust force, torque, tool wear and hole quality on glass reinforced composite. For that, they used 7 mm carbide drills with two different geometries, multifaceted (Figure 10a) and twist drill (Figure 10b) and three levels of feeds: 0.03, 0.05 and 0.07 mm/rev. It was found that multifaceted drills worn out faster than twist drills as cutting speed increases, but for both geometries tool wear increase as feed increases. Lower feeds reduced tool wear. Thrust force was affected by number of holes drilled, feed rate, cutting speed and drill geometry, due to tool wear. Delamination extension was reduced when using multifaceted drills although thrust force was higher than the values measured with twist drills.

In Rubio et al. [39] the authors compared the effect of high speed machining using an aerostatic headstock with 40,000 rpm maximum rotational speed with the effect of 4,000 rpm and 8,000 rpm. The most interesting conclusion of this work is related with the temperature build-up due to high speed machining causing matrix softening and inducing less delamination. Consequently, the effect of feed rate when machining at 40,000 rpm was not as evident as in conventional machining speeds.

Figure 10. Different drill geometries: a) multifaceted drill; b) twist drill.

2.3.2. Drilling of Carbon/Epoxy Composites

Piquet et al. [25] completed an experimental analysis of drilling damage in thin carbon/epoxy plates using special drills. The plates had quasi-isotropic properties. According to the authors, best drill material is tungsten carbide, based in economic considerations when compared with polycrystalline diamond. Peel-up delamination can be reduced with the use of a small rake angle (6°) preventing the first ply from lifting up and tearing off. A great number of cutting edges, three to six, increase the contact length between tool and part, facilitating heat removal. Point angle of 118° is necessary for the main cutting edges and then 70° for the minor cutting edges. Chisel edge dimensions have a direct relation with delamination onset, so it is necessary to be as reduced as possible. Normally the non-cutting edge represents 20% of the drill diameter. Pre-drilling neutralize the chisel edge effect, having an effect similar to the non-existence of chisel edge. Machining condition can be improved by applying a variable feed rate, combining a low machining time in the initial plies using high feeds with a reduction of axial thrust force, using low feeds to avoid delamination in the last plies of the laminate. This procedure requires the use of CNC drilling machines.

Hocheng and Puw [24] characterized the response to drilling of reinforced thermoset and thermoplastics. The specimens were epoxy, a thermoset, and ABS, a thermoplastic, with continuous carbon fibre reinforcement, and the volume fraction was 60 and 10%, respectively. Tool was a HSS drill of 5 mm diameter, cutting speeds were in the range of 2 to 63 m/min and feeds from 0.0125 to 2.4 mm/rev. Chips produced by machining were different according to the matrix material. While for carbon/ABS chips were in continuous or curling form, for carbon/epoxy the chips were discontinuous. In the authors opinion this is explained by the large elongation capability of thermoplastics under load and by the brittle nature of thermoset materials causing the chips to break earlier.

For both materials, an increase in feed rate showed an increase in thrust force, although forces involved in carbon/ABS machining were always smaller than those for carbon/epoxy drilling. Specific cutting energy, that gives a good indication of the machining effort, was greater when machining carbon/epoxy, but this could be a consequence of its rich fibre content.

Surface roughness is more dependent on the material than on cutting parameters. Carbon/epoxy drilled holes have higher values of roughness. As surface integrity of parts is concerned, it is important to avoid the use of high speeds together with low feeds, because of heat build-up that can cause matrix degradation. In the particular case of carbon/epoxy, due to its fragile behaviour, machining is more difficult leading to a more accurate feed control and tool wear to avoid or minimize the occurrence of delamination.

Machining parameters are a function of the material to be machined and tool material. As a general rule, feed shall be below 0.1 mm/rev or less, if possible. Nevertheless lower feeds tend to cause a greater amount of heat that must be dissipated. The use of CNC machines, with the possibility of variable feed regulation can improve machining conditions. Cutting speeds can vary between 25 m/min with HSS drills to 150 m/min or more, using PCD drills.

Other studies have been done on carbon/epoxy drilling, trying to obtain delamination free holes, using alternative methods of conventional drilling.

Park et al. [40] applied the helical-feed method to avoid fuzzing and delamination. Tool used for the drilling experiments was a core drill made of cast iron with bonded diamond in two shapes, straight and rounded. The latter showed the best results as no delamination was observed around drilled holes. A tungsten carbide drill was used for results comparison and it was found that drilling quality degraded and fuzzing observed as the number of drilled holes increased due to the wear of cutting edge. The use of helical-feed allows the drilling operation to be completed efficiently without any limitation of drilling depth. Tensile tests performed on specimens drilled in different conditions did not show dependency of drilling methods on test results

Persson et al. [19, 20] studied the effect of hole machining defects on strength and fatigue life of carbon/epoxy composite laminates. For that purpose they have compared two traditional machining methods using a PCD drill, a Dagger drill and the orbital drilling method. This a patented method by NOVATOR®, developed by Zackrisson, Persson and Bäcklund at the Department of Aeronautics of Kungl Tekniska Högskolan (KTH) in Sweden. The hole generation method is shown in Figure 11. The hole is machined both axially and radially by rotating the cutting tool about its own axis as well as eccentrically about a principal axis while feeding through the laminate.

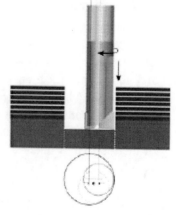

Figure 11. Hole generation using NOVATOR® orbital drilling method. Source: NOVATOR AB.

Some advantages of this method are referred by the authors in [19] when compared to traditional hole machining methods. First is the elimination of a stationary tool centre, thus reducing axial force. Second is the reduction of the risk of tool clogging, as the tool diameter is smaller than hole diameter. For this reason the cutting edges are only partially and intermittently in contact with hole surface, allowing efficient removal of cut material and efficient cooling of tool and hole surface. Third advantage is the possibility of using one tool diameter to machine holes of several diameters and fourth is the precision of the hole that is determined by tool positioning and not by tool precision itself, reducing tool costs.

Quasi-isotropic carbon/epoxy plates were drilled using the three methods and the following conclusions were drawn. Radiographs showed no damage around orbital drilled specimens, damage extended to nearly a quarter of hole radius in Dagger drilled specimens and almost equal to hole radius in PCD drilled specimens. Static testing of pin load specimens gives the highest values for orbital drilled specimens with a reduction of 2 to 3% for Dagger specimens and about 11% for PCD specimens. Fatigue testing results yielded 8 to 10% lower strengths for PCD and Dagger than orbital drilled specimens. Dagger drill, although giving fair results, has some disadvantages, in the authors' opinion. Due to its long and sharp tip it is less suitable in situations with limited space on the exit side of the laminate. Other disadvantages are related with the inability to remove chips and the relatively short tool life, around 70 holes.

Another work concerning influence of drilling on fatigue behaviour of carbon/epoxy laminates was presented by Hamdoun et al. [21]. Two kinds of specimens were produced, a 'damage-free' and another with damages intentionally introduced during drilling. The two specimens were subjected to the same compression cyclic loading. Specimens with damages showed slightly lower values of fatigue resistance. That can be explained by lower stiffness of the specimens caused by the existence of delaminations around the hole.

Murphy et al. [41] compared the performance of three different types of tungsten carbide drills, TiN (titanium nitride) and DLC (diamond like carbon) coated and an uncoated drill. Testing involved a series of consecutive holes. During these tests the thrust forces and torque were monitored, tool wear was inspected and workpiece also inspected for hole tolerance, delamination and spalling. They have verified that tool wear caused thrust force and torque to increase and unacceptable damage to the composite was found, even though the wear is small when compared with that associated to the drilling of conventional materials. When holes were measured it was found that only a small number satisfied an H8 tolerance criterion. Finally, the use of coatings was found to be of no benefit when machining carbon-epoxy laminates, reducing neither tool wear nor composite damage.

Dharan and Won [42] conducted a series of machining experiments in order to recommend an intelligent machining scheme that avoids delamination by peel-up at entrance and by push-out at exit. Study of drilling action had divided this process in three components: extrusion at chisel edge, secondary cutting at chisel edge and primary cutting at cutting lips. For metallic drilling, the portion of the thrust force acting on the chisel edge is about 50 to 60% of total thrust force. On the other hand, torque acting on chisel edge is relatively small.

The authors have divided the drilling cycle in seven steps, described herewith.

1. Approach: the drill approaches the workpiece;
2. Contact: the drill tip makes the contact;
3. Normal drilling: drilling without delamination;

4. Delamination onset: when critical thrust force is exceeded;
5. Drill breakthrough: the drill tip exits the workpiece;
6. Completion: the hole is completed;
7. Withdrawing drill: the drill must be moved backward and withdrawn from the workpiece.

An 'intelligent' system shall be able to perform steps 1 and 7 as quickly as possible and limit feed in steps 2 and 5 to avoid entry and exit delamination. During step 3, feed must be controlled to prevent delamination.

With the same objective of delamination prevention, Stone and Krishnamurthy [43] studied the implementation of a neural network thrust force controller, using Hocheng and Dharan model [23] for critical thrust force calculation. In this study, cutting forces along three axis and torque were measured with a Kistler dynamometer, although only the thrust force was of interest. As there are variations of the force signal due to drill rotation, thrust force signal was averaged over one spindle revolution. Controller act to update feed rate every three spindle revolutions allowing the system to detect the beginning and end of a revolution and carry out necessary neural network computations. The study had shown that the neural network control scheme can minimize delamination during drilling of a graphite/epoxy laminate, varying feed in order to control thrust force. For comparison, the authors also drilled some plates with constant feed and concluded that results regarding delamination were not so good.

Hocheng et al. [44] analysed the machinability of carbon reinforced plastics with different matrix material – epoxy, PEEK, ABS. The experimental observation discussed chip characteristics and the effect of cutting speed and feed on specific cutting energy and axial thrust force. Generally, an increase in feed causes an increase in thrust force and a decrease in specific cutting energy. Cutting speed effects are less visible and for the two lay-ups used in the experiments – $[0/90]_{4s}$ e $[0/\pm 45/90]_{2s}$ – it showed no influence on thrust force. Carbide drills are adequate for drilling these laminates and smaller point angles should be used when materials with less interlaminar fracture toughness are drilled, as thrust force is reduced.

In the experimental work described in [44], material with greater interlaminar fracture toughness in Mode I – G_{Ic} – was carbon/PEEK with a value of 1800 J/mm^2 and edge quality was better when compared with carbon/epoxy edges, whose G_{Ic} value is 800 J/mm^2. They also verified that lay-up of the laminate did not make a difference on hole edge quality. As a conclusion for their work, an optimal operation domain is presented, see Figure 12. Proper parameter choice shall consider the use of conservative feed and speed. An increase in feed can cause delamination and burrs, while increasing speed raises cutting thrust force and torque as well as reduces tool life.

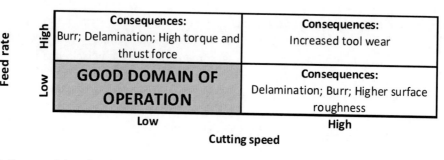

Figure 12. Suggested domain of operation.

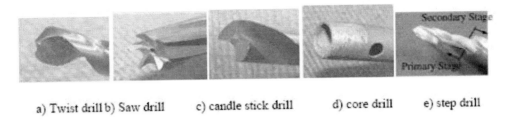

a) Twist drill b) Saw drill   c) candle stick drill   d) core drill   e) step drill

Figure 13. Drill geometries.

There is a large contribution of the work of Hocheng and Tsao for the understanding of the delamination mechanism associated with different drilling conditions like drill geometry [45, 46], the use of a core drill [47] or the influence of using an exit back-up plate on delamination depending on drill geometry [48]. In [45, 46] different drill bits are compared for drilling-induced delamination. The different drill geometries considered in these works are the twist drill, the saw drill, the candle stick drill, the core drill and the step drill, Figures 13a) to e), respectively. In [47] only core drill is studied, showing that grit size and feed rate are the most important parameters for delamination reduction and should be kept low. According to authors there are advantages in using special drill bits for composites drilling. The traditional twist drill provides a low threshold of the thrust force when compared with other geometries. From these geometries the higher threshold feed rate at delamination onset was obtained with core drill, followed by candle stick drill, saw drill and step drill [46]. The relevance of step drill geometry will be discussed in the following section.

The effects of using a backup plate on delamination are well known in composites industry. In [48] the authors had confirmed theoretically this assumption as well as experimentally, providing an analytical model that will be referred in the 'Damage Models' section.

Won and Dharan [49] conducted drilling tests on aramid and carbon fibre-reinforced composite laminates using carbide drills. During these tests, thrust force and torque were monitored to establish the contribution of chisel edge cutting force to total thrust force at different speeds. A ratio of thrust force acting on the chisel edge to total thrust force was established drilling in one step and with a pre-drilled pilot hole, whose diameter was slightly greater than the chisel edge length of the carbide drill. Results show that independently of hole diameter, chisel edge contribution to total thrust force was between 60 and 85%, increasing at higher feed rates. They also verified that trust force increases with feed rate.

Using Hocheng-Dharan delamination model for critical thrust force [23], a maximum feed of 0.145 mm/rev was found for carbon/epoxy laminates, with a 6.35 mm drill and a spindle speed of 1000 rpm. For aramid/epoxy laminates the feed would be 0.045 mm/rev for the same drill but lower spindle speed. These feeds for drilling without delamination were calculated with the following empirical relationships for thrust force, for a given feed f and a drill diameter d,

$$F_t = 40.77(fd)^{0.66} - 0.36d^2 \tag{4},$$

for carbon/epoxy laminates, and

$$F_t = 35.84(fd)^{0.50} - 0.09d^2 \qquad\qquad (5),$$

for aramid/epoxy laminates.

Calculations for maximum feed without delamination at exit side are based in Hocheng-Dharan model [23].

Enemuoh et al. [50] developed an approach for damage-free drilling of carbon fibre reinforced thermoset resins. This approach is based on a combination of Taguchi's experimental analysis technique and a multi-objective optimization criterion. The first step is used to select an optimum parameter range where the global optimum will reside while the second step will search for the global optimum within the resulting range from step one. This second step is accomplished by the use of a multi-objective optimization criterion with a nonlinear sequential quadratic programming algorithm. The authors describe an experimental procedure where this technique was applied. Four experimental responses were considered: thrust force, delamination, damage width and hole surface roughness. Experimental factors were cutting speed, feed rate and tool point angle and ANOVA was used to determine the relative significance of process factors. After an optimum range is selected, a multi-objective technique is developed. Recommendations from this work are the use of low feed, in the range of 0.02 to 0.05 mm/rev and speed from 40 to 60 m/min, to minimize delamination and have good hole surface finish.

Linbo et al. [51] presented a new vibration drilling technology using a hybrid variation parameter method, with the purpose of limiting thrust force during drilling. In vibration drilling, a piezoelectric crystal oscillator is fixed in the spindle sleeve. The application of a high voltage across the crystal produces a vibration in axis direction. In addition, the feed motion is provided by means of a servomotor driving a harmonic gearbox, which turns the spindle sleeve. Hybrid variation parameters method is a controlling strategy on vibration drilling that enables, according to numerical simulations and experimental results in carbon/epoxy laminates, the drilling of good quality holes. The applicability of this method can be extended to other laminates.

Iliescu et al. [52] developed a model in order to understand the importance of tool coating, feed and cutting speed in tool wear, thus increasing tool life without delamination onset. In their conclusions the authors pointed out some specifications that a drill must follow in order to achieve the greatest number of holes without delamination. The point angle should be able to minimize the thrust load (90°), the helix angle shall act like a corkscrew (35 to 40°), clearance angle instead of double slope flank and diamond coating. Finally a cutting speed of 170 m/min combined with a feed of 0.05 mm/rev.

High speed drilling and consequences on delamination had also deserved some attention [53]. Similarly as described above, an aerostatic headstock with 40000 rpm maximum rotational speed was compared with the effect of drilling at 4000 rpm and 8000 rpm. The effect of matrix softening as a consequence of temperature build-up seems to explain the delamination decrease when high speed machining is used. Consequently, the effect of feed rate when machining at 40000 rpm was not as evident as in conventional machining speeds.

2.3.3. Pilot Hole Drilling

As it was already referred, drill bit geometry plays an important role in thrust force and, consequently, in delamination onset and propagation. The chisel edge of a twist drill, the most frequent geometry, acts like a punch over the uncut plies of the laminate. The consequence of this punching action is delamination whenever this thrust force exceeds the interlaminar resistance of the laminate. The size of the delamination zone has been proved to be related to the thrust force developed during the drilling process [54]. The hole can be pre-drilled in order to eliminate the thrust caused by the chisel edge of twist drills. Thus, delamination hazard is significantly reduced by the minimization of the thrust force.

This drilling strategy and its influence on thrust force and delamination had been addressed by a reduced number of investigators.

In [55], Won and Dharan studied the effect of chisel edge on thrust force and also the effect of pilot hole drilling on carbon fibre reinforced composites. They have observed that the effect of chisel edge on thrust force decreases with increasing drill diameter and increases significantly with increasing feed. They also stated that thrust force acting on chisel edge, being a significant component of total thrust force, may govern the occurrence of delamination. A new empirical relationship for drills with pilot holes was determined using experimental results [55],

$$F_t = 3.50(fd)^{0.66} + 0.11d^2 \qquad\qquad (6).$$

Forces measured for specimens with pilot holes are 20 to 50% lower than those without pilot hole, with an average result of 27% reduction in data published. It is important to note that specimens drilled with pilot holes did not exhibit any delamination over the range of feed tested (0.1 to 0.7 mm/rev). A model for predicting critical thrust force for delamination onset, based on linear elastic fracture mechanics, is presented in [55]. It shows that the critical thrust force decreases slightly for specimens with pilot holes. In their study they had calculated a 9 percent reduction when the chisel edge length is approximately 0.18 of the final hole diameter.

In a preliminary conclusion, it is possible to say that a two stage drilling shall have a pilot hole diameter equal to the chisel edge length of the final drill in order to reduce potential for delamination damage [55]. A pilot hole to final hole ratio of 0.18 is well inside the optimum interval of the dimensionless chisel edge length and in the range of chisel edge lengths in standard twist drills.

Later, Tsao and Hocheng [56] took the same model but with the interest in studying the effect of chisel edge length and associated pilot hole on delamination. Two chisel edge lengths of 0.15 and 0.2 of drill diameter were considered. An optimal ratio of chisel edge length to drill diameter was derived, as a function of laminate Poisson's ratio. A thrust reduction of 25 to 50% was found when comparing results of holes with and without pilot hole, using 10 mm drills, a spindle speed of 1000 rpm and feed in the range of 0.008 to 0.012 mm/rev. In the authors' opinion, based on experimental work and models developed, the dimensionless chisel edge length should be around 0.09 to 0.2 of drill diameter. With proper settings of cutting parameters and chisel edge length or pilot hole diameter is possible to produce delamination free holes. Critical thrust force is reduced with pre-drilled hole, while axial drilling thrust is largely reduced by cancelling the chisel edge effect. Small feed rates opens a wider range of pilot hole diameters that can be used.

Durão et al. [57-59] had followed this tool concept and developed several comparative studies of step drills to conventional drills. In [57] a step drill is compared to twist and Brad drills when drilling hybrid composite laminates. Although the thrust force reduction was not evident, step drill offered the lowest delamination. In [58] a more complete comparative study is presented for the drilling of carbon/epoxy laminates. For the experimental parameters used, no advantage was found in the use of the step drill when compared with a 120° point angle twist drill. An evolution was evaluated in [59] showing a slight advantage by enabling a 2% damage reduction when using step drill when compared to conventional twist drill.

Tsao conducted different analysis on the effect of pilot hole on delamination when associated with saw drill [54] or core drill [60]. In both papers, experimental results had shown the advantages of pilot hole drilling based on thrust force reduction. This effect largely compensates the reduction of the critical thrust force for delamination onset. The key for efficient drilling lies on an appropriate ratio of pilot hole to drill diameter. A comparative study of three different step-core drills can be found in [61]. As a result of the experimental analysis conducted, Tsao concluded that the step-core-twist drill offer the lowest thrust force during drilling.

Finally, in [62], Tsao defined an optimal ratio between the first and second stage of drilling. In this work, Taguchi method was used to obtain an optimal set of parameters in the drilling of carbon fibre reinforced plates. A ratio of first to second stage drill of 0.4 should be used to minimize delamination. Moreover, this parameter resulted to be the most important in the experimental work conducted. The remaining parameters, in order of relevance to damage, were the step angle, the feed rate and the cutting speed. The step angle was defined as the angle of transition from the first to the second stage, see Figure 14.

Pilot hole drilling can be a good alternative for the reduction of thrust force, thus delamination. Tool geometry concept is essential to have good results, demonstrating the importance of a dedicated geometry for the drilling of fibre reinforced laminates.

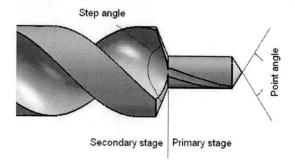

Figure 14. Step drill geometry.

2.3.4. Drilling of Aramid/Epoxy Composites

The drilling of aramid reinforced composites is more complicated and frequently causes shredding, giving a bad surface finishing to the hole. Shredded fibres have to be manually removed, increasing operation time and associated costs. In order to obtain good geometric tolerances and no shredding, it is necessary that drilling proceed in such a way that fibres are preloaded by tensile stress and then cut by shearing motion [4]. For a rotating tool, that means that the cutting edge of the drill should be C-shaped, like a Brad type drill, to cut from outside to the centre, see Figure 15.

Figure 15. Drill geometry for aramid fibre cutting.

Di Ilio et al. [63] reported the results of an experimental research on the drilling of aramid fibre reinforced plastics. For this material, parameters used were a cutting speed of 120 m/min, with feed in the range of 0.006 to 0.07 mm/rev. The authors have detected large oscillations of the thrust force during drilling, due to the inhomogeneity of the material inside single plies and poor interlaminar strength, causing piercing effects at the interfaces. Another important aspect to be taken into account is the low thermal conductivity of aramid composites, resulting in heat build-up that must be carried away by the tool, accelerating wear processes. Temperature increase at the cutting front has the effect to cause a decrease in thrust force due to the reduction of material strength. Friction forces are high, which is evidenced by torque increase, contributing to the raise of power needed for cutting as well as temperature increase.

Further work by the same authors employing a variable feed rate strategy during drilling [64] showed significant improvement in hole quality. The drill used was a Brad drill, already referred as most suitable for this material.

3. NONTRADITIONAL COMPOSITES MACHINING

Besides conventional machining, referred above, it is necessary to consider the existence of alternative machining processes. A wide variety of non-traditional machining methods are available for the machining of composites. These methods are promising as they are becoming cost effective. There are several reasons that can lead to the use of one of those machining processes, like the poor machinability of composites due to their higher hardness, abrasiveness or lower thermal resistance. It can be also the result of the need to obtain complex shapes, avoid superficial cracks and residual stresses, high accuracy or miniaturization of parts or surfaces. These processes are already known for the machining of metals or other materials. Some of them are limited in their use due to their high installation cost or low productivity. The alternative processes that are going to be briefly referred to are laser, water jet, abrasive water jet, ultrasonic, electron beam and electrochemical machining. The applicability of these processes is related with the material properties, and not all of them are suitable to reinforced epoxy composites, as the matrix is a thermoset material. The rate of material removal and cost per unit material removal rate are important considerations for the selection of the process.

3.1. Laser Machining

Laser beam techniques have already been used in industrial applications for some time. As it is a non-contact process, it allows the cut of fragile parts and as the force applied is low, clamping devices can be simplified. The basic principle of LASER – Light Amplification of Stimulated Emission of Radiation – is the application of a highly coherent light beam with a single wavelength and highly collimated [1]. High concentration of light in a very small spot produces energy densities up to 10^6 W/mm^2, causing localized material vaporization, with a reduced heat affected zone (HAZ). There are materials, however, that suffer thermal degradation by this localized heating, and therefore laser cannot be used on them.

In industrial composite cutting, two types of LASER are available. Nd:YAG laser – Figure 16 – has a wavelength of 1.06 μm, a pulse length of 200 pulses per second and an average energy of 500 W, although it is possible to reach peak values of several kW. It can be used effectively in the cutting of metallic composites without an organic resin. Those materials do not absorb the laser wavelength and can decompose [65].

Figure 16. Nd: YAG laser.

The other type is CO_2 laser, with a wavelength of 10.6 μm. With a 1500 W laser system it is possible to have a power density around 4×10^7 W/m^2. If necessary, as some materials are not effectively cut with this concentration, it is possible to reach 1.0×10^{11} W/m^2 using focusing lenses reducing beam diameter to 0.15 mm at an appropriate distance (focal distance) – Figure 17. It can operate with pulse lengths of the order of 10^{-4} pulses per second. It is effectively absorbed by most organic materials [65]. This laser system was used for cutting fibreglass sheets, with a narrow HAZ and melting the ends of the fibre, preventing fraying. However, a good exhaust system is needed, as particles resulting from vaporization create a health hazard. It is also possible the use of this laser for the cut of Kevlar®-graphite/epoxy and Kevlar®/epoxy composites. Kevlar®, known as a material difficult to cut, has good cutting behaviour by laser, leaving a small edge with burning signs that can be easily removed. Graphite showed poor cutting behaviour because of its high dissociation temperature, around 3600ºC, and thermal conductivity causing degradation at a certain distance from the cutting edge [65].

The number of papers published on this subject is reduced and are mainly concerned with the reduction of HAZ [66-68]. Herzog et al. [66] compared the damage by laser cutting with conventional milling and abrasive water jet testing specimens for tensile strength and bending. The HAZ of laser cut by pulsed Nd:YAG laser was minimal, about 0.6 mm.

Therefore, the mechanical strength of plates cut using this laser type was maximum among laser cut specimens and comparable to milling.

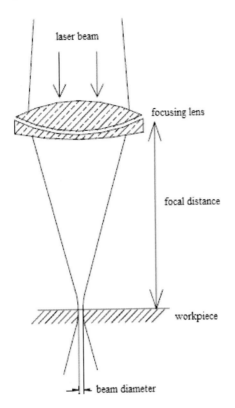

Figure 17. Use of a focusing lens for laser beam concentration.

Best results of mechanical resistance were returned by abrasive water jet cut specimens. Li et al. [67] used short pulsed UV laser to have a minimum HAZ (about 50 µm) in machining of CFRP composites. The study found that heat is easily accumulated in the material during laser processing, especially when the carbon fibres are sliced into small pieces. Mechanical strength, evaluated with bearing test, was found to be similar to that of the mechanically-drilled samples. Davim et al. [68] evaluated the effect of the processing parameters (laser power and cutting velocity) on the cut quality of polymeric materials with CO_2 laser. From the experimental results, it was evident that the HAZ increases with the laser power and decrease with the cutting velocity.

If a small slit is needed and material removal rate is not of prime importance, femtosecond laser can be a good option. This cutting technique allows for precise machining. Each pulse lasts from 50 to 1,000 femtoseconds (1 fs = 10^{-15} s). These ultrashort pulses are too brief to transfer heat or shock to the material being cut, which means that cutting, drilling, and machining can occur with virtually no damage to surrounding material. Advantages to other processes are related with high precision and the non-existence of a heat affected zone.

A short notice goes to the use of laser in ceramic turning. In this case, the light beam, with a diameter equal to cut width, focuses on the area that is going to be cut by the tool, causing an intense and localized heating. Material removal is the result of two combined effects: laser beam sublimation and softened material removal by the cutting edge of the tool.

3.2. Water Jet Machining

Water jet is an alternative method of cutting used on a wide variety of composites like reinforced polymers, metallic and non-metallic but also plywood and plastics. The jet cutting nozzle can have a diameter as small as 0.13 mm with a water pressure of 350 MPa [65].

This process can be used effectively to cut narrow slits or drill small holes. Besides other advantages, this process allows the cut to be started in any position, as long as there is care enough to avoid delamination. One possible option is to start the cut in an area that is going to be removed by the operation. Another benefit is that the cut is not affected when water jet beam suffers speed variations. The cut slit is narrow, reducing material consumption, process is easily controlled and automated with the help of an X-Y controller table and part clamping can be simplified as very low forces are imposed to the workpiece. With this method there are no burning of work surface material and no environmental pollution.

Disadvantages in this process are the high noise levels generated by air coupling into a large high-velocity volume of air/water. Because of the noise level, ear protection is needed as well as the use of noise catcher systems. Another concern is that the cutting fluid must be properly filtered and conditioned to reduce wear. The cut of composite materials can cause layer delamination, if internal pressures are generated [65].

There is a wide variety of materials, besides composites, that can be cut by this method like metallic materials, plastics, cork, rubber, wood and plywood, stone and even frozen food. Among composite materials that can be cut are those with glass, carbon, aramid or boron fibre reinforcement.

3.3. Abrasive Water Jet

Abrasive water jet – AWJ, Figure 18 – is a process very similar to water jet, the main difference being the use of an abrasive material that is introduced into the water jet after the primary jet is formed [69]. In spite of being more complex, due to the need to incorporate an abrasive feeding system, this process is well suited for the cutting of abrasive non-homogeneous materials that cause rapid tool wear when conventional cutting tools are used, for materials that can produce hazardous fumes or dusts when cut or for materials that have high hardness or that are sensitive to high temperatures. When using abrasive water jet cutting, the jet stream tends to angle away from cutting direction. This 'trailback' effect becomes more pronounced when workpiece thickness increases or if nozzle feed rate is increased. During water jet cutting, the kerf is generally wider at the water jet entrance than at the exit side [69], but this effect varies with feed rate and material.

This method is used for the cutting of metals and metal alloys, glass and metallic matrix composites, ceramic and reinforced plastics with glass, carbon or aramid fibres. When used to trim laminate composite materials, it does not cause delaminations or visible fibres beyond the cutting edges.

Ramulu and Arola [71] studied the micromechanical behaviour of both fibres and matrix of an unidirectional graphite/epoxy composite under water jet and abrasive water jet cutting conditions. In their study they have concluded that AWJ machining consists of a combination of material removal mechanisms including shearing, micromachining and erosion. Machined surface quality is better than those produced by water jet, so AWJ is a more feasible process for the cut of the material studied.

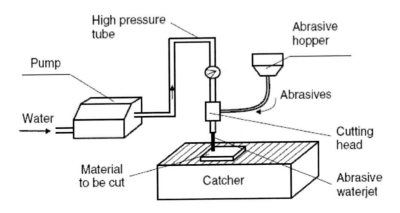

Figure 18. Schematic of an AWJ cutting system [70].

Hocheng et al. [72] studied the feasibility of AWJ milling of carbon/epoxy laminate with 64 layers. They have found that the width-to-depth ratio and lateral feed increment are two important parameters for producing acceptable surfaces. Fibre orientation does not affect the volume removal rate and surface roughness. They have found that abrasive water jet has proved to be a satisfactory machining process.

Shanmugam et al. [70] developed a semi-analytical model to predict the maximum delamination length generated by an AWJ, based on a energy conservation approach. Delamination is initiated by the shock wave impact of the water jet in the initial cutting stage when it is targeted to the material surface and can be minimized increasing water pressure and reducing the traverse speed.

Azmir and Ahsan [73] studied the cutting of glass/epoxy composite laminates. Based on experimental results and analysis of variance, it was possible to conclude that an increase of the kinetic energy of abrasive water jet machining process may produce a better quality of cuts.

3.4. Ultrasonic Machining

Ultrasonic machining is a process in which abrasives, contained in slurry, are driven at high velocity against the work by a vibrating tool. The tool has a low amplitude vibration – 0.05 to 0.125 mm – and high frequency – 20 to 30 kHz – oscillating in a direction perpendicular to the work surface and fed slowly, so that the shape of the tool is formed in the part [1]. The abrasive particles used for this process are aluminium oxide, silicon carbide, boron oxide and similar materials. According to Hocheng et al. [74], this method reduces the concerns with material damage as delaminations, fibre pull-out, micro cracking or burrs are not likely to happen when using ultrasonic machining. Besides, there are no thermal, chemical, electric or metallurgical effects on parts. This process is suitable for drilling and slitting ceramics and several composites.

3.5. Electrical Discharge Machining

The operation of electrical discharge machining – EDM – or spark machining is based on the eroding effect of an electrical spark that is generated between an electrode and the workpiece in the presence of a dielectric fluid. The spark generated produces a localized high temperature which melts and vaporizes the material to form a small crater on the workpiece surface [65]. Crater dimensions depend on machining parameters used. As electrical conductivity is necessary for the process, it is only suitable for materials that possess uniform and continuous electrical conductivity.

EDM is used for machining metal matrix and other composites that exhibit good electrical conductivity. One of the great challenges nowadays is the use of EDM to ceramic machining. Ceramics have poor machinability, due to their toughness, high hardness and abrasive characteristics, although the use of this material is growing. They also have poor electrical conductivity, making EDM use a problem that may be solved if electrical conductivity is increased.

3.6. Electron Beam Machining

Electron beam machining is a thermoelectric process, generally carried out in a vacuum, in which high speed electrons impinge on the workpiece surface, and the heat generated vaporizes the material locally [65]. Vacuum is necessary to eliminate collisions of the electrons with gas molecules. This process is considered to be micromachining, but material removal rates can reach 0.01 mg/s in metals. Extremely closed tolerances can be maintained using this process, with no heat-affected zone on the workpiece.

This process is suitable for machining small holes and cutting or slitting of most composites. On the other hand, it is too expensive and the need for a vacuum chamber increases workpiece processing time.

3.7. Electrochemical Machining

Electrochemical machining removes material from an electrically conductive workpiece by anodic dissolution. It is the reverse of electroplating. The workpiece is the anode and the tool is the cathode. Material is depleted from the anode (positive pole) and deposited onto the cathode (negative pole) in the presence of an electrolyte bath [1]. The electrolyte is usually sodium chloride mixed with water, sodium nitrate and other fluids that can chemically react with the workpiece.

This process can be used for machining complex cavities and for slitting, drilling and cutting most composites that exhibit continuous and uniform electrical conductivity. One advantage of this process is that it does not cause any thermal damage.

Finally, table 3 gives a comparison of several non-traditional processes regarding material removal rate (MRR) and specific cutting energy, with processes sorted by ascending MRR.

Table 3. Comparison of non-traditional machining processes (typical values)

Machining Process	Material Removal Rate	Specific cutting energy
	cm^3/s	GJ/m^3
Laser	0.0001	150 000
Abrasive water jet	0.0001	10 000
Electron beam	0.001	1 500
Ultrasonic	0.005	150
Electrical discharge	0.1	150
Electrochemical	1	500
Edge tool (conventional)	14	3

4. DAMAGE IN COMPOSITES

Damage in composite materials can be the consequence of several causes. Composites can be damaged during manufacture, assembly and transport or field deployment, besides expected damages during normal use in service. As composites can be used in critical applications, damages that remain undetected or that are of difficult detection can turn into a serious problem [75].

One of the possible solutions is to increase the design safety factor, over designing the part, with a penalty in weight and final cost. However, this solution does not prevent damage nor considers its possible extent, which may turn out to be wider than predicted exceeding the safety factor allowance. Unexpected ruptures during service must be avoided. Therefore, the solution most often used is to implement special damage protection methods [75].

There are a variety of damages that can be caused in composite parts during machining. These operations are normally included in the assembly phase, but one should remind that some damage can be caused prior to assembly, i.e. during parts manufacturing.

4.1. Manufacturing Damages

Composite parts manufacturing is a process in which fibres and resins are combined in a single product, using a certain processing technique. Some damages of the parts can be found after part manufacturing, like:

- uncured or not correctly cured resin;
- incorrect fibre fraction, either by excess or lack of resin;
- existence of voids;
- foreign inclusion metallic or non-metallic;
- fibre misalignment;
- ply misalignment;
- delamination;
- fibre damages;
- adhesion damages.

The possibility of occurrence of those damages is normally related with the fabrication process. Prior to be sent to the assembly phase, parts shall be conveniently inspected in order to prevent that unacceptable damaged parts are assembled. Inspection criteria shall be defined and applied either to visual inspection or non-destructive testing and damages classified in allowable, repairable and non-repairable.

In order to enable damage assessment, several types of damage have been defined, some of them related with machining:

Delamination - Separation of adjacent composite plies;

Crack - Fractures in matrix and/or fibres;

Scratch - An elongated surface discontinuity due to damage that is very small in width compared to length.

These definitions can be useful in field situations, to decide the amount of repair that must be done to correct damage and also to predict possible causes of damage.

4.2. Machining Induced Damage

Machining processes, based in the contact between a cutting edge and the part, generate stresses that are transmitted to the part. These stresses that are supported by the part can be the cause of deformation and temperature increase, that may be absorbed or not by the composite part. This is the main cause of damage existence, depending the type and amount of damage on the machining process used. In this section only drilling related damage will be referred.

Drilling is a major process used in most of the parts that are assembled in complex structures for aeronautic, aerospace, automotive, railway, wind turbines and other industries. Parts are joined either by rivets and bolts. Some parts are riveted or bolted and bonded. Bonded joints are also used, but in that case drilling is not necessary, hence bonding is not in the scope of this work. Service reliability is sensitive to machined holes quality [25].

The damages that usually can be caused by drilling composite parts are delamination, either superficial or between plies, fibre/matrix debonding, intralaminar cracks, burrs and thermal damage [6, 16, 28, 29, 40, 76]. The extent of these damages is largely dependent on drilling tools and parameters selected. Attention shall be also paid to non-circularity of holes after machining.

Delamination is the separation of adjacent composite plies. If this separation occurs between laminate outer plies it is called superficial. If this separation occurs between laminate inner plies, than delamination is internal.

Surface delamination can occur mainly in two different occasions: one is when the tip of the drill touches the part and the other is when the tip of the drill reaches the last plies of the laminate, i.e. approaches the exit side of the part. In the moment when the drill tip starts to get in contact with the material, there is a peel-up effect of the first ply along the drill main cutting edge. This kind of damage is associated with the material, the tool and the cutting conditions or developed cutting forces. This damage has the tendency to increase when the rake angle (γ) is greater and tends towards ply detachment. The chip formed tends to turn backwards with the action of the drill flank. The resulting load pushes onto the hole edge via the fibres which connect to the chip to the rest of the ply. The only force which can resist delamination is the bonding strength applied by the matrix. If the localised peel force is greater than that resistance, delamination occurs [25] – Figure 8.

Delamination at the exit side of the part is an expected consequence of the drilling itself, mainly when twist drills are used without a backup plate. The relatively large non-cutting chisel edge is its main drawback, as it has the effect of an extrusion instead of a drilling action. When the active part of the drill approaches the last laminate plies, beyond the critical thrust stress greater than the ply cohesion force, cracking forms and then spreads, in a plane perpendicular to the drill axis. Crack propagation frequently starts on a matrix-rich zone and remains on the plane. Adhesion failure in this matrix leads to bending and delamination of the remaining plies [25] – Figure 9.

Fibre/matrix pull-out is evidenced by fibres pulled out of the matrix, affecting the final value of machined surfaces superficial roughness. This damage is influenced by tool material, anisotropy and the type of loading, that is to say, according to the relative orientation between the fibres and the cutting edge. Fibres can be subjected to a compression loading as they are pushed towards the matrix, to tensile loading or to pure shear. Cutting mechanisms are different, affecting the final value of surface roughness [25].

Intralaminar cracking has its onset in the inner plies of the laminate. Normally, at the start, it has an inclination of 60° to the ply plane. The crack extends according to that direction until it reaches an interlaminar plane, turning then into a delamination, in agreement with the definition established [76].

Burrs are little portions of material broken in a corner or edged surface, but still attached to the part. Their existence is related with cutting edge roundness not allowing a clean cut. Their importance is mainly aesthetic.

Thermal damage is a consequence of friction between part and tool cutting edge, causing localised heating, which has more importance in composites cutting as cooling fluids are not recommended for these materials. An abnormally high temperature of the hole can cause local damage to the matrix, like burning or even melting, if the temperature reaches or exceeds the glass transition temperature (T_g). Low feeds increase the possibility of high temperature generation. Sometimes even fibres can be thermally affected. In a research by Caprino and Tagliaferi [76], with the aim to clarify the interactions between damage and cutting parameters, several microscopic observations showed no thermal damage in the matrix, for all feeds used from 0.0057 to 2.63 mm/rev. The authors reported a strong influence of feed in delamination, being low feeds better to reduce this damage.

If carbon reinforced composites are considered, thermal damages become more serious, due to the low thermal conductivity of carbon fibres. As some fibres are bent instead of being cut, they tend to return to the initial position, causing tightening around the drill and increasing friction. This increase in friction is responsible for added heating of the part and temperatures can reach glass transition temperature of the resin and cause matrix damage. In some cases, matrix material can stick to the drill, interrupting the drilling process and damaging the tool as well [25].

Another problem related with composites drilling is the roundness error, due to material anisotropy. During drilling the tool cutting edge orientation to the fibres varies continuously, as well as the fibre cutting mechanism. When a unidirectional laminate is considered, this orientation goes from 0° to 90°, from 90° to 180°, from 180° to 270° and from 270° to 360°, as the drill completes one revolution – Figure 19. When the orientation goes from 0° to 90°, fibres are compressed and from 90° to 180° they are under tension. From 180° to 360°, the stress cycle repeats inversely. When the orientation is equal to 90°, fibres are mostly bent and so their shrinkage by elastic deformation caused by the action of the cutting edge makes the hole narrower and become elliptical. The angle between the large diameter of the ellipse and the fibre axis is 90°. This phenomenon may be supposed identical in the case of a multidirectional ply lay-up, although ellipse main axis may have different orientations depending on material thickness.

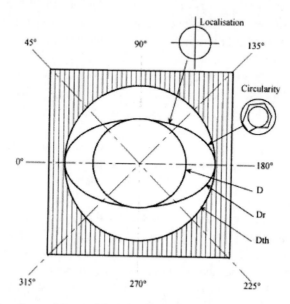

Figure 19. Hole damages observed in a unidirectional plate [25].

4.3. Damage Models

Analysis of delamination during drilling in composite materials using a fracture mechanics approach has been developed and different models presented. The models herewith referred are based in the study of carbon/epoxy laminates, although other materials, like glass/epoxy or hybrid composites are also suitable for their application. The main focus on carbon/epoxy laminates can be explained by the fragile nature of carbon fibres, when compared with glass fibres that are less troublesome in machining study. Delamination mechanisms are assumed to be modelled by linear-elastic fracture mechanics (LEFM), considering the laminate structure of composites, its high modulus of elasticity in direction 1 and the failure in delamination form.

From the known models, the one that is most referred to is the Hocheng and Dharan delamination model [23]. The authors studied the onset of delamination in two different situations: push-out at exit and peel-up at entrance. The first one is the result of the compressive thrust force that the drill exerts on the uncut plies of the laminate, whose thickness is reduced as the drill advances. At some point the loading exceeds the interlaminar bond strength of the material and delamination occurs – Figure 9.

According to authors, the applicability of LEFM to composite has been previously discussed and confirmed, provided that crack growth is collinear and the crack is in a plane of material symmetry.

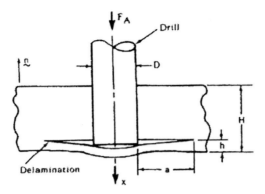

Figure 20. Circular plate model for delamination analysis [45].

As represented in figure 20, drill has a diameter D, F_A is the applied thrust force, x is the displacement, H is the thickness of the structure, h is the uncut depth under the tool and a is the assumed size of an existing crack. The equation of energy balance, from linear elastic fracture mechanics, can be written as

$$G\pi(D+2a)da = F_A dx - dU \qquad (6),$$

where G is the energy release rate per unit area and U is the stored strain energy. The correlation linking F_A, x and U together uses Timoshenko's classical plate bending theory for a circular plate with clamped ends and concentrated loads. The strain energy is given by

$$U = \frac{8\pi M x^2}{\left(a+\dfrac{D}{2}\right)^2} \qquad (7),$$

where M is the flexural rigidity of the plate given by

$$M = \frac{Eh^3}{12(1-v^2)} \qquad (8),$$

and the displacement x is expressed as

$$x = \frac{F_A\left(a+\dfrac{D}{2}\right)^2}{16\pi M} \qquad (9).$$

Substituting (7) to (9) into (6), the critical load at the onset of crack propagation can be calculated

$$F_{crit} = \pi \left[\frac{8G_{Ic}Eh^3}{3(1-\upsilon^2)} \right]^{1/2} \tag{10}.$$

To avoid delamination the applied thrust force should not exceed this value, which is a function of the material properties and uncut thickness. It can be said that a drill geometry that reduces axial thrust forces is able to reduce delamination during composite laminates drilling.

According to the authors, several simplifications are made in this model, considering the values of E and G_{Ic}, giving the results of the critical load on the conservative side.

The other mechanism of delamination, peel-up, is caused by the cutting force pushing the abraded and cut materials to the flute surface – Figure 8. The material spirals up before it is completely machined. A peeling force pointing upwards is introduced that tend to separate the upper laminas of the uncut portion held by the downward acting thrust force. Peel-up effect becomes progressively more difficult as drilling proceeds and resisting thickness becomes greater.

The authors assume that a critical peeling force, F_p*, in the axial direction is related to the horizontal critical cutting force F_c* by a *peeling factor*, k_p, defined as

$$k_p = \frac{F_C^*}{F_P^*} \tag{11}.$$

The peeling factor is a function of tool geometry and friction between tool and workpiece. The LEFM mechanism used for push-out delamination model is here assumed to be applicable. Substituting F_P for F_A and the uncut thickness by the cut thickness, the critical cutting force at the onset of delamination at entrance is

$$F_{Crit}^* = k_p \pi \left[\frac{8G_{Ic}E(H-h)^3}{3(1-\upsilon^2)} \right]^{1/2} \tag{12}.$$

The results from these models show that the cutting at entrance should be limited in order to reduce peel-up action. They also show that the thrust at exit has to be kept to a small finite value to avoid push-out delamination. Since delamination occurs only between laminas, the minimum value of F_A^*, which occurs at exit, is not zero as it is a function of last ply thickness. This value increase when drilled material has higher E and/or G_{IC} values.

Another model is presented by Lachaud et al. [77-78], for the drilling of carbon/epoxy composite laminates with twist drills. Hole entry defect does not appear on every occasion. It is related with the fibrous character of the material and with tool geometry. Circular defect is linked to the presence of an angle created by the direction of the fibres of each ply and by the direction of the cutting edge. Delamination, or hole exit defect, at the exit hole is a consequence of the cutting conditions imposed by the fact that the chisel edge of a twist drill cannot cut through the material – Figure 21. The thrust force of the twist drill, function of feed rate, may cause normal stress which is likely to open the ply interface, corresponding to a crack opening in mode I. Delamination between plies spreads beyond the hole diameter and can occur at varying depths as the drill progresses.

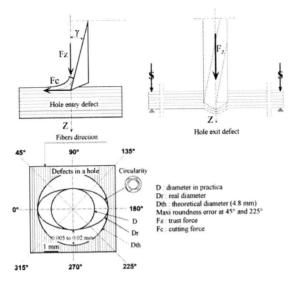

Figure 21. Different damages in a drilled composite plate [77].

In his model, Lachaud considered the existence of a normal stress perpendicular to the ply surface. To obtain the final result, the part of the plate located beneath the drill has been modelled in terms of a thin circular orthotropic plate, with radius a equal to drill radius, and clamped on the laminate surface – Figure 23. This representation does not take into account the global deflection of the plate. It is only valid for a small number of plies under the drill. Two hypotheses are considered:

- distributed load model with resultant F_z;
- point load model with the load F_z concentrated in a point.

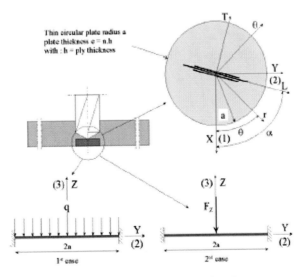

Figure 22. Analytical model with the two hypotheses represented [77].

In the distributed load model, by applying the plate theory, the equilibrium equation of a plate element is

$$\frac{\partial^2 M_{xx}}{\partial x^2} + \frac{\partial^2 M_{yy}}{\partial y^2} + \frac{2\partial^2 M_{xy}}{\partial x \partial y} = -q \qquad (13),$$

with M_{xx}, M_{yy} e M_{xy} bending moments and

$$q = \frac{F_z}{\pi a^2} \qquad (14).$$

The stress/strain law for the part of the plate in contact with the drill is

$$\begin{Bmatrix} M_{xx} \\ -M_{yy} \\ -M_{xy} \end{Bmatrix} = \begin{vmatrix} D_{11} & D_{12} & D_{16} \\ D_{12} & D_{22} & D_{26} \\ D_{16} & D_{26} & D_{66} \end{vmatrix} \begin{Bmatrix} -\dfrac{\partial^2 w}{\partial x^2} \\ -\dfrac{\partial^2 w}{\partial y^2} \\ -2\dfrac{\partial^2 w}{\partial x \partial y} \end{Bmatrix} \qquad (15).$$

The D_{ij} coefficient (bending stiffness) has been calculated by using relations of laminates theory. The terms D_{16} and D_{26} are null in the situation considered – orthotropic plate and axis-symmetrical deflection -, and the equilibrium equation (13) becomes

$$D_{11}\frac{\partial^4 w}{\partial x^4} + 2(D_{12} + 2D_{66})\frac{\partial^4 w}{\partial x^2 \partial y^2} + D_{22}\frac{\partial^4 w}{\partial y^4} = -q \qquad (16).$$

The transversal displacement w of the clamped circular plate, radius a, under uniform loading which is a solution of equation (16) is given by Timoshenko

$$w(r) = \frac{q(a^2 - r^2)^2}{64D} \qquad (17),$$

where

$$D = \frac{1}{8}(3D_{11} + 2D_{12} + 4D_{66} + 3D_{22}) \qquad (18).$$

An energetic approach based on the application of the theorem of virtual work to the equilibrium of the part of the plate affected by the drill, enables the critical drilling load to be determined.

Delamination energy Ud is derived by multiplying the critical energy release rate in mode I (G_{IC}) by the supposedly circular fracture surface

$$Ud = G_{Ic}S = G_{Ic}\pi a^2 .$$ (19).

Considering

$$D' = \frac{D_{11} + D_{12}}{2} + \frac{D_{12} + D_{66}}{3}$$ (20),

the critical value of the thrust force is

$$F_{crit} = 8\pi \left[\frac{2G_{Ic}D}{(1/3) - (D'/8D)} \right]^{\frac{1}{2}}$$ (21).

In the point load model – figure 22 – transversal displacement is given by

$$w(r) = \frac{F}{16\pi D} \left[2r^2 \ln\frac{r}{a} + (a^2 - r^2) \right]$$ (22),

by carrying out the same procedure as for the preceding model, the critical drilling axial thrust force is

$$F_{crit} = 8\pi \left[\frac{2G_{Ic}D}{1 - (D'/8D)} \right]^{\frac{1}{2}}$$ (23).

Test plates of carbon/epoxy resin with quasi-isotropic lay-up were drilled using a twist drill made of K20 tungsten carbide. Different drilled depths were chosen and a perpendicular load was applied to the non-drilled part of the plate. The drill was then used as a punch, without rotation [79]. Results from the experiments and the theoretical model were compared. The values of the distributed load model were closest to experimental results. The experimental setup is presented at Figure 23.

Zhang et al. [80] considered a different approach. In their model the shape of delamination is elliptical, even when multidirectional composites are drilled, as was revealed by drilling experiments. The ellipse has two principal directions, a and b, whose directions are aligned respectively with ply fibre direction and transverse direction.

In this model, when delamination propagates, the ellipticity ratio $a/b = \xi$ remains constant. According to LEFM theory, the energy balance equation can be written as

$$F_{crit}d\omega_0 = G_{IC}dA + dU$$ (24),

where F_C is the critical thrust force at delamination onset, $d\omega_0$ is the drill displacement, G_{IC} is the energy release rate per unit area in mode I, dA is the area of the crack and dU is the infinitesimal strain energy.

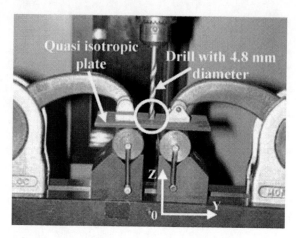

Figure 23. Experimental setup for delamination (punching) test [79].

Drill displacement and infinitesimal strain energy are given by

$$d\omega_0 = 2F_C C_3 a\, da \qquad (25),$$

$$dU = 2KaF_C^2\, da \qquad (26),$$

where C_3 and K come from mechanical characteristics of the laminate.

Substituting equations (26) and (25) into (24), the expression of critical thrust force results:

$$F_{crit} = \sqrt{\frac{\pi G_{IC}}{\xi(C_3 - K)}} \qquad (27).$$

For unidirectional composite laminates, this equation can be simplified to

$$F_{crit} = \frac{3\pi}{\xi}\sqrt{2G_{IC}D^*} \qquad (28),$$

where D^* is a result of flexure properties of the laminate.

Jain and Yang [81] developed a model starting from [23], where the delamination zone has an elliptical shape, and the critical thrust force for the onset of delamination is given by

$$F_{crit} = 3\pi\left(\frac{D_{22}}{D_{11}}\sqrt{2G_{IC}D_C^*}\right)^{1/4} \qquad (29),$$

where

$$D_C^* = 2D_{11} + \frac{2(D_{12} + 2D_{66})}{3}\sqrt{\frac{D_{11}}{D_{22}}} \qquad (30),$$

where all D's are a function of sublaminate flexural rigidity.

Sadat followed the studies by Hocheng and Dharan [23] and Jain and Yang [80] to predict delamination load [82]. In this model the critical value of thrust force for delamination onset is

$$F_{crit} = 8\pi \left[\frac{G_{IC} D}{(1/3) - (D'/8D)} \right]^{1/2} \qquad (31),$$

where

$$D = \frac{1}{8}(3D_{11} + 2D_{12} + 4D_{66} + 3D_{22})$$

$$D' = \frac{D_{11} + D_{12}}{2} + \frac{D_{12} + D_{66}}{3} \qquad (32),$$

where the D_{ij}'s are also a function of sublaminate flexural rigidity.

Jung et al. [83] presented a new formulation for the critical thrust force at delamination propagation in multidirectional laminates. Delamination zone was considered as an elliptical shape under a concentrated load with clamped boundary conditions. The load causes bending, twisting, mid-plane extension and shear of the plate. The starting relation is the energy balance equation

$$F_a dw_o = GdA + dU \qquad (33),$$

where G, dU, dA, F_a and dw_o are the energy release rate per unit area, the infinitesimal strain energy, the increase in the area of the crack, the thrust force and the infinitesimal deflection at the centre of the laminate, respectively.

Deflection is obtained in the form

$$w_o = \frac{F_a a^4}{6\pi abD'} \qquad (34),$$

where a and b are the ellipse axes, and D' is a function of the coefficients of the bending stiffness matrix.

The strain energy U for the elliptical delamination zone is obtained by

$$U = \frac{\pi w_o^2}{a^2}\left(\frac{b}{a}\right)\Delta = \frac{F_a^2 a^4}{36\pi b^2 (D')^2}\left(\frac{b}{a}\right)\Delta \qquad (35).$$

Finally, the critical thrust force comes from (33) to (35), considering the ellipticity ratio a/b at critical thrust, using G_{Ic} instead of G and considering D'_c as another function of the coefficients of the bending stiffness matrix

$$F_{crit} = 6\pi D'_c \left(\frac{a}{b}\right)^{0.25} \sqrt{\frac{G_{IC}}{(6D'_c - \Delta)}} \qquad (36).$$

If the twisting and mid-plane extensions are ignored, the results from this model agree with previous studies.

Hocheng and Tsao derived the critical thrust force as a function of drill bit [84]. For a twist drill the result is equation (10). For a saw drill, Figure 24, the critical thrust force is

$$F_{crit} = \pi \sqrt{\frac{32 G_{Ic} M}{(1 - 2s^2 + s^4)}} \qquad (37)$$

where $s = c/a$, i.e the ratio of drill radius to the assumed size of an existing crack.

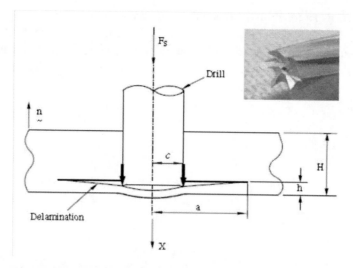

Figure 24. Circular plate model for delamination analysis of saw drill [84].

For a candle stick drill, Figure 25, the critical thrust force is

$$F_{crit} = \pi(1+\alpha)\sqrt{\frac{32 G_{Ic} M}{1 + \alpha^2(1 - 2s^2 + s^4)}} \qquad (38)$$

where α is the ratio of the central concentrated force to the peripheral circular force.

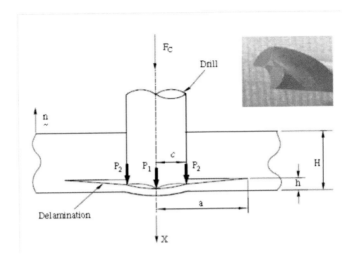

Figure 25. Circular plate model for delamination analysis of candle stick drill [46].

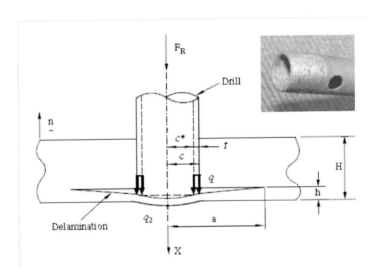

Figure 26. Circular plate model for delamination analysis of core drill [45].

For a core drill, Figure 26, the model becomes more complex because of the peculiar geometry of the drill, with a distributed load over an annulus

$$F_{crit} = \pi \left\{ \frac{32 G_{Ic} M}{1 - [(2 - 2\beta + (3\beta^2/2)) + (4(1 - \beta^2)/\beta(2 - \beta))\ln(1 - \beta)]s^2 + [((2 - 4\beta + 5\beta^2 - 3\beta^3 + \beta^4)/2) + (2(1 - \beta^2)(2 - 2\beta + \beta^2)/\beta(2 - \beta))\ln(1 - \beta)]s^4} \right\}^{1/2}$$

(39)

where β is the ratio between thickness and radius of core drill.

Equation (39) can be expressed in a simpler way if uniform load over the tool diameter is considered [45]

$$F_{crit} = \pi\beta(2-\beta)\sqrt{\frac{32G_{Ic}M}{\{[1-(1-\beta)^4]-(1/2)s^2[1-(1-\beta)^6]\}}} \qquad (40)$$

If an exit backup plate is used to avoid delamination, the effect will be a greater critical thrust force, as demonstrated in the work of Tsao and Hocheng [48]. For the drilling with saw drill with backup the critical thrust force is calculated

$$F_{crit} = \frac{\pi}{2s^2|\ln s|}\sqrt{32G_{Ic}M} \qquad (41)$$

If a core drill is used the critical thrust force with backup is

$$F_{crit} = \pi\sqrt{\frac{32G_{Ic}M}{(k_8 + k_{23} + k_{24}\ln s + k_{25}\ln^2 s)s^4}} \qquad (42)$$

and k_i are derived from tool radius, crack radius and core drill thickness.

Won and Dharan [55] determined quantitatively the effect of chisel edge and pilot hole in composite laminates drilling. The results showed a large reduction in the thrust force when a pilot hole is present, removing the chisel edge contribution. From the experimental results they have formulated nonlinear relationships between drilling forces and drilling parameters.

For specimens without pilot holes this relationship is

$$F = 40.77(fd)^{0.66} - 0.36d^2 \qquad (43).$$

For specimens with pilot holes it becomes

$$F = 3.50(fd)^{0.66} + 0.11d^2 \qquad (44).$$

A delamination model for predicting the critical thrust force for specimens with predrilled pilot holes is then presented. The starting model is equation (6) and LEFM and the final expression is equivalent to equation (45).

They concluded that the potential for delamination can be reduced considerably if drilling is divided in two stages: first with a pilot hole whose diameter is equal to the chisel edge width of the final hole; secondly with the final diameter drill. This strategy divides the total thrust force into two stages.

Tsao and Hocheng [56] also studied the effect of pre-drilling in delamination, showing that the existence of a pre-drilled pilot hole can reduce significantly the occurrence of this damage. A model based in LEFM was presented – Figure 27 – and the final result equivalent to the model presented in [56], with an exception on the consideration of a new variable – ζ – to represent the ratio between pilot hole to final hole diameter.

In this model, the pilot hole is selected equal to the chisel length of the drill, in order to eliminate the disadvantage of the chisel-induced thrust force and avoid the threat of create large delamination by large pre-drilled hole. Considering $2b$ as the diameter of pilot hole and d the drill diameter, the critical thrust force at the onset of crack propagation with pre-drilled pilot hole is

$$F_{crit} = \frac{4\pi}{1-\upsilon}\left\{\frac{G_{IC}Eh^3\left[(1-\upsilon)+2(1+\upsilon)\zeta^2\right]^2}{3(1+\upsilon)\left[2(1-\upsilon)(1+2\upsilon^2)-(12-4\upsilon+3\upsilon^2+3\upsilon^3)\zeta^2-8(1+3\upsilon)\zeta^2\ln\zeta\right]}\right\}^{1/2}$$
(45),

where $\zeta = 2b/d$. The differentiation of equation (45) with respect to ζ and letting the result equal to zero, gives the value of ζ^* at the minimum of the critical thrust force

$$\zeta^* = \frac{\partial F_A^*}{\partial \zeta} = \zeta^2\left(-16+16\upsilon+26\upsilon^2-12\upsilon^3-6\upsilon^4-16\ln\zeta-64\upsilon\ln\zeta-48\upsilon^2\ln\zeta\right)+\ln\zeta+$$
$$(24(8+16\upsilon-24\upsilon^2)-8\upsilon+3\upsilon^2-19\upsilon^4)=0$$
(46)

This value is a function of Poisson's ratio v. When $v = 0.3$, ζ is found equal to 0.1176. The higher the value of v, the larger is ζ^*, see Figure 28.

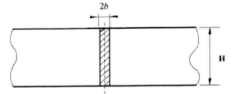

Pre-drilled hole of diameter $2b$

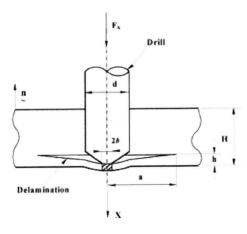

Figure 27. Delamination model for pre-drilled holes [56].

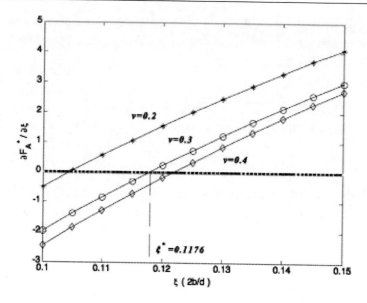

Figure 28. Critical ratio of chisel edge length to drill diameter with various Poisson's ratio [56].

For a twist step drill, Figure 29, the critical thrust force at the onset of delamination is [45]

$$F_{crit} = \pi[1-(i\xi)^2]\sqrt{\frac{32G_{Ic}M}{\{[1-(i\xi)^4]-(1/2)s^2[1-(i\xi)^6]\}}} \qquad (47)$$

where $i=1-n$, being n the number of sequential increment of secondary lips in action and ξ the divided thickness of the last lamina according to n.

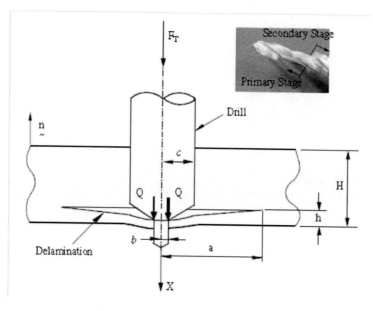

Figure 29. Circular plate model for delamination analysis of step drill [46].

In a similar way Tsao [85] derived the thrust force of the saw drill with pilot hole at the onset of delamination

$$F_{crit} = \pi \sqrt{\frac{32G_{Ic}M}{(1-4s^2+5s^4-2s^6-4s^2\ln s+\frac{(1-\eta^2)s^2(1-s^2)(2\ln s+1-s^2)}{[(1+\upsilon)\eta^2+(1-\upsilon)]^2})}} \tag{48}$$

where η is the ratio of the pilot hole radius to crack radius.

Finally, for the thrust force of the core drill with pilot hole [86]

$$F_{crit} = \pi \sqrt{\frac{32G_{Ic}M}{(k_3-k_7+\frac{k_7k_9}{2}-k_{11})-\frac{(1-\upsilon)k_7k_{10}}{2}+k_5\ln\frac{[1+\eta(1+\gamma)]s}{2}+[1+\eta(1+\gamma)]^2k_4s^2+[1-\eta^2(1+\gamma)^2][1-\frac{k_9}{2}+\frac{(1-\upsilon)k_{10}}{2}]k_8s^2}} \tag{49}$$

where γ is the ratio between pilot hole to inner uncut portion of core drill.

For all tool geometries similar results indicate that although the critical thrust force is reduced with pilot hole, the reduction of thrust force during second stage drilling is large enough to compensate this effect. Thus, it is possible to drill at higher feed rates without delamination damage.

5. DAMAGE EVALUATION

After laminate holes are drilled, it is important to establish criteria that can easily compare the delamination extension caused by different processes, even though they can only be applied to composites with the same lay-up regarding orientation and number of plies. Damage extension can be evaluated through nondestructive testing, using, for example, a tool maker's microscope [37, 39, 53, 87], ultrasound [88], acoustic emission [89], radiography [90, 91] – Figure 30a), C-Scan [92] – Figure 30b) or computerized tomography (CT) [92-94] – Figure 30c), in order to obtain images representing the hole surrounding areas that can be further analyzed.

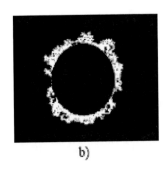
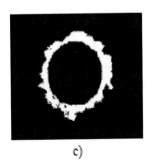

Figure 30. Damage evaluation: a) radiography; b) C-Scan [92]; c) computerized tomography [92].

The second step is the measurement of the damaged diameter or area in order to be able to calculate a factor that expresses the damaged region – Figure 31. Chen [95] presented a comparing factor that enables the evaluation and analysis of delamination extent in laminated composites. That ratio was called the *Delamination Factor* F_d and it was defined as the quotient between the maximum delaminated diameter D_{max} and the hole nominal diameter D,

$$F_d = D_{max}/D \tag{50}$$

In the experimental work presented in [95], the author examined the effects of tool geometry and cutting parameters as well as tool wear on delamination factor. Two types of drills were used: a carbide drill and a HSS drill with 5 mm diameter. Damage zone was evaluated by using X-ray non-destructive inspection. Results showed a near-linear relationship between the delamination factor and average thrust forces for both drill materials. The author also concluded that thrust force increased when drill point angle increases and that helix angle did not have a significant effect on this force. Tool flank wear causes an increase of delamination factor, as thrust force increases with tool wear.

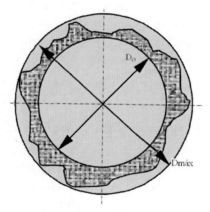

Figure 31. Measurement of maximum delaminated diameter.

Although feed rate has a strong influence on thrust force, cutting speed did not show a significant effect on that force. Finally, he has noticed the absence of built up edge during carbon/epoxy machining.

Tsao and Hocheng [96] evaluated the delamination factor in the use of three different drills, computing the results with images obtained from ultrasonic C-Scan of drilled holes. In the experimental work, Taguchi's method and ANOVA were used in order to establish the influence of the different factors involved. At the end, it was found that feed rate, as well as drill diameter, have the largest contribution on drilling performance, but the use of an adequate tool can reduce delamination.

Davim et al. [87, 97] studied the effect of drilling parameters on delamination factor, based on Taguchi's method and ANOVA. The authors concluded that feed rate has the greater influence on specific cutting pressure and thrust force, and when feed rate increases also the damage increases. The same effect is verified when cutting speed is higher, but is not so significant. Brad drill produces less delamination, both at entrance and exit sides.

Mehta et al. [98] have suggested a different ratio with the same purpose, named Damage Ratio (D_{RAT}), defined as the ratio of Hole Peripheral Damage Area (D_{MAR}) to Nominal Drilled Hole Area (A_{AVG}), i.e.,

$$D_{RAT} = D_{MAR}/A_{AVG} \qquad (51)$$

This hole damage evaluation method is based on the existence of damage images from C-Scan and pixel counting of the digitized damage area, as described in [98], or from digitized radiographs [99]. In the study referred there, the authors assessed hole quality executed with three different types of drill. Statistical analysis showed that drill type was the factor having the greatest influence on hole quality. Other conclusions from this study were that the use of a backup plate only act to reduce the push-out delamination and roughness did not appear to be a reliable parameter on hole quality measurement. Based on the results of Damage Ratio observed, a value of peripheral damage around two should not be exceeded.

One limitation of Chen's criterion is related to situations when the delamination involved isn't round, but presents breaks and cracks. In such cases, the values of the delaminated area are more appropriated for the damage quantification. Based on this, Davim et al. [100] presented a novel approach known as the adjusted delamination factor $- F_{da}$,

$$F_{da} = \alpha \frac{D_{max}}{D} + \beta \frac{A_{max}}{A}, \qquad (52)$$

where A_{max} is the area corresponding to the maximum delaminated diameter and A_o the nominal hole area. In this new criterion, the first term is the conventional delamination factor and a second term is added, taking into account the damaged area contribution and the parameters α and β are used as weights, being their sum always equal to 1 (one).

6. MECHANICAL TESTS

The main purpose of hole making in any plate is to have the possibility to assemble the part with others in a more complex structure. In service, parts will be subjected to efforts that will cause stress at the hole surrounding area where a screw, bolt or rivet was placed. It is important for design engineers to know what the load carrying capacity of a connection is. In order to acquire the indispensable knowledge and establish safety coefficients, mechanical tests are necessary. From ASTM standards, two test involving reinforced composites plates with a drilled hole can be found: Bearing Test [101] and Open-hole tensile strength [102].

Bearing test is used to determine the bearing response of multidirectional polymer matrix composite laminates reinforced by high modulus fibres. Although the test standard describes two procedures, A and B, procedure A is more widely used [57, 59, 103-105] for every reinforced composites, glass fibre, carbon fibre or even hybrid composites [99]. Figure 32 shows schematically the test assembly. In Figure 33 the bearing test failure modes according to ASTM standard are schematically presented. This procedure uses a single fastener and is recommended for basic material evaluation and comparison. It has to be kept in mind that damage due to hole preparation will affect strength results. So, this test can be useful to compare different hole machining methods. Results of the bearing strength are usually correlated with delamination extension. Greater delamination corresponds to lower bearing strength.

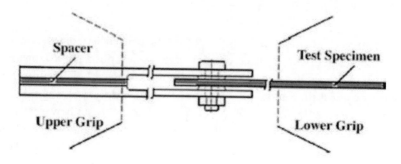

Figure 32. Bearing test assembly [106].

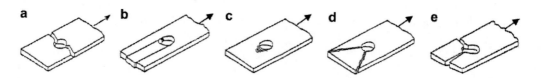

Figure 33. Bearing test failure modes according to ASTM D5961/D5961M-05: a) lateral (net-tension); b) shear-out; c) bearing; d) tearout; e) cleavage [106].

Another kind of test is the pin bearing test, which was used by Khashaba et al. [14] to evaluate the effect of feed and speed on bearing strength of drilled hole. Test fixture is represented in Figure 34. The conclusions were similar to those of bearing test, showing that higher thrust force will cause higher delamination, thus reducing bearing strength.

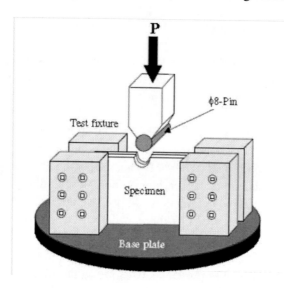

Figure 34. Pin bearing test fixture [14].

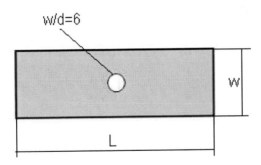

Figure 35. Open-hole test specimen.

Aktas [107] investigated both the static and dynamic bearing strengths of a carbon epoxy laminates. For static tests the author adopted ASTM D953 [108]. The fatigue tests were conducted according to ASTM STP 749 [109]. His work was focused on the effect of the stacking sequence.

Open-hole tensile strength test is intended to evaluate the effect of the presence of a hole in the middle of the plate, Figure 35. However, test description refers the need of an adequate hole preparation in order to avoid influence on results. It is possible that the existence of delamination has the opposite effect as it can blunt the stress concentration caused by the presence of the hole, thus increasing the load-carrying capacity of the specimen. This test is referred in the work of Rao et al. [105] where a correlation between delamination extension and notched tensile strength was found and in a Taguchi analysis conducted by Kishore et al. [110]. In this work, the authors concluded that the drilling induced damage affects the residual tensile strength. Both works were carried out on glass fibre reinforced epoxy composites. A final reference to the review work published by Thoppul et al. [106] comparing relevant mechanical test methods and standards used in the experimental characterization and strength prediction of mechanically fastened joints in fibre reinforced plastics. Open-hole test and bearing test are described as well as their relevance and factors that can affect the final result like joint design, washer size, stacking sequence or initial clamping pressure. Generally, these factors and their influence on bearing strength need to be further investigated.

Fatigue tests to evaluate the effect of machining damage on carbon/epoxy composite laminates were carried out by Persson et al. [20, 21]. For these tests the authors used a servo hydraulic fatigue testing machine and a frequency of 5 Hz, amplitude of 66% of the static strength of drilled specimens and a load ratio (R) of zero, meaning no compression load. From the results of the fatigue test showed that, according to diverse damage extension, a difference of 8 to 10% in fatigue strength was verified. Identical conclusion was possible if fatigue life is used as evaluation factor, showing that holes with less delamination increase fatigue life of the composite plate.

CONCLUSION

This chapter was concerned on the machining of polymeric matrix composite materials. From the diverse operations that are common in order to achieve the final shape of a workpiece, drilling is by far the most usual, due to the need of parts assembly.

New problems arise during the machining of composite materials. Unlike metal machining, where there are a vast number of handbooks, there are few publications addressing composites machining. Experience acquired from metal cutting cannot be directly transplanted to fibre reinforced plastics. The reinforcement fibres are very abrasive leading to rapid tool wear and deterioration of machined surfaces. During drilling operation, the tool comes across matrix and reinforcement materials, alternatively. These two constituents are, by definition of a composite, different in nature and in properties. So, the machining of these materials deals with unique requirements on tool geometry and abrasion resistance.

It is generally accepted that the fibre type, content and structure are the governing factors in tool and cutting parameters selection. Thrust forces are to be kept low as they are associated to delamination, tool wear and general hole quality.

Delamination is referred as the most critical type of damage that can occur during a drilling operation as it can be responsible for a severe loss in load carrying capacity of a composite plate. Thus, good hole drilling processes are essential, based on a knowledge of reinforcement used, ply stacking sequence and fibre orientation. A large range of cutting speeds and feeds can be found in different studies as a function of the process factors, showing that there is still a long way to go until quality and reliability are combined with satisfactory tool life. In order to avoid delamination, feed rate must be as low as possible as it appears to be the most important factor on delamination onset. The importance of cutting speed appears to be small when compared to feed.

Another important factor is tool geometry. A dedicated tool geometry, able to drill with lower thrust force, will induce less damage on the composite. Endless work can be found on this particular aspect regarding drill bit geometry. It is possible to find work published on twist drills, saw drills, candle stick drills, core drills, Brad drills, Dagger drills, orbital drilling, step drills, pilot hole drilling and other alternatives. Step drill and pilot hole drilling reduce central material removal cancelling the chisel edge effect of a twist drill bit. Thus, delamination formed in the first stage drilling is removed during second stage drilling. Tool material seems to be a closed issue. Depending on material and number of holes, carbide drills or PCD drills should be used. Finally, it is noteworthy that conclusions that are valid for a certain type of composite, as a result of fibre nature and orientation may not be valid if some material parameter is changed.

It is recognized that much experimental and theoretical work must be carried out before optimization of the machining conditions for composite materials can be accomplished.

Nowadays, production of higher quality holes in polymeric matrix composites, with damage minimization, still offers a challenge to everyone involved in this industry.

REFERENCES

[1]   Groover, M. P., Fundamentals of Modern Manufacturing : Materials, processes, and systems, Prentice Hall, New York, 1996, pp 543-611.

[2]   Abrate, S., Composites Engineering Handbook, P. K. Mallick, Marcel Dekker, New York, 1997; pp 777-809.

[3]   Wern, C. W.; Ramulu, M.; Shukla, A. *Experimental Mechanics* 1994, pp 33 – 41.

Machining of Fibre-Reinforced Composites 435

[4] Boldt, J. A.; Chanani, J. P. Engineered Materials Handbook, ASM Intern. *Handbook Committee*, 1987, Vol.1, pp 667-672.

[5] Koplev, A.; Lystrup, Aa.; Vorm, T. *Composites*, 1983, 14, pp 371-376.

[6] Wang, X. M.; Zhang, L. C. *Int. J. Machine Tools Manufacture*, 2003, 43, pp 1015-1022.

[7] Klocke, F.; Koenig, W.; Rummenhoeller, S.; Wuertz, C. Machining of Ceramics and Composites, Ed Marcel Dekker, New York, 1998, pp 249-266.

[8] Krishnamurthy, R.; Santhanakrishnan, G.; Malhotra, S. K. Proc Machining Composite Materials Symposium, *ASM Materials* Week 1992, pp 139-148.

[9] Sreejith, P.S.; Krishnamurthy, R.; Malhotra, S. K.; Narayanasamy K. *J. Materials Processing Technology*, 2000, 104, pp 53-58.

[10] El-Sonbaty, I.; Khashaba, U. A.; Machaly, T. Composite Structures 2004, 63, pp 329-338.

[11] Mohan, N.S.; Ramachandra, A.; Kulkarni, S.M. *Composite Structures* 2005, 71, pp 407–413.

[12] Sardiñas, R.Q.; Reis, P.; Davim, J.P. *Composites Science Technology* 2006, 66, pp 3083–3088.

[13] Gaitonde, V.N.; Karnik, S.R.; Mata, F.; Davim, J.P. *J. Composite Materials* 2007, 43, pp 725-739.

[14] Khashaba, U.A.; El-Sonbaty, I.A.; Selmy, A.I.; Megahed, A.A. *Composites A* 2010, 41, pp 391–400.

[15] Khashaba, U.A.; El-Sonbaty, I.A.; Selmy, A.I.; Megahed, A.A. *Composites A* 2010, 41, pp 1130-1137.

[16] Chandrasekharan, V.; Kapoor, S. G.; DeVor, R. E. *J. Engineering Industry* 1995, 117, pp 559-570.

[17] Langella, A.; Nele, L.; Maio, A. *Composites A* 2005, 36, pp 83-93.

[18] Fernandes, M.; Cook, C. *Int. J. Machine Tools Manufacture* 2006, 46, pp 70–75.

[19] Persson, E.; Eriksson, I.; Zackrisson, L. *Composites A* 1997, 28, pp 141-151.

[20] Persson, E.; Eriksson, I.; Hammersberg, P. *J. Composite Materials* 1997, 31, pp 383-408.

[21] Hamdoun, Z.; Guillaumat, L.; Lataillade, J. L. Proc ECCM 11, Rhodes, Greece, May 2004.

[22] Hocheng, H.; Tsao, C.C. *J. Materials Processing Technology* 2005, 167, pp 251–264.

[23] Hocheng, H.; Dharan, C. K. H. *J Engineering Industry* 1990, 112, pp 236-239.

[24] Hocheng, H.; Puw, H. Y. *Int. J. Machine Tools Manufacture* 1992, 32, pp 583-592.

[25] Piquet, R.; Ferret, B.; Lachaud, F.; Swider, P. *Composites A* 2000, 31, pp 1107-1115.

[26] Abrão, A.M.; Faria, P.E.; Campos Rubio, J.C.; Reis, P.; Davim, J.P. *J. Materials Processing Technology* 2007, 186, pp 1–7.

[27] Schulze, V.; Becke, C.; Weidenmann, K.; Dietrich, S. *J. Materials Processing Technology* 2011, 211, pp 329–338.

[28] Tagliaferri, V.; Caprino, G.; Ditterlizzi, A. *Int. J. Machine Tools Manufacture* 1190, 30, pp 77-84.

[29] Bongiorno, A.; Capello, E.; Copani, G.; Tagliaferri, V. Proc ECCM-8, Naples, Italy 1998 Vol. II, pp 525-532.

[30] Khashaba, U. A. *J. Composite Materials* 1996, 30, pp 2042-2055.

[31] Khashaba, U. A. *Composite Structures* 2004, 63, pp 313-327.

[32] Khashaba, U.A.; Seif, M.A.; Elhamid, M.A. *Composites A* 2007, 38, pp 61–70.

[33] Davim, J. P.; Reis, P.; António, C. C. *Composites Science Technology* 2004, 64, pp 289-297.

[34] Velayudham, A.; Krishnamurthy, R.; Soundarapandian, T. *Int. J. Machine Tools Manufacture* 2005, 45, pp 399-496.

[35] Ramkumar, J.; Malhotra, S. K.; Krishnamurthy, R. *J. Materials Processing Technology* 2004, 152, pp 329-332.

[36] Aoyama, E.; Nobe, H.; Hirogaki, T. *J. Materials Processing Technology* 2001, 118, pp 436-441.

[37] Palanikumar, K.; Campos Rubio, J.; Abrão, A.; Esteves, A.; Davim, J.P. *J Reinforced Plastics Composites* 2008, 27, pp 1615-1623.

[38] Lin, S-C.; Shen, J-M.; *J. Composite Materials* 1999, 33, pp 827-851.

[39] Campos Rubio, J.; Abrao, A.M.; Faria, P.E.; Esteves Correia, A.; Davim, J.P. *Int. J. Machine Tools Manufacture* 2008, 48, pp 715–720.

[40] Park, K. Y.; Choi, J. H.; Lee, D. G. *J. Composite Materials* 1995, 29, pp 1988-2002.

[41] Murphy, C.; Byrne, G.; Gilchrist, M. D. *Proc Instn Mech Engrs* 2001, 216 Part B, pp 143-152.

[42] Dharan, C. H. K.; Won, M. S. *Int. J. Machine Tools Manufacture* 2000, 39, pp 415-426.

[43] Stone, R.; Krishnamurthy, K. *Int. J. Machine Tools Manufacture* 1996, 36, pp 985-1003.

[44] Hocheng, H.; Puw, H. Y.; Yao, K. C. *Proc Machining Composite Materials Symp, ASM Materials Week* 1992, pp 127-138.

[45] Hocheng, H.; Tsao, C.C. *J. Materials Processing Technology* 2003, 140, pp 335-339;

[46] Hocheng, H.; Tsao, C.C. *Int. J. Machine Tools Manufacture* 2006, 46, pp 1403–1416.

[47] Tsao, C.C.; Hocheng, H. *J. Materials Processing Technology* 2007, 192–193, pp 37–40.

[48] Tsao, C.C.; Hocheng, H. *Int. J. Machine Tools Manufacture* 2005, 45, pp 1261-1270.

[49] Won, M. S.; Dharan, C. H. K. *Trans ASME J. Manufacturing Science Engineering* 2002, 124, pp 778-783.

[50] Enemuoh, E. U.; El-Gizawy, A. S.; Okafor, A. C. *Int. J. Machine Tools Manufacture* 2001, 41, pp 1795-1814.

[51] Linbo, Z.; Lijiang, W.; Xin, W. *Composites A* 2003, 34, pp 237-244.

[52] Iliescu, D.; Gehin, D.; Gutierrez, M.E.; Girot, F. *Int. J. Machine Tools Manufacture* 2010, 50, pp 204–213.

[53] Campos Rubio, J.C.; Abrão, A.M.; Faria, P.E.; Correia, A.E., Davim, J.P. *J. Composite Materials* 2008, 42, pp 1523-1532.

[54] Tsao, C.C. *Int. J. Machine Tools Manufacture* 2007, 47, pp 2172–2176.

[55] Won, M. S.; Dharan, C. H. K. *Trans. of ASME J. Manufacturing Science Engineering* 2002, 124, pp 242-247.

[56] Tsao, C. C.; Hocheng, H. *Int. J. Machine Tools Manufacture* 2003, 43, pp 1087-1092.

[57] Durão, L.M.P.; Magalhães, A.G.; Marques, A.T.; Baptista, A.M.; Figueiredo, M. *Materials Science Forum* 2008, 587-588, pp 706-710.

[58] Durão, L.M.P.; Gonçalves, D.J.S.; Tavares, J.M.R.S; de Albuquerque, V.H.C.; Vieira, A.A.; Marques, A.T. *Composite Structures* 2010, 92, pp 1545-1550.

[59] Durão, L.M.P.; Marques, A.T.; Magalhães, A.G.; Silva, J.F.; Tavares, J.M.R.S. *Composites Science Technology* 2009, 69, pp 2376–2382.

Machining of Fibre-Reinforced Composites 437

[60] Tsao, C.C. *Int. J. Machine Tools Manufacture* 2006, 46, pp 1653–1661.

[61] Tsao, C.C. *Materials Design* 2008, 29, pp 1740–1744.

[62] Tsao, C.C. *Int. J. Advanced Manufacturing Technology* 2008, 36, pp 11–18.

[63] Di Ilio, A.; Tagliaferri, V.; Veniali, F. *Int. J. Machine Tools Manufacture* 1991, 31, pp 155-165.

[64] Di Ilio, A.; Tagliaferri, V.; Veniali, F. *Proc Machining Composite Materials Symposium, ASM Materials Week* 1992, pp 199-203.

[65] Sadat, A. B., "Machining of composites", *International Encyclopaedia of Composites*, 1991, Vol. 3, pp 95-102.

[66] Herzog, D.; Jaeschke, P.; Meier, O.; Haferkamp, H. *Int. J. Machine Tools Manufacture* 2008, 48, pp 1464– 1473.

[67] Li, Z.L.; Zheng, H.Y.; Lim, G.C.; Chu, P.L.; Li, L. *Composites A* 2010, 41, pp 1403–1408.

[68] Davim, J.P.; Barricas, N.; Conceição, M.; Oliveira, C. *J. Materials Processing Technology* 2008, 198, pp 99–104.

[69] Korican, J. Engineered Materials Handbook, *ASM Int. Handbook Committee* 1987, Vol.1, pp 673-675.

[70] Shanmugam, D.K.; Nguyen, T.; Wang, *J. Composites A* 2008, 39, pp 923–929.

[71] Ramulu, M.; Arola, D. *Composites* 1993, 24, pp 299-308.

[72] Hocheng, H.; Tsai, H. Y.; Shiue, J. J.; Wang, B. *Trans ASME J Manufacturing Science and Engineering* 1997, 119, pp 133-141.

[73] Azmir, M.A.; Ahsan, A.K. *J. Materials Processing Technology* 2009, 209, pp 6168–6173.

[74] Hocheng, H.; Tai, N. H.; Liu, C. S. *Composites A* 2000, 31, pp 133-142.

[75] Strong, A. B. *International Encyclopaedia of Composites*, 1991, Vol. 2, pp 1-6.

[76] Caprino, G.; Tagliaferri, V. *Int. J. Machine Tools Manufacture* 1995, 35, pp 817-829.

[77] Lachaud, F.; Piquet, R.; Collombet, F.; Surcin, L. *Composite Structures* 2001, 52, pp 511-516.

[78] Piquet, R.; Lachaud, F.; Ferret, B.; Swider, P. Mécanique. Industriel 2000, 1, pp 105-111.

[79] Zitoune, R.; Collombet, F. *Composites: Part A* 2007, 38, pp 858–866.

[80] Zhang, L-B.; Wang, L-J.; Liu, X-Y. *Proc Instn Mech Engrs* 2001, 215 Part B, pp 135-146.

[81] Jain, S.; Yang, D. C. H. *J. Engineering Industry* 1993, 115, pp 398-405.

[82] Cit. by Kim, D. W. PhD thesis, Univ. of Washington, 2002.

[83] Jung, J. P.; Kim, G. W.; Lee, K. Y. *Composite Structures* 2005, 68, pp 391-397.

[84] Tsao, C.C. *Int. J. Machine Tools Manufacture* 2007, 47, pp 1132-1138.

[85] Tsao, C.C. *Int. J. Machine Tools Manufacture* 2007, 47, pp 2172–2176.

[86] Tsao, C.C. *Int. J. Machine Tools Manufacture* 2006, 46, pp 1653–1661.

[87] Davim, J.P.; Reis, P. Materials Design 2003, 24, pp 315-324.

[88] Hosur, M.V.; Chowdhury, F; Jeelani, S. *J. Composite Materials* 2007, 41, pp 2195-2212.

[89] Jong, H. J. *J. Composite Materials* 2006, 40, pp 37-69.

[90] de Albuquerque, V.H.C.; Tavares, J.M.R.S.; Durão, L.M.P. *J. Composite Materials*, 2010, 44, pp 1139-1159.

438   Luís Miguel Durão

[91] Johnson, W. S.; Treasurer, P.; Woodruff, G. W. *J. Composite Materials* 2008, 42, pp 2143-2161.

[92] Tsao, C.C.; Hocheng, H. Int. J. Machine Tools Manufacture 2005, 45, pp 1282–1287.

[93] Wang, L.B.; Frost, J.D. Voyiadjis, G.Z.; Harman, T.P. *Mechanics Materials* 2003, 35, pp 777–790.

[94] Schilling, P.J.; Karedla, B.P.R.; Tatiparthi, A.K.; Verges, M.A.; Herrington, P.D. *Composites Science Technology* 2005, 65, pp 2071–2078.

[95] Chen, W. C. *Int. J. Machine Tools Manufacture* 1997, 37, pp 1097-1108.

[96] Tsao, C. C., Hocheng, H. *Int. J. Machine Tools Manufacture* 2004, 44, pp 1085-1090.

[97] Davim, J. P.; Reis, P. Proc VI Congr Ibero-Americano de Engenharia Mecânica, Coimbra 2003, pp 1079-1084.

[98] Mehta, M.; Reinhart, T. J.; Soni, A. H. *Proc. Machining Composite Materials Symposium*, ASM Materials Week 1992, pp 113-126.

[99] Durão, L. M. P.; Magalhães, A. G.; Tavares, J. M. R. S.; Marques, A. T. Electronic Letters Computer Vision Image Analysis (ELCVIA) 2008, 7(2), pp 11-21.

[100] Davim, J.P.; Campos Rubio, J.C.; Abrão, A.M. *Composites Science Technology* 2007, 67, pp 1939-1945.

[101] ASTM D 5961M-01, *ASTM INTERNATIONAL*, USA 2001.

[102] ASTM D 5766M-07, *ASTM INTERNATIONAL*, USA 2007.

[103] Wang, Y. *J. Composite Materials* 2002, 36, pp 2199-2216.

[104] Aktas, A. *J Reinforced Plastics Composites* 2007, 26, pp 555-564.

[105] Rao, B.S.; Rudramoorthy, R.; Srinivas, S.; Rao, B.N. *Materials Science Engineering A* 2008, 472, pp 347–352.

[106] Thoppul, S.D.; Finegan, J.; Gibson, R.F. *Composites Science Technology* 2009, 69, pp 301–329.

[107] Aktas, A. *Composite Structures* 2005, 67, pp 485-489.

[108] ASTM D 953- 10, *ASTM INTERNATIONAL*, USA 2010.

[109] ASTM STP 749, Joining of Composite Materials, Ed. K. Kedward, *ASTM INTERNATIONAL*, USA, 1981.

[110] Kishore, R.A.; Tiwari, R.; Dvivedi, A.; Singh, I. *Materials Design* 2009, 30,pp 2186–2190.

In: Fiber–Reinforced Composites
Editor: Qingzheng Cheng

ISBN: 978-1-61470-303-7
© 2012 Nova Science Publishers, Inc.

Chapter 14

ANN APPLICATIONS IN MACHINING OF FIBER-REINFORCED COMPOSITES

Devi K. Kalla[1], Jamal Sheikh-Ahmad[2] and Janet Twomey[1]*
[1]Department of Industrial and Manufacturing Engineering,
Wichita State University, Wichita, KS 67260, USA
[2]Mechanical Engineering Program, The Petroleum Institute,
Abu Dhabi, UAE

ABSTRACT

Composite materials are inhomogeneous in nature and the study of their behavior during secondary manufacturing is rather difficult due to this complex nature. During machining of fiber reinforced composites, the basic mechanisms of chip formation and the resulting effects on machinability are greatly influenced by the material composition and architecture. The chip formation mode, type of chip produced, cutting force and surface quality are influenced by fiber type, fiber orientation, and cutting tool geometry. The advancement of the cutting tool against alternating hard fibers and soft matrix phases generates cyclic cutting forces and nonuniform surface characteristics. In addition, brittle failure of the fibers and the fiber-binder interfaces induce surface and subsurface cracks and delamination. Thus, the experimental study of machining of composites often involves the analysis and modeling of complex phenomena relying on noisy data that is highly nonlinear and lacking normality. Such data may not be easily analyzed by rigid mathematical models.

Artificial Neural Network (ANN) methods have emerged as a powerful and flexible approximating and predictive tool that is capable of self-learning and adapting to complex data sets. The advantages of ANN over mathematical methods are that it requires no explicit mathematical solutions or restrictive assumptions, and it is faster than other algorithms because of the use of parallel computing. ANN has been successfully applied in the analysis and modeling of machining fiber reinforced composites. This includes the prediction of tool wear, surface roughness, cutting forces and delamination. ANN has also been utilized in devising process control techniques for preventing and

---

* Corresponding author present address: Industrial and Manufacturing Engineering Department, Wichita State University, Box 35, Wichita, KS, USA. E-mail: devi.kalla@wichita.edu.

minimizing delamination. Special types of ANN structures and training methods were developed specifically to address the type of data obtained from composites machining. Significant improvements in the model predictive capabilities were achieved when ANN methods are compared to multiple regression methods. In this chapter we review recent advancements in neural network techniques and their application in machining FRP composites. Future trends of this modeling technology are also discussed.

Keywords: Fiber-reinforced composites, Artificial Neural Network, machining, bootstrap committee networks, cutting forces, specific cutting energy, delamination, surface roughness, tool wear.

1. INTRODUCTION

Fiber-reinforced polymers (FRPs) have attracted a great deal of interest in modern industries due to superior properties such as high specific strength and stiffness, low weight and directional properties compared to conventional metallic materials. All these promising properties make these FRPs interesting potentials for a wide variety of application in several fields of engineering such as aerospace, automobiles, space research, transportation etc. Although FRPs are produced as near net shapes, the machining has to be carried out in the final stage of assembly for the required finish. The FRP machining is completely different from that of metals and the theory and experience from metals cannot be directly applied. Due to inhomogeneity of FRP composites, they do not machine by plastic deformation in a similar way to ductile metals, producing a uniform chip. They rather machine by fracture of the matrix and the reinforcement fibers in a powdery-like chip. The chip formation mode, type of chip produced, cutting forces and surface quality are influenced by fiber type, fiber orientation, and cutting tool geometry. However, it has been reported that the strong anisotropy and inhomogeneity of FRP introduces many specific problems in machining, such as fiber pullout, delamination, surface damage, burrs and burning.

Inspired by the neural architecture and operation of the brain, an ANN approach is been used as a powerful mathematical tool for a wide variety of complex scientific and engineering problems. ANN is proven to be a promising field of research in predicting experimental trends and has become increasingly popular as they can often solve problems much faster compared to other approaches with the additionally ability to learn [1]. A neural network has several features, which make it suitable for material modeling. It has learning capability and can acquire, store and retrieve knowledge. The learning capabilities allow neural networks to be directly trained with the results of experiments. Neural networks, which can be defined as massively parallel computational models for knowledge representation and information processing, have unique learning capabilities that can be used in learning complex nonlinear relationships, and they are also noise and fault tolerant and truly adaptive systems. These types of characteristics are found to be suitable for the material modeling task, which often deals with material behavior that contains nonlinearity, path dependency, and uncertainty.

2. ANN PREDICTION METHODS

ANN prediction methods are models that try to capture the functional relationship between a set of input (predictor) variables and a set of output (response) variables. ANN models develop a mapping from the input variables to the output variables through an iterative learning process with little or no prior assumption of model form. A multi-layered feed-forward neural network has been proven to be suitable and effective as material models in computational mechanics. The multi-layered neural network is the most widely applied neural network, which has been utilized in most of the research related to the composites [2]. A back propagation algorithm can be used to train these multilayer feed-forward networks with differentiable transfer functions to perform function approximation, pattern association and pattern classification [3].

Typically, statistical (nonparametric regression) prediction and neural network prediction are measured in terms of a mean squared error (MSE) of a the targeted function, $y=g(x)$, and of the ANN approximation function, $\hat{f}(T_n,x)$, respectively. The goal is to minimize this MSE, which consists of the sum of two terms, $bias^2$ and $variance$. Since the dependent function $g(x)$ is usually unknown, $g(x)$ is often defined as the expected value of the output y corresponding to any input x, $E[y|x]$. Therefore, the mean squared distance or error between $\hat{f}(T_n,x)$ and $E[y|x]$ is given by:

Mean squared error $bias^2 \rightarrow$ - the squared error between the expected

$$E[(E[y|x] - \hat{f}(T_n,x))^2] = \hat{f}(T_n,x) \text{ and the targeted function:}$$

$$(E[\hat{f}(T_n,x)] - E[y|x])^2 + variance \rightarrow \text{ - variance of approximating function } \hat{f}(T_n,x)$$
error:

$$E[|\hat{f}(T_n,x) - E[\hat{f}(T_n,x)]|^2]. \tag{1}$$

From preceding expression, we can notice that the dilemma has been formed according to: first, nonparametric models require large amount of data to insure convergence and to control variance. The problem with large amount of data is slow convergence rate, and large sample sizes are rarely found in the real world. Leaving us to deal with small sample sizes, in which the estimators may become too dependent on the particular sample resulting in high variance of the estimators. Second, since real world models are unknown and difficult to identify, a good model that fits data well and causes significant reduction in bias is hard to come up. However, a highly parameterizing model may fit data too close and result in variance increase.

In order to reduce both bias and variance while guarantee convergence, a statistical nonparametric estimator (including ANN estimator) requires good model structure and large numbers of observations. The following methodologies try to reduce both bias and variance.

2.1. Bootstrapping

The bootstrap method of error estimation was originated by Efron [4]. It gives the nonparametric maximum likelihood estimate of the excess error of a prediction model that corrects for the bias of the apparent error. Unlike cross-validation data sets, bootstrap data sets are generated by resampling $\hat{F}$ with replacement, whereas cross-validation does it without replacement. Bootstrap samples are produced as follows: Let $\hat{F}$ be the empirical distribution function where random sample T_n is drawn with equal probability mass $\frac{1}{n}$ on t_1, $t_2, \ldots, t_n$; $t_i = (x_i, y_i)$. And, let T_n^* be a random sample of size n drawn *iid* with replacement from $\hat{F}$, where t_i^* is a single random observation. If an observation is taken more than once, probability mass $\frac{1}{n}$ times the number of being selected is assigned to that observation. The true error estimation is obtained through the independent bootstrap training sets T^{*1}, T^{*2}, ... , T^{*B}; where B is the "total number of bootstrap samples", each created as previously described. The prediction model $\hat{f}(T^{*b}, x_i)$ is constructed according to each T^{*b}. The first term of the bootstrap estimate of true error is the resubstitution error of the application model. The second term is the difference of the average error over the original sample and the average error over the bootstrap sample (mean apparent error), summed over all bootstrap validation models divided by B. The bootstrap error estimate is given as following [5]:

$$\hat{Err}_{BOOT} = \frac{1}{n}\sum_{i=1}^{n} L[y_i, \hat{f}(T_n, x_i)] + \frac{1}{B}\sum_{b=1}^{B}\left(\frac{1}{n}\sum_{i=1}^{n} L[y_i, \hat{f}(T^{*b}, x_i)] - \frac{1}{n}\sum_{i=1}^{n} L[y_i^*, \hat{f}(T^{*b}, x_i^*)]\right) \quad (2)$$

For ordinary bootstrap (described in this section), B validation model plus one application model (trained on all original n observations) are constructed; where B value traditionally starts from a minimum of 20.

2.2. Train-and-Test Methodology

The train-and-test method is the most commonly employed method of neural network evaluation. It obtains an estimate of true error from an independent set of data not used to construct the prediction model. This method is often referred to as "cross validating the network"; where the *test set error* represents the true error of the prediction model. Subdividing a given sample of size n into two sub-samples such that $n_1 + n_2 = n$. One sample of size n_1 is used to build the prediction model $\hat{f}(T_{n_1}, x)$, and another sample of size n_2 is used to validate the trained network. The train-and-test estimate of true error is the error over the validation set of size n_2 [5]:

$$\hat{Err}_{T/T} = \frac{1}{n_2}\sum_{j=n_1+1}^{n} L[y_j, \hat{f}(T_{n_1}, x_j)] \quad (3)$$

Train-and-test method, unlike the resubstitution and resampling, does not use all available n data for constructing the application network. It constructs the network from a reduced data set of size n_1; where only one model is built.

2.3. Committee Networks

It is believed that a combination of many different predictors may improve overall predictions. In recent years, a committee network has been one of the focused subjects in the ANN community. A committee network (CN) is composed of several individually trained neural networks. These individual networks can be trained either on the same data set or different subsets of data. The outputs of the individual networks are usually combined according to some rules to produce the final output of the CN. Common approaches such as simple averaging or weighted averaging have been used. The networks in the committee can be either trained using the same data set or different subsets of data. The outputs of these committee members are normally combined using a fuser, which is established from a certain rule such as the simple average and weighted average. Figure 1 demonstrates a typical committee network.

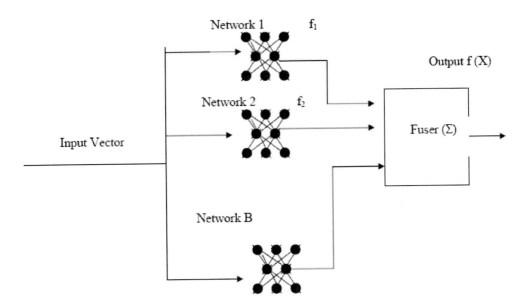

Figure 1. Block Diagram of a Committee Network.

The committee's output is the weighted sum of the committee member's outputs, which is written as

$$\hat{f}(X) = \sum_{i=1}^{B} \alpha_i f_i(X), \tag{4}$$

where $f_i(X)$ is the output from committee member i, and α_i is the corresponding combination-weight, $i = 1,2,.......,B$. The terms α_i are chosen such that the error of the committee is minimized. The simplest way to determine α_i is to use the simple average method (SAVG), which assumes that all the committee members are equally important. The committee's output based on the SAGV is written as

$$\hat{f}(X)=1/B(\sum_{i=1}^{B}f_i(X))$$

(5)

Some researchers have demonstrated that CN can perform better than a single network. Twomey and Smith [5] considered CN's derived from the resampling validation networks. They showed that using a CN approach by fusing all the resampling validation networks outperformed the single network that was developed using reduced data set.

Parmanto [7] indicates that neural network models can inherently suffer through high variance, where two possibilities can occur. First, some neural network models may perform better than others on different unseen data or cases. As it is known that the network model changes significantly with a slightly change in training set; as the result, different models will perform differently on different unseen data. Second, with excessive complexity in neural network models, different models may obtain different useful information about the actual system. From the two preceding reasons of Parmanto [7], one best (selected) model may not perform well on all varieties of unseen data; or it may not contain all useful information about the true model. In this research, it is impractical to use only one best model to predict the future outputs since the experimented data is sparse. Therefore, the committee methodology is used to combine the various performances and information of different prediction models together.

2.4. Bias-Variance Dilemma and the Committee

The concept behind building an accurate predictor is to be able to optimally balance the tradeoff between bias and variance (that contributing to the final prediction) to reduce the total error. According to Geman, Bienenstock, and Doursat, [6] the mean squared error of a statistical or ANN estimator can be broken down into two parts: bias^2 and variance, given as following.

$$E[(\hat{f}(T_n,x)-E[y\,|\,x])^2]=\underbrace{(E[\hat{f}(T_n,x)]-E[y\,|\,x])^2}_{"bias^2"}+\underbrace{E[(\hat{f}(T_n,x)-E[\hat{f}(T_n,x)])^2]}_{"variance"},$$

(6)

where $\hat{f}(T_n,x)$ is the empirical estimator built from the training set $T_n=\{(x_1, y_1),..., (x_n, y_n)\}$; and $E[y|x]$ is the true estimator that we try to emulate. "n" empirical estimator $\hat{f}(T_n,x)$ is believed to be a biased estimator of the true estimator $E[y|x]$, if (based on the entirely possible population of T_n) the mean of $\hat{f}(T_n,x)$ is different from the mean of $E[y|x]$.

However, the empirical estimator $\hat{f}(T_n, x)$ is considered to yield high variance, if (based on giving a different training set of T_n) an output of $\hat{f}(T_n, x)$ significantly differs from other outputs of $\hat{f}(T_n, x)$. From equation (6), we can notice that both bias and variance can give the empirical estimator poor performance. Typically, decreasing the prediction error is usually achieved by reducing variance caused by an excessive number of parameters in the ANN predictor [7]. Parmanto [7] viewed the committee predictor as a linear combination of different ANN network predictors; where bias^2 and variance of the committee can be formulated as following:

$$Bias^2[\hat{f}_{com}(T, x)] = \left(\left\{ \sum_{k=1}^{K} \alpha_k E[\hat{f}_k(T_k, x)] \right\} - E[y \mid x] \right)^2 \tag{7}$$

and the variance: $Var[\hat{f}_{com}(T, x)] = E\left[\left(\sum_{k=1}^{K} \alpha_k \hat{f}_k(T_k, x) - \sum_{k=1}^{K} \alpha_k E[\hat{f}_k(T_k, x)] \right)^2 \right]$

$$= \sum_{k=1}^{K} \alpha_k^2 \left(E[\hat{f}_k^2(T_k, x)] - (E[\hat{f}_k(T_k, x)])^2 \right)$$

$$+ \sum_{k \neq k'} \alpha_k \alpha_{k'} \left(E[\hat{f}_k(T_k, x) \hat{f}_{k'}(T_{k'}, x)] - (E[\hat{f}_k(T_k, x)] E[\hat{f}_{k'}(T_{k'}, x)]) \right) \tag{8}$$

[7] Indicates that if the simple average decision combining is used to form the committee, then the first term of eq (6) becomes:

$$\sum_{k=1}^{K} \alpha_k (E[\hat{f}_k(T_k, x)]) = \frac{1}{K} \sum_{k=1}^{K} E[\hat{f}_k(T_k, x)] = E[\hat{f}(T, x)],$$

which exactly makes eq. (6) equal to the bias^2 term in eq. (7). As the result, the bias^2 term of the committee predictor is equal to the bias^2 term of a single predictor;

$Bias^2[\hat{f}_{com}(T, x)] = Bias^2[\hat{f}(T, x)]$. For the weighted average decision combining under the constraint of $\sum_{k=1}^{K} \alpha_k = 1$ without the constant term, the same result is still applied [7].

2.5. Bootstrap Committee

Let B be the total number of bootstrap samples. In bootstrap error estimation procedure [5]: n items are randomly sampled with replacement from the original data set of n points for each resampling process to create a (bootstrap) resampled set of data. The left-out data points are the items that have not been selected. One item or data point can be randomly selected more once. Repeat the resampling process B times. Therefore, B resampled sets of training data (size of n) can be used to construct B network members or predictors of the committee based on a particular decision-making method (simple averaging, voting, or weighted averaging), and then test on the original data set (size of n).

Efron et al. [8] suggested that the number of bootstrap replicates (B) should be between 20 and 200 in order to obtain a good, stable error prediction for the parameter estimation problem. Since the number of networks in the committee may become very large, to reduce the computational time it is necessary to control the training time and the number of networks for the optimal overall performance. From Twomey et al. [5] study on the estimation of the ANN prediction error using bootstrap method, the optimal number of bootstrap samples is between 20 and 30. There is no significant improvement in the generalization performance after 30.

The two studies by [1] and [9] demonstrated the application of ANN to the critical issue of delamination in drilling FRPs. The advantages to a neural network approach are well documented – the ability to model complex non-linear, multi-dimensional functional relationships without any prior assumptions about the nature of the relationships.

3. ANN APPLICATIONS IN MACHINING FRPS

For materials research, a certain amount of experimental results is always needed first to develop a well performing neural network, including architecture, training functions, training algorithms and other parameters, followed by training process and evaluation method. After a network has learned to solve the problems based on these datasets, new data from the same knowledge domain can then be put into the trained network, in order to output realistic solutions. The process of creating ANNs for materials research can, therefore be summarized in terms of the following stages:

1. Database collection: analysis and preprocessing of the data.
2. Training of the neural network: this includes the choice of it architecture, training functions, training algorithms and parameters of the network.
3. Test of the trained network: to evaluate the network performance.
4. Use of the trained ANNs for simulation and prediction.

The greatest advantage of ANNs is its ability to model complex non-linear, multi-dimensional functional relationships without any prior assumptions about the nature of the relationships, and the network is built directly from the experimental data. However, the limitations of the ANN method are as follows:

1. Training data of the database should have close relationship with the predicting parameters.
2. Sufficient training data for complex ANNs are necessary.

The objective of predictive modeling in machining in general is to estimate the machining process outcomes, within a given set of process parameters, in order to evaluate whether these outcomes are desirable. The ultimate goal of such an exercise is to dynamically control the machining process by continuously adjusting the process parameters so that the outcomes are always in close proximity to the desired targets. To be able to reach this goal, it is required that the modeling schemes used are capable of capturing the nonlinear nature of the process, reasonably fast, robust and readily implementable in the control schemes of commercially available CNC machine tools. Our recent review of the literature have shown a large activity in the area of predictive modeling of machinability of FRPs using ANN [1, 9, 10-27], but very few works extend over the entire range of prediction and dynamic control [10]. In the sections below we will review the most significant works on predictive machinability as it applies to fiber reinforced composites.

Machinability studies are used to assess the ease or difficulty with which materials can be machined under a given set of conditions which include, in addition to the work material itself, the cutting tool, machine tool, machining operation and cutting conditions. Machinability is often assessed by criteria such as tool wear or tool life, cutting forces, power consumption, and surface finish. Thus, good machinability correlates with small amounts of tool wear, low cutting forces and good surface finish. Machinability may also be assessed by the type of chips produced and the cutting temperatures generated. However, it is also realized that some of these criteria are interrelated. High cutting forces lead to higher power consumption and higher cutting temperatures. Tool wear is also affected by the cutting temperatures, and the cutting forces are influenced by tool wear. Therefore, it is preferred to assess machinability in terms of desirable outcomes of the machining process, which include low tool wear, low cutting forces and good surface finish.

Fiber-reinforced polymers have their unique characteristics that affect their machinability in different ways than metals. The properties of FRP materials are primarily determined by the physical properties of the fibers and the matrix, fiber volume fraction and fiber orientation or architecture. The low strength of fiber-matrix interface and interlaminate bonding makes disbonding or delamination a likely and serious damage that may take place during machining. This type of damage is often associated with drilling because of the tendency of the drill point to separate the laminates in a multiply composite structure. However, delamination is also common in edge trimming, milling, grinding and abrasive waterjet machining. Because of the risk of reducing part strength due to this internal damage [16], delamination may account for a high percent of part rejection in the aircraft industry [28]. Therefore, delamination is the most commonly used criterion in assessing machinability of FRPs, and the evaluation of drilling delamination has received the most attention in published works included in this review [1, 9, 10-18].

3.1. Delamination Damage

Delamination is an intrinsic problem in drilling layered materials because the drill bit feed motion and the resulting thrust force act normal to the stacking plane of the composite material. This tends to separate the plies along the weak matrix interface in between. Delamination leaves cracks between the plies in the drilled component, which may severely affect their structural integrity and long-term reliability and hence a decision may be made for its rejection. Since drilling is usually done at the assembly stages of finished components, such decisions to reject or accept based on machining quality are very costly and must be avoided. This requires the implementation of process monitoring and control strategies that will help in minimizing delamination.

Drilling induced delamination occurs at the entry and exit surfaces of the drilled component as illustrated schematically in Figure 2. These two modes of damage are called peel-up and push-out delamination. Two different mechanisms are responsible for delamination on each side of the laminate. Peal-up delamination occurs by sliding the pierced plies up the flutes of the drill similar to the action of a power screw. This tends to separate and severely bend the surface plies leading to its fracture under Mode III loading. In push-out delamination the drill point exerts compressive force on the uncut plies below causing them to bend elastically. As the drill approaches the exit the number of uncut plies supporting it reduces and the resistance to bending decreases. At a critical thickness the bending stress becomes greater than the interlaminar strength between the plies and an interlaminar crack is initiated around the hole. Further pushing down by the drill point causes the crack to propagate and the flexural rigidity of the supporting plies becomes weaker. This leads to fracturing the material below the drill point as the chisel edge proceeds exiting the laminate. The fracture of the bottom surface plies occurs by both Mode I and Mode III fracture.

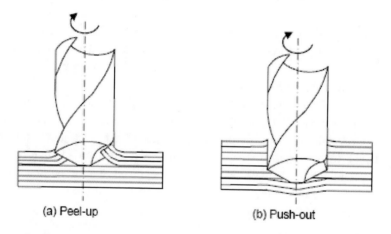

Figure 2. Schematic of (a) peel-up delamination at entry and (b) push-out delamination at exit.

Delamination damage at the surface plies is shown as spalling that extends beyond the hole diameter as illustrated in Figure 3. A delamination factor commonly used to measure the extent of delamination damage at the surface and is defined as the ratio of the maximum diameter of the damage zone, D_{max}, to the hole diameter, D,

$$F_d = \frac{D_{max}}{D} \qquad (9)$$

However, this factor does not capture the irregular shape of the delamination zone caused by spalling of fibers and hence an adjusted delamination factor was proposed according to the equation [29].

$$F_{adj} = \alpha \frac{D_{max}}{D} + \beta \frac{A_{max}}{A} \qquad (10)$$

where A_{max} is the area of damage and A is the nominal area of the hole. The coefficients α and β are weights which sum is equal to 1.

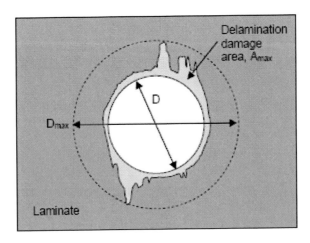

Figure 3. Schematic representation of the delamination factor.

A linear elastic fracture mechanics model for predicting the onset of exit delamination due to the applied thrust force was introduced by Hocheng and Dharan [30]. In this model the laminate consists of a number of isotropic plies and crack growth is assumed to be coplanar and is confined to the plane of the crack. The uncut thickness h ahead of the drill point is modeled as an isotropic circular plate clamped on its contour to the cut portion of the laminate, which is assumed to be rigid. As the drill cuts downward, this plate is deformed elastically by the action of the applied thrust force at its center. Crack propagation occurs when the thrust force reaches a critical value,

$$F_{cr} = \pi \sqrt{\frac{8 G_{IC} E h^3}{3(1-v^2)}} \qquad (11)$$

where G_{IC} is the critical strain energy release rate in plane strain conditions and Mode I fracture. The modulus of elasticity of the isotropic ply is considered to be the highest modulus of a unidirectional laminate, E_{11}. It is apparent that the thrust force at the onset of delamination is independent of drill diameter when a concentrated load is assumed. The concentrated load condition is best applied to twist drills with narrow chisel point. This model was later extended to other drill point geometries by introducing a geometry factor Φ into the equation [31]. Φ is typically greater than 1.0 and represents the extent of thrust force distribution to the periphery of the drill bit. This results in a higher thrust force for the onset of delamination. For a twist drill the value of Φ is 1.0, while it is approximately 1.4 for a core drill, 1.14 for a saw drill and 1.06 for a candlestick drill.

$$F_{cr} = \Phi \cdot \pi \sqrt{\frac{8G_{IC}Eh^3}{3(1-v^2)}} \tag{12}$$

The thrust force in drilling is influenced greatly by the drill diameter and feed rate, and to a lesser extent by the drill point geometry and spindle speed. Because of the direct relationship between thrust force and delamination, the efforts to control delamination damage are generally classified into two approaches:

- studies that aim at preventing delamination through selecting process parameters that result in a thrust force that is less than the critical thrust force as predicted by the equations above, and
- studies that aim at minimizing delamination through selecting process parameters that result in minimum delamination factor.

As the thrust force can be easily monitored on-line, the first approach is more readily adaptable to strategies of dynamic process monitoring and control. The second approach is mostly a post-mortem type and is more difficult to provide feedback on-line. It is nevertheless very useful in establishing process trends and control limits suitable for statistical process control.

3.1.1. ANN Thrust Force Monitoring and Control

Stone and Krishnamurthy [10] have developed one of the earliest models for the prediction and control of delamination by means of monitoring and controlling the thrust force. In this system, two neural networks were implemented. A neural identifier network was used to model and predict the thrust force dynamically. A neural controller network was used to control the drilling process in order to keep the thrust force within limits that are predetermined by the analytical equations of linear elastic fracture mechanics [30,31]. The network type used was a recurrent neural network with a recursive least-squares training algorithm. This type of network was adopted for improving performance and reducing training time. Two hidden layers with two neurons each were used and no effort was made to optimize the network architecture. The training of the network was a two-step process. First, the neural identifier was trained to predict the thrust force resulting from the drilling process. Then the trained neural identifier was used to modify the neural controller by back propagating the process errors through the trained neural identifier and updating the neural controller weights such that error was minimized. The inputs to the neural identifier were current and previous command feed rate and the previous measured thrust force. The output was the current thrust force. The inputs to the neural controller were the previous feed rate and difference between the desired thrust force and current thrust force. The output was the current feed rate. The neural controller was tested on AS4/3501-6 graphite-epoxy laminate approximately 6.35 mm thick, a 6.65 mm drill and spindle speed of 652 rpm. The laminate consisted of 48-plies in quasi-isotropic arrangement $(0^\circ, \pm 45^\circ, 90^\circ)$. It was shown that for the most part the controller was able to maintain the thrust force within 5% of the desired value. The robustness of the system was tested by changing some process parameters and the results were satisfactory.

The effect of drill diameter, feed rate and spindle speed on the thrust force for different drill point geometries was investigated by Tsao [11] and Tsoa and Hocheng [12]. Radial basis function network (RBFN) was used in these studies. RBFN network has only one hidden layer with a large number of neurons and Gaussian function is the most common transformation function used in the hidden neurons. The number of hidden neurons and other network parameters were determined by trial and error. The drill point geometries investigated were the twist drill, core drill, center core drill [11] and candlestick drill [12]. The composite laminate was approximately 5 mm thick and was fabricated from woven carbon fiber fabric with stacking sequence $[0/90]_{12s}$. These studies showed that the drill diameter is the most important factor affecting the thrust force, followed by feed rate. The spindle speed effect was insignificant. It was also found that RBFN is an effective modeling technique for predicting the thrust force. The average absolute error of the model predictions was from 0.6% [2] to less than 2% [3].

Athijayamani et al. [13] developed a multilayer feed-forward network with back propagation training algorithm for predicting the thrust force and torque in drilling natural fiber hybrid composites.

The inputs to the network were feed rate, spindle speed and drill diameter and the outputs were thrust force and torque. By trial and error an optimum network architecture with two hidden layers, five neurons each was selected and trained. The prediction results of the network were very satisfactory with an average absolute error of 1.32% for thrust force and 4.21% for torque predictions. Figure 4 demonstrates the adequacy of the predictive model for 10 data points used for testing the network.

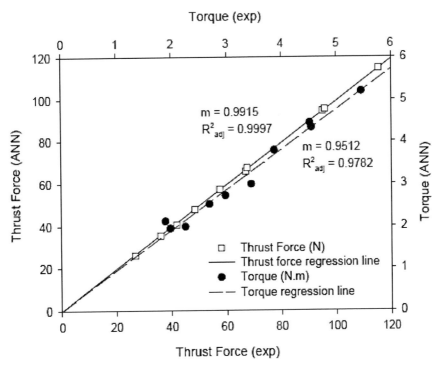

Figure 4. Comparison of experimental and ANN predicted values of thrust force and torque in drilling roselle/sisal hybrid composite [13].

3.2.1. Delamination Damage Prediction

Several studies have investigated delamination damage in drilling composites and proposed neural network models for on-line and off-line prediction of the damage [1, 9, 14-18]. These studies can be divided into two groups with respect the method by which the optimum neural network architecture was determined:

- one group used design of experiments techniques to identify the factors that affect network performance and then identify the most appropriate network for the job. This includes input parameters and hidden neurons [1,9,14]
- the other group relied mainly on trial and error mainly to determine the number of hidden layers and the number of neurons in each [15-18].

Enemuoh et al. [14] used design of experiments to select the most informative signals and most appropriate network architecture to predict delamination damage and surface roughness in drilling AS4/PEEK carbon fiber composites. A number of neural networks were designed and trained with different sets of input sensors according to a predetermined orthogonal array. Table 1 below shows the different signals and network architectures used in the experiment. The neural networks implemented had two hidden layers with varying number of neurons as specified by the orthogonal array. The neuron transfer functions were log sigmoid in the first layer and tangent sigmoid in the second layer. The output layer has two neurons, one for delamination and the other for surface roughness. Resilient back propagation training algorithm was implemented for fast convergence. Variance analysis of the network error showed that network architecture has the most significant influence on network performance (accounts for 73% of variance), followed by cutting speed (14.6%) and thrust force (10.4%). Feed rate, tool material, acoustic emission and point geometry had little contribution to the network error (less than 0.84%).

Table 1. Sensor fusion orthogonal array [14]. 1 means a parameter is included in input array, 0 means parameter is not included. Network architecture identification numbers indicate number of input signals, number of neurons in hidden layers and number of output signals, respectively

Trial	Cutting Speed (V)	Drill Material (M)	Drill Point Angle (G)	Feed Rate (F)	Acoustic Emission (AE)	Thrust Force (Fz)	Network Architecture (AR)	RMSE
1	1	1	1	1	1	1	6,3,2	0.1400
2	1	1	1	0	0	0	3,5,2	0.0330
3	1	0	0	1	1	0	3,5,2	0.0097
4	1	0	0	0	0	1	2,3,2	0.1524
5	0	1	0	1	0	1	3,5,2	0.0111
6	0	1	0	0	1	0	2,3,2	0.3107
7	0	0	1	1	0	0	2,3,2	0.3100
8	0	0	1	0	1	1	3,5,2	0.0599

Analysis of main effects of input signals on network performance further indicated that the only sensors that could significantly reduce network error are thrust force, cutting speed, feed rate and tool material. The best performing network architecture was identified as (4,5,2). Even though this architecture was not included in the orthogonal array, a new network was built, trained and tested and results of the prediction of delamination and surface roughness were satisfactory (R = 0.994 and 0.853, respectively).

The significant contribution of this work is that it points out the importance of proper fusion of sensors and neural networks for improving model predictions. However, the selection of signals building the model may have not been an ideal one. It has been pointed out in the literature that drill diameter plays an important role in inducing delamination, but this parameter was left out of the experimental design. Also amble evidence in the literature indicates the strong relationship between feed rate and drill point geometry as inputs and the thrust force as output in the drilling process. Using all three parameters as inputs in the model may have been redundant and may as well explain the little contribution of feed rate and drill point geometry to the network error.

In a later study Enemuoh and El-Gizawy [1] further investigated different network architectures for the purpose of identifying the optimum architecture for predicting delamination damage and surface roughness. The variables included in the design of experiment included the number of hidden layers, the number of neurons in the hidden layer, the neuron activation function and the training algorithm. Analysis of variance of network error indicated that the number of hidden layers has the most significant contribution to network error (53.8%) followed by activation function (26.7%) and learning algorithm (19.4%). The number of neurons had very little influence on network error (0.1%). An optimum architecture was identified as one with 2 hidden layers, 12 neurons each. The network implements log-sigmoid and linear transfer functions and resilient back propagation learning algorithm. The inputs to the network were spindle speed, tool geometry, feed rate and thrust force. The outputs were delamination length and width and hole roughness. The network was trained and tested in a drilling operation of BMS 8-256 carbon fiber reinforced epoxy and the model results were satisfactory. The regression factors for delamination length, width and surface roughness were 0.902, 0.848 and 0.737, respectively.

In a similar study Dini [9] considered different network architectures for predicting delamination damage both at entry and exit when drilling glass fiber reinforced polyester. Multilayered feed forward networks with input layer, hidden layer and output layer has been adopted and trained with Levemberg-Marquardt back-propagation learning algorithm. The number of input, hidden and output neurons was varied and the network performance was examined. Drilling experiments were conducted according to factorial design and statistical analysis of the results was carried out to determine the most important input parameters influencing peel-up and push out delamination. It was found that thrust force and feed rate have the most significant influence on peel-up delamination, while drill diameter, thrust force and feed rate have the most significant influence on push out delamination. Cutting speed had no significant effect on delamination on either side of the laminate. Therefore, two neural networks were built, one for peel-up delamination and the other for push out delamination. Furthermore, two variations of each network were introduced, the first uses an analogue output neuron for the delamination factor and the second provides 4 digital output neurons for indicating the level of damage. The input and hidden number of neurons was varied from 2 to 6 and the effect of this variation of network performance was investigated. It was determined that the best network architectures for analog output are 4,3,1 and 4,2,1 for peel-up and push-down delamination, respectively. The best network architectures for digital output were 5,5,4 and 5,6,4 for peel-up and push-down delamination, respectively. Even though statistical analysis of experimental data indicated that only two inputs are significant for peel-up delamination and three inputs are significant for push-out delamination, the recommended networks have 4 and 5 inputs. It was unclear from this study what additional inputs were used.

Neural networks were also used to predict delamination damage [15] and residual tensile strength [16] after drilling glass fiber reinforced plastic laminates. Three layered, feed forward network with back propagation learning was used. The neurons in the input layer were determined by the process parameters. The output layer has one neuron. The number of neurons in the hidden layer was adjusted to optimize network performance. The process parameters were spindle speed, drill point geometry, feed rate and drill diameter [15]. In the second study only the first three parameters were used. This works demonstrated the effective of neural networks as post-mortem modeling techniques and the network performance was satisfactory (absolute error in most cases was less than 5%). However, not effort was made to investigate the effect of process parameters on drilling damage. On the other hand, the study conducted by Karnik et al. [17] employed results of the neural network model to analyze the effects of process parameters on delamination damage at hole entry (peel-up delamination). This study considered spindle speed, feed rate and drill point angle as input parameters in the high speed drilling of carbon fiber-reinforced polymer composites. Spindle speeds were varied in three levels from 4,000 rpm to 40,000 rpm and feed rates were varied in four levels from 1,000 mm/min to 9,000 mm/min. Three levels of drill point angle were also used (85, 115 and 130°). The study used a multilayer feed forward ANN, trained using error back-propagation learning algorithm to model the drilling process and build a database for direct effect analysis of the input parameters on delamination factor. This analysis revealed that delamination is sensitive to all three input parameters. A combination of high spindle speed and low levels of feed rate and drill point angle was determined to be the most appropriate for minimizing drilling damage. This is perhaps due to the fact that cutting temperatures are higher at higher cutting speeds, which results in softening the matrix and lowering the thrust force. But the most obvious reason for this is that such a combination results in smaller chip thickness and thus lower cutting forces.

De Albuquerque [18] presented a novel application of ANN in the image analysis and measurement of delamination. Unlike all previous works which used more or less manual systems for measuring delamination damage, the study in [18] used a combination of radiographic imaging and ANN based image recognition to identify and measure the damage zone around the drilled hole. The network topography consisted of three layers for input, hidden neurons and output. The input layer consisted of three neurons for the color components R, G and B of a pixel. The output layer had three neurons for class assignment of the pixel being analyzed into a hole area, damaged area or non-damaged area. The network is trained by presenting it with representative pixels from the three classes of image area with identification of each class. In the prediction stage, the network is presented with unidentified pixels and class assignment is made. Pixel by pixel the network is capable of constructing a digitized image of the drilled hole consisting only of three areas: hole, damaged and un-damaged. This digitized image is then readily available for the computation of each area and quantification of the delamination damage.

ANN Applications in Machining of Fiber-Reinforced Composites

3.2. Cutting Forces and Specific Cutting Energy

Cutting forces, specific cutting energy (or cutting pressure) and machining power are interrelated quantities that can be used interchangeably to assess machinability. In an orthogonal turning operation, for example, the machining power is related to the cutting force by the equation,

$$P_m = F_c \cdot v \tag{13}$$

where P_m is machining power, F_c is the cutting (principal) force and v is the cutting speed. Specific cutting energy is defined as the energy required to remove a unit volume of the work material in unit time, that is,

$$K_c = \frac{P_m}{Q_w} = \frac{F_c}{f \cdot d} \tag{14}$$

where K_c is the specific cutting energy (or cutting pressure) in the direction of the cutting force, $Q_w = vfd$ is the material removal rate, f is the feed rate and d is the depth of cut (or undeformed chip thickness). Similarly, a cutting pressure coefficient in the feed (thrust) direction is determined as,

$$K_t = \frac{F_t}{f \cdot d} \tag{15}$$

where F_t is the feed force. Thus, if one of these basic quantities and the cutting geometry are known, the other two quantities can be determined.

Cutting forces and specific cutting energies are commonly used to assess machinability because they greatly influence machining quality. High cutting forces can cause delamination of the composite material. Higher cutting forces and higher machining power also lead to higher cutting temperatures, which is a known cause of accelerated tool wear. Higher cutting temperature may also lead to deterioration of the epoxy matrix in the fiber reinforced polymer composite, which in turn would result in lowering the strength of the composite. Therefore, it is desirable to machine composites at low cutting forces and low machining power levels.

The works on prediction of cutting forces and machining power using ANN in composites are very limited. Karnik et al. [19] studied the machinability in turning of carbon/PEEK composites. A multilayer feed-forward network using error back-propagation algorithm was used to model the machining process. The inputs to the network were cutting tool material (PCD, K10), workpiece material (unreinforced PEEK, PEEK/30%CF and PEEK/30%GF), cutting speed and feed rate. The outputs were specific cutting energy and machining power. Model predictions were very useful in the analysis of the effects of input parameters and their interactions on machining power and specific cutting energy. It was found that machining power increases with an increase in cutting speed and an increase in feed rate for all tool-workpiece combinations. Thus, the lowest machining power is consumed at the lowest levels of cutting speed and feed rate. The effect of cutting speed and feed rate on machining power can be readily explained by equation (13), as an increase in cutting speed will directly increase the machining power while an increase in feed rate will result in an increase in undeformed chip thickness, and hence an increase in the cutting force. The specific cutting energy was influenced mainly by the feed rate where lower values of the specific cutting energy were associated with high feed rates (i.e. large undeformed chip thickness). It was also found that the effect of the interaction between feed rate and cutting speed on specific cutting energy is more significant for the K10 tool than the PCD tool. With the aid of this analysis it was determined that machinability of PEEK/30%CF and PEEK/30%GF are best performed with K10 tool and PCD tool, respectively, at a cutting speed of 125 m/min.

Antonio et al. [20] introduced genetic learning with an elitist strategy as a method of improving the speed of convergence and performance of the neural network. This method uses supervised learning process in which the weights and biases of hidden and output neurons are used as design variables. A fitness function is used in an evolutionary search process to identify the best members of the set of design variables to be used in the next generation group and regulization functions are used to accelerate the convergence of the learning process. Typical operators of genetic algorithm were applied to the population: selection, crossover, implicit mutation and replacement by similarity are used to support the evolutionary learning process of the network. The results of this study showed the advantage of network optimization with genetic learning. Figure 5 shows a comparison between the predictions of genetically optimized NN and a normal feed-forward, error back-propagation supervised learning network in machining PEEK and reinforced PEEK composites. Even though the two network topographies and input data sets may have not been completely identical, the figure highlights the benefit of genetic algorithm learning in improving network performance.

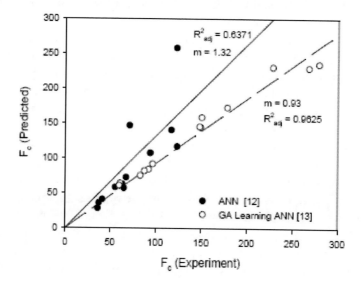

Figure 5. Comparison between the predictions of multilayer feed-forward error back-propagation network with and without genetic algorithm learning in machining PEEK, PEEK/GF30 and PEEK/CF30 [19] and PEEK and PEEK/GF30 [20].

The machinability of CFRP in orthogonal milling [21] and helical milling [22] was also investigated recently using ANN models. Unlike the previous studies which utilized a single neural network to model the machining process, these works implemented a committee of neural networks (CN) to capture the highly complex and nonlinear data that is generated in milling. Committee networks were employed because of demonstrated smoothing capabilities needed for the high degree of noise exhibited by the data. The idea behind committee networks is that a composite system of networks will be better than the best of the networks that went into the composite system. A committee of 20 networks having identical architectures, training parameters and initial conditions was implemented. The multi-layer perceptron trained by back-propagation was the basis of the committee members. Decorrelation of the network errors was achieved through different 20 Bootstrap training sets. The inputs to the committee members were fiber orientation angle and undeformed chip thickness. The output was the specific cutting energy. Separate committees were built and trained for K_c and K_t.

The data for training the committee network was generated from milling experiments of unidirectional CFRP laminates. Four unidirectional laminate were used, namely $0°$, $45°$, $90°$ and $135°$. The depth of cut, spindle speed, feed rate and cutting tool geometry were held constant. Unlike turning, in milling both undeformed chip thickness and fiber orientation vary continuously with the cutter engagement angle. Therefore, both cutting forces and specific cutting energy are strong functions of the fiber orientation angle and chip thickness [21]. Figure 6 shows the variation of specific cutting energies K_c and K_t with fiber orientation and undeformed chip thickness for the four laminates used in the experiments. The figures also show the predictions the committee networks. It is evident from these figures that CN is capable of capturing the highly complex data generated by milling. Correlation analysis of the data revealed an adjusted correlation coefficient $R^2_{adj} = 0.945$ for the predicted K_c and $R^2_{adj} = 0.964$ for K_t, where an adjusted correlation coefficient of 1.0 means a perfect fit.

The CN model predictions of specific cutting energies provided an orthogonal cutting data based that can be used to determine the cutting forces in composite machining operations other than orthogonal milling. Prediction of the cutting forces at this stage requires analysis of the cutting geometry in order to determine the instantaneous undeformed chip area and the fiber orientation angle. Once these quantities are determined for a cutting operation, then the instantaneous cutting and thrust forces are calculated by multiplying the undeformed chip area by the specific cutting energy from the CN orthogonal milling data base. Figure 7 shows a comparison of experimental and predicted cutting forces in the helical milling of $0°$ unidirectional CFRP laminate [22]. The model predictions in this study were found to be in good agreement with the experimental values. An average absolute error was less than 10% for the three force components predicted.

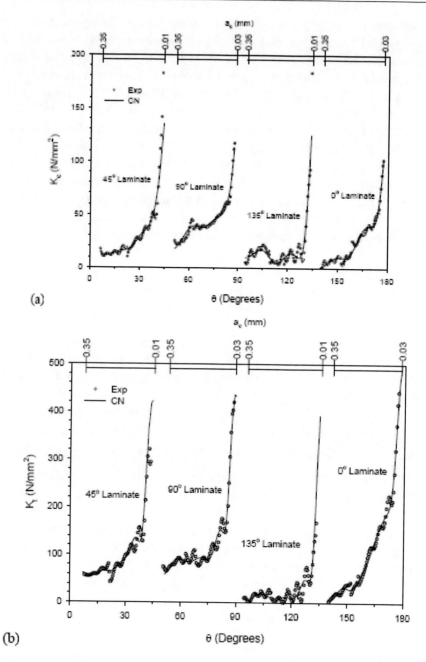

Figure 6. Variation of specific cutting energies (a) K_c and (b) K_t with fiber orientation angle (□) and undeformed chip thickness (a_c) in orthogonal milling of CFRP.

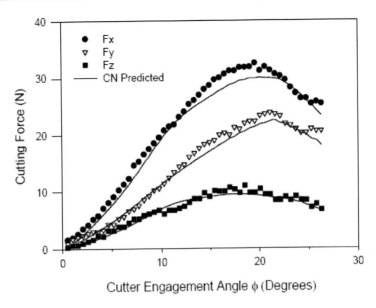

Figure 7. Comparison between experimental and predicted forces for 0° laminate at depth of cut of approximately 1.0 mm.

3.3. Surface Roughness

Surface roughness refers to micro-geometrical irregularities of the machined surface with respect to a mean line and is typically a function of the cutting tool geometry, cutting speed, feed rate, fiber type and content, fiber orientation and matrix type and content. The reliability of machined components, especially of high strength applications, is critically dependent on the quality of the surfaces produced by machining. Surface roughness also greatly influences the manufacturing cost because of the costly and time consuming finishing process that may follow. Surface roughness may be described by one of many average roughness height measures such as arithmetic average height, R_a, peak to valley height, R_{max}, root mean square height, R_q and ten-point average height R_z. However, it has been shown that R_{max} and R_z are the most sensitive when machining inhomogeneous materials such as FRPs. This is due to the presence of pulled out fibers, loose fibers and cracks.

Surface roughness in drilling composites was investigated in [1,12,14]. The network topographies implemented in these studies were described earlier in section 3.2. R_a was used to characterize the surface roughness of the drilled hole in these works. It was generally found that network predictions of surface roughness were not as good as those of the thrust force. For example, a regression coefficient of $R = 0.737$ was reported for model predictions of surface roughness while $R = 0.902$ was reported for model predictions of the thrust force [1]. Similar results were also reported in [12] and [14]. This might be attributed to the fact that surface roughness data by nature has more variance and are more complex than thrust force data and that R_a is not the most sensitive factor to represent surface topography. Furthermore, it was shown in [12] that neural networks are more capable of modeling surface roughness than multiple regression. The mean absolute error of the neural network predictions was seven times less than that of multiple regression (see Table 2).

Bagci and Isik [23] investigated the influence of process parameters on surface roughness in the turning of GFRP bars. The process parameters studied were cutting speed, depth of cut and feed rate. Surface roughness was represented by the arithmetic mean R_a. The network topography used was (3,9,10,1). It was found that ANN model is a good predictive tool for surface roughness. But when the prediction results were compared with response surface model results it was found that the latter provided better predictive capability (see Table 2). Feed forward networks with supervised learning were also used in the modeling of surface roughness in the CNC face milling of GFRP [24]. The training data for the networks were obtained by design of experiments and the input parameters were cutting speed, feed rate, depth of cut and mean light intensity and the output signal was R_a. The optimum network topography was determined as (4,23,1). Mean light intensity was used as another indicator of surface roughness and it was measured by an image analysis software on surface areas of the sample were roughness measurements were made. It was found that the light intensity input to the network, along with process parameters, has greatly improved the performance of the models.

Multilayer feed forward networks with back-propagation and Bayesian regularization training algorithm were used in the prediction of kerf width and surface roughness in abrasive waterjet machining of Kevlar composites [25]. The input process parameters were waterjet pressure, abrasive flow rate and quality level and the responses were top kerf width, bottom kerf width and surface roughness. Quality level is a dimensionless quantity that is used to describe the surface roughness of the upper, middle and lower zones of the kerf. Quality levels of 3, 4 and 5 corresponded to smooth surface with striations marks, striation fee smooth surface, and very smooth surface, respectively. The optimum network topography was found by trial and error to be (3,9,3). The training and testing data was generated by design of experiments and regression analysis was also applied to model the process. NN model predictions of surface roughness were satisfactory, but of less accuracy than linear regression (see Table 2).

The regression factor for NN was R = 0.947 as compared to R = 0.970 for linear regression. The authors attributed this less satisfactory performance to the added complexity of the network introduced by including three responses. To overcome this, a separate network was constructed and implemented for modeling surface roughness alone. The topography of the simplified network was (3,3,1) and its regression factor was R = 0.992.

Table 2. Comparison between mean absolute error for statistical and neural networks models in machining fiber reinforced composites. Input parameters: D = drill diameter, f = feed rate, N = spindle speed, d = depth of cut in turning and V = cutting speed, AWJP = abrasive waterjet pressure, AFR = abrasive flow rate, QL = quality level

Source	Input Parameters	Response	Analytical Model	AM Avg ABS Error (%)	ANN Avg ABS Error (%)
[11]	D, f, N	Thrust force	Response surface method	11	0.6
[12]	D, f, N	Thrust force	Taguchi and multi-variable regression	8.7	0.83

	D, f, N	Surface roughness	Taguchi and multi-variable regression	10.1	1.46
[13]	D, f, N	Thrust force	Multi-variable regression	6.24	1.32
	D, f, N	Torque	Multi-variable regression	24.04	4.21
[23]	V, d, f	Surface roughness	Response surface method	1.72	4.51
[25]	AWJP, AFR, QL	Surface roughness	Linear regression	6.59	3.07

3.4. Tool Wear

The study of tool wear in machining composites did not receive any significant attention in comparison to the studies of delamination or surface roughness, or as in the case of metal cutting where a substantial number of studies have been conducted in the past two decades (see for example the review by Sick [26]). This is perhaps because tool wear is not viewed as an important machinability indicator as delamination. Tool wear is also more difficult to measure on-line without stopping the machining process and indirect assessment of tool wear is more useful for on-line monitoring. In addition, tool wear may not have been an issue in machining composites because PCD tooling is utilized most of the time to resist the abrasive wear that dominates this field. Segreto and Teti [27] utilized ANN based pattern recognition to analyze acoustic emission signals in the orthogonal machining of unidirectional glass and carbon fiber reinforced polymers and sheet molding compound. The acoustic emission signals were analyzed by extracting spectrum parameters (features) using linear predictive analysis. The extracted AE spectrum feature vector was then used as an input to train the ANN network. The output of the network was a coded value for tool identification: 0 = new tool and 1 = worn tool. A three layer feed-forward back-propagation network was utilized in this study. The number of neurons in the input layer was equal to the number of elements in the signal features vector (n = 4, 8 or 16). The number of neurons in the hidden layer was a multiple of the number of neurons on the input layer (1n, 2n or 4n). It was shown that the ANN success rate in identifying the state of the tool was high for all three composite materials (over 70%). However, the highest success rates were obtained by tailoring the NN to each composite material. For GFRP the 16-16-1 network exhibited a success rate of 97%, for CFRP the 4-8-1 network exhibited a success rate of 92% and for SMC the 4-4-1 network exhibited a success rate of 90%.

4. COMPARISON WITH REGRESSION METHODS

Since its inception, neural network methods were conceived as the new alternative modeling techniques to the more traditional and well established statistical and analytical methods. Their ability to learn complex and highly nonlinear behaviors made them of interest in various applications in engineering, life sciences and social sciences. However, the application of ANN in modeling is not always justifiable. They usually require more development time than traditional statistical methods for which commercial software packages have been available for many years and have been extensively tested and improved. Depending on level of complexity, computing time of ANN may also be longer, despite the fact that they heavily use parallel processing. Overall, the choice of ANN over statistical methods involves a trade-off between cost and performance.

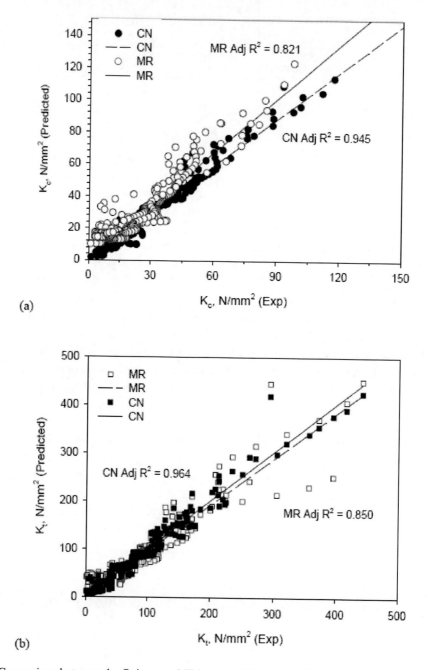

Figure 8. Comparison between the fitting capabilities of multiple regression and committee network models in fitting (a) K_c and (b) K_t in milling CFRP.

A comparison of the performance of ANN and statistical methods in machining fiber reinforced composites is shown in Table 2. The performance level of the modeling technique is represented here in terms of the mean absolute error. It can be seen that with the exception of [23] and [25], statistical methods produce an error that is between 5 and 18 times that of ANN methods. Other evidence in the literature also suggests that ANN methods provide better fitting of the data presented to it. The goodness of fitting may be measured by the correlation factor R^2 or adjusted R^2, where a value of 1.0 means perfect correlation. Figure 8 shows the correlation between specific cutting energies predicted by multiple regression and committee networks on one hand and the experimental values on the other hand. It is clear that CN models provide much better correlation of the data than MR, even though the mean absolute errors in both cases are somewhat similar [21].

CONCLUDING REMARKS AND FUTURE TRENDS

ANN has been successfully applied in the analysis and modeling of machining fiber reinforced composites. This chapter presents a review of some of the applications of artificial neural network in machining FRPs: prediction of cutting forces, delamination damage, surface roughness and tool wear. The literature shows that artificial neural networks have a lot of potential to offer for application in modeling FRPs machining processes. The following conclusions can be drawn from the extensive review on ANN applications in machining Fiber-reinforced composites:

1. Most of the applications of NN were in modeling of the machining process for the purpose of identifying optimum process parameters. Very few studies extended their efforts beyond prediction and into controlling the machining process.
2. An ANN is ideally suited for predicting complex fiber reinforced composite problems because, like its biological counterparts, it can learn, and therefore can be trained to find solutions.
3. Multilayer feed forward error back propagation network topography was the most commonly used to mode machining of FRPs. Different transfer functions were implemented, which include log-sigmoid, tan-sigmoid and linear. The optimum topography of the network was either determined by trial and error or by design of experiments.
4. When compared to regression methods, performance of neural networks was generally better.
5. Prediction ability of surface roughness is improved by including other quality input parameters such image light intensity and quality level.
6. Hybrid models that combine NN with other modeling techniques such as genetic algorithms and fuzzy logic may provide significant improvements in the model performance.

REFERENCES

[1] Enemuoh, U.G., and El-Gizawy, A.S. (2003). Optimal Neural Network Model for Characterization of Process-Induced Damage in Drilling Carbon Fiber Reinforced Epoxy composites. *Machining Science and Technology,* 7 (3), 389-400.

[2] Ghaboussi, J., Garret, J. H., and Wu, X. (1991). Knowledge-Based modeling of material behavior with neural networks. *Journal of Engineering Mechanics Division,* ASCE, 117 (1), 132-153.

[3] Attoh-Okine, N.O. (1999). Analysis of learning rate and momentum term in backpropagation neural network algorithm trained to predict pavement performance. *Advances in Engineering Software,* 30, 291-302.

[4] Efron, B. "The Jacknife, the Bootstrap, and other resampling plans", *SLAM NSF-CBMS*. Monograph 38, 1982.

[5] Twomey, J. M., and Smith, A. E. (1995). Committee networks by resampling. Intelligent neural engineering systems through artificial networks. *ASME Press*, 5, 153-158.

[6] Geman, S., Bienenstock, E., and Doursat, R. (1992). Neural networks and the bias/variance dilemma. *Neural Computation*, 4, 1-58.

[7] Parmanto, B. (1995). Agitating dissent: Methods for improving performance of a neural network committee by error decorrelation. PhD Thesis, Department of Information Science, University of Pittsburgh, PA.

[8] Efron, B., and Tibshirani, R. (1986). Bootstrap methods for standard errors, confidence interval and other measures of statistical accuracy. *Statistical Science,* 1, 54-77.

[9] Dini, G. (2003). On-Line prediction of delamination in drilling of GFRP by using neural network approach". *Machining Science and Technology*, 7 (3), 295-314.

[10] Stone, R., and Krishnamurthy, K. (1996). A neural network thrust force controller to minimize delamination during drilling of graphite-epoxy laminates. *International Journal of Machine Tools and Manufacture*, 36 (9), 985-1003.

[11] Tsao, C.C. (2008). Comparison between response surface methodology and radial basis function network for core-center drill in drilling composite materials. *International Journal of Advanced Manufacturing Technology,* 37 (11-12), 1061-1068.

[12] Tsao, C.C., and Hocheng, H. (2008). Evaluation of thrust force and surface roughness in drilling composite material using Taguchi analysis and neural network. *Journal of Materials Processing Technology*, 203 (1-3), 342-348.

[13] Athijayamani, A., Natarajan, U., and Thiruchitrambalam M. (2010). Prediction and comparison of thrust force and torque in drilling of natural fibre hybrid composite using regression and artificial neural network modeling. *International Journal of Machining and Machinability of Materials*, 8 (1-2), 131-145.

[14] Enemuoh, U.E., El-Gizawy, A.S., and Okafor, C.A. (1999). Neural network based sensor fusion for on-line prediction of delamination and surface roughness in drilling AS4/PEEK composites. Proceedings of NAMRC XXVII, Paper MS99-187, Berkeley, California, May 25-28, 1999.

[15] Mishra, R., Malik, J., and Singh, I. (2010). Prediction of drilling-induced damage in unidirectional glass-fibre-reinforced plastic laminates using an artificial neural network

Proceedings of the Institution of Mechanical Engineers, Part B: Journal of Engineering Manufacture, 224 (5), 733-738.

[16] Mishra, R., Malik, J., Singh, I., Davim, J. Paulo. (2010). Neural network approach for estimating the residual tensile strength after drilling in uni-directional glass fiber reinforced plastic laminates. *Materials and Design*, 3 (6), 2790-2795.

[17] Karnik, S.R., Gaitonde, V.N., Rubio, J. C., Correia, A. E., Abrão, A.M., Davim, J. P. (2008). Delamination analysis in high speed drilling of carbon fiber reinforced plastics (CFRP) using artificial neural network model. *Materials and Design*, 29 (9), 1768-1776.

[18] De Albuquerque, V.C., Tavares, J. M. R. S., Durão, L. M. P. (2010). Evaluation of delamination damage on composite plates using an artificial neural network for the radiographic image analysis. *Journal of Composite Materials*, 44 (9), 1139-1159.

[19] Karnik, S.R., Gaitonde, V.N., Mata, F., Davim, J. Paulo. (2008). Investigative study on machinability aspects of unreinforced and reinforced PEEK composite machining using ANN model. *Journal of Reinforced Plastics and Composites*, 27 (7), 751-768.

[20] Antonio, C., Davim, J., Lapa, V. (2008). Artificial neural network based on genetic learning for machining of polyetheretherketone composite materials. *International Journal of Advanced Manufacturing Technology*, 39 (11), 1101-10.

[21] Sheikh-Ahmad, J., Twomey, J., Kalla, D., and Lodhia, P. (2007). Multiple regression and committee neural network force prediction models in milling FRP. *Machining Science and Technology*, 11 (3), 391-412.

[22] Kalla, D., Sheikh-Ahmad, J., and Twomey, J. (2010). Prediction of cutting forces in helical end milling fiber reinforced polymers. *International Journal of Machine Tools and Manufacture*, 50 (10), 882-891.

[23] Bagci, E., and Isik, B. (2006). Investigation of surface roughness in turning unidirectional GFRP composites by using RS methodology and ANN. *International Journal of Advanced Manufacturing Technology*, 31 (1-2), 10-17.

[24] Alexandrakis, S., Benardos, P., Vosniakos, G.-C., and Tsouvalis, N. (2008). Neural surface roughness models of CNC machined glass fibre reinforced composites. *International Journal of Materials and Product Technology*, 32 (2-3), 276-294.

[25] Shukla, M., Tambe, P. B. (2010). Predictive modelling of surface roughness and kerf widths in abrasive water jet cutting of Kevlar composites using neural network. *International Journal of Machining and Machinability of Materials*, 8 (1-2), 226-246.

[26] Sick, B. (2002). On-line and indirect tool wear monitoring in turning with artificial neural networks: A review of more than a decade of research. *Mechanical Systems and Signal Processing*, 16 (4), 487-546.

[27] Segreto, T., Teti, R., Tool condition monitoring in composite materials machining through neural network processing of acoustic emission. In IPROMS 2007 - Virtual International Conference on Innovative Production Machines and Systems, 2-13 July 2007. http://www.iproms.orgTechnology Conference, Atlanta, GA, USA, 1982, pp. 471–483.

[28] Wong, T.L., Wu, S.M., and Croy, G.M. An analysis of delamination in drilling composite materials. Proceedings of 14th SAMPE Technology Conference, Atlanta, GA, USA, 1982, 471-483.

[29] Davim, J.P., Rubio, J.C. and Abrao, A.M. (2007). A Novel Approach Based on Digital Image Analysis to Evaluate the Delamination Factor after Drilling Composite Laminates, *Composites Science and Technology*, 67 (9), 1939-1945.

[30] Hocheng, H., Dharan, C.K.H., (1990). Delamination during drilling in composite laminates. *Journal of Engineering for Industry*, 112, 236-239.

[31] Hocheng, H., Tsao, C.C., (2003). Comprehensive analysis of delamination in drilling of composite materials with various drill bits. *Journal of Materials Processing Technology*, 140, 335-339.

In: Fiber–Reinforced Composites
Editor: Qingzheng Cheng

ISBN: 978-1-61470-303-7
© 2012 Nova Science Publishers, Inc.

Chapter 15

INITIAL VERSUS FINAL FRACTURE OF FIBER-REINFORCED COMPOSITES, ANALYZED VIA ACOUSTIC EMISSION

Nadia Ereifej[*] and Mohammad Al-Rababah
School of Dentistry, The University of Jordan, Amman, Jordan

ABSTRACT

The aim of this chapter was to compare fracture resistance of composite resin-based materials using three-point bending in conjunction with acoustic emission (AE) analysis.

Samples were prepared of dimensions of (25 x 2 x 2 mm), according to ISO 4049 standard. Non-reinforced composite samples (NRC) were prepared using Synergy composite (Coltene, Altstätten-Switzerland), while in unidirectional (UFRC) and multidirectional fiber-reinforced groups (MFRC), samples were prepared by incorporating unidirectional EverStick and multidirectional fibers (Stick Tech Ltd Oy, Turku, Finland), respectively. Eight samples were prepared per group, placed at a universal testing machine and loaded at a crosshead speed of 0.2 mm/min. An AE system (Physical Acoustics Corporation, New Jersey, USA) was used to detect audible signals produced while the specimens were loaded. Flexural strength (σ) was calculated at initial and final fracture. One-Way ANOVA was used to detect differences in fracture strength and AE amplitudes between the groups. Paired t-tests were used to detect differences between the flexural strength at initial and final failure of each group. Correlation and regression analyses were used to detect the correlations between initial and final failure strength in each group.

Two major sets of acoustic signals were detected demarcating initial and final failure. Initial fracture strengths and AE amplitudes were significantly lower than those at final fracture in all groups (p<0.05). Initial fracture strength of UFRC (171.4 MPa) was significantly higher than that of MFRC (125.1 MPa) and NRC (88.3 MPa). Final fracture strength of UFRC was also significantly higher (197.7 MPa) than the rest of the groups; MFRC (150.5 MPa) and NRC (108.8 MPa). Initial and final fracture strengths were significantly correlated (r= 0.983).

---

[*] Corresponding author. Department of Prosthodontics, School of Dentistry, The University of Jordan Amman, 11942, Jordan Tel: 00962797379525 e.mail: nadia_ereifej116@hotmail.com

It was concluded that fiber-reinforcement with unidirectional or multidirectional glass fibers improves the fracture resistance of composite resin materials and monitoring acoustic signals released during fracture can reveal significant information regarding the fracture process.

1. INTRODUCTION

1.1. Fiber-Reinforced Materials

Fiber-reinforced composites (FRC) have been used in industry for a long time in the construction of boats, buses and windmills [1]. The use of fibers to reinforce dental resins was described more than 40 years ago [2] and fibers are now being used to reinforce denture bases, prosthodontic frameworks used in implants, fixed prostheses, periodontal splints, orthodontic appliances [3] and endodontic posts [4-7].

Fibers are thin flexible cylindrical structures that usually have fewer defects than bulk materials and they make the restoration more resistant to fracture. If fracture occurs, they act to hold the fractured fragments together [8]. The positive effects of using fibers on the mechanical and physical properties of dental materials were proven [9]. Fibers act as crack stoppers providing strength and rigidity to the composite and the surrounding resin acts to protect the fibers and fix their arrangement so that reinforcement is supported [10].

FRCs can be classified according to different criteria, including the type of fibers: carbon, Kevlar, polyethelene or glass [1, 11], fiber orientation: unidirectional, braided or woven [5, 10, 12], the presence of pre-impregnation: non-impregnated fibers are prepared in the laboratory or at chairside while pre-impregnated fibers are impregnated with the resin by the manufacturer [13] and method of construction: manually or by vacuum- forming [14].

FRCs are easy to handle and posses good mechanical and physical strengths that might reach seven times the strength of conventional composites while maintaining their flexibility [15, 16]. They are recommended where excellent aesthetics are required, to reduce wear of opposing teeth and when adhesive bonding is possible [17]. Prostheses constructed using FRC have good aesthetics, simple design and fabrication and good durability. However, proper case selection is very important in these situations [18-20].

1.1.1. Mechanical Properties of FRC

Long - term serviceability of a dental material is an important factor that should be taken into consideration. Restorations are subjected to complex types of loads intra-orally, and the resistance of the material to fracture under stresses is an important factor that determines its longevity and clinical performance.

Various studies investigated the effect of fiber-reinforcement on the strength of composites. According to Freilich et al. (1997), FRCs had a comparable strength to alloys used in fixed partial dentures and higher strength than non-reinforced composites [21]. Loose et al. (1998) measured higher strength for FRC material than In-Ceram material [22]. According to Behr et al. (1999), FRC bridges have greater strength than the maximum masticatory load expected in clinical situations [23] and Bae et al. (2004) reported significantly higher fatigue strength for FRC than non-fiber-reinforced composites [24]. More recently, the review undertaken by Van Heumen et al. (2008) concluded that fibers do enhance the flexural strength of composites [25]. Furthermore, Ereifej et al. (2009) showed that the addition of fibers enhances the edge strength of resin-composite materials and that the presence of these fibers prevented the extension of marginal cracking through the whole sample thickness and therefore resulted in a less catastrophic and a relatively repairable pattern of failure [26].

However, some FRC systems were not proven to have superior fracture resistance than non-reinforced systems [27]. Kolbeck et al. (2002) showed that FiberKor/Conquest Sculpture had fracture strength that is significantly below the values needed to withstand occlusal loads [28]. Furthermore, Pereira et al. in 2003 showed that reinforcement of the hybrid composite with polyethene fibers did not increase the flexural strength of the composite [29]. Similar findings were reported by Behr et al. (2003) who showed that fiber reinforcement does not increase the fracture strength of single molar crowns [30].

1.1.2. Variables Affecting Mechanical Properties of FRCs

Mechanical properties vary between different types of FRC materials depending on many parameters including the following:

- Amount and diameter of fibers: Fracture strength of FRCs was found to increase as the diameter and volume fraction of fibers incorporated increase [5, 31]. Therefore, unidirectional fibers of the diameters of 20-30 μm, and fiber weight fraction of 37-42% were recommended for dental use [12, 31].
- Impregnation of fibers with the matrix: Nonimpregnated composites have more voids increasing their water absorption and reducing their mechanical strength. Also, poor impregnation reserves oxygen which inhibits polymerisation, decreases the strength and increases residual monomer content that is irritant to oral tissues [32].
- Properties of fibers versus properties of matrix polymer [31, 32]: Modern FRC systems consist mainly of bis-GMA polymers in addition to Polymethyl methacrylate (PMMA) in some of them. While fibers are the components that bear the tensile forces, the resin-matrix is the component that withstands and absorbs the compressive stresses. Due to the difference in elastic moduli, stresses normally develop at the fiber/matrix interfaces and structural flaws in the matrix also weaken the structure. Therefore, improving the matrix properties and bonding between the fibers and the matrix is essential to improve the strength of FRC structure [6].
- The type of fibers used: Behr et al. (2000) and Kolbeck et al. (2002) showed that fracture resistance of crowns made of glass FRC was higher than those of polyethelene FRC crowns [33, 34]. These results were supported by Ellakwa et al. (2002) who showed that the use of silane-containing wetting agent increased the strength of composite significantly when reinforced by glass-fibers [35].
- Fiber orientation or architecture: Theoretically, if a material is to be loaded in one direction, unidirectional fiber-reinforcement should provide the ultimate reinforcement compared to 50% reinforcement provided by bidirectional or multidirectional fibers. In the clinical situations, however, load is applied in multiple directions and it would be logical to assume that multidirectional fibers would provide better strength enhancement [36]. Goldberg et al. (1998) indicated that FRC with unidirectional fibers had the highest flexural strength [37]. Similar results were

reported by Chong et al. (2003) who suggested that debonding of fibers from the matrix in the unidirectional fiber-reinforced materials resulted in strengthening of the material by means of crack deflection and fiber bridging [38]. Chong and Chai undertook a similar study later in 2003 and concluded that bidirectional FRC material has a significantly higher probability of fracture than unidirectional FRC [39]. On the other hand, Karbhari et al. (2007) reported that the highest fracture strength in FRCs was achieved when braided fibers rather than unidirectional fibers were used. According to those authors, cracking in the unidirectional FRC occurred longitudinally and unconstrained due to lack of transverse reinforcement in addition to the possibility of delamination at matrix/fiber level, while braided or woven samples had more strain energy due to the interlocking nature of the fibers that allowed local points of intersection and sliding and acted as crack arrestors [12].

- Inner surface layer: The mechanical properties of FRC can be enhanced either by using pre-impregnated fibers beneath the composite resin restorative material or by reinforcing the composite material with fiber impregnation [36]. The highest fracture strength was found by researchers when the fiber frameworks were placed at the tensile side or the base of the composite [40, 41].

- Cementation material and technique: The development of adhesive systems has revolutionized many aspects of restorative dentistry and allowed more conservative tooth preparations [42]. The results of previous studies by suggested that that adhesive cementation is recommended to minimise the risk of fracture of FRC restorations [43-45].

- Variation in preparation design: A chamfer finish line and more parallel tooth preparation pattern than ceramic restorations were recommended for FRC restorations to increase the fracture strength of such restorations [46, 47].

1.2. Mechanical Testing Methods of Dental Materials

Increasing demands for restorative dental materials that combine improved aesthetics and optimal physical and chemical properties have led to the introduction of various types of dental ceramic and FRC materials [8, 48]. Despite the continuous improvement in the mechanical properties of these materials, fracture is still one of the most common causes of their failure [49-51]. Fracture occurs when a particular stress limit is exceeded [52]. Therefore, the load-bearing capacity of a particular material is an important criterion influencing the performance and serviceability of this material in the clinical situations [53].

Various testing methods have been used to investigate the mechanical strength of dental materials [54]. Both compressive and tensile stresses must be applied on specimens during testing. Three-point, four-point bending tests and biaxial flexural tests are examples [55]. Strength is not an inherent property of a particular material. It is dependent on the condition of the tested material and it is very sensitive to the defects present within the material as well as interfacial stresses. Therefore, testing conditions should be very carefully controlled since the results can be affected by factors like stressing rate, stress concentration, specimen size and failure mode. Storage of the specimens in water and cyclic fatigue loading should also be applied [56-60].

Alternatives to conventional testing methods have often been sought and therefore some techniques were suggested. Electronic speckle pattern interferometry, based on observing the changes in the distribution of speckles (grains) on the surface of the specimen under loading using suitable laser irradiation, has been used previously to investigate the fracture resistance of fiber-reinforced and non-reinforced composite prostheses and it was found that fiber-reinforcement enhances the strength of composite materials [61].

Finite element analysis (FEA) is based on modeling complex structures by dividing them into small elements with simple geometric shapes, assigning them different values according to the material properties they represent and applying loading conditions and parameters. Using special softwares, stresses can be calculated at each element and stress distributions in the sample under investigation can be analyzed [62, 63]. FEA has been commonly used in dentistry. Different materials, including ceramics, composites and metals, were investigated [64]. A combination of FEA and experimental analyses is usually recommended for complete investigations [65-67].

Photoelasticity and R-T(2D) are other experimental stress analysis methods that are used to study the mechanical problems of elasticity, based on the analysis of interference fringe patterns, representing the stress states, in model structures under loading. From these patterns, the stress distributions for different materials can be seen, facilitating the identification of fracture-risk areas and locations of crack initiation [68].

1.2.1. Acoustic Emission Technique

1.2.1.1. The Concept and Basic Set Up of the Acoustic Emission Technique

Acoustic emissions (AE) are the stress waves induced by the sudden internal stress redistribution of the materials caused by changes in the internal structure [68]. The concept of AE was first proposed by Joseph Kaiser in the early 1950s. When an object is subjected to external stimuli, areas with flaws develop high stress energies. Seeking a more stable position, some of this energy is released in the form of a pressure wave or sound which occurs long before the object starts to fail. These waves are generated by any type of elastic or plastic deformation, dislocation and microcracking, inter-crystal fracture, impurity fractures and phase transformation. They are generated from both ductile and brittle materials and can be affected by many variables like the material nature and geometry, loading mode, temperature and other environmental conditions [69-72].

AE signals are classified into two categories; continuous AE in which signals of low-amplitude are continuously released and superimposed on one another while burst type signals are released as distinct high amplitude events [73, 74]. AE of the second category were used in most studies investigating mechanical properties and fracture processes of biomaterials [73].

The general arrangement for AE testing is shown in Figure 1 and is comprised of the following:

1. Acoustic sensors: These contain broadband piezoelectric transducers which convert the mechanical energy carried by the elastic wave into accessible electronic signals [75]. As AE waves hit the broadband transducer, part of them is converted to an oscillating voltage [76, 77]. The active element of a piezoelectric transducer is a thin disc of piezoelectric material that is metal-coated on each surface for electrical

contact and mounted in a metal cylinder. Different sizes and shapes of sensors are available depending on the object being tested. The number of sensors used in a test varies and the more transducers used, the higher the accuracy of the results obtained. Sensors should be placed on a flat surface of the specimen coated with a thin film of a greasy acoustic couplant for better signal transduction [71, 77, 78].
2. Preamplifiers: As the output voltage of an AE sensor is usually very small, millivolts or less, preamplifiers are needed to enlarge the signal so that it is acceptable for further analysis. The gain of the preamplifier, which measures the ability of the amplifier to increase signal amplitude, is calculated as the mean ratio of the signal output to the signal input in dB. This can also be adjusted according to the test performed [78].
3. Cables: These connect the different components of the system together such as those connecting the sensors to preamplifiers and the preamplifiers to the PCI card [78].
4. A PC-based data acquisition board (PCI card) with AE application software: Through these, the settings of AE data acquisition and display are adjusted, progression of data acquisition is monitored, analysis of the acoustic signals perceived is performed and data regarding the different AE characteristics are displayed. Furthermore, the PCI card contains a frequency filtering system, in addition to those provided in the amplifiers, to select the appropriate bandpass and help eliminate unavoidable background noises [77-80].

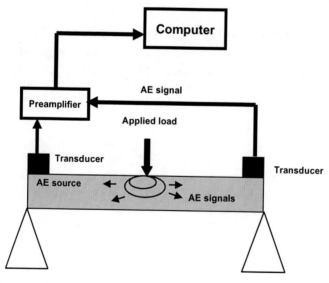

Figure 1. The basic set up for AE acquisition.

In order to minimize the effects of background noise, a threshold value below which signals are not recorded or recorded but not subjected to further analysis is set [81]. Each time a signal exceeds the predetermined threshold an event is recorded [72, 77].

Upon receiving of the AE signals, these are analyzed by the special software through which measurable AE features can be obtained and used as part of the AE signal description. Several AE features can be acquired as shown in Figure 2. The most useful signal characteristics are [77, 78, 81, 82]:

- Peak amplitude: The largest voltage peak in the AE signal waveform, expressed in dB.
- Duration: The time from the first threshold crossing till the end of the last threshold crossing in milliseconds.
- Rise time: The time from an AE signal first threshold crossing to its peak in milliseconds.
- Absolute energy: The true energy measure of the AE event (in AttoJoules).
- The counts of hit: The number of the times the AE signal crossed the detected threshold.

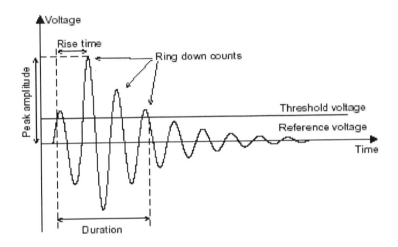

Figure 2. Basic AE signal features.

In the majority of studies that applied AE technology in fracture testing, one or more AE features were used to analyze the fracture behaviour of materials. AE signal characteristics most commonly used are counts and peak signal amplitudes [73, 83]. Some studies included the duration of signals and their energy [84, 85]. The greatest difficulty in analyzing AE based on amplitude is that the structure of the specimen can considerably disturb the amplitude of the signals, depending on the structure and geometry of the sample and must be understood to properly interpret AE signals.

1.2.1.2. Advantages and Disadvantages

AE technique offers the advantages of being a non-stop technique which can monitor the condition of the material under investigation throughout the test without the need to stop it to visually examine the specimens [81]. It is used to detect deformation, fatigue and fracture of tissues and materials including metals, composites and ceramics. It can detect failure initiation, the initial site of damage, damage propagation, measure accurately the maximum flexural strength of the material and detect the mechanism of failure at multiple sites even subsurface and non-visible ones. It is considered as an early warning of the failure of the material, regarded as having high sensitivity as it registers even small cracks, as long as a noise is generated and signals are emitted [69, 86, 87].

However, the technique has some limitations including the following:

1. The technique is most suitable for simple homogenous objects and its applicability with complex non-homogenous structures is far more complicated.
2. Ideally, background noise should be eliminated. Nevertheless, this is not always possible.
3. As acoustic waves propagate through an object they are often subjected to distortion and attenuation, affecting the features of AE signals obtained.
4. Although the previous limitations can be overcome by using modern AE equipment, those are more expensive, time consuming, need larger memory space, advanced training and still are in need of further development [79, 88, 89].
5. AE technique is meant to predict failure. Nevertheless, detection of acoustic signals suggests that damage has already occurred [87].
6. Errors might result during computation processes such as the algorithms, round off, and truncation in addition to errors related to using fixed threshold values [90].
7. The stress distributions, nucleation and coalescence of microcracks, extension of main cracks and the way propagation of elastic stress waves released influences AE signals are still unknown using the AE technique. Combinations of AE and FEA is therefore suggested [75].

1.2.1.3. Applications of AE Technique in Dentistry

One of the major advantages of the AE technique is that it can detect initial failure well before being observed using other techniques like micrographs [91]. AE release coincides with the onset of cracking and starts well before or at the initiation of cracking detected optically. Therefore, the AE technique can be used as an early warning for the damage of materials [85].

Acoustic signals are usually first emitted at the end of the elastic region just before the yield point is reached. These signals, related to friction noises and grating between the fractured surfaces, are of long duration and low amplitude (55-65 dB). Further loading causes more deformation which emits few signals of moderate amplitude (up to 80 dB) that increase slowly until final fracture occurs accompanied by the release of high amplitude (up to 100 dB) burst type acoustic signals released at high rates [73, 83, 85, 87, 92, 93].

Investigating the AE release form can provide more in depth insight of the fracture process, specifically clarifying the mode of fracture. Determining the sequence of failure and fracture mode can indicate the different properties of the phases of materials helping eventually designing materials with optimum resistance against the specific mode of failure [94]. Furthermore, AE technology has also been used to locate the site of failure initiation in materials. This is usually done by placing two sensors at the sides of the sample, of a material in which the speed of the wave is known, and calculating the difference in time of arrival of the signals from the 2 sensors [84, 90, 95].

In dentistry, composite structures often exhibit local failures, referred to as damage, before rupture into two or more pieces. Progression from initial local failure to rupture is defined as damage accumulation. Early damage can reduce the life of a material and sub-threshold cracking allows for water ingress into composite structures. Therefore, understanding the lower failure thresholds may assist in setting performance criteria and prosthetic design [96].

AE technology has been commonly applied in the investigation of fracture behavior of dental composite materials. Acoustic signals can be released from composites by plastic deformation, crack growth, matrix cracking or debonding and fracture of fillers [75, 77, 97].

Kondo et al. (1985) compared AE signal patterns in conventional and microfilled composites. The study showed that fracture of conventional composites released burst-type signals of high amplitude and short duration resulting from matrix-reinforcement debonding while the more ductile microfilled composites generated low amplitude signals with longer durations related to their plastic deformation. It was concluded that material characteristics, such as the molecular weight, cross-linking structure, particle size, shape and distribution are all involved in the fracture mechanism of composite materials under loading [74].

Kim et al. (1991) conducted a study to investigate the fracture toughness of 3 commercially available dental composites using an AE technique. The study revealed that composites having different types of filler generate different AE activities during testing and that AE activity increased as the fracture toughness of the material increased [98].

In 1994, Kim et al. also used AE analysis to investigate fracture behavior of dental composites with regular and well-defined spherically-shaped filler particles. The study showed that AE activity differs according to the filler content of the composite and that debonding of the filler from the matrix generated microcracks that emitted relatively low-amplitude acoustic signals at low load levels [70]. Similar results were found more recently by Kim et al. (2000) concerning composite resins containing prepolymerised particle fillers. It was found that AE signals were released at loads close or equal to the maximum loading force. Matrix-filler separation generated signals of low AE amplitude while fracture of filler particles themselves produced signals of high amplitude [99].

More recently, Kim et al. (2002) conducted a similar study investigating AE released during fracture toughness testing of resin-composites containing irregular-shaped fillers. The results showed that the morphology of the filler influenced the fracture toughness of composites and therefore the amplitude, the number of AE signals released and the load at which signals were first released [97].

Vallittu (2002) used AE technique to investigate fracture propagation patterns between different groups of glass FRC veneers used in the repair of ceramo-metal crowns in fracture testing. AE analysis showed that increasing the amount of fibers in the material was accompanied with an increase in the acoustic signals emitted [100].

Alander et al. (2004) analyzed AE signals produced during flexural strength testing of FRC materials. According to AE analysis, initial failure occurred at values of loading that were 19-32% lower than the final failure loads. Low- amplitude signals were associated with matrix cracking while high amplitude signals indicated fiber-cracking [86].

Fennis et al. (2005) investigated the fracture resistance of FRC cusp-replacing restorations by means of AE analysis. According to AE analysis, there was no significant difference in the loads at which the first acoustic signals were emitted among the groups, but where fibers were not used, failure was more catastrophic and fewer acoustic signals were recorded [101].

Duray et al. (1996) investigated acoustic signals released from dental composites upon laser heating. The results showed evenly spaced spikes or bursts of AE signals when heating started. Each burst corresponded to an event such as matrix cracking, filler particle fracture or matrix-filler interfacial debonding. The pattern and the intensity of the AE were different according to the quantity of the filler in the composites [77]. Lee et al. (2000) used the same technique in experimental dental composites with different filler contents. It was found that the acoustic signals counts decreased as filler content of the composite increased and that aging in ethanol resulted in loosening of the composite structure and therefore decreased the amplitudes of the released AE signals [102].

Although they haven't used AE equipment, many authors detected audible signals and used them as an indication for the initial failure of composites [7, 10, 96].

In addition to dental composites, AE technique was used to analyze the fracture process of many other dental materials including porcelains [80], ceramics [103-106], titanium alloys [107], gypsum-bonded materials [108], denture base materials [76], Ti_6Al_4V oral implant [109] and dental amalgam and glass-ionomer materials [110].

1.3. Aims and Objectives

The aim of this chapter was to investigate the fracture resistance of different fiber-reinforced resin-composite materials.

The specific objectives were:

1. To use AE technique during flexural strength testing of these materials.
2. To identify signs of initial failure as determined by AE technique.
3. To compare the values of loading at initial and final failure of the materials tested using AE analysis.
4. To assess the effect of fiber orientation on the fracture strength of these materials.

The Null hypotheses were:

1- AE can not detect initial fracture of FRC materials.
2- There are no significant differences in initial and final fracture strength values or amplitudes of released AE signals among the different materials.
3- Fiber orientation does not affect the fracture strength of composites.

2. METHODS AND MATERIALS

2.1. Specimen Preparation

Stainless steel moulds (25.0 ± 2.0 mm) x (2.0 ± 0.1 mm) x (2.0 ± 0.1 mm), shown in Figure 3, were used for sample preparation according to ISO 4049 standard Table 1 shows the different groups tested in this chapter, their composition and their manufacturers.

For the preparation of Group 1 samples (NRC), the stainless steel moulds were first coated with Vaseline to facilitate removal of the samples after curing. Synergy composite was packed into the mould using a stainless steel carver slightly overfilling the moulds to allow for finishing and polishing procedures. A3 shade was used for all the specimens. The first layer of composite was packed first, cured for 40s, followed by the second layer on top of which a polyethelene sheet was placed followed by a 49N weight to flush the excess of the material. Samples were then cured for 40 s with a halogen high-intensity curing unit (optilux 501; SDS Kerr, Danbury, Conn; light intensity 800 mW/cm², wavelength range 400-505 nm) following the manufacturers' recommendations.

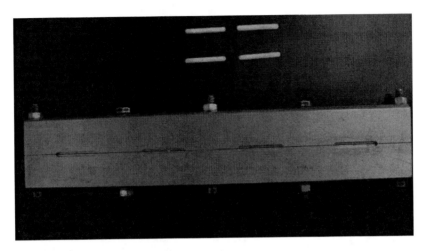

Figure 3. The mould used in the construction of the samples.

Table1. The different groups investigated in this chapter

Group	Composition	Manufacturers
1: Non-reinforced Composite (NRC)	Synergy: Packable nano-filled resin-composite	Coltene, Altstätten-Switzerland
2: Unidirectional Fiber-reinforced Composite (UFRC)	Unidirectional EverStick Fiber/Synergy	Stick Tech Ltd Oy, Turku, Finland/ Coltene, Altstätten-Switzerland
3: Multidirectional Fiber-reinforced Composite (MFRC)	Multidirectional EverstickNet/ Synergy	Stick Tech Ltd Oy, Turku, Finland/ Coltene, Altstätten-Switzerland

In order to determine the amount of fibers to be included for the preparation of Group 2 and 3 specimens, Group 1samples were weighed using an electronic balance, the average weight was determined and 40% of the weight of the specimens of Everstick unidirectional fibers and EverstickNet multidirectional fibers were weighed for each specimen of groups 2 and 3 [111].

According to the manufacturers, Everstick and EverstickNet fibers are E-glass fibers that are composed of calcium-boroaluminosilicate glass of nominal composition 56% SiO_2, 14% Al_2O_3, 22% CaO, 6% B_2O_3 (by weight) in addition to fractional percentages of Na_2O, MgO, TiO_2, Fe_2O_3 and fluoride [11]. For Group 2 samples (UFRC), Everstick glass fibers were wetted with StickResin (StickTech, Turku, Findland) first, condensed at the bottom of the mould, cured for 40s followed by the application of the resin-composite material as described for group1. Similar procedures were used for Group 3 (MFRC), except that StickNet multidirectional fibers were used instead. Fibers were placed at the bottom part of the specimens as it was found previously to optimize the reinforcing effect of fibers [25, 111].

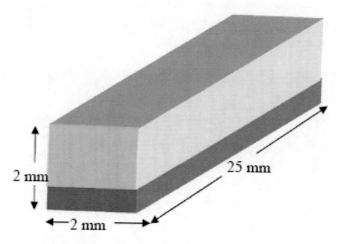

Figure 4. Specimen configuration used in this chapter.

Specimens were visually inspected for voids and in case any defects were found, they were discarded and new ones were prepared. Specimens were finished and polished and their dimensions were verified using a digital micrometer (Digimatic, Mitutoyo Corp., Niles, Il, USA) with an accuracy of 0.01 mm at three locations along the rectangular bar specimens. Ten specimens were prepared per group (n=10) and stored in distilled water at 37 ± 1 °C until the time of testing. Figure 4 shows the configuration of the specimens used in this study.

2.2. The Zwick Universal Testing Machine and AE Machine

Load was applied on the specimens using a Zwick universal testing machine (Zwick GmbH and Co., Ulm, Germany) shown in Figure 5. Specimens were placed at the lower part of the machine while being loaded by the upper apparatus of the machine. A load cell of 20 kN was used.

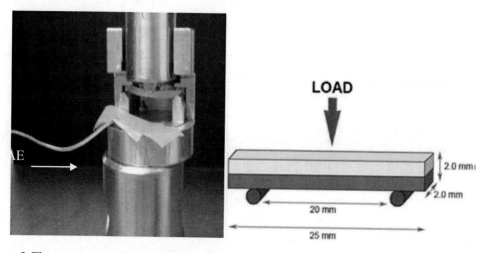

Figure 5. The test set up used in this chapter. (5.a): The specimen placed at the zwick machine with an AE sensor attached. (5.b): The 3-point bending assembly

An AE system (Physical Acoustics Corporation, New Jersey, USA) was used to detect audible signals produced while the specimens were being loaded. The system consisted of a sensor, containing a transducer that converts the acoustic waves into electrical signals that pass then through a preamplifier, of 40 dB gain and band pass of 100 kHz to 2 MHz, then to a computer that has special software used to analyze the perceived signals. Data obtained were graphically represented where the amplitude of acoustic signals (in dB), the counts (number of hits) and the absolute energy (in AttoJoules) were plotted against test time (s).

2.3. Fracture Testing of the Specimens

Specimens were placed in the Zwick machine, as shown in Figure 5, having their ends supported with the 20 mm distance between the supports. The AE sensor was placed at the loading platform of the zwick machine. The surface on which the sensor was attached was first coated with Vaseline, the sensor was placed onto the surface afterwards and then an adhesive tape was wrapped around them to secure the sensor in place. Both machines were set so that the collection of AE signals started at the same time as the loading apparatus of the Zwick machine contacted the specimen surface and started the loading process. As the specimens were being loaded, the acoustic signals were continuously collected by the sensor and consequently the number, energy and amplitude of the signals were plotted against time on the computer display. The other computer screen was connected to the Zwick machine and therefore the force applied (N) was plotted against time.

Loading was performed at a crosshead speed of 0.2 mm/min. The samples for all the groups were tested until catastrophic failure. As the specimen fractured, a drop of 20% of the maximum loading force was measured causing the Zwick machine to stop the loading process automatically. The AE machine was stopped manually immediately afterwards. All specimen preparation and loading procedures were performed by one person for standardization.

The data obtained from the AE machine and those obtained from the Zwick machine were used to analyze the loads at which initial cracking and final failure occurred. Flexural strength (σ) values were first calculated according to the following equation:

$$\sigma = 3Fl/2bh^2$$

where F is load to failure in Newtons, l is the length of the specimens in mm, b is the width at the centre of the specimens in mm, h is the height at the centre of the specimens in mm.

Two graphs were initially prepared, using Sigmaplot software (V 10.0), for each sample tested; flexural strength (MPa) - time (Sec) and AE amplitude (dB) - time (Sec). The graphs were then superimposed and a graph of strength and AE amplitude versus time was obtained for each sample. This is illustrated in Figure 6.

The graphs were used in conjunction with the data of time, flexural strength and acoustic signal amplitudes. Since a threshold of 50 dB was determined before as a value of acoustic signals at which cracks occurring while loading are detectable microscopically [105], the value of loading at which acoustic signals $\geq$ 50 dB were first recorded was considered as the loading value of initial fracture. The value of final catastrophic fracture was that at which a drop of 20% of load was detected and at which the Zwick machine stopped the loading process. This usually coincided with the release of very high AE signals.

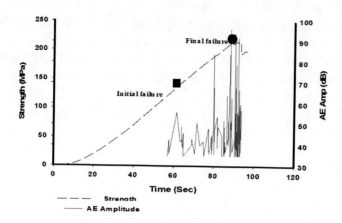

Figure 6. An example of strength-AE amplitude-time graph for one of the samples.

One-Way ANOVA followed by Games-Howell post-hoc tests were used to detect the differences in fracture strength values and amplitudes of AE signals between the groups. A series of paired t-tests were used to detect the statistical differences between the flexural strength at initial and final failure of each group as well as differences between AE signals at initial and final failure. Correlation and regression analyses were used to detect the correlations between the initial and final failure strengths and AE at initial and final failures.

3. RESULTS

Two major sets of acoustic signals were detected demarcating initial and final failure loads. Figures 7.a – 7.c are examples of Strength-AE amplitude-Time plots of each of the three groups investigated.

Table 2. Summary of the results of this chapter

Group	Initial fracture force (MPa)	Final fracture force (MPa)	% of Initial to final force	Initial fracture amplitude (dB)	Final fracture amplitude (dB)
NRC	88.3 (14.9) [A]	108.8 (17.7) [a]	81.2 %	56.6 (12.7) [E]	95.8 (8.2) [e]
UFRC	171.4 (8.1) [B]	197.7 (10.0) [b]	86.7 %	67.8 (5.9) [F]	91.0 (8.5) [e]
MFRC	125.1 (14.2) [C]	150.5 (12.1) [c]	83.1 %	59.8 (8.3) [EF]	86.9 (10.2) [e]

* Similar letters indicate statistically similar groups.

As evident from the graphs above, initial and final fracture points could be clearly identified. Therefore, values of force and AE amplitudes were extracted for all the specimens. Table 2 shows the means and standard deviations of initial and final fracture strength (MPa), the percentage of initial fracture to the final fracture strength and AE amplitudes (dB) at initial as well as at final fracture for all materials. These data are also represented in Figures 8 and 9.

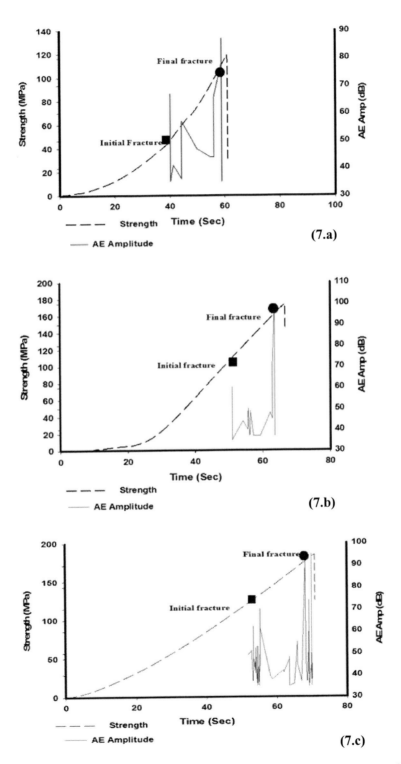

Figure 7. Graphs of strength (MPa) and AE amplitude (dB) *versus* time (s) of a representative specimen for each Group. (a): NRC, (b): UFRC, (c): MFRC, (■): strength at initial fracture, (●): strength at final fracture.

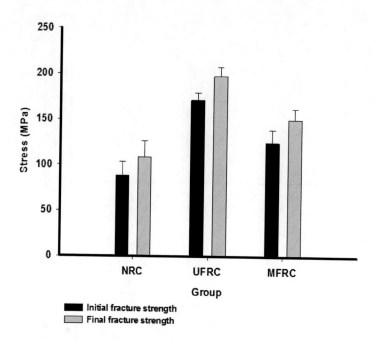

Figure 8. Initial failure and final fracture strengths for the different groups.

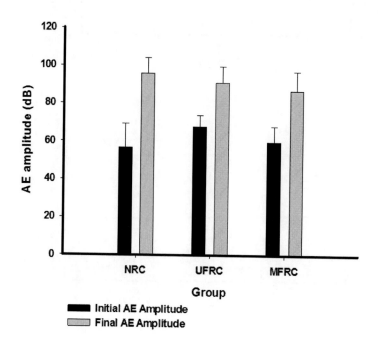

Figure 9. AE amplitude at initial and final failure of the different groups.

NRC had the lowest initial fracture values (88.3 MPa), followed by MFRC (125.1 MPa) while UFRC was the strongest (171.4 MPa). The differences between the three groups were significant. Mean final fracture strength of NRC (108.8 MPa) was significantly lower than the rest of the groups, followed by MFRC (150.5 MPa) while UFRC (197.7 MPa) had the highest values. Initial fracture force values were always significantly lower than final fracture values with the highest difference among UFRC samples (26.3 MPa) and the lowest in NRC (20.5 MPa).

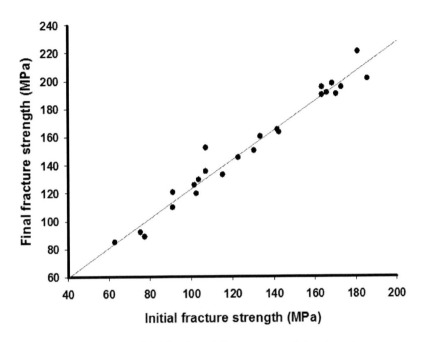

Figure 10. The correlation between initial and final fracture strength (r = 0.983).

The initial fracture AE amplitude was significantly lower than the final fracture AE amplitude in all the groups. Comparing these values between the different groups, significantly higher values of initial fracture AE amplitude were found for UFRC than those of NRC which had statistically similar values to those of MFRC while the values of AE amplitude of the final fracture were similar.

Correlation and regression analyses were performed to detect any relationship between initial fracture strength and final fracture strength. This is illustrated in Figure 10. A positive significant correlation was found ($p < 0.05$), with a high r value of 0.983.

The statistical analysis performed revealed no significant correlation between AE amplitude at initial failure and AE at final failure, no correlation between initial fracture strength and AE amplitude at initial fracture or final fracture strength and AE at final fracture ($p > 0.05$).

AE release patterns after crack initiation and before final fracture varied among the specimens of the different groups. In NRC, most of the specimens had limited AE activity during crack propagation. In fiber-reinforced specimens, higher AE activity was recorded during the fracture process as cracks propagated through the different material layers. More frequent acoustic signals were observed in UFRC than MFRC specimens.

Although not analyzed microscopically, the patterns of failure varied among the groups, particularly between the NRC and the other two groups. Fracture of NRC specimens seemed to be catastrophic as specimens broke completely into pieces while those reinforced by fibers fractured less catastrophically as the fibers seemed to keep the broken parts held within the specimens.

4. DISCUSSION

In this chapter, the AE technique was employed in combination with flexural strength testing of three different resin-composite based groups. The technique was used to detect the time at which failure was initiated and therefore determine the load that caused microcrack initiation. Monitoring acoustic signals released upon further loading revealed information about further cracking and failure propagation. Finally, AE results combined with those of the loading machine were used to determine the values of loading at which final fracture occurred. Failure processes start as soon as the first microcrack is initiated. In clinical service, further loading and intra-oral fatigue conditions propagate cracks and ultimately lead to final fracture [112, 113]. The presence of these microcracks weakens the structure of the material and results in its failure even at lower stress levels than those at which they were initiated. Therefore, recording the loading values at which initial microcracks occur might be the more accurate way of measuring the strength of a particular material bringing the analysis closer to the real clinical situation [110]. The ability of the AE technique to detect initial fracture and final failure was demonstrated and therefore the first null hypothesis was rejected. Furthermore, statistically significant differences were found between the different materials tested regarding values of initial and final fracture strength and the amplitudes of the released acoustic signals. Thus the second null hypothesis was also rejected. Finally, fiber orientation or architecture was found to affect the flexural strength of FRC groups and hence the third null hypothesis was also rejected.

Although the use of FRC is growing, their clinical behavior is not fully understood. Their mechanical behavior is complex and can range from isotropic to anisotropic and can be affected by many variables including the volume, the direction and location of fibers [25]. Therefore, future research would continue investigating the fracture mechanics and behavior of such materials.

4.1 Flexural Strength Testing and AE Set Up

Initial failure represents a more reliable indicator for performance than final failure point as the former sets off a chain of damage processes which might reduce the structure lifespan. It represents the first damage of the least resistant phase. Therefore, postponing the initial failure is essential [7]. Reinforcement strategies that only raise the final failure and not the initial failure simply elongate the damage accumulation phase. When low fracture thresholds are present, fatigue loading and corrosive oral environment may elicit further damage and failure below the expected limit. Therefore, reinforcement and design techniques that elevate the initial failure may be more valuable for dental prostheses and restorations [96, 110].

Flexural strength is often a reliable indicator of structural performance for brittle dental materials including composites. Composites are subjected to flexural loading during function. Flexural strength measures the mechanical properties of the material at both compression and tension acting together [10]. Three-point bending tests simulate the loading of an overlying bridge construction, or a fixed partial denture. A standard three-point bending test has been published by the international Standards Organization (ISO 4049) and has been used extensively since then [25]. However, while simple to perform, flexural testing only assesses the performance of the material under single loading condition while the stresses applied clinically are far more complicated and occur in multidirectional patterns [12].

A crosshead speed of 0.2 mm/min was selected for loading as it was thought to be more conducive to AE data collection and was also previously used in the literature for fracture testing of dental materials [103, 110].

During fracture testing, the sensor was placed at the base of the machine close to the specimen itself as recommended by the instrument manufacturers. It was quite difficult to place the sensor on the specimens as it is was relatively small for the sensors to be placed on and wrapped around with adhesive tapes. In addition, the transducers are better placed at flat surfaces for optimum contact and there was a chance that the transducers could be damaged by the loading component if they were placed at the specimens. The position of the transducer was standardized in this way such that both the transducer and the specimens were placed in the same exact position every time the test was undertaken. If they had been placed at the specimens, the position of the sensor relative to the loading points could have varied. The transducer was placed as close as possible to the specimens and the instrument surfaces on which the transducer was placed were metallic. Sound waves travel quickly enough through metals, so the delay in receiving the signal was negligible.

4.2 Flexural Strength and Fracture of Samples

Direct resin composite restorations are commonly used as small restorations and not recommended for large restorations with high occlusal stresses. Unidirectional continuous fibers will be difficult to adapt as substructures of fillings, short random fibers or bidirectional fibers can be used for such purposes and were found previously to significantly reinforce resin-composite materials, especially if used in sufficient thicknesses [114]. The flexural strengths of all groups in this study fulfilled the minimum requirement specified by ISO 4049 which states a value of 50 MPa. However, although the effect of fiber reinforcement on enhancing the strength of the resin-composite used was found significant, it is unclear whether such restorations can withstand the complex intra-oral conditions.

Analysis of the results showed that initial fracture, detected by the AE technique, always started at lower loading values (13.3-18.8%) than the final loading values. This indicated that failure actually started earlier than the final fracture point and that loading machines recording only forces at catastrophic failure have overestimated the fracture strength of tested materials. This agreed with Alander et al. (2004) who investigated the acoustic signals emitted during flexural strength testing of different types of FRCs and found that AE signals were first released at values of loading that are 19-32% lower than the final failure points and this difference was statistically significant [86]. Other studies investigating fracture of materials using similar techniques have also shown that acoustic signals were first released, as a result of formation of subcritical cracks, at values of loading lower than the final failure points [109, 115]. However, a general tendency for the values of final fracture force to increase as the values of initial fracture force increased was observed in this chapter and since the correlation was strong (r = 0.983), it can be concluded that initial failure point can be used to predict the final failure one.

Loading forces at initial failure varied significantly among the groups. UFRC had the highest values followed by MFRC while NRC was the weakest. Initial crack formation was found previously to be greatly dependent on the stress distribution of the whole structure, which in turn is controlled by many material and geometrical variables [110]. While thickness was maintained constant among the different samples, it seems evident that the inclusion of fibers in the samples had a positive effect on their flexural strength. The highest resistance for crack initiation by UFRC specimens can be explained by the superior mechanical properties, as fibers placed at the tension side of the samples acted to delay crack initiation, especially that fracture was proven to be initiated at the tension side of the samples in such a test geometry. The highest flexural strength of UFRC can be explained by the optimum fiber orientation of these samples relative the direction of load application. In this chapter, UFRC had the highest strength as the load direction was perpendicular to the fibers. This group had the highest AE amplitude and more frequent AE during fracture propagation.

The ratio between initial to final failure can provide an insight into the failure of complex structures. A high initial to final failure ratio indicates minimal damage accumulation while a low ratio indicates that a large portion of damage occurs after initiation and before rupture. In general it is desirable to have a high initial failure load and long damage accumulation phase to avoid instantaneous failure [96]. In this chapter, in addition to the fact that it was evident that fracture was less catastrophic in the fiber-reinforced groups, the ratio initial/final failure was the lowest in the NRC group proving the effect of the addition of fibers on improving the fracture resistance and preventing catastrophic failure of composites.

AE release patterns during crack propagation were different among the different groups. In NRC, most of the specimens had limited AE activity during crack propagation indicating fewer events occurring within the specimens at fracture [71]. This can be explained by the structure of these specimens. Those samples were composed of composite material only and since no fibers were included, fewer microcracks were expected to occur. Composites undergo plastic deformation before they break catastrophically. This plastic deformation is a relatively quiet process and few AE signals, above the preset threshold, are expected to be recorded. Plastic deformation of a material usually results in emission of AE signals with reduced amplitude and number [76]. On the other hand, in FRC groups, higher AE activity was recorded during the fracture process as cracks propagated through the different layers of the specimens or at the interfaces between the different components. When fibers fractured, AE signals with higher frequency and amplitude can be recorded. The fact that the fibers of these samples were manually layered adds to the complexity of their structure and might have induced more defects within their structure. Construction-induced flaws might have increased AE activity of these multilayered samples. Two Signal levels were identified in the FRC groups. The low amplitude one is believed to be associated with the matrix cracking while the high amplitude might indicate fiber-cracking [86].

Fewer signals were recorded during fracture process of the MFRC samples than UFRC ones. The fibers in the woven design are divided equally in the longitudinal and transverse directions which gives the material orthotropic mechanical properties. Therefore, MFRC specimens might be more efficient in impediment of crack propagation and thus fewer fracture events and acoustic signals are produced. MFRC might be suitable in cases where multidirectional reinforcement of the restoration is needed and the direction of loading is difficult to predict [10].

Final failure points were clearly identified among the different specimens. The loading machine was programmed to stop when a drop of 20% of the maximum load was recorded. At final fracture of the NRC specimens, parts of the specimens were chipped off and the loss of maximum contact between the loading component and the samples reduced the loading force and thus loading process was stopped. This fracture was accompanied by the release of AE signals of high amplitude. In the UFRC and MFRC groups, the fractured fragments of the specimens were kept in place by the underlying reinforcing fibers. The more flexible nature of those samples resulted in a gradual rather than a sudden drop of the loading force, as observed in the FRC group. This can be attributed to the presence of fibers that seemed to break gradually. This agrees with the results of Turkaslan et al. (2009) who detected initial failure of FRC during testing by observing a sharp drop in the loading curve. Initial failure signifies the beginning of the damage which usually occurs by microcracking of the matrix. This is usually followed by the release of a part of the strain energy as a wave that propagates through the structure. In FRC, as the crack propagates from fiber to another, the fibers absorb the energy of the wave propagation thus decreasing the energy of the final failure making the final failure less catastrophic in the form of localised chippings and limited fractures. The fibers convert the fracture of composites from instantaneous to an elongated process [36].

From a clinical point of view, keeping the fractured pieces linked together may prevent swallowing or aspiration of the displaced pieces. Separation of the fibers from the matrix can occur due to inadequate bonding between the fibers and the resin-composite or due to the difference in the elastic modulus between them [116]. No separation was observed in the MFRC probably due to the fact that fibers were bidirectionally oriented at 90 degrees to each other, therefore less fracture lines are expected and this was proven by the presence of fewer AE signals collected in this group [117]. This agrees with Fennis et al. who found that where fibers were not used, failure was more catastrophic than FRCs and fewer acoustic signals were recorded [101]. Better bonding between the fibers and the veneering composite can be achieved by silanization or sandblasting of the fibers [118].

Mean AE amplitudes at initial fracture ranged between 56.6 and 67.8 dB and were higher in UFRC and MFRC groups, while those at final failure ranged between 86.9 and 95.8 dB. AE amplitudes at initial failure were found to be significantly lower than those at final fracture within all the groups. Signals released at initial failure are believed to be caused by small cracks while those at final failure were caused by more catastrophic fracture that involved, in some cases in NRC, total chipping of one or two fragments of the specimens. Consequently, it can be concluded that AE amplitude might be an indication of the severity of the fracture. This agrees with previous studies that indicated that plastic deformation of the matrix and fibers produces generally continuous signals of low amplitudes while cracking of resin matrix, reinforcing fibers or debonding of the fibers generates burst-type signals of higher amplitudes [74, 77, 93]. According to Alander et al. the low amplitude signals of FRCs is usually associated with the matrix cracking and the high amplitude signals indicate fiber-cracking [86].

Watanabe et al. (2001) found a positive linear relationship between the AE activity, in terms of cumulative counts and both the load for initiation of acoustic signals and the final failure load [87]. In the present chapter, AE amplitude was analyzed instead of the counts of the signals. Although it was evident that AE of final fracture was significantly higher than AE at initial failure, no significant correlation was found neither between AE amplitude at initial failure and initial failure strength nor between AE at final failure and final fracture point. Differences between the two studies can be related to the differences in the materials used. Watanabe et al. used bone samples in their study while the samples in the present chapter were resin-based composites. For such structures, fracture mechanism could be more complicated in which several factors can be involved.

One of the limitations of this chapter is that testing was performed in dry conditions and no thermo-mechanical cycling or aging were performed. Because E-glass fibers that are used in this chapter contain alkali and earth alkali ions and boron oxide (B_2O_3) which are reactive to water, hydrolytic effects are expected to decrease the physicochemical and mechanical properties of the E-glass FRC systems[11]. Water can destroy the fiber-polymer matrix bond which can result in an irreversible reduction of the strength of FRC and also a reversible reduction in the mechanical properties of FRC can be caused by the plasticization of the polymer matrix by water molecules [116, 119]. Furthermore, fatigue is a mode of failure whereby cracks are induced by subjecting a material or structure to repeated subcritical loads that leads eventually to failure which makes fatigue resistance one of the most important material properties of clinical relevance [41]. Under the influence of cyclic loading, microcracks start to develop at structural flaws as the initial sign of failure. With continuous loading and due to the effect of stress concentration at these defects, microcracks grow and start to join to form larger cracks serving as entrance to oral fluids and bacteria which accelerate the failure process. Although the presence of fibers act as barriers in the direction of crack propagation preventing immediate failure and prolonging the fracture process especially if the fibers are placed at the tension side of the restorations [41], future testing should focus on the evaluation of the effects of these factors on the strength of such materials.

Another limitation of this chapter is that in 3-point bending, load is applied only in one direction, which makes the reinforcing effect of unidirectional fibers superior. Clinically, occlusal forces are applied in various directions. Therefore, future studies should be directed to design the samples and the mode of load application to be more clinically relevant [25, 41].

Finally, it can be concluded that failures in the fiber-reinforced groups are less catastrophic and take longer duration to occur. The addition of fibers to composites improves the fracture resistance of these materials.

REFERENCES

[1] Freilich MA, Meiers JC, Duncan JP and Goldberg AJ, Fibre-reinforced composites in clinical dentistry. 2000: Quintessence publishing.

[2] Goldberg AJ and Burstone CJ. The use of continuous fiber reinforcement in dentistry. *Dent. Mater.*, 1992; 8, 192-202.

[3] Cacciafesta V, Sfondrini MF, Lena A, Scribante A, Vallittu PK and Lassila LV. Flexural strengths of fiber-reinforced composites polymerized with conventional light-curing and additional postcuring. *Am. J. Orthod. Dentofacial Orthop.*, 2007; 132, 524-7.

[4] Kelly JR. Developing meaningful systematic review of CAD/CAM reconstructions and fiber-reinforced composites. *Clin. Oral. Impl. Res.*, 2007; 18, 205-17.

[5] Kanie T, Arikawa H, Fuji K and Ban S. Mechanical properties of woven glass fiber-reinforced composites. *Dent. Mater. J.*, 2006; 25, 377-81.

[6] Seefeld F, Wenz HJ, Ludwig K and Kern M. Resistance to fracture and structural characteristics of different fiber reinforced post systems. *Dent. Mater.*, 2007; 23, 265-71.

[7] Schlichting LH, de Andrada MA, Vieira LC, de Oliveira Barra GM and Magne P. Composite resin reinforced with pre-tensioned glass fibers. Influnece of prestressing on flexural properties. *Dent. Mater.*, 2010; 26, 118-25.

[8] Brown D. Fibre-reinforced materials. Dent Update, 2000; 27, 442-8.

[9] Krause WR, Park SH and Straup RA. Mechanical properties of BIS-GMA resin short glass fibre composites. *J. Biomed. Mater. Res.*, 1989; 23, 1195-211.

[10] Eronat N, Candan U and Turkun M. Effects of glass fiber layering on the flexural strength of microfill and hybrid composites. *J. Esthet Restor Dent.*, 2009; 21, 171-81.

[11] Meric G, Dahl JE and Ruyter IE. Physicochemical evaluation of silica-glass fiber reinforced polymers for prosthodontic applications. *Eur. J. Oral. Sci.*, 2005; 113, 258-64.

[12] Karbhari V and Strassler H. Effect of fiber architechture on flexural charactersitics and fracture of fiber-reinforced dental composites. *Dent. Mater.*, 2007; 23, 960-8.

[13] Kumbuloglu O, Ozcan M and User A. Fracture strength of direct surface-retained fixed partial dentures: Effect of fiber reinforcement versus the use of particulate filler composites only. *Dent. Mater. J.*, 2008; 27, 195-202.

[14] Rosenstiel SF, Land MF and Fujimoto J, Contemporary fixed prosthodontics. 3rd edition ed. 2002, St Louis: Mosby.

[15] Freilich MA, Karmaker AC, Burstone CJ and Goldberg AJ. Development and clinical applications of light polymerized fiber-reinforced composite. *J. Prosthet Dent.*, 1998; 80, 311-8.

[16] Giordano R. Fibre reinforced composite resin systems. *Gen. Dent.*, 2000; 48, 244-9.

[17] Freilich MA, Duncan JP, Meiers JC and Goldberg AJ. Preimpregnated, fibre-reinforced prosthese. Part I. Basic rationale and complete coverage and intracoronal fixed partial denture designs. *Quintessence Int.*, 1998; 29, 689-96.

[18] Meiers JC, Duncan JP, Freilich MA and Goldberg AJ. Preimpregnated, fiber-reinforced prostheses. Part II. Direct applications: splints and fixed partial dentures. *Quintessence Int*, 1998; 29, 761-8.

[19] Freilich MA and Meiers JC. Fibre-reinforced composite prostheses. *Dent. Clin. North Am.*, 2004; 48, 545-62.

[20] Goldberg AJ and Freilich MA. An innovative pre-impregnated glass fibre for reinforcing composites. *Dent. Clin. North Am.*, 1999; 43, 127-33.

[21] Freilich MA, karmaker AC, Burstone CJ and Goldberg AJ. Flexure strength of fibre-reinforced composites designed for prosthodontic application [abstract 999]. *J. Dent. Res*, 1997; 76, 138.

[22] Loose M, Rosentritt M, Leibrock A, Behr M and Handel G. In vitro study of fracture strength and marginal adaptation of fibre-reinforced composite versus all ceramic fixed partial dentures. *Eur. J. Prosthodont Restor Dent.* 1998; 6, 55-62.

[23] Behr M, Rosentritt M, Leibrock A, Schneider-Feyrer S and Handel G. In Vitro study of fracture strength and marginal adaptation of fibre-reinforced adhesive fixed partial inlay dentures. *J. Dent.*, 1999; 27, 163-8.

[24] Bae JM, Kim KN, Hattori M, Hasegawa K, Yoshinari M, Kawada E and Oda Y. Fatigue strengths of particulate filler composites reinforced with fibres. *Dent. Mater. J.*, 2004; 23, 166-74.

[25] van Heumen CC, Kreulen CM, Bronkhorst EM, Lesaffre E and Creugers NH. Fiber-reinforced dental composites in beam testing. *Dent. Mater.*, 2008; 24, 1435-43.

[26] Ereifej N, Silikas N and Watts DC. Edge strength of indirect restorative materials. *J. Dent.*, 2009; 37, 799-806.

[27] Brunton PA, Cattell P, Burke FJ and Wilson NH. Fracture resistance of teeth restored with onlays of three contemporary tooth-colored resin-bonded restorative materials. *J. Prosthet. Dent.*, 1999; 82, 167-71.

[28] Kolbeck C, Rosentritt M, Behr M, Lang R and Handel G. In vitro examination of the fracture strength of 3 different fiber-reinforced composite and 1 all-ceramic posterior inlay fixed partial denture systems. *J. Prosthodont*, 2002; 11, 248-53.

[29] Pereira CL, Demarco FF, Cenci MS, Osinaga PW and Piovesan EM. Flexural strength of composites: influences of polyethelene fibre reinforcement and type of composite. *Clin Oral Investig*, 2003; 7, 116-9.

[30] Behr M, Rosentritt M, Latzel D and Handel G. Fracture resistance of fiber-reinforced vs. non-fiber-reinforced composite molar crowns. *Clin. Oral. Investig*, 2003; 7, 135-9.

[31] Obukuro M, Takahashi Y and Shimizu H. Effect of diameter of glass fibers on flexural properties of fiber-reinforced composites. *Dent. Mater. J.*, 2008; 27, 541-8.

[32] Behr M, Rosentritt M, Lang R and Handel G. Flexural properties of fibre-reinforced composite using a vacuum /pressure or a manual adaptation manufacturing process. *J. Dent.*, 2000; 28, 509-14.

[33] Behr M, Rosentritt M, Latzel D and Kreisler T. Comparison of three types of fiber-reinforced composite molar crowns on their fracture resistance and marginal adaptation. *J. Dent.*, 2001; 29, 187-96.

[34] Kolbeck C, Rosentritt M, Behr M, Lang R and Handel G. In vitro study of fracture strength and marginal adaptation of polyethelene-fiber-reinforced-composite versus glass-fiber-reinforced-composite fixed partial dentures. *Clin. Oral Investig.*, 2002; 29, 668-74.

[35] Ellakwa AE, Shortall AC and Marquis PM. Influence of fiber type and wetting agent on the flexural properties of an indirect fiber reinforced composite. *J. Prosthet. Dent.*, 2002; 88, 485-90.

[36] Turkaslan S, Tezvergil-Mutluay A, Bagis B, Vallittu PK and Lassila LV. Effect of fiber-reinforced composites on the failure load and failure mode of composite veneers. *Dent. Mater. J.*, 2009; 28, 530-6.

[37] Goldberg AJ, Freilich MA, Haser KA and Audi JH. Flexure properties and fibre architecture of commercial fibre-reinforced composites [abstract 967]. *J. Dent. Res,* 1998; 77, 226.

[38] Chong KH and Chai J. Strength and mode of failure of unidirectional and bidirectional glass fiber-reinforced composite materials. *Int. J. Prosthodont.*, 2003; 16, 161-6.

[39] Chong KH and Chai J. Probability of failure of veneered glass fibre-reinforced composites and glass infiltrated alumina with or without zirconia reinforcement. *Int. J. Prosthodont.*, 2003; 16, 487-92.

[40] Behr M, Rosentritt M, Sikora MI, Karl P and Handel G. Marginal adaptation and fracture resistance of adhesively luted glass fiber- composite reinforced molar crowns with different inner crown surfaces. *J. Dent.*, 2003; 31, 503-8.

[41] Keulemans F, Palav P, Aboushelib MN, van Dalen A, Kleverlaan CJ and Feilzer AJ. Fracture strength and fatigue resistance of dental resin-based composites. *Dent. Mater.*, 2009; 25, 1433-41.

[42] McCabe JF and Walls AW, Applied Dental Materials. 8th ed. 1998: Blackwell Science.

[43] Behr M, Rosentritt M, Mangelkramer M and Handel G. The influence of different cements on the fracture resistance and marginal adaptation of all-ceramic and fibre-reinforced crowns. *Int. J. Prosthodont.*, 2003; 16, 538-42.

[44] Lehmann F, Eickemeyer G and Rammelsberg P. Fracture resistance of metal- free composite crowns- Effects of fiber reinforcement, thermal cycling and cementation technique. *J. Prosthet. Dent.*, 2004; 92, 258-64.

[45] Behr M, Rosentritt M, Ledwinsky E and Handel G. Fracture resistance and marginal adaptation of conventionally cemented fiber-reinforced composite three unit FPDs. *Int. J. Prosthodont.*, 2002; 15, 467-72.

[46] Cho L, Song HY, Koak JY and Heo SJ. Marginal accuracy and fracture strength of ceromer/fiber-reinforced composite crowns: Effect of variations in preparation design. *J. Prosthet. Dent.*, 2002; 88, 388-95.

[47] Cho L, Choi JM, Yi YJ and Park CJ. Effect of finish line variants on marginal accuracy and fracture strength of ceramic optimized polymer/fibre-reinforced composite crowns. *J. Prosthet. Dent.*, 2004; 91, 554-60.

[48] McLaren EA and White SN. Glass-infiltrated zirconia/alumina-based ceramic for crowns and fixed partial dentures. *Pract Periodontics Aesthet Dent*, 1999; 11, 985-94.

[49] Walton JN, Gardner FM and Agar JR. A survey of crown and fixed partial denture failures: length of service and reasons for replacement. *J. Prosthet. Dent.*, 1986; 56, 416-21.

[50] Oden A, Andersson M, Krystek-Ondracek I and Magnusson D. Five-year clinical evaluation of Procera AllCeram crowns. *J. Prosthet. Dent.*, 1998; 80, 450-6.

[51] El-Mowafy O and Brochu JF. Longevity and clinical performance of IPS-Empress ceramic restorations--a literature review. *J. Can. Dent. Assoc.*, 2002; 68, 233-7.

[52] Lawn BR, Strength and reliability, in Fracture of brittle solids, Davis EA and Ward IM, Editors. 1993, Cambridge University Press.

[53] Darvell BW, Mechanical testing in Materials Science for Dentistry, 7th edition, Darvell BW, Editor. 2002, Darvell, BW: Hong Kong. 1-35.

[54] Miller A, Long J, Miller B and Cole J. Comparison of the fracture strengths of ceramometal crowns versus several all-ceramic crowns. *J. Prosthet. Dent.*, 1992; 68, 38-41.

[55] Ban S and Anusavice KJ. Influence of test method on failure stress of brittle dental materials. *J. Dent. Res*, 1990; 69, 1791-9.

[56] Kelly JR. Perspectives on strength. *Dent. Mater.*, 1995; 11, 103-10.

[57] Kelly JR. Clinically relevant approach to failure testing of all-ceramic restorations. *J. Prosthet. Dent.*, 1999; 81, 652-61.

[58] Kelly JR. Dental ceramics: current thinking and trends. *Dent. Clin. North Am.*, 2004; 48, 513-30.

[59] Rosentritt M, Plein T, Kolbeck C, Behr M and Handel G. In vitro fracture force and marginal adaptation of ceramic crowns fixed on natural and artificial teeth. *Int. J. Prosthodont.*, 2000; 13, 387-91.

[60] Attia A and Kern M. Fracture strength of all-ceramic crowns luted using two bonding methods. *J. Prosthet. Dent.*, 2004; 91, 247-52.

[61] Gohring TN, Zappini G, Mayer J and Zehnder M. Glass-fiber frameworks for fixed partial dentures: laser-interferometrical in vitro analysis. *Quintessence Int.*, 2004; 35, 668-75.

[62] Gungor MA, Kucuk M, Dundar M, karaoglu C and Artunc C. Effect of temperature and stress distribution on all-ceramic restorations by using a three-dimensional finite element analysis. *J. Oral Rehabil*, 2004; 31, 172-8.

[63] Selna LG, Shillingburg HT, Jr. and Kerr PA. Finite element analysis of dental structures--axisymmetric and plane stress idealizations. *J. Biomed. Mater. Res.*, 1975; 9, 237-52.

[64] Mackerle J. Finite element modelling and simulations in dentistry: a bibliography 1990-2003. *Comput Methods Biomech. Biomed. Engin*, 2004; 7, 277-303.

[65] Morin DL, Douglas WH, Cross M and DeLong R. Biophysical stress analysis of restored teeth: experimental strain measurement. *Dent. Mater.*, 1988; 4, 41-8.

[66] Morin DL, Cross M, Voller VR, Douglas WH and DeLong R. Biophysical stress analysis of restored teeth: modelling and analysis. *Dent. Mater.*, 1988; 4, 77-84.

[67] Lang LA, Wang RF, Kang B and White SN. Validation of finite element analysis in dental ceramics research. *J. Prosthet. Dent.* 2001; 86, 650-4.

[68] Kou W, Kou S, Liu H and Sjogren G. Numerical modeling of the fracture process in a three-unit all-ceramic fixed partial denture. *Dent. Mater.*, 2007; 23, 1042-9.

[69] Kohn DH. Acoustic emission and nondestructive evaluation of biomaterials and tissues. *Crit. Rev. Biomed. Eng*, 1995; 23, 221-306.

[70] Kim KH, Park JH, Imai Y and Kishi T. Microfracture mechanisms of dental resin composites containing spherically-shaped filler particles. *J. Dent. Res*, 1994; 73, 499-504.

[71] Lee SY, Lin CT, Wang MH, Tseng H, Huang HM, Dong DR, Pan LC and Shih YH. Effect of temperature and flux concentration on soldering of base metal. *J. Oral Rehabil*, 2000; 27, 1047-53.

[72] Nicholls PJ and Berg E. Acoustic emission properties of callus. *Med. Biol. Eng. Comput,* 1981; 19, 416-8.

[73] Wright TM, Vosburgh F and Burstein AH. Permanent Deformation of Compact-Bone Monitored by Acoustic-Emission. *J. Biomechanics*, 1981; 14, 405-409.

[74] Kondo S, Ohkawa S, Hanawa T, Sugawara T and Ota M. Evaluation of conventional and microfilled composite resins using an acoustic emission technique. *Dent. Mater. J.,* 1985; 4, 81-7.

[75] Lee SY, Chiang HC, Lin CT, Huang HM and Dong DR. Finite element analysis of thermo-debonding mechanism in dental composites. *Biomaterials*, 2000; 21, 1315-26.

[76] Lin CT, Lee SY, Tsai TY, Dong DR and Shih YH. Degradation of repaired denture base materials in simulated oral fluid. *J. Oral Rehabil*, 2000; 27, 190-8.

Initial versus Final Fracture of Fiber-Reinforced Composites ...   493

[77] Duray SJ, Lee SY, Menis DL, Gilbert JL, Lautenschlager EP and Greener EH. Laser acoustic emission thermal technique (LAETT): a technique for generating acoustic emission in dental composites. *Dent. Mater.*, 1996; 12, 13-8.

[78] Corporation PA, PCI-2 Based AE system user's manual. 2nd ed. 2004, New Jersey Physical Acoustics Corporation.

[79] Qi G, Pujol J and Fan Z. 3-D AE visualization of bone-cement fatigue locations. *J. Biomed. Mater. Res.*, 2000; 52, 256-60.

[80] Asaoka K, Yoshida K and Sakamaki K. Effect of Transient Stress on Acoustic-Emission Behavior during Firing of Dental Porcelain. *J. Mater. Sci.*, 1992; 27, 3118-3122.

[81] Jeffers JR, Browne M and Taylor M. Damage accumulation, fatigue and creep behaviour of vacuum mixed bone cement. *Biomaterials*, 2005; 26, 5532-41.

[82] Roques A, Browne M, Thompson J, Rowland C and Taylor A. Investigation of fatigue crack growth in acrylic bone cement using the acoustic emission technique. *Biomaterials*, 2004; 25, 769-78.

[83] Wells JG and Rawlings RD. Acoustic emission and mechanical properties of trabecular bone. *Biomaterials*, 1985; 6, 218-24.

[84] Roy C and Elghorba M. Monitoring Progression of Mode-Ii Delamination during Fatigue Loading through Acoustic-Emission in Laminated Glass-Fiber Composite. *Polym Composites*, 1988; 9, 345-351.

[85] Kohn DH, Ducheyne P and Awerbuch J. Acoustic emission during fatigue of porous-coated Ti-6Al-4V implant alloy. *J. Biomed. Mater. Res.*, 1992; 26, 19-38.

[86] Alander P, Lassila LV, Tezvergil A and Vallittu PK. Acoustic emission analysis of fiber-reinforced composite in flexural testing. *Dent. Mater.*, 2004; 20, 305-12.

[87] Watanabe Y, Takai S, Arai Y, Yoshino N and Hirasawa Y. Prediction of mechanical properties of healing fractures using acoustic emission. *J. Orthop. Res.*, 2001; 19, 548-53.

[88] Akkus O. Comments on acoustic emission visualization of bone cement fatigue locations. *J. Biomed. Mater. Res.*, 2002; 59, 398-9.

[89] Geng RS. Modern acoustic emission technique and its application in aviation industry. *Ultrasonics*, 2006; 44: E1025-9.

[90] Qi G, Li JH, Mann KA, Mouchon WP, Hamstad MA, Salehi A and Whitten SA. 3D real time methodology monitoring cement failures in THA. *J. Biomed. Mater. Res. A*, 2004; 71A, 391-402.

[91] Hasegawa K, Takahashi HE, Koga Y, Kawashima T, Hara T, Tanabe Y and Tanaka S. Mechanical properties of osteopenic vertebral bodies monitored by acoustic emission. *Bone*, 1993; 14, 737-43.

[92] Berkovits A and Fang DN. Study of Fatigue-Crack Characteristics by Acoustic-Emission. *Eng. Fract. Mech.*, 1995; 51, 401-8.

[93] Narisawa I and Oba H. An Evaluation of Acoustic-Emission from Fiber-Reinforced Composites .1. Acoustic-Emission Interpretation of Epoxy Matrix and Model Composites Containing Glass-Beads, Carbon and Glass-Fibers. *J. Mater. Sci.*, 1984; 19, 1777-86.

[94] Aggelis DG, Barkoula NM, Matikas TE and Paipetis AS. Acoustic emission monitoring of degradation of cross ply laminates. *J. Acoust Soc. Am.*, 2010; 127, 246-51.

[95] Sachse W and Kim KY. Quantitative Acoustic-Emission and Failure Mechanics of Composite-Materials. *Ultrasonics*, 1987; 25, 195-203.

[96] Dyer SR, Sorensen JA, Lassila LV and Vallittu PK. Damage mechanics and load failure of fiber-reinforced composite fixed partial dentures. *Dent. Mater.*, 2005; 21, 1104-10.

[97] Kim KH and Okuno O. Microfracture behaviour of composite resins containing irregular-shaped fillers. *J. Oral. Rehabil*, 2002; 29, 1153-9.

[98] Kim KH, Park JH, Imai Y and Kishi T. Fracture-Toughness and Acoustic-Emission Behavior of Dental Composite Resins. *Eng. Fract. Mech.*, 1991; 40, 811-9.

[99] Kim KH, Kim YB and Okuno O. Microfracture mechanisms of composite resins containing prepolymerized particle fillers. *Dent. Mater. J.*, 2000; 19, 22-33.

[100] Vallittu PK. Use of woven glass fibres to reinforce a composite veneer. A fracture resistance and acoustic emission study. *J. Oral. Rehabil.*, 2002; 29, 423-9.

[101] Fennis WM, Tezvergil A, Kuijs RH, Lassila LV, Kreulen CM, Creugers NH and Vallittu PK. In vitro fracture resistance of fiber reinforced cusp-replacing composite restorations. *Dent. Mater.*, 2005; 21, 565-72.

[102] Lee SY, Lin CT, Keh ES, Pan LC, Huang HM, Shih YH and Cheng HC. Laser-induced acoustic emissions in experimental dental composites. *Biomaterials*, 2000; 21, 1399-408.

[103] Peterson IM, Pajares A, Lawn BR, Thompson VP and Rekow ED. Mechanical characterization of dental ceramics by hertzian contacts. *J. Dent. Res.*, 1998; 77, 589-602.

[104] Dong XD and Darvell BW. Stress distribution and failure mode of dental ceramic structures under Hertzian indentation. *Dent. Mater.* 2003; 19, 542-51.

[105] Ereifej N, Silikas N and Watts DC. Initial versus final fracture of metal-free crowns, analyzed via acoustic emission. *Dent. Mater.*, 2008; 24, 1289-95.

[106] Yi YJ and Kelly JR. Effect of occlusal contact size on interfacial stresses and failure of a bonded ceramic: FEA and monotonic loading analyses. *Dent. Mater.*, 2008; 24, 403-9.

[107] Kim KH, Choi MY and Kishi T. Fracture analysis of cast pure Ti and Ti-6Al-4V alloy for dental use. *Biomed. Mater. Eng.*, 1997; 7, 271-6.

[108] Kim KH, Asaoka K and Yoshida K. Acoustic emission signals from gypsum-bonded dental casting molds during thermal processing. *Dent. Mater. J.*, 1998; 17, 11-20.

[109] Schrooten J and Helsen JA. Adhesion of bioactive glass coating to Ti6A14V oral implant. *Biomaterials*, 2000; 21, 1461-9.

[110] Wang Y and Darvell BW. Failure mode of dental restorative materials under Hertzian indentation. *Dent. Mater.*, 2007; 23, 1236-44.

[111] Vakiparta M, Yli-Urpo A and Vallittu PK. Flexural properties of glass fiber reinforced composite with multiphase biopolymer matrix. *J. Mater. Sci. Mater. Med.*, 2004; 15, 7-11.

[112] Wolfart S, Ludwig K, Uphaus A and Kern M. Fracture strength of all-ceramic posterior inlay-retained fixed partial dentures. *Dent. Mater.*, 2007; 23, 1513-20.

[113] Attia A, Abdelaziz KM, Freitag S and Kern M. Fracture load of composite resin and feldspathic all-ceramic CAD/CAM crowns. *J. Prosthet. Dent.*, 2006; 95, 117-23.

[114] Garoushi S, Lassila LV, Tezvergil A and Vallittu PK. Load bearing capacity of fiber-reinforced and particulate filler composite resin combination. *J. Dent.*, 2006; 34, 179-84.

[115] Ozcan M, Breuklander MH and Vallittu PK. The effect of box preparation on the strength of glass fiber-reinforced composite inlay-retained fixed partial dentures. *J. Prosthet. Dent.*, 2005; 93, 337-45.

[116] Al-Darwish M, Hurley RK and Drummond JL. Flexure strength evaluation of a laboratory processed fiber-reinforced composite resin. *J. Prosthet. Dent.*, 2007; 97, 266-70.

[117] Stiesch-Scholz M, Schulz K and Borchers L. In vitro fracture resistance of four-unit fiber-reinforced composite fixed partial dentures. *Dent. Mater.*, 2006; 22, 374-81.

[118] Kolbeck C, Rosentritt M, Behr M, Schneider S and Handel G. Fracture strength and bond capacities of electron irradiated fiber reinforced composites. *Dent. Mater.*, 2007; 23, 1529-34.

[119] Drummond JL and Bapna MS. Static and cyclic loading of fiber-reinforced dental resin. *Dent. Mater.*, 2003; 19, 226-31.

In: Fiber–Reinforced Composites
Editor: Qingzheng Cheng

ISBN: 978-1-61470-303-7
© 2012 Nova Science Publishers, Inc.

Chapter 16

ANALYSIS OF DYNAMIC LOADING OF LONG FIBER-REINFORCED COMPOSITES

*Carlos Santiuste[1], Xavier Soldani[2], Jorge López-Puente[1] and Henar Miguélez[2]**

[1]Department of Solid Mechanics and Structural Analysis
[2]Department of Mechanical Engineering
Universidad Carlos III of Madrid, Avda. Universidad 30, 28911, Madrid (Spain)

ABSTRACT

Structural applications of LFRP composites could involve dynamic loading during the service life of the component; for instance, the impact of foreign objects such as birds against components of aircrafts during service. On the other hand, LFRP composite components are usually made to the final size of the desired product however some machining operations are needed, in general, to achieve work-piece requirements. Cutting operations of composite can be considered a dynamic problem and usually involve damage in the component.

Experimental work is needed to understand phenomena involved during dynamic loading states and it is also required to validate models. Numerical modeling based on Finite Element Analysis (FEA) is a powerful tool used to simulate dynamic loading of LFRP composites, allowing the analysis of different parameters influent on the process. The use of numerical models validated with experimental results is of crucial importance for designing, giving an alternative to costly and complex experiments.

This paper is focused on the analysis of dynamic loading of LFRP composites including both impact and machining studies. Previous work of the authors is reviewed and summarized in the paper together with other contributions obtained from the scientific literature.

---

* Email: mhmiguel@ing.uc3m.es

1. INTRODUCTION

The use of Long Fiber-Reinforced Polymer (LFRP) composite in different industrial sectors has been continuously increasing in the last decades, due to the excellent properties of this family of materials. The combination of high strength and specific stiffness, fatigue and corrosion resistance and low density of LFRP composite makes them suitable for high responsibility applications in aircraft, automotive, naval and sport components [1].

LFRP composites are an extensive group of materials including three main families of composites based on glass, carbon and or aramid fibers (mainly Kevlar) in a polymeric matrix (commonly denoted GFRP, CFRP and AFRP/KFRP respectively). Long fibers can be unidirectional (all fibers parallel to each other) or woven into a fabric or cloth. Glass fiber reinforced plastics (GFRP) are the most commonly used materials not only due to their high specific mechanical properties also because of the competitive cost. Carbon fiber-reinforced plastics (CFRP) and aramid fiber (commonly Kevlar) reinforced plastics (AFRP) present better specific strength, higher specific stiffness and lighter weight. However these composites are used in high performance applications because of their elevated cost. CFRP composites are widely used in structural components in aircrafts. AFRP is used instead of CFRP when higher strength, lightness and toughness are required, for instance for personal protections [2].

The use of these materials in different mobile systems in transport applications justifies the interest of understanding their behaviour under dynamic loading. The possibility of suffering an impact of a foreign object is elevated in such applications. For example the collision between an aircraft structure and a bird; or the impact of metallic aero-engine small pieces separated from the original component against the external components of the airplane made of composite; or the ballistic hazard in military or civil aircraft; illustrate the interest of understanding and preventing the onset of damage and failure of composite under dynamic loading [3].

On the other hand, this family of materials are also subjected to dynamic loading during the last stages of the component manufacture. Although the components are manufactured close to the final shape, they commonly require some machining operations, mainly milling and drilling, previously to the final assembly, in order to achieve dimensional specifications. Machining can be considered a dynamic process involving high cutting speed, and extreme contact conditions at the interface tool-chip [4].

Experimental research concerning dynamic loading is crucial for safety designing composite structural components and also for the proper definition of machining processes previously to the final assembly of the component. However experimental tests are always costly in economical terms and time consuming. Moreover, sometimes it is not possible to test the component used in the current applications due to its elevated cost, and in consequence, the tests should be defined in a simplified way. Machining experiments involve also other problems such as the attachment of the work piece to the machine-tool or the health hazards associated to the inhalation/contact with skin of the fiber.

In the context of high responsibility applications of composite materials, it is important to develop modelling tools able to simulate dynamic processes. These tools, once validated with experimental results, offer the possibility to complete experimental studies, avoiding technical problems and elevated cost. The accuracy of these models depends on different factors, being one of the most important the proper selection of the damage model of the composite.

This paper focuses on the study of dynamic processes in LFRP composites, mainly impact loading and machining. Both experimental work and theoretical models are commented. Modelling deals with some analytical efforts and especially with finite element models, a versatile technique commonly used in designing processes. The objective of the paper is presenting a brief review of current trends, available techniques and strategies concerning the study of impact loading and cutting processes of composites. Results summarized in the work refer mainly to carbon and glass composites.

The paper is structured as follows. After a brief introduction presented in this section, damage models commonly used in scientific literature are summarized in the second section. The third section deals with studies of impact processes affecting composites, presenting both low and high velocity impacts. Composite cutting, with attention to modelling techniques, is presented in the next section. Finally conclusions of the work are included in the last section.

2. Damage Modeling

Modeling of damage is one of the most important aspects for the accurate simulation of impact loadings and machining operations of LFRP composites. Among the different techniques available to predict composite damage, the failure-criteria approach has demonstrated its accuracy in both static and dynamic loading states. Many sets of failure criteria can be found in the literature [5,6].

The analysis of the different models available for long fiber composites and the identification of damage model parameters, have motivated an international exercise developed between 1998 and 2004 [7] comparing different failure criteria under static conditions. Despite of the research efforts, a consensus on selecting the best criteria to predict the composite failure has not been reached, especially in events involving out-of-plane loads.

No comparable study has been performed under dynamic loading. The most common failure criteria used in the dynamic conditions can be categorized in simple criteria which use an equation to predict the global failure of the laminate, and complex criteria involving several failure mechanisms (matrix cracking, matrix crushing, fiber failure, delamination, etc).

Although the complex failure of composite materials needs to distinguish between different failure modes, simple criteria have been used in several works to model the impact behavior [8,9] and the machining processes of composite materials [10].

The failure of composite materials cannot be predicted using Von Mises yield criterion that apply only to isotropic materials. The failure of anisotropic materials has been determined for many years by means of Hill criterion, which is an extension of Von Mises criterion to anisotropic materials. Tsai-Hill criterion is based on the application of the strength parameters present in Hill criterion to the critical strength values in three orthogonal directions formulating a criterion for LFRP composite, see Eq. 1 (X_1 and X_2 are the longitudinal and transversal strength respectively and S is the shear strength).

$$\frac{\sigma_1^2}{X_1^2} - \frac{\sigma_1 \sigma_2}{X_1^2} + \frac{\sigma_2^2}{X_2^2} + \frac{\sigma_{12}^2}{S^2} \geq 1 \qquad (1)$$

Tsai-Wu criterion [11] is a modification of Tsai-Hill criterion taking into consideration different values for tensile and compressive strength. The formulation of Tsai-Wu criterion is the same given in Eq. 1, but the values of the longitudinal and transversal strengths are dependent on the tensile or compressive stress state.

These models do not allow predicting the different failure modes characterizing composite materials: fiber failure, matrix failure, and fiber-matrix interface. This limitation has motivated the formulation of more realistic failure criteria. The most common sets of criteria used in the analysis of composite material in dynamic conditions are those due to Hashin [12] and Hou [13], constituting a three-dimensional version of Chang-Chang criteria [14].

Both Hashin and Hou formulations consider different failure mechanism and equations. Four failure modes (fiber failure, matrix cracking, matrix crushing and delamination) are accounted in Hou criteria, considering quadratic interaction between stresses, see Table 1. Hashin formulation considers also four failure modes: tensile and compression failure of fiber and matrix, however delamination is not considered. A quadratic interaction between the components of the stress vector associated with the failure plane governs each mode. Several variations for the Hashin criterion [15,16] are presented in the scientific literature.

Table 1. Hou and Hashin failure-criteria formulation

Failure mode	Hashin Formulation	Hou formulation
Fibre tension	$d_{ft}^2 = \left(\dfrac{\sigma_{11}}{X_T}\right)^2 + \left(\dfrac{\sigma_{12}}{S_L}\right)^2$	$d_f^2 = \left(\dfrac{\sigma_{11}}{X_T}\right)^2 + \left(\dfrac{\sigma_{12}^2 + \sigma_{13}^2}{S_L{}^2}\right)$
Fibre compression	$d_{fc}^2 = \left(\dfrac{\sigma_{11}}{X_c}\right)^2$	
Matrix cracking	$d_{mt}^2 = \left(\dfrac{\sigma_{22}}{Y_T}\right)^2 + \left(\dfrac{\sigma_{12}}{S_L}\right)^2$	$d_{mt}^2 = \left(\dfrac{\sigma_{22}}{Y_T}\right)^2 + \left(\dfrac{\sigma_{12}}{S_L}\right)^2 + \left(\dfrac{\sigma_{23}}{S_T}\right)^2$
Matrix crushing	$d_{mc}^2 = \left(\dfrac{\sigma_{22}}{2S_T}\right)^2 + \left[\left(\dfrac{Y_c}{2S_T}\right)^2 - 1\right]\left(\dfrac{\sigma_{22}}{S_L}\right)$ $+ \left(\dfrac{\sigma_{12}}{S_L}\right)^2$	$d_{mc}^2 = \dfrac{1}{4}\left(\dfrac{-\sigma_{22}}{S_T}\right)^2 + \left(\dfrac{Y_c\sigma_{22}}{4S_T{}^2}\right) - \dfrac{\sigma_{22}}{Y_c}$ $+ \left(\dfrac{\sigma_{12}}{S_L}\right)^2$
Delamination		$d_{del}^2 = \left(\dfrac{\sigma_{33}}{Z_T}\right)^2 + \left(\dfrac{\sigma_{23}}{S_T}\right)^2 + \left(\dfrac{\sigma_{13}}{S_L}\right)^2$

Parameters in Table 1 are the following: σ_{11}, σ_{22}, and σ_{33}, are the stresses in longitudinal, transverse and through-the-thickness direction respectively; σ_{12}, σ_{23}, and σ_{13}, are the shear stresses; X_T and X_C are the tensile and compressive strengths in longitudinal direction; Y_T and Y_C are the tensile and compressive strengths in the transverse direction; Z_T is the tensile strength in the through-thickness direction; S_L is the longitudinal shear strength; S_T is the transverse shear strength, (failure occurs when d_{ij} reaches the value 1).

The formulation of both Hou and Hashin criteria is explained below:

- Fiber tension: Both Hou and Hashin consider a combination of axial stress and shear stresses, with a quadratic interaction of σ_{11}, σ_{12} and σ_{13}. A weighting factor is sometimes considered in order to model the interaction between shear and tensile stress [17].

- Fiber compression: Hou considers a similar formulation as tensile fiber-breakage criterion, while Hashin considers just linear dependence on fiber-direction stress, σ_{11}.
- Matrix cracking and matrix crushing: the stresses in transverse direction, σ_{22}, σ_{23}, and σ_{12} are considered by the Hou criteria. Hashin criteria also include the stresses in the normal direction, σ_{33} and σ_{13}. The effect of these stresses is included in the delamination criterion in the Hou formulation. The impossibility of determining the plane of failure, leads to the use of a quadratic interaction between stress invariants considered in the Hashin matrix-failure formulation. The stress σ_{11} is not considered since any possible plane of failure is parallel to the fibers. The linear term resulting from the interaction between invariants is maintained in Hashin matrix crushing criterion, but it disappears in the matrix cracking criterion.
- Delamination: the delamination criterion proposed by Kim and Soni [18] and modified by Brewer and Lagace [19] was modified by Hou assuming that delamination was avoided by compressive through-the-thickness stresses ($\sigma_{33}<0$).

Continuum damage mechanics theory is a different approach to model the behavior of composite materials. This family of models has become a common approach to model the behavior under static and dynamic conditions for fiber reinforced plastic materials in the last decade.

Strength -based failure criterions can be used to predict the ultimate failure, but they cannot represent with enough accuracy the evolution of the damage due to the accumulation of damage as result of different damage mechanism. This fact is of great importance for damage tolerance analysis of structures made with composite materials. In addition those families of models have less mesh size sensibility, since the characteristic element length is part of the model equation. One of the most cited models is the LaRC, which has different versions; this model has been developed by the NASA agency [20].

3. IMPACT LOADING OF COMPOSITES

Long fiber-reinforced composite materials have been increasingly used in components for vehicles in marine, automotive, aviation and space industries due to their excellent mechanical properties. The impact between any of those composite structures and foreign objects is critical and usually unpredicted. In composite structures, impacts induce internal damage often difficult to detect by visual inspection. Internal damage could cause severe reductions in strength and are also susceptible to grow under service loads. Composites are vulnerable to impact processing, especially when out of the plane of laminate loads are involved, due to brittleness of the polymeric phase and the poor translaminar properties.

Therefore, it is necessary to improve the understanding of the effects of foreign object impacting on composite structures to prevent future damage during service, even from the earliest stages of design process. Concerns about the effect of impacts on the performance of composite structures have been a critical factor limiting the use of these materials. Thus, the problem of impact has received considerable attention in the literature.

During the life of a composite structure, impacts by foreign objects can be expected to occur during service, manufacturing and maintenance operations. Two regimes of dynamic loading are usually considered, high and low-velocity impact.

For instance, high-velocity impact occurs during aircraft take-offs and landings, when stones and other small debris from the runway are propelled at high velocities by the tires. These impacts, characterized by small mass but high impact velocity can perforate the structure, compromising its structural integrity. According to the comprehensive review of Abrate [21] high velocity impacts are defined as those where the ratio between impact velocity and the velocity of compressive waves propagating through the thickness is larger than the maximum strain to failure in that direction. This implies that damage is generated during the first few travels of the compressive wave through the thickness when overall plate motion is not yet established. Thus, high -velocity impact is a phenomenon controlled by wave propagation, and is essentially independent of boundary conditions, whereas a low-velocity impact is highly influenced by the boundary conditions.

Low-velocity impacts can occur during the manufacturing process or during maintenance when tools can be dropped on the structure. In this case, impact velocities are small but the mass of the projectile is larger. Abrate [21] considers low-velocity impacts when stress wave propagation through the thickness of the specimen plays no significant role. As soon as the projectiles enter in contact with the target, a compressive wave, a shear wave, and Rayleigh waves propagate outward from the impact point. Compressive and shear waves reach the back face and reflect back. After many reflections through the thickness of the laminate, the plate motion is established, and damage is induced after plate motion establishment.

The study of impact on composite structures involves many different topics, including contact mechanics, structural dynamics, strength, stability, fatigue, damage mechanics, and micromechanics. Impacts are simple events involving complex phenomena. Understanding the process of impact damage initiation and growth and identifying the governing parameters are important for the development of models for damage prediction.

Although experimental studies provide essential information, since impact phenomena depend on numerous parameters, understanding its influence on impact behavior requires a broad test program, which is time consuming and expensive. To reduce both cost and time, it is possible to use validated theoretical modeling. Both analytical [22, 23] and numerical [24,25] approaches have been used to analyze the dynamic response of composites.

The main advantage of analytical models is the quick analysis of the influence of different parameters. However, the complexity of the physical phenomena, which includes dynamic structural behaviour and loading, contact, friction, damage and failure, often results in an oversimplification of the problem and limits the utility of analytical approach. The main drawback of this methodology is the limited applicability of a given model to a short range of impact conditions (type of FRP and projectile, impact regime).

The numerical approach by means of FE analyses is a more flexible and powerful alternative to the analytical formulations. The possibility of modelling the constitutive behaviour of each material at local (element) level improves the capacity of simulation of complex structures under seemingly complex external loads and boundary conditions. The advantages of FE analysis include the possibility to easily modify the structure geometry, the ability to model complex geometric shapes and complex boundary conditions, and the relatively simple implementation of failure criteria and damage evolution models.

On the other hand, the use of FEM in the analysis of impact behavior of composite structures presents some difficulties including numerical instabilities due to the extremely distorted elements during impact event and the definition of contacts between projectile and composite plate which lead to high computational cost. Details of the numerical models used for simulation of impact can be found in references [24,25].

3.1. High Velocity Impact

The behavior of composite materials under high velocity impact has been widely analyzed in the last decades. The first experimental studies of the behavior of composite materials against ballistic impact were made in the 60s, mainly by the military agencies; all the documentation generated in those investigations has been remained secret until the 80s, when several studies were published. An example of this work is the article of [26], focusing on the damage induced in carbon/epoxy laminates for a wide range of impact velocities, from 10 to 500 m/s, the effect being measured by ultrasonic non-destructive inspection. This last study was one of the earliest works showing the effects of the localization of the damage and its extension for different laminate thicknesses. Fuji et al. [27], verified for CFRPs specimens that the extension of damage decreases with impact velocity. The effect of stacking sequence on the ability of the laminate to dissipate the kinetic energy of the projectile was studied in [28]. The differences in the response of CFRP laminate manufactured with woven or tape plies were discussed in the work due to [29]. Tape laminates were found to be less effective against impact than woven laminates, both below and above the ballistic limit, due to the different flexural behavior of its plies. Tanabe et al., [30] analyzed the behavior of carbon-reinforced plastics damaged by the impact of a steel sphere. The stress and stress-time histories on the specimens were strongly affected by the type of carbon fibre used and by the fibre/matrix interface strength. Hour et al. [31] carried out high-velocity impact tests up to perforation on stitched and unstitched panels. The damage was more reduced in the case of stitched samples than in the unstitched samples. However, the presence of stitches lowered the ballistic limit. Hammond et al. [32] studied high velocity impact on carbon-fibre-reinforced polymers and performed an extensive work to measure in-plane and out-of-plane deformation of the sample during impact, the residual velocity of the projectile, as well as microscopic and macroscopic damage. Herzsberg and Weller [33], fired steel spherical projectiles at post buckled carbon/epoxy laminates, stitched and unstitched, to study an eventual catastrophic failure due to impact damage.

Investigations in the range of high velocity impacts on composite laminates, usually focus on the measurement of two variables, the residual velocity in the case of perforation and the extension of the damage in the composite laminate. The ballistic limit, defined as the minimum impact velocity needed to induce perforation, is important when the composite laminate is used as a protection barrier for important devices such as oil pipes in aircrafts against different kinds of debris. The second variable is also of great importance because depending on the damage area, the stress redistribution along the structure could lead to a catastrophic failure. The localization of the damage around the impact point depends strongly on the laminate architecture. Tape laminates which usually are used in a quasi-isotropic stacking sequence present larger damage area, compared to woven laminates; this larger damage area in the first type of laminates appears because of the different flexure stiffness in the through-thickness direction.

In the following paragraphs, an explanation of the evolution of both variables is carried out; this explanation is first analyzed for velocities below impact limit, and then above this value.

Impacts Below Ballistic Limit

Impact velocities below the ballistic limit are high enough to enforce a strongly transient response, in opposition to quasi-static conditions, with little global structural response. Figure 1 left shows the cross-section of a woven laminate (2.2 mm thickness) impacted at 60 m/s; the white ellipsis marks the damaged zone with some delamination due to stress wave reflection (spalling effect). As the impact velocity increases (figure 1 right), the damage also increases; is easy to observe a crushed zone at the impacted face and fibre tensile failure of the fibres in the bottom plies; a conical-shaped shear plug started to form. In addition, small interlaminar cracks were found, progressively longer as they approached the back surface of the laminate. The maximum damage inflicted corresponds to the ballistic limit velocity.

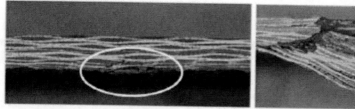

Figure 1. cross-section of a woven laminate impacted at 60 m/s (left) and 100 m/s (right).

Tape laminates shows different behavior when impacted at velocities below the ballistic limit. The damage extension is much larger as before mentioned; figure 2 shows the cross-section of a tape laminate (2.2 mm thickness) impacted at 60 m/s. Different failure mechanism could be observed, delamination matrix crushing and matrix cracking.

Figure 2. cross-section of a tape laminate impacted at 60 m/s.

At this range of velocities all the kinetic energy of the projectile is absorbed by the laminate generating damage; thus it is possible to state that the damaged area is proportional to the square of the velocity. In order to obtain an accurate value for the damaged area C-Scan technique is commonly used. Figure 3 shows two C-Scan images of impacted composite laminates at velocities around 100 m/s; the dark area correspond to the zone where the laminate is damaged by any mechanism, delamination, matrix cracking, matrix crushing or fiber failure. It is observed the larger damaged area in the case of tape laminate.

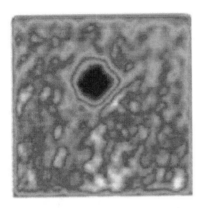 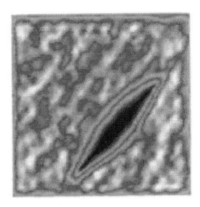

Figure 3. C-Scan image; woven laminate impacted at 100 m/s (left) and tape laminate impacted at 95 m/s (right).

Impacts above Ballistic Limit

In the velocity range above ballistic limit the damage extension decreases with velocity. Although the decrement is not easily appreciable in the cross-sections, depicted in figure 4 and 5, in those images it is possible to distinguish a conical-shaped shear-plug failure, oriented to the impact direction.

This conical hole changes and tends to a cylindrical shape parallel to the projectile trajectory as impact velocity increases; in addition some interlaminar cracks appear close to the crater but their area is much smaller than that observed in specimens impacted at velocities below ballistic limit.

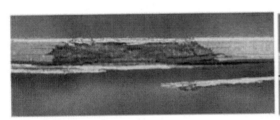 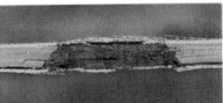

Figure 4. cross-section of a tape laminates impacted at 200 m/s (left) and 500 m/s (right).

 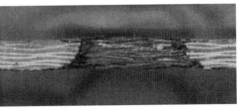

Figure 5. cross-section of a woven laminate impacted at 200 m/s (left) and 500 m/s (right).

The damage area measured with the C-Scan also accorded with these trends, the damage extent decreases with velocity. At high velocities the main mechanism developed by the laminate to absorb the energy of the projectile is the momentum transfer and scant delamination is observed. As velocity impact decreases the characteristic time of penetration process increases involving local bending phenomena and thus delamination. As observed in the images of the cross sections, the hole produced by a high-velocity impact has the shape of the intersection between the laminate and the trajectory of the projectile, and damage is strongly localized around this orifice.

Figure 6 shows C-Scan images of impacted laminates at velocities around 400 m/s. The white circle that appears approximately in the center of the laminate is the hole made by the projectile. It is clear that the damage extent is much smaller in the woven laminate, as before mentioned.

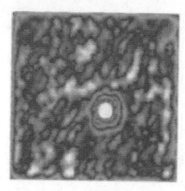 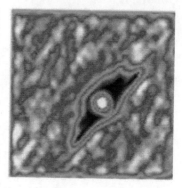

Figure 6. C-Scan image; woven laminate impacted at 400 m/s (left) and tape laminate impacted at 410 m/s (right).

Resume of Results

A very common graph that is usually used in ballistic problems, and that allows to easily visualize the laminate capability of protection is the impact velocity vs. residual velocity graph (figure 7).

Using this graph the ballistic limit is easily determined around 125 m/s for both laminate types; in addition the differences between the two architectures are very small in terms of residual velocity, and hence in energy absorbed by the laminate when penetration occurs. This behavior is explained because the main absorption mechanism is the linear momentum transfer, and that mechanism depends only on laminate density, which is almost the same. At high velocities the relation between the impact and residual velocity is linear, which is also due to the fact that the main absorption mechanism is the linear momentum transfer.

The damage induced in the laminate versus the impact velocity is plotted in figure 8. Differences between the two architectures are significant. Below ballistic limit the curves of both architectures have quadratic dependence with impact velocity, being the curve for the tape laminate higher because it is easier delaminated. As the impact velocity increases, above ballistic limit, the damage decreases more or less linear with the inverse of the impact velocity. Those curves tend asymptotically to the area of the projectile.

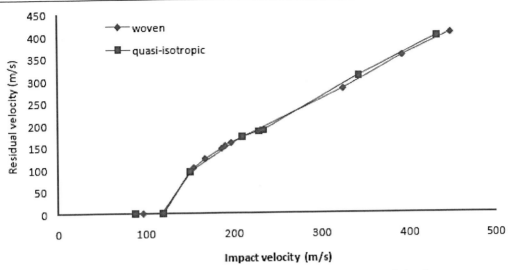

Figure 7. Impact velocity vs. residual velocity for both woven and quasi-isotropic laminates.

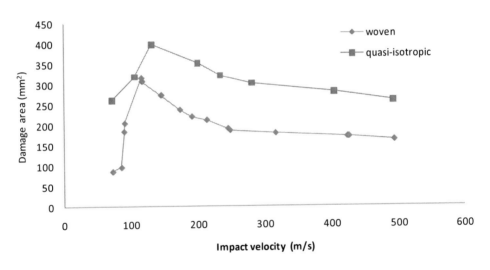

Figure 8. Damage area vs. residual velocity for both woven and quasi-isotropic laminates.

3.2. Low Velocity Impact

Low-velocity impacts or the drop of minor objects, such as tools during assembly or maintenance operation can produce damage on composite structures. These impacts are particularly dangerous because they can drastically impair the mechanical behavior of the structure after impact while exhibit little or no visible damage. The studies on composite behavior under low-velocity impact are focused on the damage inflicted by the impact. This damage can be related to several variables as damaged area, absorbed energy or residual strength.

Low velocity impact devices are mainly pendulum type (Charpy, Izod) and the drop-weight type. The pendulum is usually used to measure the in-plane impact resistance, but it has limitations concerning the size of specimen, impact direction, and boundary conditions. On the other hand, the drop-weight type, which better represents situations of falling objects or thrown debris, enables to measure the impact behaviour of a whole structure for the out-of-plane impact loads. Also, the drop-weight test machine test can be easily adapted to various situations by changing the boundary conditions or impactor shapes. Considering these benefits, the drop-weight test has been the main experimental device used in low-velocity impact on composite materials.

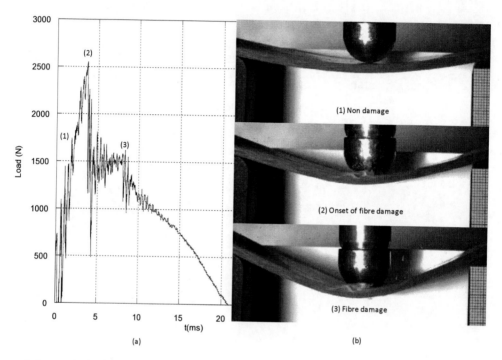

Figure 9. Low velocity impact on glass/polyester composite beams. a) Contact force history. b) Damage evolution.

The main variable of a low-velocity impact test is the contact force exerted by the impactor, Figure 9. The contact force can be used to determinate the onset of the damage in the laminate [34]. However, damage cannot be quantitatively predicted by the analysis of the contact force.

Absorbed energy can be directly related to the damaged originated on the composite. The higher is the absorbed energy, the higher the damaged inflicted on the composite. Figure 10 shows the evolution of the absorbed energy in a low-velocity impact test on a glass/polyester composite beam [34]. The impactor kinetic energy is transferred to the composite specimen from the initiation of contact until it is completely stopped. This means that the total kinetic energy has been absorbed by the composite structure in terms of elastic energy storage in the laminate and dissipative damage mechanisms. The elastic energy absorbed by the specimen is transferred back to the impactor which bounces with a post-ricochet kinetic energy. However, the post-ricochet kinetic energy of the impactor is lower than the impact energy, the energy transferred from the impactor to the composite plate is absorbed by the failure mechanisms activated. Thus, each failure mode, i.e., fiber failure and/or delaminations is related with the absorption of a fraction of impact energy. Therefore, the extension and type of failure mode activated will affect the absorbed energy values. However, the amount and the type of failure mechanisms activated depend on many factors as mass and velocity of the impactor (impact energy), geometry of the impactor, type of fiber and matrix, and stacking sequence. The experimental analysis of the influence of these factors is time and cost consuming thus the use of numerical models is a powerful tool to analyze the failure mechanisms of composites under low-velocity impacts [25,35].

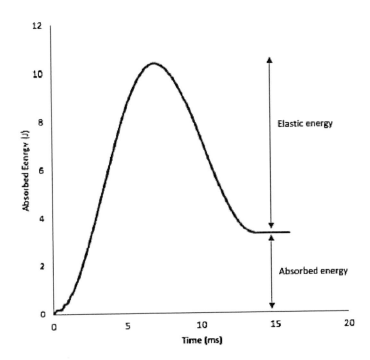

Figure 10. Low velocity impact on glass/polyester composite beams. Evolution of absorbed energy vs. time.

The absorbed energy is dependent on the impact energy. One of the main objectives of a low-velocity impact study is the estimation of the critical impact energy producing significant damage on the specimen. FE models have been used by several authors to predict the absorbed energy in low-velocity impacts [36,37].

Figure 11 shows the absorbed energy predicted by a FE model including Hou criteria and the experimental results obtained in low-velocity impact tests on composite sandwich beams [36].

Damage area is an important parameter for estimation of the residual strength of the impacted specimen and is mainly dominated by delamination failure. Several researchers have reported that delaminations occur mainly at interface between plies with different fiber orientations [21]. Figure 12 shows damage area as a function of impact energy on glass/vinylester plates with two configurations: woven and quasi-isotropic laminates [38].

All the plies in the woven plates have the same fiber orientation thus the extension of delaminations is lower than in quasi-isotropic plates where two adjacent plates have different fiber orientation. To model this phenomenon at the interface between adjacent plies cohesive elements have been using by several authors in FE modeling [39,40].

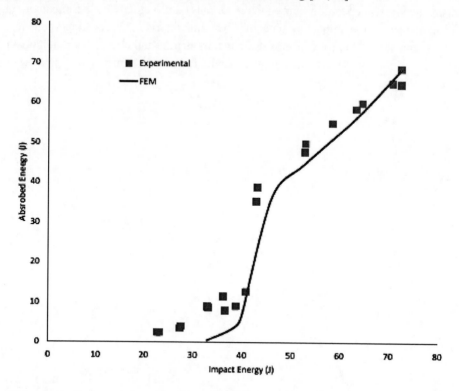

Figure 11. Low velocity impact on composite sandwich beams. Absorbed energy vs impact energy.

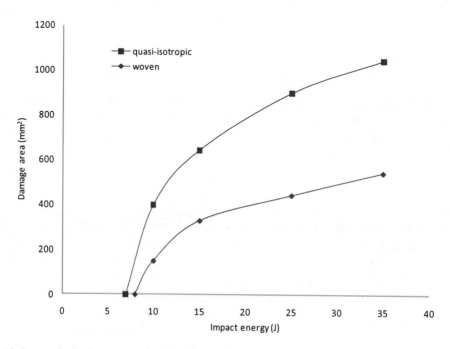

Figure 12. Low velocity impact on glass/vinylester plates. Damaged area vs impact energy.

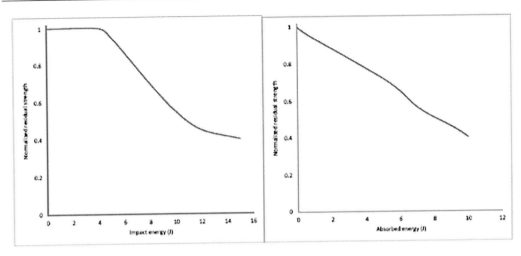

Figure 13. Low velocity impact on glass/polyester composite plates. Left: normalized residual strength vs impact energy. Normalized residual strength vs absorbed energy.

The estimation of the residual mechanical properties of the impacted laminates is of most concern since they can be significantly reduced by damage mechanisms. Studies dealing with the estimation of residual properties usually focus on compression after impact behaviour [41,42] and also on residual flexural properties [34,43]. Normalized flexural residual strength as a function of impact energy is shown in Figure 13 for low-velocity impacts on glass/polyester plates. There is a threshold energy from which the residual strength is reduced as the impact energy increases and the damage mechanisms are activated. Figure 13 also shows the normalized flexural residual strength as a function of absorbed energy. There is a linear relationship between absorbed energy and residual strength thus the residual mechanical properties can be directly related to the absorbed energy. Wang et al. used Hashin failure criteria to predict the residual tensile strength after low-velocity impacts on carbon fiber composite laminates [44].

4. MACHINING PROCESSES

Turning, and especially drilling and milling are the common technologies for the conventional machining of LFRP [45]. Most works dealing with machining of this family of materials focus on glass and carbon/polymeric matrix composites, however reduced information is available about machining of aramid composites. The properties of the fibers and matrix influence the machinability of FRPs that also depends on fiber orientation and fiber volume fraction. During cutting glass and carbon fibers break in a brittle manner under bending stresses, while aramid fibers undergo shearing fracture under high deformation bending and tear under tensile loading [2].

Turning is commonly performed to achieve final dimensions and surface requirements of rotation-symmetric components. This process is characterized by an almost constant contact between tool and chip: quasi-continuous cut exists during turning of FRP, despite of oscillations in stress caused by the different cutting behavior of the fibers and the matrix.

Composite milling (also known as routing or trimming) is an end-machining operation, thus low quantity of material is removed during machining. It is usually applied to the contour of the component to achieve final dimensional and superficial requirements. Modeling current three dimensional turning of milling processes involves such complexity, even in the case of metal cutting, that most information is obtained from the modelization of orthogonal cutting. This is a simplified process that allows obtaining valuable information about the mechanisms of chip formation, contact characteristics and cutting forces.

Drilling is a common operation to machine the holes needed for assembly composite components. The use of composite material in high responsibility applications is often limited due to delamination induced during machining (see figure 14). In fact, in the aerospace sector, a significant percentage of rejected parts during the assembly process is due to delamination induced during drilling [46]. Drilling is a complex process, thus models described in literature simplify the process in order to relate cutting forces with the problem of delamination.

In the following subsections, modeling of composite machining, mainly orthogonal cutting and some attempts to model drilling process are presented. The section does not include tool characteristics or cutting conditions, the wide variety of these technological parameters for each type of composite can be found in some specialized references previously included (see for instance [2,45].

Figure 14. Drilling induced defects in CFRP due to inappropriate cutting parameters (by courtesy of Prof. M. Marcos, Laboratory of Manufacturing, Universidad de Cádiz, Spain).

4.1. Orthogonal Cutting of Composite

As was explained previously machining processes of composite are very complex not only due to the characteristics of the operation (turning, drilling or milling), but also because of the anisotropy of the workpiece. This fact has motivated the development of numerical works focused on the study of orthogonal cutting of LFRP composites and also the development of experiments to validate the models.

Orthogonal cutting of composite assumes that the cutting edge of the tool is orthogonal to the direction of the cutting velocity. The boundary conditions and geometry of the FE models commonly used to simulate orthogonal cutting is illustrated in figure 15.

The shape of the composite, usually thin walled, complicates the execution of the tests, commonly performed at low cutting speeds achieved after complex positioning process in machine-tool, see for instance the works [47-49] focused on orthogonal cutting of Glass and Carbon FRP composites, those have been widely used to validate numerical models in scientific literature.

FE method has been used to develop macro and micro approaches to simulate orthogonal cutting of FRP composites.

Macro-mechanical approaches consider anisotropic homogeneous material and unidirectional fiber orientation is commonly assumed. Arola and Ramulu [50] developed one of the earliest FE studies of orthogonal cutting of unidirectional composites based on maximum stress and Tsai-Hill failure criteria.

Ramesh et al. [51] implemented a failure mechanism based on fiber-matrix cracking following the previous work in [52]. The authors analyzed the influence of fiber orientation for different LFRP composites and demonstrated that dependence of the load needed to induce failure on the fiber orientation.

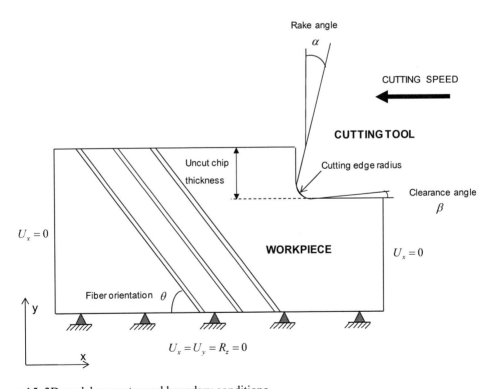

Figure 15. 2D model geometry and boundary conditions.

Tsai Hill criterion was considered in the 2-D analysis developed by Mahdi and Zhang [10] to simulate composite cutting. They reproduced an equivalent homogeneous material predicting cutting forces of LFRP composite as a relation to the fiber orientation. Three-dimensional approach simulated a composite cell based on perfectly bonded constituents (fiber and matrix).

The same failure criterion was considered in [53]. Sensitivity of chip formation mechanism and damage generation on the fibre orientation was confirmed.

Arola et al. [55] explored the influence of tool geometry on both cutting forces and subsurface damage in orthogonal machining of unidirectional LFRPs.

Lasri et al. [56] considered Hashin, Maximum stress and Hoffman failure criteria in combination with the material property degradation rule for the free surface failure analysis during machining of the composite workpiece.

Numerical simulation based on finite element technique has been used to analyze orthogonal cutting of LFRP composites using both macro and micro mechanical approaches [57,58].

In references [47,59] micro-macro approaches were combined to model the orthogonal cutting process, obtaining good accuracy in the prediction of cutting forces.

Micro mechanical modeling of both glass and carbon fiber reinforced polymer composite has been presented in [60]. The influence of the fiber orientation on the failure mechanisms has been reported and the de-cohesion phenomena in matrix-fiber interface was analyzed.

A new approach of modeling of orthogonal cutting of carbon LFRP using discrete element method (DEM) was presented in a recent work [61]. To simulate the material, solid link between particles were created; this link was considered as elastic whose ends are glued on each particle. The forces were calculated not only on the basis of the particles centers distance but also according to the distance between grip points. The behavior of the composite was characterized by a set of macroscopic parameters representing the behavior of fiber and matrix components. The micro data (contact force, link force) had to be deduced from the macro-scale. Validation of the numerical model was carried out by comparing cutting forces and predicted chip morphology with experimental observations.

2D Modelling

The authors have presented a recent numerical work focused on the mechanism of chip formation of both glass and carbon LFRP composites. The numerical model is based in the scheme of figure 16, details can be found in reference [62].

Dynamic explicit analysis was carried out in the commercial Finite Element code ABAQUS/Explicit using plane stress, quadrilateral, linearly interpolated, elements, with reduced integration and automatic hourglass control (CPS4R in ABAQUS/Explicit notation [63].

Geometrical and material characteristics were chosen according to the experimental results in reference [48] used for model validation. In order to analyze complete chip formation long cutting length was simulated equal to 2 mm, large enough to reach steady state conditions. Contact at the interface chip/tool was modeled by using surface – node surface contact available in ABAQUS/Explicit with a constant coefficient of friction equal to 0.5 as proposed in ([56]. The tool was assumed to be rigid. Both workpiece materials, glass and carbon FRP were modeled with an elastic behavior up to failure, taking into account the anisotropy of the material (see details about material properties in [62]). The damage initiation criteria for fiber reinforced composites are based on Hashin theory (see section 2).

The aim of the work was analyzing the differences between both materials: cutting forces, chip morphology and damage were compared for GFRPs and CFRP (see Figure 16, fiber orientation equal to 45°).

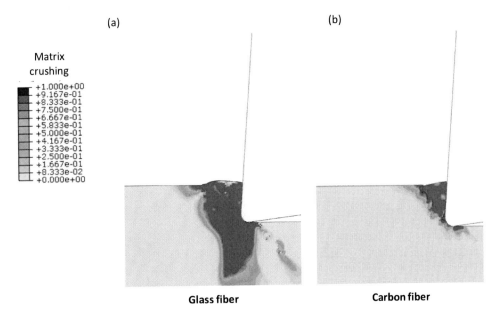

Figure 16. Matrix crushing damage predicted for complete chip formation of GFRP and CFRP.

Simulations of orthogonal cutting of GFRPs showed that the damaged zone was widely extended both beneath and in front of the tool tip. However numerical predictions of damage of CFRP showed much smaller affected zone, being restricted to the vicinity of chip. In this case, the damage was initiated close to the tool tip, evolved towards the region located between the primary shear zone and the interface chip-tool. This damaged zone is removed with the chip segmentation. This behavior shows an important benefit for CFRPs materials commonly and is in accordance with experimental observations in [64] showing better surface quality when turning carbon FRPs.

The different behavior of GFRPs and CFRPs could be partially explained with the level of energy needed to complete breakage of the elements. This parameter has strong influence on the simulation of the material behavior. The statement of low level of energy leads to the erosion of the element just after the damage onset, while high level of energy is related with elevated deformation of the element before total breakage. Material stiffness should be decreased after the onset of damage, this reduction of the stiffness coefficients is controlled by damage variables ranging between zero (corresponding to undamaged state) and one (related with fully damage state for the mode corresponding to the considered damage variable). The evolution of damage after the initiation is governed by the fracture energy dissipated G_I. This approach is based on the model proposed by Camanho and Dávila ([39] for inter-laminar delamination using cohesive elements.

CFRPs present catastrophic failure with weak progression of the damage, thus the elements suffer negligible deformation, breaking just after the damage onset limiting the damage extension on the workpiece [43]. However GFRPs could be considered more ductile composite materials exhibiting progressive failure: after the onset of damage in an element it is deformed up to the level of energy allowed.

The energy needed for complete breakage of the element is related with the mechanical behavior of the composite and could not be considered a purely numerical parameter. In a recent work of the authors it was shown that the damage distribution strongly depends on the level of energy needed for breakage. The value of energy considered for GFRP composite was ranged from the nominal value to lower and higher values (200, 400 and 600 J/m^2).

The highest value of the energy leads to wide extension of damage and high deformation without chip segmentation, reminding to the mechanisms observed in metal cutting. When the energy decreases, the damage extension tends to be reduced in a small area close to the tool and beneath the machined surface [65], see figure 17.

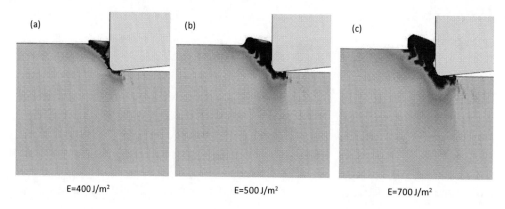

Figure 17. Evolution of chip morphology with the energy for element breakage (reference material, CFRP, fibre angle 45°).

3D Modeling

Even the simple case of orthogonal cutting could improve its accuracy when the 3D approach is implemented. With the aim of analyzing the validity of the assumptions of the 2D approach a 3D model was developed using the commercial Finite Element code ABAQUS/Explicit. Dynamic explicit analysis was carried out using 8-node brick elements with reduced integration (C3D8R in ABAQUS/Explicit notation [63], see figure 18.

It was necessary to develop a VUMAT subroutine based on the Hou failure criteria [13] including a procedure to degrade material properties to model the workpiece behavior.

The stresses at each integration point in the composite are computed with the user subroutine when the load is applied. The failure criteria are checked and when it is reached, the material properties are degraded according to the mode of failure. The stresses are reduced to reproduce the elastic-properties degradation in the damaged elements depending on the failure mode. Thus fiber failure results in complete collapse of the material at that point ($\sigma_{11}= \sigma_{22}= \sigma_{33}= \sigma_{12}= \sigma_{23}= \sigma_{13}= 0$), whereas delamination eliminate the possibility of supporting stresses in the normal direction ($\sigma_{33}= \sigma_{23}= \sigma_{13}= 0$).

An element erosion criterion, based in maximum strain criteria was implemented in the VUMAT subroutine in order to avoid numerical problems due to distorted elements. After each time increment the longitudinal strains (ε_{11}, ε_{22} and ε_{33}) are evaluated, and the element is removed if one of the strains exceeds a threshold value.

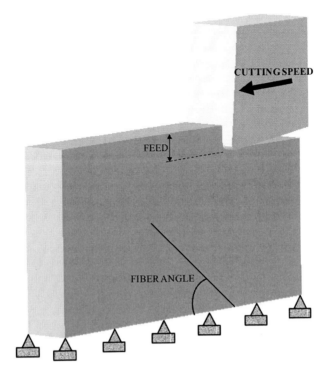

Figure 18. Scheme of 3D model for orthogonal cutting of composite.

As was considered in the 2D approach, the tool was assumed to be rigid. Interaction between workpiece and tool was modeled by using the algorithm surface–node surface contact available in ABAQUS/Explicit. A constant coefficient of friction equal to 0.5 at the tool/workpiece interface was assumed. Other model characteristics were similar to those implemented in the 2D approach.

As was described previously plane stress state is commonly assumed in modeling of orthogonal cutting of composite. However out-of-plane stresses could play an important role during machining. Not only experimental results have shown that the damage varies along the thickness also numerical simulations performed by the authors have shown the same trend. Differences in the damage area in external and internal plies can be seen in figure 19.

Another reason to develop 3D models of orthogonal cutting is that damaged area is strongly dominated by delamination in dynamic processes. This phenomenon is produced by out-of-plane stresses, thus a 3D model is required to simulate the evolution of damage.

An important limitation of the 2D approach is also the impossibility of simulating non unidirectional laminates with a given ply sequence, widely used in industry. These laminates are also the most sensitive to delamination damage that is not commonly observed in unidirectional laminates.

Further improvement in the 3D modeling of composite cutting could be reached with cohesive elements. Some problems were found by the authors when using cohesive formulation in ABAQUS/explicit. This commercial code gives the possibility of using cohesive interaction to model the interface between plies. The main problem encountered when using this approach to model the laminate is the interaction between the workpiece and the tool. The use of cohesive interaction is not allowed in combination with the type of contact algorithm (surface/node surface) required for modeling the chip formation and removal. Thus it is not possible the use of cohesive interaction in composite cutting. Current difficult found in the process is the elevated computational cost of the simulations, that up to date is not optimized.

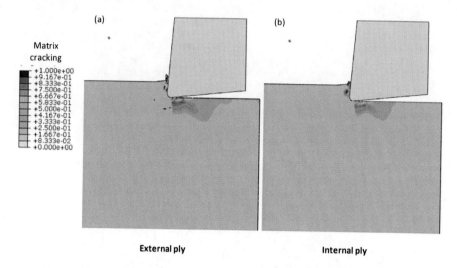

Figure 19. Matrix cracking damage in the external and internal ply predicted with 3D model.

4.2. Drilling

As was explained previously damage delamination due to drilling is a common defect causing rejection of the final component in aeronautical applications. It is possible to observe two different mechanisms of delamination when drilling FRP composites: peel-up at entrance and push-out at exit. The material spirals up the drill flute due to the acting cutting force, prior to be completely machined in the first case, resulting this phenomenon in a peeling force which combined with the thrust force, tries to separate the upper lamina from the uncut part of workpiece. This mechanism can be reduced with a moderate feed rate.

It is possible to observe the second mode, push-out delamination, in the interlaminar zone. This phenomenon depends on the matrix properties and it is weakly influenced by the fiber nature. It is commonly admitted that this mode of damage depends on the compressive trust force exerted by the drill tip on the uncut laminate plies that could exceed the interlaminar bonding strength [66]. The thrust force should remain lower than the critical value inducing delamination. Also the tool geometry is related with the development of delamination. For instance special drills acting on the distribution of thrust force towards the periphery rather than the center of the hole could diminish the delamination damage.

Main types suitable for composite drilling are twist, Brad, Dagger and step (see for instance [67]) and other special core geometries (see [68]) have been analyzed. The optimal results depend not only on the drill geometry but also on the definition of the corresponding optimal cutting parameters. The optimum selection of the drill point geometry is crucial to ascertain the minimum drilling induced damage and subsequently the maximum residual tensile strength [69].

On the other hand, different authors have focused attention on the estimation of critical thrust force related with the onset of delamination. Linear elastic fracture mechanics approach has been applied in different analytical models for critical thrust force prediction. Hocheng and Dharan [70] developed one of the most referred models. The critical thrust force corresponding to the initiation of delamination (F_{crit}) (Eq. 1) depends on the properties of the unidirectional laminate: the elastic modulus (E_1), the Poisson ratio (v_{12}), the interlaminar fracture toughness in mode I (G_{Ic}) and the uncut laminate thickness (h):

$$F_{crit} = \pi \left[\frac{8G_{Ic}E_1 h^3}{3(1-v_{12}^2)} \right]^{1/2} \tag{2}$$

Similar analysis was developed by Lachaud et al. [71] for small number of uncut plies. The uncut plate was modeled as a thin orthotropic plate clamped on the laminate surface and it was assumed normal stress acting in the ply surface. Both distributed and localized load were considered for the estimation of resultant critical thrust force, see Eq. 3 and in Eq. 4 respectively (being D and D_0 parameters calculated using laminates theory).

$$F_{crit} = 8\pi \left[\frac{G_{Ic}D}{1/3 - D'/8D} \right]^{1/2} \tag{3}$$

$$F_{crit} = 8\pi \left[\frac{2G_{Ic}D}{1 - D'/8D} \right]^{1/2} \tag{4}$$

The Finite Element Method (FEM) has been used to predict, stresses, strains and delamination onset in the workpiece in drilling operations. Simplified models were developed due to the difficulty of simulate drill rotation and feed and establish both damage and erosion criteria at the elements. Thus most works in scientific literature model drilling as a punching process on the laminate [72]. It is assumed that the drill point acts as a punch that pierces through the laminate (see Figure 14). This is the approach used by Durao et al. [73] and Singn et al. [74] to study GFRP drilling, showing the influence of the drill point angle in the induced damage.

CONCLUSIONS

Dynamic loading of composites has been briefly reviewed including experimental works and modeling approach. It is important to improve understanding of dynamic processes because composites are susceptible to suffer impact of a foreign object or induced damage during machining. High responsibility applications of composite materials involve elevated loads and extreme service operations those could lead to the progression of the damage and catastrophic failure of the structural component. One of the objectives of the design process should be preventing the failure of the component subjected to different hazards.

Experimental work is a valuable research field for the study of mechanisms involved in dynamic loading of composites. It was included a section dealing with this topic, including both low and high velocity impact ranges.

Modelling of dynamic processes could help in the achievement of the design objectives allowing the evaluation of difficult to measure parameters. Models should be validated through the comparison with experimental work. Among the modeling efforts in the field of composites there are a significant number of research work focusing on finite element modeling. The versatility of this technique allows the development of useful models reproducing dynamic loading of composites.

The development of accurate FE models strongly depends on the selection of proper damage model. Different damage models are available for composite however no consensus about the optimal criteria has been reached. Main formulations developed in scientific literature have been summarized in the work. Most of these models have been proved its ability to predict damage in different applications.

Although a wide variety of experimental works concerning machining of composites are available in the literature, modeling of current machining operations of composites is still a poorly developed topic. Even in the case of the simplest case of simulation of orthogonal cutting, few works are found in scientific literature. The results obtained in modeling orthogonal cutting of composite have been included, with attention to the influence of numerical parameters. Limitations of this approach are related with the assumption of two dimensional modeling in plane stress conditions, some considerations to the importance of stating three dimensional models are included. No information about tool characteristics or cutting conditions is included in the paper, but some specialized references are provided.

The paper does not try to be exhaustive in the recompilation of information in both broad fields of impact processes and machining. On the contrary it pretends to present together two types of processes those are commonly presented in separated way, while they can be studied with similar approaches. The development of the processes at high velocity, the implementation of damage models, the establishment of erosion criteria in FE modeling or the trend to generate delamination are common characteristics for both impact and machining requiring similar approaching strategies.

ACKNOWLEDGEMENTS

The authors are indebted for the financial support of this work, to the Ministry of Science and Technology of Spain (under Projects DPI2008-06746 and DPI2010-15123).

REFERENCES

[1] W.H. Zhong, Major Trends in Polymeric Composites Technology in Composite Materials Research Progress, Nova Publishers, 2008.

[2] R. Teti, *CIRP Annals - Manufacturing Technology*, Volume 51, Issue 2, 2002, Pages 611-634.

[3] R. Zaera, Ballistic Impacts on Polymer Matrix Composites, Composite Armor, Personal Armor in Impact Engineering of Composite Structures, Abrate, Serge (Ed.) Springer, 2011.

[4] A.M. Abrão, J. Paulo Davim, J.C. Campos Rubio and P.E. Faria, Department of Mechanical Engineering, University of Minas Gerais, Pampulha, Brazil, and others, The Surface Integrity of Composites Laminates Subjected to Drilling, pp. 439-451 in Composite Laminates: Properties, Performance and Applications, Nova Publishers, 2010.

[5] Nahas NM. Survey of failure and post-failure theories of laminated fiber-reinforced composites. *J. Compos. Tech. Res.* 1986; 8(4):138-153.

[6] Paris F. A study of failure criteria of fibrous composite materials. Technical report: NASA-cr210661, 2001.

[7] World-Wide Failure Exercise: *Composite Science and Technology*, vols. 58, 62 and 64.

[8] Prakash Jadhav, P. Raju Mantena, Ronald F. Gibson, Energy absorption and damage evaluation of grid stiffened composite panels under transverse loading, *Composites: Part B* 37 (2006) 191–199.

[9] Luise Kärger, Jens Baaran, Jan Teßmer, Efficient simulation of low-velocity impacts on composite sandwich panels, *Computers and Structures*, Volume 86, Issue 9, May 2008, Pages 988-996.

[10] Mofid Mahdi, Liangchi Zhang A finite element model for the orthogonal cutting of fiber-reinforced composite materials, *Journal of Materials Processing Technology, Volume 113, Issues 1-3, 15 June 2001, Pages 373-377].*

[11] S.W. Tsai and E.M. Wu, A general theory of strength for anisotropic materials. *J. Comp. Mater.* 5 (1971), pp. 58–80.

[12] Hashin Z. Failure criteria for unidirectional fiber composites. *J. Appl. Mech.* 1980, 47(2): 329-334.

[13] Hou JP, Petrinic N, Ruiz C, Hallett SR. Prediction of impact damage in composite plates. *Compos. Sci. Tech.* 2000; 60 (2): 273-28.

[14] Chang F, Chang K. A progressive damage model for laminated composites containing stress concentrations. *J. Compos. Mater.* 1987; 21:834-55.

[15] S. Chan, Z. Fawaz, K. Behdinan and R. Amid, Compos. Struct.77, 466–74 (2007).

[16] Z. Hashin and A. Rotem, *J. Compos. Mater.*7, 448–64 (1973)).

[17] (D. Zangani, M. Robinson and A.G. Gibson, *Appl. Compos. Mater.*15, 139–156 (2008).).

[18] R.Y. Kim and S.R. Soni in K. Kawata, S. Umekawa and A. Kobayashi, eds., Composites'86: recent advances in Japan and the United States, Proc. Japan-U.S. CCM-III, Tokyo, 1986, pp. 341–350.

[19] J.C. Brewer and P.A. Lagace, Quadratic stress criterion for initiation of delamination *J. Compos. Mater.*, 22,1141–55 (1988).

[20] Dávila C, Camanho P. Failure criteria for FRP laminates in plane stress. Tech. Rep., NASA, Langley Research Center, Hampton, VA, NASA/TM-2003-212663; November 2003.

[21] Abrate S. "Impact on composite structures". CambridgeUniversity Press. 1998.

[22] Santiuste C, Sanchez-Saez S, Barbero E. Application of the flexibility influence function method in the dynamic analysis of composite beams. *International Journal of Solids and Structures* 44 (2007) 4795–4809.

[23] J. López-Puente, R. Zaera, C. Navarro. An analytical model for high velocity impacts on thin CFRPs woven laminated plates. *International Journal of Solids and Structures, Volume 44*, Issue 9, 1 May 2007, Pages 2837-2851.

[24] J. López-Puente, R. Zaera, C. Navarro. Experimental and numerical analysis of normal and oblique ballistic impacts on thin carbon/epoxy woven laminates. *Composites Part A: Applied Science and Manufacturing*, Volume 39, Issue 2, February 2008, Pages 374-387.

[25] Santiuste C, Sanchez Saez S, Barbero E. A comparison of progressive-failure criteria in the prediction of the dynamic bending failure of composite laminated beams. *Composite Structures* 2010; 92: 2406-2414.

[26] Cantwell, W.J., 1988. Influence of target geometry on the high velocity impact response of CFRP. *Compos. Struct.* 10, 247–265.

[27] Fuji, K., Aoki, M., Kiuchi, N., Tasuda, E., 2002. Impact perforation behavior of cfrps using high-velocity stell sphere. *Int. J. Impact Eng.* 27, 497–508.

[28] Will, M.A., Franz, T., Nurick, G.N., 2002. The effect of laminate stacking sequence of CFRP filament wound tubes subjected to projectile impact. *Compos. Struct.* 58, 259–270.

[29] López-Puente, J., Zaera, R., Navarro, C., 2002. The effect of low temperatures on the intermediate and high velocity impact response of cfrps. *Compos. Part B: Eng.* 33 (8), 559–566.

[30] Tanabe Y, Aoki M, Fujii K, Kasano H, Yasuda E. Fracture behavior of cfrps impacted by relatively high-velocity steel sphere. *Int. J. Impact. Eng.* 2003;28:627–42).

[31] Hosur MV, Vaidya UK, Ulven C, Jeelani S. Performance of stitched/unstitched woven carbon/epoxy composites under high velocity impact loading. *Compos. Struct.* 2004;64:455–66.

[32] Hammond RI, Proud WG, Goldrein HT, Field JE. High-resolution optical study of the impact of carbon–fibre reinforcedpolymers with different lay-ups. *Int. J. Impact. Eng.* 2004;30:69–86.

[33] Herzsberg I, Weller T. Impact damage resistance of buckled carbon/epoxy panels. *Compos. Struct.* 2006;73:130–7.

[34] Santiuste C, Sanchez Saez S, Barbero E. Residual flexural strength after low-velocity impact in glass/polyester composite beams. *Composite Structures* 2010; 92: 25-30.

[35] Volnei Tita, Jonas de Carvalho, DirkVandepitte. Failure analysis of low velocity impact on thin composite laminates: Experimental and numerical approaches. *Composite Structures*, Volume 83, Issue 4, June 2008, Pages 413-428.

[36] Inés Ivañez, Carlos Santiuste, Sonia Sanchez-Saez, FEM analysis of dynamic flexural behaviour of composite sandwich beams with foam core, *Composite Structures*, Volume 92, Issue 9, August 2010, Pages 2285-2291.

[37] Davies GAO, Hitchings D, Wang J. Prediction of threshold impact energy for onset of delamination in quasi-isotropic carbon/epoxy composite laminatesunder low-velocity impact. *Compos. Sci. Technol.* 2000;60(1):1–7.

[38] S.K. García-Castillo, S. Sanchez-Saez, E. Barbero, C. Navarro. Comportamiento frente a impacto de placas de vidrio/vinilester. MATCOMP 05. Congreso nacional de materials compuestos. Valencia, 2005.

[39] Camanho, P.P., Davila, C.G., 2002. Mixed-mode decohesion finite elements for the simulation of delamination in composite materials. NASA/TM-2002-211737.

[40] N. Hu, Y. Zemba, T. Okabe, C. Yan, H. Fukunaga, A.M. Elmarakbi. A new cohesive model for simulating delamination propagation in composite laminates under transverse loads. *Mechanics of Materials*, Volume 40, Issue 11, November 2008, Pages 920-935.

[41] Sánchez-Sáez S, Barbero E, Zaera R, Navarro C. Compression after impact of thin composite laminates. *Composite Science and Technology* 2005; 65(13): 1911-1919.

[42] Schubel PM, Luo J-J, Daniel IM. Impact and post impact behaviour of composite sandwich panels. *Composites Part A: Applied Science and Manufacturing* 2007; 38: 1051-1057.

[43] Rotem A. Residual flexural strength of FRP composite specimens subjected to transverse impact loading. *SAMPE Journal* 1988; 24: 19-25.

[44] Shi-Xun Wang, Lin-Zhi Wu, Li Ma, Low-velocity impact and residual tensile strength analysis to carbon fiber composite laminates, *Materials and Design*, Volume 31, Issue 1, January 2010, Pages 118-125.

[45] F. Girot, L. N. Lopez de Lacalle, A. Lamikiz, D. Iliescu and E. Gutiérrez, Machinability aspects of polymer matrix composites, in Machining Composite Materials, Edited by J. Paulo Davim, Wiley (2009).

[46] U.A. Khashaba, Delamination in drilling polymeric composites: a review, in J.P. Davim, ed., Drilling of composite materials, Nova Science Publishers Inc, Aveiro, 2009.

[47] Nayak D, Bhatnagar N, Mahajan P. Machining studies of uni-directional glass fiber reinforced plastic (UD-GFRP) composites part 1: effect of geometrical and process parameters. *Mach. Sci. Technol.* 2005;9:481–501.

[48] Bhatnagar N, Nayak D, Singh I, Chouhan H, Mahajan P. Determination of machining-induced damage characteristics of fiber reinforced plastic composite laminates. *Mater. Manuf. Process* 2004;19(6):1009–23.

[49] Wang XM, Zhang LC. An experimental investigation into the orthogonal cutting of unidirectional fiber reinforced plastics. *Int. J. Mach. Tools Manuf.* 2003;43:1015–22].

[50] Arola, D; Ramulu, M. Orthogonal cutting of fiber-reinforced composites: a finite element analysis, *International Journal of Mechanical Sciences* 1997; 39: 597-613.

[51] Ramesh, M.V; Seetharamu, K.N; Ganesan, N; Shivkumar, MS. Analysis of Machining of FRPs using FEM, *International Journal of Machine Tools and Manufacture* 1998; 38: 1531-1549)).

[52] R. Varizi, M.D. Olson, D.L. Anderson, Finite element analysis of fibrous composite structures: a plasticity approach, *Computers and Structures* 44 (1992) 103-116.

[53] Ali Mkaddem, Mohamed El Mansori, Finite element analysis when machining UGF-reinforced PMCs plates: Chip formation, crack propagation and induced-damage, *Materials and Design* 30 (2009) 3295–3302.

[54] Arola, D; Sultan, M.B; Ramulu, M. Finite element modeling of edge trimming fiber-reinforced plastics. Transactions of the ASME, *Journal of Engineering Materials and Technology* 2002; 124: 32-41.

[55] Arola, D; Sultan, M.B; Ramulu, M. Finite element modeling of edge trimming fiber-reinforced plastics. Transactions of the ASME, *Journal of Engineering Materials and Technology* 2002; 124: 32-41.

[56] Lasri L, Nouari M, El Mansori M. Modelling of chip separation in machining unidireccional FRP composites by stiffness degradation concept. *Compos. Sci. Technol.* 2009;69:684–92.

[57] Mahdi, M; Zhang, L. An adaptive three – dimensional finite element algorithm for the orthogonal cutting of composite materials, *Journal of Materials Processing Technology* 2001; 113: 368-372.

[58] Mkaddem, A; Demirci, I; El Mansori, M. A micro–macro combined approach using FEM for modelling of machining of FRP composites: Cutting forces analysis, *Composites Science and Technology* 2008; 68: 3123–3127.

[59] Wang X.M; Zhang L.C. An experimental investigation into the orthogonal cutting of unidirectional fiber reinforced plastics. *International Journal of Machine Tools and Manufacture* 2003; 43: 1015-1022.).

[60] Venu Gopala Rao, G; Mahajan, P; Bhatnagar, N. Micro-mechanical modeling of machining of FRP composites- Cutting force analysis. *Composites Science and Technology* 2007; 67: 579-593.

[61] D. Iliescu, D. Gehin, I. Iordanoff, F. Girot, M.E. Gutiérrez. A discrete element method for the simulation of CFRP cutting. *Composite Science and Technology*, 2009, doi:10.1016/j.compscitech.2009.09.007.

[62] Santiuste C., Soldani X., Miguélez H., Machining FEM model of long fiber composites for aeronautical components, Composite Structures 92, 2010, 691–698).

[63] Hibbit, Karlsson and Sorensen, *ABAQUS user's manual v.6.4-1*, ABAQUS Inc., Rhode Island; 2003.

[64] Santhanakrishnan G, Krishnamurthy R, Malhotra SK (1988), Machinability characteristics of fibre reinforced plastic composites., *J. Mech. Work Technol.* 17:195–204.

[65] X. Soldani, C. Santiuste, A.Muñoz-Sánchez and M.H. Miguélez, *Composites part A,* (Under review).

[66] Luís Miguel P. Durão, Daniel J.S. Gonçalves, João Manuel R.S. Tavares, Victor Hugo C. de Albuquerque , A. Aguiar Vieira, A. Torres Marques, Drilling tool geometry evaluation for reinforced composite laminates, *Composite Structures* 92 (2010) 1545–1550).

[67] Antonio T. Marques, Luís M. Durão, António G. Magalhães, João Francisco Silva, João Manuel R.S. Tavares, Delamination analysis of carbon fibre reinforced laminates: Evaluation of a special step drill, *Composites Science and Technology* 69 (2009) 2376–2382.

[68] C.C. Tsao, Investigation into the effects of drilling parameters on delamination by various step-core drills, *Journal of Materials Processing Technology* 206 (2008) 405–411.

[69] R.A. Kishore, R. Tiwari, A. Dvivedi, I. Singh, Taguchi analysis of the residual tensile strength after drilling in glass fiber reinforced epoxy composites, *Materials and Design* 30 (2009) 2186–2190.

[70] Hocheng H, Tsao CC. The path towards delamination-free drilling of composite materials. *J. Mater. Process Technol.* 2005;167:251–64.

[71] Lachaud F, Piquet R, Collombet F, Surcin L. Drilling of composite structures.*Compos Struct* 2001;52:511–6.).

[72] Durão LMP, de Moura MFSF, Marques AT. Numerical prediction of delamination onset in carbon/epoxy composites drilling.*Eng. Fract. Mech.* 2008;75:2767–78).

[73] L.M.P. Durao, M.F.S.F. de Moura, A.T. Marques, Numerical simulation of the drilling process on carbon/epoxy composite laminates, *Composites: Part A* 37 (2006) 1325–1333.

[74] Singh, N. Bhatnagar, P. Viswanath, Drilling of uni-directional glass fiber reinforced plastics: Experimental and finite element study, *Materials and Design* 29 (2008) 546–553.

In: Fiber–Reinforced Composites
Editor: Qingzheng Cheng

ISBN: 978-1-61470-303-7
© 2012 Nova Science Publishers, Inc.

Chapter 17

A MICROMECHANICAL MODEL AND REINFORCING DISTRIBUTION OPTIMISATION IN FIBRE-REINFORCED MATERIALS

Roberto Brighenti[*]

Department of Civil and Environmental Engineering and Architecture
University of Parma, Viale G.P. Usberti 181/A, 43100 Parma, ITALY

ABSTRACT

The use of composite materials has known a crescent interest in the last decades in many application fields due to their desirable mechanical characteristics such as high specific strength and stiffness, high fracture and fatigue resistance, high wear resistance, high damping durability performance, low thermal coefficient, and so on. For the above mentioned reasons composite materials replace or strategically compliment other traditional structural materials.

The extensive use of advanced materials such as composite materials, requires to describe, with an appropriate accuracy, their overall mechanical behaviour to correctly assess the safety level of structural components in the design process. In the present chapter the problem of the description of the macroscopic mechanical characteristics of such a class of materials is considered. The behaviour of fibre-reinforced composites (FRC) is examined through the formulation of a micromechanical-based model. The macro constitutive equations for such a class of materials, composed by a matrix phase, for which an elastic-plastic behaviour is eventually allowed, and a fibre-reinforcing phase, is obtained through a micro mechanical model which takes into account the possibility of an imperfect bond between the matrix and the fibres. A two-parameters mechanical model, obtained from energetic considerations and by considering the evolution of the shear stress distribution along a single fibre during the loading process, is formulated to determine the entity of the debonding and its mechanical influence, from a macroscopic point of view, on the composite material. The detailed aspects of the mechanical model are presented and discussed and its implementation in a 2D FE code is finally illustrated.

---

[*] Phone +39 0521 905910 - Fax +39 0521 905924 - e-mail: brigh@unipr.it

In the second part of the chapter the problem of the optimal content distribution of fibres in a fibre-reinforced composites, in order to maximise or minimise a given objective function and by assuming some suitable constraints, is investigated by using a biological-based procedure known as Genetic Algorithm (GA).

The evolution process simulated by the developed algorithm, can be performed in order to get the maximisation or minimisation of some mechanical desired performance of the structure (stiffness, compliance, peak stress, etc) by keeping constant the total fibres content (optimal constrained problem).

The proposed mechanical model with the optimisation algorithm is finally used in some numerical simulations in order to assess its reliability in material design composition with respect to some expected optimal performance, and quantitative comparisons – in term of the improvements with respect to classical homogeneously distributed fibres situations – is finally illustrated and quantified.

Keywords: Fibre-reinforced composite, micromechanical model, structural optimization, Genetic Algorithm.

NOTATION

A_{ij}, B_{ij}	random matrices for crossover, with the property $A_{ij} + B_{ij} = 1$
A_f^p	cross section area of the fibres belonging to the p-th fibre phase
$\mathbf{b}$	body force vector field
$\mathbf{C}_m(\mathbf{x})$	elastic tensor of the matrix
$\mathbf{C'}_m(\mathbf{x})$	tangent elastic tensor of the matrix
$\mathbf{C}_{eq}$	homogenised elastic tensor of the composite
$\mathbf{C'}_{eq}$	tangent homogenised elastic tensor of the composite
$\mathbf{C}_f$	elastic tensor of the homoheneous material made only of fibres
d	characteristic microscopic length
D	characteristic macroscopic length
$D_{f,ij}$, $D_{f,ij}$ *	$N \times M$ matrix of the fibre density in the FE model before and after crossover, respectively
$D_{f,ij}^L$, $D_{f,ij}^K$	matrices having all the columns composed by the two better individuals, respectively, i.e. $D_{f,ij}^L = D_{f,iL}, D_{f,ij}^K = D_{f,iK}$,
E_f	fibre Young modulus
$E_{f,p}$	Young modulus of the p-th fibre phase
E_m	Young modulus of the matrix
E_t	post-yielding stress-strain slope of the matrix
$f_{t,f}$	fibre tensile strength

$F(\sigma, k_1, ..., k_n) = 0$ yield function

$F_c(\eta_c(\mathbf{x}))$, objective function for compliance and maximum stress optimisation, respectively
$F_s(\eta_s(\mathbf{x}))$

G_m shear modulus of the matrix

H hardening parameter for an uniaxial stress state

$\mathbf{i}$ unit vector parallel to the generic fibre axis

$k_1, ..., k_n$ hardening parameters

$\bar{k}$ stiffness of the fibre-matrix interface

$2L_{ad}^p$ non-dobonded length of the fibres belonging to the p-th fibre phase

$2L_f^p$ length of the fibres belonging to the p-th fibre phase

$2L_f^p$ length of the fibres belonging to the p-th fibre phase

M, N number of individuals and number of finite elements in the model, respectively

N_c, N_m crossover point and elitism point, respectively

P_c, P_m probability for crossover and elitism, respectively

$P_{m,g,crossover}$, $P_{m,g,elitism}$ probability of mutation for individuals generated by crossover or by elitism from the previous population, respectively

REV   Reference Elementary Volume

$s(\varepsilon_f^m)$ sliding function such that: $[\![\varepsilon_{f-m}]\!] = \varepsilon_f^m \cdot [1 - s(\varepsilon_f^m)]$

$s(\overline{\varepsilon_f^m})$ mean value of the sliding function along a single fibre

$\tilde{\mathbf{u}}$ virtual displacement field

V volume of the composite REV

V_m volume of the matrix phase present in the REV

$V_{f,p}$ volume of the p-th fibre phase fraction present in the R.E.V.

$\mathbf{x}$ position vector

$\boldsymbol{\varepsilon}, \boldsymbol{\sigma}$ strain and stress tensors, respectively

$\dot{\boldsymbol{\varepsilon}}^{el}, \dot{\boldsymbol{\varepsilon}}^{pl}$ elastic and plastic part of the strain rate tensor, respectively

$\tilde{\boldsymbol{\varepsilon}}, \dot{\tilde{\boldsymbol{\varepsilon}}}$ virtual strain and strain rate fields, respectively

$[\![\varepsilon_{f-m}]\!]$ strain jump between the fibre and the matrix (parallel to fibre axis) in the case of imperfect bond

ε_f uniaxial fibre strain

ε_f^m uniaxial matrix strain measured at the location and in the fibre direction

$\overline{\varepsilon_f^m}$ matrix mean strain along the fibre measured in its direction

$\dot{\lambda}$ plastic multiplier

$\dot{\boldsymbol{\sigma}}, \dot{\boldsymbol{\sigma}}_f, \dot{\boldsymbol{\sigma}}_{eq}$	generic stress rate tensor, stress rate tensor at the location of the generic fibre belonging to the *p-th* phase, stress rate tensor in the equivalent material, respectively
$\chi_p(\mathbf{x})$	point function denoting the presence of a fibre, belonging to the *p-th* phase, at the location $\mathbf{x}$
$\kappa(\mathbf{x})$	point function denoting the presence of the matrix at the location $\mathbf{x}$
$\mu = V_m / V$	REV matrix volume fraction
$\eta(\mathbf{x}) = V_f(\mathbf{x})/V$	fibre reinforcing phase distribution in the matrix material
$\eta'(\mathbf{x})$	normalised fibre distribution design function, defined in the range $0 \le \eta'(\mathbf{x}) \le 1$
$\overline{\eta}$	uniform fibre content distribution in the composite
$\eta_p = V_p / V$	REV fibre volume fraction of the *p-th* fibre phase, $p = 1,..,q$
$\eta_c(D_{f,c})$, $\eta_s(D_{f,s})$	fibre distributions, encoded by the matrix $D_{f,c}$ and $D_{f,s}$, respectively, for compliance and maximum stress optimisation
τ_i	fibre-matrix interface shear stress
τ_{au}	fibre-matrix interface shear strength
τ_{fu}	friction shear stress at the fibre-matrix interface
$\Gamma = \Gamma_t \cup \Gamma_u$	boundary of the solid
Γ_t	portion of the boundary on which tractions are prescribed
Γ_u	portion of the boundary on which displacements are prescribed
ξ	random number

1. INTRODUCTION

The use of composite materials in many application fields has known a crescent interest in the last decades due to their important mechanical characteristics such as high fracture and fatigue resistance, high wear resistance, durability performance, high damping durability performance, low thermal coefficient, and so on which are easily obtainable by such a class of materials at an acceptable economic cost, making their use convenient with respect to other traditional structural materials. The term 'composite materials' is quite general and indicates a very wide class of composed (or multiphase) materials: among them laminated materials, particles-reinforced, long or short fibre-reinforced materials, etc can be mentioned. The extensive use of composite materials has determined the necessity to describe with an appropriate accuracy - from the engineering point of view - their overall mechanical behaviour to correctly assess the safety level in the design of structural components. In order to obtain suitable mechanical model for such materials, various approaches can be used such as micromechanical models (physically based approach, [1-3]), homogenisation models (mathematically based approach, [4, 5]), etc.

From a mathematical point of view, the problem of the equilibrium of a multi-phase material involves a sets of equations, characterised by spatially variable mechanical parameters which are related to the mechanical properties of the individual phases of the composite material, that are not easily solvable. Such a boundary-value problem is rather complex and an approximate solution, for example obtained by considering material averaged properties, should be acceptable for practical applications.

Furthermore it must be considered that, since the material is composed by two or more phases, the formulation of the mechanical phases-interaction behaviour must be also taken into account, leading to a very convoluted problem [6, 7]. From the above consideration it can be argued as, even if the single components behaves in a linear elastic fashion, the composite material can show non-linear stress-strain relationship due, for instance, to the imperfect bounds between the constituents: in fact it originates an irreversible phenomenon due to the energy dissipation involved in the slip relative displacements. The perfect bonds between different phases is seldom verified and damage phenomenon at the matrix-inclusion interfaces heavily affect the whole behaviour of the composite material.

In the present paper, after a brief overview on the mechanical model formulated in the literature for several types of composite materials, the attention is focused on short fibre-reinforced composites (FRC) with are assumed, for sake of generality, to be characterised by an elastic-plastic matrix and by a dispersion of fibre having an imperfect bond with respect to the matrix material. The incremental equilibrium equation is established and solved for simple cases. The imperfect bonds is studied by considering the sliding between the matrix and the fibre, obtained by using the shear stress distribution evolution along a single fibre during the loading process applied to the composite.

A sliding parameter is introduced in order to take into account the effective bond between the matrix and the fibres: an imperfect bond between the phases, which is responsible of a non linear overall composite behaviour, is modelled and quantified through energetic considerations.

The obtained mechanical constitutive law has been implemented in a F.E. code and numerical simulation are performed in order to assess the reliability of the proposed model.

Since in fibre-reinforced materials the mechanical behaviour strongly depends on the amount, density and orientation of the fibres, the spatial distribution of the reinforcing phase inside the matrix can be used as a *"design variable"* to optimise the mechanical performance for this class of materials (stiffness, compliance, weight, etc.) or a particular response quantities (energy density, peak stress, displacements, etc.) in order to reduce the *cost to performance* ratio.

The above task can be considered as a minimum or maximum optimum search constrained problem. Such an optimisation process related to the spatial fibre distribution in a fibre-reinforced composites (FRC) structure – formulated by taking into account the effective bound between the matrix and the fibres – is considered. By using a biological-based algorithm approach, known as Genetic Algorithm (GA), some extreme conditions related to a desired objective function can be approximately imposed. Different optimal conditions are considered in order to demonstrate the versatility of the adopted evolutionary approach in order to reduce as much as possible the *cost-effectiveness* ratio.

The proposed mechanical model with optimisation is finally employed to perform simple numerical simulation in order to assess its reliability in finding the optimal fibre volume fraction distribution in FRC, which is assumed as a design spatial variable function, with respect to different objective functions.

2. MECHANICAL MODELLING OF COMPOSITE MATERIALS

Recent trends have shown as composite materials can be successfully applied in several fields ranging from small and simple structures up to large-scale structures in the mechanical, aerospace, mechanical, power, transportation, civil engineering and sports fields, by replacing other traditionally employed structural materials.

The use of composite materials in challenging structural applications requires the knowledge of their macroscopic behaviour which must be precisely known along the design process.

In order to develop a suitable mechanical model, the composite material must be observed from both the microscopic as well as the macroscopic point of view.

The microscopic point of view considers the behaviour and the individual characteristics of the various constituents of the composite such as the matrix material, the reinforcing fibres, the matrix-fibre interface etc., whereas at the macroscopic level the composite material is treated as a single entity. A physically-based micromechanical model must considers both the local and the global aspects of the composite by taking into account the mechanical and geometrical characteristics of the reinforcing phases in the composite [8] and its interaction with the matrix material.

The development of micromechanical models for composites materials has been widely investigated in the literature; it is worth mention here the pioneering research by Hashin [9] and by Hashin and Rosen [10] that considered the composite as a material with embedded inclusions made of spherical particles for which the shear and bulk moduli of macroscopically isotropic composites were determined.

Other researches were conducted by Hill [11, 12] and Budiansky [13] that considered a composite with rigid inclusions embedded in an incompressible matrix. In particular Hill [11] studied the problem of two isotropic media forming perfect bond with arbitrary volume fractions of the constituents.

A variational approach has been used by Hashin and Shtrikman [14] to determine the upper and lower limits for the effective elastic properties of multiphase materials, while Eshelby [15] considered the problem of an ellipsoidal inclusion in an infinite matrix.

Russel [16] considered arbitrary cross-section slender elastic inclusions, oriented in a preferred direction, embedded in an infinite elastic medium characterized by boundary conditions corresponding to a constant uniform strain. For such a case he was able to determine both the Poisson's ratio and the bulk modulus of the composite by varying the inclusion volume fraction.

More recently Kalamkarov and Liu [17] proposed a work energy-based approach for multiphase fibre-matrix composite material in order to obtain its characteristic mechanical properties at the a so-called mesostructure scale.

As stated above, the global formulation of the macroscopic mechanical behaviour of a composite materials, reflects the mechanical aspects at a microscopic scale. Usually the microscopic scale is of the same order of magnitude as the size or spacing distribution of the reinforcements; on the other hand the macroscopic scale has typically a characteristic dimension with an order of magnitude similar to a characteristic dimension of the composite structure. The consideration of these two scales at the same time leads to a very complex problem to be solved and represent a very difficult task in the mathematical modelling of material science.

To solve this class of problems, the method of asymptotic homogenization has been developed to decouple the microscopic and the macroscopic variations in order to allow each single problem to be solved independently or sequentially; the mathematical bases of such an asymptotic homogenization approach can be found for example in Sanchez-Palencia [18], Kalamkarov [8, 19, 20] and Cioranescu and Paulin [21].

3. EQUILIBRIUM PROBLEM IN A COMPOSITE MATERIAL

As is well-known the equilibrium of a body can be mathematically stated as follows:

$$
\begin{aligned}
div\ \boldsymbol{\sigma} + \mathbf{b} &= \mathbf{0} &&\text{in } B \\
\mathbf{n} \cdot \boldsymbol{\sigma} &= \mathbf{t} &&\text{on } \Gamma_t \\
\mathbf{u} &= \mathbf{u}_0 &&\text{on } \Gamma_u
\end{aligned}
\tag{1}
$$

where $\mathbf{u}, \boldsymbol{\sigma}, \mathbf{b}, \mathbf{n}$, are the displacement field, the stress tensor, the body forces and the unit outward normal to the boundary, respectively, while $\mathbf{t}, \mathbf{u}_0$ represent the prescribed traction and displacements on the portion Γ_t and Γ_u of the boundary (Figure 1a) where traction and displacements are prescribed.

The linearised strain-displacement equation is given by the relation:

$$
\varepsilon = \mathrm{sym}(\mathrm{grad}\,\mathbf{u}) = \mathrm{sym}[\nabla \otimes \mathbf{u}] = \frac{1}{2}\left[(\mathrm{grad}\,\mathbf{u}) + (\mathrm{grad}\,\mathbf{u})^T\right]
\tag{2}
$$

where ε is the strain tensor and the symbols "∇" and "$\otimes$" stand for gradient and tensor operator product, respectively.

The incremental elastic constitutive relation, written in a specific point inside the material, can be written as:

$$
\dot{\boldsymbol{\sigma}}(\mathbf{x}) = \mathbf{C}'(\mathbf{x}):\dot{\varepsilon}(\mathbf{x}) = \mathbf{C}'(\mathbf{x}):\left[\mathrm{sym}(\mathrm{grad}\,\dot{\mathbf{u}})\right] = \mathbf{C}'(\mathbf{x}):\left[\mathrm{sym}(\nabla \otimes \dot{\mathbf{u}})\right]
\tag{3}
$$

where the dot denotes an increment when applied to the stress or strain tensors, while the apex indicates the fourth-order elastic tangent tensor, $\mathbf{C}'(\mathbf{x})$; the location of the point inside the body is identified by the position vector $\mathbf{x} = (x_1, x_2, x_3)$. In a composite material the latter tensor presents a high variability with the position vector (Figure 1).

The heterogeneous (composite) material is supposed to be composed by one matrix phase, denoted by the subscript m, and by q different fibre phases embedded in the matrix (Figure 1b).

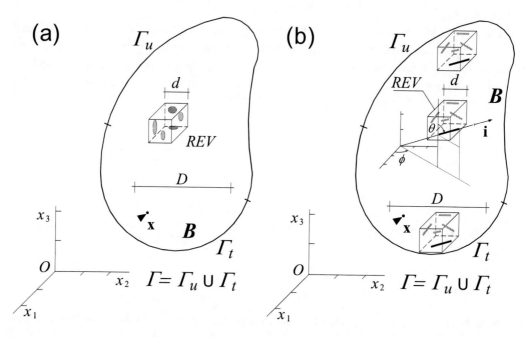

Figure 1. Body made of a generic composite material (a) and fibre-reinforced composite material (b).

We assume that the fibres belonging to the same phase p have the same length $2L_f^p$ and cross section A_f^p, while their orientation are randomly distributed in the solid angle, so the assumption that each direction is equally represented, can be made. The direction of a single fibre can be identified by the unit vector $\mathbf{i}$ parallel to its axis; using the polar angles ϕ, θ (Figure 1b), such vector can be written as:

$$\mathbf{i} = \{i_1 \ i_2 \ i_3\} = \{\sin\theta \cdot \cos\phi \ \ \sin\theta \cdot \sin\phi \ \ \cos\theta\} \tag{4}$$

A fundamental hypothesis for the mechanical model being developed is the assumptions that each fibre phase is homogeneously distributed inside the matrix and that a reference elementary volume (REV), having a characteristic length d, has the same average composition - and consequently the same mechanical properties - of the composite. Moreover the characteristic length D of the body is assumed to be much more greater than the microscopic characteristic length d, e.g. $d/D << 1$; this implies that the composite is macroscopically homogeneous.

By considering the REV, the following volume fractions of each component can be defined:

$$\mu = V_m / V \qquad \text{matrix volume fraction}$$

$$\eta_p = V_{f,p} / V \qquad \text{fibre volume fraction of the } p\text{-th fibre phase} \qquad (5)$$

where $V, V_m, V_{f,p}$ are the REV volume, the matrix and the $p-th$ fibre phase volume present in the REV, respectively. It is trivial to observe that:

$$V = V_m + \sum_{p=1}^{q} V_{f,p} \qquad (6)$$

or equivalently, in dimensionless form, $1 = \mu + \sum_{p=1}^{q} \eta_p$.

4. A MICRO-MECHANICAL MODEL FOR FIBRE-REINFORCED MATERIALS

4.1. Equivalent Homogeneous Material

Under the previous hypotheses it is possible to determine the averaged properties of the composite material; this goal can be obtained by equating the virtual power w' (work rate), evaluated in the composite material, with those in the homogenised equivalent one [2, 7]. By introducing a generic virtual displacement field $\widetilde{\mathbf{u}}$, and the corresponding strain tensor $\widetilde{\varepsilon}$ and strain rate tensor $\dot{\widetilde{\varepsilon}}$, defined as follows:

$$\widetilde{\mathbf{u}}, \quad \widetilde{\varepsilon} = \mathrm{sym}(\mathrm{grad}\ \widetilde{\mathbf{u}}) = \mathrm{sym}[\nabla \otimes \widetilde{\mathbf{u}}], \quad \dot{\widetilde{\varepsilon}} = \left[\mathrm{sym}(\nabla \otimes \dot{\widetilde{\mathbf{u}}})\right] \qquad (7)$$

the stated condition can be written, for a generic composite volume V, equal or greater than the REV volume, as:

$$w' = \underbrace{\int_V \kappa(\mathbf{x}) \cdot \dot{\widetilde{\varepsilon}} : \sigma\, dV + \sum_{p=1}^{q} \int_V \chi_p(\mathbf{x}) \cdot \dot{\widetilde{\varepsilon}}_f \cdot \sigma_f\, dV}_{\text{composite's work rate}} = \underbrace{\int_V \dot{\widetilde{\varepsilon}} : \sigma_{eq} dV}_{\text{homogenized material's work rate}} \qquad (8)$$

In Eq. (8) $\dot{\tilde{\varepsilon}}_f$, σ_f are the virtual strain rate and the stress in a fibre belonging to a given *p-th* phase, respectively, while the scalar functions $\kappa(\mathbf{x})$, $\chi_p(\mathbf{x})$ defined as:

$$\kappa(\mathbf{x}) = \begin{cases} 1 & \text{if } (\mathbf{x}) \in V_m \\ 0 & \text{if } (\mathbf{x}) \notin V_m \end{cases} \quad \text{and} \quad \chi_p(\mathbf{x}) = \begin{cases} 1 & \text{if } (\mathbf{x}) \in V_{f,p} \\ 0 & \text{if } (\mathbf{x}) \notin V_{f,p} \end{cases} \tag{9}$$

identify the location of the point $\mathbf{x}$ in the matrix or in the reinforcing phase, respectively.

The fibre stress can be evaluated by mean of its elastic modulus $E_{f,p}$ multiplied by the fibre strain ε_f, while the stress tensor of the homogenised equivalent material $\boldsymbol{\sigma}_{eq}$ can be obtained by mean of the equivalent elastic tensor $\mathbf{C}_{eq}$ multiplied by the actual strain tensor ε, if a linear elastic behaviour for the fibres and the equivalent homogenised material, are assumed:

$$\sigma_f = E_{f,p} \cdot (\mathbf{i} \otimes \mathbf{i}) : \varepsilon, \qquad \boldsymbol{\sigma}_{eq}(\mathbf{x}) = \mathbf{C}_{eq}(\mathbf{x}) : \varepsilon \tag{10}$$

The above relation (10_1) is written by observing as the strain measured in the fibre direction can be written as $\varepsilon_f = (\mathbf{i} \otimes \mathbf{i}) : \varepsilon$ and analogously for the virtual $\tilde{\varepsilon}_f$ and virtual strain rate, $\dot{\tilde{\varepsilon}}_f$, $\tilde{\varepsilon}_f = (\mathbf{i} \otimes \mathbf{i}) : \tilde{\varepsilon}$, $\dot{\tilde{\varepsilon}}_f = (\mathbf{i} \otimes \mathbf{i}) : \dot{\tilde{\varepsilon}}$.

By inserting Eqs (9, 10) in the expression of the virtual work rate (8) we get:

$$w' = \int_V \kappa(\mathbf{x}) \dot{\tilde{\varepsilon}} : \mathbf{C}_m(\mathbf{x}) : \varepsilon \, dV + \sum_{p=1}^{q} \int_V \chi_p \cdot (\mathbf{i} \otimes \mathbf{i}) : \dot{\tilde{\varepsilon}} \cdot E_{f,p} \cdot (\mathbf{i} \otimes \mathbf{i}) : \varepsilon \, dV = \int_V \dot{\tilde{\varepsilon}} : \mathbf{C}_{eq}(\mathbf{x}) : \varepsilon \, dV$$

$$\tag{11}$$

where ε indicates the matrix strain tensor.

Thus the previous equality can be rewritten:

$$\int_V \dot{\tilde{\varepsilon}} : \left\{ \kappa(\mathbf{x}) \cdot \mathbf{C}_m(\mathbf{x}) + \sum_{p=1}^{q} \chi_p(\mathbf{x}) \left[E_{f,p} \cdot \mathbf{Q} \otimes \mathbf{Q} \right] \right\} : \varepsilon \, dV = \int_V \dot{\tilde{\varepsilon}} : \mathbf{C}_{eq}(\mathbf{x}) : \varepsilon \, dV \tag{12}$$

where the second-order tensor $\mathbf{Q} = \mathbf{i} \otimes \mathbf{i}$ has been introduced.

It is trivial to observe as, since the equivalent material is macroscopically homogeneous (at least at the scale of the REV), the elastic tensor $\mathbf{C}_{eq}(\mathbf{x})$ can be assumed to be constant with respect to the position vector $\mathbf{x}$, i.e. $\mathbf{C}_{eq}(\mathbf{x}) = \mathbf{C}_{eq}$.

By comparing the first and the second member of Eq. (12), the equivalent tangent elastic tensor can be recognised; since we assume $\mathbf{C}_{eq}(\mathbf{x}) = \mathbf{C}_{eq}$, its mean value in the volume V can be obtained:

A Micromechanical Model and Reinforcing Distribution Optimisation ...	537

$$\mathbf{C}_{eq} = \frac{1}{V} \cdot \int_V \left\{ \kappa(\mathbf{x}) \cdot \mathbf{C}_m + \sum_{p=1}^{q} \chi_p(\mathbf{x}) \cdot E_{f,p} \cdot [\mathbf{Q} \otimes \mathbf{Q}] \right\} dV = \mu \cdot \mathbf{C}_m + \sum_{p=1}^{q} \eta_p E_{f,p} \cdot \int_V \mathbf{Q} \otimes \mathbf{Q} \, dV$$

(13)

where the definition of the fibre and matrix volume fraction (5):

$$\mu = \frac{1}{V} \cdot \int_V \kappa(\mathbf{x}) dV = \frac{V_m}{V} \quad , \qquad \eta_p = \frac{1}{V} \cdot \int_V \chi_p(\mathbf{x}) dV = \frac{V_{f,p}}{V}$$

(14)

has been used.

The tangent elastic properties of the materials must be used to calculate the increment of stresses in the matrix, $\dot{\sigma}$, in the fibre, $\dot{\sigma}_f$, and in the equivalent material, $\dot{\sigma}_{eq}$, when an incremental situation is considered:

$$\dot{\sigma}(\varepsilon) = \mathbf{C'}_m(\varepsilon) : \dot{\varepsilon} \quad , \qquad \dot{\sigma}_f(\varepsilon_f^m) = \left(\frac{d\sigma_f}{d\varepsilon_f} \cdot \frac{d\varepsilon_f}{d\varepsilon_f^m} \right)_p \dot{\varepsilon}_f \quad , \qquad \dot{\sigma}_{eq}(\varepsilon) = \mathbf{C'}_{eq}(\varepsilon) : \dot{\varepsilon} \quad (15)$$

where $\mathbf{C'}_m$, $\mathbf{C'}_{eq}$ are the tangent elastic tensor of the matrix and of the equivalent material, respectively. For sake of simplicity all the equations can be written with reference to the actual matrix strain tensor ε.

In the following the hypothesis of a single fibre phase presents in the composite will be made in order to simplify the notation, so the summation over the index p will be omitted.

Remembering Eq. (14) and indicating with ε_f^m and ε_f the strain in the matrix measured in the fibre direction and the actual fibre strain, respectively, the tangent stiffness tensor $\mathbf{C'}_{eq}(\varepsilon)$ of the equivalent homogenised material can be obtained:

$$\mathbf{C'}_{eq} = \frac{1}{V} \cdot \int_V \left[\kappa(\mathbf{x}) \cdot \mathbf{C'}_m + \chi(\mathbf{x}) \cdot \left(\frac{d\sigma_f}{d\varepsilon_f^m} \right) \cdot \mathbf{Q} \otimes \mathbf{Q} \right] dV \cdot = \mu \cdot \mathbf{C'}_m + \eta \left(\frac{d\sigma_f}{d\varepsilon_f^m} \right) \cdot \int_V \mathbf{Q} \otimes \mathbf{Q} \, dV \quad (16)$$

$$\mathbf{C'}_{eq} = \mu \cdot \mathbf{C'}_m + \eta \cdot \mathbf{C'}_f$$

To evaluate the tangent stiffness tensor $\mathbf{C'}_{eq}$, the last integral in Eq. (16) can be calculated on a hemisphere, having a generic volume V_{em} (greater than the REV volume in order to be sufficiently representative of the macroscopic mechanical behaviour of the composite) and divided by the hemisphere volume itself, in order to represent all the possible fibre orientations in the composite:

$$\frac{1}{V_{hem}} \int_{V_{hem}} \mathbf{Q} \otimes \mathbf{Q} \, dV = \int_0^R \int_0^{2\pi} \int_0^{\frac{\pi}{2}} (\mathbf{Q} \otimes \mathbf{Q}) r \, d\phi \, r \sin\theta \, d\theta \, dr =$$

$$= \frac{R^3}{3} \frac{1}{\frac{2}{3}\pi R^3} \int_0^{2\pi} \int_0^{\frac{\pi}{2}} (\mathbf{Q} \otimes \mathbf{Q}) d\phi \sin\theta \, d\theta = \frac{1}{2\pi} \int_0^{2\pi} \int_0^{\frac{\pi}{2}} (\mathbf{Q} \otimes \mathbf{Q}) d\phi \sin\theta \, d\theta \qquad (17)$$

In the case of a fibre phase homogeneously distributed and oriented in the matrix, the contribution $\mathbf{C}_f$ of the fibre to the elastic tensor $\mathbf{C}_{eq}$ of the composite is isotropic (if the matrix material is isotropic too), and the integral (17) can be simply evaluated; as an example some of the corresponding terms of the elasticity tensor $\mathbf{C}_f$ are reported below:

$$C^f_{1111} = \frac{1}{5}\eta E_f, \quad C^f_{1122} = \frac{1}{15}\eta E_f, \quad C^f_{1133} = \frac{1}{15}\eta E_f, \quad C^f_{1112} = C^f_{1123} = C^f_{1113} = 0$$
$$C^f_{1212} = \frac{1}{15}\eta E_f, \quad \ldots\ldots\ldots \qquad (18)$$

where E_f is the elastic modulus of the fibres.

As an example in Figure 2 the dimensionless composite elastic modulus (Figure 2a) and Poisson's ratio (Figure 2b) against fibre volume fraction η for different values of the Poisson's ratio of the matrix v_m and ratios E_m/E_f between the matrix and fibre Young modules are displayed. All the above relations are obtained under the hypothesis of a perfect bound between the fibres and the matrix, i.e. the actual strain in a fibre can be simply written from the matrix strain tensor ε as $\varepsilon_f^m = \varepsilon_f = (\mathbf{i} \otimes \mathbf{i}) : \varepsilon$.

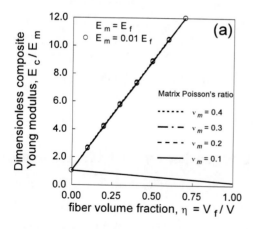
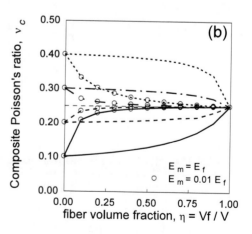

Figure 2. Dimensionless composite elastic modulus E_c (a) and its Poisson's ratio v_c (b) against the fibre volume fraction η for different values of the Poisson's ratio of the matrix v_m and Young's modulus ratios E_m/E_f.

4.2. Matrix-Fibre Debonding

When an imperfect bound between the reinforcing fibre and the matrix takes place, a strain jump, $[\![\varepsilon_{f-m}]\!]$, can be assumed to exists at the interface. Such a strain jump can be written in the following form:

$$[\![\varepsilon_{f-m}]\!] = \varepsilon_f^m - \varepsilon_f = [(\mathbf{i} \otimes \mathbf{i}):\varepsilon] \cdot (1 - s(\varepsilon_f^m)) = \varepsilon_f^m \cdot (1 - s(\varepsilon_f^m)) \quad (19)$$

where ε_f^m represents the matrix strain evaluated along the fibre direction and the scalar function s measures the local (i.e. at a given position along the fibre) *"degree of sliding"* between the fibre and the matrix.

When $s = 0$ the sliding is complete, i.e. $[\![\varepsilon_{f-m}]\!] = \varepsilon_f^m$, and no shear stress is transferred between the matrix and the fibre; in such a case the fibre-matrix interface has not stiffness and the fibre doesn't bear any stress, that is to say the composite material behaves as an elastic material, with inclusions of voids having the shape of the fibres. When $s = 1$ the interface is perfect, i.e. no strain jump occurs, $[\![\varepsilon_{f-m}]\!] = 0$; this case represents the maximum shear stress transfer between the matrix and the fibre, and the reinforcing phase carries the maximum possible load, giving the maximum contribution to the composite load bearing capacity (Figure 3a).

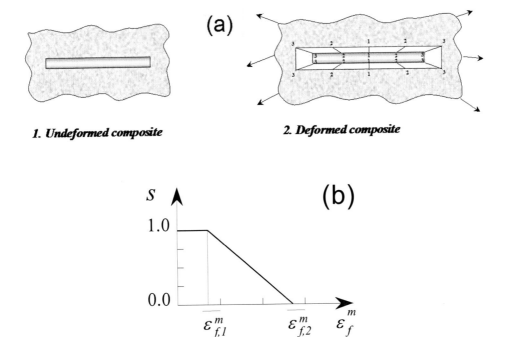

Figure 3. Imperfect bounds between the fibre and matrix (a). A simplified relation assumed to describe strain jump at the interface (b).

540 Roberto Brighenti

In order to refer the actual fibre strain ε_f to the matrix strain ε when the sliding is present, it can be written:

$$\varepsilon_f = \varepsilon_f^m - [\![\varepsilon_{f-m}]\!] = \varepsilon_f^m \cdot [1 - (1 - s(\varepsilon_f^m))] = [(\mathbf{i} \otimes \mathbf{i}) : \varepsilon] \cdot s(\varepsilon_f^m) = \varepsilon_f^m \cdot s(\varepsilon_f^m) \quad (20)$$

It is reasonable to assume that the value of the function s depends on the amount of the matrix strain evaluated in the fibre direction, ε_f^m, as the used notation indicates, $s(\varepsilon_f^m)$.

A very simple and reasonable relation for $s(\varepsilon_f^m)$ can be assumed to be, for instance, a piecewise linear dependence between s and ε_f^m as shown in Figure 4b; such a relation requires the introduction of two characteristic strains, $\overline{\varepsilon_{f,1}^m}$, $\overline{\varepsilon_{f,2}^m}$, that define the matrix strain value at which debonding takes place ($\overline{\varepsilon_{f,1}^m}$) and the matrix strain value at which debonding is complete ($\overline{\varepsilon_{f,2}^m}$), i.e. $s = 0$. Nevertheless the quantification of $s(\varepsilon_f^m)$ can be made on the basis of physically-based concepts as illustrated in the following.

As stated above, the sliding function is defined locally along the fibre; by assuming the hypothesis of very short fibres, it reasonable to consider s as constant along the fibre; its constant value can be assumed to be equal to the mean value of $s(\varepsilon_f^m)$ along the fibre as described below.

Since the strains are evaluated with reference to the matrix strains, even if the case of linear elastic fibre material (with Young's modulus E_f), the stress-matrix strain relation for the fibre is non-linear when the sliding takes place; the elastic tangent modulus of the fibre with respect to the matrix strain, $E_f(\varepsilon_f^m) = d\sigma_f / d\varepsilon_f^m$, can thus be calculated as follows:

$$E_f(\varepsilon_f^m) = \frac{d}{d\varepsilon_f^m}\left[E_f \cdot \overbrace{\left(\varepsilon_f^m - [\![\varepsilon_{f-m}]\!]\right)}^{\varepsilon_f = s(\varepsilon_f^m)\cdot\varepsilon_f^m} \right] = E_f \cdot \left[s(\varepsilon_f^m) + \varepsilon_f^m \cdot \frac{ds(\varepsilon_f^m)}{d\varepsilon_f^m} \right] \quad (21)$$

By considering the above relation, the tangent stiffness tensor $\mathbf{C'}_{eq}(\varepsilon)$ of the equivalent homogenised material becomes now (see Eq. (16)):

$$\begin{aligned}
\mathbf{C'}_{eq} &= \mu \cdot \mathbf{C'}_m + \eta \left(\frac{d\sigma_f}{d\varepsilon_f^m} \right) \cdot \int_V \mathbf{Q} \otimes \mathbf{Q}\, dV = \\
&= \mu \cdot \mathbf{C'}_m + \eta_p \cdot E_f \cdot \left[s(\varepsilon_f^m) + \varepsilon_f^m \cdot \frac{ds(\varepsilon_f^m)}{d\varepsilon_f^m} \right]_p \cdot \int_V \mathbf{Q} \otimes \mathbf{Q}\, dV
\end{aligned} \quad (22)$$

In order to write explicitly the relation $s(\varepsilon_f^m)$ the stress transfer between the matrix and the fibre must be considered.

4.3. Equilibrium of the Fibre-Matrix System

In order to determine the relation $s(\varepsilon_f^m)$, a cylinder of material around a single fibre can be taken into account; the equilibrium equation along the fibre direction can be written (Figure 5a):

$$\frac{df(x)}{dx} = -c \cdot \tau_i = -c \cdot \bar{k} \cdot (u_R - u_r) \qquad (23)$$

where $c, \tau_i, f(x), \bar{k}, u_R, u_r$ are the fibre perimeter, the shear stress at the interface, the total axial force sustained by the fibre in the fibre section located at the abscissa x, (Figure 4a) the stiffness of the fibre-matrix interface, the displacement measured in the fibre direction far from the fibre and at the interface (which is assumed to coincide with the displacement measured on the fibre axis due to the negligibility of the fibre radius with respect to the radius of the cylinder of material under study), respectively, Figure 5a.

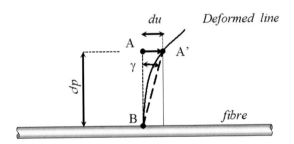

Figure 4. Scheme of the shear deformation around a single fibre.

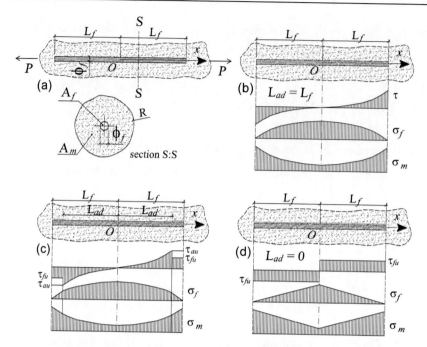

Figure 5. Cylinder of matrix material around a single fibre under tension (a). Stress distributions around a single fibre in the pre-debonding stage (b), during the debonding stage (c) and during the complete debonding stage (d).

The stiffness $\bar{k}$ of the fibre-matrix interface can be calculated by writing the shear strain and stress at a generic distance R from the fibre centre as a function of the shear interface stress τ_i:

$$\tau(\rho) = G_m \cdot \gamma(\rho) = G_m \cdot \frac{du}{d\rho} = \tau_i \frac{r}{\rho} \tag{24}$$

where G_m, r and ρ indicate the matrix shear modulus, the fibre radius and a generic radius measured from the fibre centre, respectively and u is the displacement parallel to the fibre direction (Figure 4). The last equality in Eq. (24) is written by considering the longitudinal equilibrium of the hollow cylinder defined by the internal radius r and by the external radius ρ, $2\pi \cdot r \cdot \tau_i \cdot dx = 2\pi \cdot \rho \cdot \tau(\rho) \cdot dx$ (Figure 5a).

By expressing the interface stress τ_i from the above relation we have: $(d\rho/\rho)\tau_i = G_m \cdot du/r$. Finally by integrating the above relation between r and $\rho = R$ (where we suppose that the displacements u are known, $u_r = u(r), u_R = u(R)$), we get:

$$\int_r^R \frac{d\rho}{\rho} \cdot \tau_i = \frac{G_m}{r} \cdot \int_{u(r)}^{u(R)} du \rightarrow \tau_i = \underbrace{\frac{G_m}{r} \cdot \frac{1}{\ln(R/r)}}_{\bar{k}} \cdot (u_R - u_r) = \bar{k} \cdot (u_R - u_r) \tag{25}$$

By considering the following relations:

$$\frac{du_r(x)}{dx} \approx \varepsilon_f(x) = \frac{\sigma_f(x)}{E_f} = \frac{\sigma_f(x)}{E_f} \cdot \frac{A_f}{A_f} = \frac{f(x)}{E_f \cdot A_f} \tag{26}$$

and $\dfrac{du_R(x)}{dx} \approx \varepsilon_m(x) = \dfrac{\sigma_m(x)}{E_m}$

and deriving Eq. (23) with respect to x we get [22]:

$$\frac{d^2 f(x)}{dx^2} + c \cdot \overline{k} \cdot \left(\frac{\sigma_m(x)}{E_m} - \frac{f(x)}{E_f \cdot A_f} \right) = 0 \tag{27}$$

or equivalently:

$$\frac{d^2 f(x)}{dx^2} = -c \cdot \overline{k} \cdot \left(\frac{P - f(x)}{E_m \cdot A_m} - \frac{f(x)}{E_f \cdot A_f} \right) \tag{28}$$

since $\underbrace{\sigma_f(x) \cdot A_f}_{f(x)} + \underbrace{\sigma_m(x) \cdot A_m}_{m(x)} = P$, where P is the total force sustained by the cylinder of composite under study (Fig. 5) that, for each value of the co-ordinate x, is the sum of the load carried by the fibre ($f(x)$) and by the matrix ($m(x)$).

By making the positions $\alpha = \dfrac{1}{E_m \cdot A_m} + \dfrac{1}{E_f \cdot A_f}$, $\beta = \sqrt{c \cdot \overline{k} \cdot \alpha}$, $F = \dfrac{P}{\alpha \cdot E_m \cdot A_m}$,

Eq. (28) can be rewritten in the following form:

$$\frac{d^2 f(x)}{dx^2} - \beta^2 \cdot f(x) + \beta^2 \cdot F = 0 \tag{29}$$

By integrating the previous second order differential equation and by imposing the boundary conditions at the fibre ends, $f(x = \pm L_f) = 0$ (the fibre is assumed to have length $2 L_f$), the solution of Eq. (29) becomes:

$$f(x) = F \cdot \left[1 - \frac{\cosh(\beta \cdot x)}{\cosh(\beta \cdot L_f)} \right] \tag{30}$$

while the shear stress $\tau_i = \tau(x)$ at the interface and the force sustained by the matrix can be calculated as:

$$\tau(x) = -\frac{1}{c} \cdot \frac{df(x)}{dx} = \frac{F \cdot \beta}{c} \cdot \left[\frac{\sinh(\beta \cdot x)}{\cosh(\beta \cdot L_f)} \right] \quad \text{and}$$

$$m(x) = P - f(x) = F \cdot \left\{ \alpha \cdot E_m \cdot A_m - \left[1 - \frac{\cosh(\beta \cdot x)}{\cosh(\beta \cdot L_f)} \right] \right\} \tag{31}$$

respectively.

4.4. Sliding Function Evaluation

In order to obtain the mean value $s(\overline{\varepsilon_f^m})$ of the sliding function $s(\varepsilon_f^m)$ along the fibre (which relates the mean matrix deformation in the fibre direction, $\overline{\varepsilon_f^m}$ and the mean fibre deformation), we can determine the actual elastic energy stored in the fibre and by imposing such a value to be equal to the average energy due to a constant stress distribution, $\overline{\sigma_f}$, along the fibre. The actual elastic density energy w in the fibre is:

$$w = \frac{1}{2 \cdot V_f} \cdot \int_{V_f} \sigma_f(x) \cdot \varepsilon_f(x) dV_f = \frac{1}{2 \cdot V_f} \cdot \frac{A_f}{E_f} \cdot \int_{-L_f}^{L_f} \sigma_f^2(x) \, dx = \frac{1}{4 \cdot L_f \cdot E_f} \cdot \int_{-L_f}^{L_f} \sigma_f^2(x) \, dx \tag{32}$$

where the fibre volume is written as $V_f = 2 \cdot L_f \cdot A_f$, while the mean fibre elastic energy density $\overline{w}$ corresponding to a constant fibre stress, $\overline{\sigma_f}$, is:

$$\overline{w} = \frac{1}{2} \cdot \overline{\sigma_f} \cdot \overline{\varepsilon_f} = \frac{1}{2} \cdot E_f \cdot \left[s\left(\overline{\varepsilon_f^m}\right) \cdot \overline{\varepsilon_f^m} \right]^2 \tag{33}$$

where $\overline{\varepsilon_f} = s\left(\overline{\varepsilon_f^m}\right) \cdot \overline{\varepsilon_f^m}$ and the mean fibre stress is written as $\overline{\sigma_f} = E_f \cdot \left[s\left(\overline{\varepsilon_f^m}\right) \cdot \overline{\varepsilon_f^m} \right]$.

In the above relations the mean value of the matrix strain $\overline{\varepsilon_f^m}$ measured in the fibre direction has been used; it can be computed as:

$$\overline{\varepsilon_f^m} = \frac{1}{2L_f} \cdot \int_{-L_f}^{L_f} \varepsilon_f^m(x) dx = \frac{1}{2L_f} \cdot \int_{-L_f}^{L_f} [(\mathbf{i} \otimes \mathbf{i}) : \varepsilon(x)] dx \tag{34}$$

The (constant) value of $s\left(\overline{\varepsilon_f^m}\right)$ can be explicitly written by solving Eq. (33):

$$s\left(\overline{\varepsilon_f^m}\right) = \frac{1}{\overline{\varepsilon_f^m}} \cdot \sqrt{\frac{2 \cdot \overline{w}}{E_f}} \tag{35}$$

and by assuming $w = \overline{w}$, we get the value sought:

$$s\left(\overline{\varepsilon_f^m}\right) = \frac{1}{\overline{\varepsilon_f^m} \cdot E_f} \cdot \sqrt{\frac{\int_{-L_f}^{L_f} \sigma_f^2(x)\, dx}{2L_f}} = \frac{1}{\dfrac{E_f}{2L_f} \cdot \int_{-L_f}^{L_f} \varepsilon_f^m(x)\, dx} \cdot \sqrt{\frac{\int_{-L_f}^{L_f} \sigma_f^2(x)\, dx}{2L_f}} \tag{36}$$

with

$$\varepsilon_f^m(x) = m(x)/(E_m \cdot A_m) = [P - f(x)]/(E_m \cdot A_m) = F\left\{\alpha - \frac{1}{E_m \cdot A_m}\left[1 - \frac{\cosh(\beta \cdot x)}{\cosh(\beta \cdot L)}\right]\right\} \tag{37}$$

In the following, Eq. (36) will be calculated by using the proper fibre stress law distribution in the pre-, post- and ultimate stage.

i) Pre-Debonding Stage

In the pre-debonding stage the condition $\tau_{i,\max} = \tau_i\left(x = \pm L_f\right) < \tau_{au}$ the stress in the fibre can be expressed by using $f(x)$ obtained from Eq. (30): $\sigma_f(x) = f(x)/A_f$ (Figure 5b).

In this case the constant value of the sliding function evaluated through the mean fibre strain:

$$s\left(\overline{\varepsilon_f^m}\right) = \frac{1}{\overline{\varepsilon_f^m} \cdot E_f} \cdot \sqrt{\frac{\int_{-L_f}^{L_f}\left[\dfrac{F}{A_f}\cdot\left(1 - \dfrac{\cosh(\beta \cdot x)}{\cosh(\beta \cdot L_f)}\right)\right]^2 dx}{2L_f}} = \frac{F}{\sqrt{2}\cdot\overline{\varepsilon_f^m}\cdot E_f \cdot A_f}\cdot\sqrt{\frac{\int_{-L_f}^{L_f}\left(1 - \dfrac{\cosh(\beta \cdot x)}{\cosh(\beta \cdot L_f)}\right)^2 dx}{L_f}} \tag{38}$$

and by solving the integral under the square root, the sliding function $s\left(\overline{\varepsilon_f^m}\right)$ in the pre-debonding stage can be finally evaluated:

$$s\left(\overline{\varepsilon_f^m}\right) = \frac{F}{\sqrt{2}\cdot\overline{\varepsilon_f^m}\cdot E_f \cdot A_f}\cdot\sqrt{2 + \frac{1}{L_f \cosh^2(\beta \cdot L_f)}\cdot\left(L_f + \frac{\sinh(2\beta \cdot L_f)}{2\beta}\right) - \frac{4 \cdot \sinh(\beta \cdot L_f)}{L_f \beta \cdot \cosh(\beta \cdot L_f)}} \tag{39}$$

with the mean matrix strain evaluated parallel to the fibre direction, $\overline{\varepsilon_f^m}$, which can be analytically determined:

$$\overline{\varepsilon_f^m} = \frac{1}{2L_f} \cdot \int_{-L_f}^{L_f} \varepsilon_f^m(x)\, dx = \frac{1}{2L_f} \cdot \int_{-L_f}^{L_f} F\left\{ \alpha - \frac{1}{E_m \cdot A_m}\left[1 - \frac{\cosh(\beta \cdot x)}{\cosh(\beta \cdot L_f)}\right]\right\} dx =$$

$$= F \cdot \left[\alpha - \frac{1}{E_m \cdot A_m} + \frac{\sinh(\beta \cdot L_f)}{\beta \cdot L_f \cdot E_m \cdot A_m \cosh(\beta \cdot L_f)}\right]$$

(40)

ii) Debonding Initiation

When the shear stress transmitted from the matrix becomes equal to the ultimate interface shear stress τ_{au} at fibre the fibre extremities, $\tau_i(x = \pm L_f) = \tau_{au}$, the debonding phenomenon starts. As in the previous case, since the sliding function must be expressed with respect to $\overline{\varepsilon_f^m}$, it is necessary to obtain such a quantity corresponding to this situation.

By using Eq. (31₁), the debonding initiation condition can be identified:

$$\left|\tau_i(\pm L_f)\right| = \frac{F \cdot \beta}{c} \cdot \left[\frac{\sinh(\beta \cdot L_f)}{\cosh(\beta \cdot L_f)}\right] = \frac{F \cdot \beta}{c} \cdot \tanh(\beta \cdot L_f) = \tau_{au} \Rightarrow F_{deb} = \frac{\tau_{au} \cdot c}{\beta \cdot \tanh(\beta \cdot L_f)}$$

(41)

where the subscript '*deb*' stands for debonding; F_{deb} is related (through the relation $F_{deb} = P_{deb}/(\alpha \cdot E_m \cdot A_m)$) to the minimum value of the total force applied to the single-bar reinforced cylinder under study, that causes the starting of the debonding phenomenon.

The corresponding mean strain in the matrix, measured in the fibre direction, becomes:

$$\overline{\varepsilon_{f,deb}^m} = \frac{1}{2L_f} \cdot \int_{-L_f}^{L_f} \varepsilon_f^m(x)\, dx = \frac{\tau_{au} \cdot c}{\beta \cdot \tanh(\beta \cdot L_f)} \cdot \left(\frac{\alpha \cdot \beta \cdot L_f \cdot E_m \cdot A_m - \beta \cdot L_f + \tanh(\beta \cdot L_f)}{\beta \cdot L_f \cdot E_m \cdot A_m}\right)$$

(42)

iii) Partial Debonding Stage

When the debonding phenomenon occurs in two limited zones (with length $L_f - L_{ad}$ each) located at the fibre extremities, it can be assumed that in such zones the shear stress is constant and equal to the characteristic interface friction shear stress τ_{fu} (Figure 5c). By integrating the differential Eq. (29) with the boundary conditions $f(x = \pm L_{ad}) = c \cdot \tau_{fu} \cdot (L_f - L_{ad})$ (Fig. 5c), we get:

$$f(x) = \begin{cases} \sigma_{f,deb} \cdot A_f = c \cdot \tau_{fu} \cdot \left(L_f - |x|\right) & \text{for } -L_f \leq x \leq -L_{ad} \text{ and } L_{ad} \leq x \leq L_f \\ \sigma_f(x) \cdot A_f = F \cdot \left[1 + \dfrac{c \cdot \tau_{fu} \cdot \left(L_f - L_{ad}\right) - F}{F} \cdot \dfrac{\cosh(\beta \cdot x)}{\cosh(\beta \cdot L_{ad})}\right] & \text{for } -L_{ad} \leq x \leq L_{ad} \end{cases}$$

$$(43)$$

and the shear stress at the matrix-fibre interface can be written:

$$\tau_i(x) = -\frac{1}{c} \cdot \frac{df(x)}{dx} = \begin{cases} \tau_{fu} & \text{for } -L_f \leq x \leq -L_{ad} \text{ and } L_{ad} \leq x \leq L_f \\ -\beta \cdot \left[\tau_{fu} \cdot \left(L_f - L_{ad}\right) - \dfrac{F}{c}\right] \cdot \dfrac{\sinh(\beta \cdot x)}{\cosh(\beta \cdot L_{ad})} & \text{for } -L_{ad} \leq x \leq L_{ad} \end{cases}$$

$$(44a)$$

Since $\tau_i\left(x = \pm L_{ad}\right) = \tau_{au}$, we can determine the value of the corresponding $F = F_{deb}(L_{ad})$ as:

$$F_{deb}(L_{ad}) = \frac{\tau_{au} \cdot c}{\beta \tanh(\beta \cdot L_{ad})} + \tau_{fu} \cdot c \cdot \left(L_f - L_{ad}\right)$$

$$(44b)$$

The axial force in the fibre and the interface shear stress distribution in the non-debonded zone can be written:

$$f(x) = \begin{cases} \sigma_{f,deb} \cdot A_f = c \cdot \tau_{fu} \cdot \left(L_f - |x|\right) & \text{for } -L_f \leq x \leq -L_{ad} \text{ and } L_{ad} \leq x \leq L_f \\ \sigma_f \cdot A_f = F_{deb}(L_{ad}) \cdot \left[1 + \dfrac{c \cdot \tau_{fu} \cdot \left(L_f - L_{ad}\right) - F_{deb}(L_{ad})}{F_{deb}(L_{ad})} \cdot \dfrac{\cosh(\beta \cdot x)}{\cosh(\beta \cdot L_{ad})}\right] = F_{deb}(L_{ad}) \cdot \left[1 - M \cdot \dfrac{\cosh(\beta \cdot x)}{\cosh(\beta \cdot L_{ad})}\right] & \text{for } -L_{ad} \leq x \leq L_{ad} \end{cases}$$

$$(45a)$$

and

$$\tau_i(x) = -\frac{1}{c} \cdot \frac{df(x)}{dx} = \begin{cases} \tau_{fu} & \text{for } -L_f \leq x \leq -L_{ad} \text{ and } L_{ad} \leq x \leq L_f \\ \beta \cdot \left[-\tau_{fu} \cdot \left(L_f - L_{ad}\right) + \dfrac{F_{deb}(L_{ad})}{c}\right] \cdot \dfrac{\sinh(\beta \cdot x)}{\cosh(\beta \cdot L_{ad})} & \text{for } -L_{ad} \leq x \leq L_{ad} \end{cases}$$

$$(45b)$$

By taking into account Eqs (34, 35), the mean sliding function $s\left(\overline{\varepsilon_f^m}\right)$ in this stage can be obtained as follows:

$$s\left(\overline{\varepsilon_f^m}\right) = \frac{1}{\overline{\varepsilon_f^m} \cdot E_f} \cdot \sqrt{\frac{\displaystyle\int_{-L_f}^{L_f} \sigma_f^2(x)\, dx}{2L_f}} = \frac{1}{\overline{\varepsilon_f^m} \cdot E_f} \cdot \sqrt{\frac{\displaystyle\int_{-l}^{l} \sigma_f^2(x)\, dx + 2\displaystyle\int_{l}^{L_f} \sigma_{f,deb}^2(x)\, dx}{2L_f}} =$$

$$= \frac{F_{deb}}{\sqrt{2} \cdot \overline{\varepsilon_f^m} \cdot E_f \cdot A_f} \cdot \sqrt{\frac{2L_{ad} + \dfrac{M^2}{\cosh^2(\beta \cdot L_{ad})} \cdot \left(L_{ad} + \dfrac{\sinh(2\beta \cdot L_{ad})}{2\beta}\right) - \dfrac{4M \cdot \sinh(\beta \cdot L_{ad})}{\beta \cdot \cosh(\beta \cdot L_{ad})} + \dfrac{2}{3} \cdot \dfrac{\pi^2 d^2 \tau_{fu}^2}{F_{deb}^2}\left(L_f - L_{ad}\right)^3}{L_f}}$$

$$(46)$$

where in the previous relation the following terms can be acknowledged:

$$\int_{-L_{ad}}^{L_{ad}} \sigma_f^2(x)\ dx = \frac{F^2}{A_f^2}\cdot\left[2L_{ad}+\frac{M^2}{\cosh^2(\beta\cdot L_{ad})}\cdot\left(L_{ad}+\frac{\sinh(2\beta\cdot L_{ad})}{2\beta}\right)-\frac{4M\cdot\sinh(\beta\cdot L_{ad})}{\beta\cdot\cosh(\beta\cdot L_{ad})}\right]$$

(47)

and $2\int_{L_{ad}}^{L_f} \sigma_{f,deb}^2(x)\ dx = \dfrac{2}{3}\cdot\dfrac{\pi^2 d^2\tau_{fu}^2}{A_f^2}\left(L_f-L_{ad}\right)^3$

In which we indicate with M the following quantity: $M=\dfrac{c\cdot\tau_{fu}\cdot\left(L_{ad}-L_f\right)+F_{deb}(L_{ad})}{F_{deb}(L_{ad})}$. In this stage the matrix strain function evaluated in the fibre direction can be obtained:

$$\varepsilon_f^m(x)=\frac{m(x)}{E_m\cdot A_m}=\begin{cases}F_{deb}(L_{ad})\cdot\alpha+\dfrac{c\cdot\tau_{fu}\cdot\left(L_{ad}-L_f\right)}{E_m\cdot A_m}+\dfrac{c\cdot\tau_{fu}\cdot\left(|x|-L_{ad}\right)}{E_m\cdot A_m} & \text{for } -L_f\le x\le-L_{ad}\ \text{ and }\ L_{ad}\le x\le L_f\\[3mm] F_{deb}(L_{ad})\left\{\alpha-\dfrac{1}{E_m\cdot A_m}\left[1-\dfrac{c\cdot\tau_{fu}\cdot\left(L_{ad}-L_f\right)+F_{deb}(L_{ad})}{F_{deb}(L_{ad})}\cdot\dfrac{\cosh(\beta\cdot x)}{\cosh(\beta\cdot L_{ad})}\right]\right\} & \text{for } -L_{ad}\le x\le L_{ad}\end{cases}$$

(48)

while its mean value along the fibre is:

$$\overline{\varepsilon_f^m}=\frac{1}{2L_f}\cdot\int_{-L_f}^{L_f}\varepsilon_f^m(x)\,dx=\frac{1}{2L_f}\cdot\left(\int_{-L_{ad}}^{L_{ad}}\varepsilon_f^m(x)\,dx+2\int_{L_{ad}}^{L_f}\varepsilon_f^m(x)\,dx\right)=$$

$$\overline{\varepsilon_f^m}=F_{deb}(L_{ad})\cdot\left[\alpha-\frac{L_{ad}}{L_f\cdot E_m\cdot A_m}\right]+\frac{\sinh(\beta\cdot L_{ad})}{L_f\cdot E_m\cdot A_m}\cdot\left[\left(\frac{c\cdot\tau_{fu}\cdot\left(L_{ad}-L_f\right)+F_{deb}(L_{ad})}{\beta\cdot\cosh(\beta\cdot L_{ad})}\right)\right]+$$

$$+\frac{c\cdot\tau_{fu}(L_{ad}-L_f)\cdot L_{ad}}{L_f\cdot E_m\cdot A_m}+\frac{c\cdot\tau_{fu}}{L_f\cdot E_m\cdot A_m}\cdot\left(\frac{L_f^2+L_{ad}^2}{2}-L_f\cdot L_{ad}\right)$$

(49)

In the present case, once the fibre and interface parameters L_f and τ_{fu},τ_{au} are known, the solution of the problem can be achieved by determining first the distance L_{ad}, at which dobonding starts, by using Eq. (45b) and setting $\tau_i(L_{ad})=\tau_{au}$. From Eq. (49) the value of $F_{deb}(L_{ad})$ can be expresses in function of $\overline{\varepsilon_f^m}$. Finally substituting the result obtained for $F_{deb}(L_{ad})$ in Eq. (46) the constant value of the sliding function $s\!\left(\overline{\varepsilon_f^m}\right)$ can be obtained.

iv) Complete Debonding Stage

When the debonding phenomenon reaches the centre of the fibre ($L_{ad}=0$) the shear stress transmitted from the matrix to the fibre is everywhere equal to the friction shear stress τ_{fu} (Figure 5d).

In this stage the forces carried by the fibre at a generic distance x from the fibre midpoint is expressed by:

$$f(x) = c \cdot \tau_{fu} \cdot \left(L_f - |x|\right) \qquad \text{for } -L_f \leq x \leq L_f \qquad (50)$$

while the shear stress at the interface assumes a constant value, $\tau_i(x) = \tau_{fu}$.

The sliding function $s\left(\varepsilon_f^m\right)$ in this stage can be finally obtained simply by setting $L_{ad} = 0$ in Eq. (46):

$$s\left(\varepsilon_f^m\right) = \frac{1}{\varepsilon_f^m \cdot E_f} \cdot \sqrt{\frac{2\int_0^{L_f} \sigma_f^2(x)\, dx}{2L_f}} = \sqrt{\frac{1}{3}} \cdot \frac{c \cdot \tau_{fu} \cdot L_f}{A_f} \cdot \frac{1}{E_f \cdot \varepsilon_f^m} \qquad (51)$$

4.5. Elasto-Plastic Behaviour of the Matrix

In order to generalize the matrix mechanical behaviour, it can assumed to behave as an elastic-plastic material; in incremental form, the well-known elastic-plastic constitutive equations can be expressed by [23, 24]:

$$\dot{\varepsilon} = \dot{\varepsilon}^{el} + \dot{\varepsilon}^{pl} = \mathbf{C}_m^{-1} \cdot \dot{\boldsymbol{\sigma}} + \dot{\varepsilon}^{pl} \qquad (52)$$

where $\dot{\varepsilon}^{el}$, $\dot{\varepsilon}^{pl}$ indicate the elastic and plastic part of the strain rate tensor, respectively, while $\dot{\boldsymbol{\sigma}}$ is the stress rate tensor.

Assuming that the elastic behaviour is defined by a yield function $F(\sigma, k_1, ..., k_n) = 0$ with $k_1, ..., k_n$ n hardening parameters, we have the elastic behaviour when: i) $F(\sigma, k_1, ..., k_n) < 0$, while plastic flow occurs when ii) $F(\sigma, k_1, ..., k_n) = 0$ and $F'(\sigma, k_1, ..., k_n) = 0$; when iii) $F(\sigma, k_1, ..., k_n) = 0$ and $F'(\sigma, k_1, ..., k_n) < 0$ we have elastic unloading.

The yielding condition for a perfectly plastic or a hardening material can be written:

$$F(\sigma, k_1, ..., k_n) = 0 \quad \text{and} \quad \begin{cases} dF = \dfrac{\partial F_0}{\partial \sigma} d\sigma = 0 & \text{perfectly plastic material} \\[3mm] dF = \dfrac{\partial F}{\partial \sigma} d\sigma + \dfrac{\partial F}{\partial k_h} dk_h = 0 & \text{hardening material} \end{cases} \qquad (53)$$

A simple standard associative flow rule is assumed to described the plastic strain increment:

$$\dot{\varepsilon} = \dot{\varepsilon}^{el} + \dot{\varepsilon}^{pl} = \mathbf{C}_m^{-1} : \dot{\boldsymbol{\sigma}} + \dot{\lambda}\frac{\partial F}{\partial \sigma} \quad \text{with} \quad \dot{\lambda} = \frac{\mathbf{f} : \mathbf{C}'_m : \dot{\varepsilon}}{H + \mathbf{f} : \mathbf{C}'_m : \mathbf{f}} \qquad (54)$$

where $\mathbf{f} = \partial F / \partial \sigma_{ij}$, H is the hardening parameter, while $\dot{\lambda} \geq 0$ if $F = 0$ and $dF = 0$, $\dot{\lambda} = 0$ if $F = 0$ and $dF < 0$ or $F < 0$. The plastic multiplier $\dot{\lambda}$ can be calculated by eliminating the stress rate tensor from the equations $dF(\sigma, k_1,...., k_n) = 0$ and the first of Eq. (54).

Yielding functions depending upon a single yielding parameter k are considered. The yielding parameter k will be identified with the uniaxial yield stress of the material σ_0, $F(\sigma, k) = F(\sigma, \sigma_0) = 0$. The evolution of the yielding parameter k is written in the simple form: $\sigma_0 = \sigma_Y + H\varepsilon^{pl}$ where H is the hardening parameter that can be related to the post-yielding stress-strain slope E_t, as $H = E_t / (1 - E_t / E_m)$ and ε^{pl} is an uniaxial equivalent plastic strain evaluated in the matrix. All the above relations are calculated by considering the stress state in the matrix depurated from the stress state carried by the fibre phase.

4.5. F.E. Implementation

The proposed model has been implemented in a F.E. program in order to verify its capability in predicting the behaviour of fibre-reinforced composite materials characterised by an elasto-plastic matrix reinforced with fibres having the possibility of debonding. The main characteristic of the composite mechanical behaviour is the calculation of the mean sliding function $s\left(\overline{\varepsilon_f^m}\right)$ at every Gauss integration point. In order to obtain such a function, the mean matrix strain measured in the fibre direction, $\overline{\varepsilon_f^m}$, is needed. Since the main assumption is the random distribution of the fibres in the matrix material, i.e. each direction of the solid angle is equally represented, the mean matrix strain can be simply obtained by computing the hydrostatic strain from the strain tensor of the matrix, $\overline{\varepsilon_f^m} = \mathrm{tr}\varepsilon / 3$. In the case of fibres not uniformly oriented in every direction, $\overline{\varepsilon_f^m}$ can be obtained by knowing the fibres orientation distribution in the solid angle and by performing the corresponding weighted mean matrix strain evaluation.

4.6. Numerical Application

The first example considers a FRC material with random-oriented uniformly dispersed fibres; the mechanical and geometrical characteristics of the composite are assumed as follows: matrix Young modulus $E = 10\,\mathrm{GPa}$, matrix Poisson's ratio equal to $\nu = 0.33$, fibres (volume fraction ratio equal to $\eta = V_f / V = 5\%$, length and diameter equal to $2L_f = 24mm$ and $\phi_f = 0.5mm$, respectively) with elastic modulus equal to $E_f = 100\,\mathrm{GPa}$ and ultimate matrix-fibre interface shear stress equal to $\tau_{au} = 1\,\mathrm{MPa}$.

A uniaxial monotonic stress state is assumed to be applied to the FRC.

In Fig. 6 the sliding function values (Fig. 6a) are represented against the mean matrix strain $\overline{\varepsilon_f^m}$ measured along the fibre direction; as can be observed the sliding function generally shows a decreasing behaviour (for small values of the τ_{au}/τ_{fu} ratio an initial increasing can be noted); the sliding function tends to assume nearly constant values for high strain levels (when the debonding phenomenon is diffused along the whole length of the fibres). Such constant values depend on the friction shear stress value τ_{fu}: the lower such a value, the lower the asymptotic sliding function.

In Fig. 6b the half fibre non-debonded length L_{ad} is displayed against the mean matrix strain measured in the fibre direction; it can be observed as the debonding phenomenon is more pronounced for high values of the τ_{au}/τ_{fu} ratio, i.e. low friction shear stress values facilitate the fibre-matrix relative sliding as the low values of the parameter L_{ad} indicate.

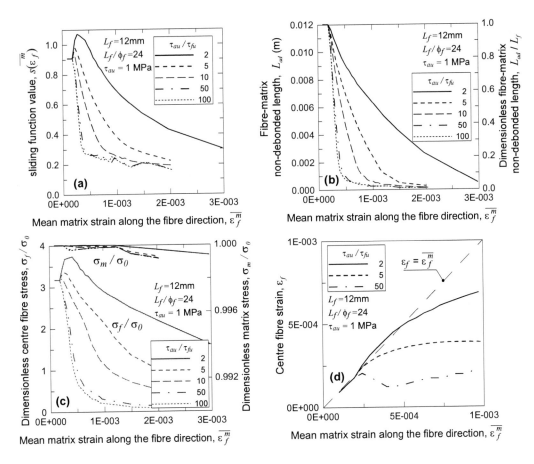

Fig. 6. Fibre-reinforced material with random distributed fibres: sliding function values (a), fibre-matrix non debonded length (b), dimensionless fibre and matrix stresses (c) and and fibre strain (d) against the vertical matrix strain for different τ_{au}/τ_{fu} ratios.

In Fig. 6c the fibre and matrix dimensionall stresses (obtained by dividing the current FRC stress by the remote applied stress σ_0) are represented against the mean matrix strain; it can be observed as the dimensionless stress carried by the fibres decreases by increasing the main matrix strain $\overline{\varepsilon_f^m}$ due to the progressive increasing of the fibre debonded length Such a trend is much more pronounced for high τ_{au}/τ_{fu} values. On the other hand the stress carried by the matrix material is practically unaffected by the presence of the reinforcing fibres due to the limited fibre volume fraction adopted. Finally in Fig. 6d the centre fibre strain is plotted against the mean matrix strain; it can be noticed as the fibre strain initially increases and subsequently tends to a constant value as much as the fibre debonding increases as is expected for a constant friction shear stress along the whole fibre length.

The next example considers an elastic-plastic unreinforced and fibre-reinforced thick-walled cylinder under internal pressure; a metal matrix, obeying the Mises criterion, is assumed to be reinforced with a random distribution of fibres stiffer than the matrix. Two different F.E. models are considered, both discretised through 4-nodes and 8-nodes elements: the plane strain and the axisymmetric one (Figure 7). The mechanical parameters of the composite under study are reported in Figure 8. The maximum value of the applied internal pressure (reached in ten equal load steps) is equal to 180 MPa and the obtained inner face radial displacement against the load factor is displayed in Figure 8. As can be noted the final displacement decreases by increasing the fibre volume fraction with respect to the unreinforced case, for both the plane strain and the axisymmetric models that present similar results.

The sliding function patterns $s\left(\overline{\varepsilon_f^m}\right)$ for the axisymmetric model, evaluated by assuming the friction shear stress equal to $\tau_{fu} = 0.1 \cdot \tau_{au}$ and $\tau_{fu} = 0.5 \cdot \tau_{au}$, is reported in Figure 8; in proximity of the inner face the mean value of the sliding function is greater (dark zones) with respect to its values in proximity of the outer face, because of the constraint effect given by the compressive radial stress that limits the occurrence of the debonding phenomenon to in the composite material.

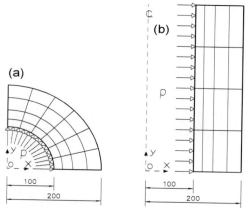

Figure 7. Elasto-plastic fibre-reinforced cylinder under internal pressure: model with 24 elements in plane strain condition (a), and model with 16 elements in axisymmetric condition (b) (dimensions in millimetres).

A Micromechanical Model and Reinforcing Distribution Optimisation ...

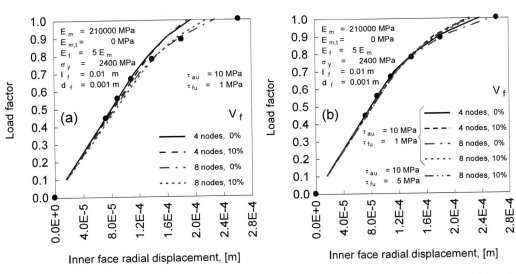

Figure 8. Radial displacement of the inner face of the cylinder in the case of plane strain model (a) and axisymmetric model (b) for unreinforced and fibre-reinforced materials. Dots represent the theoretical solution for an unreinforced material (ref. [24]).

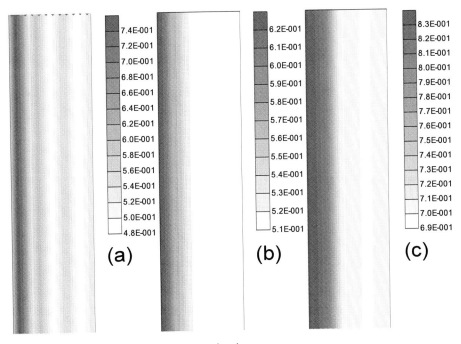

Figure 9. Contour map of the sliding function $s\left(\overline{\varepsilon_f^m}\right)$ at the final load level for the axisymmetric model: (a) 4 nodes elements with $\tau_{fu} = 0.1 \cdot \tau_{au}$; (b) 8 nodes elements with $\tau_{fu} = 0.1 \cdot \tau_{au}$; (c) 8 nodes elements with $\tau_{fu} = 0.5 \cdot \tau_{au}$.

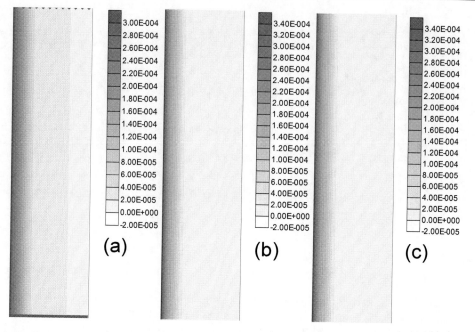

Figure 10. Contour map of the hoop plastic deformation at the final load level for the axisymmetric model: (a) 8 nodes elements, without fibres; (b) 8 nodes elements with $V_f = 10\%$ and $\tau_{fu} = 0.1 \cdot \tau_{au}$; (c) 8 nodes elements with $V_f = 10\%$ and $\tau_{fu} = 0.5 \cdot \tau_{au}$.

Furthermore it is possible to note that the fibres produce with respect to the unreinforced case (apart the reduction of the final displacements), a limitation of the zone size where the plastic deformations take place while their maximum value does not change significantly (Figure 10).

Other applications will be discussed in the next sections, where the fibre content distribution will be considered as a design variable of composite structural elements in order to get some desired optimised mechanical performances.

5. FIBRES DISTRIBUTION OPTIMISATION BY USING GENETIC ALGORITHM

5.1. Introduction to Optimisation

The optimisation of design parameters in applied engineering problems is usually a complex and difficult task and its solution through conventional techniques [25], such as the gradient approach, is often very hard. In the last decades there has been an increasing interest in algorithms simulating the natural evolutionary process of life [26-29] known as Genetic Algorithms (GAs).

To optimise one or more design parameters with respect to a chosen objective function for the problem under study, a random, stochastic, global search optimisation method known as Genetic Algorithm (GA) is used. The GA is a biologically-inspired optimisation method that enhance iteratively the current solution by using genetic concepts.

The use of GA, instead of - for example - classical techniques such as the response surface or gradient-based techniques [30, 31], allows to easily tackle problems with multiple minima or problems characterised by non convexity properties, without the risk to miss global optimum and avoiding numerical instability as often happen for classical optimum search methods. The GA does not require any mathematical-based special theory and can handle any kind of objective function; it simply makes use of basic concepts such as generation of random population, selection, crossover and mutation to get a new population usually characterised by a better performance with respect to the previous one. The outlined procedure can be iteratively repeated until a desired tolerance is reached and the sought optimal condition can be approximately achieved.

The GA has been successfully applied to the solution of several problems such as structural [30, 31] and topology [32, 33] optimisation, material identification [33, 34], optimal design of composite structures or materials [35-40] and even non-structural problems have also been successfully solved by this technique [41].

5.2. Basic Concepts of the GA

In the GA method, a random guessed population representing a set of design parameters is generated; the generated populations must be adequately numerically encoded in order to allows computational operations. In the present study a *real encoding* technique [28] has been used, i.e. a real number representation of the design vector defining the fibre volume content distribution has been employed. Other encoding technique, such as *binary encoding* [28], can be used.

The fibre volume fraction distribution can be conveniently represented by the elements of a rectangular matrix (in the present study it will be indicated with $D_{f,ij}$, which contains the fibre distribution concentration values in the FRC); its *j-th* column contain the design parameters ($i = 1,...,N$) belonging to the *j-th* individual with $j = 1,...,M$ (which are also usually called *chromosomes* in GA theory), where M is the total number of individuals of the population. A suitable objective function must be chosen for the problem under study to represent the design performance for each population, in order to determine its fitness, with the final aim to find out the better ones among the given individuals.

In order to iteratively solve the stated optimisation problem, a *crossover strategy* must be defined to generate a new population from the previous one; in the present paper the scheme described in the following is used.

Starting from an initial population (for instance generated by a random numbers algorithm generator), the fitness for each chromosome belonging to population must be evaluated; the two individuals (namely L and K, which are supposed to be located in the *l-th* and *k-th* columns of $D_{f,ij}$, i.e. $D_{f,iL}, D_{f,iK}$, respectively) that have the best fitness with respect to the objective function, are employed to generate a new populations $D_{f,ij}$ *

(*offspring*) by using genetic concepts such as crossover.

The new population is obtained as follows:

$$D_{f,ij}* = A_{ij} \cdot D_{f,ij}^L + B_{ij} \cdot D_{f,ij}^K \quad \text{with probability} \quad P_c \leq 1$$
$$D_{f,ij}* = D_{f,ij} \quad\quad\quad\quad\quad\quad \text{with probability} \quad (1 - P_c) \leq 1 \tag{55}$$

where P_c represents the probability that the new population is generated by the so-called *continuous crossover* [26] with the values of individual index $1 \leq j \leq M$ randomly generated with a constant probability algorithm.

In Eq. (55$_1$), $D_{f,ij}^L, D_{f,ij}^K$ are matrices having all the columns composed by the two better individuals $D_{f,iL}, D_{f,iK}$ respectively, i.e. $D_{f,ij}^L = D_{f,iL}, D_{f,ij}^K = D_{f,iK}$, while A_{ij} and B_{ij} are matrices of 0 and 1 only, with the property: $A_{ij} + B_{ij} = 1$. In order to form such a matrices, a $N \times M$ matrix of random numbers is generated. The matrix A_{ij} (or B_{ij}) is obtained as follows: if the value of the random number in position (i, j) is above a given crossover value, its corresponding value in the matrix A_{ij} is set to 1, otherwise it is set to 0.

Furthermore, in order to not lose the good chromosomes of the better individuals (apart L and K), some other individuals are simply transferred to the new population (*elitism*, Eq. (55$_2$)) [28].

Alternatively to *continuous crossover*, a randomly generated number, a so-called *crossover point* N_c ($1 < N_c < N$), can be defined to perform the crossover operation: in such a case the new population $D_{ij}*$ receives part of the chromosomes from the first parent L and the remaining from the second parent K:

$$D_{f,ij}* = \begin{cases} D_{f,ij}^L & \text{if } i \leq N_c \\ D_{f,ij}^K & \text{if } i > N_c \end{cases} \quad \text{with probability} \quad P_c \leq 1 \quad\quad \text{crossover}$$
$$D_{f,ij}* = D_{f,ij} \quad\quad\quad\quad\quad \text{with probability} \quad (1 - P_c) \leq 1 \quad \text{elitism} \tag{56}$$

An alternative to the above approach is the so-called *direction-based crossover* [28] which is able to assure that the offspring individuals are better than their parents:

$$D_{f,ij}* = \xi \cdot (D_{f,ij}^L - D_{f,ij}^K) + D_{f,ij}^L \quad \text{with probability} \quad\quad P_c \leq 1 \quad \text{crossover}$$
$$D_{f,ij}* = D_{f,ij} \quad\quad\quad\quad\quad\quad\quad\quad \text{with probability} \quad (1 - P_c) \leq 1 \quad \text{elitism}$$
$$\tag{57}$$

where ξ is a random number lying in the range $(0, 1)$ while the assumption that to $D_{f,ij}^L$ corresponds a better fitness than $D_{f,ij}^K$, is made.

In order to explore a design space as wide as possible, the *mutation* concept is used; it can be reasonable to change the obtained offspring by partially modifying such a new obtained population. By defining a *probability of mutation* $P_m \leq 1$, a certain number of individuals of

$D_{f,ij}$* (with the given probability P_m) are randomly redefined in a number N_m ($1 \leq N_m \leq N$) of their design parameters (genes). The *mutation* procedure is performed (once the crossover operation has been done), on individuals randomly chosen from the new population; for each individual chosen for mutation, the number of genes - to be randomly mutated - is assumed to be equal to $P_{m,g,crossover}$, $P_{m,g,elitism}$ for individuals generated by crossover or by elitism from the previous population, respectively.

The application of the described genetic-based algorithm for the determination of the optimum fibre reinforcing phase distribution in the matrix material, quantified through the spatial scalar variable $\eta(\mathbf{x}) = V_f(\mathbf{x})/V$, is straightforward. Since the problem under study is characterised by a fixed total fibre content, $V_{f,tot} = V \cdot \overline{\eta}$ (where $\overline{\eta}$ is the theoretical uniformly distribution fibre content in the composite), it must be regarded as a constrained optimisation problem (non-linear programming problem, [28]).

As stated above, for the problem under study a continuous representation of the design fibre density function $\eta(\mathbf{x})$, "encoded" in a matrix of real numbers $D_{f,ij}$, is used. Such a function must obviously obey to the stated constraint condition. For sake of computational simplicity, a normalised fibre distribution design function, $\eta'(\mathbf{x})$, defined in the unit range $0 \leq \eta'(\mathbf{x}) \leq 1$, is employed; the relationship between $\eta(\mathbf{x})$ (which is the optimisation function having physical meaning) and $\eta'(\mathbf{x})$ (the fictitious fibre volume content) can be simply obtained:

$$\eta(\mathbf{x}) = \frac{V_{f,tot}}{V'} \cdot \eta'(\mathbf{x}) \text{ where } V' = \int_V \eta'(\mathbf{x})dV \tag{58}$$

The normalised fibre distribution function $\eta'(\mathbf{x})$ is used by the genetic algorithm optimisation procedure as the computational design variable.

As can be deduced from the above introduction, the use of Genetic Algorithm to optimise the fibre content distribution in a composite structure, is quite simple and attractive and allows the possibility to choice any kind of objective function to get the desired performance, such as those explained in the following.

5.3. Fibre Distribution Content Optimisation in FRC

As a first example, the optimal fibre distribution which allows to get the minimum compliance of the structural component under study is considered; the compliance can be quantified through the maximum displacement (irrespective of the considered direction i) u_i^{max} in the structure, which must be minimised. The corresponding dimensionless objective function F_c to be minimised can be written:

$$F_c(\eta_c(\mathbf{x})) = \frac{u_i^{max}(\eta_c(\mathbf{x}))}{u_{i,V_f=0}^{max}} \tag{59}$$

where $u_i^{max}(\eta_c(\mathbf{x}))$ is the maximum displacement in the structure which depends on the design function $\eta_c(\mathbf{x})$ or on the corresponding dimensionless counterpart $\eta_c'(\mathbf{x})$. As can be noted the objective function is defined as the maximum displacement normalised with respect to the corresponding maximum displacement of the unreinforced structure, $u_{i,V_f=0}^{max}$. In the above equation the subscript c stands for *compliance*.

Another useful required performance to be minimised as much as possible can be assumed to be the maximum principal stress value s_I in the FRC solid under study: analogously to the previous case the optimal fibre distribution $\eta_s(\mathbf{x})$, to which corresponds such a condition, can be defined through the following dimensionless objective function to be minimised:

$$F_s(\eta_s(\mathbf{x})) = \frac{\sigma_I^{max}(\eta_s(\mathbf{x}))}{\sigma_{I,V_f=0}^{max}} \tag{60}$$

where $\sigma_I^{max}(\eta_s(\mathbf{x}))$ is the maximum principal stress which depends on the design function $\eta_s(\mathbf{x})$ or on the corresponding dimensionless counterpart $\eta_s'(\mathbf{x})$. As can be noted the objective function is defined as the maximum stress normalised with respect to the corresponding maximum stress of the unreinforced structure, $\sigma_{I,V_f=0}^{max}$. In the above equation the subscript s stands for *stress*.

The two above stated optimal conditions can be mathematically written as:

$$\text{search for } D_{f,c}(\mathbf{x}): \quad \min[F_c(\eta_c(\mathbf{x}))] = \min\left[F_c(\eta_c(D_{f,c}))\right] \text{ is reached}$$

$$\text{search for } D_{f,s}(\mathbf{x}): \quad \min[F_s(\eta_s(\mathbf{x}))] = \min\left[F_s(\eta_s(D_{f,s}))\right] \text{ is reached} \tag{61}$$

where $\eta_c(D_{f,c})$ and $\eta_s(D_{f,s})$ indicate the fibre volume distributions, encoded by the matrix $D_{f,c}$ and $D_{f,s}$, respectively, which satisfy the required optimal conditions.

The proposed FRC mechanical model, together with the presented GA, is implemented in a F.E. program in order to verify its capability in predicting the optimal fibre distribution with respect to the desired objective functions.

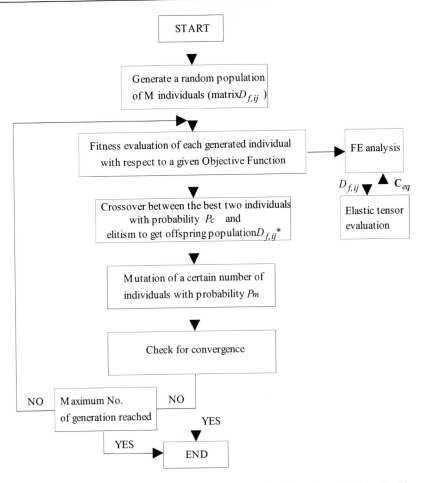

Figure 11. Flow-chart of the genetic algorithm implemented in a FE code applied to the fibre distribution optimisation problem.

As previously stated, the genetic algorithm applied to the optimal fibre distribution in fibre-reinforced composites, is implemented by using a *real encoded* design matrix $D_{f,ij}$ which values corresponds to the fibre volume content in every finite element of the discretised structures. By using a *direction-based crossover* and *elitism*, with a defined amount of *mutation* - which can occur both in offspring or in individuals obtained through the elitism operation - the optimisation procedure is performed until a required error tolerance between two subsequent solution is satisfied or the maximum number of generations is reached. The flow-chart of the implemented optimisation procedure is shown in Figure 11.

5.4. Numerical Tests

In the present section the optimum fibre distribution is determined for very simple structures in order to assess the applicability of the procedure and the convergence of the method in reaching the criterion of maximum fibre exploitation i.e. the lowest cost to reinforcing material ratio.

The obtained results are discussed and some qualitative and quantitative considerations are made.

No comparisons with literature results are considered, since, in the author's knowledge, similar studies (apart the research field of the so-called Functionally Graded Materials (FGM) which deals with the optimum composition of a binary mixture of particulate materials, typically metal-ceramics materials and studies related to the optimum fibre orientation distribution [42]) have not been published.

i) Three-Point Bending of a Wall-Beam

The first case examined deals with a three-point bending problem of a plane stress wall-beam with length/height and thickness/height ratios equal to 5 and equal to 0.3, respectively; the matrix material is assumed linear elastic, characterised by an elastic modulus $E_m = 20000\,MPa$ and Poisson's ratio $v_m = 0.1$ while the fibre-matrix Young modulus ratio is assumed equal to $E_m/E_f = 1/10$. The total fibre content is set to be equal to $\bar{\eta} = 10\%$ of the total beam volume V, $\bar{\eta} = V_f/V = 0.1$, while the fibres are geometrically characterised by length and diameter equal to $2L_f = 2 \cdot 10^{-2}$, $\phi_f = 5 \cdot 10^{-4}$ m, respectively with an aspect ratio $r_f = 2L_f/\phi_f = 40$. The adopted fibre distribution is characterised to be equally oriented in every direction of the space. The considered mechanical characteristics correspond to a fibre-reinforced concrete-like beam with a dispersion of a steel reinforcing phase (Figure 12a).

The first optimisation problem searches the optimal fibre distribution, for the given total fibre content, in order to minimise the maximum displacement of the beam (which occurs at the central point where the concentrated load is applied) in the vertical direction. The genetic algorithm described above is characterised by an initial population of 30 individuals, $M = 30$, probability of crossover equal to $P_c = 0.70$, probability of mutation equal to $P_m = 0.30$ and probability which defines the number of genes to be mutated for each individual equal to $P_{m,g,crossover} = 0.1$ and $P_{m,g,elitism} = 0.60$, respectively. In Figure 12b the adopted F.E. mesh (100 four-noded elements is used) is displayed together with a qualitative plot of the deformed shape of half of the structure, due to the symmetry which characterise the problem.

In Figure 12c the objective function $F_c\big(\eta_c(\mathbf{x})\big) = u_i^{max}\big(\eta_c(\mathbf{x})\big)/u_{i,V_f=0}^{max}$ is plotted against the number of generations while in Figure 12e the corresponding obtained final fibre volume fraction distribution pattern is displayed.

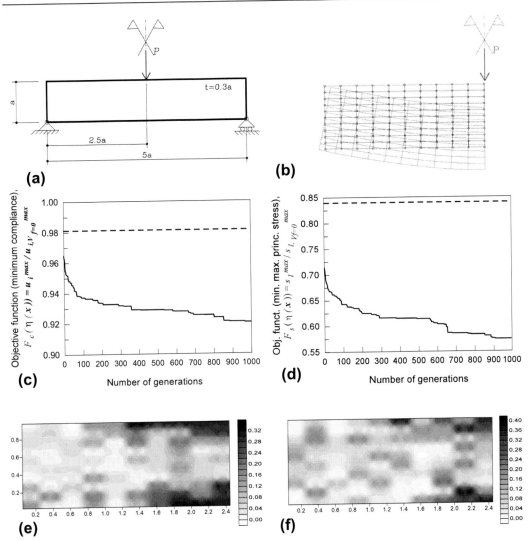

Figure 12. Three-point bending of a wall-beam (a); F.E. mesh and qualitative deformed shape (b). Objective function values against the number of generations to minimise the maximum displacements (c) or the maximum principal stress (d) and the corresponding fibre distribution patterns, respectively (e, f). Values corresponding to the uniformly reinforced material are reported (dashed lines).

As can be observed the fibres tend to concentrate at the top and at the bottom side of the central part of the beam in order to increase the moment of inertia of the cross sections which undergo the biggest values of the bending moment. The improvement with respect to the uniformly reinforced material (for the same total amount of fibres) is equal to about 6.3% after 1000 generations. The problem at this number of iteration has not reached the final convergence but some qualitative information can be obtained; the algorithm has been stopped just after 1000 generations to contain computational cost (note that fitness, i.e. the F.E. solution with $\mathbf{C}_{eq}$ assessment, has been evaluated 30.000 times).

The second optimisation problem searches for the better fibre distribution that allows to minimise the maximum principal stress in the structure. The parameters defining the genetic algorithm are kept the same as in the previous case; in Figure 12d the objective function

$F_s(\eta_s(\mathbf{x})) = s_I^{max}(\eta_s(\mathbf{x}))/s_{I,V_f=0}^{max}$ is plotted against the number of generations, while in Figure 12f the corresponding fibre volume fraction distribution pattern is displayed. Now the fibres tend to concentrate in the very near top and bottom part of the central cross section where the maximum tensile and compressive stresses takes place; in this way the matrix' material undergo a lower stress field which is mainly "absorbed" by the reinforcing phase. The improvement with respect to the uniformly reinforced material is equal to about 32% after 1000 generations.

ii) Square Plate with a Central Hole under Tension

The second case considers a thin plane stress square plate with a central hole under uniform tension, characterised by the relative dimensions reported in Figure 13a. The mechanical characteristics of the materials and the parameters of the GA are exactly the same as in the first problem. Due the double axes of symmetry which characterise the problem, only a quarter of the plate is modelled. The two different objective functions considered are, as in the previous case, the minimisation of the displacement which occurs at the central top point of the plate (i.e. minimum compliance) and of the maximum principal stress which occurs near the hole where the stress concentration phenomenon takes place.

In Figure 13b the F.E. mesh (72 four-noded elements are used) adopted is displayed together with a qualitative plot of the deformed shape of the modelled portion of the structure. In Figure 13c the objective function of the first optimisation problem, $F_c(\eta_c(\mathbf{x})) = u_i^{max}(\eta_c(\mathbf{x}))/u_{i,V_f=0}^{max}$, is plotted against the number of generations, while in Figure 13e the corresponding final fibre distribution pattern is displayed. The optimisation of the compliance give rise to a fibres' distribution concentrated mainly in the left side of the weak section of the plate where the hole is located; in this way a sort of uniform horizontal cross section stiffness is obtained along the height of the structure. The improvement with respect to the uniformly reinforced material is equal to about 3.5% after 1000 generations.

In Figure 13d the objective function, $F_s(\eta_s(\mathbf{x})) = s_I^{max}(\eta_s(\mathbf{x}))/s_{I,V_f=0}^{max}$, is plotted against the number of generations for the second optimisation problem while Figure 13f the corresponding obtained fibre distribution pattern is displayed. As can be noted, the fibres tend to concentrate around the stress concentration area, just on the right of the hole, as could be expected. The present optimisation algorithm leads to an improvement, with respect to the uniformly reinforced material, equal to about 60% after 1000 generations. It should be noted that the fibre concentration after the optimisation procedure, reaches very high and unrealistic values for a practical FRC material: nevertheless the problem could be simply constrained by limiting the maximum fibre content per unit volume. In the present simulations such a constraint has not been considered.

A Micromechanical Model and Reinforcing Distribution Optimisation ... 563

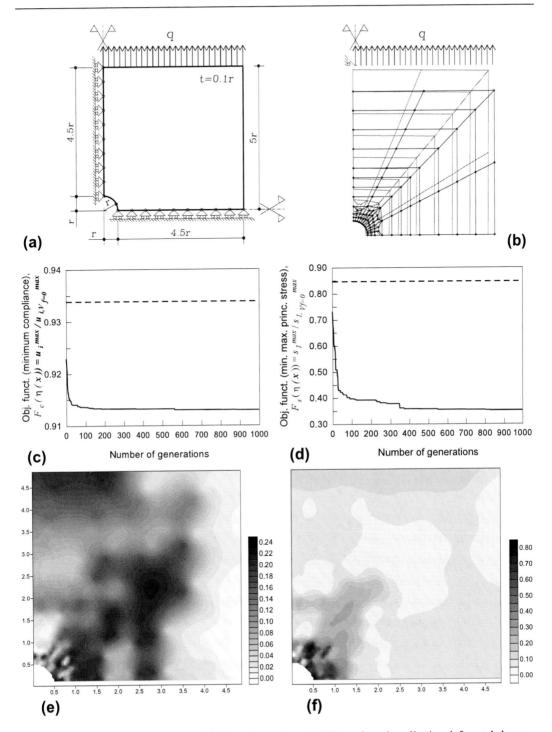

Figure 13. Square plate with a central hole under tension (a); FE mesh and qualitative deformed shape (b). Objective function values against the number of generations to minimise the maximum displacements (c) or the maximum principal stress (d) and the corresponding fibre distribution patterns, respectively (e, f). Values corresponding to the uniformly reinforced material are reported (dashed lines).

iii) Bending of a Notched Cantilever Beam

The last considered example deals with the optimisation of the fibre distribution in a plane stress notched cantilever beam (characterised by the relative dimensions reported in Figure 14a) under a concentrated load at its free end. The objective functions described above are considered. The mechanical characteristics of the materials are exactly the same as in the previously described problems.

The first optimisation problem is the minimisation of the maximum displacement (equivalent to the minimum compliance condition), which occurs at the free end of the beam and of the maximum principal stress which occurs near the notched zone where the stress concentration phenomenon occurs.

In Figure 14b the FE mesh adopted (84 four-noded elements are used) is displayed together with a qualitative plot of the deformed shape of the structure. In Figure 14c the objective function of the first optimisation problem, $F_c(\eta_c(\mathbf{x})) = u_i^{max}(\eta_c(\mathbf{x}))/u_{i,V_f=0}^{max}$, is plotted against the number of generations, while in Figure 14e the corresponding fibre distribution pattern is displayed.

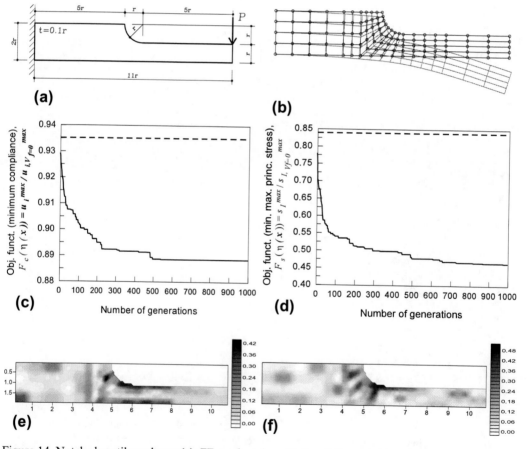

Figure 14. Notched cantilever beam (a); FE mesh and qualitative deformed shape (b). Objective function values against the number of generations to minimise the maximum displacements (c) or the maximum principal stress (d) and the corresponding fibre distribution patterns, respectively (e, f). Values corresponding to the uniformly reinforced material are reported (dashed lines).

Similarly to the case of the wall-beam, the fibres tend to concentrate at the top and at the bottom side of the central part of the beam (where the reduced section is present) in order to increase the moment of inertia of the reduced cross sections which is much more weak than the plain cross section located after the filled zone. The improvement with respect to the uniformly reinforced material is equal to about 5% after 1000 genetic algorithm loops.

In Figure 14d, the objective function $F_s\left(\eta_s(\mathbf{x})\right) = s_I^{max}\left(\eta_s(\mathbf{x})\right)/s_{I,V_f=0}^{max}$ is plotted against the number of generations while the corresponding final fibre distribution pattern for the second optimisation problem is displayed in Figure 14e. In order to reduce the maximum principal stress (which takes place near to the lower zone of the linking zone), the fibres must concentrate in this area in order to help the matrix material in bearing the stress field. The improvement with respect to the uniformly reinforced material, is equal to about 46% after 1000 iterations. Once again the maximum fibre content per unit volume is not constrained to be lower than a given value.

5.5. Discussion and Further Developments

The mechanical model developed for fibre-reinforced materials with reinforcement distribution optimisation, is characterised by a wide and straightforward applicability to the design of fibrous composite structural elements. The reinforcing phase is assumed to be uniformly distributed in the 3-D space but the fibre orientation can be simply considered also to follows preferential directions. A probability distribution function for fibre orientation description can also be considered and introduced to evaluate the integral of Eq. (17):

$$\int_{SolidAngle} P_\varphi(\varphi) \cdot P_\theta(\theta) \cdot \mathbf{Q} \otimes \mathbf{Q} d\phi \, d\theta = \int_0^\pi \int_0^\pi P_\varphi(\varphi) \cdot P_\theta(\theta) \cdot (\mathbf{Q} \otimes \mathbf{Q}) d\varphi \, d\theta \qquad (62)$$

with $\int_0^\pi P_\varphi(\varphi)d\varphi = 1$ and $\int_0^\pi P_\theta(\theta)d\theta = 1$, where $p_\varphi(\varphi), p_\theta(\theta)$ are the probability distribution functions that describe the fibres orientation in the space.

If fibre shapes different form the cylindrical one have to be considered (such as hooked, wrinkled, etc. fibres), this can be done in the present mechanical model by defining an equivalent fibre length and by properly setting the limit shear stress τ_{au} and the friction shear stress τ_{fu} values.

Fibre failure is another crucial aspect which has not been explicitly considered in the present mechanical model. Since fibre breaking occurs by producing a subdivision of the initial fibre into two parts having approximately the same length (in fact the fibre failure can be reasonably assumed to occur in the most stressed section located at the fibre middle point, see Figure 5), the mechanical model allows also to account for such a phenomenon.

From the above consideration, once the actual fibre tensile stress σ_f reaches the fibre tensile strength $f_{t,f}$, i.e.:

$$\sigma_f = E_f \cdot \varepsilon_f = E_f \cdot \varepsilon_f^m \cdot s(\varepsilon_f^m) = E_f \cdot tr\, \boldsymbol{\varepsilon} \cdot s(tr\, \boldsymbol{\varepsilon}) \geq f_{t,f} \qquad (63)$$

the fibre length $2L_f$ must be simply updated to L_f for the F.E. for which such a condition is fulfilled.

In such a way the load bearing capacity of the fibres is lowered due to their reduced length with respect to the initial length. In fact from Eqs (29, 30) it can be observed as the axial force $f(x)$ along the fibre is maximum at the midpoint of the fibre and equal to zero at the extremities while in the central region of the fibre we have a nearly constant value for $f(x)$ as much as the fibre length to cross section size ratio increases. The fibre breaking phenomenon can occur several times as the load level applied to the composite increases.

Strain localisation in the matrix, which develops as shear bands in ductile materials or as cracks in brittle materials, has not been considered. Such a phenomenon requires the introduction of displacements discontinuity in the matrix with implies a different F.E. formulation [43-47]. Nevertheless for fibre-reinforced composite materials characterised a sufficiently high fibre volume fraction, it is reasonable to expect a reduction of the brittle behaviour even for fragile matrix materials; such a situation can frequently prevent the crack formation (at least as a single dominant crack) and avoids the necessity to numerically describe such a strong discontinuity of the displacement field. Nevertheless also if a displacement discontinuity is introduced, the present model is able to model correctly the mechanical fibre behaviour through the computation of the corresponding matrix mean volumetric strain, $tr\, \boldsymbol{\varepsilon}$ [47], which simply takes into account the effects of the displacement jump due to the crack opening. The crack bridging effect due to the cohesive fracture behaviour of the matrix must be also taken into account in such cases [48], while the crack bridging effect due to the fibre is already represented by the present model which considers the stress and strain values in the matrix depurated from the fibre contribution.

CONCLUSIONS

In the last decades the use of composite materials in many engineering fields has known a crescent interest due to their better mechanical characteristics – such as fracture and fatigue resistance, wear, durability performance, and so on – with respect to traditional materials. The use of high performance materials requires to accurately describe their mechanical behaviour in the design process in order to correctly assess the safety level of structural components.

The present chapter has been devoted to the development of a micromechanical model for fibre-reinforced materials. In particular the formulation of a mechanical phases-interaction model has been considered in order to describe the imperfect bonds between the matrix and the fibre (quantified through a scalar sliding parameter evaluated by mean of energetic considerations), and an elastic-plastic constitutive model has been considered for the matrix material.

Being the fibre-reinforced composite materials characterised by several useful parameters – such as the amount, density, orientation and the spatial distribution of the reinforcing fibres inside the matrix – it is reasonable to use such parameters as *"design variables"* in order to optimise the mechanical performance of structural elements made of this class of materials (stiffness, compliance, weight, energy density, peak stress, maximum displacements, etc.) in order to reduce the *cost effectiveness* ratio. Such an optimisation problem related to the spatial density of the fibre distribution in a fibre-reinforced composites structure (FRC) has been considered. By using an evolutionary biological-based approach known as Genetic Algorithm (GA), the optimisation of the fibre volume fraction distribution has been obtained in order to fulfil some extreme conditions related to desired objective functions.

Different optimal conditions are considered with the aim to maximise or minimise a particular mechanical behaviour (such as the compliance, the stiffness, the stress concentration factor, etc.) in order to get the maximum fibre exploitation.

The proposed mechanical model with optimisation, implemented in a 2-D F.E. code, has finally been employed to perform simple numerical simulation in order to assess its reliability in the description of the proper mechanical behaviour of such class of materials and in the optimal fibre content distribution evaluation, assumed as a design spatial variable function in FRC.

ACKNOWLEDGEMENTS

The author gratefully acknowledges the research support for this work provided by the Italian Ministry for University and Technological and Scientific Research (MIUR).

REFERENCES

[1] Hori M; S. Nemat-Nasser. On two micromechanics theories for determining micro-macro relations in heterogeneous solids. *Mech. of Mat.*, 1999, 31, 667–682.

[2] Kalamkarov AL; Liu HQ. A new model for the multiphase fibre-matrix composite materials. *Compos. Part B*, 1998, 29B, 643–653.

[3] Kalamkarov AL; Liu HQ; McDonald DO. Experimental and analytical studies of smart composite reinforcement. *Compos. Part B*, 1998, 29B, 21–30.

[4] Balendran B; Nemat-Nasser S. Bounds on elastic moduli of composites. *J. Mech. Phys. Sol.*, 1995, 43, 1825–1853.

[5] Hassani B; Hinton E. A review of homogenization and topology optimization: I-homogenization theory for media with periodic structure. *Comput. and Struct.*, 1998, 69, 707–717.

[6] Bisegna P; Luciano R. Bounds on the overall properties of composites with debonded frictionless interfaces. *Mech. of Mat.*, 1998, 28, 23–32.

[7] Kalamkarov AL; Kolpakov AG. Design problems for the fibre-reinforced composite materials. *Compos. Part B*, 1996, 27B, 485–492.

[8] Kalamkarov AL. Composite and reinforced elements of construction, Wiley, Chichester, UK, 1992.

[9] Hashin Z. The elastic moduli of heterogeneous materials. *J. Appl. Mech.* 1962, 29, 143.

[10] Hashin Z; Rosen BW. The elastic moduli of fibre-reinforced materials. *J. Appl. Mech.*, 1964, 31, 223–232.

[11] Hill R. Elastic properties of reinforced solids. *J. Mech. Phys. Sol.*, 1963, 11, 357–372.

[12] Hill R. A self-consistent mechanics of composite materials. *J. Mech. Phys. Sol.*, 1965, 13, 213–222.

[13] Budiansky B. On the elastic moduli of some heterogeneous materials, *J. Mech. Phys. Sol.*, 1965,13, 223–227.

[14] Hashin Z; Shtrikman S. A variational approach to the theory of elastic behavior of multiphase materials. *J. Mech. Phys. Sol.* 1963, 11, 127–140.

[15] Eshelby JD. The determination of the elastic field of an ellipsoidal inclusion, and related problems. *Proc. R. Soc. Lond.*, 1957, A241, 376–396.

[16] Russel WB. On the effective moduli of composite materials: effect of fibre length and geometry at dilute concentrations. *Z. Angew. Math. Phys.*, 1973, 24, 581–600.

[17] Kalamkarov AL; Liu HQ. A new model for a multiphase fibre-matrix composite materials. *Compos. Part B*, 1998, 29B, 643–653.

[18] Sanchez-Palencia E. Non-homogeneous media and vibration theory, Springer-Verlag, Berlin, 1980.

[19] Kalamkarov AL; Georgiades AV. Micromechanical modeling of smart composite structures. *Smart Mat. Struct.*, 2002, 11, 423–434.

[20] Kalamkarov AL; Georgiades AV. Asymptotic homogenization models for smart composite plates with rapidly varying thickness: part I-theory. *Int. J. Multiscale Comput. Engng*, 2004, 2, 133–148.

[21] Cioranescu D; Paulin JSJ. Homogenization of reticulated structures, Springer-Verlag, New York, 1999.

[22] Ngollè A; Péra J. Microstructural based modelling of the elastic modulus of fibre-reinforced cement composites. *Adv. Cem. Based Mat.*, 1997, 6, 130–137.

[23] Hodge PG; White GN. A quantitative comparison of flow and deformation theories of plasticity. *J. Appl. Mech.*, 1950, 17, 180–184.

[24] Owen DRJ; Hinton E. Finite elements in plasticity: theory and practice, Pineridge Press Ltd, Swansea, U.K. (1980).

[25] Haftka RT; Grandhi RV. Structural shape optimization - a survey. *Comp. Meth. Appl. Mech. Engng.*, 1986, 57, 91–106.

[26] Goldberg DE. Genetic algorithms in search, optimization, and machine learning. MA: Addison-Wesley Publishing Company inc., 1989.

[27] Davis L. (ed.) Handbook of genetic algorithms. New York: Van Nostrand Reinhold, 1991.

[28] Gen M; Cheng R. Genetic algorithms and engineering design. New York: John Wiley and Sons, 1996.

[29] Pham DT, Karaboga D. Intelligent optimisation techniques. London: Springer-Verlag Ltd, 2000.

[30] Gantovnik VB; Anderson-Cook CM; Gürdal Z, Watson LT. A genetic algorithm with memory for mixed discrete-continuous design optimization. *Comput. and Struct.*, 2003, 81, 2003–2009.

[31] Wieghardt K; Hartmann D; Leimbach KR. Interacting shape optimisation of continuum structures. *Engng Struct.*, 1997, 19, 325–331.

[32] Cappello F, Mancuso A. A genetic algorithm for combined topology and shape optimisations. *Comput.-Aided Des.*, 2003, 35, 761–769.

[33] Liu GR; Chen SC. A novel technique for inverse identification of distributed stiffness factor in structures. *J. Sound and Vibration*, 2002, 254, 823–835.

[34] Zohdi TI. Constrained inverse formulations in random material design. *Comput. Meth. Appl. Mech. Engng.*, 2003, 192, 3179–3194.

[35] Sadagopan D; Pitchumani R. Application of genetic algorithms to optimal tailoring of composite materials. *Compos. Sci. and Tech.*, 1998, 58, 571–589.

[36] Muc A; Gurba W. Genetic algorithm and finite element analysis in optimization of composite structures. *Compos. Struct.*, 2001, 54, 275–281.

[37] Cho JR; Ha DY. Volume fraction optimization for minimizing thermal stress in Ni-Al2O3 functionally graded materials. *Mat. Sci. and Engng A*, 2002, A334, 147–155.

[38] Surendranath H; Bruck HA; Gowrisankaran S. Enhancing the optimization of material distributions in composite structures using gradient architecture. *Int. J. Sol. Struct.*, 2003, 40, 2999–3020.

[39] Eduardo MRF; Marcos MS; Romildo DTF; José LDA; Nelson FFE. Optimisation of mass concrete construction using genetic algorithms. *Comput. and Struct.*, 2004, 82, 281–299.

[40] El Kadi H. Modeling the mechanical behavior of fibre-reinforced polymeric composite materials using artificial neural networks - A review. *Compos. Struct.*, 2006, 73, 1–23.

[41] Freeman J. Simulating neural networks with mathematics. New York: Addison-Wesley, 1994.

[42] Duvaut G; Terrel G; Léné F., Verijenko VE. Optimization of fiber reinforced composites. *Compos. Struct.*, 2000, 48, 83–89.

[43] Belytschko T; Fish J; Engelmann BE. A finite element with embedded localization zones. *Comp. Meth. Appl. Mech. Engng*, 1988, 70, 59–89.

[44] Dvorkin EN; Assanelli AP. 2D finite elements with displacement interpolated embedded localization lines: The analysis of fracture in frictional materials. *Comp. Meth. Appl. Mech. Engng*, 1991, 90, 829–844.

[45] Alfaiate J; Simone A; Sluys LJ. Non-homogeneous displacement jumps in strong embedded discontinuities. *Int. J. Sol. Struct.*, 2003, 40, 5799–5817.

[46] Oliver J; Cervera M; Manzoli O. Strong discontinuities and continuum plasticity models: the strong discontinuity approach. *Int. J. Plasticity*, 1999, 15, 319–351.

[47] Brighenti R. Fracture behaviour of brittle fibre-reinforced solids by a new FE formulation. *Compos. Struct.*, 2009, 91, 324–336.

[48] Hillerborg A; Modéer M; Peterson PE. Analysis of crack formation and crack growth in concrete by means of fracture mechanics and finite elements. *Cem. Concrete Res.*, 1976, 6, 773–782.

In: Fiber–Reinforced Composites
Editor: Qingzheng Cheng

ISBN: 978-1-61470-303-7
© 2012 Nova Science Publishers, Inc.

Chapter 18

APPLICATION OF FRP COMPOSITES FOR REHABILITATION IN CIVIL ENGINEERING: EXPERIMENTS AND ADVANCED COMPUATIONAL METHODS

B. R. Kim[*], S. K. Ha, S. Na and H. K. Lee

Department of Civil and Environmental Engineering,
Korea Advanced Institute of Science and Technology (KAIST)
373-1 Guseong-dong, Yuseong-gu, Daejeon 305-701, South Korea

ABSTRACT

Fiber-reinforced polymer (FRP) composites are advanced composites created by placing extremely strong and stiff fibers in a resin matrix. FRP has rapidly gained recognition as a superior construction material, sparking a revolution in the civil engineering society. Thus far, the majority of FRP applications in the field of civil engineering have been limited to the strengthening and retrofitting of structures (Zhao and Zhang, 2007). FRP composite materials have also been used to construct transportation infrastructures in numerous countries (USA, Germany, England, etc.). FRP composites have increasingly replaced conventional construction materials such as steel and concrete in civil engineering. As a result, technology of FRP composites has continued to advance, allowing for the promotion of knowledge and the creation of new materials (Karbhari and Zhao, 2000; Van Den Einde et al., 2003). In this chapter, the application of FRP composites for rehabiliation in civil engineering will be reviewed.

Keywords: Fiebr-reinforced polymer (FRP) composite, Carbon fiber-reinforced plastic (CFRP), Sprayed fiber-reinforced polymer (Sprayed FRP) coating, Rehabilitation, Civil infrastructures

---

[*] E-mail address: bong-ida@kaist.ac.kr

1. INTRODUCTION

Composites are new systems with enhanced properties gained from the advantages of each of the constituents. Because they offer improved performance over conventional materials, these composites are increasingly being considered as attractive engineering materials in a wide variety of fields (Kaw, 1997; Peters, 1998; Kim and Lee, 2010). In particular, *fiber-reinforced polymer (FRP) composites* are an advanced composite made from very strong, stiff structural fibers and an environmentally resistant polymer resin matrix. The fibers carry the load in predesigned directions while the polymer resin matrix transfers the stress between the fibers and protects them from the environment (Matthews and Rawlings, 1994; Reddy, 1997; Herakovich, 1998; Berthelot, 1999; Karbhari and Zhao, 2000; Bakis et al., 2002; Kim and Lee, 2010). FRP composites have recently been demonstrated as an excellent alternative to conventional materials because of the following competitive advantages: high specific strength/stiffness, excellent fatigue resistance, outstanding corrosion resistance, potentially high overall durability, controllable thermal properties, and tailored properties (Herakovich, 1998; Karbhari and Zhao, 2000; Karbhari and Seible, 2000; Tong et al., 2002; Teng et al., 2004; Shrivastava et al., 2009). In spite of these advantages, FRP composites were initially limited to industries related to aerospace, due to higher upfront material cost and a lack of familiarity with the general use of standards (Karbhari and Zhao, 2000; Bakis et al., 2002). However, after many attempts to lower the cost of FRP composites and expand their application, they have recently become one of the most important advanced materials in many research fields: automotive, military, infrastructure, medical, and electronic applications (Herakovich, 1998; Bakis et al., 2002).

In the field of civil engineering, the application of FRP composite has been evolving very rapidly in recent years (Neale, 2000; Bakis et al., 2002; Karbhari et al., 2003; Sen, 2003; Rizkalla et al., 2003; Cheng and Kabhari, 2006; Pendhari et al., 2008; Hollaway, 2010). One of the principal challenges facing many societies all over the world is the need to renew and upgrade civil infrastructures in order to meet more stringent design requirements (Triantafillou, 1998; Karbhari and Zhao, 2000; Karbhari and Seible, 2000; Lopez-Anido and Naik, 2000; Lee et al., 2008a). Thus, there has been considerable interest in the use of new and emerging materials and technologies which can facilitate the functionality, efficiency, and durability of civil infrastructures (Karbhari and Seible, 2000). In particular, FRP composites have revolutionized the civil engineering society by rapidly gaining recognition as a superior construction material. In civil engineering, FRP composites have increasingly replaced conventional construction materials such as steel and concrete. As a result, the technology of FRP composites has continued to advance, allowing for the promotion of knowledge and the creation of new materials (Karbhari and Zhao, 2000; Van Den Einde et al., 2003).

Up until now, in civil engineering, FRP composites have primarily been used for the rehabilitation of civil infrastructures (Karbhari and Zhao, 2000; Zhao and Zhang, 2007). *Rehabilitation* (including *repair*, *retrofitting*, and *strengthening*) in the civil engineering field means a general operation to meet the required specific performance level of existing civil infrastructures (Karbhari and Zhao, 2000; Van Den Einde et al., 2003; Al-Salloum and Almusallam, 2003). The differentiation of repair, retrofitting, and strengthening is given below.

"Repair is to fix a structural or functional deficiency such as a crack or a severely degraded structural component, retrofitting is specifically used as related to the seismic upgrade of facilities, such as in the case of the use of composite jackets for the confinement of columns, and strengthening is specifically used to those cases wherein the addition or application of the composite would enhance the existing designed performance level, as would be the case in attempting to increase the load rating (or capacity) of a bridge deck through the application of composites to the deck soffit." (Karbhari and Zhao, 2000).

Conventional materials and construction techniques (e.g., externally bonded steel plates, steel or concrete jackets, and external post tensioning) have thus far been used in the rehabilitation of civil infrastructures (Mukherjee and Joshi, 2002; Nezamian and Setunge, 2007). These conventional rehabilitations using various types of steel and concretes have been reliable and quantifiable, but they have many drawbacks including corrosion, deterioration of the bond, and the difficulty of handling heavy materials (Bakis et al., 2002; Lee et al., 2005, 2008a). These deficiencies have led to the development of alternative rehabilitation methods such as FRP composites (Nanni, 1995; Barboni et al., 1997; Midwater, 1997; Hollaway and Mays, 1999; Bakis et al., 2002; Karbhari and Li, 2002; Mufti, 2003; Sen, 2003; Anania et al., 2005; Hollaway and Teng, 2008; Lee et al., 2008a).

In this chapter, the application of FRP composites for rehabilitation in civil engineering will be reviewed. Specifically, this chapter focuses on research on FRP composites for rehabilitation in civil engineering society. This chapter is organized as follows. The general introduction of FRP composites including their constituents (fibers and polymer matrix) and manufacturing method is summarized in Section 2. In Section 3, a variety of research on FRP composites for rehabilitation are introduced. In particular, research on the application of FRP composite systems for flexural, shear, and seismic strengthening of reinforced-concrete (RC) structures are reviewed in this section. Finally, Section 4 introduces the advanced computational methods for FRP composite systems, focusing primarily on micromechanical analysis. A series of studies on the characteristic behavior of FRP composite systems using micromechanical analysis are reviewed.

2. FIBER-REINFORCED POLYMER COMPOSITES

2.1. Fibers and Polymer Matrix Materials

Fiber-reinforced polymer (FRP) composites are the most common advanced composites. FRP composites consist of a polymer matrix reinforced with high performance fibers. Figure 1 shows the classification of fibers and the polymer matrix in FRP composites. Fibers used in FRP composites can be largely classified into two groups: synthetic (man-made) fibers and natural fibers. *Synthetic fibers* are most widely used today, due to their low costs. They include glass, carbon, aramid, etc. (Lubin, 1969; Parratt, 1972; Mallick, 1993; Herakovich, 1998; Mazumdar, 2002). Glass fiber (or fiberglass) is a material with extremely fine fibers of glass which are usually 524μm in diameter (Mazumdar, 2002). There are several types of glass fibers with different mechanical and chemical properties which can be used for a variety of purposes. E-glass (or electric glass) is the most common type of glass fiber; it offers good tensile strength, compressive strength, stiffness and electrical properties at a relatively low cost (Lubin, 1969; Bakis et al., 2002). However, its impact resistance is relatively poor. C-glass (or chemical glass) is the most resistant to chemical attack. C-glass fibers are mainly used as an outer layer surface tissue or laminate for chemical pipes and water tanks (Lubin, 1969; Jang, 1994). S-glass has a higher elastic modulus and more strength than E-glass. These properties are achieved by a smaller filament diameter, which results in higher interlaminar shear strength and wet-out properties (Lubin, 1969). Above these, there are also A-, D-, L-, and M-glass fibers (Lubin, 1969). Carbon fiber (also known as graphite fiber or carbon graphite) has the highest specific stiffness of any commercially available fiber, with a filament diameter of approximately 5~7 μm for most types (Shalin, 1995). Even though it has lower impact strength than either aramid or glass, carbon fiber has very high compression and tensile strength, as well as high resistance to corrosion, creep, and fatigue (Galhano et al., 2005). Depending on the modulus band, carbon fibers are usually grouped by their various properties (Meier, 1992). These bands are referred to as: high strength (HS, less than 265 GPa), intermediate modulus (IM, 265~320GPa), high modulus (HM, 320~440GPa), and ultra high modulus (UHM, roughly 440 GPa) (Meier, 1992). Aramid fiber is an aromatic polyamide which is a class of man-made heat-resistant fibers (Shalin, 1995). Aramid fiber which is also called "Kevlar," has very high specific strength and good impact resistance, making it ideal for aerospace and military applications (Czarnecki et al., 1980). However, aramid's compressive strength is similar to E-glass and it may degrade slowly under ultraviolet light. It is also significantly more expensive than glass fibers (Zahr and Riewald, 1989). Along with glass fiber, carbon fiber, and aramid fiber, other synthetic fibers include metal fiber, silicon carbide fiber, alumina fiber, boron fiber, polyethylene fiber, quartz fiber, and silica fiber (Lubin, 1969; Parratt, 1972; Mallick, 1993; Herakovich, 1998; Mazumdar, 2002). All of these fibers are mainly used as reinforcements for FRP composites. Detailed explanation of each fiber can be found in the relevant literature.

Unlike synthetic fibers, *natural fibers* are made from natural ingredients, such as an animal, vegetable, and mineral source (Saheb and Jog, 1999; Möller and Popescu, 2009). Natural fibers can be grouped as seed-hair fibers (e.g., coir, cotton, kapok, and milk weed floss), bast fibers (e.g., jute, flax, hemp, kenaf, ramie, roselle, and urena), and leaf fibers (e.g., banana, sisal, abaca, henequen, pineapple, cantala, mauritius, and phormium) (Rowell, 1995; Saheb and Jog, 1999). Recently, natural fibers have attracted more attention from scientists and engineers because they offer advantages such as low-cost, low density, and high specific properties (Gram, 1983; Canovas et al., 1992; Berhane, 1994; Gauthier et al., 1998; Gassan and Bledzki, 1999; Luo and Netravali, 1999; Saheb and Jog, 1999; Tripathy et al., 1999; Toledo et al., 2000; Wambua et al., 2003; Ramakrishna and Sundararajan, 2005; Silva, 2008). For a variety of natural fibers, details can be found in the relevant literature.

The polymer matrix resins used in FRP composites can be classified into two classes: thermoset polymer and thermoplastic polymer (Herakovich, 1998). *Thermoset polymer resins* are generally liquid at room temperature, but they harden by curing (heating). The curing process is permanent and irreversible (Tung and Dynes, 1982). Applying heat to a cured thermoset polymer resin will not return it back to the liquid state. These types of resins are high in durability, wear and chemical resistance, even under extreme environments (Kamal, 1973). Thermoset polymer resins are usually cheap and easily processed. In addition, they are usually stronger and better suited for higher temperature applications, as compared to thermoplastics (Astrom, 1997). The most typical thermoset polymer resins are epoxy and polyester. Polyester has been used (in particular, with glass fibers) for the longest period in the widest range of structures because it is inexpensive, lightweight, somewhat resistant to environmental exposure, and has a useful temperature range up to 100°C (Herakovich, 1998). Epoxy has also been widely used as a matrix for advanced composites. Its advantages include resistance to moisture, creep, and fatigue, as well as excellent adhesive properties and low shrinkage during cure (Herakovich, 1998). The other thermoset polymer resins include vinylester, phenolics, cyanate esters, bismaleimides, polyimides, etc.

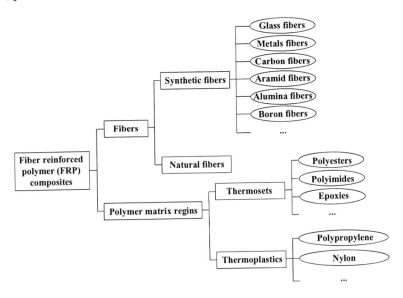

Figure 1. Classification of fibers and polymer matrix in FRP composites.

Thermoplastic polymer resins are usually solid at room temperature, but soften at high temperatures. This allows the setting process to be reversible (Johnson, 1983). The viscosity for this type of resin is generally much higher compared to thermoset polymer resins, which means that it is difficult to make them flow through the fibers (Gibson, 1992). However, they offer several advantages such as greater fracture toughness, longer raw material shelf life, recycling possibilities, and a safer working environment due to the absence of organic solvents during the hardening process exist (Rijswijk and Bersee, 2007). The thermoplastic polymer resins used for composites include polypropylene, polyvinyl chloride, nylon, polyurethane, poly-ether-ether-ketone (PEEK), polyphenylene sulfide (PPS), poly-ether-imide (PEI), and polysulfone (Johnson, 1983; Gibson, 1992; Rijswijk and Bersee, 2007). Details of the advantages, drawbacks, and applications of a variety of polymer matrix resins can be found in the relevant literature.

2.2. Manufacturing Methods for FRP Composites

The manufacturing methods for FRP composites can be classified as open mold process (e.g., spray lay-up, hand (or wet) lay-up, filament winding, sheets molding compound, and expansion tool molding), closed mold process (e.g., compression molding, vacuum bag molding, pressure bag molding, autoclave molding, resin transfer molding, and reaction injection molding), and continuous process (e.g., pultrusion and braiding) (Mallick, 1993;

Rudd et al., 1997; Herakovich, 1998; Rosato et al., 2001; Mazumdar, 2002; Advani and Sozer, 2003; Biron, 2004; Campbell, 2004). In these manufacturing methods, liquid composite molding (LCM) technologies such as resin transfer molding and reaction injection molding have been widely used in polymer composite manufacturing (Mohan et al., 1998; Shojaei et al., 2003). The main idea of LCM is to rapidly inject either thermosetting or thermoplastic resin into reinforcement with minimum voids and cycle time (Boyajian, 1993; Rudd et al., 1997; Biron, 2004; Campbell, 2004). In general, according to the resin systems, LCM can be divided into two categories: resin transfer molding (RTM) with a thermal activated resin system, and structural reaction injection molding (SRIM) with a mixing activated resin system (Frederick and Phelan, 1997; Karlsson and Åströn, 1997; Ngo and Tamma, 2001). *Resin transfer molding (RTM)* is a closed mold low-pressure process which can fabricate composites ranging in size from small to large and in complexity from simple, low-performance to complex, high-performance parts (Johnson, 1993; Potter, 1997; Mohan et al., 1998; Shojaei et al., 2003; Harper, 2009). Since the RTM process is quite popular for manufacturing complex shaped components which require good surface finishing, a considerable amount of past and present research has been conducted on this process (Kendall et al., 1992). Several alternatives to the RTM process include injection compression molding (I/CM), thermal expansion resin transfer molding (TERTM), and vacuum-assisted resin transfer molding (VARTM). Technically, the process of *structural reaction injection molding (SRIM)* is similar to RTM, with the exception of the resin systems (Ngo and Tamma, 2001). A detailed comparison (including resin formulation, resin reactivity, mould type, mould material, impregnation, cycle time, and part size) between RTM and SRIM processes can be found in Karlsson and Åströn (1997). Similar injection molding processes include reaction injection molding (RIM), reinforced reaction injection molding (RRIM), vacuum-assisted resin injection molding (VARI) and ultimately reinforced thermoset resin injection molding (URTRI) (Boyajian, 1993). In order to manufacture composite materias of high quality, it is essential to have expertise in cure kinetics, mold design, and resin movement through the perform (Long et al., 1998). Details of the manufacturing methods for FRP composites can be found in the relevant literature.

3. APPLICATION OF FRP COMPOSITE SYSTEMS FOR REHABILITATION

3.1. Overview

FRP composites have been extremely attractive in the civil engineering field as materials for repair and strengthening materials (Shahawy et al., 1996; Toutanji et al., 2006; Karbhari, 2007). Among FRP composites, *carbon fiber-reinforced plastic (CFRP) composites* such as sheets and plates are ideal materials which are very strong, lightweight, and resistant to chemicals (Shahawy et al., 1996; Li et al., 2006; Karbhari, 2007). In the civil engineering field, CFRP composites can be used to effectively enhance the flexural, shear, and seismic performances of RC structures in terms of structural load-carrying capacity, stiffness, and ductility under static, cyclic, and fatigue loading conditions, while also improving their environmental durability (Li et al., 2006). However, in order to effectively enhance the load-carrying capacity of the strengthened RC structure with CFRP composites, the interfacial stresses between the CFRP composites and concrete must be transferred (Teng et al., 2002). Typical methods using CFRP sheets and plates are sensitive to unevenness on the surface of the RC beam soffit, and such unevenness can lead to the debonding of composites in an interfacial zone (Karbhari, 2007). In addition, there are a lot of significant physical differences in an interfacial zone between CFRP composites and concrete (Boyd, 2000). Moreover, it has been noted that CFRP composites, unlike reinforcing steel, are anisotropic, having different properties in the longitudinal and transverse directions of the fibers. This anisotropy is also apparent in relation to the different coefficients of thermal expansion in the different directions of fibers (Karbhari, 2007).

In order to overcome some of the problems (e.g., bond ability) between the adhesive and FRP composites, as well as the anisotropic properties of CFRP composites, a newly developed method for using a *sprayed fiber-reinforced polymer (sprayed FRP) coating* to repair and strengthen deteriorated concrete was introduced at the University of British Columbia (Banthia et al., 1996; Banthia and Boyd, 2000; Boyd, 2000). This method makes it relatively easy to achieve good bond ability between the adhesive and FRP composites, as the adhesive and the matrix resin consist of a single entity and thus only require one bond (Boyd, 2000). Using sprayed FRP coating with discontinuous fibers can solve the problem of anisotropy, since the fibers in the sprayed FRP coating will be dispersed randomly throughout a matrix (Caibal et al., 2003). In this method, the base resin and activator resins (e.g., epoxy, polyester, and vinylester resins) are transferred to a spray gun, which then sprays the resin system stream as a single stream (Banthia et al., 1996; Banthia, 2002; Lee and Hausmann, 2004; Lee, 2004; Lee et al., 2008a). Simultaneously, the fibers (e.g., glass, carbon, and aramid) are fed into a chopper unit which is fixed on the spray gun, such that one or two strands of fibers at a time are pulled between a rubber roller and a steel roller, with several rolling blades placed at equal distances (Banthia et al., 1996; Banthia, 2002; Lee and Hausmann, 2004; Lee, 2004; Lee et al., 2008a). The two streams of chopped fibers and resin system are mixed in the air and applied onto the surface to concrete (Banthia et al., 1996; Banthia, 2002; Lee and Hausmann, 2004; Lee, 2004; Lee et al., 2008a). Details of this application of sprayed FRP coating can be found in Boyd (2000), Young and Harries (2004), Ross et al. (2004), Lee and Hausmann (2004), Lee (2004), Boyd et al. (2008), and Lee et al. (2008a).

In this section, a variety of studies on FRP composites which focus on CFRP composites (in sheet and plate shape) and sprayed FRP coating for the strengthening of RC structures are introduced. Details of the strengthening of RC structures bonded with CFRP composites (e.g., sheets or plates) and sprayed FRP coating on flexural, shear, and seismic performances are reviewed.

3.2. Strengthening of RC Structures Using FRP Composites in Civil Engineering

3.2.1. Flexural Strengthening of RC Structures Using FRP Composites

Traditional methods including external post-tensioning and the bonding of steel plates have commonly been used for the flexural strengthening of RC structures in civil engineering

(Meier, 1995; Takeda et al., 1996; Arduini and Nanni, 1997; Garden and Hollaway, 1998; Grace et al., 1999; Nguyen et al., 2001; Teng et al., 2002; Ashour et al., 2004). However, in recent years, the bonding of CFRP sheets or plates has become an attractive option as a flexural strengthening material for RC structures (Meier, 1995; Takeda et al., 1996; Arduini and Nanni, 1997; Garden and Hollaway, 1998; Grace et al., 1999; Nguyen et al., 2001; Teng et al., 2002: Rabinovitch and Frostig, 2003; Ashour et al., 2004). The application of CFRP composites for the flexural strengthening of RC structures was first introduced in the mid-1980s at the Swiss Federal Laboratory for Materials Testing and Research (EMPA) (Meier et al., 1993). In recent years, as CFRP composites have been used more extensively in civil engineering for strengthening RC structures, the number of worldwide projects using FRP composites to improve structural strength has exploded (Teng et al., 2002; Ashour et al., 2004; Toutanji et al., 2006). As a result, the last decade has seen a vast quantity of research on flexural strengthening using CFRP composites (e.g., Ritchie et al., 1991; Triantafillou and Plevris, 1992; Meier, 1995; Shahawy et al., 1996; Takeda et al., 1996; Arduini and Nanni, 1997; Garden and Hollaway, 1998; GangaRao and Vijay, 1998; Grace et al., 1999; Nguyen et al., 2001; Rabinovitch and Frostig, 2003; Ashour et al., 2004; Barros and Fortes, 2005; Toutanji et al., 2006). This section is concerned with the flexure performances of RC structures which have been strengthened with CFRP composites (e.g., CFRP sheets or plates) and the wet lay-up method using unstressed FRP plates on the soffit of RC structures.

A series of experimental tests were carried out to investigate the effectiveness of external strengthening using FRP plates made with glass, carbon, and aramid fibers. In the test conducted by Ritchie et al. (1991), the stiffness increased from 17 to 99 % and the ultimate strength increased from 40 to 97% with RC beams strengthened with FRP plates. Shahawy et al. (1996) observed and comprehensively tested the flexural behavior of RC rectangular beams strengthened with CFRP laminates. In this experimental test, the first crack load, cracking behavior, deflections, serviceability loads, ultimate strength, and failure modes were observed (Shahawy et al., 1996). The results generally indicated that the flexural strength of beams strengthened with CFRP laminate was significantly increased (Shahawy et al., 1996). Arduini and Nanni (1997) investigated how the strengthening performance of precracked and un-precracked RC beams strengthened with CFRP sheets was affected by several variables, including CFRP material systems, concrete surface preparations, RC cross sections, and the number and location of CFRP plies. The results showed that externally bonded CFRP sheets resulted in significant improvements in ultimate load capacity and flexural stiffness, though major attention to the extent length of the CFRP sheets was necessary to avoid extent peeling (Arduini and Nanni, 1997). Twenty-four reinforced concrete beams strengthened with carbon fiber wraps were experimentally tested by GangaRao and Vijay (1998) under bending loading. Four different carbon wrap configurations were applied to RC beams, and an evaluation and comparison of the experimental data was conducted (GangaRao and Vijay, 1998). The static responses of all the concrete beams were evaluated in terms of strength, stiffness, compatibility between wrap and concrete, and other key variables governing the percentage increase in the ultimate strength (GangaRao and Vijay, 1998). The behaviors of RC beams strengthened with various types of FRP laminates were reported by Grace et al. (1999). Five different strengthening systems using CFRP/glass fiber-reinforced plastic (GFRP) composites were applied for the tests (Grace et al., 1999). The experimental data showed that FRP vertical layers forming a U-shape around the cross section of a beam could significantly reduce deflections and the increase load-carrying capacity of the strengthened

RC beams (Grace et al., 1999). Existing reinforced concrete structures using externally bonded composite materials were experimentally investigated by Rabinovitch and Frostig (2003). In particular, Rabinovitch and Frostig (2003) focused on the stress concentration near the edge of the carbon/epoxy FRP strips, the failure modes triggered by the edge effects, and the method for preventing stress concentration. The results showed that a significant enhancement in the serviceability and strength of the RC beams was observed and the various edge designs efficiently controlled the brittle failure modes of the strengthened RC beams (Rabinovitch and Frostig, 2003). Ashour et al. (2004) performed a series of tests on sixteen RC beams strengthened with different arrangements of internal steel bars and external CFRP laminates. Key variables included the length, thickness, position and form of the CFRP laminates (Ashour et al., 2004). Three failure modes of beams with external CFRP laminates were observed from this experimental test (Ashour et al., 2004). The results showed that the ductility of all of the strengthened RC beams was reduced compared with the un-strenghtened control beam (Ashour et al., 2004). A total of eight RC beams, including one control beam and seven beams strengthened with CFRP sheets using an inorganic epoxy, were tested and evaluated by Toutanji et al. (2006). The results of a four-point bending test showed that the load-carrying capacity of the beams increased with the number of layers of carbon fiber sheets, while the ductility of the strengthened beams was greatly reduced compared to the control beam (Toutanji et al., 2006). In addition, a moment deflection model is developed and presented by Toutanji et al. (2006), and good agreement between experimental and predicted results was reported.

More recently, several studies on flexural strengthening of RC structures using sprayed FRP coating have been carried out (e.g. Boyd, 2000; Young and Harries, 2000; Ross et al., 2004; Lee and Hausmann, 2004; Lee, 2004; Boyd et al., 2008; Lee et al., 2008a). In order to evaluate how the application of sprayed FRP coating on concrete specimens affected the flextural behavior of strengthened beams, Young and Harries (2000) conducted a series of experimental tests. They reported that key parameters including fiber loading and coating thickness affected the flexural performance of RC beams strengthened with sprayed FRP coating, and they identified potential real-world applications for this method (Young and Harries, 2000). Young and Harries (2000) also provided a brief background of the sprayed FRP coating, and reported that the sprayed FRP coating was efficient to enhance the strength, stiffness, and toughness of retrofitted concrete specimens. In order to evaluate the influence of sprayed FRP coating parameters on the performance of RC beams, aseries of flexural bending tests was carried out by Lee and Hausmann (2004) on both damaged (precracked) and undamaged RC beams strengthened with the sprayed FRP coating. The tests were also intended to assess the feasibility of sprayed FRP coating for repair and strengthening materials (Lee and Hausmann, 2004).

Figures 2 and 3 show a skilled person applying the sprayed FRP coating to concrete specimens and the set-up for three-point bending tests for measuring the applied load and the deflection of the coated beams, respectively. The key variables in the experimental tests were coating thickness, fiber length, fiber types, and fiber loading in volume (Lee and Hausmann, 2004). The experimental results showed that the sprayed FRP coating can substantially increase the load capacity, ductility, and energy-absorbing capacity strength and that this method is effective for strengthening and repairing damaged RC beams (Lee and Hausmann, 2004).

Reprinted from Composite Structures, Vol. 63, H.K. Lee and L.R. Hausmann, Structural repair and strengthening of damaged RC beams with sprayed FRP, pp. 201209, Copyright (2004), with permission from Elsevier.

Figure 2. A skilled person applies the sprayed FRP coating to concrete specimens (Lee and Hausmann, 2004).

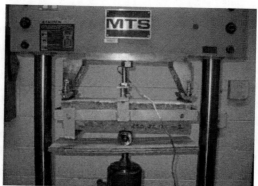

Reprinted from *Composite Structures*, Vol. 63, H.K. Lee and L.R. Hausmann, Structural repair and strengthening of damaged RC beams with sprayed FRP, pp. 201209, Copyright (2004), with permission from Elsevier.

Figure 3. Set-up for three-point bending test to measure the applied load and the deflection of RC beams with the sprayed FRP coating (Lee and Hausmann, 2004).

An experimental study carried out by Lee (2004) reported on the effectiveness of anchorage in concrete beams strengthened with the sprayed FRP coating. The anchorage is created by embedding the ends of the sprayed FRP coating into slits which have been cut into the concrete, enabling the coating layer and anchors to act as a single body (Lee, 2004). A series of three-point bending tests were carried out on the notched concrete specimens strengthened with sprayed FRP coating, as shown in Figure 4. The flexural performances of control specimens were compared with those of specimens strengthened with the sprayed FRP coating, both with and without anchorage (Lee, 2004). The experimental results showed that the bond ability between the coating layer and the applied face was sufficient enough to allow the loads to be transferred from the specimen to the strengthening system (Lee, 2004). Another series of experimental flexural tests were carried out by Boyd et al. (2008) on three AASHTO Type II girders, in order to assess the feasibility of the sprayed FRP coating as a strengthening method for impact damaged girders. For the experimental study, an undamaged girder was tested as a control specimen, while two other specimens were tested for flexure after being subjected to simulated impact damage at mid-span (Boyd et al., 2008). The study showed that the specimens strengthened with sprayed FRP coating reached the target rehabilitation goal of 95% of the original undamaged girder strength (Boyd et al., 2008).

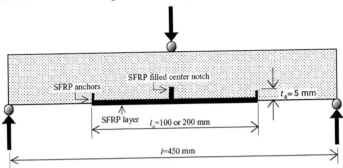

The final, definitive version of this paper has been published in <Journal of Reinforced Plastics and Composites>, Vol. 23/Issue 12, August/2004 by <<SAGE Publications Ltd.>>/<<SAGE Publications, Inc.>>, All rights reserved. © [as appropriate].

Figure 4. A schematic of three-point bending tests on notched specimens with sprayed FRP coating anchors (Lee, 2004).

3.2.2. Shear Strengthening of RC Structures Using FRP Composites

CFRP composites such as CFRP sheets and plates have recently attracted attention for use in shear strengthening of RC structures (e.g., Arduini et al., 1994; Araki et al., 1997; Sato et al., 1997a, 1997b; Mitsui et al., 1998; Alexander, 1996; Fanning and Kelly, 1999; Khalifa and Nanni, 2000; Kachlakev and McCurry, 2000; Zhang et al., 2004; Islam et al., 2005; Anil, 2006; Bencardino et al., 2007; Jayaprakash et al., 2008; Lee et al., 2011). Teng et al. (2002) summarized the typical FRP shear-strengthening schemes, including bonding FRP only to the sides of an RC structure, bonding FRP U jackets to both the sides and the tension face, and wrapping FRP around the whole cross-section of a RC structure.

Mitsui et al. (1998) conducted experimentals on the shear strengthening capabilities of RC beams strengthened by having CFRP sheets externally bonded to both sides of the beam. Six RC beams, including specimens that were crack-damaged by pre-loading and then repaired by injecting epoxy resin into the cracks, were tested under an asymmetric bending load (Mitsui et al., 1998). The results showed that the ultimate shear strength of the strengthened beam was higher than that of the control beam, and that the shear-strengthening effect of CFRP sheets can be accurately predicted by the truss model analysis (Mitsui et al., 1998). A series of experimental tests carried out by Khalifa and Nanni (2000) evaluated the shear performance of T-section RC beams strengthened with CFRP sheets. Key variables used in this study were strengthening schemes, amount of CFRP sheets, 90°/0° ply combination, and end anchor (Khalifa and Nanni, 2000). The tests showed that externally bonded CFRP sheets can significantly enhance the shear strength of the beam (Khalifa and Nanni, 2000). Design approaches using both the ACI code (ACI code, 1995) and Euro code (Eurocode, 1992) formats were carried out to predict the shear capacity of the strengthened T-section RC beams (Khalifa and Nanni, 2000). A series of four-point bending tests were conducted by Kachlakev and McCurry (2000) to assess the feasibility of using FRP laminates such as CFRP and GFRP sheets as structural strengthening materials. The results showed that the RC beams strengthened with both the GFRP and CFRP sheets should well exceed the static demand moment capacities (Kachlakev and McCurry, 2000). Anil (2006) carried out an experimental program on strengthening shear-deficient RC beams by an external bonding of CFRP straps. For this study, a total of six RC beams with a T-section were tested under cyclic loading, and key variables included width of CFRP straps, arrangements of straps, and anchorage technique (Anil, 2006). In Anil (2006)'s tests, all of the CFRP arrangements significantly improved the strength and stiffness of the specimens. In addition, Anil (2006) compared his experimental results with analytical approaches by using the ACI-440 Committee report (ACI Committee 440, 1996). Experimental tests on four RC beams designed both with and without an external anchorage system were carried out by Bencardino et al. (2007) in order to clarify the structural performance of shear-deficient RC beams strengthened with CFRP laminates. In this study, the strains, cracking, load capacity, and failure modes were measured during four-point bending tests to systematically analyze deformability, strength, and failure processes (Bencardino et al., 2007). The results showed that a well-defined designed anchorage system can allow the brittle shear failure of a strengthened specimen to be transferred to the ductile failure (Bencardino et al., 2007). Jayaprakash et al. (2008) performed a series of experimental tests on the shear-strengthening capacity and modes of failure of RC rectangular beams strengthened with bi-directional CFRP composites. Key variables used in the study included the longitudinal tensile reinforcement ratio, shear span to effective depth ratio (a/d), spacing of CFRP strips, and amount and orientation of CFRP strips (Jayaprakash et al., 2008). Jayaprakash et al. (2008) found that CFRP strips can significantly improve the shear capacity of RC shear specimens, and they compared their experimental results with theoretical results.

There have thus far been only limited studies to investigate the shear behavior and performance of deep beams shear strengthened with FRP composites (Zhang et al., 2004; Islam et al., 2005). Zhang et al. (2004) carried out a series of tests to assess the feasibility of using externally bonded CFRP composites to improve the shear capacity of deep beams. In addition, Zhang et al. (2004) developed new design equations for shear strengthening the deep beams with CFRP composites. Islam et al. (2005) performed a series of tests to improve the strengthening performance of structurally deficient deep beams using an externally bonded FRP system. Their results showed that the FRP system dramatically slowed the growth of critical diagonal cracks and improved the load-carrying capacity of the deep beam (Islam et al., 2005). Recently, another series of experimental tests were carried out by Lee et al. (2011) to investigate the behavior and performance of reinforced concrete T-section deep beams strengthened in shear with CFRP sheets, as shown in Figure 5. A total of fourteen RC T-section deep beams were designed to be deficient in shear with a shear span-to-effective depth ratio (a/d) of 1.22 (Lee et al., 2011). During the experimental tests, failure modes were observed as shown in Figure 6. Lee et al. (2011) used the load-deflection curves to address the effects of key variables on the shear performance of the strengthened deep beams. In addition, a series of comparative studies between the experimental data and theoretical results were carried out to evaluate the shear strength of deep beams strengthened with CFRP sheets (Lee et al., 2011).

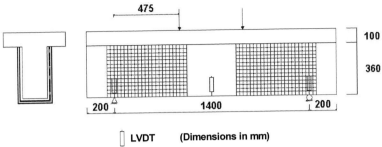

Reprinted from Composite Structures, Vol. 93, Lee, H.K., Cheong, S.H., Ha, S.K. and Lee, C.G., Shear behavior and performance of RC T-section deep beams externally strengthened with CFRP sheets, pp. 911922, Copyright (2011), with permission from Elsevier.

Figure 5. Scheme of test set-up and strengthening methods (Lee et al., 2011).

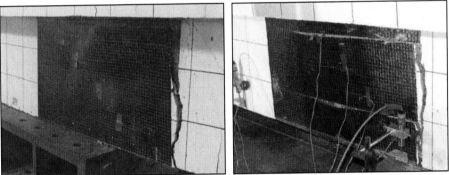

Reprinted from Composite Structures, Vol. 93, Lee, H.K., Cheong, S.H., Ha, S.K. and Lee, C.G., Shear behavior and performance of RC T-section deep beams externally strengthened with CFRP sheets, pp. 911922, Copyright (2011), with permission from Elsevier.

Figure 6. Partial delamination failure in a triangular shape and rupture failure of CFRP sheets (Lee et al., 2011).

Recently, there have been a few studies on the feasibility of using the sprayed FRP coating to strengthen the shear of RC structures (e.g., Banthia, 2002; Soleimani et al., 2007; Lee et al., 2007a). At the University of British Columbia, Banthia (2002) conducted a series of tests to assess the new repair method using the sprayed FRP coating. The sprayed FRP coating was applied to eleven girders on a single-span bridge near the town of Duncan on Vancouver Island (Banthia, 2002). Three channels—a control specimen strengthened with the sprayed FRP coating, and a specimen strengthened with a continuous wrap system—were subjected to a three-point loading test using four large hydraulic jacks as load actuators (Banthia, 2002). The results showed that both strengthening methods enhanced the stiffness and the load-carrying capacity, thereby asserting the effectiveness of the sprayed FRP coating (Banthia, 2002). The use of the sprayed GFRP as a potential technique for enhancing the impact resistance of RC beams was investigated by Soleimani et al. (2007). RC beams with a small number of stirrups as shear reinforcement were strengthened with different configurations and thicknesses of the sprayed GFRP on two or three sides of the RC beams (Soleimani et al., 2007). Under both static and impact loading, the RC beams strengthened with the sprayed GFRP coating were observed to have a higher load-carrying capacity and much greater energy absorption capacity compared to control beams under both static and impact loading (Soleimani et al., 2007). Lee et al. (2007a) investigated the flexural and shear strengthening performance of damaged RC structures strengthened with sprayed FRP coating. Lee et al. (2007a) found that both the flexural and shear strengthening performances of RC beams bonded with the sprayed FRP coating were reasonably similar to those of the FRP sheets, verifying that the sprayed FRP coating could perform as a strengthening scheme for existing RC structures.

3.2.3. Seismic Strengthening of RC Structures Using FRP Composite

Many civil infrastructures have been severely damaged in recent earthquakes (e.g., the Whittier Narrouws (1987), Loma Prieta (1989), and Northridge (1994) earthquakes in the USA, and the Kobe (1995) earthquake in Japan), highlighting the need for retrofitting and strengthening of seismically insufficient structures (Teng et al., 2002). In the mid-1980s, Katsumata et al. (1987, 1988) first proposed the use of FRP composite materials for the retrofitting of existing RC columns to improve seismic resistance. Matsuda et al. (1990) identified two retrofit methods for RC bridge columns using FRP composites: strength-oriented retrofit and ductility-oriented retrofit. In Matsuda et al. (1990)'s study, the FRP plates were used longitudinally to increase the flexural strength of the RC columns, while the FPR composites were bonded like hoops around the RC columns to enhance their ductility, as suggested by Katsumata et al. (1987, 1988). Practical applications of the CFRP sheets and plates for seismic retrofitting and strengthening of the existing RC columns have attracted a lot of attention in recent years (e.g., Katsumata et al., 1987, 1988; Ballinger et al., 1993; Seible et al., 1997; Kobatake, 1998; Ogata and Osada, 2000; Ye et al., 2003; Balsamo et al., 2005; Yalcin et al., 2008; Promis et al., 2009; Ozcan et al., 2008; Li et al., 2009; Ozcan et al., 2010).

In response to the Kobe earthquake, Ogata and Osada (2000) focused on the seismic retrofit of RC piers in their study. They conducted experimental tests using CFRP sheets to strengthen high, hollow circular RC piers with cut-off from longitudinal reinforcement and variable wall thickness along the height (Ogata and Osada, 2000). Ogata and Osada (2000) found that the seismic performance of the retrofitted piers was verified, and they discussed a non-linear earthquake response analysis through a multi-degree-of-freedom system. A new seismic retrofit system for RC columns was developed by Seible et al. (1997). The jacketing system for RC columns consisted of continuous carbon fiber prepreg tows wound onto existing circular or RC columns, with variable jacket thickness along the column height (Seible et al., 1997). Seible et al. (1997) carried out large-scale bridge column model tests using the carbon jacket designs and found that steel-shell jacketing was effective in providing the desired inelastic design deformation capacity levels. Ye et al. (2003) conducted experimental tests under constant axial load and lateral cyclic load to evaluate the seismic performance of RC columns strengthened with CFRP sheets. In this study, eight specimens with varying damage levels were used to investigate the ductility enhancement provided by the CFRP sheets (Ye et al., 2003). Ye et al. (2003) examined the strain development and distribution in the CFRP, and concluded that a key factor of CFRP sheets. They also introduced an equivalent transversal reinforcement index (Ye et al., 2003). A series of experimental works were conducted by Yalcin et al. (2008) on the retrofitting of RC columns using CFRP sheets to enhance structural strength and ductility capacities. In order to reflect outdated construction design, the experiment's key variables were plain rebars, inadequate splicing lengths, and insufficient stirrup spacing (Yalcin et al., 2008). The results showed that the retrofitting method using CFRP sheets was effective for RC columns with continuous rebars (Yalcin et al., 2008). In addition, the strength, ductility, and energy dissipation capacity of the columns retrofitted with CFRP sheets were significantly improved in the columns which were wrapped with CFRP sheets after welding dowel rebars with longitudinal rebars near the splicing region.

Promis et al. (2009) performed experimental tests to evaluate how CFRP composites might affect the mechanical and energetic performance of RC short columns under combined compressive and flexural loading. In order to ensure the shear failure of the RC specimens, eight short columns with insufficient transverse reinforcements were tested. Seven of the specimens were strengthened either continuously or discontinuously with CFRP/GFRP composites (Promis et al., 2009).

Promis et al. (2009) calculated damage indices to investigate the influence of the nature and geometry of the composite reinforcement. The flexural behavior and performance of strengthened rectangular RC columns with CFRP composites and insufficient confining steel were studied by Ozcan et al. (2010). In this study, five specimens of typical deficient building columns with low concrete strength were tested under cyclic displacement excursions with a constant axial load (Ozcan et al., 2010).

The effect of CFRP anchor dowels and the FRP confinement ratio on the ultimate drift ratio of the specimens was observed, with the results showing that an increase in the confinement ratio can improve ultimate drift capacities of the strengthened columns (Ozcan et al., 2010). It was also found that an appropriate configuration of CFRP anchor dowels showed a better seismic behavior compared to the control column without any dowels (Ozcan et al., 2010).

To date, limited studies have been conducted to determine the feasibility of using the sprayed FRP coating on RC columns as a seismic strengthening material (e.g., Furuta et al., 2002; Lee et al., 2007b).

Furuta et al. (2002) evaluated the performance of RC columns which had been repaired and strengthened with the sprayed FRP composites. The fundamental behaviors and performance of the strengthened RC structures were observed under asymmetrical loading, with the results showing that the sprayed FRP coating and the FRP sheets displayed similar shear strengthening performance (Furuta et al., 2002). An anchor system using sprayed FRP slits on a concrete surface was used to assess whether such slit could be expected to provide mechanical bearing (Furuta et al., 2002).

Lee et al. (2007b) investigated the structural performance of six 2/3-scaled RC columns strengthened with sprayed FRP coating through the pseudo-static reversed cyclic load under a constant axial load. In this study, four specimens were strengthened with sprayed FRP coating comprising different combinations of fibers (i.e., carbon or glass) and resins (i.e., epoxy or vinylester) (Lee et al., 2007b).

In order to assess the structural performance of the strengthened RC columns, a specimen strengthened with carbon fiber sheets and a control specimen were tested (Lee et al., 2007b). The results showed that the shear strengths and ductility capacities of the columns strengthened with the sprayed FRP coating were significantly enhanced compared to those of the control specimen (Lee et al., 2007b).

4. ADVANCED COMPUTATIONAL METHODS FOR FRP COMPOSITE SYSTEMS

4.1. Overview

The effective use of advanced composite systems such as FRP requires a reliable knowledge of their characteristic behavior. Such knowledge can be gained through experiemtnal techniques and an extensive analytical and numerical evaluation of their mechanical properties (Biolzi et al., 1994). The reason analytical and numerical evaluations of advanced composite systems are so important is because they are considerabley more economic and expedient than laboratory and field testing (Lee et al., 2005). As a general numerical technique for solving a variety of practical engineering problems, the finite element method (FEM) has shown great capabilities in predicting the behavior and evaluating the performance of conventional materials (Biggs et al., 2000). However, predicting the performance of structural systems composed of advanced materials such as FRPs is a rather complex process, due to their complicated damage constitutive behavior (Lee et al., 2005). Therefore, it is necessary to determine how to make predictions for advanced materials in order to evaluate the performance of new structure systems.

The problem of predicting the effective properties of advanced materials has been a key subject of interest in the mechanical field (Biolzi et al., 1994). In particular, the problem of predicting the effective properties must be considered from a microscopic perspective to enable prediction of physical phenomena such as damage and failure of advanced materials (Kwon and Berner, 1995; Kawagai et al., 2006; Kim and Lee, 2010). The physical properties of advanced materials such as FRP composites typically vary from point to point (Yan, 2003). Hence, the stress and strain fields within the material element are also not uniform at the microscopic scale (Nemat-Nasser and Hori, 1993). Therefore, a characterization model based on the homogenization process has been widely used for simplification of the advanced composite systems (Ju and Chen, 1994a; Kalamkarov and Georgiades, 2002; Lee and Kim, 2010). Here, *homogenization process* can be interpreted as finding a homogeneous reference material that is energetically equivalent to a heterogeneous material, such as FRP composites (Böhm, 1998; Kim and Lee, 2010). The most straightforward application of the homogenization process is the characterization of the advanced materials, and analysis of this kind has been performed by many researchers, who have examined rule of mixture method (Jones, 1975; Tsai and Hahn, 1980), variational principle methods (Hashin, 1962, 1972; Hashin and Strikman, 1962a, 1962b, 1963; Hashin and Rosen, 1964; Hill, 1964; Walpole, 1966a, 1966b, 1969; Ponte Castañeda, 1991; Willis, 1991), the self-consistent method (Hershey, 1954; Hill, 1965; Budiansky, 1965; Budiansky and O'Connell, 1976; Christensen and Lo, 1979; Krajcinovic and Sumarac, 1989; Ju, 1990), the differential scheme (Roscoe, 1973; McLaughlin, 1977; Hashin, 1988), and the Mori-Tanaka method (Mori and Tanaka, 1973; Taya and Chou, 1981; Benveniste, 1986; Weng, 1990; Qiu and Weng, 1990). Extensive details and in-depth analysis of the homogenization process can be found in the relevant literature.

4.2. Micromechanical Analysis for FRP Composites

The theoretical foundation of micromechanics is based on *Eshelby's equivalent inclusion method* (Eshelby, 1957, 1959, 1961), which is a celebrated method for solving the problem of stress and strain distributions in homogenous media containing a subregion that spontaneously changes its shape or size (Böhm, 1998). According to Eshelby (1957, 1959, 1961), when an elastic homogeneous ellipsoidal inclusion in an infinite linear elastic medium is subjected to a uniform strain ε^*, uniform stress and stain states are induced in the constrained inclusion, and the uniform strain in the constrained inclusion ε_c can be expressed as (Mura, 1987; Nemat-Nasser and Hori, 1993; Böhm, 1998):

$$\varepsilon_c = \mathbf{S} : \varepsilon^* \tag{1}$$

where S is referred to as Eshelby's tensor, and the uniform strain ε^* can also be called the "eigenstrain" (stress-free transformation strains) (Mura, 1987; Collini, 2005; Maranganti and Sharma, 2005; Chen, 2006; Kim and Lee, 2010; Lee and Kim, 2010). The detailed expressions of Eshelby's tensor for various shapes are well documented in Mura (1987) and Nemat-Nasser and Hori (1993).

A new micromechanical framework using the *ensemble-volume average method* was proposed by Ju and Chen (1994a, 1994b) for obtaining the effective mechanical properties of advanced composites. For the ensemble-volume average method (as well as the Eshelby's equivalent inclusion method), a renormalization procedure is employed to avoid the conditional convergence problem (Ju and Chen, 1994a, 1994b; Kim and Lee, 2010; Lee and Kim, 2010). The effective elastic stiffness tensor C_* of the advanced composites can be finally expressed from the ensemble-volume average method as (Ju and Chen, 1994a, 1994b)

$$\mathbf{C}_* = \mathbf{C}_0 \cdot \left[\mathbf{I} + \sum_{r=1}^{N} \left\{ \phi_r (\mathbf{A}_r + \mathbf{S}_r)^{-1} \cdot \left[\mathbf{I} - \phi_r \mathbf{S}_r \cdot (\mathbf{A}_r + \mathbf{S}_r)^{-1} \right]^{-1} \right\} \right] \tag{2}$$

where $\mathbf{A}_r = (\mathbf{C}_r - \mathbf{C}_0)^{-1} \cdot \mathbf{C}_0$, the subscript r signifies the r-th phase, "$\cdot$" is the tensor multiplication, and I denotes the fourth-rank identity tensor. In addition, ϕ_r is the volume fraction of r-phase, and $\mathbf{C}_r$ and $\mathbf{C}_0$ are the stiffness tensor of the r-phase and the matrix, respectively. Details of the expansion process for the ensemble-volume averaged micromechanical equation in an elastic medium can be found in Ju and Chen (1994a, 1994b). The ensemble-volume averaged micromechanical framework has been widely used for the prediction of advanced composites. In particular, the computational simulation of advanced composite systems based on the micromechanical framework is introduced and reviewed in this chapter.

4.3. Computational Simulation of Advanced Composite Systems

The primary purpose of the computation simulation is to develop a reliable and accurate computational model for simulating the performance of advanced composite systems. In particular, the rehabilitation performance of CFRP and sprayed FRP is discussed in this chapter. The overall procedure for simulating the performance of advanced composite systems is shown in Figure 7. First, constitutive models are developed for both CFRP and sprayed FRP to consider a phenomenon of dominant damage. The developed constitutive model is then implemented into the finite element program using a user-supplied material (UMAT) subroutine to numerically characterize the mechanical and damage behavior of the CFRP and sprayed FRP. Finally, the predicted results from the simulation are compared with the relevant experimental results for the validation and verification of the proposed constitutive models for CFRP and sprayed FRP.

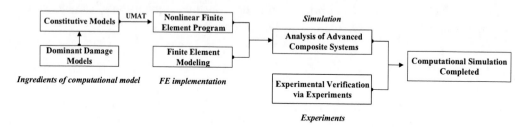

Figure 7. Procedure for simulating the performance of advanced composite systems.

4.3.1. Simulation of Performance of CFRP System

For the CFRP system, the constitutive model for unidirectional fiber-reinforced composites proposed by Liang et al. (2006) is reviewed in this section. The unidirectional fiber-reinforced composites are considered as an initially perfectly bonded composite consisting of an elastic matrix (phase 0) and aligned continuous fibers (phase 1) (Liang et al., 2006; Lee and Kim, 2007). The diameters of the continuous fibers are small enough relative to their length that the fibers can be assumed to be elastic circular cylindrical inclusions (Liang et al., 2006). As loadings or deformations are applied, some fibers are partially debonded and microcracks are nucleated (Martin, 1995; Liang et al., 2006). The partially debonded fiber (phase 2) is replaced by an equivalent, perfectly bonded fiber that possesses transversely isotropic moduli (Zhao and Weng, 1996, 1997; Ju and Lee, 2000). The microcracks (phase 3) can be regarded as the limiting case of aligned spheroidal voids with the aspect ratio $\alpha \to 0$, which is modeled as oblate spheroids (Liang et al., 2006).

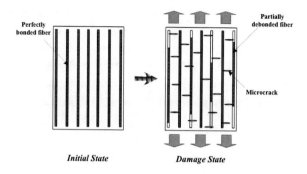

Figure 8. Schematic description of the unidirectional fiber-reinforced composites subjected to loadings or deformations.

The nucleation of microcracks is simulated by adopting a continuum damage model proposed by Karihaloo and Fu (1989). Figure 8 shows the schematic description of the unidirectional fiber-reinforced composites subjected to loadings or deformations. Based on the effective elastic stiffness tensor C* shown in Eq. (2), in conjunction with the stiffness transformation by Herakovich (1998), the stiffness tensor of the unidirectional fiber-reinforced composites with the fibers oriented off-axis $\bar{C}$ can be expressed as (Liang et al., 2006; Lee and Kim, 2007).

$$\sigma = \bar{C}:\varepsilon = \begin{bmatrix} \bar{C}_{11} & \bar{C}_{12} & \bar{C}_{13} & 0 & 0 & \bar{C}_{16} \\ \bar{C}_{12} & \bar{C}_{22} & \bar{C}_{23} & 0 & 0 & \bar{C}_{26} \\ \bar{C}_{13} & \bar{C}_{23} & \bar{C}_{33} & 0 & 0 & \bar{C}_{36} \\ 0 & 0 & 0 & \bar{C}_{44} & \bar{C}_{45} & 0 \\ 0 & 0 & 0 & \bar{C}_{45} & \bar{C}_{55} & 0 \\ \bar{C}_{16} & \bar{C}_{26} & \bar{C}_{36} & 0 & 0 & \bar{C}_{66} \end{bmatrix} : \varepsilon \qquad (3)$$

where the components of $\bar{C}$ are given in Liang et al. (2006). Refer to Liang et al. (2006) and Lee and Kim (2007) for more details on the effective elastic stiffness tensor for unidirectional fiber-reinforced composites.

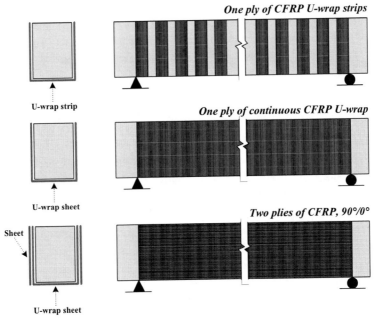

Reprinted from Composites Part B: Engineering, Vol. 39, H.K. Lee, B.R. Kim, S.K. Ha, Numerical evaluation of shear strengthening performance of CFRP sheets/strips and sprayed epoxy coating repair systems, pp. 851862, Copyright (2008), with permission from Elsevier.

Figure 9. Schematic descriptions of CFRP systems (Lee et al., 2008b).

Several simulations of CFRP systems using the micromechanics-based constitutive model were carried out by Lee et al. (2008b, 2008c). The performance of reinforced-concrete

(RC) beams with CFRP systems was numerically investigated according to the procedure shown in Figure 9. Lee et al. (2008b, 2008c) chose to use Khalifa and Nanni (2002)'s experiments on RC beams with CFRP systems as a benchmark for their study. In particular, Khalifa and Nanni (2002) investigated the shear performance of RC beams designed with shear deficiencies; the beam fails suddenly without sufficient warning and diagonal shear cracks are considerably wider than the flexural cracks (Khalifa and Nanni, 2002; Nawy, 2008). The experimental performance tests of three-type CFRP systems shown in Figure 9 were carried out by Khalifa and Nanni (2002): one-ply of CFRP U-wrap strips, one-ply of continuous CFRP U-wrap, and two-plies of CFRP.

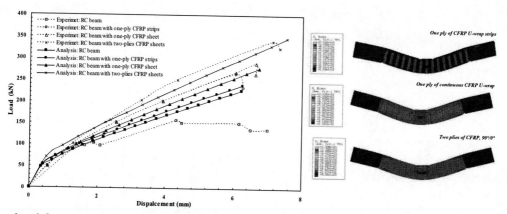

Reprinted from *Composites Part B: Engineering*, Vol. 39, H.K. Lee, B.R. Kim, S.K. Ha, Numerical evaluation of shear strengthening performance of CFRP sheets/strips and sprayed epoxy coating repair systems, pp. 851862, Copyright (2008), with permission from Elsevier.

Figure 10. (a) The comparison of p-u curves between the prediction (Lee et al., 2008b) and the experiment (Khalifa and Nanni, 2002) in the RC beam with CFRP systems, and (b) the deformed shape of the RC beam with CFRP systems through the numerical simulation (Lee et al., 2008b).

The RC beams with these CFRP systems were analyzed in studies by Lee et al. (2008b, 2008c) to show numerically how shear deficient RC beams can be effectively strengthened with CFRP systems. Figure 10 compares the numerical predictions with the experimental load-displacement (p-u) curves of the RC beams with the three CFRP systems during the four-point bending test (cf. Lee et al., 2008b). Figure 10 shows that the numerical predictions based on the micromechanics-based constitutive model are in strong quantitative agreement with the experimental data (Khalifa and Nanni, 2002). This figure also confirms that the RC beam with two-plies of CFRP system exhibits the most superior shear performance (Lee et al., 2008b). Additional series of numerical results for RC beams with the CFRP systems can be found in Lee et al. (2008c). Lee et al. (2008c) also carried out a numerical simulation for the flexural strengthening performance of the CFRP strips and sheets with the shear strengthening performance. The RC beam with CFRP sheets was shown to be capable of sustaining larger loads and absorbing more energy at equal deflection in comparison with the RC beam with the CFRP strips (cf. Lee et al., 2008c). It is noted that the slight differences between the numerical predictions calculated by Lee et al. (2008b, 2008c) and the experimental data (Khalifa and Nanni, 2002) is partly due to other damage mechanisms between the CFRP systems and the RC beam such as debonding which were not considered by Lee et al. (2008b, 2008c). Details of the numerical procedure of the RC beams with the CFRP systems, and more detailed results including the figures can be found in Lee et al. (2008b, 2008c).

Other studies on the analytical and numerical methods for externally bonded CFRP systems have been carried out by several researchers. Arduini et al. (1997) formed an analytical model considering the influences of concrete confinement in the compression zone and the tensile softening properties of concrete in order to more accurately predict crack propagation and the failure mechanism of the flexural member. According to the FRP type, Arduini et al. (1997) investigated thickness, bonded length, and various failure types ranging from FRP rupture to concrete shear failure. Buyle-Bodin et al. (2002) conducted a non-linear FE analysis for prediction of the flexural behavior of RC beams with CFRP plates. They noted that the possibility of brittle failure needs to be considered in spite of the accuracy of their non-linear FE analysis (Buyle-Bodin et al., 2002). Kishi et al. (2005) proposed a numerical analysis method for investigating the load-carrying behavior of RC beams which failed in the CFRP sheets peel-off mode. This study showed that the dominant cracks were discretized according to the crack patterns obtained from the experiments, and it was therefore recommended that the development of a general method for deciding the location of the discrete cracks is necessary (Kishi et al., 2005). Li et al. (2006) performed an FE analysis of RC beams with CFRP sheets in order to investigate the effect of CFRP thickness and length on the failure load and ductility. Pham et al. (2006) carried out an FE study of the bond characteristic between CFRP and concrete. An interface crack model (cf. Hillerborg et al. (1976) and Cervenka et al. (1998)) based on nonlinear fracture mechanics was used for simulation of the debond failures in retrofitted beams. As applied successfully to simulate intermediate span debond of a retrofitted beam, the ultimate load and CFRP strain distributions predicted by Pham et al. (2006) correlated closely with the experimental results. Kotynia et al. (2008) implemented a 3D nonlinear EF analysis of RC beams strengthened in flexure with various externally bonded CFRP. Prior to the numerical study, Kotynia et al. (2008) carried out experimental work to characterize the mode of failure by intermediate crack debonding for all of the specimens. In addition, they conducted an FE analysis which predicts the CFRP debonding mode of failure observed from their experimental work (Kotynia et al., 2008). Details of the analytical and numerical methods for the externally bonded CFRP systems can be found in the referenced literature.

4.3.2. Simulation of Performance of Sprayed FRP System

For the sprayed FRP system, the constitutive model for aligned and randomly oriented discontinuous fiber-reinforced composites proposed by Lee and Simunovic (2000, 2001) is reviewed in this section. The aligned discontinuous fiber-reinforced composites are considered to consist of a matrix (phase 0) and the aligned discontinuous fibers (phase 1) (Lee and Simunovic, 2000, 2001). As loadings or deformations are applied, some fibers are partially debonded (phase 2) and microcracks are nucleated (phase 3). With the help of Eshelby's tensor for an ellipsoidal inclusion, the effective elastic stiffness tensor C_* of aligned discontinuous fiber-reinforced composites can be expressed as (Lee and Simunovic, 2000, 2001):

$$\mathbf{C}_* = \tilde{F}_{ijkl}(l_1, l_2, l_3, l_4, l_5, l_6) \tag{4}$$

where a transversely isotropic fourth-rank tensor $\tilde{\mathbf{F}}$ is defined as

$$\tilde{F}_{ijkl}(\iota_m) = \iota_1 \tilde{n}_i \tilde{n}_j \tilde{n}_k \tilde{n}_l + \iota_2(\delta_{ik}\tilde{n}_j\tilde{n}_l + \delta_{il}\tilde{n}_j\tilde{n}_k + \delta_{jk}\tilde{n}_i\tilde{n}_l + \delta_{jl}\tilde{n}_i\tilde{n}_k) \\ + \iota_3 \delta_{ij}\tilde{n}_k\tilde{n}_l + \iota_4 \delta_{kl}\tilde{n}_i\tilde{n}_j + \iota_5 \delta_{ij}\delta_{kl} + \iota_6(\delta_{ik}\delta_{jl} + \delta_{il}\delta_{jk}) \tag{5}$$

in which $\tilde{n}$ denotes the unit vector and δ_{ij} signifies the Kronecker delta. The parameters of ι_m (m=16) are given in Lee and Simunovic (2000, 2001).

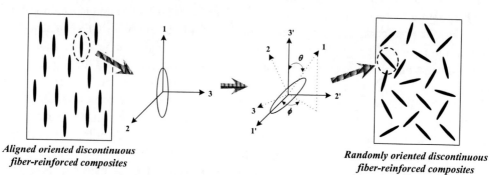

Aligned oriented discontinuous fiber-reinforced composites

Randomly oriented discontinuous fiber-reinforced composites

Reprinted from *Composites Part B: Engineering*, Vol. 31, H.K. Lee, S. Simunovic, Modeling of progressive damage in aligned and randomly oriented discontinuous fiber polymer matrix composites, pp. 7786, Copyright (2000), with permission from Elsevier.

Figure 11. Coordinate designation between an aligned oriented discontinuous fiber and randomly oriented discontinuous fiber (cf., Lee and Simunovic, 2000).

In order to obtain the effective elastic stiffness tensor $\mathbf{C}_*$ of randomly discontinuous fiber-reinforced composites, the coordinate designation between an aligned discontinuous fiber and a randomly oriented discontinuous fiber (shown in Figure 11) was introduced (Lee and Simunovic, 2000). The averaging process over all orientations of governing constitutive field equations was conducted by Lee and Simunovic (2000). The averaging process for all possible orientations can be expressed as (Lee and Simunovic, 2000):

$$\subset \cdot \supset \equiv \int_0^\pi \int_0^\pi (\cdot) P(\theta,\phi) \sin\theta \, d\theta \, d\phi \tag{6}$$

where $P(\theta,\phi)$ is the probability density function. Finally, assuming the overall uniform distribution of strains, the effective elastic tensor $\subset \mathbf{C}_* \supset$ of the randomly discontinuous fiber-reinforced composites can be obtained as noted in Lee and Simunovic (2000). Refer to Lee and Simunovic (2000, 2001) for more details on the effective elastic stiffness tensor for aligned and randomly oriented discontinuous fiber-reinforced composites.

Several simulations of sprayed FRP systems using micromechanics-based constitutive model were carried out by Lee et al. (2003, 2005) and Ha et al. (2009). In particular, Lee et al. (2005) implemented numerical studies on damaged reinforced concrete beams and bridge superstructures coated with sprayed FRP. For this study, a series of numerical three-point bending tests were conducted on undamaged and damaged (pre-cracked) RC beams both with and without sprayed FRP (Lee et al., 2005). The predicted results were compared with the typical load-displacement curves obtained from the experiments found in Lee and Hausmann (2004). Figure 12 compares the prediction (Lee et al., 2005) and the experimental results (Lee and Hausmann, 2004) for the load-displacement of undamaged and damaged (pre-cracked) RC beams with sprayed FRP. A series of parametric studies on the thickness of sprayed FRP coating, damage parameters, and mesh sensitivity were also conducted by Lee et al. (2005).

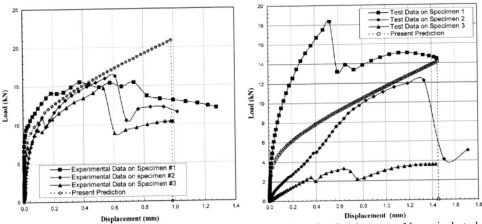

Reprinted from *Engineering Structures*, Vol. 27, H.K. Lee, G. Avila, C. Montanez, Numerical study on retrofit and strengthening performance of sprayed fiber reinforced polymer, pp. 14761487, Copyright (2005), with permission from Elsevier.

Figure 12. The comparison of load-displacement between the prediction (Lee et al., 2005) and the experiment (Lee and Hausmann, 2004): (a) undamaged RC beams with sprayed FRP and (b) damaged (pre-cracked) RC beams with sprayed FRP.

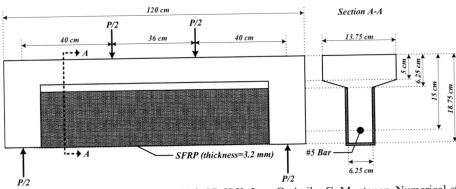

Reprinted from *Engineering Structures*, Vol. 27, H.K. Lee, G. Avila, C. Montanez, Numerical study on retrofit and strengthening performance of sprayed fiber reinforced polymer, pp. 14761487, Copyright (2005), with permission from Elsevier.

Figure 13. The dimension of the T-beam analyzed by Lee et al. (2005) and the schematic of the four-point bending test on the T-beam with sprayed FRP coating.

In addition, in order to study the benefits that sprayed FRP can provide to a deteriorated bridge superstructure, and to show how a shear deficient RC beam can be strengthened with sprayed FRP, Lee et al. (2005) carried out a numerical simulation on the T-beam bridge superstructure whose shear strengthening was well analyzed by Kaliakin et al. (1996). The original experiments by Kaliakin et al. (1996) were on T-beams bonded with woven composite fabrics. In order to show the superior characteristics of sprayed FRP, Lee et al. (2005) carried out the numerical simulations on T-beams with sprayed FRP coating instead of woven composite fabrics. The dimension of the T-beam analyzed by Lee et al. (2005) and the schematic of the four-point bending test on the T-beam with sprayed FRP coating are shown in Figure 13.

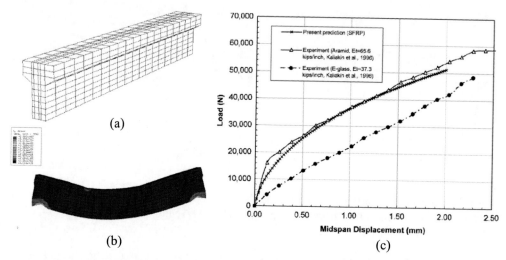

Reprinted from *Engineering Structures*, Vol. 27, H.K. Lee, G. Avila, C. Montanez, Numerical study on retrofit and strengthening performance of sprayed fiber reinforced polymer, pp. 14761487, Copyright (2005), with permission from Elsevier.

Figure 14. (a) The FE mesh of 3D T-beam model, (b) the deformed shape and von Mises effective stress during the four-point bending test, and (c) the predicted load-displacement curve on the sprayed FRP coated T-beam (Lee et al., 2005).

Figure 14 shows the FE mesh of the 3D T-beam model, the deformed shape and von Mises effective stress during the four-point bending test, and the predicted load-displacement curve on the sprayed FRP coated T-beam, as well as the experimental results obtained by Kaliakin et al. (1996). The figure shows that the T-beam coated with sprayed FRP is able to sustain a larger load and absorb more energy at equal displacements in comparison with the T-beam bonded with woven composite fabrics with low modulus per unit width (Lee et al., 2005). However, the FRP-treated T-beam sustains virtually the same load and absorbs the same amount of energy as the T-beam bonded with woven composite fabrics with high modulus per unit width (Lee et al., 2005). Details of the numerical prediction of RC beams with sprayed FRP systems and more detailed results, including the figures, can be found in Lee et al. (2003, 2005) and Ha et al. (2009).

There have been limited studies on the analytical and numerical methods for sprayed FRP coating systems. Ross et al. (2004) conducted an FE analysis of deteriorated reinforced concrete channel beam highway bridges which were retrofit with glass fiber reinforced polymer (GFRP) fabric and a GFRP spray material. Ross et al. (2004) compared and calibrated their experimental results at non-retrofit beam element and GFRP element developing stages. Silva et al. (2008) carried out a Monte Carlo simulation on shear capacity of walls retrofitted by rapid-setting sprayed polyurea reinforced GFRP grids. Details of the analytical and numerical methods for sprayed FRP systems can be found in the referenced literature.

SUMMARY

A series of studies on the application of FRP composite systems for rehabilitation in civil engineering have been reviewed in this chapter. All over the world, the renewal of civil infrastructures has been recognized as one of the most principal challenges facing many societies. Thus, there has been considerable interest in the use of new and emerging materials and technologies to facilitate the functionality, efficiency durability of civil infrastructures. In the field of civil engineering, FRP composites have rapidly gained recognition as a superior construction material, and the application of FRP composites has been evolving very rapidly in recent years.

In the civil engineering field, a high load-carrying capacity is always necessary in order for civil infrastructures to accommodate higher performance than they were originally designed for. This constant need and difficulty can be resolved through the successful application of superior rehabilitation systems. As introduced in this chapter, FRP composite systems such as CFRP and sprayed FRP can be used to effectively enhance the flexural, shear and seismic performances of RC structures in structural load-carrying capacity, stiffness, and ductility under static, cyclic, and fatigue loading conditions, while protecting them from the environmental. However, there are still some limitations associated with FRP composite systems, including bonding problemS and application orientation. Therefore, in order to encourage and convince the civil engineering community to strengthen and retrofit civil infrastructures with FRP composite systems, more experiments are needed to show the effectiveness of the rehabilitation methods and to improve these FRP composite systems. More reliable methods of analyzing the expected results also need to be developed before moving to the field and repairing structures with the novel FRP composite systems.

REFERENCES

[1] ACI Committee 318 (1995), *Building Code Requirements for Structural Concrete: (ACI 318-95); and Commentary (ACI 318R-95)*, American Concrete Institute, Detroit.

[2] ACI Committee 440 (1996), *State-of-the-Art Report on Fiber Reinforced Plastic (FRP) Reinforcement for Concrete Structures*, American Concrete Institute, Detroit.

[3] Advani, S.G., Sozer, E.M. (2003), *Process Modeling in Composite Manufacturing*, Marcel Dekker, New York.

[4] Al-Salloum, Y.A., Almusallam, T.H. (2003), *Journal of King Saud University*, 16: 1-21.

[5] Alexander, J.G.S. (1996), *Proceedings of Annual Conference of Canaian Society for Civil Engineering*, 2: 167-178.

[6] Anania, L., Badalà, A., Failla, G. (2005), *Construction and Building Materials*, 19: 55-61.

[7] Anil, Ö. (2006), *Cement and Concrete Composites*, 28: 638-649.

[8] Araki, N. Matsuzaki, Y., Nakano, K., Kataoka, T. (1997), *Non-Metallic (FRP) Reinforcement for Concrete Structures, Proceedings of the Third International Symposium*, Sapporo, Japan, pp. 512-522.

[9] Arduini, M., D'Ambrish, A., Di Tommaso, A. (1994), *Infrastructure: New Materials and Methods of Repair, Proceedings of the Materials Engineering Conference 804*, ASCE, New York, USA, pp. 123-130.

[10] Arduini, M., Nanni, A (1997), *ASCE Journal of Composites for Construction*, 1: 63-70.

[11] Arduini, M., Di Tommaso, A., Nanni, A. (1997), *ACI Structural Journal*, 94: 363-370.

[12] Ashour, A.F., El-Refaie, S.A., Garrity, S.W. (2004), *Cement and Concrete Composites*, 26: 765-775.

[13] Astrom, B.T. (1997), *Manufacturing of polymer composites*, Chapman and Hall Publisher, London.

[14] Bakis, C.E., Bank, L.C., Brown, V.L., Cosenza, E., Davalos, J.F., Lesko, J.J., Machida, A., Rizkalla, S.H., Triantafillou, T.C. (2002), *ASCE Journal of Compoistes for Construction*, 6: 73-87.

[15] Ballinger, C., Maeda, T., Hoshijima, T. (1993), *Proceedings of the International Symposium on Fiber-Reinforced-Plastic Reinforcements for Concrete Structures* (Editors: Nanni, A. and Dolan, C.W.), American Concrete Institute, Farmington Hills, Michigan, USA.

[16] Balsamo, A., Colombo, A., Manfredi, G., Negro, P., Prota, C. (2005), *Engineering Structures*, 27: 769-780.

[17] Banthia, N., Yan, C., Nandakumar, N. (1996), *Proceedings of the 2nd International Conference on Advanced Composite Materials*, August, Montreal, Que.

[18] Banthia, N., Boyd. A.J. (2000), *Canadian Journal of Civil Engineering*, 27: 907-915.

[19] Banthia, N. (2002), *Proceedings of 1st International Conference on Structural Health Monitoring*, Winnipeg: ISIS Canada Corporation, pp. 135-144.

[20] Barboni, M., Benedetti, A., Nanni, A. (1997), *ASCE Journal of Composites for Construction*, 1: 168-174.

[21] Barros, J.A.O., Fortes, A.S. (2005), *Cement and Concrete Composites*, 27: 471-480.

[22] Bencardino, F., Spadea, G., Swamy, R.N. (2007), *Construction and Building Materials*, 21: 1997-2006.

[23] Benveniste, Y. (1986), *Mechancis Research Communications*, 13: 193-201.

[24] Berhane, Z. (1994), *Materials and Structures*, 27: 347-352.

[25] Berthelot, J.M. (1999), *Composite Materials: Mechanical Behavior and Structural Analysis*, Springer-Verlag New York, Inc.

[26] Biggs, R.M., Barton, F.W., Gomez, J.P., Massarelli, P.J., McKeel, W.T. (2000), *Finite element modeling and analysis of reinforced-compoiste bridge decks*, Final Report: Virgina Transportation Research Council, September, VTRC 01-R4.

[27] Biolzi, L., Castellani, L., Pitacco, I. (1994), *Journal of Materials Science*, 29: 2507-2512.

[28] Biron, M. (2004), *Thermosets and Composites: Technical Information for Plastics Users*, Elsevier Ltd.

Application of FRP Composites for Rehabilitation in Civil Engineering 597

[29] Böhm, H.J. (1998), *A Short Introduction to Basic Aspects of Continuum Micromechanics*, ILSB Report/ILSB-Arbeitsbericht 206.

[30] Boyajian, P.G. (1993), *Manufacturing and design of Textile performs for composite materials*, M.S.thesis, Dept. of Civil and Environ. Eng., MIT.

[31] Boyd, A.J. (2000), *Rehabilitation of Reinforced Concrete Beams with Sprayed Glass Fiber Reinforced Polymers*, Ph. D. dissertation, The University of British Columbia.

[32] Boyd. A.J., Liang, N., Gree, P., and Lammert, K. (2008), *Construction and Building Materials*, 22: 411-416.

[33] Budiansky, B. (1965), *Journal of the Mechanics and Physics of Solids*, 13: 223-227.

[34] Budiansky, B., O'Connell, R.J. (1976), *International Journal of Solids and Structures*, 12: 81-97.

[35] Buyle-Bodin, F., David, E., Ragneau, E. (2002), *Engineering Structures*, 24: 1423-1429.

[36] Caibal, M.R., Obera, I., Murata, H., Niwa, J. (2003), *JSPS Core University Program on Environment Engineering*, pp.1-27.

[37] Campbell, F.C. (2004), *Manufacturing Processes for advanced composites*, Elsevier Ltd.

[38] Canovas, M.F., Selva, N.H., Kawiche, G.M. (1992), *Materials and Structures*, 25: 417-422.

[39] Cervenka, J., Kishen, J.M.C., Saouma, V.E. (1998), *Engineering Fracture Mechanics*, 6: 95-107.

[40] Cheng, L., Karbhari, V.M. (2006), *Progress in Structural Engineering and Materials*, 8: 143-154.

[41] Christensen, R.M., Lo, K.H. (1979), *Journal of the Mechanics and Physics of Solids*, 27: 315-330.

[42] Collini, L. (2005), *Micromechanical Modeling of the Elasto-plastic Behavior of Heterogenous Nodular Cast Iron*, Ph.D. Dissertation, Universita' degli Studi di Parma.

[43] Cooke, T.F. (1991), *Journal of the American Ceramic Society*, 74: 2959-2978.

[44] Czarnecki, L., White, J.L. (1980), *Journal of Applied Polymer Science*, 25: 1217-1244.

[45] Dhingra, A.K. (1980), *Philosophical Transactions of the Royal Society of London. Series A, Mathematical and Physical Sciences*, 294: 411-417.

[46] Eshelby, J.D. (1957), *Proceedings of the Royal Society London A*, 241: 376-396.

[47] Eshelby, J.D. (1959), *Proceedings of the Royal Society London A*, 252: 561-569.

[48] Eshelby, J.D. (1961), *Progress in Solid Mechanics* (Editors: Senddon, N.I. and Hill, R.), Amsterdam, North-Holland.

[49] Eurocode (1992), *Design of Concrete Structures, No. 2*, European Committee for Standardization, Lausanne.

[50] Fanning, P., Kelly, O. (1999), *Proceedings of the Eighth International Conference on Structural Faults and Repair* (Editor: Forde, M.C.), Engineering Technics Press, Edinburgh, UK.

[51] Furuta,T., Kanakubo, T., Nemoto, T., Takahashi, K., Itoh, K., Minamihara, H. (2002), *downloaded from http://www.quakewrap.com/frp%20papers/Sprayed-Up-FRP-Strengthening-For-Reinforced-Concrete-Beams.pdf.*

[52] Frederick, R., Phelan, J.R. (1997), *Polymer Composites*, 18: 460-476.

[53] Galhano, G.Á., Valandro, L.F., de Melo, R.M., Scotti, R., Bottino, M.A. (2005), *Journal of Endodontics*, 25: 209-211.

[54] GangaRao, H.V.S., Vijay, P.V. (1998), *ASCE Journal of Structural Engineering*, 124: 3-10.

[55] Garden, H.N., Hollaway, L.C. (1998), *Composite Structures*, 42: 175-88.

[56] Gassan, J., Bledzki, A. (1999), *Polymer Composites*, 20: 604-611.

[57] Gauthier, R., Joly, C., Compas, A., Gaultier, H., Escoubes, M. (1998), *Polymer Composites*, 19: 287-300.

[58] Gibson, A.A. (1992), *Composites Manufacturing*, 3: 223-233.

[59] Grace, N.F., Sayed, G.A., Soliman, A.K., Saleh, K.R. (1999), *ACI Structural Journal*, 96: 865-874.

[60] Gram, H.E. (1983), *Durability of Natural Fibres in Concrete*, CBI Research No. 1-83, Swedish Cement and Concrete Research Institute, Stockbolm.

[61] Ha, S.K., Yang, B.J., Lee, H.K. (2009), *Journal of the Computational Structural Engineering Institute of Korea*, 22: 579-585. (in Korean)

[62] Harper, A. (2009), *Reinforced Plastics*, 53: 30-33.

[63] Hashin, Z. (1962), *ASME Journal of Applied Mechanics*, 29: 143-150.

[64] Hashin, Z. (1972), *Theory of Fiber Reinforced Materials*, Technical Report (NASA).

[65] Hashin, Z. (1988), *Journal of the Mechanics and Physics of Solids*, 36: 719-734.

[66] Hashin, Z., Rosen, B. (1964), *ASME Journal of Applied Mechanics*, 31: 223-232.

[67] Hashin, Z., Strikman, S. (1962a), *Journal of the Mechanics and Physics of Solids*, 10: 335-342.

[68] Hashin, Z., Strikman, S. (1962b), *Journal of the Mechanics and Physics of Solids*, 10: 343-352.

[69] Hashin, Z., Strikman, S. (1963), *Journal of the Mechanics and Physics of Solids*, 11: 127-140.

[70] Hassan, N., Rizkalla, S., Hassan, T. (2003), *Progress in Structural Engineering and Materials*, 5: 16-28.

[71] Herakovich C.T. (1998), *Mechanics of fibrous composites*, John Wiley and Sons, New York.

[72] Hershey, A. (1954), *ASME Journal of Applied Mechanics*, 21: 236-240.

[73] Hill, R. (1964), *Journal of the Mechanics and Physics of Solids*, 12: 199-212.

[74] Hill, R. (1965), *Journal of the Mechanics and Physics of Solids*, 13: 189-198.

[75] Hillerborg, A., Modeer, M., Petersson, P.E. (1976), *Cement and Concrete Research*, 6: 773-782.

[76] Hollaway, L.C. (2010), *Construction and Building Materials*, 24: 2419-2445.

[77] Hollaway, L.C., Mays, G.C. (1999), *Strengthening of Reinforced Concrete Structures: using externally-bonded FRP composites in structural and civil engineering* (Editors: Hollaway, L.C. and Leeming, M.B.), pp. 83-134, CRC-Press.

[78] Hollaway, L.C., Teng, J.G. (2008), *Strengthening and Rehabilitation of Civil Infrastructures Using Fiber-Reinforced Polymer (FRP) Composites*, CRC press.

[79] Islam, M.R., Mansur, M.A., Maalej, M. (2005), *Cement and Concrete Composites*, 27: 413-420.

[80] Jang, B.Z. (1994), *Advanced Polymer Composites: Principles and Applications*, ASM International, Materials Park , OH 44073-0002, USA.

[81] Jayaprakash, J. Samad, A.A.A., Abbasovich, A.A., Ali, A.A.A. (2008) *Construction and Building Materials*, 22: 1148-1165.

[82] Johnson, C.F. (1993), *Engineered materials handbook. Vol. 1-Composites*, ASM International, pp. 564-568.

[83] Johnson, R.O., Burlhis, H.S. (1983), *Journal of Polymer Science: Polymer Symposia*, 70: 129-143.

[84] Jones, R.M. (1975), *Mechanics of Composite Mateirals*, McGraw-Hill, New York.

[85] Ju, J.W. (1990), *International Journal of Solids and Structures*, 27: 227-258.

[86] Ju, J.W., Chen, T.M. (1994a), *Acta Mechanica*, 103: 103-121.

[87] Ju, J.W., Chen, T.M. (1994b), *Acta Mechanica*, 103: 123-144.

[88] Ju, J.W., Lee, H.K. (2000), *Computer Methods in Applied Mechanics and Engineering*, 183: 201-222.

[89] Kachlakev, D., McCurry, D.D. (2000), *Composites Part B: Engineering*, 31: 445-452.

[90] Kalamkarov, A.L., Georgiades, A.V. (2002), *Smart Materials and Structures*, 11: 423-434.

[91] Kaliakin, V.N., Chajes, M.J., Januszka, T.F. (1996), *Composites Part B: Engineering*, 27: 235-244.

[92] Kamal, M.R. (1973), *Polymer Engineering and Science*, 13: 59-64.

[93] Karbhari, V.M., Seible, F. (2000), *Applied Composite Materials*, 7: 95-124.

[94] Karbhari, V.M., Zhao, L. (2000), *Computer Methods in Applied Mechanics and Engineering*, 185: 433-454.

[95] Karbhari, V.M., Li, Y. (2002), *Proceedings of the Tenth U.S.-Japan Conference on Composite Materials* (Editor: Chang, F.K.), pp.292-300, DEStech Publications, Inc.

[96] Karbhari, V.M., Chin, J.W., Hunston, D., Benmokrane, B., Juska, T., Morgan, R., Lesko, J.J., Sorathia, U., Reynaud, D. (2003), *ASCE Journal of Compoistes for Construction*, 7: 238-247.

[97] Karbhari, V.M. (2007), *Durability of composites for civil structural applications*, Woodhead Publishing.

[98] Karihaloo, B.L., Fu, D. (1989), *European Journal of Mechanics-A/Solids*, 8: 373-384.

[99] Karlsson, K.F., Åströn, B.T. (1997), *Composites Part A: Applied Science and Manufacturing*, 28: 97-111.

[100] Katsumata, H., Kobatake, Y., Takeda, T. (1987), *Proceedings of the Seminar on Repair and Retrofit of Structures, Workshop on Repair and Retrofit of Existing Structures*, US-Japan Panel on Wind and Seismic Effects, UJNR, Tsukuba, Japan.

[101] Katsumata, H., Kobatake, Y., Takeda, T. (1988), *Proceedings of the Ninth World Conference on Earthquake Engineering*, Tokyo-Kyoto, Japan, Vol. 7, pp. 517-522.

[102] Kaw, A.K. (1997), *Mechanics of Composite Materials*, CRC-Press.

[103] Kawagai, M., Sando, A., Takano, N. (2006), *Modelling and Simulation in Mateirals Science and Engineering*, 14: 53-69.

[104] Kendall, K.N., Rudd, C.D., Owen, M.J., Middleton, V. (1992), *Composites Manufacturing*, 3: 235-249.

[105] Khalifa, A., Nanni, A. (2000), *Cement and Concrete Composites*, 22: 165-174.

[106] Khalifa, A., Nanni, A. (2002), *Construction and Building Materials*, 16: 135-146.

[107] Kim, B.R., Lee, H.K. (2010), *Compoiste Laminates: Properties, Performance and Applications* (Editors: Doughett, A. and Asnarez, P.), pp. 201-232, Nova Science Publishers, Inc.

[108] Kishi, N., Zhang, G., Mikami, H. (2005), *ASCE Journal of Composites for Construction*, 9: 507-514.

[109] Kobatake. Y. (1998), *Advanced Composite Materials*, 7: 1-22.

[110] Kotynia, R., Baky, H.A., Neale, K.W. (2008), *ASCE Journal of Composites for Construction*, 12: 190-201.

[111] Krajcinovic, D., Sumarac, D. (1989), *Journal of the Mechanics and Physics of Solids*, 56: 51-62.

[112] Kwon, Y.W., Berner, J.M. (1995), *Engineering Fracture Mechanics*, 60: 233-253.

[113] Lee, H.K. (2004), *Journal of Reinforced Plastics and Composites*, 23: 1285-1300.

[114] Lee, H.K., Avila, G., Schadler, D. (2003), *KSCE Journal of Civil Engineering*, 7: 675-682.

[115] Lee, H.K., Avila, G., Montanez, C. (2005), *Engineering Structures*, 27: 1476-1487.

[116] Lee, H.K., Cheong, S.H., Ha, S.K., Lee, C.G. (2011), *Composite Structures*, 93: 911-922.

[117] Lee, H.K., Hausmann, R.L. (2004), *Composite Structures*, 63: 201-209.

[118] Lee, H.K., Hausmann, R.L., Seaman, W.C. (2008a), *Journal of Reinforced Plastics and Composites*, 27: 1269-1286.

[119] Lee, H.K., Kim, B.R. (2007), *Composite Science and Technology*, 67: 2221-2230.

[120] Lee, H.K., Kim, B.R. (2010), *Advances in Polymer Composites (Volume I): Macro and Micro Composites* (Editors: Thomas, S., Joseph, K., Malhotra, S.K., Goda, K., and Sreekala, M.S.), Wiley-VCH Verlag GmbH and Co. KGaA, submitted for publication.

[121] Lee, H.K, Kim, B.R., Ha, S.K. (2008b), *Composites Part B: Engineering*, 39: 851-862.

[122] Lee, H.K., Ha, S.K., Afzal, M. (2008c), *Structural Engineering and Mechanics*, 30: 247-261.

[123] Lee, H.K., Simunovic, S. (2000), *Composites. Part B: Engineering*, 31: 77-86.

[124] Lee, H.K., Simunovic, S. (2001), *International Journal of Solids and Structures*, 38: 875-895.

[125] Lee, K.S., Son, Y.S., Lee, M.S. (2007a), *Journal of the Korea Concrete Institute*, 19: 421-431.

[126] Lee, K.S., Byeon, I.H., Lee, M.S. (2007b) *Korea institute for Structural Maintenance Inspection*, 11: 132-142.

[127] Li, J., Gong, J., Wang, L. (2009), *Construction and Building Materials*, 23: 2653-2663.

[128] Li, L.J., Guo, Y.C., Liu, F., Bungey, J.H. (2006), *Construction and Building Materials*, 20: 901-909.

[129] Liang, Z., Lee, H.K., Suaris, W. (2006), *International Journal of Solids and Structures*, 43: 5674-5689.

[130] Long, A.C., Blanchard, C.D., Smith, R.P. (1998), *Composites Part A: Applied Science and Manufacturing*, 29: 847-854.

[131] Lopez-Anido, R.A., Naik, T.R. (2000), *Emerging Materials for Civil Infrastructure-State of the Art*, ASCE.

[132] Lubin, G. (1969), *Handbook of Fiberglass and Advanced Plastics Composites*, Krieger Publishing Company, Huntington, New York.

[133] Luo, S., Netravali, A. (1999), *Polymer Composites*, 20: 367-378.

[134] Mallick, P.K. (1993), *Fiber-Reinforced Composites: Materials, Manufacturing, and Design*, CRC Press.

[135] Maranganti, R., Sharma, P. (2005), *Handbook of Theoretical and Computational Nanotechnology, Nanotechnology Book Series, Vol. 2* (Editors: Rieth, M. and Schommers, W.), American Scientific Publishers.

[136] Martin, R.H. (1995), *Composite Materials: Fatigue and Fracture*, Fifth volume, ASTM Publication,

[137] Matsuda, T., Sato, T., Fujiwara, H., Hinashida, N. (1990), *Proceeding of 1st US-Japan Workshop on Seisic Retrofit of Bridges*, Public Works Institute, Ministry of Construction, Tsukuba, Japan, pp. 356-374.

[138] Matthews, F.L., Rawlings, R.D. (1994), *Composite Materials: Engineering and Science*, Chapman and Hall.

[139] Mazumdar, S.K. (2002), *Composite Manufacturing: Materials, Product, and Process Engineering*, CRC Press.

[140] McLaughlin, R. (1977), *International Journal of Engineering Science*, 15: 237-244.

[141] Meier, U. (1992), *Structural Engineering International*, 25: 7-12.

[142] Meier, U. (1995), *Construction and Building Materials*, 9: 341-351.

[143] Meier, U., Deuring, M., Meier, H., Schwegler, G. (1993), *Fibre-Reinforced-Plastic (FRP) Reinforcement for Concrete Structures: Properties and Applications* (Editor: Nanni, A.), Elsevier Science, Amsterdam, The Netherlands.

[144] Midwater, K.R. (1997), *Construction Repair*, 11: 5-8.

[145] Mitsui, Y., Murakami, K., Takeda, K., Sakai, H. (1998), *Composite Interfaces*, 5: 285-295.

[146] Mohan, R.V., Tamma, K.K., Shires, D.R., Mark, A. (1998), *Advances in Engineering Software*, 29: 249-263.

[147] Möller, M., Popescu, C. (2009), *Sustainable Solutions for Modern Econmies* (Editor: Höfer, R.), Royal Society of Chemistry.

[148] Mori, T., Tanaka, K. (1973), *Acta Metallurgica*, 21: 571-574.

[149] Mufti, A.A. (2003), *Construction and Building Materials*, 17: 379-387.

[150] Mukherjee, A., Joshi, M.V. (2007), *Indian Concrete Journal*, 76: 496-502.

[151] Mura, T. (1987), *Micromechanics of Defects in Solids*, 2nd edition, Martinus Nijhoff Publishers, Dordrecht.

[152] Nanni, A. (1995), *Concrete International*, 17: 22-26.

[153] Nawy, E.G. (2008), *Reinforced Concrete: A Fundamental Approach*, 6th edition, Prentice Hall.

[154] Neale, K.W. (2000), *Progress in Structural Engineering and Materials*, 2: 133-138.

[155] Nemat-Nasser, S., Hori, M. (1993), *Micromechanics: Overall Properties of Heterogenous Materials*, Elsevier Publishers, Amsterdam.

[156] Nezamian, A., Setunge, S. (2007), *ASCE Journal of Compoistes for Construction*, 11: 531-544.

[157] Ngo, N.D., Tamma, K.K. (2001), *International Journal for Numerical Methods in Engineering*, 50: 1559-1585.

[158] Nguyen, D.M., Chan, T.K., Cheong, H.K. (2001), *ASCE Journal of Composites for Construction*, 5: 12-17.

[159] Ogata, T., Osada, K. (2000), *Cement and Concrete Composites*, 22: 17-27.

[160] Ozcan, O., Binici, B., Ozcebe, G. (2008), *Engineering Structures*, 30: 1632-1646.

[161] Ozcan, O., Binici, B., Ozcebe, G. (2010), *Engineering Structures*, 32: 964-973.

[162] Parratt, N.J. (1972), *Fibre-Reinforced Materials Technology*, Van Nostrand Reinhold Co., London and New York.

[163] Pendhari, S.S., Kant, T., Desai, Y.M. (2008), *Composite Structures*, 84: 114-124.

[164] Peters, S.T. (1998), *Handbook of Composites*, 2nd edition, Chapman and Hall.

[165] Pham, H.B., Al-Mahaidi, R., Saouma, V. (2006), *Composite Structures*, 75: 145-150.

[166] Ritchie, P.A., Thomas, D.A., Lu, L.W., Connely, G.M. (1991), *ACI Structural Journal*, 88: 490-500.

[167] Ponte Castañeda, P. (1991), *Journal of the Mechanics and Physics of Solids*, 39: 45-71.

[168] Potter, K. (1997), *Resin Transfer Moulding*, Chapman and Hall.

[169] Qiu, Y.P., Weng, G.J. (1990), *International Journal of Engineering Science*, 289: 1121-1137.

[170] Rabinovitch, O., Frostig, Y. (2003), *Composites Part B: Engineering*, 34: 663-677.

[171] Ramakrishna, G., Sundararajan, T. (2005), *Cement and Concrete Composites*, 27: 547-553.

[172] Reddy, J.N. (1997), *Mechanics of Laminated Composite Plates: Theory and Analysis*, CRC-Press.

[173] Rijswijk, K.V., Bersee, H.E.N. (2007), *Composites Part A: Applied Science and Manufacturing*, 38: 666-681.

[174] Rosato, D.V., Rosato, D.V., Rosato, M.G. (2001), *Plastics Design Handbook*, Springer-Verlag.

[175] Roscoe, R. (1973), *Rheologica Acta*, 12: 404-411.

[176] Ross, S., Boyd, A., Johnson, M., Sexsmith, R., Banthia, N. (2004), *Journal of Bridge Engineering*, 9: 66-74.

[177] Rowell, R.M. (1995), *Polymers and Other Advanced Materials: Emerging Technologies and Business Opportunities* (Editors: Prasad, P.N., Mark, J.E., Ting, T.J.), Plenum Press.

[178] Promis, G., Ferrier, E., Hamelin, P. (2009), *Composite Structures*, 88: 367-379.

[179] Rudd, C.D., Long, A.C., Kendall, K.N., Mangin, C.G.E. (1997), *Liquid Moulding Technologies: Resin Transfer Moulding, Structural Reaction Injection Moulding and Related Processing Techniques*, Woodhead Publishing Ltd.

[180] Saadatmanesh, H., Malek, A.M. (1998), *ASCE Journal of Composites for Construction*, 2: 158-164.

[181] Saheb, D.N., Jog, J.P. (1999), *Advances in Polymer Technology*, 18: 351-363.

[182] Sato, Y., Ueda, T., Kakuta, Y., Ono, S. (1997a), *Non-Metallic (FRP) Reinforcement for Concrete Structures, Proceedings of the Third International Symposium*, Sapporo, Japan, pp. 499-505.

[183] Sato, Y., Katsumata, H., Kobatake, Y. (1997b), *Non-Metallic (FRP) Reinforcement for Concrete Structures, Proceedings of the Third International Symposium*, Sapporo, Japan, pp. 507-513.

[184] Seible, F., Priedstley, M.J.N., Hegemier, G.A., Innamorato, D. (1997), *ASCE Journal of Composites for Construction*, 1: 52-62.

[185] Sen, R. (2003), *Progress in Structural Engineering and Materials*, 5: 99-113.

[186] Shahawy, M.A., Arockiasamy, M., Beitelman, T., Sowrirajan, R (1996), *Composites Part B: Engineering*, 27: 225-233.

[187] Shalin, R.E. (1995), *Polymer Matrix Composites*, Chapman and Hall.

[188] Shojaei, A., Ghaffarian, S.R., Karimian, S.M.H. (2003), *Polymer Composites*, 24: 525-544.

[189] Shrivastava, R., Gupta, U., Choubey, U.B. (2009), *International Journal of Recent Trends in Engineering*, 1: 89-93.

[190] Silva, F.A. (2008), *Composites Science and Technology*, 68: 3438-3443.

[191] Silva, P.E., Yu, P., Nanni, A. (2008), *ASCE Journal of Composites for Construction*, 12: 405-415.

[192] Soleimani, S.M., Banthia, N., Mindess, S. (2007), *Advances in Construction Materials*, 3: 279-286.

[193] Takeda, K., Mitsui, Y., Murakami, K., Saki, H., Nakamura, M. (1996), *Composites Part A: Applied Science and Manufacturing*, 27: 981-987.

[194] Taya, M., Chou, T.W. (1981), *International Journal of Solids and Structures*, 17: 553-563.

[195] Teng, J.G., Chen, J.F., Smith, S.T., Lam, L. (2002), *FRP Srengthened RC Sructures*, John Wiley and Sons, Ltd.

[196] Teng, J.G., Lam, L., Chen, J.F. (2004), *Progress in Structural Engineering and Materials*, 6: 173-184.

[197] Toledo, R.D., Scrivener, K., England, G.L., Ghavami, K. (2000), *Cement and Concrete Composites*, 22: 127-143.

[198] Tong, L., Mouritz, A.P., Bannister, M.K. (2002), *3D Fiber Reinforced Polymer Composites*, Elsevier Scinece Ltd.

[199] Toutanji, H., Zho, L., Zhang, Y. (2006), *Engineering Structures*, 28: 557-566.

[200] Triantafillou, T.C. (1998), *Progress in Structural Engineering and Materials*, 1: 126-134.

[201] Triantafillou, T.C., Plevris, N. (1992), *Materials and Structures*, 25: 201-211.

[202] Tripathy, S., Mishra, S., Nayak, S. (1999), *Polymer Composites*, 20: 62-71.

[203] Tsai, S., Hahn, T. (1980), *Introduction to Composite Mateirals*, CRC-Press.

[204] Tumialan, G., Belarbi, A., Nanni, A. (1999), *Department of Civil Engineering, Centre for Infrastructure Engineering Studies, Report No. CEIS 99-01*, University of Missouri-Rolla, USA.

[205] Tung, C.Y.M., Dynes, P.J. (1982), *Journal of Applied Polymer Science*, 27: 569-574.

[206] Van Den Einde, L., Zhao, L., Seible, F. (2003), *Construction and Building Materials*, 17: 387-403.

[207] Walpole, L.J. (1966a), *Journal of the Mechanics and Physics of Solids*, 14: 151-162.

[208] Walpole, L.J. (1966b), *Journal of the Mechanics and Physics of Solids*, 14: 289-301.

[209] Walpole, L.J. (1969), *Journal of the Mechanics and Physics of Solids*, 17: 235-251.

[210] Wambua, P., Ivens, J., Verpoest, I. (2003), *Composites Science and Technology*, 63: 1259-1264.

[211] Weng, G.J. (1990), *International Journal of Engineering Science*, 28: 1111-1120.

[212] Willis, J.R. (1991), *Journal of the Mechanics and Physics of Solids*, 39: 73-86.

[213] Yalcin, C., Kaya, O., Sinangil, M. (2008), *Construction and Building Materials*, 22: 295-307.

[214] Yan, C. (2006), *On Homogenization and De-homogenization of Composite Materials*, Ph.D. Dissertation, Drexel University.

[215] Ye, L.P., Zhang, K., Zhao, S.H., Feng, P. (2003), *Construction and Building Materials*, 17: 499-506.

[216] Young, S.C., Harries, K.A. (2000), *An Investigation of Properties and Procedure of Spray Layed-Up Fiber Reinforced Polymer Materials for Concrete Rehabilitation*, Report No.ST00-03, University of South Caronia, pp. 1-57.

[217] Zahr, G.E., Riewald, P.G. (1989), *Plastics Compounding*, 12: 60-65.

[218] Zhang, Z., Hsu, C.T.T., Moren, J., (2004), *ASCE Journal of Composites for Construction*, 8: 403-414.

[219] Zhao, Y.H., Weng, G.J. (1996), *International Journal of Plasticity*, 12: 781-804.

[220] Zhao, Y.H., Weng, G.J. (1997), *International Journal of Solids and Structures*, 34: 493-507.

[221] Zhao, X.L., Zhang, L. (2007), *Engineering Structures*, 29: 1808-1823.

In: Fiber–Reinforced Composites
Editor: Qingzheng Cheng

ISBN: 978-1-61470-303-7
© 2012 Nova Science Publishers, Inc.

Chapter 19

FLEXIBLE FIBER COMPOSITES FOR SPACE STRUCTURES

Leri Datashvili[*] and Horst Baier

Institute of Lightweight Structures (LLB), Technische Universität München (TUM),
Boltzmannstr. 15, D-85747 Garching, Germany

ABSTRACT

Flexible fiber composites are those in which fibres reinforce (highly) flexible matrix materials like urethane, silicone or similar. In this chapter basics of flexible fiber composites are given focusing on carbon fiber-reinforced silicone (CFRS) composites. Specific methods are addressed for manufacturing, testing, and analysis of unidirectional CFRS laminate as well as CFRS of both biaxially and triaxially woven fabric (TWF). A method of homogenization of material properties of TWF flexible composites is given in detail.

For this a derivation procedure is developed and discussed for a 9 x 9 ABD stiffness matrix of an equivalent laminate, which is based on micromechanical finite element analysis of a repeated unit cell of the composite. Using an effective beam finite element model, 3D deformability of the cell is fully accounted for. The resulting properties include tensile and flexural as well as transverse shear stiffnesses, which are particularly needed for characterizing the flexible fiber composites. It is concluded that TWF composites behave as transversely (quasi) isotropic materials according to the calculated homogenized properties.

After addressing possible space applications like large deployable reflectors and large membrane structures, a focus is given on shape variable space structures using the flexible fiber composites as reflecting morphing skins. The prospective of the use of CFRS in shape morphing structures opens new challenging application direction for this material.

---

[*] Email: Datashvili@llb.mw.tum.de

NOMENCLATURE

$[ABD]$ = Stiffness matrix of a laminate
$[A]$ = tensile and shear stiffness sub-matrix of [ABD] matrix
$[D]$ = bending stiffness sub-matrix of [ABD] matrix
$[B]$ = coupling stiffness sub-matrix of [ABD] matrix
$[U]$ = Displacement matrix of all boundary nodes of the repeated unit cell (RUC)
$[F]$ = Resultant force and moment matrix of all boundary nodes of the RUC
l_x = Width of the RUC
l_y = Height of the RUC
E_f, E_m = Modulus of elasticity of fiber and matrix respectively
E_a = E-Modulus in a direction
G_{ab} = Shear-Modulus in a b direction
v_{ab} = Poisson ratio in a b direction
n_b = Number of boundary nodes of the RUC
S = Vector of thermal forces and moments per unit length
α_1 = Coefficient of thermal expansion (CTE) for 1 or x direction
α_2 = CTE for 2nd or y direction
α_3 = CTE for 3rd or z direction
α_9 = CTT – thermal twisting coefficient
ε = Vector of strains of the RUC
κ = Vector of curvatures of the RUC

INTRODUCTION

Flexible fiber composites are those in which fibers or fiber fabrics reinforce (highly) flexible matrix materials. Aramid, glass, carbon, steel or other fibres are used for reinforcement of flexible matrices like urethane or silicone elastomers. Space structures, where such flexible composites can be applied, are usually of large size and light weight. These are then required to be stowed for transportation/launch into a space orbit. The use of flexible composites leads to the ease of packaging providing the possibility of high elastic shape variation maintaining the high dimensional stability and integrity of the structures.

The focus in this chapter is given to the investigations of carbon fiber reinforced silicones (CFRS). Results, methods and tools developed during these investigations are similarly valid for other types of fibers, fabrics and stiff matrix materials. The CFRS as a space structural member takes the origin [22] as an alternative to metal knitted meshes, which fold and deploy very easily, making up the reflecting surface of large deployable reflectors (LDR) (Figure 1). In addition, CFRS (and FRS generally) has received more and more attention for space structural applications. These applications range LDRs, morphing/reconfigurable surfaces of reflectors, large gossamer space structures, deployable membrane-truss structures, direct radiating and reflect arrays, and structurally integrated hinges for foldability among others.

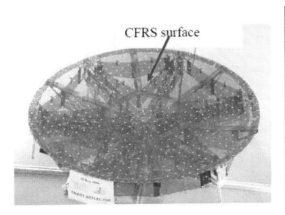 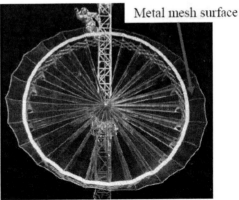

Figure 1. CFRS and metal mesh reflecting surfaces: SMART, LLB/TUM (left) and Georgian Mesh Reflector.

As the material properties of FRS, especially stiffness, are drastically different, simulation, design, manufacturing and testing of FRS composites have to be carried out differently compared to traditional composite materials. For example, isotropic properties of the stiff matrix composites are not present for flexible matrix composites in some cases (see section 3.1). During the design phase (local) buckling and wrinkling of the flexible composite structure has to be taken into account for much lower loads. In most cases there is no compression load bearing capability in structures made of flexible composites.

Most of the studies in FRS analysis and technology have been performed on triaxially woven carbon fiber fabric (TWF) reinforced silicone composite. Biaxially woven fabric (BWF) as well as unidirectional (UD) carbon fiber inlays have also been used for flexible composite laminates and were therefore investigated as well. Material characterization test results for mechanical and thermo-elastic properties as well as for radio frequency (RF) properties are discussed below. The tests have been correlated to the simulation results and appropriate analysis methods as well as tools have been developed. The use of this material allows for easy stowing and deployment. Because of the carbon fibers dominate the properties of CFRS, the CFRS is characterized with a very low thermal expansion coefficient (CTE) with no moisture absorption or micro-cracking effects even under severe environmental cycles. These points together with the RF properties indicate high dimensional stability and multi functionality capabilities of this material.

Fundamentals of the design, simulation, testing and manufacturing issues for flexible composites are discussed in this chapter showing different application directions. For example, in case of the carbon fiber TWF flexible composites, a number of applications have been investigated, predominately laboratory models of LDRs. These investigations resulted in the creation of novel concepts of LDRs, new simulation methods and tools, as well as to well-established manufacturing processes.

Some other applications include reconfigurable reflector antenna surfaces, which are based on the bending and shear flexibility of such composites. Triaxially woven carbon fiber reinforced silicone (TWF-CFRS) and biaxially woven fabric CFRS (BWF-CFRS) reconfigurable surfaces have been investigated and promising results have been obtained. These investigations are discussed in this chapter showing lab model test results and correlation to the simulation results. Final assessment of the surfaces is performed based on the RF analyses results.

Challenges and advantages of flexible fiber composite structures will receive an increasing attention because of satisfactory of environmental, design and cost requirements of the highly flexible deployable shape variable space large lightweight structures.

CARBON FIBER-REINFORCED SILICONE COMPOSITE MATERIAL

Reasons and Motivation for Creating Such a Material

Large deployable reflectors use metal knitted meshes as a reflecting surface (RS). Metal mesh is easy to transform as well as to fold and deploy due to its flexibility. The mesh surface becomes distorted due to the "pillow effect" because of the need to be tensed in an operational state [1]. Having evaluated the pillow effect of meshes, one can conclude that it is not possible to fully eliminate it. In the best case, it can be reduced close to a theoretical limit being faceting approximation error of the paraboloid. This in turn depends on the facet size: the larger the facet size the larger the deviations are.

On the other hand, carbon fiber reinforced plastic (CFRP) shell reflecting surfaces are used for small reflectors for high accuracy and dimensional stability. These shells are stiff and therefore their use is very limited in large deployable and/or reconfigurable reflectors.

A new composite reflecting material, combining the advantages of above mentioned reflecting surfaces, has been created and developed at the Institute of Lightweight Structures (LLB) of Technische Universität München (TUM) in the frame of Shell Membrane Antenna Reflector Technology (SMART) [2]. The reflectivity of the material is based on carbon fibers, which reinforce a soft silicone matrix. This material, on the one hand, has very good in-plane stiffness properties and low thermal expansion, on the other hand, it can be transformed and/or reconfigured into different shapes as well folded and deployed easily for large surfaces. This allows for the consideration of the reflecting surface material as a membrane. In addition, the following advantages are acquired:

- The necessity for tensioning, which is present in metal mesh LDRs, is fully eliminated. Therefore, the pillow effect is fully eliminated, as well. This is achieved by assigning a small bending stiffness to the RS, allowing consideration of the reflecting surface material as a shell.
- The reflector structure is free (unloaded) of the mesh tensioning forces. Therefore high accuracy and stiffness can be achieved with lower mass compared to the same structure with a mesh RS.

Mechanical, thermo-elastic and RF properties of the CFRS reflecting surface flexible composite material and its manufacturing peculiarities are discussed in the following.

Materials for CFRS Composite

A CFRS reflecting surface is composed of two materials: carbon fibers and a silicone elastomer. Study of fiber reinforced silicones has been performed on UD, biaxially and triaxially woven fabrics as reinforcement materials. In a particular case, investigations were focused on a triaxially woven carbon T300 fiber fabric of SK-802 style from SACASE Adtech, Japan, which was used to reinforce a space qualified silicone elastomer S 690 from Wacker Silicones, Germany. A piece of the TWF-CFRS made of these two materials is shown in Figure 2 with fiber and material directions indicated and basic dimensions given. Basic geometry parameters of the TWF are also shown in the figure. This TWF has dry fabric thickness of 0.13 mm and dry fiber tow width of 0.9 mm.

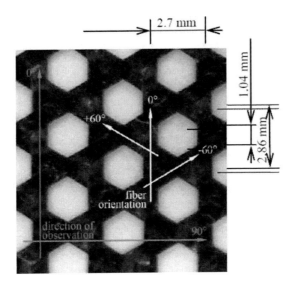

Figure 2. T300 fiber triaxially woven fabric reinforced silicone S 690.

Carbon fiber TWF has a number of attractive mechanical and RF properties promoting the research with further applications. Namely,

- large size TWF composite is characterized with transversely (quasi) isotropic mechanical properties as well as in-plane quasi-isotropic thermo-elastic and RF behavior
- possibility of the use of a single layer as a composite reinforcement due to the mentioned isotropicity
- very low aerial density of a single layer TWF composite due to its lattice (porous) structure and, therefore, ultra lightweight designs are possible
- Porosity helps in decreasing the response on acoustic loads and through height thermal gradients in a structure

The silicone S 690 is a development of the company Wacker Silicones, Germany. The most important and attractive properties of the silicone S 690 for the use in space reflectors and generally in space transformable structures are

- low glass transition temperature (Tg) below -105°C,
- high flexible above this temperature
- low outgassing and volatile content
- no moisture uptake
- resistivity to ultraviolet radiation, electron radiation and atomic oxygen
- wide range of service temperatures (-200 to over 250°C)
- possibility of curing at room temperature
- extremely low shrinkage during curing

The resulting composition satisfies stiffness and folding/deploying basic requirements of large space structures. In addition, it features a very low CTE (measured and calculated about -0.36e-6/°C), quasi-isotropic thermal behavior and no pillow effect since it does not require pretension when used as a double curved shell-membrane component of those structures.

Manufacturing Peculiarities of a CFRS

Since prepreg materials with silicone elastomers are not (yet) available, a wet lamination process had to be adapted for manufacturing of CFRS flexible composites.

A reliable manufacturing process has been developed at LLB enabling production of the CFRS composite with remarkably consistent quality. Moreover, in case of the TWF, it is ensured that hexagonal holes of the fabric remain free of silicone. A volume fraction of fibers is controlled within 3% for manufacturing of different specimens or components, which is still a well acceptable value since silicone provides insignificant influence on the properties of the final composite.

As with any other composite material not made out of prepregs, the manual manufacturing process for a single layer of CFRS consists of the following basic steps:

- Cutting of a reinforcing material: Hand laminating with relatively high viscosity matrix material makes it particularly important to care for the fiber directions in unidirectional laminates and biax or triax fabrics. To prevent the de-orientation of the fibers, support frames can be applied to the reinforcing material before cutting.
- Application of primer to the reinforcing material for reliable bonding of fibres and silicone: Standard primer materials are preferred solutions for the use in priming, but, if high flexibility of the final composite is required, special highly flexible primers shell be used. For the elastomer S 690, the G 790 primer is standard, while a specially created G 481 series is a very flexible primer preventing fibers from breaking during transformation.
- Applying of a matrix material and impregnation of the reinforcing material: At this step fiber volume is tuned, which, as practice shows, falls in the range of 40% with total of 3% variation bandwidth for S 690 silicone.
- Curing of the composite: Curing at room conditions gives an advantage of manufacturing of large sized doubly curved structures and structural elements on significantly cheap molds.

The attention in the manufacturing shall be paid to ensuring good fiber impregnation and uniform distribution of the silicone on both sides of the fabric as well as over the surface.

Another point, which has to be taken into account particularly for flexible composite materials with UD inlays, consists in the following: Usually UD fiber inlays are delivered with a support grid made of e.g. polyethylene (PE) fibers (Figure 3). One has to remove this PE grid before processing in order to ensure dimensional stability and integrity of the composite under the loaded conditions.

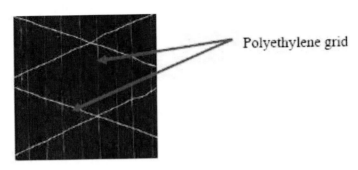

Figure 3. Unidirectional fiber material with polyethylene grid.

As a last step, a post-cure treatment of CFRS in an ethylene atmosphere has to be performed. Such a treatment helps to reduce a compression set of the silicone from about 40% to about 6% (Figure 4), giving the best long term elastic behavior of the silicone.

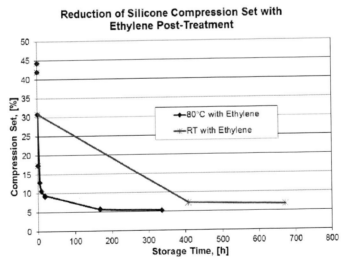

Figure 4. Compression Set improvement with the Ethylene post-treatment of Silicone.

As the graphs in Figure 4 show, the compression set of the silicone improves significantly after storage of the cured silicone in the Ethylene atmosphere. Improvement of the compression set is even better and faster if storing in Ethylene atmosphere and at higher temperatures at the same time.

Mechanical and Thermo-Elastic Behavior of the CFRS

High flexibility of the CFRS calls for the development of the special methods and tools for testing as well as for the analysis of thermo- mechanical properties for its characterization. Use of standard strain measurement techniques is complicated and sometimes not possible due to the very low stiffness of the silicone: it is much lower than of the strain gauges and attachment of extensometers to the CFRS surface is not reliable. It is best to use a 3D strain measurement system e.g. ARAMIS [23]. In addition, attention has to be paid to the size of the specimens correlating all the measurement results to the theoretical (numerical) predictions.

Due to the low stiffness of the flexible matrix and relatively high matrix content, the properties of CFRS have to be predicted either by solid finite elements or beam/shell elements with transverse shear capabilities. Calculation of the in-plane mechanical properties of the UD-CFRS composite laminate material can be also successfully done using the classical laminate theory (CLT) [12], [13], [16] taking into account transverse shear flexibility for out-of-plane (bending, twist) properties. In addition, triaxially and biaxially woven fabrics composites require the use of homogenization methods for calculating accurate properties of these materials for further use in large structures analysis.

Same is valid for the measurement of thermo-elastic properties of CFRS composite. Because of high flexibility of the matrix, standard measurement methods cannot be used therefore a new method has been developed [2] and used for the measurements, results of which are discussed below.

Results of the tests and micromechanical analyses for TWF composites are compared and discussed in the following deriving useful homogenized properties. Micromechanical modeling and analysis method is explained further as an effective homogenization method.

Basics Considerations in TWF Composite Mechanical Properties

A complex geometry of the triaxial weave makes it more of a structure than a material, and the need to find an effective material law for modeling of large size structures makes the analysis very challenging. In addition, instead of conventional stiff matrices like an epoxy is (TWF-CFRP), having a flexible silicone rubber as a matrix material (TWF-CFRS) makes the analysis even more challenging.

Accounting for these challenges, several different techniques of micromechanical modeling have been examined using beam, shell or solid elements. Solid elements give better possibility to realize the triax fabric geometry compared to the shell elements, while beam elements are simpler giving the possibility of modeling and analysis of relatively large size structural components in an effective computational time.

In a solid element finite element (FE) model of the TWF composite the crossing tows are directly modeled with the joint interface forming a "micro-laminate". In the beam models of TWF (or BWF) composites the crossing tows have to be modeled separately from each other due the line (beam) element geometry. Then accurate connection modeling and calculating of the tow properties from that micro-laminate of the composite is necessary for the beam FE model (micro-laminate phenomenon is addressed in details in section 3.3 for the beam FE model).

For TWF composites with stiff matrix materials, several different possibilities (coupling, rigid beam) were used for tow-to-tow connection modeling, e.g. [6], which turns the beam FE model to the quasi-2D deformable model. Besides, tows were not assigned properties from the "micro-laminate". Therefore, the solid element modeling had to be used for predicting the thermo-elastic properties.

For the composite with a flexible matrix however, the importance of accurate modeling of the tow and the connection element increases drastically. Such modeling is necessary even for prediction of the mechanical properties of the composite. With accurate modeling of the connection, the beam FE model turns into a 3D deformable model, Therefore, only using accurately defined tow properties at crossing interface and flexible/elastic beams as connection elements between tows a good correlation to the sample test results has been achieved (Figure 5). In the figure the comparison of the FE analysis results to the tensile test results is performed based on deformed pictures and the values of elastic modulus in the transverse to the fiber direction (90° direction of the material). The sample size in these tests is 100 × 200 mm.

It will be shown later that such a modeling approach enables accurate prediction of the thermo-elastic properties along the mechanical properties of TWF composites using the same beam element model. It allows further application of the higher order plate theory as compared to Kirchhoff plate theory, e.g. inclusion of transverse shear capabilities into the homogenized material model.

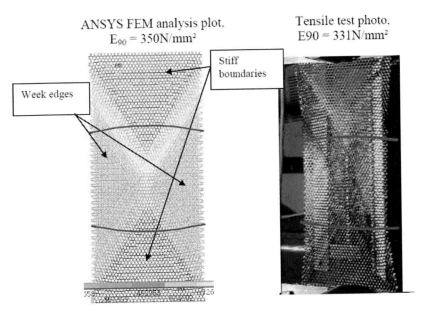

Figure 5. Triax CFRS FEM and lab tensile tests for 90° direction (vertical in the figure).

In-plane mechanical properties in fiber direction of the specimen are not remarkably influenced by the size of the specimen. On the contrary, properties in transverse to the fiber direction are drastically dependent to the size of the specimen; therefore the deeper investigations were carried out for that. From all the analysis performed on a finite size specimen level, the following have been concluded for the test direction in transverse to the fiber direction (Figure 5):

- Mechanical behavior of TWF composites is influenced by
- a free edge effect which in turn is a function of the width of the specimen
- a fixed boundary effect which in turn depends on the length of the specimen.
- The free edge effect is being reduced with increasing the width since contribution of the weakness of the free edges to the total cross section stiffness decreases. Therefore the specimen with higher width shows higher stiffness.
- The fixed boundary influence is decreased at long specimen sizes, so that the influence, therefore the stiffness, decreases with length increase.
- Both effects become reduced with the increasing the size of specimens. Therefore the aspect ratio (length to width ratio, L/w) dependence disappears. This dependence is not at all expressed for larger ratios above the tangent of the angle between of the inclined fibres and fixed edge (60° in this case).

Tensile test results of TWF-CFRS single ply specimens of finite size (Figure 6) also show that it exhibits an anisotropic behavior irrespective of a quasi-isotropic behavior of TWF-CFRP single ply specimens of finite size.

The modulus of elasticity $E_{0°}$ in the fiber direction is much higher than orthogonal to that direction - $E_{90°}$ (see also Figure 2). This is the case even for the variation of the width up to 1 m with a fixed aspect ratio of the specimens.

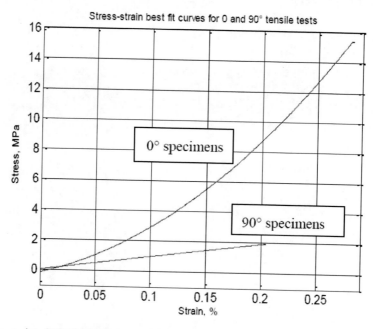

Figure 6. Stress-strain of TWF-CFRS, test results.

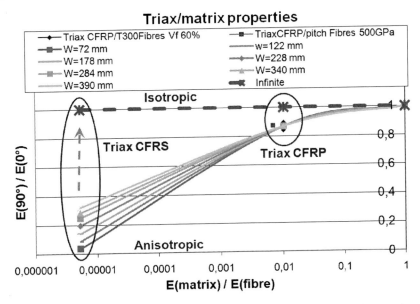

Figure 7. In-plane mechanical properties of TWF-CFRP and TWF-CFRS, tests and FE analyses results.

The test results and FE analyses results for flexible as well as stiff matrix materials and increasing size with the fixed aspect ratio are summarized in Figure 7. It shows relations between the ratio of $E_{90°}$ to $E_{0°}$ of a ply (vertical axis in the figure) and the ratio of a matrix modulus of elasticity to a fiber modulus of elasticity (horizontal axis). Results of calculations and tests of TWF-CFRS and TWF-CFRP materials are given in the figure.

In-plane anisotropy of a finite size TWF-CFRS ply is very easy to detect from the figure. The stiffness orthogonal to fiber direction ($E_{90°}$) increases with the increase of specimen width. The increase of $E_{90°}$ is shown with the dashed vertical arrow in Figure 7 taking into account that the stiffness in fiber direction - $E_{0°}$ is almost constant. Therefore with the increase of the width the in-plane properties of triax CFRS go to the isotropic direction which is $E_{90°}/E_{0°}=1$. A TWF-CFRP ply shows quasi-isotropic properties already at very low width of a test article (Figure 7).

Changing the analysis from finite to the infinite size, both materials show in-plane isotropic behavior. This conclusion is a result of the linear analysis performed on the unit cell of TWF flexible composite with periodic boundary conditions (see section 3.3). An indication of in-plane isotropic properties is plotted with a horizontal dashed line in Figure 7.

Tests and FE analyses confirmed that, when dealing with finite size of the TWF-CFRS structure, boundary conditions and its geometry have to be treated carefully and anisotropic behavior has to be taken into account. Although in some special cases like circular or quadratic large structures (e.g. large reflector surfaces) and statically determined boundaries, it can be also considered isotropic for small strains.

Thermo-Elastic Properties of CFRS

A set of measurements were performed aiming at the characterization of CFRS material coefficient of thermal expansion (CTE) in different fiber orientations. A method of the CTE measurement developed at LLB [2] is based on the use of non-standard single layer rolled tube or stacked specimens fixed with ropes. It has enabled measurements of the CFRS composite material in a standard vertical push rod dilatometer discussed hereafter.

Coefficient of Thermal Expansion of the TWF-CFRS

Specimens of three different directions (0°, 45° and 90°) were cut from TWF-CFRS single layer for CTE measurements and thermo-elastic characterization of the composite. Thermal deformations graph, which represents the average of all specimens' measurement results (Figure 8), shows the following:

- Two characteristic ranges are observed with different CTE: below glass transition temperature (Tg ≈ -105°C) of silicone and above Tg.
- In the first temperature range the CTE equates to $3.86 \pm 0.25 \times 10^{-6}/°C$. Since all specimens are stiff in this range due to stiffening of silicone, a silicone influence on the resulting CTE is significant. Specimens with higher silicone volume show higher CTE.
- The second range is characterized with nearly no influence of silicone because of its flexibility. The resulting CTE is negative ($-0.36 \pm 0.08 \times 10^{-6}/°C$) and is in the range of the fibers' CTE.
- The effective CTE of TWF-CFRS is calculated as $0.224 \times 10^{-6}/°C$

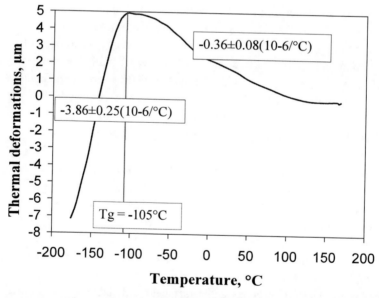

Figure 8. Thermal deformation and CTE of TWF-CFRS in different temperature ranges.

Comparison of the measured CTEs in different fiber directions shows a quasi-isotropic thermo-elastic behavior of the triax CFRS. Using the FE micromechanical model explained in section 3.3, a CTE of TWF-CFRS was predicted as -0.48e-6/°C, which matches very well the measurement results and confirms the quasi-isotropic thermal behavior of the TWF composite.

Coefficient of Thermal Expansion of UD-CFRS Quasi-Isotropic Laminate

One of the promising results has been observed when measuring the CTE of quasi-isotropic laminate (~40% of fiber volume) constructed of UD-CFRS layers. As it is shown in Figure 9 the CTE of the laminate is almost zero above the glass transition temperature of silicone.

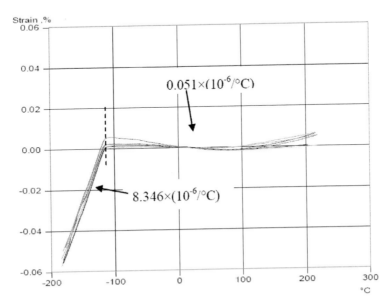

Figure 9. Thermal strain vs. temperature and CTE of q/i specimens of the UD CFRS in different temperature ranges.

The influence of the silicone and off-axis fibres in the quasi-isotropic laminate is much higher than in the TWF-CFRS composite below of the glass transition temperature, therefore CTE equates to 8.3 ppm/°C.

CTE measurement result above the glass transition temperature indicates that creation of the highly accurate and dimensionally stable surfaces is feasible using quasi-isotropic laminates of the UD-CFRS.

Homogenization of Mechanical and Thermo-Elastic Properties of TWF Composites via Micromechanical Modeling and Analysis of a Repeated Unit Cell

General Discussions and Assumptions

The problem of modeling and analysis of TWF composites with stiff matrices has been investigated intensively according to literature, [5] - [8]. A micromechanical modeling method of the repeated unit cell (RUC) was used, which was introduced and discussed in detail in [9]. Using the RUC model and Kirchhoff's plate assumptions homogenized properties in terms of an [ABD] stiffness matrix were calculated. Two different modeling methods are applied in the papers for TWF composites: beam FE modeling and solid FE modeling. Solid element FE models are more representative. They can be used for prediction of mechanical as well thermo-elastic properties. Beam FE models are simple and fast, but proper modeling is required to predict all the mentioned properties.

Using the beam micromechanical model, prediction of thermo-elastic properties of TWF composites were not possible [6] and use of solid element FE models was practiced. In the beam FE models of RUC a rigid (coupled) tow to tow connection was used, turning the model into a quasi 2D deformable model. This way the modeling was simplified but the inter-influence of the tows and thickness change influence was lost (see section 3.1) leading to nonrealistic thermo-elastic properties. Because of this, local bending stiffening was also not considered at the cross-over area of the tows.

In fact, a single ply of the TWF composite consists of many tow-to-tow connections which are angle-ply antisymmetric [12] "micro-laminates" and dominate the definition of the resulting TWF composite mechanical and thermo-elastic properties.

Here are some general assumptions which have to be stated prior to the analysis:

- Material principal directions are aligned with a tow and transverse to this tow and will be called $0°$ (x) and $90°$ (y) directions respectively ((see also Figure 2)). The 3rd (z) direction is orthogonal to the first two directions.
- The behavior of a triax composite material is nonlinear. In deriving the homogenized material properties, only small strains are considered and therefore, the material behavior can be assumed linear. Then the results should be correlated to the tests in a very small starting strain range.
- The repeated unit cell will be modeled with beam FE elements allowing 3D deformation. Nine load cases will be applied for determining the material properties in the following six force and three moment cases: three tensions, three shears, two bending and one twist load cases. These kinds of loadings will allow accounting for transverse shear and 3rd direction strains.
- According to the assumed linearity, linear analysis can be performed for calculation of the [ABD] stiffness matrix, which can reasonably represent the TWF composite material.

Based on the "micro-laminate" phenomenon the following conclusions can already be made using the CLT and the related assumptions:

- A TWF composite on a microstructure level consists of many asymmetric two layers micro-laminates with layers of the following orientations: $0°$ and $+60°$, $0°$ and $-60°$, $+60°$ and $-60°$. These micro-laminates are constructed by the three tows of $0°$, $-60°$ and $+60°$ orientations. In their own coordinate system all of them can be considered as antisymmetric $±θ$ angle ply micro-laminates. The existence of an equal number of differently oriented antisymmetric $±θ$ angle ply micro-laminates makes the composite in-plane (quasi) isotropic.
- Since tows always interchange positions going from up to bottom and from bottom to up the overall bending of micro-laminates is completely stabilized compared to the infinite antisymmetric angle ply laminate. In fact, interface planes of tow-contacts are not bent at all in case of equal axial forces in the tows. Instead twisting of the

composite takes place as expected from a [B] matrix of the antisymmetric angle ply laminate.

- An interchange of the positions of the tows, where each tow is free alone, is responsible for the transverse shear flexibility of TWF composites together with the inter-laminar shear flexibility of the micro-laminates.
- The previous point is also proved from the following: although the 0/-60° and 0/+60° antisymmetric micro-laminates have fully populated [ABD] matrices, if considering them in series, they have sign-inverted tensile-bending coupling terms so they compensate for overall bending, while the tensile-twisting coupling terms are not compensated and are summed up. In addition, the micro-laminates of -60°/+60° directions have only the twisting coupling terms since they are of $\pm\theta$ orientation. Therefore, only expansion-twisting coupling is expressed globally. Same can be easily explained for shear and bending coupling
- In-plane shear and tensile coupling is compensated in the same way for the micro-laminates of tows of 0/60 and 0/-60 directions. This is the reason that in a tensile test an overall shear is not present. The same is valid for bending and twisting coupling: when bending the specimen no twisting will take place because they are compensated locally. Instead, a global shear will occur.
- These discussions are valid for linear interpretation of the TWF composite (small strain). In reality the TWF composite material has an in-equally expressed nonlinear behavior in only some directions.
- As for the thermal behavior of the TWF composite, there is thermal twisting additionally to the in-plane thermal expansion, similarly to classical antisymmetric $\pm\theta$ angle ply laminates.

For example, comparing the [ABD] stiffness matrix of an antisymmetric angle ply laminate to the 6×6 [ABD] matrix of TWF composite from [5], one can observe an identical population of the matrices (with only a difference in sign antisymmetry observed in the [B] sub-matrix of TWF composite). CTE vectors of both are populated similarly as well having bi-axial in-plane expansions and twisting non-zero members only (see also section 3.3.5).

The discussions above have been taken into account and a beam FE model has been established for the RUC of TWF composites. This model avoids the common modeling over-simplifications and enables acquiring both the homogenized mechanical and thermo-elastic properties requiring significantly less computational time compared to solid element FE models.

Repeated Unit Cell of the TWF-CFRS

For all the modeling tasks ANSYS finite element software is used and a beam model has been established in it. The established beam FE model [17] is fully parametric and allows simulation of a composite with any style of TWF, of different resin type (stiff/flexible), fiber materials, filament amount in a tow, cell size, ply thickness and tow width, being input parameters of the model.

The beam FE model of a unit cell consists of series of beam elements arranged in sinusoidal shape from one cross point to the other combining a tow (Figure 10). 0° tows are oriented parallel to x axis therefore y axis corresponds to the material 90° direction which always appears between the tows. Therefore a cell height is defined along y axis while a cell width is defined along x axis.

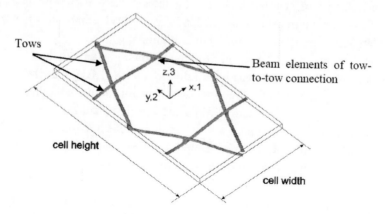

Figure 10. FE unit cell of triax composite.

Quadratic beam elements (BEAM189 in ANSYS) are used for modeling the tows. This element type is selected for its transverse shear capabilities according to the Timoshenko beam theory. It gives the possibility of accurate representation of the transverse shear stiffness (flexibility) particularly needed for the flexible matrix materials.

Note that, in contrast to the beam models found in the literature, in which the tow-to-tow connection is assumed rigid, the connection has been modeled by the same element type as for the tow itself. This gives an increased possibility to model the triax with beam elements as it permits to properly specify the stiffness and thermal-expansion characteristics of the connection. Importantly, 3D deformability of the RUC is maintained.

The modeled cross-section of the tows and its coordinate system can be seen in Figure 11. The cross-section has an adaptive (parameterized) shape going from diamond shape via approached lens shape to a rectangular shape. It is introduced for simulating the tows with higher or lower fiber fractions enabling a change of the tow cross section area. As it is denoted in Figure 11 the parameterized points are used to vary the cross-sectional area without changing the tow width or height, which are (measured) input geometry parameters.

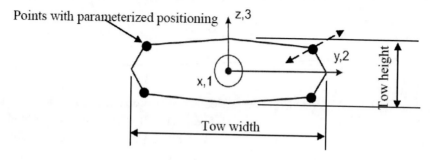

Figure 11. Tow cross-section with tow coordinate system.

Figure 12. Two different cross-sections indicated by different colours (expanded cross-sections of the beam FE model).

Two types of cross sections with different areas are defined in the model (Figure 12) using the same adaptive shape principle. One of the sections uses high fiber volume and represents tow at contact areas while the other is used to model tows at free regions with higher resin content.

Another important feature of the model is a possibility of modeling of the tow with a layered cross-section. The need of the layered modeling was identified when analyzing the micro-sections of the TWF composites having relatively high fiber volume at tow-contact interfaces and lower volume far from that interfaces. Influence of this effect is significant if resin variation takes place differently on either side of the composite. If the resin variation is symmetric then the influence of the effect on homogenized TWF composite properties is expressed mainly in bending properties, for example in TWF-CFRS.

The fact that antisymmetric angle-ply laminate properties (Figure 13) are calculated for tow-to-tow connections and assigned to the connecting beam as well as to the tow beams makes the modeling different from any existing in the literature beam models. Namely, such an approach enables the simulation of mechanical and thermo-elastic properties in a single run of the unit cell beam model.

The connection beam length is set equal to half the thickness of the TWF composite ply. It corresponds to the sum of the half thickness of the upper tow and the half of the lower tow.

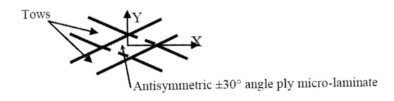

Figure 13. Tow cross over area with a coordinate system.

As mentioned, in crossover ranges, the tows form micro-laminates. Therefore, they are stiffened as compared to the free areas. To account for this behavior, that cannot directly be modeled using beam models, the crossover area was stiffened using "multi point constraint" MPC184 elements with the point-in-plane key option For each crossover area, a separate plane was defined. This plane was fixed to the center node of the crossover area and was left free for rotation in space to follow the node. All the nodes falling inside of the effective interface of tows keep their distance to this plane. The whole tow could still rotate, strain and displace. Bending stiffening of tows is shown in Figure 14 schematically. The tow coordinate system is shown in the figure as well.

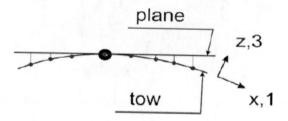

Figure 14. Model of the bending stiffening of the tow crossover area.

Determination of the Properties of the Tow at Free Area

Having measured input geometry and mechanical properties of the tow material, some missing material properties of the tow have to be calculated. These will be assigned to the tow beams and especially to the connection beams. The calculation of all the necessary material properties is performed in three steps:

- Using the mixture rule and other well-known formulas for calculating the tow properties at the free area as of a single ply of a UD composite material
- Calculating tow properties at the crossing area from the antisymmetric angle ply micro-laminate
- Assigning of the latter properties also to the connection beam but rotated into right coordinate system

A set of formulas has been selected and used in the model described here based on comparing many different available formulas for calculating the UD composite material properties. The selected formulas correlate best to the test results as well as to those from literature.

The extensional modulus along the tow x-axis E_1 and the Poisson's ratio v_{12} have been obtained from the rule of mixture using the known fiber and matrix properties:

$$E_1 = \varphi_f E_{1,f} + (1-\varphi_f)E_m \tag{1}$$

$$v_{12} = \varphi_f v_{12,f} + (1-\varphi_f)v_m \tag{2}$$

Transverse extensional modulus E_2 has been determined using the Halpin-Tsai semi empirical equation [15]

$$E_2 = E_3 = E_m \frac{1+\xi\eta\varphi_f}{1-\eta\varphi_f} \tag{3}$$

where

$$\eta = \frac{E_{2,f} - E_m}{E_{2,f} + \xi E_m} \tag{4}$$

and the parameter $\xi=2$ is a measure of reinforcement of the composite that depends on the fiber geometry, packing geometry, and loading conditions.

Similarly, for the shear modulus G_{12}, the Halpin-Tsai semi-empirical relation [15] has been used.

$$G_{12} = G_{13} = G_m \frac{\left(G_{12,f} + G_m\right) + \varphi_f\left(G_{12,f} - G_m\right)}{\left(G_{12,f} + G_m\right) - \varphi_f\left(G_{12,f} - G_m\right)} \tag{5}$$

The shear modulus G_{23} is obtained by solving the following quadratic equation [14] (used also in [6])

$$\left(\frac{G_{23}}{G_m}\right)^2 A + \left(\frac{G_{23}}{G_m}\right) B + C = 0 \tag{6}$$

where

$$\begin{aligned}
A = {} & 3\varphi_f\left(1-\varphi_f\right)^2\left(\frac{G_{12,f}}{G_m} - 1\right)\left(\frac{G_{12,f}}{G_m} + \zeta_f\right) \\
& + \left[\left(\frac{G_{12,f}}{G_m}\right)\zeta_m + \zeta_m\zeta_f - \left(\left(\frac{G_{12,f}}{G_m}\right)\zeta_m - \zeta_f\right)\left(\varphi_f\right)^3\right] \\
& \times \left[\zeta_m\varphi_f\left(\frac{G_{12,f}}{G_m} - 1\right) - \left(\left(\frac{G_{12,f}}{G_m}\right)\zeta_m + 1\right)\right]
\end{aligned} \tag{7}$$

$$B = -6\varphi_f\left(1-\varphi_f\right)^2\left(\frac{G_{12,f}}{G_m}-1\right)\left(\frac{G_{12,f}}{G_m}+\zeta_f\right)$$

$$+\left[\left(\frac{G_{12,f}}{G_m}\right)\zeta_m+\left(\frac{G_{12,f}}{G_m}-1\right)\varphi_f+1\right]$$

$$\times\left[\left(\zeta_m-1\right)\left(\frac{G_{12,f}}{G_m}+\zeta_f\right)-2\left(\varphi_f\right)^3\left(\left(\frac{G_{12,f}}{G_m}\right)\zeta_m-\zeta_f\right)\right] \tag{8}$$

$$+\left(\zeta_m+1\right)\varphi_f\left(\frac{G_{12,f}}{G_m}-1\right)\left[\frac{G_{12,f}}{G_m}+\zeta_f+\left(\left(\frac{G_{12,f}}{G_m}\right)\zeta_m-\zeta_f\right)\left(\varphi_f\right)^3\right]$$

$$C = 3\varphi_f\left(1-\varphi_f\right)^2\left(\frac{G_{12,f}}{G_m}-1\right)\left(\frac{G_{12,f}}{G_m}+\zeta_f\right)$$

$$+\left[\left(\frac{G_{12,f}}{G_m}\right)\zeta_m+\left(\frac{G_{12,f}}{G_m}-1\right)\varphi_f+1\right] \tag{9}$$

$$\times\left[\frac{G_{12,f}}{G_m}+\zeta_f+\left(\left(\frac{G_{12,f}}{G_m}\right)\zeta_m-\zeta_f\right)\left(\varphi_f\right)^3\right]$$

with

$$\zeta_m = 3 - 4v_m, \text{ and } \zeta_f = 3 - 4v_{12,f} \tag{10}$$

Having shear modulus G_{23} Poisson's ratiov_{23} can be computed from

$$G_{23} = \frac{E_2}{2\left(1-v_{23}\right)} \tag{11}$$

The CTE of the tow in the longitudinal direction can be accurately calculated with many different formulas. As for the transverse thermal expansion of the tow, several CTE formulas were compared in [10]. It has been shown that the formulas by Rosen and Hashin were most reliable correlating best to the test results. Therefore, both CTEs were calculated using these formulas in the present analysis. Superscripts f and m denote fiber and matrix respectively, terms with an overbar refer to effective composite properties and terms with a hat refer to volume average composite properties in the formulas (12) to (19) given below.

The calculation procedure for both longitudinal and transverse CTE values of the UD composite according to [11] is as follows:

$$\alpha_1 = \hat{\alpha}_1 + \left(\overline{S}_{11}-\hat{S}_{11}\right)\cdot\left[\left(\alpha_1^f-\alpha_1^m\right)\cdot P_{11}+\left(\alpha_2^f-\alpha_2^m\right)\cdot 2P_{12}\right]$$
$$+ \left(\overline{S}_{12}-\hat{S}_{12}\right)\cdot\left[\left(\alpha_1^f-\alpha_1^m\right)\cdot 2P_{12}+\left(\alpha_2^f-\alpha_2^m\right)\cdot 2\left(P_{22}+P_{23}\right)\right] \tag{12}$$

$$\alpha_2 = \hat{\alpha}_2 + \left(\overline{S}_{12} - \hat{S}_{12}\right) \cdot \left[\left(\alpha_1^f - \alpha_1^m\right) \cdot P_{11} + \left(\alpha_2^f - \alpha_2^m\right) \cdot 2P_{12}\right]$$
$$+ \left(\overline{S}_{22} - \hat{S}_{22}\right) \cdot \left[\left(\alpha_1^f - \alpha_1^m\right) \cdot P_{12} + \left(\alpha_2^f - \alpha_2^m\right) \cdot \left(P_{22} + P_{23}\right)\right] \tag{13}$$

where S is the symmetric compliance tensor

$$S_{11} = 1/E_1, \ S_{22} = S_{33} = 1/E_2 \tag{14}$$

$$S_{12} = S_{13} = -\nu_{12}/E_1, \ S_{23} = -\nu_{23}/E_2 \tag{15}$$

$$S_{66} = S_{55} = 1/G_{12}, \ S_{44} = 1/G_{23} \tag{16}$$

P is the symmetric modified stiffness tensor

$$P_{ab}\left(S_{bc}^f - S_{bc}^m\right) = I_{ac} \tag{17}$$

and I is the unit tensor. The required elements of P can be written as

$$P_{11} = \left(A_{22}A_{22} - A_{23}A_{23}\right)/D_A$$
$$P_{12} = \left(A_{12}A_{23} - A_{22}A_{12}\right)/D_A$$
$$P_{22} = \left(A_{11}A_{22} - A_{12}A_{12}\right)/D_A$$
$$P_{23} = \left(A_{12}A_{12} - A_{11}A_{23}\right)/D_A \tag{18}$$

where A_{ab} is defined as $\left(S_{ab}^f - S_{ab}^m\right)$ and D_A is given by

$$D_A = \left[A_{11}\left(A_{22}A_{22} - A_{23}A_{23}\right) + 2A_{12}\left(A_{12}A_{23} - A_{22}A_{12}\right)\right] \tag{19}$$

Tow twist and fiber undulation (if any) influence on the matrix dominated properties of flexible matrix TWF composites have to be taken into account. Namely transverse shear modulus and transverse CTE values are affected.

Now, since all the properties of the tow have been determined for the free area, we need to proceed for determining tow properties at crossover area.

Properties of the tows at cross over area and of the connection beam

Mechanical and thermo-elastic properties of the tows at crossover area and the connection beam have to be determined from the antisymmetric angle ply micro-laminate formed by the tows. They follow similar rules based on inter-influence of tows in the micro-laminate. The properties have been calculated by setting up the [ABD] matrix of this micro-laminate. Connection beam properties have been obtained from setting the layers as $\pm\theta$ oriented, while for the tow the layers have been set with the 0 and 60° directions. The layer properties are calculated with the same formulas as before, taking into account the increased tow fiber volume fraction and possible layered cross sections in the crossover range. The corresponding [ABD] matrices are calculated and inversed into the compliance matrix $[a]$:

$$[a] = \begin{bmatrix} [A] & [B] \\ [B]^T & [D] \end{bmatrix}^{-1}$$

(20)

Now the properties of the micro-laminate (tow at cross over area) can be calculated as well as assigned to the connection beam. The x-direction extensional modulus of the laminate $E_{1,lam}$ corresponds to the z-direction extensional modulus $E_{3,con}$ in the connection beam coordinate system (t being laminate thickness).

$$E_{3,con} = E_{1,lam} = \frac{1}{a_{11}t}$$

(21)

The y-direction extensional modulus of the laminate $E_{2,lam}$ corresponds to the y-direction extensional modulus $E_{2,con}$ in the connection coordinate system.

$$E_{2,con} = E_{2,lam} = \frac{1}{a_{22}t}$$

(22)

The x direction extensional modulus of the connection beam $E_{1,con}$ is the transverse extensional modulus of the tow at free area from (3)

$$E_{1,con} = E_2$$

(23)

$v_{12,lam}$ and G_{12} can be obtained from

$$v_{12,lam} = \frac{-a_{12}}{a_{11}}$$

(24)

$$G_{12,lam} = \frac{E_{2,lam}}{2(1+v_{12})}$$

(25)

The thermo-elastic properties of the antisymmetric angle ply micro-laminate are computed similarly using the formulas from CLT. The connection is again assumed as a 0/60 laminate. The corresponding layer CTEs, which have to be calculated for the UD tow at crossover area from (12) and (13), are

$$\alpha_{1,0^\circ} = \alpha_1$$

$$\alpha_{1,60^\circ} = \alpha_1 \sin^2(60^\circ) + \alpha_2 \cos^2(60^\circ)$$

The CTE of the micro-laminate in 0° direction, which is a longitudinal CTE for the tow at crossover area, is given by

$$\alpha_{1,\mathrm{mod}} = \frac{\alpha_{1,60^\circ} E_{60^\circ} + \alpha_{1,0^\circ} E_1}{E_{60^\circ} + E_1} \tag{26}$$

where E_1 is the 0° E-Modulus and E_{60° is the 60° E-Modulus of the tow

$$E_{60^\circ} = E_1 \cos^2(60) + E_2 \sin^2(60)$$

As for the axial coefficient of thermal expansion of the connection beam, the transverse CTE α_2 of the tow at crossover area is taken as calculated from equation (13).

Homogenization Analysis Procedure and Results

The analysis method used in this chapter is based on the direct micromechanics method from [9], which has been adapted to triaxial weaves in [5] and [6]. The method has been further expanded and adapted to the beam model discussed above which accounts for the 3D deformable plate. Namely, the method has been expanded to account for the non-rigid tow-to-tow connections both inside the cell as well as at the boundary. With this expansion, a total 20 boundary nodes are defined from which 12 nodes are coupled pairwise for the in-plane periodic boundary conditions and 8 nodes are internal nodes at cross-over areas and are constrained pairwise as well. The derivation procedure of thermo-elastic properties is also incorporated into the method. Below the basic steps are outlined, which have to be performed for calculation of homogenized mechanical and thermo-elastic properties according to the improved method.

A repeated unit cell of the single ply TWF composite (section 3.3.2), which is constrained via periodic boundary conditions in in-plane directions, can deform in all directions. Only an in-plane bending and XZ as well as YZ direction twists do not take place because of infinite stiffness coming out from the periodicity conditions and linearity of the analysis. The homogenized material properties are calculated as a 9 × 9 [ABD] stiffness matrix (in contrast to the Kirchhoff plate assumptions in [5], [6], which deal with 6 × 6 [ABD] matrix). Cross sections originally normal to the mid plane can rotate relative to the mid-surface and change the shape. Small strains are assumed for the linearization of the material behavior of the TWF composite. The forces and moments per unit length are related to the three expansion strains, three shear strains and three out-of-plane curvatures according to equations (27, 28). Note that, 3rd, 4th and 5th columns are omitted in the [B] and [D] matrices and same number rows – in the $[B]^T$ and [D] matrices.

$$\left\{ \begin{array}{c} [N] \\ [M] \end{array} \right\} = \left[\begin{array}{cc} [A] & [B] \\ [B]^T & [D] \end{array} \right] \bullet \left\{ \begin{array}{c} [\varepsilon] \\ [\kappa] \end{array} \right\} \tag{27}$$

$$\left\{ \begin{array}{c} N_x \\ N_y \\ N_z \\ N_{yz} \\ N_{zx} \\ N_{xy} \\ M_x \\ M_y \\ M_{xy} \end{array} \right\} = \left[\begin{array}{cccccc} A_{11} & A_{12} & A_{13} & A_{14} & A_{15} & A_{16} \\ A_{21} & A_{22} & A_{23} & A_{24} & A_{25} & A_{26} \\ A_{31} & A_{32} & A_{33} & A_{34} & A_{35} & A_{36} \\ A_{41} & A_{42} & A_{43} & A_{44} & A_{45} & A_{46} \\ A_{51} & A_{52} & A_{53} & A_{54} & A_{55} & A_{56} \\ A_{61} & A_{62} & A_{63} & A_{64} & A_{65} & A_{66} \\ B_{11} & B_{21} & B_{31} & B_{41} & B_{51} & B_{61} \\ B_{12} & B_{22} & B_{32} & B_{42} & B_{52} & B_{62} \\ B_{16} & B_{26} & B_{36} & B_{46} & B_{56} & B_{66} \end{array} \begin{array}{ccc} B_{11} & B_{12} & B_{16} \\ B_{21} & B_{22} & B_{26} \\ B_{31} & B_{32} & B_{36} \\ B_{41} & B_{42} & B_{46} \\ B_{51} & B_{52} & B_{56} \\ B_{61} & B_{62} & B_{66} \\ D_{11} & D_{12} & D_{16} \\ D_{21} & D_{22} & D_{26} \\ D_{61} & D_{62} & D_{66} \end{array} \right] \bullet \left\{ \begin{array}{c} \varepsilon_x \\ \varepsilon_y \\ \varepsilon_z \\ \varepsilon_{yz} \\ \varepsilon_{zx} \\ \varepsilon_{xy} \\ \kappa_x \\ \kappa_y \\ \kappa_{xy} \end{array} \right\} \tag{28}$$

The RUC model is used for the nine sequential analyses in a single ANSYS run. Each analysis is performed for a single strain or curvature as a unit load with the other eight being zeros. The tenth loading is a unit temperature for calculating the in-plane and out-of-plane thermo-elastic properties in the same FE analysis.

The [ABD] matrix is then calculated using the virtual work method. Two matrices of deformations [U] and forces/moments [F] are filled up after the FEM analysis of the 9 loading cases for all boundary nodes.

$$[ABD] = \frac{1}{l_x l_y} [U]^T \cdot [F] \tag{29}$$

[U] is a $9 \times (n_b \cdot 6)$ matrix, where $n_b = 20$ is the number of boundary nodes. The rows contain the values of the three displacements and three rotations for each boundary node. [F] is also a $9 \times (n_b \cdot 6)$ matrix. The rows contain the three reaction forces and three reaction moments for each boundary node. The nine rows represent nine loading cases in each of the two matrices.

The resulting [ABD] matrix of the TWF-CFRS with the materials applied (see section 2.2) is given in Table 1. The calculated stiffness matrix shows that the infinite size TWF composite is transversely (quasi) isotropic material. Here the term "quasi" is mentioned because of the slight difference and low couplings between the transverse shears and very low couplings of them to the rest. In-plane and bending isotropic properties are also easy to detect from the table. In contrast to conventional laminates, a single ply of the TWF composite – which cannot form a symmetric laminate with the given weave – shows isotropic coupling behavior as well. This is a consequence of the B matrix sign asymmetry: $B_{16}=-B_{61}=B_{62}=-B_{26}$, [5].

Flexible Fiber Composites for Space Structures 629

Table 1. 9 × 9 [ABD] matrix of the TWF-CFRS with T300 fibres and S 690 silicone

1084.14	1053.12	8.854	0	-0.00036	0	0	0	-0.00158
1053.12	1084.14	8.854	0	-0.00038	0	0	0	0.00158
8.854	8.854	0.310	0	0	0	0	0	0
0	0	0	1952.23	-0.1543	0	0	0	0
-0.00036	-0.00038	0	-0.1543	1952.04	0	0	0	0
0	0	0	0	0	15.508	0.00158	-0.00158	0
0	0	0	0	0	0.00158	1.1117	0.3703	0
0	0	0	0	0	-0.00158	0.3703	1.1117	0
-0.00158	0.00158	0	0	0	0	0	0	0.3707

It can be concluded from Table 1 that extending the [A] matrix to 6 × 6 in [ABD] matrix calculations has an effect to the calculated mechanical properties of TWF-CFRS and allows making a wide range of conclusions. Transverse shear stiffness per unit length equates to 1952 N/mm, it is in the finite range and shall be accounted for. The effective z-direction stiffness is low and corresponding stresses can be neglected. Instead, the related strains shall be associated to transverse Poisson's effect and can be easily obtained from the analysis. Transverse Poisson's ratios for the TWF-CFRS are in the same range as the in-plane one (about 0.9). In Table 1 the not-bold-written terms, which are couplings of transverse shear stiffnesses in between each other and to the in-plane tensile stiffnesses are rather low and can be neglected. Because of these, it is found that the TWF composite material is a transversely isotropic material. The bending stiffness [D] matrix of the TWF-CFRS is slightly overestimated and requires further effort for understanding possible reasons.

Further, the above discussions point that TWF composites with flexible matrix materials are better to be considered with transverse shear properties that should be derived from the 3D deformable RUC micromechanical model. In contrast, 2D deformable plate assumptions are accurately characterizing the stiff matrix TWF composite material mechanical properties. Although, 3D deformability of the RUC model is necessary for derivation of the thermo-elastic properties of TWF composites with stiff as well as flexible matrices.

Effective thermo-elastic properties of the TWF composites are determined from the unit temperature loading case of the FE analysis of the RUC. All opposing in-plane boundary nodes of the RUC are coupled in all DOFs. This is equivalent to setting all strains and curvatures from equation (28) to 0, maintaining the periodic boundary conditions. Temperature caused forces and moments at boundary nodes, which are located on each positive axes side of the RUC, are summated separately and thermal force and moment vector per unit length S is calculated in this loading case.

The CTE vector can now be calculated by multiplying the inverted [ABD] matrix to the vector S, [18].

$$(\alpha_i) = [ABD]^{-1} \cdot [S] \tag{30}$$

where $i = 1\ldots9$.

CTE calculation results are given in a vector below:

$$
\begin{pmatrix} \alpha_1 \\ \alpha_2 \\ \alpha_3 \\ \alpha_9 \end{pmatrix} = \begin{bmatrix} -0.4864 \\ -0.4847 \\ 44.398 \\ -0.703 \times (1/mm) \end{bmatrix} \times 10^{-6} (1/{}^\circ C)
\tag{31}
$$

The 0 terms are omitted in the vectors above. The non-zero results are CTEs α_1, α_2 and α_3 for expansions in three principal directions and a coefficient of thermal twist (CTT), α_9, for thermally induced twisting that describe the thermo-elastic behavior of the TWF composite.

Population of the above vector as well as the population of the resulting [ABD] stiffness matrix of TWF composite (not considering the 3rd direction) is identical to the population of the respective vector and matrix of an antisymmetric angle ply conventional laminate. Another important result of these investigations is the low coupling ratio between in-plane and out-of-plane stiffness properties (Table 1) as compared to the stiff matrix composites from [17], [6] and [5].

The thermal twist coefficient is also much lower compared to the respective value of the stiff matrix composites. These facts facilitate further design of the highly accurate reflecting surfaces of space antenna reflectors using fiber composites with flexible matrix materials.

RADIO-FREQUENCY CHARACTERISTICS OF THE CFRS REFLECTING MATERIAL

Radio-frequency (RF) characterization of the CFRS has been performed so far on the sample level. Investigations show that RF properties of the CFRS are mainly defined by fibers and no significant change of the properties occurs due to introducing the flexible silicone matrix instead of conventional stiff matrices. RF tests were performed in the following three types of measurement facilities for the frequencies up to 8 GHz:

- Free space measurements of fairly large samples in a compensated compact test range
- Waveguide measurements of small samples
- Free space measurements of samples put close to the aperture of a standard gain horn antenna

The measurement results obtained with the three test setups are summarized in Table 2 and compared to the performance requirements given in the same table.

Flexible Fiber Composites for Space Structures

Table 2. RF measurement results

Subject / RF-Requirement	TWF-CFRS
Reflection loss <0.1 dB	0.05 … 0.13
Depolarization <-40dB	<-47 …- 42
Amplitude variation for 2 orth. Polarizations < 0.02 dB	< 0.02
Phase variation for 2 orthogonal polarizaions < 2°	< 0.5°
Transmission coefficient < -20 dB	- 33 … - 21

All requirements are fulfilled for TWF CFRS samples for frequencies up to 8 GHz. The TWF CFRS reflecting surface shows excellent polarization behavior which means, that all requirements concerning depolarization, amplitude and phase variations for orthogonal polarizations are completely fulfilled with some additional margins facilitating the use of the material with selected composite up to 15 GHz.

Coating of the surface is being studied currently which aims for improvement of the RF properties of CFRS enabling the high frequency use of it with the above discussed materials. The applicable frequency can be also increased using high modulus fiber fabrics and closed weave fabrics (e.g. BWF) in the CFRS.

DEPLOYMENT AND FOLDING OF STRUCTURES FOR SPACE USE WITH THE CFRS FLEXIBLE COMPOSITE

Several structures and structural components were manufactured at the LLB lab and tested for verification of the expected behavior of CFRS composite material. These models include membrane rib structures, reflecting surfaces for different type of deployable space reflectors, structurally integrated hinges for rigid rotating parts of structures.

Since CFRS composite materials are characterized with relatively low shear stiffness a combination of CFRS with CFRP was investigated for tension and shear stiff membrane structures. As a result of these particular investigations a hybrid membrane material was created for use as a tension membrane. The material has been named CFRS/P. An example of CFRS/P hybrid material was manufactured and tested for the stiffening rib of the SMART reflector [1-4]. In this hybrid material, a carbon fiber fabric reinforces zones with flexible (silicone) and stiff (plastics - epoxy, cyanate ester etc.) matrix materials in series. The hybrid CFRS/P material exhibits high tensile stiffness and high dimensional stability (C-fibers), high flexibility for stowing/deploying (CFRS zones) and high in-plane shear stiffness (CFRP zones). In Figure 15 the deployed and folded membrane ribs are shown installed in a tensile test setup. One can see in this figure that in the folded state the stiff matrix zones remain planar, while the flexible matrix zones bend and function like hinges.

This hybrid material with flexible and stiff matrix materials is a promising lightweight candidate for building tensioned membrane structures where stiffness and accuracy in a specific direction is provided by the membrane, while the deployment and compression load bearing are performed by a deployable truss structure. This is demonstrated further in Figure 16.

In this figure the membrane CFRS/P ribs are installed and tensioned for providing stiffness and accurate profile of the reflecting surface of SMART. Deployment and tension is provided by a radially deployable pantographic truss structure. Thus, axial stiffness of such a structure is combined by a tensioned membrane and compressed truss while transverse stiffness and in-plane shear stiffness is mainly provided by the tensioned membrane. The folded hybrid ribs, which are mounted on either side of radially installed pantograph trusses, are also shown in Figure 16.

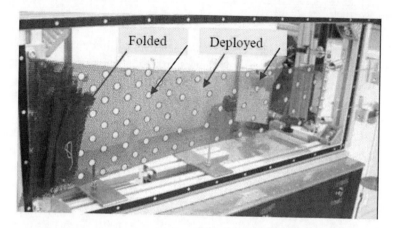

Figure 15. Hybrid membrane ribs for SMART, deployed and folded.

Figure 16. Hybrid membrane ribs of SMART back structure, folded and deployed.

The completed SMART reflector model includes TWF-CFRS reflecting surface. An assembly of the hybrid membrane ribs and the CFRS reflecting surface can be easily folded and deployed despite the orthogonal bonding. The folded, deploying and deployed states of the structure are demonstrated in Figure 17.

Figure 17. Deployment of SMART with Hybrid membrane ribs and CFRS reflecting surface.

In this figure the membrane ribs, which cannot be seen in this figure, are folded similarly as it is shown in the left photo of Figure 16 and the reflecting surface CFRS material is folded on the top of the folded ribs. The partly deployed reflector structure, shown in the middle figure, demonstrates the hybrid ribs being pulled by a deployment and tensioning mechanism (radial pantograph trusses) of the reflector. The ribs themselves transform the reflecting surface from a folded state into the operational double curved tension-free shape (right of Figure 17). Note that on the top of the reflector structure an auxiliary cross is mounted which is used as a coordinate system for shape photogrammetric measurements and is not part of the reflector structure.

The reflector concept investigated under SMART [3] was easily scaled up to the 7.5 m × 5 m elliptical aperture of centrally mounted reflector for the Multi-Application Purpose SAR (MAPSAR) mission (Figure 18).

Usage of the TWF-CFRS surface provides high accuracy and dimensional stability of the MAPSAR reflector fulfilling mission requirements in terms of stowed volume for Taurus class launchers as well as accuracy.

This reflector concept can be further scaled up to 15-20 m. If a side attachment of reflector is requested, a deployable ring supported CFRS RS is a preferable solution [21] instead of radially supporting ribs.

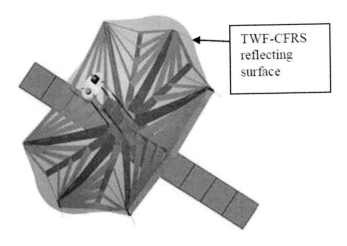

Figure 18. SMART for MAPSAR.

Another example of the use of the CFRS reflecting surface is given in Figure 19. Deployable radial ribs support CFRS RS in the demonstrated reflector. Deployment force is provided by springs which rotate each radial rib thus forming an umbrella type deployment mechanism.

Figure 19. Deployed and folded laboratory model of medium size LDR with TWF-CFRS reflecting surface.

Such a reflector concept is very promising for applications requiring medium sized LDRs (maximum 8 m diameter) because of its simplicity, reliability and functionality. Accuracies and reflecting properties are fulfilled for X-band RF with further prospective of higher RF as briefly outlined in section 4.

SHAPE MORPHING REFLECTING CFRS SKINS FOR SPACE RECONFIGURABLE ANTENNA REFLECTORS

Reconfigurable space reflectors are requested for telecommunication systems needing to change the coverage during in-orbit operation. Such a change might be needed several times per life time of the antenna. This means then that several different surface shapes must be achievable with the same reflector. In addition, thinking about the series of the reflectors for various missions, in which totally other shapes might be necessary, one concludes that reflecting skins of highly variable shapes are required for this. Recent investigations in shape reconfigurable reflectors for Ku band targeted applications have shown high challenge and feasibility of the use of the CFRS as a shape morphing skin of reflectors [19], [20].

Mechanical Requirements for Reconfigurable Reflectors

Mechanical design requirements to the reconfigurable reflectors can be split in several parts depending on their structural components. These are requirements to the reflecting surface, to the actuators and their placement, to the boundary conditions of surface, actuators as well as their interface. Most relevant to the material of CFRS are the following requirements to the surface to be fulfilled by the shape morphing skins:

Flexible Fiber Composites for Space Structures — 635

- Low tensile modulus in particular if surface edge is not free
- Low (even close to zero) in-plane shear modulus to allow wrinkle free reshaping (see also next)
- High bending stiffness in combination to the previous two properties and with a certain ratio to the shear modulus [19]

These requirements define a case of shear morphing with bending for the reflecting skin. In addition, the reflecting material shall be stable to temperature variations, have low RF losses, and shall be free of passive intermodulation as well as satisfying other space environmental requirements. Generally, it is impossible to find an available material satisfying the listed (mechanical) requirements. Therefore one has to design such a surface material. Taking plane weave triaxially or biaxially woven fabric CFRS materials, one can construct a laminate suitable for reconfiguration. The BWF-CFRS layer is about two orders of magnitude better than the TWF-CFRS layer in terms of bending to shear stiffness ratio, but does not satisfy the stiffness ratio requirement with a single layer. To achieve an acceptable stiffness ratio, the BWF-CFRS concept needs to have a laminate with only several layers due to the very low in-plane shear modulus. The TWF-CFRS concept needs many more layers due to the high in-plane shear modulus. Lower shear modulus (with the fixed stiffness ratio) results in lower required actuator forces. Therefore, layers of lower shear modulus (e.g. BWF-CFRS) are more desirable for designing and building the morphing skin of the reflector. As for the other mechanical requirements, statically indeterminate and over-determinate fixation of the surface to the actuators must be avoided. Further, a rotation-free interface design is needed. The free surface boundary of the reflector has been shown in [19] to best fulfill the shape-morphing surface requirements.

Testing of Shape-Variable Surfaces of Reconfigurable Reflectors

Mechanical and thermo-elastic properties of a single layer CFRS have been measured analyzed and discussed in previous sections. In terms of the mechanical behavior of both TWF and BWF-CFRS multilayer flexible laminates, CFRS multilayer laminate material properties do not remain constant with increasing the number of layers. Instead, the laminate tensile modulus, shear modulus and flexural stiffness are functions of the number of layers in the laminate caused by a very low tensile modulus of the matrix material and that the laminate thickness does not change linearly with the change of number of layers because of the weave structure.

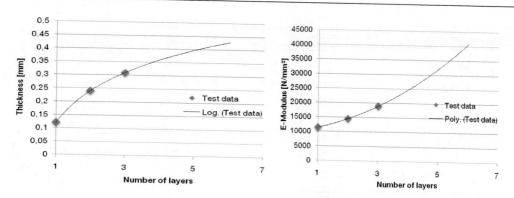

Figure 20. Laminate thickness (left) and E-modulus vs. number of layers.

The measured results of the thickness and tensile modulus of the BWF-CFRS are plotted in Figure 20. Nonlinear dependences of the thickness and elastic modulus of the biax are clearly observed from the figure. These observations are taken into account for the FE modeling of the CFRS surface concepts.

A breadboard of reconfigurable reflector was built at LLB for correlation of the FE model and verification purposes. The reflective surface was made of two continuous CFRS layers and patches on the back side around the actuator connection points. The patches increased locally the bending stiffness of the reflecting surface. The laboratory model is equipped with 19 manual linear actuators, Figure 21.

The deformed shapes were measured with a photogrammetry system from AICON and compared to the simulated shape. The maximum actuator displacements applied are in the range of ±5 mm with the RMS of 0.2 mm of the shape differences.

Figure 21. Laboratory model of CFRS surface reflector concept with 19 actuators.

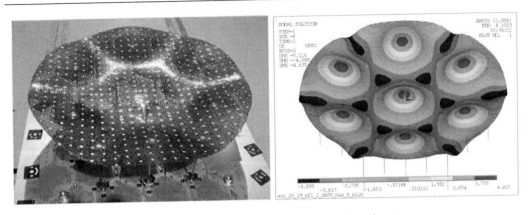

Figure 22. Reshaping results comparison according to the FE analysis and test.

Results of the FE Investigations

Prior to FE investigations of the shape, three target shapes have been defined for reconfiguration. These are European coverage, Brazil coverage and Conus coverage [20]. A main reflector of an offset earth deck antenna configuration as well as its sub-reflector has been studied for reconfiguration of all of the three coverages.

These three missions are considered as being representative of a typical scenario of reconfiguration. Target shapes have been used in the mechanical optimization procedure aiming all of them simultaneously in the FE analysis of the main reflector. The surface was constructed according to the designed material in both the FE and the lab models. For accounting for the transverse shear flexibility of the CFRS multilayer surface, pure silicone layers were used between the CFRS layers in the FE model. The thickness of the silicone layers was adapted from the overall amount of silicone. Then the CFRS material properties were taken for the respectively increased fiber content. The target and achieved shapes of the BWF-CFRS surface concept as well as the differences between these two shapes are plotted in Figure 23 for the European coverage as an example.

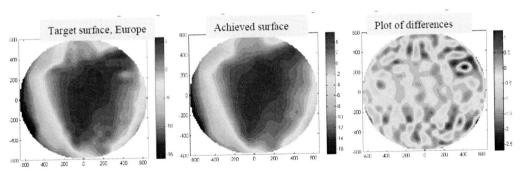

Figure 23. BWF-CFRS shape analysis results for European coverage.

Accuracy results are given in Table 3 for both CFRS concepts of the surface. As it can be seen from the table, BWF-CFRS performs slightly better. It also performs better according to levels of actuator forces (Table 4). The actuator forces in BWF-CFRS concept are about the half of that of the TWF-CFRS concept.

Table 3. RMS results of the CFRS concepts

RMS (mm)	Europe
TWF-CFRS	0.46
BWF-CFRS	0.43

Table 4. Actuator forces of the concepts for European coverage

Actuator forces (N)	European Coverage			
	Max	Min.	abs. mean	Std
TWF-CFRS with 91 actuators	18.1	-25.2	8.57	10.3
BWF-CFRS with 91 actuators	12	-11.6	3.91	4.8

The achieved RMS results for different technologies, as well as the co-polarization degradations, are compared in [19] and [20]. For European coverage the best RF performance is achieved with the CFRS technology. Multidisciplinary design optimization, simultaneously considering RF and mechanical designs will further improve results of reshaping of the reflector surfaces with the CFRS or other technologies.

CONCLUSIONS

A new flexible fiber composite material CFRS (carbon fiber-reinforced silicone) has been created and characterized for space structural applications. A major use of this material is in space deployable reflectors as a doubly curved reflecting surface. It can also be successfully used in large tensed membrane structures, while it can be easily folded and deployed. Further it can be also used as a connecting part of rigid members performing a function of structurally integrated hinges in foldable/deployable structures. An extended micromechanical numerical analysis method of homogenization of mechanical and thermo-elastic properties of the TWF composite material has been developed. It additionally accounts for transverse shear stiffnesses and thickness change. Calculations are performed using a 3D deformable beam FE model of a repeated unit cell of the material with periodic boundary conditions. The use of CFRS composites in morphing surfaces of reconfigurable reflectors will meet the need of the telecommunication industry for in-orbit variable coverage. Investigations have shown feasibility and high interest of the use. The CFRS reflecting surface material shows promising results in all characterization tests for mechanical, thermo-elastic, environmental and RF properties defined by space applications.

ACKNOWLEDGMENTS

Investigations in the CFRS material have been supported by the ESA ESTEC in a number of projects under supervision of Dr. Cyril Mangenot, Dr. Julian Santiago-Prowald, and Dr. Lucio Scolamiero among others.

The LLB staff and workshop are gratefully acknowledged for the provided support and assistance.

The following companies are acknowledged for their help and support in different tests and manufacturing processes: Wacker Silicones, Germany, with Dr. T. Sixt, Munich University of Applied Sciences, Germany, with Prof. D. Fasold, Thales Alenia Space, France, with Dr. Philippe Lepeltier, among others.

REFERENCES

[1]   L. Datashvili, H. Baier, J. Schimitschek, M. Lang, M. Huber, "High Precision Large Deployable Space Reflector Based on Pillow-Effect-Free Technology", *Proceedings 48th AIAA/ASME/ASCE/AHS/ASC Structures, Structural Dynamics, and Materials Conference*, 23 - 26 May 2007, Honolulu, Hawaii, USA.

[2]   L. Datashvili et. al, "Technical Assessment of High Accuracy Large Space Borne Reflector Antenna (TAHARA)", *Final Report* on Contract Change Notice 3 on ESTEC/LLB Contract number 16757/02/NL/LvH/bj.

[3]   H. Baier, L. Datashvili, N. Nathrath, "The Deployable Precision Shell-Membrane Reflector SMART", *Proc. European Conference on Antennas and Propagation,* Eucap 2006, Nice, France, Nov. 2006.

[4]   Datashvili, L., "Multifunctional and Dimensionally Stable Flexible Fibre Composites for Space Applications", *Acta Astronautica*, Vol. 66, numbers 7-8 April-May , pp. 1081-1086, 2010.

[5]   T. Aoki, K. Yoshida, "Feasibility Study of Triaxially Woven Fabric Composite for Deployable Structures", Paper AIAA-2007-1811, *Proceedings of the 48th AIAA/ASME/ASCE/AHS/ASC Structures, Structural Dynamics, and Materials Conference*, Honolulu, Hawaii, 2007.

[6]   A.B.H. Kueh and S. Pellegrino, "ABD Matrix of Single-Ply Triaxial Weave Fabric Composites", Paper AIAA-2007-2161, *Proceedings of the 48th AIAA/ASME/ASCE/AHS/ASC Structures, Structural Dynamics, and Materials Conference*, Honolulu, Hawaii, 2007.

[7]   Obst and G. Palermo, "Modelling of Triaxial Woven Fabrics for Antenna Reflectors", *Proceedings of the European Conference on Spacecraft Structures, Materials and Mechanical Testing 2005 (ESA SP-581).* 10-12 May 2005, Noordwijk, The Netherlands.

[8]   S. C. Quek, A. M. Waas et. al., "Analysis of 2D triaxial flat braided textile composites", *Internation Journal of Mechanical Sciences* Vol. 45 pp 1077-1096, Elsevier Ltd., USA, 2003.

[9]   R. L. Karkkainen, B. V. Sankar, "A direct micromechanics method for analysis of failure initiation of plain weave textile composites", *Journal of Composites Science and Technology,* Volume 66, Issue 1, January 2006, Pages 137-150.

[10] David E. Bowles, Stephen S. Tompkins, "Prediction of Coefficients of Thermal Expansion for Unidirectional Composites", *Journal of Composite Materials*, 1989; 23; 370.

[11] Rosen, B. W. and Z. Hashin. "Effective Thermal Expansion Coefficients and Specific Heats of Composite Materials", *International Journal of Engineering Science,* Volume 8, Issue 2, February 1970, Pages 157-173.

[12] R. M. Jones, "Mechanics of composite materials", Scripta Book Company, Washington D.C., USA, 1975.

[13] C. T. Herakovich, "Mechanics of fibrous composites", John Wiley and Sons, Inc., New York, USA, 1998.

[14] R. M. Christensen, K.H. Lo, "Solutions for effective shear properties in three phase sphere and cylinder models*", J. Mech. Phys. Solids* Vol. 27 pp. 315-330, Pergamon Press Ltd, UK, 1979.

[15] M. Daniel and O. Ishai, "Engineering Mechanics of Composite Materials", Oxford University Press, 2006.

[16] H. Schuermann, „Konstruieren mit Faser-Kunststoff-Verbunden", Springer, 2007.

[17] L. Datashvili, H. Baier, L. da Rocha-Schmidt, "Multi-scale Analysis of Structures Made of Triaxially Woven Fabric Composites with Stiff and Flexible Matrix Materials", *Proceedings of the 52th AIAA/ASME/ASCE/AHS/ASC Structures, Structural Dynamics, and Materials Conference*, Denver, 2011.

[18] ANSYS 11.0 Software Documentation, ANSYS, Inc.

[19] L. Datashvili, H. Baier, B. Wei, J. Hoffmann, E. Wehrle, L. Schreider, C. Mangenot, J. Santiago-Prowald, L. Scolamiero, J.-C. Angevain, "Mechanical Investigations of in-space-Reconfigurable Reflecting Surfaces ", *Proceedings of the 32th ESA Antenna Workshop*, 2010, Noordwijk.

[20] L. Schreider, I. Faro, S. Depeyre, V. Belloeil, L. Datashvili, H. Baier, J. Hoffmann, B. Wei, "Potential Membrane Technologies for in Orbit Reconfigurable Reflector ", *Proceedings of the 32th ESA Antenna Workshop*, 2010, Noordwijk.

[21] L. Datashvili, "Review and Evaluation of the Existing Designs / Technologies for Space Large Deployable Apertures", *International Scientific Conference on Advanced Lightweight Structures and Reflector Antennas*, 14 - 16 October 2009, Tbilisi, Georgia.

[22] Datashvili L., Baier H., "The Method of Manufacturing of the Deployable Space Antenna Reflector And The Deployable Reflector Made By This Method", Georgian Patent GE P 2003 3005 B, Patent Application N3382/01-01 13.04.2001.

[23] Aramis 3D strain measurement system: http://www.gom.com/de/messsysteme/systemuebersicht/aramis.html.

INDEX

1

1,2-methylenedioxybenzene (MDOB), 254

3

3D deformability, xvi, 605, 620, 629

A

Adhesion, 40, 42, 43, 52, 53, 139, 251, 252, 259, 280, 290, 294, 297, 306, 310, 316, 415, 494
aramid fiber, vii, 1, 3, 12, 498, 511, 574, 578
Aramid fiber, 12, 574
Artificial Neural Network, xiv, 439, 440
Atmospheric pressure plasma, 290, 292, 293, 299
ATR-FTIR, 255, 261, 262, 263, 264, 289, 300, 303, 305, 306

B

Bearing, 431, 432
Biaxially, 607
Boron fiber, 13
Brittleness, 29, 320

C

Carbides, 18
carbon, vii, xvi, 6, 7, 11, 13, 15, 19, 20, 21, 22, 34, 43, 49, 59, 76, 78, 246, 248, 323, 326, 451, 452, 453, 454, 461, 465, 511, 514, 523, 574, 576, 578, 584, 585, 605, 606, 607, 608, 609, 631, 638
carbon XE "carbon" fiber, vii, xvi, 6, 7, 11, 13, 15, 19, 20, 21, 22, 34, 43, 49, 59, 76, 78, 246, 248, 323, 326, 451, 452, 453, 454, 461, 465, 511, 514, 523, 574, 576, 578, 584, 585, 605, 606, 607, 608, 609, 631, 638

Carbon fiber, 3, 11, 19, 21, 498, 571, 574, 609
Carbon fibres, xi, 251, 252, 280
Carbon matrix XE "matrix" composites, 21
Carpet fiber, 381
cellulose, 105
Cellulose, 7, 8, 51, 53, 58, 64, 67, 85, 86, 87, 92, 101, 102, 103, 143, 145, 147, 156
cellulose fibre, 105
Ceramic fiber, 12
Ceramic matrix XE "matrix" composite, 19
Characterisation, 88, 144
Characterization, 37, 38, 39, 40, 50, 52, 85, 87, 94, 95, 96, 165, 279, 335, 336, 464
Coir, 53, 60, 67, 68, 76
Compocasting, 17
compression, 576
Compression molding, 32, 35
Contact angle, 258, 302
Cotton, 6, 8, 68, 86
Crack bridging, 323
Crack deflection, 322, 328
Cutting forces,, 455

D

Damage, 215, 216, 403, 413, 414, 416, 429, 430, 431, 447, 452, 464, 493, 494, 499, 507, 508, 509
Delamination, 229, 259, 395, 398, 402, 411, 414, 415, 416, 418, 421, 423, 427, 430, 434, 447, 448, 452, 465, 466, 493, 500, 501, 523, 524
Differential scanning calorimetry, 45, 146
differential scheme, 586
Double cantilever beam (DCB), 303
Drilling, 160, 388, 395, 397, 399, 404, 406, 414, 448, 453, 464, 466, 512, 518, 521, 523, 524, 525

Dynamic loading, 519

E

eigenstrain, 587
ensemble-volume average method, 587
Epoxy, x, 25, 50, 53, 54, 79, 117, 118, 121, 217, 244, 245, 259, 397, 399, 406, 464, 493, 575
Epoxy resin, x, 117, 118, 217

F

Fatigue, v, 51, 123, 159, 187, 212, 215, 216, 401, 433, 490, 493, 601
fiber, xvi, 572, 577, 578, 605, 606
Fiber, vii, xvi, 1, 3, 5, 14, 21, 33, 40, 42, 43, 50, 51, 52, 53, 54, 55, 57, 60, 63, 67, 68, 69, 76, 85, 86, 87, 89, 101, 140, 319, 338, 344, 345, 348, 350, 358, 361, 366, 376, 378, 379, 382, 439, 440, 447, 463, 464, 467, 468, 469, 476, 477, 몸490, 493, 497, 498, 500, 501, 571, 573, 595, 596, 597, 598, 600, 603, 605, 608
Fiber-reinforced composites, vii, 1, 338, 358, 440, 463, 468
fiber XE "fiber" -reinforced polymer XE "polymer" (FRP), 572
Fibres, 50, 52, 53, 139, 140, 161, 251, 283, 392, 415, 554, 598
Filament winding, 31
Final fracture strength, xiv, 467
Finite Element Analysis (FEA), xv, 497
Finite element model, 492, 523, 596
Flammability, v, 105, 109, 120
Flax, 6, 9, 44, 51, 52, 53, 55, 60, 64, 67, 68, 71, 79, 87, 89, 139
Flexible fiber XE "fiber" composites, xvi, 605, 606
Flexural strength, xiv, 121, 337, 467, 479, 485, 488, 490
Fracture resistance, 247, 269, 490, 491
Fracture toughness, 18, 122, 336, 353, 356

G

glass fiber, 2, 6, 7, 11, 12, 13, 19, 24, 34, 37, 41, 42, 43, 49, 60, 61, 62, 67, 71, 76, 79, 81, 82, 84, 126, 216, 453, 454, 465, 468, 477, 488, 489, 490, 491, 494, 495, 523, 524, 525, 573, 574, 578, 594
Glass fiber, 12, 498, 573
glass fiber XE "fiber" -reinforced plastic (GFRP) composite, 578
Glass fibre reinforced polyester (GFRP), xi, 297

Glassy carbon, 252, 265
Gliding arc, 299, 301, 316

H

Hemp, 6, 9, 60, 67, 68, 72, 87
homogenization process, 586
Hyperbranched polymer, 247

I

ImageJ, 143, 147
Impact toughness, 119
Inclusion, 239
Initial fracture strengths, xiv, 467
injection compression XE "compression" molding (I/CM), 576
Injection molding, 32
Interfacial shear XE "shear" strength, 40

J

Jute, 6, 7, 8, 37, 50, 51, 52, 53, 54, 55, 60, 64, 67, 68, 69, 73, 74, 85, 86, 139, 140

K

Kenaf, 6, 9, 60, 68, 73, 84, 88

L

LFRP composite, xv, 497, 498, 499, 512, 513, 514
liquid composite molding (LCM), 576

M

Machining, 387, 388, 391, 399, 400, 407, 410, 411, 412, 413, 414, 435, 436, 437, 438, 439, 446, 464, 465, 498, 511, 523, 524
matrix, 15, 16, 19, 21, 22, 49
Matrix, 13, 14, 15, 17, 20, 21, 22, 40, 42, 43, 50, 51, 52, 54, 55, 139, 162, 246, 475, 493, 500, 501, 515, 518, 520, 539, 541, 549, 573, 602, 639, 640
Mechanical properties, 27, 36, 68, 85, 123, 143, 165, 350, 354, 385, 469, 489, 493
Mechanical testing, 491
Metal matrix XE "matrix" composite, 15, 16
Metallic fiber, 13
micromechanics, 122, 502, 567, 587, 589, 590, 592, 627, 639

Index

Microstructure, 50, 244, 335, 337, 338, 344, 348, 356, 357, 359, 360
Mori-Tanaka method, 586
Morphology, 102, 156, 245

N

Nanocomposites, v, 91, 92, 95, 105, 108, 123, 245
Nanocrystals, 102
Nanofibers, 50, 91, 96
Nanofibres, 227
natural fiber, vii, viii, 1, 6, 7, 10, 14, 42, 43, 49, 57, 58, 59, 60, 61, 62, 63, 64, 66, 67, 68, 69, 70, 71, 72, 73, 74, 76, 77, 78, 80, 81, 82, 83, 89, 122, 123, 127, 451, 573, 574
Natural fiber, viii, 4, 6, 57, 58, 62, 63, 66, 69, 71, 83, 84, 89, 574
Natural fibers, viii, 4, 6, 57, 58, 62, 63, 66, 69, 71, 83, 84, 574
Natural fibres, 106, 144
Nettle, 68, 72
Nitrides, 18

O

OES, 256, 270, 271, 300, 302, 308

P

Permeability, 379
Phase separation, 221, 247, 249
Plasma polymerization, 255, 267
Plasma torch, 258, 259
Plasma treatment, xi, 251, 298, 314
Polyester, 23, 54, 55, 85, 140, 297, 316, 574
Polylactic Acid, 26
polymer, 572, 574, 575, 577
Polymer matrix XE "matrix" composites, 22, 49
Polymer pyrolysis, 20
Polypropylene, 27, 33, 37, 38, 39, 42, 46, 47, 48, 49, 50, 51, 52, 53, 54, 55, 89, 139, 140, 143, 144, 145, 147
Polypropylene composites, 89
Powder metallurgy, 16
Pultrusion, 33, 51

R

Ramie, 6, 64, 67, 68, 74, 88
reaction injection molding (RIM), 576

Recycled cellulose XE "cellulose" fibre XE "cellulose fibre" s (RCF), 105
rehabilitation, 298, 572, 573, 581, 587, 595
reinforced reaction injection molding (RRIM), 576
Reinforcing agent, 4
repair, 414, 475, 576, 577, 579, 580, 583, 589, 590
resin transfer molding (RTM), 576
retrofitting, 571, 572, 573, 584
rule of mixture method, 586

S

self-consistent method, 586
SEM, ix, xii, 113, 114, 115, 133, 143, 146, 147, 152, 153, 155, 222, 223, 224, 226, 227, 228, 231, 233, 239, 240, 241, 242, 300, 303, 304, 320, 334, 335, 336, 339, 340, 341, 342, 345, 346, 347, 349, 355, 359, 360
shear, 40
Short-fibre composites, 143
SiC fibers, xii, 20, 319, 320, 326, 327, 328, 333, 334, 337, 340, 347, 349, 357, 358, 359, 364
Silk, 10, 37, 50, 51, 52
Sisal, 6, 52, 54, 60, 64, 67, 68, 74, 88, 89
Soil, xii, 88, 369, 371, 384, 385, 386
Specific cutting energy, 455
sprayed fiber XE "fiber" -reinforced polymer XE "polymer" (sprayed FRP) coating, 577
Squeeze casting, 17
Strength, 21, 52, 53, 71, 72, 73, 74, 75, 76, 108, 111, 470, 480, 484, 485, 490, 491, 501
strengthening, xvi, 5, 323, 364, 470, 571, 572, 573, 576, 577, 578, 579, 580, 581, 582, 583, 584, 585, 589, 590, 593, 594
Structural applications, xv, 497
structural reaction injection molding (SRIM), 576
Surface energy, 87, 144
Surface roughness, 393, 398, 400, 459, 460
synthetic fiber, vii, 1, 4, 7, 11, 12, 13, 14, 43, 49, 59, 61, 66, 68, 76, 370, 574

T

TEM, xii, 320, 336, 340, 341, 342, 343, 346, 361, 365
Thermal analysis, 45
Thermal degradation, 122
Thermogravimetric analysis, 45
Thermoplastic, 25, 51, 52, 102, 139, 159, 218, 219, 220, 227, 243, 244, 575
thermoplastic polymer XE "polymer" resin, 575
Thermoplastics, 26, 52, 85, 86, 139, 156

Thermoset, 21, 23, 244, 574
thermoset polymer XE "polymer" resin, 574, 575
Three-point bending, 485, 561
Tool wear, 396, 447, 461
Toughening, x, 123, 217, 218, 219, 224, 227, 234, 244, 248, 322, 324, 325, 351, 356

U

ultimately reinforced thermoset resin injection molding (URTRI), 576
Ultrasonic irradiation, 311, 312, 313

V

vacuum, 576
vacuum XE "vacuum" -assisted resin injection molding (VARI), 576

vacuum XE "vacuum" -assisted resin transfer molding (VARTM), 576
variational principle method, 586

W

Whisker, 51, 319, 342, 346, 354
Whiskers, 327, 328, 364
Wool, 10

X

XPS, 254, 258, 261, 263, 264, 265, 266, 269, 273, 274, 276, 277, 278, 280, 281, 283, 284, 285, 287, 289, 290, 293, 300, 303, 306, 307, 309, 310, 311, 312, 313, 314, 317
X-μCT analysis, 143, 153, 155